Function Keys

- F1 Displays online help.
- F2 Toggles text & graphics windows.
- F3 Toggles object snap.
- F4 Toggles tablet mode.
- F5 Cycles through isoplanes.
- F6 Cycles coordinate display.
- F7 Toggles grid display.
- F8 Toggles ortho mode.
- F9 Toggles snap mode.
- F10 Toggles polar tracking.
- F11 Toggles object snap tracking.

- Ctrl F4 Closes drawing.
- Ctrl F6 Switches to next drawing.
- Alt F8 Runs VBA.
- Alt F11 Opens VBA IDE.

Other Keys

- Esc Cancels commands and grips.
- Delete Deletes selected objects.
- Executes and repeats commands.

Command Prefixes

- **'** Specifies transparent command:
  `From point: 'zoom`
- **'?** Provides context-sensitive help:
  `Command: line '?`
- **~** Forces display of file dialog box:
  `Command: -insert ~`
- **-** Forces display on command line:
  `Command: -mtext`
- **+** Prompts for tab number:
  `Command: +options`
- **.** Forces use of undefined command:
  `Command: .line`
- **_** Forces English cmd in int'l version:
  `Command: _line`
- **multiple** Automatically repeats command:
  `Command: multiple circle`

- **(** Begins AutoLISP function.
- **)** Ends AutoLISP function:
  `Radius: (/ 3.2 2.0)`
- **$(** Begins Diesel macro.

... Copies to Clipboard.

- Ctrl Shift C Copies with basepoint.
- Ctrl D Cycles coordinate display.
- Ctrl E Cycles through isoplanes.
- Ctrl F Toggles object snap mode.
- Ctrl G Toggles grid display.
- Ctrl H Toggles pick style.
- Ctrl K Displays Hyperlinks dialog box.
- Ctrl L Toggles ortho mode.
- Ctrl N Opens new drawing.
- Ctrl O Opens drawing files.
- Ctrl P Displays Plot dialog box.
- Ctrl Q Quits AutoCAD.
- Ctrl R Cycles through viewports.
- Ctrl S Saves drawing.
- Ctrl Shift S Displays Save As dialog box.
- Ctrl T Toggles tablet mode.
- Ctrl U Toggles polar tracking.
- Ctrl V Pastes from Clipboard.
- Ctrl Shift V Pastes with insertion point.
- Ctrl X Cuts to Clipboard.
- Ctrl Y Redoes last undo.
- Ctrl Z Undoes last command.
- Ctrl 0 Toggles cleanscreen mode.
- Ctrl 1 Toggles **Properties** window.
- Ctrl 2 Toggles **DesignCenter** window.
- Ctrl 3 Toggles **Tool Palettes** window.
- Ctrl 4 Toggles **SheetSet Manager**.
- Ctrl 5 Toggles **Info Palette** window.
- Ctrl 6 Toggles **dbConnect Manager**.
- Ctrl 7 Toggles **Markup Set Manager**.

Mouse Buttons

(***Sketch*** *command buttons.)*

① Pick objects (*raises, lowers pen*).

② Displays cursor menu (*draws line*).

③ Selects (*records sketch*).

Ctrl ② Displays object snap cursor menu.

THE

Illustrated AutoCAD® 2005 Quick Reference

Ralph Grabowski

autodesk® Press

THOMSON
DELMAR LEARNING™

Australia • Canada • Mexico • Singapore • Spain • United Kingdom • United States

autodesk® Press

The Illustrated AutoCAD® 2005 Quick Reference
Ralph Grabowski

Vice President, Technology and Trades SBU:
Alar Elken

Editorial Director:
Sandy Clark

Senior Acquisitions Editor:
James DeVoe

Senior Development Editor:
John Fisher

Editorial Assistant:
Katherine Bevington

Marketing Director:
Dave Garza

Channel Manager:
Fair Huntoon

Marketing Coordinator:
Casey Bruno

Production Director:
Mary Ellen Black

Production Manager:
Andrew Crouth

Production Editor:
Tom Stover

Art & Design Specialist:
Mary Beth Vought

Cover Image:
Getty Images, Inc.

Printed in the United States
1 2 3 4 5 XXX 07 06 05 04

For more information contact Delmar Learning, Executive Woods, 5 Maxwell Drive, PO Box 8007, Clifton Park, NY 12065-8007

Or find us on the World Wide Web at www.delmarlearning.com

ISBN: 1-4018-8366-4

About This Book

The Illustrated AutoCAD 2005 Quick Reference presents concise facts about all commands found in AutoCAD 2005 and earlier. The clear format of this reference book demonstrates each command, starting on its own page, illustrated by over 500 figures. Each command includes one or more of the following: command line options, dialog box options, shortcut menu options, related commands, related toolbar icons, and related system variables. Plus:

- All variations of commands, such as the **View**, **-View**, and **+View** commands.
- Dozens of AutoCAD 2005 commands and system variables not documented by Autodesk.
- Icons, such as ☑ and ☐, that indicate default settings of check boxes in dialog boxes.
- "Quick Start" mini-tutorials that help you get started quicker.
- Over 100 definitions of acronyms and hard-to-understand terms.
- Nearly 1,000 context-sensitive tips.
- All system variables in Appendix A, including those not listed by the **SetVar** command.
- Obsolete commands that no longer work in AutoCAD 2005 in Appendix B.
- External commands that operate outside of AutoCAD in Appendix C.
- Express Tool command names in Appendix D.

The name of each command shown in mixed upper and lower case to help you understand the name, which is often condensed. For example, the name of the **VpClip** command is short for "ViewPort CLIP." Each command includes all alternative methods of command input:

- Alternate command spelling, such as **Donut** and **Doughnut**.
- ' (the apostrophe prefix) indicating transparent commands, such as **'Blipmode**.
- All aliases, such as **L** for the **Line** command.
- Pull-down menu picks, such as **Draw** ↳ **Construction Line** for the **XLine** command.
- Control-key combinations, such as CTRL+E for the **Isoplane** toggle.
- Function keys, such as F1 for the **Help** command.
- Alt-key combinations, such as ALT+TE for the **Spell** command.
- Table menu coordinates, such as **M2** for the **Hide** command.

The version or release number indicates when the command first appeared in AutoCAD, such as **Ver. 1.0**, **Rel. 14**, or **2005** — useful when working with older versions of AutoCAD.

Thank you for reviewing this book: Dr. Ejike Charles Igboegwu, Ivy Tech State College, East Chicago, Indiana; John Knapp, Metropolitan Community College, Omaha, Nebraska; Phil Kreiker, Looking Glass Microproductions, Loveland, Colorado; Alex Lepeska, Renton Technical College & Pierce College, Renton, Washington; and Brian Matthews, North Carolina State University, Raleigh, North Carolina.

Special thanks to Stephen Dunning for his keen copy editing, and to Phil Kreiker for his accurate technical editing. *Soli Deo Gloria!*

Ralph Grabowski
Abbotsford, British Columbia, Canada
March 22, 2004
Contact: grabowski@telus.net

Table of Contents

Indicates the command is new to AutoCAD 2005.

' Indicates the command is transparent.

Appendices

'About

Rel.12 Displays the AutoCAD version and serial numbers, ownership information, and copyright notices.

Command	Alias	Ctrl+	F-key	Alt+	Menu Bar	Tablet
about	...	...	...	**HO**	**Help** ↳**About**	...

Command: about

Displays dialog box:

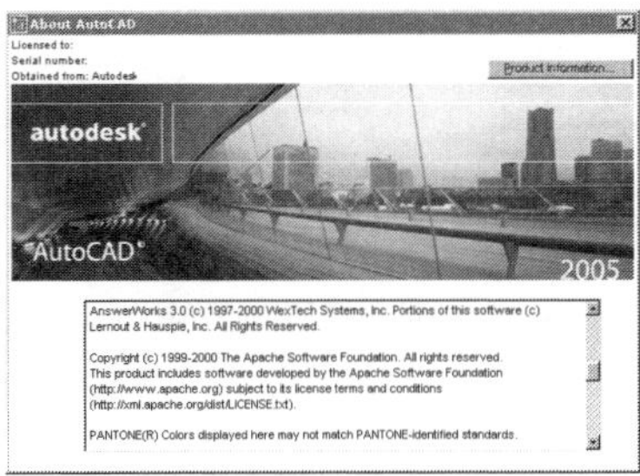

Shown on the splash screen is the Yokohama Ferry Terminal in Japan.

DIALOG BOX OPTIONS

x dismisses dialog box; alternatively, press **ESC**.

Product Information displays the Product Information dialog box:

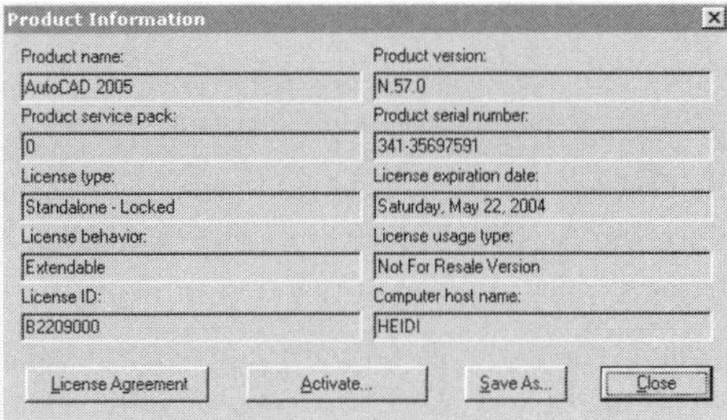

License Agreement opens the computer's default word processor, and then displays the Autodesk Software License Agreement document (*license.rtf*).

Activate runs the Product Activation wizard (*new to AutoCAD 2005*).

Save As displays the Save As dialog box, which records the information displayed above.

Close returns to AutoCAD.

RELATED COMMANDS

Properties displays information about selected objects.

Status displays information about the drawing and environment.

RELATED SYSTEM VARIABLES

_PkSer displays the AutoCAD software serial number.

_Server displays the network authorization code.

AcisIn

Rel. 13 Imports *.sat* files into drawings, and creates 3D solids, 2D regions, and bodies.

Command	Alias	Ctrl+	F-key	Alt+	Menu Bar	Tablet
acisin	...	...	...	IA	Insert ↳ACIS File	...

Command: acisin

Displays Select ACIS File dialog box. Select a .sat file, and then click ***Open.***

DIALOG BOX OPTIONS

Cancel dismisses the dialog box.

Open opens the selected *.sat* file.

RELATED COMMANDS

AcisOut exports solid objects — 3D solids, 2D regions, and bodies — to *.sat* files for import into ACIS-aware CAD software.

AmeConvert converts AME v2.0 and v2.1 solid models and regions into solids.

RELATED FILE

.sat is the ASCII format of ACIS model files; short for "save as text."

TIPS

- See the **Open** command for options related to the Select ACIS File dialog box.
- When system variable **FileDia** is turned off (set to 0), this command prompts:

  Enter SAT file name:

  Enter the tilde (~) to force the display of the dialog box.
- "ACIS" comes from the first names of the original developers, "Andy, Charles, and Ian's System." In Greek mythology, Acis was the lover of the goddess Galatea; when Acis was killed by the jealous Cyclops, Galatea turned the blood of Acis into a river.
- ACIS is the solids modeling technology from the Spatial Technologies division of Dassault Systemes, and is used by numerous 3D CAD packages. As of AutoCAD 2004, Autodesk is using its own ACIS-derived solids modeler, called ShapeManager.

AcisOut

Rel.13 Exports AutoCAD 3D solids, 2D regions, and bodies in *.sat* format.

Command	Alias	Ctrl+	F-key	Alt+	Menu Bar	Tablet
acisout	...	...	...	**FE**	**File**	...
				↳ACIS	**↳Export**	
					↳ACIS	

Command: acisout

Select objects: *(Select one or more objects.)*

Select objects: *(Press* ENTER *to end object selection.)*

*Displays Create ACIS File dialog box. Enter a name for the .sat file, and then click **Save**.*

DIALOG BOX OPTIONS

Cancel dismisses the dialog box.

Save saves the selected objects as a *.sat* file.

RELATED COMMANDS

AcisIn imports *.sat* files, and creates 3D solids, 2D regions, and bodies.

StlOut exports solid models in STL format for use by stereolithography devices.

3dsOut exports solid models as 3D faces for import into 3D Studio software.

RELATED FILE

***.sat** is the ASCII format of ACIS model files.

TIPS

- **AcisOut** exports objects that are 3D solids, 2D regions, and bodies only.
- When system variable **FileDia** is turned off (set to 0), this command prompts:

  Select objects: *(Select one or more objects.)*

  Select objects: *(Press* **ENTER.***)*

  Enter SAT file name <Drawing1.sat>: *(Type a file name.)*
- The *.sat* files can be read by other ACIS-based CAD programs.

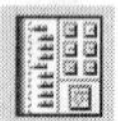

'AdCenter / 'AdcClose

2000 Opens amd closes the DesignCenter window (*short for Autocad Design.*).

Command	Aliases	Ctrl+	F-key	Alt+	Menu Bar	Tablet
adcenter	**adc** **content**	**2**	**...**	**TG**	**Tools** **↳DesignCenter**	**X12**
adcclose	**...**	**2**	**...**	**TG**	**Tools** **↳DesignCenter**	**X12**

Command: adcenter

Displays DesignCenter window:

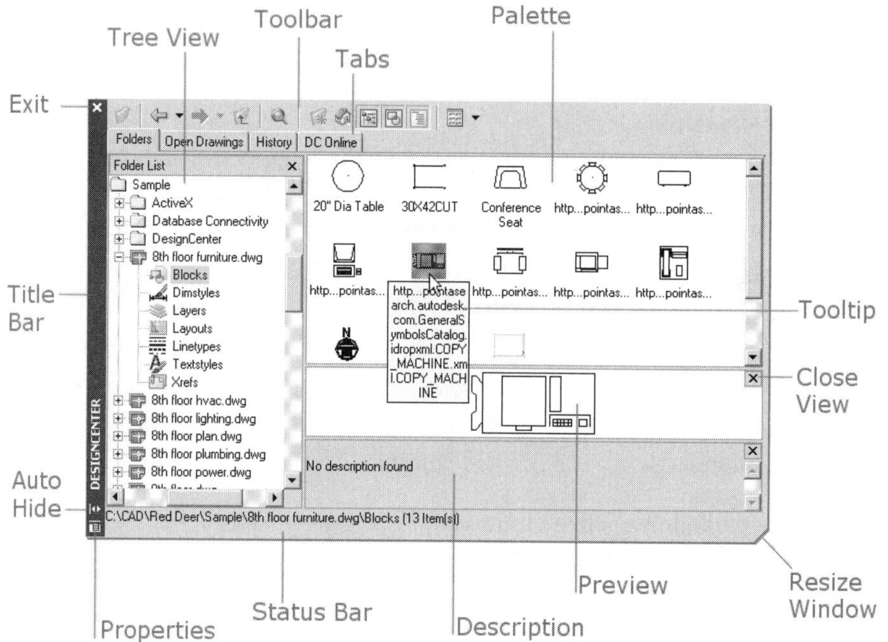

Command: adcclose

Closes the DesignCenter window:

TAB OPTIONS

Folders displays the content found in the the local computer, as well as networked computers.

Open Drawings displays the content found in drawings currently open in AutoCAD.

History displays the drawings previously opened.

DC Online displays the content available from Autodesk via the Internet.

TOOLBAR OPTIONS

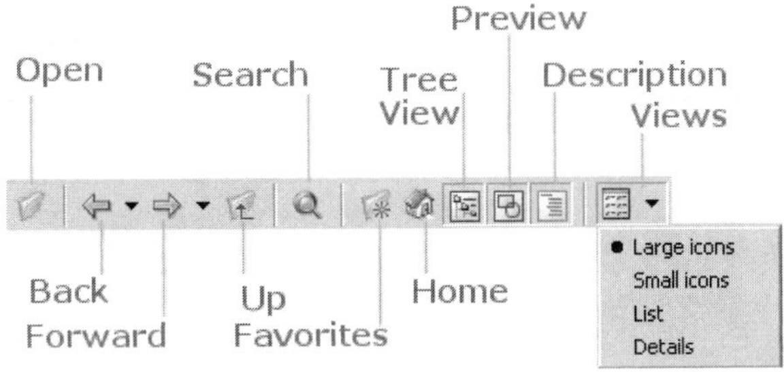

Open displays the **Open** dialog box to open the following vector and raster file types: *.dwg*, *.bil*, *.bmp*, *.cal*, *.cg4*, *.dib*, *.flc*, *.fli*, *.gif*, *.gp4*, *.ig4*, *.igs*, *.jpg*, *.mil*, *.pat*, *.pcx*, *.png*, *.rlc*, *.rle*, *.rst*, *.tga*, and *.tif*.

Back returns to the previous view.
Forward goes to the next view.
Up moves up one folder level.

Search opens the Search dialog box to search for AutoCAD files (on the computer) and the following objects (in drawings): blocks, dimstyles, drawings, hatch patterns, layers, layouts, linetypes, text styles, table styles, and xrefs.

Favorites displays the files in the *\documents and settings\<username>\favorites\autodesk* folder.
Home displays the files in the *\autocad 2005\sample\designcenter* folder.
Tree View Toggle hides and displays the Folders and Open Drawings tree views.
Preview toggles the display of the preview image of *.dwg* and raster files.
Description toggles the display of the description area.

View changes the display format of the palette area.

RELATED COMMAND

AdcNavigate specifies the initial path for DesignCenter.

RELATED SYSTEM VARIABLE

AdsState specifies whether DesignCenter is open.

TIPS

- Use DesignCenter to keep track of drawings and parts of drawings, such as block libraries.
- You can drag blocks and other drawing parts from the DesignCenter into the drawing. As of AutoCAD 2005, you can also share table styles.
- The DesignCenter window can switch between floating and docked by right-clicking and selecting the option from the menu.
- CTRL+2 opens and closes the DesignCenter each time you press the keys.

AdcNavigate

2000 Specifies the initial path for the DesignCenter to access content.

Command	Alias	Ctrl+	F-key	Alt+	Menu Bar	Tablet
adcnavigate	...	...	...	...	...	...

Command: adcnavigate

Opens DesignCenter, if not already open. Press ENTER *a second time.*

Enter pathname <>: *(Enter a path, and then press* ENTER.*)*

COMMAND LINE OPTION

Enter pathname specifies the path, such as *c:\program files\acad2005\sample.*

RELATED COMMAND

AdCenter opens the DesignCenter window.

TIPS

- AutoCAD uses the path specified by **AdcNavigate** to locate content displayed by DesignCenter's Desktop option.
- You can enter the path to a file, folder, or network location:

  Example of a folder path: `c:\program files\autocad 2005\sample`
  Example of a file path: `c:\design center\welding.dwg`
  Example of a network path: `\\downstairs\project`

Ai_Box

Draws 3D boxes as surface objects (*undocumented command*).

Command	Alias	Ctrl+	F-key	Alt+	Menu Bar	Tablet
ai_box	...	...	...	...	...	...

Command: ai_box
Specify corner point of box: *(Pick a point.)*
Specify length of box: *(Pick a point.)*
Specify width of box or [Cube]: *(Pick a point, or type* **C***.)*
Specify height of box: *(Pick a point.)*
Specify rotation angle of box about the Z axis or [Reference]: *(Specify the rotation angle, or type* **R***.)*

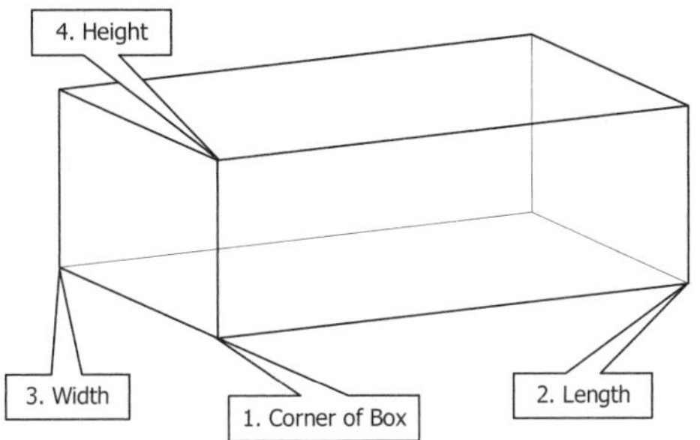

COMMAND LINE OPTIONS

Corner of box specifies the initial corner of the box.

Length specifies the length of one side of the box along the x-axis.

Cube creates a cube based on the **Length**.

Width specifies the width of the box, along the y-axis.

Height specifies the height of the box, along the z-axis.

Rotation angle specifies the angle the box rotates about the z-axis.

Reference prompts you to pick two points that represent the new angle.

TIPS

- This command creates 3D surface objects of rectangular boxes and cubes; to draw a solid model, use the **Box** command.
- When specifying the **Width** and **Height**, move the cursor back to the **Corner of box** point; otherwise the box may have a different size than you expect.
- The box is made of a single polymesh; **Explode** converts each side to an independent 3D face object.
- If no z coordinate is specified, then the base of the box is drawn at the current setting of the **Elevation** system variable.

*The following apply to the **Ai_** surface model commands:*

RELATED SYSTEM VARIABLES

SurfU specifies the surface mesh density in the m-direction.

SurfV specifies the surface mesh density in the n-direction.

TIPS

- You *cannot* perform Boolean operations (such as intersect, subtract, and union) on 3D surface objects.
- You cannot convert 3D surface objects into 3D solid objects.
- To convert 3D solid objects into 3D surface objects, export with the **3dsOut** command; then import with the **3dsIn** command.
- Types of 3D objects drawn by the **Ai_** series of commands:

Command	Object
Ai_Box	Rectangular boxes, and cubes.
Ai_Dish	Bottom half of a sphere.
Ai_Dome	Top half of a sphere.
Ai_Cone	Pointy cones, and truncated cones.
AI_Mesh	Non-planar polyface meshes.
Ai_Pyramid	Pyramids, truncated pyramids, tetrahedrons, truncated tetrahedrons, and roof shapes.
Ai_Sphere	Spheres.
Ai_Torus	Tori (donuts).
Ai_Wedge	Wedges.

- Mesh m and n sizes (**SurfU** and **SurfV**) are limited to values between 2 and 256. The meaning of these varies with the type of surface model you create. You have to vary these and re-create the model to determine their particular effects.
- Each of these surface shapes is made of a single polygon mesh; use the **Explode** command to convert faces to independent 3Dface objects.
- You can use the **Hide**, **Shade**, and **Render** commands on these surface model shapes.

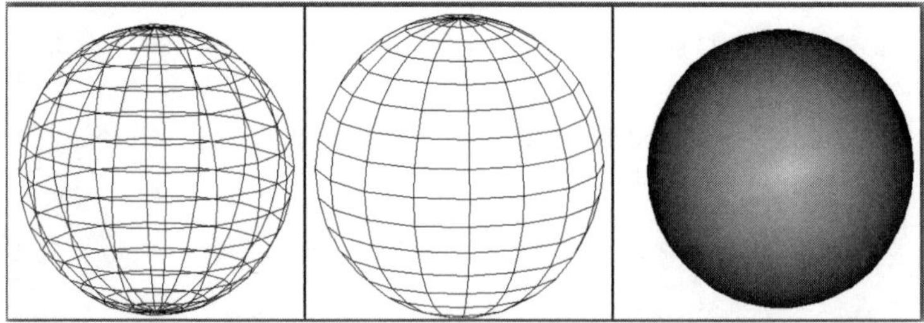

- The bases of boxes, cones, dishes, domes, meshes, and pyramids are drawn at the current setting of the **Elevation** system variable — unless you specify the z coordinate.
- The centers of spheres and tori are at the current elevation, unless you specify the zcoordinate.

'Ai_CircTan

Draws circles tangent to three points (*undocumented command*).

Command	Alias	Ctrl+	F-key	Alt+	Menu Bar	Tablet
ai_circtan	...	...	...	...	...	...

Command: ai_circtan
Enter Tangent spec: *(Pick an object.)*
Enter second Tangent spec: *(Pick an object.)*
Enter third Tangent spec: *(Pick an object.)*

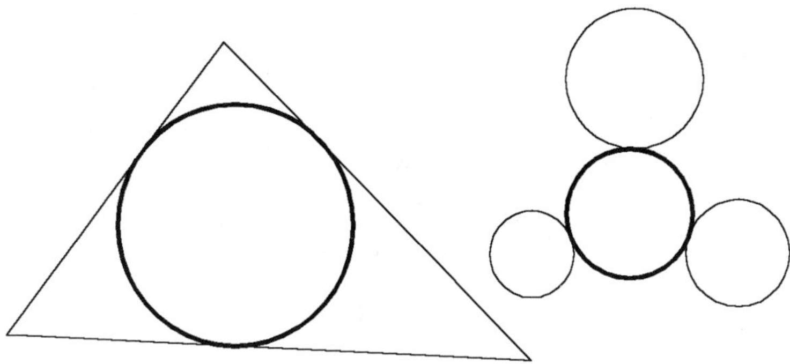

A circle drawn tangent to three lines (left) and three circles (right).

COMMAND LINE OPTION

Enter Tangent spec picks the objects to which the circle will be made tangent.

TIPS

- This command is meant for use in toolbar and menu macros.
- If the circle cannot be drawn between three tangents, AutoCAD complains, "Circle does not exist."
- This command is an alternative to using the **Circle** command's **3P** option with the TANgent object snap.

Ai_Cone

Draws 3D cones as surface objects (*undocumented command*).

Command	Alias	Ctrl+	F-key	Alt+	Menu Bar	Tablet
ai_cone	...	...	...	...	...	...

Command: ai_cone

Specify center point for base of cone: *(Pick a point.)*

Specify radius for base of cone or [Diameter]: *(Pick a point, or type **D**.)*

Specify radius for top of cone or [Diameter] <0>: *(Specify the radius, or type **D**; enter **0** for a pointy top.)*

Specify height of cone: *(Pick a point.)*

Enter number of segments for surface of cone <16>: *(Enter a number between 3 to 255, or press ENTER.)*

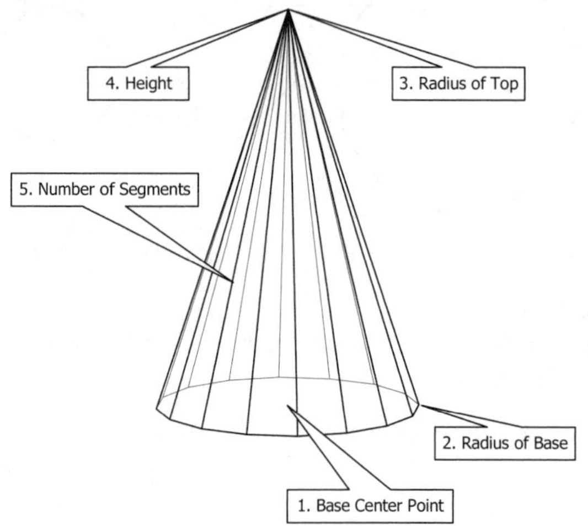

COMMAND LINE OPTIONS

Base center point specifies the center point of the base of the cone.

Diameter of base specifies the diameter of the base.

Radius of base specifies the radius of the base.

Diameter of top specifies the diameter of the top of the cone.

Radius of top specifies the radius of the top of the cone; 0 = cone with a point.

Height specifies the height of the cone.

Number of segments specifies the number of "lines" that define the curved surface of the cone; default = 16.

TIPS

- The **Ai_Cone** command creates "pointy" and truncated cones; see figure.
- The base of the cone is drawn at the current setting of the **Elevation** system variable, unless a z coordinate is specified.

'Ai_Custom_Safe, *etc.*

Accesses Autodesk support Web pages (*undocumented commands*).

Commands	Alias	Ctrl+	F-key	Alt+	Menu Bar	Tablet
ai_custom_safe	...	...	...	**HRC**	**Help** **↳Online Resources** **↳Customization**	...
ai_product_support				**HRP**	**Help** **↳Online Resources** **↳Product Support**	
ai_product_support_safe						
ai_training_safe				**HRT**	**Help** **↳Online Resources** **↳Training**	

Command: ai_custom_safe

_.browser Enter Web location (URL) <http://www.autodesk.com> http://www.www.autodesk.com/developautocad

Opens Web browser.

COMMAND LINE OPTIONS

None.

TIPS

- The commands access the following Web pages:

Command	Web Page Accessed
ai_custom_safe	www.autodesk.com/developautocad
ai_product_support	www.autodesk.com/autocad-support
ai_product_support_safe	www.autodesk.com/autocad-support
ai_training_safe	www.autodesk.com/autocad-training

- These commands are meant for use in menu and toolbar macros.

AiDimPrec

Changes the precision displayed by existing dimensions (*undocumented command*).

Command	Alias	Ctrl+	F-key	Alt+	Menu Bar	Tablet
aidimprec	...	...	...	...	...	...

Command: aidimprec
Enter option [0/1/2/3/4/5/6] <4>: *(Enter a digit.)*
Select objects: *(Select one or more dimensions; non-dimensions are ignored.)*
Select objects: *(Press* ENTER.*)*

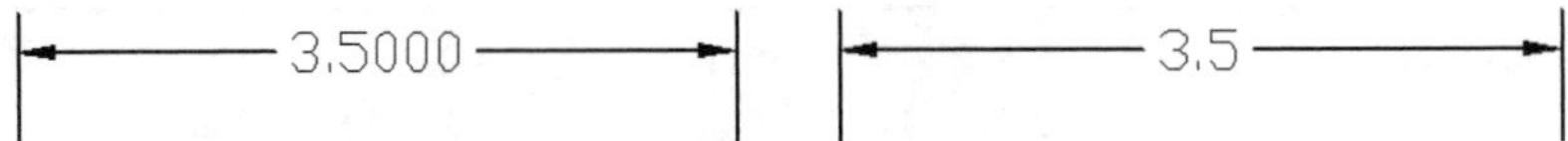

Before (left) and after (right) applying ***AiDimPrec = 1*** *to a decimal dimension.*

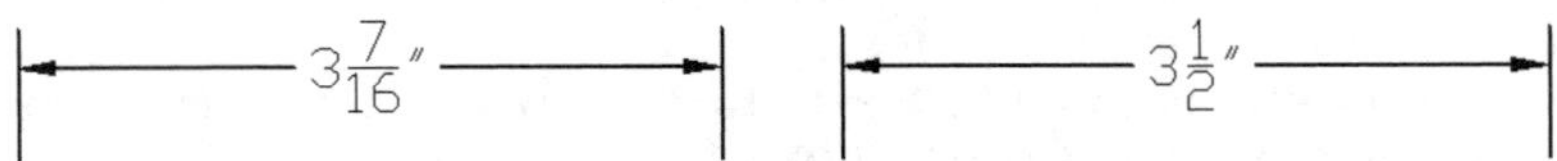

Before (left) and after (right) applying ***AiDimPrec = 1*** *to a fractional dimension.*

COMMAND LINE OPTIONS

Enter option specifies the precision (number of decimal places, or fractional equivalent); enter a number between 0 and 6.

Select objects selects one or more dimensions.

TIPS

- This command allows you to retroactively change the precision displayed by selected dimensions. It is used by the right-click shortcut menu: right click a dimension, and then select **Precision** from the shortcut menu.
- Zero to six decimal places can be specified; fractional units are rounded to the nearest fraction:

AiDim	Prec Architectural Units
0	Rounded to the nearest unit.
1	1/2"
2	1/4"
3	1/8"
4	1/16"
5	1/32"
6	1/64"

- *Caution!* Because **AiDimPrec** rounds off dimensions, it can create false values. The dimension line below measures 3.4375", but setting **AiDimPrec** to 0 rounds down to 3".

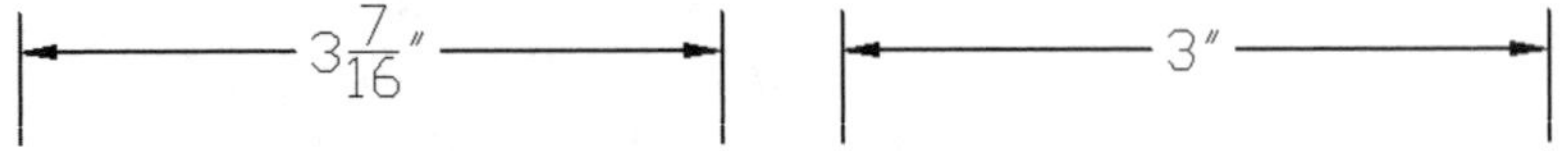

Applying ***AiDimPrec = 0*** *to a 3 $^7/_{16}$" dimension.*

AiDimStyle

Saves and applies preset dimension styles (*undocumented command*).

Command	Alias	Ctrl+	F-key	Alt+	Menu Bar	Tablet
aidimstyle	...	...	...	...	...	...

Command: aidimstyle
Enter option [1/2/3/4/5/6/Other/Save] <1>: *(Specify option, then press* ENTER.*)*
Select objects: *(Select one or more dimensions.)*
Select objects: *(Press* ENTER.*)*

COMMAND LINE OPTIONS

Enter option specifies a predefined dimension style, numbered 1 through 6.

Other applies a named dimension style to selected dimension(s).

Save saves the style of the selected dimension(s).

Select objects selects one or more dimensions.

Other option
Enter option [1/2/3/4/5/6/Other/Save] <1>: o

Displays dialog box after selecting objects:

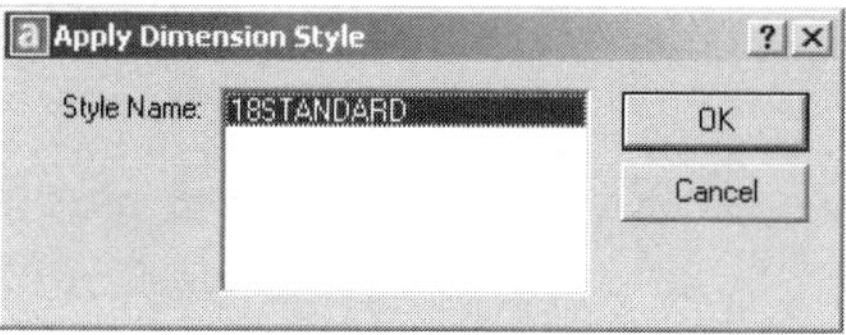

Save option
Enter option [1/2/3/4/5/6/Other/Save] <1>: s

Displays dialog box after selecting exactly one dimension:

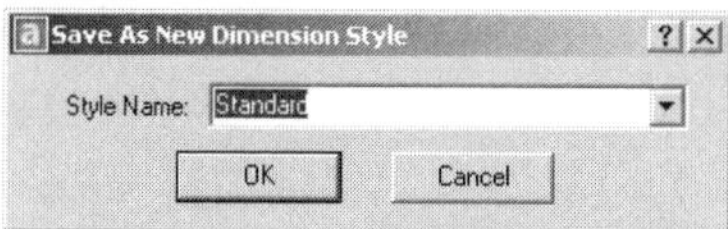

OK saves style to the selected dimension style name.

TIPS

- This command quickly applies and saves dimensions styles. It is used by the right-click shortcut menu: select a dimension, right click, and then select **Dim Style** from the shortcut menu.
- *Caution!* The **Save** option overwrites existing dimstyles; it does not create new style names.

Ai_Dim_TextAbove/Center/Home

Moves dimension text relative to dimension lines (*undocumented commands*).

Command	Alias	Ctrl+	F-key	Alt+	Menu Bar	Tablet
ai_dim_textabove	...	...	...	...	...	...
ai_dim_textcenter						
ai_dim_texthome						

Command: ai_dim_textabove
Select objects: *(Select one or more dimensions.)*
Select objects: *(Press* ENTER.*)*

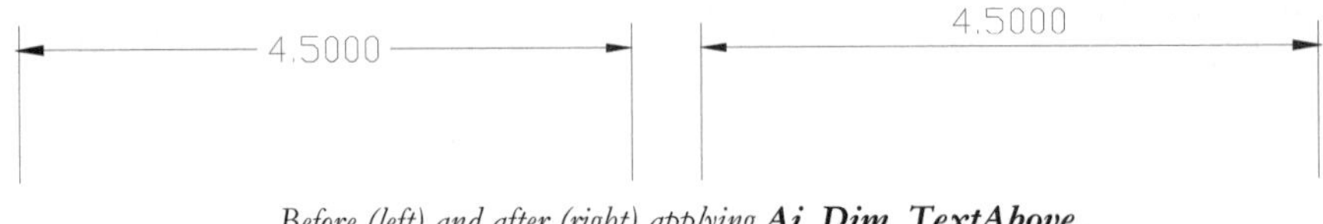

Before (left) and after (right) applying ***Ai_Dim_TextAbove*** .

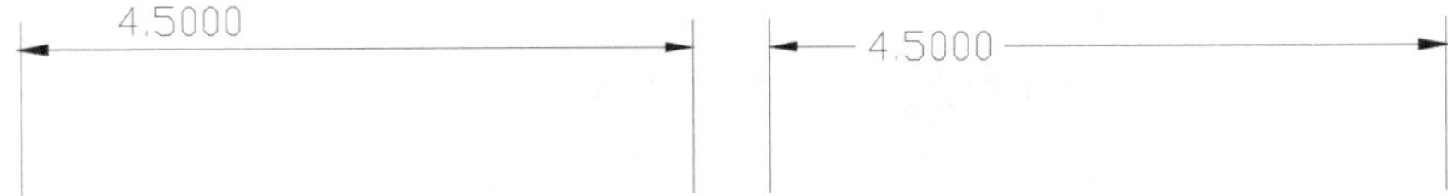

Before (left) and after (right) applying ***Ai_Dim_TextCenter***.

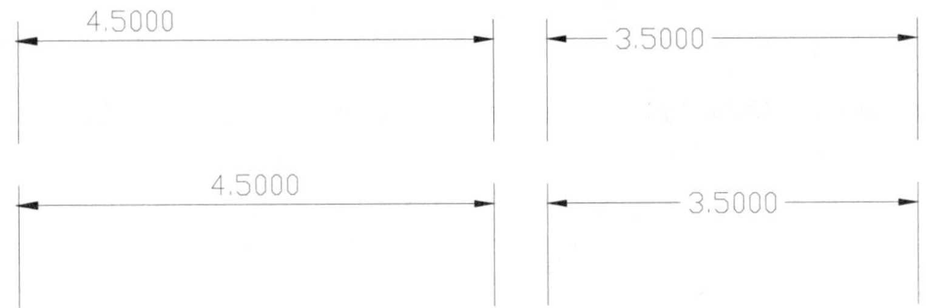

Before (top) and after (bottom) applying ***Ai_Dim_TextHome***.

COMMAND LINE OPTION

Select objects selects one or more dimensions.

TIPS

- Right click a selected dimension, and then select **Dim Text Position** from the shortcut menu:
  Ai_Dim_TextAbove makes dimensions compliant with JIS dimensioning quickly.
  Ai_Dim_TextCenter centers text vertically on the dimension line, but not horizontally.
  Ai_Dim_TextHome centers text horizontally on the dimension line, but not vertically.
- Use the **DimTEdit** command to align text to the left, center, or right on horizontal dimensions.

AiDimTextMove

Moves the location of dimension text (*undocumented command*).

Command	Alias	Ctrl+	F-key	Alt+	Menu Bar	Tablet
aidimtextmove	...	...	...	...	...	...

Command: aidimtextmove
Enter option [0/1/2] <2>: *(Enter an option, and then press* **ENTER.***)*
Select objects: *(Select one dimension.)*
Select objects: *(Press* **ENTER.***)*

Before (left) and after (right) applying ***AiDimTextMove = 0*** *to dimension text.*

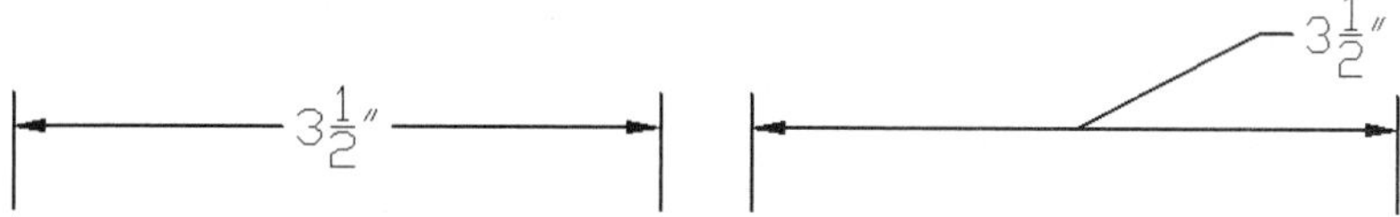

Before (left) and after (right) applying ***AiDimTextMove = 1*** *to dimension text.*

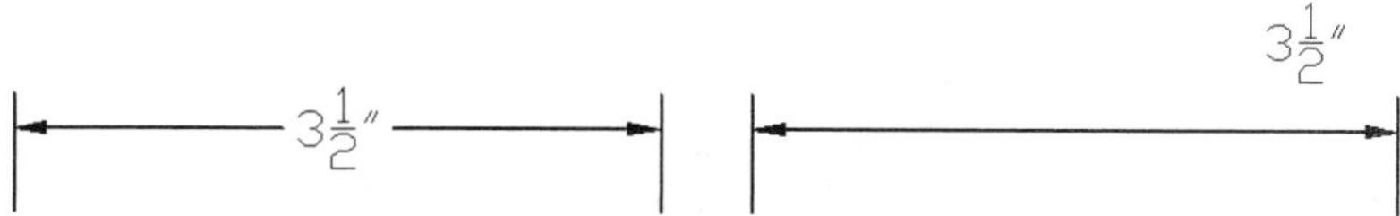

Before (left) and after (right) applying ***AiDimTextMove = 2*** *to dimension text.*

COMMAND LINE OPTIONS

Enter option specifies the style of text movement:

Option	Meaning
0	Text moves with the dimension line.
1	Adds a leader to the moved text.
2	Text moves independent of dimension line and leader (default).

Select objects selects one or more dimensions.

TIPS

- This command allows you retroactively to change the position of dimension text. It is used by the right-click shortcut menu: right click a selected dimension, and then select **Dim Text Position** from the shortcut menu.
- Although the command allows you to select more than one dimension, it operates on the first-selected dimension only.

Ai_Dish / Ai_Dome

Draws the bottom and top halves of 3D spheres as surface models (*undocumented commands*).

Command	Alias	Ctrl+	F-key	Alt+	Menu Bar	Tablet
ai_dish	...	...	...	...	...	...
ai_dome						

Command: ai_dish
Specify center point of dish: *(Pick a point.)*
Specify radius of dish or [Diameter]: *(Pick a point, or type* **D***.)*
Enter number of longitudinal segments for surface of dish <16>: *(Enter a value between 3 and 255, or press* **ENTER***.)*
Enter number of latitudinal segments for surface of dish <8>: *(Enter a value between 3 and 255, or press* **ENTER***.)*

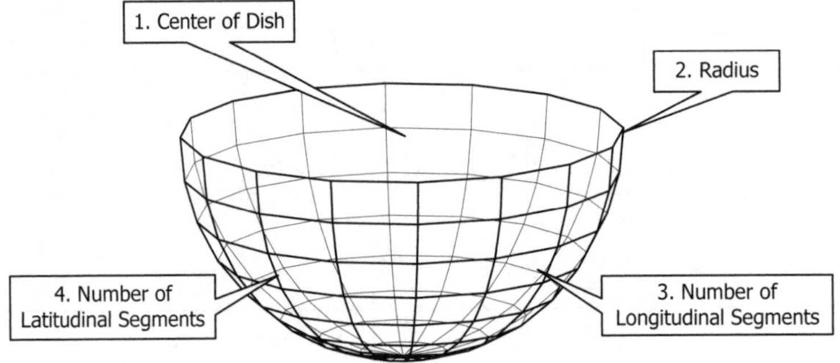

COMMAND LINE OPTIONS

Center of dish specifies the center of the dish's base.

Diameter specifies the diameter of the dish.

Radius specifies the radius of the dish.

Number of longitudinal segments specifies the number of "lines" that define the curved surface in the vertical direction; default = 16.

Number of latitudinal segments specifies the number of "lines" that define the curved surface in the horizontal direction; default = 8.

TIPS

- The **Ai_Dish** command draws the bottom half of a sphere; **Ai_Dome** draws the top half.
- The "base" of the dish and dome is drawn at the current setting of the **Elevation** system variable; the dish is drawn downward (in the negative z-direction), and the dome is drawn upward (in the positive z-direction) — unless a z coordinate is provided.

Ai_Fms

Switches to layout mode, and then to floating model space (*short for Floating Model Space; undocumented command*).

Command	Alias	Ctrl+	F-key	Alt+	Menu Bar	Tablet
ai_fms	...	...	...	...	...	...

Command: ai_fms

Switches to the last active layout, then to the first floating model viewport.

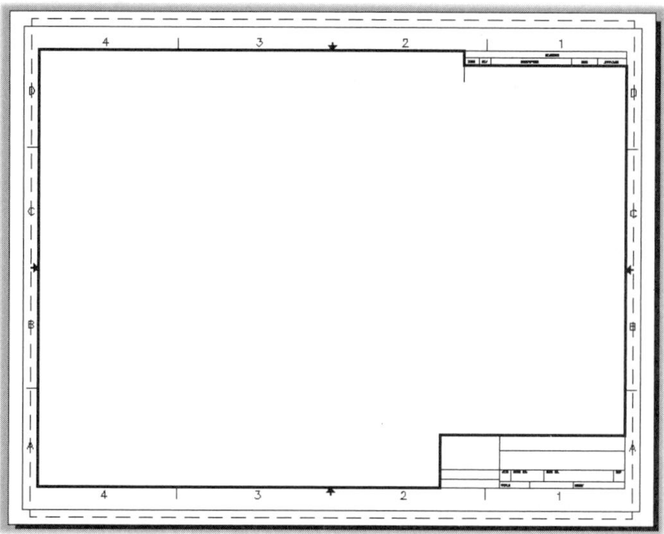

The heavy border indicates the currently-active floating viewport in model space.

COMMAND LINE OPTIONS

None.

TIPS

- This command combines two commands: **Tilemode 0** followed by **MSpace**.
- This command is meant for use with menu and toolbar macros.
- *Warning!* If there are no model space viewports in the last active layout, this command leaves you in the **MView** command with **Undo** Auto Off.

Ai_Mesh

Draws non-planar polyface meshes as surface models (*undocumented command*).

Command	Alias	Ctrl+	F-key	Alt+	Menu Bar	Tablet
ai_mesh	...	...	...	...	...	...

Command: ai_mesh
Specify first corner point of mesh: *(Pick point 1.)*
Specify second corner point of mesh: *(Pick point 2.)*
Specify third corner point of mesh: *(Pick point 3.)*
Specify fourth corner point of mesh: *(Pick point 4.)*
Enter mesh size in the M direction: *(Specify a number.)*
Enter mesh size in the N direction: *(Specify a number.)*

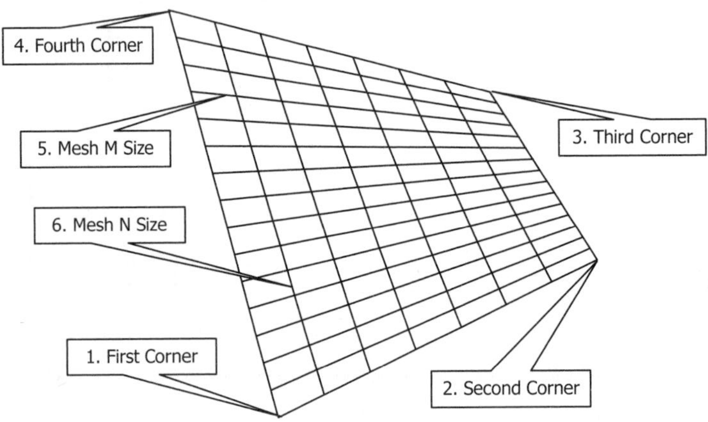

COMMAND LINE OPTIONS

First corner specifies the location of the mesh's first corner.

Second corner specifies the location of the mesh's second corner.

Third corner specifies the location of the mesh's third corner.

Fourth corner specifies the location of the mesh's last corner.

Mesh size M direction specifies the number of horizontal "lines" that define the mesh's surface.

Mesh size N direction specifies the number of vertical "lines" that define the mesh's surface.

TIP

- Use the **.xy** filter first to specify the x,y coordinate, followed by the z coordinate, as follows:

  First corner:.xy
  of *(Pick a point.)*
  need Z: *(Specify z.)*

'Ai_Molc

Changes the current layer to the one on which the selected object is located (*short for Make Object Layer Current; undocumented command*).

Command	Alias	Ctrl+	F-key	Alt+	Menu Bar	Tablet
ai_molc	...	...	...	...	...	...

Command: ai_molc
Select object whose layer will become current: *(Pick an object.)*

COMMAND LINE OPTION

Select object selects a single object.

RELATED COMMANDS

Layer displays the Layer Properties Manager dialog box.

LayerP reverts to the previous layer.

MatchProp matches the properties of one object to other objects.

RELATED SYSTEM VARIABLE

CLayer holds the name of the current layer.

TIPS

- This command is activated by the **Make Object's Layer Current** button on the toolbar.
- Merely clicking an object displays its layer on the Object Properties toolbar without changing the current layer.

Ai_Pyramid

Draws 3D pyramids as surface models (*undocumented command*).

Command	Alias	Ctrl+	F-key	Alt+	Menu Bar	Tablet
ai_pyramid	...	...	...	...	...	...

Command: ai_pyramid
Specify first corner point for base of pyramid: *(Pick point 1.)*
Specify second corner point for base of pyramid: *(Pick point 2.)*
Specify third corner point for base of pyramid: *(Pick point 3.)*
Specify fourth corner point for base of pyramid or [Tetrahedron]: *(Pick point 4, or type* **T***.)*

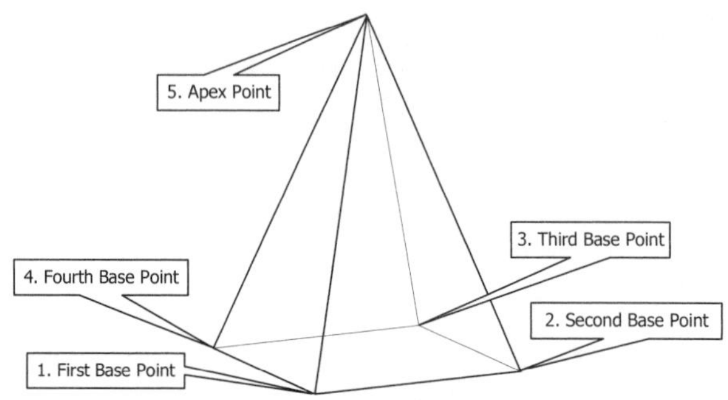

COMMAND LINE OPTIONS

First base point specifies the location of the pyramid's first base point.

Second base point specifies the location of the pyramid's second base point.

Third base point specifies the location of the pyramid's third base point.

Fourth base point specifies the location of the pyramid's last base point.

Tetrahedron draws a pyramid with triangular sides.

Ridge specifies a ridge-top for the pyramid; see figure below.

Top specifies a flat-top for the pyramid; see figure below.

Apex point specifies a point for the pyramid's top.

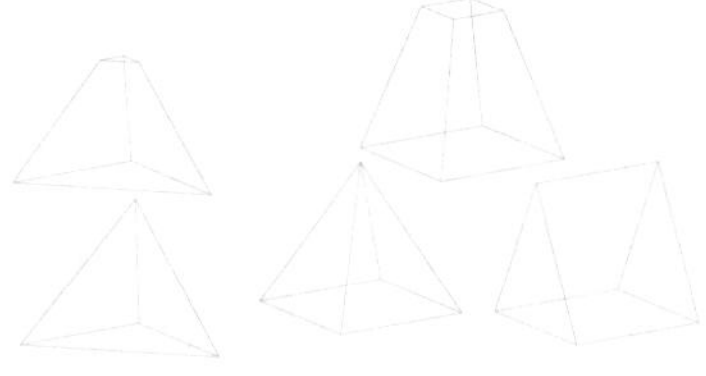

TIPS

- To draw a 2D pyramid, enter no z-coordinate for the **Ridge**, **Top**, and **Apex point** options.
- Use the **.xy** filter to specify the z-coordinate for the **Ridge**, **Top**, and **Apex** point options.

'Ai_SelAll

Selects all objects in drawings (*undocumented command*).

Command	Alias	Ctrl+	F-key	Alt+	Menu Bar	Tablet
'ai_selall	...	a	...	...	...	...

Command: ai_selall
Selecting objects...done.

COMMAND LINE OPTIONS

None.

TIPS

- This command is meant for use in menu macros and toolbars.
- Use the CTRL+A shortcut to select all objects in the drawing, other than those on frozen layers.
- The command opposite to selecting all is **(ai_deselect)**, an AutoLISP routine.

Ai_Sphere

Draws 3D spheres as surface models (*undocumented command*).

Command	Alias	Ctrl+	F-key	Alt+	Menu Bar	Tablet
ai_sphere	...	...	...	...	...	...

Command: ai_sphere
Specify center point of sphere: *(Pick point 1.)*
Specify radius of sphere or [Diameter]: *(Pick point 2, enter a radius, or type* **D***.)*
Enter number of longitudinal segments for surface of sphere <16>: *(Enter a value.)*
Enter number of latitudinal segments for surface of sphere <16>: *(Enter a value.)*

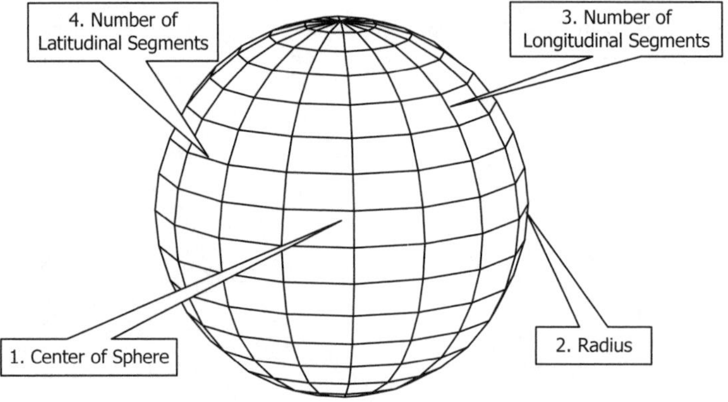

COMMAND LINE OPTIONS

Center of sphere specifies the center point of the sphere.

Diameter specifies the diameter of the sphere.

Radius specifies the radius of the sphere.

Number of longitudinal segments specifies the number of "lines" that define the curved surface in the vertical direction; default = 16.

Number of latitudinal segments specifies the number of "lines" that define the curved surface in the horizontal direction; default = 16.

Ai_Torus

Draws 3D tori as surface models (*undocumented command*).

Command	Alias	Ctrl+	F-key	Alt+	Menu Bar	Tablet
ai_torus	...	...	...	...	...	...

Command: ai_torus
Specify center point of torus: *(Pick point 1.)*
Specify radius of torus or [Diameter]: *(Pick point 2, specify the radius, or type* **D***.)*
Specify radius of tube or [Diameter]: *(Pick point 3, specify the radius, or type* **D***.)*
Enter number of segments around tube circumference <16>: *(Press* ENTER*.)*
Enter number of segments around torus circumference <16>: *(Press* ENTER*.)*

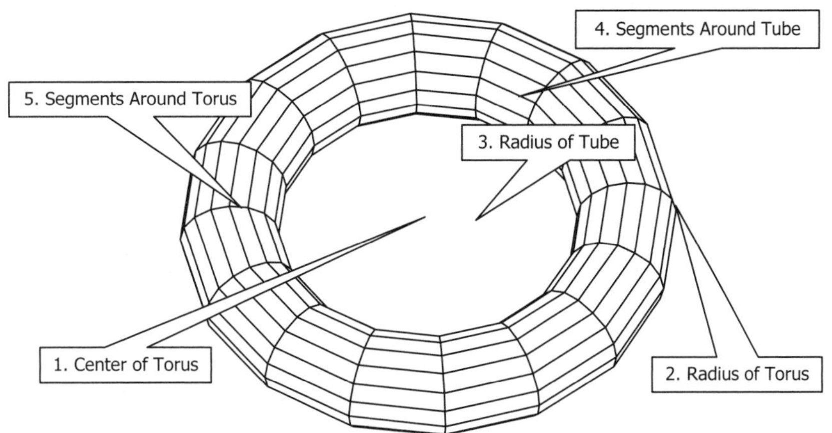

COMMAND LINE OPTIONS

Center of torus specifies the center point of the torus.

Diameter of torus specifies the diameter of the torus as measured across the center of the tube.

Radius of torus specifies the radius of the torus, as measured from the center of the torus to the center of the tube.

Diameter of tube specifies the diameter of the tube cross-section.

Radius of tube specifies the radius of the tube cross-section.

Segments around tube circumference <16> specifies the number of faces defining the curved surface; default = 16, range = 3 to 255.

Segments around torus circumference <16> specifies the number of face defining the curved surface; default = 16, range = 3 to 255.

TIPS

- The **Ai_Torus** command draws donut shapes.
- The tube diameter cannot exceed the torus radius.

Ai_Wedge

Draws 3D wedges as surface models (*undocumented command*).

Command	Alias	Ctrl+	F-key	Alt+	Menu Bar	Tablet
ai_wedge	...	...	...	...	...	...

Command: ai_wedge
Specify corner point of wedge: *(Pick point 1.)*
Specify length of wedge: *(Pick point 2.)*
Specify width of wedge: *(Pick point 3.)*
Specify height of wedge: *(Pick point 4.)*
Specify rotation angle of wedge about the Z axis: *(Enter a rotation angle, or type* **0** *for no rotation.)*

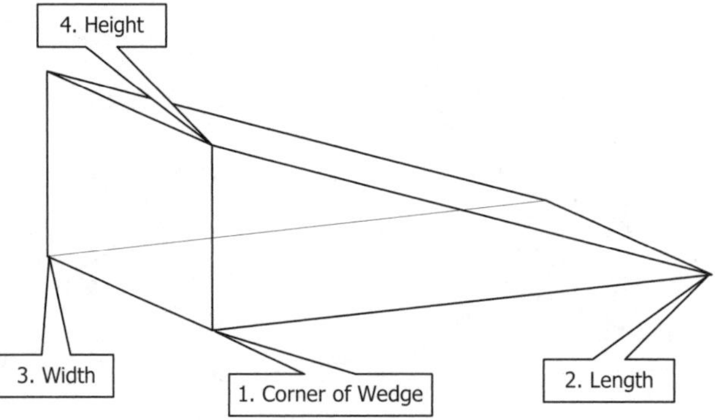

COMMAND LINE OPTIONS

Corner of wedge specifies the corner of the wedge's base.

Length specifies the length of the wedge's base.

Width specifies the width of the wedge's base.

Height specifies the height of the wedge.

Rotation angle specifies the angle the box rotates about the z-axis.

TIPS

- The point you pick for the **Corner of wedge** option determines the taller end of the wedge.
- When specifying the **Width** and **Height** options, move the cursor back to the **Corner of wedge** point; otherwise the wedge may have a different size than you expect.
- You can use the **.xy** point filter to specify the height.

Align

Rel.12 Moves, transforms, and rotates objects in three dimensions.

Command	Alias	Ctrl+	F-key	Alt+	Menu Bar	Tablet
align	**al**	...	...	**M3L**	**Modify** **↳3D Operation** **↳Align**	**X14**

Command: align
Select objects: *(Select one or more objects to be moved.)*
Select objects: *(Press* ENTER*.)*
Specify first source point: *(Pick a point.)*
Specify first destination point: *(Pick a point.)*
Specify second source point: *(Pick a point.)*
Specify second destination point: *(Pick a point, or press* ENTER*.)*
Specify third source point or <continue>: *(Pick a point, or press* ENTER*.)*
Specify third destination point: *(Pick a point.)*

COMMAND LINE OPTIONS

First point moves object in 2D or 3D when one source and destination point are picked.

Second point moves, rotates, and scales object in 2D or 3D when only two source and destination points are picked.

Third point moves and object in 3D when three source and destination points are picked.

Continue option
Scale objects based on alignment points? [Yes/No] <N>: *(Type* **Y** *or* **N***.)*

Specifies that the distance between the first and second soure points are to be scaled to the distance between the first and second destination points.

RELATED COMMANDS

Mirror3d mirrors objects in three dimensions.

Rotate3d rotates objects in three dimensions.

TIPS

- Enter the first pair of points to define the move vector (distance and direction):

  Specify first source point: *(Pick a point.)*
  Specify first destination point: *(Pick a point.)*
  Specify second source point: *(Press* ENTER*.)*

- Enter two pairs of points to define a 2D (or 3D) transformation, scaling, and rotation:

Points	Alignment Defined
First	Base point for alignment.
Second	Rotation angle.
Third	Aligns planes defined by source and destination points.

- The third pair defines the 3D transformation.

AmeConvert

Rel.13 Converts PADL solid models and regions created by AME v2.0 and v2.1 (*AutoCAD Releases 11 and 12*) to ShapeManager solid models.

Command	Alias	Ctrl+	F-key	Alt+	Menu Bar	Tablet
ameconvert	...	...	...	...	...	...

Command: ameconvert
Select objects: *(Select one or more objects.)*
Processing Boolean operations.

COMMAND LINE OPTION

Select objects selects AME objects to convert; ignores non-AME objects, such as the ACIS solids produced by AutoCAD Release 13 through 2002, and ShapeManager solids from AutoCAD 2004 and 2005.

RELATED COMMAND

AcisIn imports ACIS models from an *.sat* file.

TIPS

- After conversion, the AME model remains in the drawing in the same location as the solid model. Erase, if necessary.
- AME holes may become blind holes in the solid model.
- AME fillets and chamfers may be placed higher or lower in the solid model.
- Once the Release 12 PADL drawings is converted to an AutoCAD 2005 solid model, it cannot be converted back to PADL format.
- This command ignores objects that are neither AME solids nor regions.
- Old AME models are stored in AutoCAD as anonymous block references.

DEFINITIONS

ACIS — solids modeling technology used by AutoCAD in Release 13 through 2002.

AME — short for "Advanced Modeling Extension," the solids modeling module used by AutoCAD in Releases 10 through 12.

PADL — short for "Parts and Description Language," the solids modeling technology used by AutoCAD Releases 10 through 12.

ShapeManager — the solids modeling technology used by AutoCAD 2004 and 2005.

'Aperture

V. 1.3 Sets the size (in pixels) of the object snap target height, or box cursor.

Command	Alias	Ctrl+	F-key	Alt+	Menu Bar	Tablet
aperture	...	...	...	...	...	...

Command: aperture
Object snap target height (1-50 pixels) <10>: *(Enter a value.)*

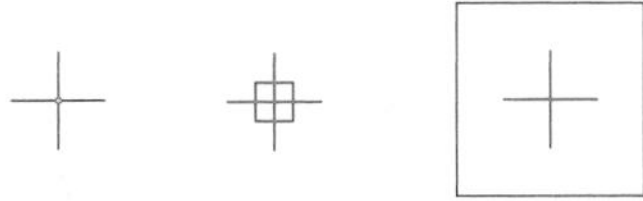

Aperture size = 1 (left), 10 (center), and 50 pixels (right).

COMMAND LINE OPTION

Height specifies the height of the object snap cursor's target.

RELATED COMMAND

Options allows you to set the aperture size interactively (Drafting tab).

RELATED SYSTEM VARIABLE

Aperture contains the current target height, in pixels:

Aperture	Meaning
1	Minimum size.
10	Default size, in pixels.
50	Maximum size.

TIPS

- AutoCAD has two similar-looking cursors: the *osnap* cursor appears only during object snap selection; the *pick* cursor appears anytime AutoCAD expects object selection. The size of both cursors can be changed; to change the size of the *pick* cursor, use the **Pickbox** command.
- By default, the box cursor does not appear. Nevertheless, it determines how close you have to be at an object for AutoCAD to "snap" to it.
- To display the box cursor, use the **Options** command, select the Drafting tab, then enable Display AutoSnap aperture box.
- Use the **Options** command to change the size of the aperture visually: select the Drafting tab, and then move the Aperture Size slider.

AppLoad

Rel.12 Creates a list of LISP, VBA, ObjectARx, and other applications to load into AutoCAD (*short for APPlication LOADer*).

Command	Alias	Ctrl+	F-key	Alt+	Menu Bar	Tablet
appload	**ap**	...	...	**TL**	**Tools** ↳**Load Applications**	**V10**

Command: appload

Displays dialog box:

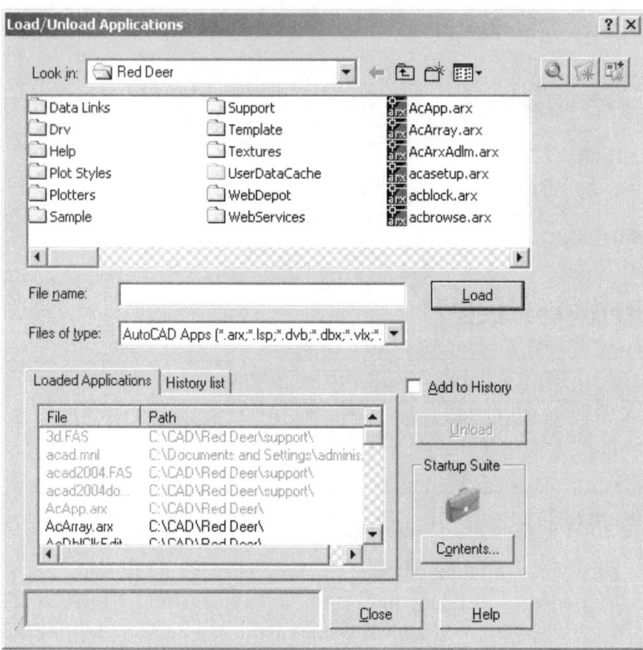

DIALOG BOX OPTIONS

Look in lists the names of drives and folders available to this computer.

File name specifies the name of the file to load.

Files of type displays a list of file types:

Filetype	Meaning
ARX	objectARX.
DVB	Visual Basic for Applications (VBA).
DBX	objectDBX.
FAS	FASt load autolisp.
LSP	autoLiSP.
VLX	Visual Lisp eXecutable.

Load loads all or selected files into AutoCAD.

Loaded Applications displays the names of applications already loaded into AutoCAD.

History List displays the names of applications previously saved to this list.

☐ **Add to History** adds the file to the history tab.

Unload unloads all or selected files out of AutoCAD.

Close exits the dialog box.

Startup Suite options

Contents displays dialog box:

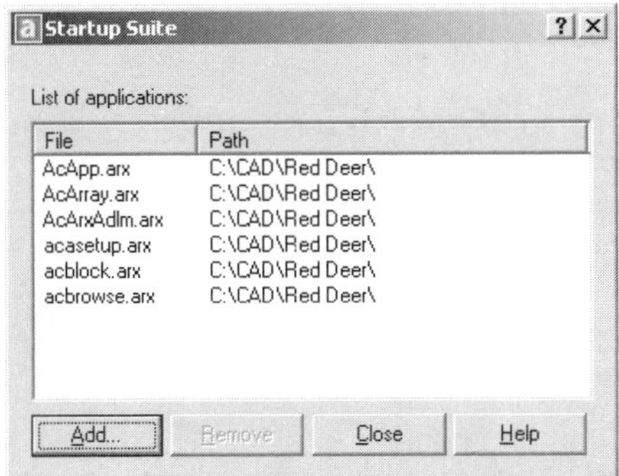

List of applications lists the file names and paths of applications to be automatically loaded each time AutoCAD starts. AutoLISP related files (*.lsp*, *.fas*, and *.vlx*) are loaded whenever a drawing is loaded or a new drawing created. All others are loaded when AutoCAD starts.

Add displays the Add File to Startup Suite dialog box; allows you to select one or more application files.

Remove removes the application from the list.

Close returns to the Load/Unload Applications dialog box.

RELATED COMMANDS

Arx lists ObjectARX programs currently loaded in AutoCAD.

VbaLoad loads VBA applications.

RELATED AUTOLISP FUNCTIONS

(load) loads an AutoLISP program.

(autoload) predefines commands to load related AutoLISP programs.

TIPS

- Use **AppLoad** when AutoCAD does not automatically load a command.
- ObjectARX, VBA, and DBX applications are loaded immediately; FAS, LSP, and VLX files are loaded after this dialog box closes.
- This command was a transparent command in earlier versions of AutoCAD.
- You can drag files from Windows Explorer into the **Loaded Applications** list.
- The *acad2005doc.lsp* file establishes autoloader and other utility functions, and is loaded automatically each time a drawing is opened; the *acad2005.lsp* file is loaded only once per AutoCAD session. Use the **AcadLspAsDoc** system variable to control whether these files are loaded with AutoCAD.

Arc

V. 1.0 Draws 2D arcs of less than 360 degrees, by eleven methods.

Command	Alias	Ctrl+	F-key	Alt+	Menu Bar	Tablet
arc	**a**	...	...	**DA**	**Draw** ↳**Arc**	**R10**

Command: arc

Specify start point of arc or [CEnter]: *(Pick a point, or enter the* **CE** *option.)*

Specify second point of arc or [CEnter/ENd]: *(Pick a point, or enter an option.)*

Specify end point of arc: *(Pick a point.)*

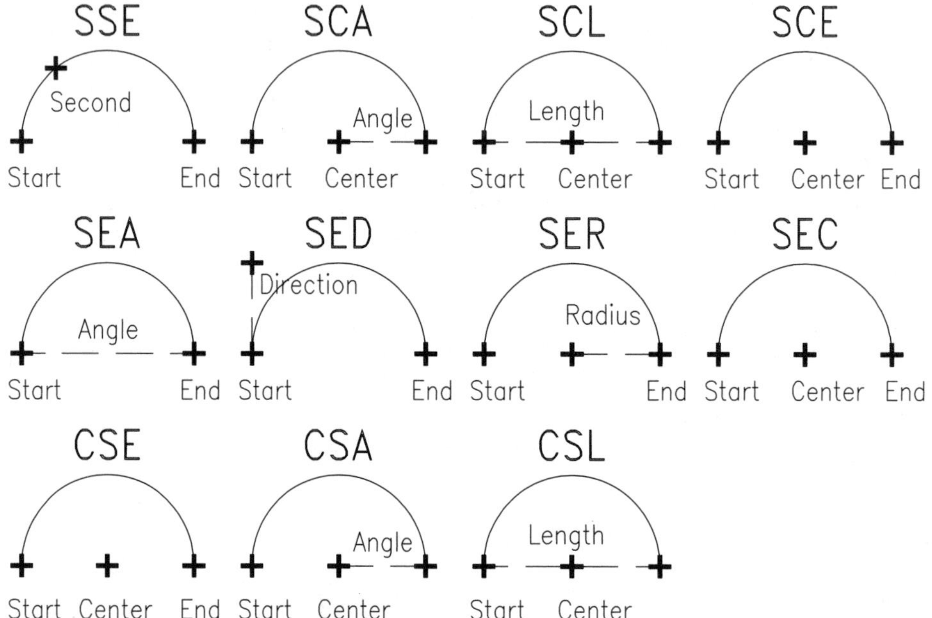

COMMAND LINE OPTIONS

SSE (start, second, end) arc options

Start point indicates the start point of a three-point arc.

Second point indicates a second point anywhere along the arc.

Endpoint indicates the end point of the arc.

SCE (start, center, end), SCA (start, center, arc), and SCL (start, center, length) options

Start point indicates the start point of a two-point arc.

Center indicates the arc's center point.

Angle indicates the arc's included angle.

Length of chord indicates the length of the arc's chord.

Endpoint indicates the arc's end point.

SEA (start, end, angle), SED (start, end, direction), SER (start, end, radius), and SEC (start, end, center) options

Start point indicates the start point of a two-point arc.

End indicates the arc's end point.

Center point indicates the arc's center point.

Angle indicates the arc's included angle.

Direction indicates the tangent direction from the arc's start point.

Radius indicates the arc's radius.

CSE (center, start, end), CSA (center, start, angle), and CSL (center, start, length) options

Center indicates the center point of a two-point arc.

Start point indicates the arc's start point.

Endpoint indicates the arc's end point.

Angle indicates the arc's included angle.

Length of chord indicates the length of the arc's chord.

Continued Arc option

ENTER continues the arc tangent from endpoint of last-drawn line or arc.

RELATED TOOLBAR ICONS

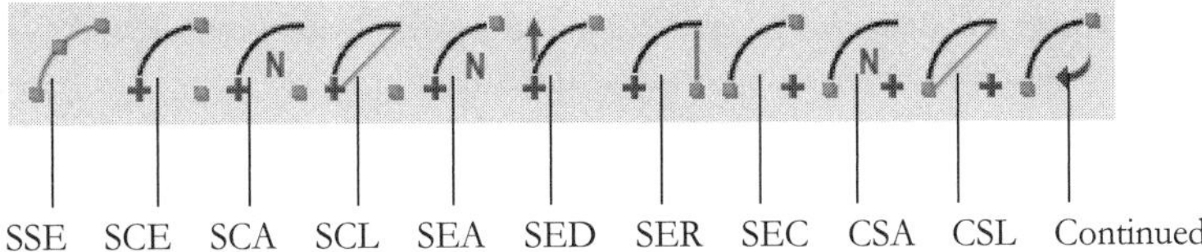

SSE SCE SCA SCL SEA SED SER SEC CSA CSL Continued

RELATED COMMANDS

Circle draws an "arc" of 360 degrees.

Ellipse draws elliptical arcs.

Polyline draws connected polyline arcs.

ViewRes controls the roundness of arcs.

RELATED SYSTEM VARIABLE

LastAngle saves the included angle of the last-drawn arc (read-only).

TIPS

- To start an arc precisely tangent to the end point of the last line or arc, press **ENTER** at the 'Specify start point of arc or [CEnter]:' prompt.
- You can drag the arc only during the last-entered option.
- Specifying an x,y,z-coordinate as the starting point of the arc draws the arc at the z-elevation.
- In most cases, it is easier to draw a circle, and then use the **Trim** command to convert the circle into an arc.
- The components of an AutoCAD arc:

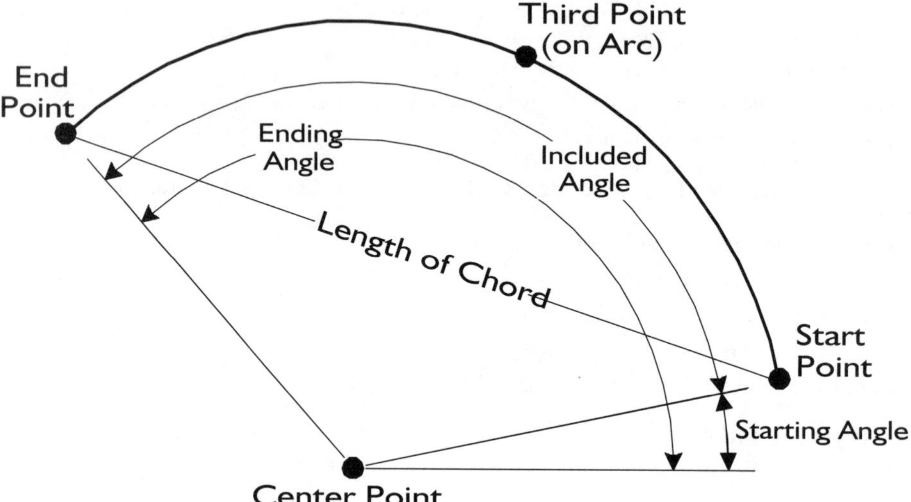

- When the chord length is positive, the minor arc is drawn counter clockwise from the start point; when negative, the major arc is drawn counter clockwise.

Archive

2005 Packages together all files related to the current sheet set.

Command	Alias	Ctrl+	F-key	Alt+	Menu Bar	Tablet
archive	...	...	...	...	...	...
-archive						

Command: archive

When no sheet sets are open, AutoCAD complains, "No Sheet Set is Open," and terminates the command. (To open a sheet set, use the ***OpenSheetset*** *command.)*

When at least one sheet set is open, AutoCAD displays the following dialog box:

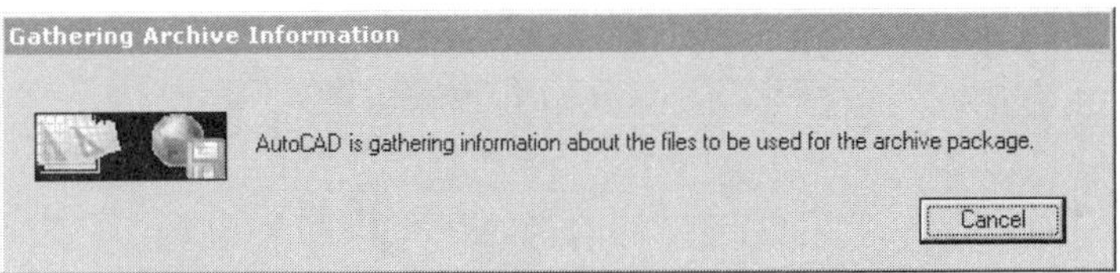

After a moment, AutoCAD displays the next dialog box:

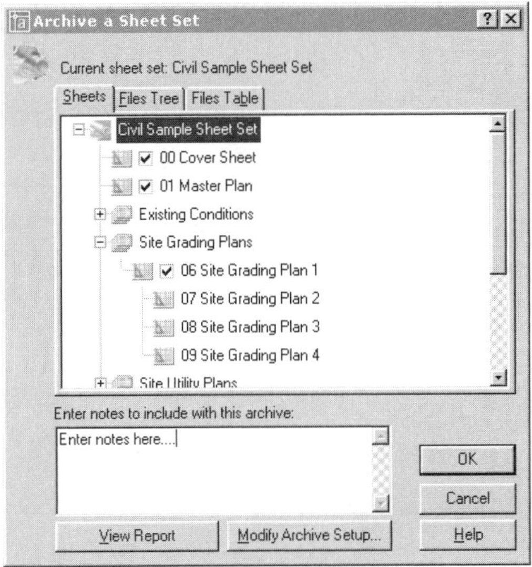

DIALOG BOX OPTIONS

Sheets tab displays sheets included with archive.

Files Tree tab displays names of drawing and support files, grouped by category.

Files Table tab displays file names in alphabetical order.

Enter notes to include with this archive provides space for entering notes.

View Report displays the View Archive Report dialog box.

Modify Archive Setup displays the Modify Archive Setup dialog box.

Files Tree tab

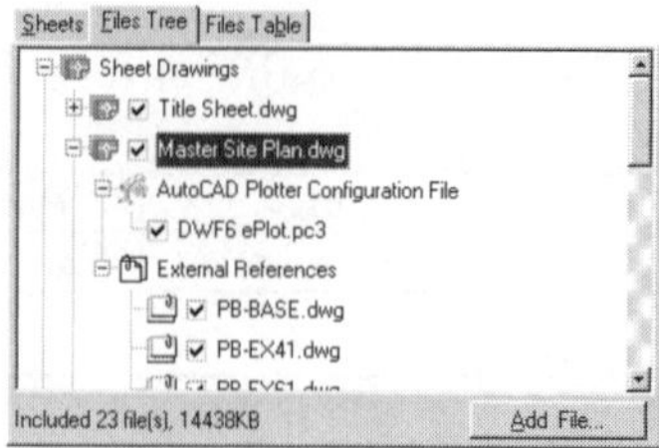

☑ File is included in archive.

☐ File excluded from archive.

Add File displays the Add File to Archive dialog box, which adds files to the archive set.

Files Table tab

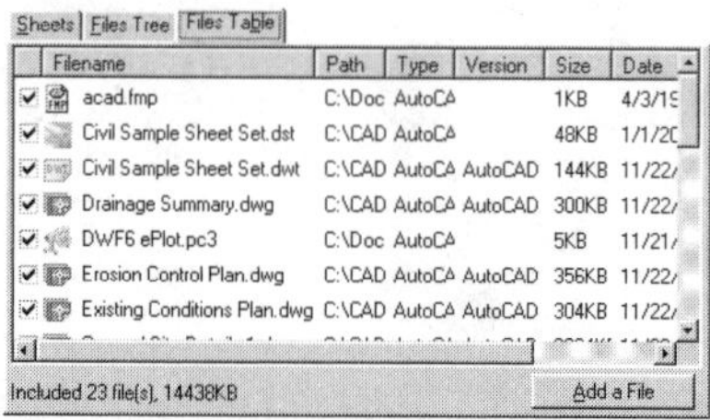

View Report dialog box

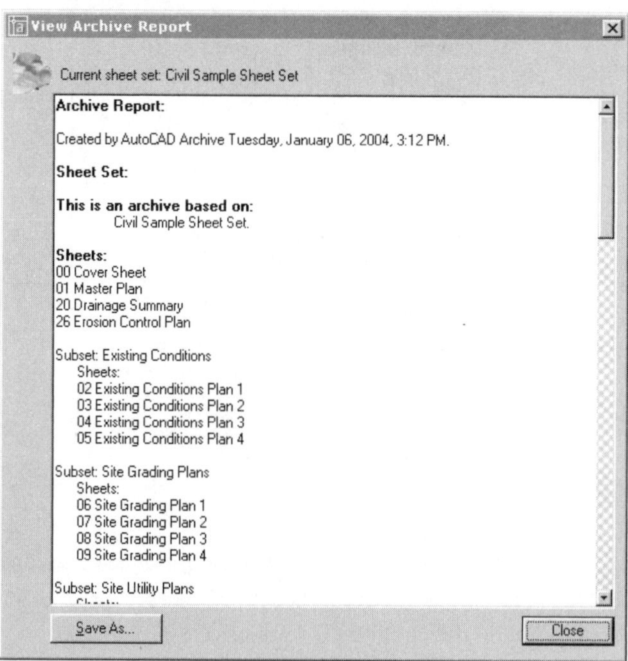

Save As saves the report in *.txt* format (plain ASCII text).

Close closes the dialog box, and then returns to the previous dialog box.

Modify Archive Setup dialog box

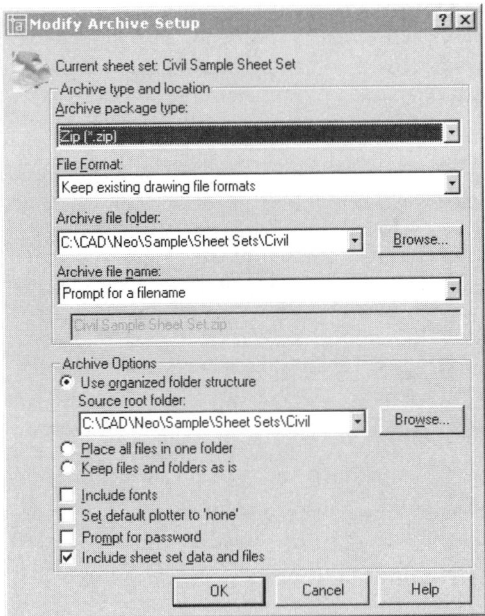

Archive Type and Location options

Archive Package Type options:

Folder (set of files) archives uncompressed files in new and existing folders. This is the best option when archiving to CD or FTP.

Self-Extracting Executable (*.exe) archives files in a compressed, self-extracting file. Uncompress by double-clicking the file.

Zip (*.zip) (*default*) archives files as a compressed ZIP file. Uncompress the file using the PkZip or WinZip programs. The is the best option when sending by email.

File Format options

Keep existing drawing file formats archives files in their native format, including AutoCAD 2005.

AutoCAD 2004/LT 2004 Drawing Format archives files in AutoCAD 2004 format.

AutoCAD 2000/LT 2000 Drawing Format archives files in AutoCAD 2000 format. *Caution*: some objects and properties not found in AutoCAD 2000 may be changed or lost.

Archive File Folder specifies the location in which to archive the files. When no location is specified, the archive is created in the same folder as the *.dst* file.

Archive File Name options

Prompt for a File Name prompts the user for the file name.

Overwrite if Necessary uses the specified file name, and overwrites the existing file of the same name.

Increment File Name if Necessary uses the specified file name, and appends a digit to avoid overwriting the existing file of the same name.

Archive Options

⊙**Use Organized Folder Structure** preserves the folder structure in the archive, but makes the changes listed below; allows you to specify the name of the folder tree. The option is unavailable when saving archives to the Internet. Autodesk reminds you of the following:

Relative paths remain unchanged.

Absolute paths inside the folder tree are converted to relative paths; absolute paths outside the folder tree are converted to "No Path," and are moved to the folder tree.

A *Fonts* folder is created when font files are included in the archive.

A *PlotCFG* folder is created when plotter configuration files are included.

A *SheetSets* folder is created for sheet set support files; the *.dst* sheet set data file is placed in the root folder.

○**Place All Files in One Folder** locates all files in a single folder.

○**Keep Files and Folders As Is** preserves the folder structure in the archive. The option is unavailable when saving archives to the Internet.

☐ **Include Fonts** includes all *.ttf* and *.shx* font files in the archive.

☐ **Set Default Plotter to 'None'** resets the plotter to None for all drawings in the archive.

☐ **Prompt for Password** displays a dialog box for specifying a password for the archive; not available when the Folder archive type is selected.

☑ **Include Sheet Set Data (DST) File** includes the *.dst* sheet set data file with the archive.

-ARCHIVE Command

Command: -archive

Sheet Set name or [?] <Sheet Set>: *(Enter a name or* **?***.)*

Enter an option [Create archive package/Report only] <Create>: *(Type* **c** *or* **r***.)*

Gathering files ...

Sheet Set name specifies the name of the sheet set to archive.

? lists the names of sheet sets open in the current drawing.

Create archive package creates the archive in ZIP format.

Report only displays the Save Report File As dialog box for saving the archive report as a text file; also prompts you for a Transmital Note.

RELATED COMMANDS

eTransmit creates an archive of only the current drawing and its support files.

SheetSet creates and controls sheet sets.

TIPS

- This command works only when at least one sheet set is open in the current drawing.
- Archive files can become large when they hold many drawing files.

Area

V. 1.0 Calculates the area and perimeter of areas, objects, and polylines.

Command	Alias	Ctrl+	F-key	Alt+	Menu Bar	Tablet
area	**aa**	...	...	**TQA**	**Tools** ↳**Inquiry** ↳**Area**	**T7**

Command: area
Specify first corner point or [Object/Add/Subtract]: *(Pick a point, or enter option.)*
Specify next corner point or press ENTER for total: *(Pick a point, or press* ENTER.*)*

Sample response:

Area = 1.8398, Perimeter = 6.5245

COMMAND LINE OPTIONS

First corner point specifies the first point to begin measurement.

Object selects the object to be measured.

Add switches to add-area mode.

Subtract switches to subtract-area mode.

ENTER indicates the end of the area outline.

RELATED COMMANDS

Dist returns the distance between two points.

Id lists the x,y,z coordinates of a selected point.

MassProp returns surface area, and so on, of solid models.

RELATED SYSTEM VARIABLES

Area contains the most recently-calculated area.

Perimeter contains the most recently-calculated perimeter.

TIPS

- At least three points must be picked to calculate an area; AutoCAD "closes the polygon" with a straight line before measuring the area.
- You can specify 2D x,y coordinates or 3D x,y,z coordinates.
- The **Object** option returns the following information:

Object	Measurement Returned
Circle, ellipse	Area and circumference.
Planar closed spline	Area and circumference.
Closed polyline, polygon	Area and perimeter.
Open objects	Area and length.
Region	Net area of all objects in region.
2D solid	Area.

- Areas of wide polylines are measured along center lines; closed polylines must have one closed area only.
- This command does not measure the volume of solid objects; use **MassProp** command.

Array

V. 1.3 Creates 2D linear, rectangular, and polar arrays of objects.

Commands	Aliases	Ctrl+	F-key	Alt+	Menu Bar	Tablet
array	ar	...	...	MA	Modify ↳Array	V18
-array	-ar					

Command: array

Displays dialog box:

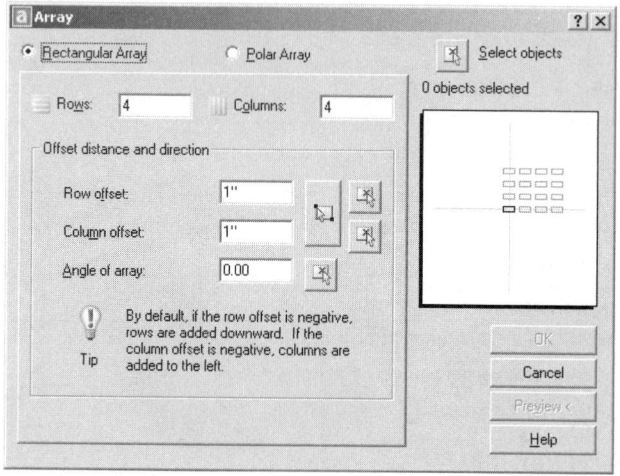

DIALOG BOX OPTIONS

⊙ **Rectangular Array** displays the options for creating a rectangular array.

○ **Polar Array** displays the options for creating a circular array.

Select objects dismisses the dialog box temporarily, so that you can select the objects in the drawing:

Select objects: *(Select one or more objects.)*
Select objects: *(Press* ENTER.*)*
Press Enter, or right-click to return to the dialog box.

Preview dismisses the dialog box temporarily, so that you can see what the array will look like.

Rectangular Array options

Rows specifies the number of rows; minimum=1, maximum=100,000.

Columns specifies the number of columns; minimum=1, maximum=100,000.

Row offset specifies the distance between the centerlines of the rows; use negative numbers to draw rows in the negative x-direction (to the left).

Column offset specifies the distance between the centerlines of the columns; use negative numbers to draw rows in the negative y-direction (downward).

Angle of array specifies the angle of the array, which "tilts" the x and y axes of the array.

Polar Array options

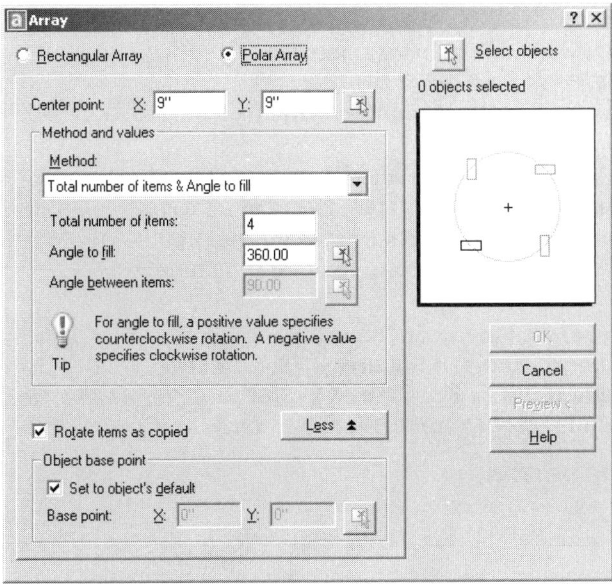

Center point specifies the center of the polar array.

Method specifies the method by which the array is constructed:

- Total number of items and angle to fill.
- Total number of items and angle between items.
- Angle to fill and angle between items.

Total number of items specifies the number of objects in the array; minimum=2.

Angle to fill specifies the angle of "arc" to construct the array; min=1 deg; max=360 deg.

Angle between items specifies the angle between each object in the array.

Rotate items as copied

☑ Objects are rotated so that they face the center of the array.

☐ Objects are not rotated.

More displays additional options for constructing a polar array.

More options

Set to the object's default:

☑ Uses the default base point of the object, as follows:

Object	Default base point
Arc, circle, ellipse	Center point.
Polygon, rectangle	First vertex.
Line, polyline, 3D polyline, ray, spline	Start point.
Donut	Start point.
Block, mtext, text	Insertion point.
Xline	Midpoint.
Region	Grip point.

☐ Allows you to specify the base point. If you are not rotating objects, you should select a base point to avoid unexpected results.

-ARRAY Command

Command: -array
Select objects: *(Select one or more objects.)*
Select objects: *(Press* **ENTER**.*)*
Enter the type of array [Rectangular/Polar] <R>: *(Type* **R** *or* **P**.*)*

Rectangular options
Enter the number of rows (---) <1>: *(Enter a value, or press* **ENTER**.*)*
Enter the number of columns (|||) <1>: *(Enter a value, or press* **ENTER**.*)*
Enter the distance between rows or specify unit cell (---): *(Enter a value.)*
Specify the distance between columns (|||): *(Enter a value.)*

Polar options
Specify center point of array or [Base]: *(Select one or more objects, or enter* **B**.*)*
Enter the number of items in the array: *(Enter a value.)*
Specify the angle to fill (+=ccw, -=cw) <360>: *(Enter a value, or press* **ENTER**.*)*
Rotate arrayed objects? [Yes/No] <Y>: *(Enter* **Y** *or* **N**.*)*

COMMAND LINE OPTIONS

R creates a rectangular array of the selected object.

P creates a polar array of the selected object.

B specifies base point of objects, and the center point of the array.

Center point specifies the center point of the array.

Rows specifies the number of horizontal rows.

Columns specifies the number of vertical columns.

Unit cell specifies the vertical and horizontal spacing between objects.

RELATED COMMANDS

3dArray creates a rectangular or polar array in 3D space.

Copy creates one or more copies of the selected object.

MInsert creates a rectangular block array of blocks.

RELATED SYSTEM VARIABLE

SnapAng determines the default angle of a rectangular array.

TIPS

- To create a rectangular array at an angle, use the **Rotation** option of the **Snap** command.
- Rectangular arrays are drawn upward in the positive x-direction, and to the right right in the positive y-direction; to draw the array in the opposite directions, specify negative row and column distances.
- Polar arrays are drawn counterclockwise; to draw the array clockwise, specify a negative angle.

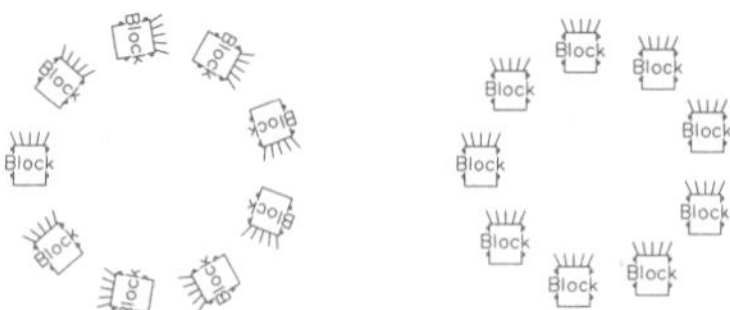

Nine-item polar arrays — rotated (left) and unrotated (right).

- For linear arrays, enter 1 for the number of rows or columns, or use **Divide** or **Measure**.

Arx

Rel.13 Displays information regarding currently loaded ObjectARX programs.

Command	Alias	Ctrl+	F-key	Alt+	Menu Bar	Tablet
arx	...	...	...	...	...	...

Command: arx

Enter an option [?/Load/Unload/Commands/Options]: *(Enter an option.)*

COMMAND LINE OPTIONS

? lists the names of currently loaded ObjectARX programs.

Load loads the ObjectARX program into AutoCAD.

Unload unloads the ObjectARX program out of memory.

Commands lists the names of commands associated with each ObjectARX program.

Options options

CLasses lists the class hierarchy for ObjectARX objects.

Groups lists the names of objects entered into the "system registry."

Services lists the names of services entered in the ObjectARX "service dictionary."

RELATED COMMAND

AppLoad loads LISP, VBA, ObjectDBX, and ObjectARX programs via a dialog box.

RELATED AUTOLISP FUNCTIONS

(arx) lists currently loaded ObjectARX programs.

(arxload) loads an ObjectARX application.

(autoarxload) predefines commands that load the ObjectARX program.

(arxunload) unloads an ObjectARX application.

RELATED FILE

.arx are objectARX program files.

TIPS

- Use the **Load** option to load external commands that do not seem to work.
- Use the **Unload** option of the **Arx** command to remove ObjectARX programs from AutoCAD to free up memory.

Removed Commands

The following ASE (AutoCAD SQL Extension) commands were removed from AutoCAD 2000: **AseAdmin**, **AseExport**, **AseLinks**, **AseRows**, **AseSelect**, and **AseSqlEd**. They were replaced by **DbConnect**.

'Assist / 'AssistClose

2000i Opens and closes the Info Palette window, which provides real-time assistance.

Command	Alias	Ctrl+	F-key	Alt+	Menu Bar	Tablet
assist	...	5	...	HA	Help ↳Info Palette	...
assistclose	...	5	...	...	Tools ↳Info Palette	...

Command: assist

Displays window, which updates as you enter commands (unless Lock is turned on):

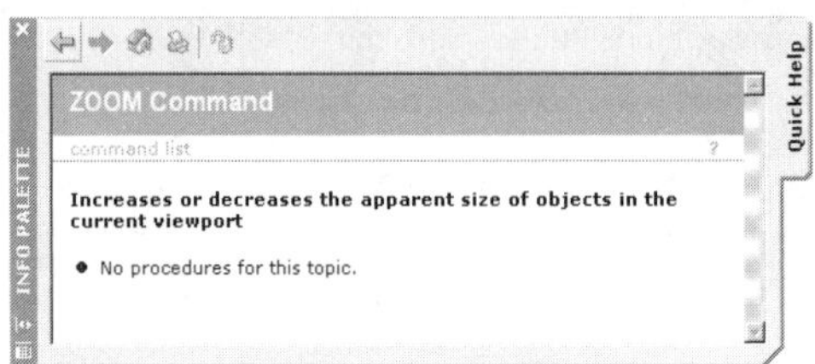

Command: assistclose

Closes window.

TOOLBAR OPTIONS

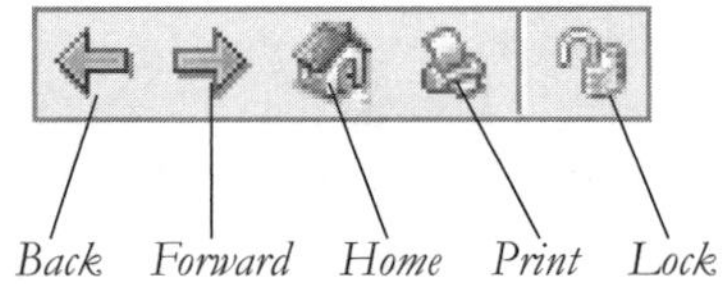

Back Forward Home Print Lock

SHORTCUT MENU OPTIONS

Home displays the "home" page.

Back displays the previous topic.

Forward displays the next topic, if the Back option has been used.

Print prints the topic.

Lock prevents contents from updating when a different command is started.

RELATED COMMAND

Help displays online help in a window.

RELATED SYSTEM VARIABLE

AssistState reports whether the Info Palette is active.

TIPS

- This menu provides real-time help for commands, dialog boxes, and system variables. It received a new user interface and a new name ("Info Palette") with AutoCAD 2005.
- For better performance, turn off the Info Palette with the **AssistClose** command during scripts, AutoLISP routines, and VBA macros.

AttachURL

Rel.14 Attaches hyperlinks to objects and areas.

Command	Alias	Ctrl+	F-key	Alt+	Menu Bar	Tablet
attachurl	...	...	...	...	...	...

Command: attachurl
Enter hyperlink insert option [Area/Object] <Object>: *(Type* **A** *or* **O***.)*
Select objects: *(Select one or more objects.)*
Select objects: *(Press* ENTER*.)*
Enter hyperlink <current drawing>: *(Enter an address.)*

COMMAND LINE OPTIONS

Area creates rectangular hyperlinks by specifying two corners of a rectangle.

Object attaches hyperlinks after you selecting one or more objects.

Enter hyperlink requires you to enter a valid hyperlink.

Area options

First corner picks the first corner of rectangle.

Other corner picks the second corner of rectangle.

RELATED COMMANDS

Hyperlink displays a dialog box for adding a hyperlink to an object.

SelectUrl selects all objects with attached hyperlinks.

TIPS

- The hyperlinks placed in the drawing can link to *any* other file: another AutoCAD drawing, an office document, or a file located on the Internet.
- Autodesk recommends that you use the following URL (uniform resource locator) formats:

File Location	Example URL
Web Site	**http://***servername***/***pathname***/***filename***.dwg**
FTP Site	**ftp://***servername***/***pathname***/***filename***.dwg**
Local File	**file:///***drive***:/***pathname***/***filename***.dwg**
or	**file:////***localPC***/***pathname***/***filename***.dwg**
Network File	**file://***localhost***/***drive***:/***pathname***/***filename***.dwg**

- The URL (hyperlink) is stored as follows:

Attachment	URL
One object	Stored as xdata (extended entity data).
Multiple objects	Stored as xdata in each object.
Area	Stored as xdata in a rectangular object on layer URLLAYER.

- The **Area** option creates a layer named URLLAYER with the default color of red, and places a rectangle object on this layer; do not delete the layer.

AttDef

V. 2.0 Defines attribute modes and prompts (*short for ATTribute DEFinition*).

Commands	Aliases	Ctrl+	F-key	Alt+	Menu Bar	Tablet
attdef	**att**	...	...	**DKD**	**Draw**	...
	ddattdef				↳**Block**	
					↳**Define Attributes**	
-attdef	**-att**					

Command: attdef

Displays dialog box:

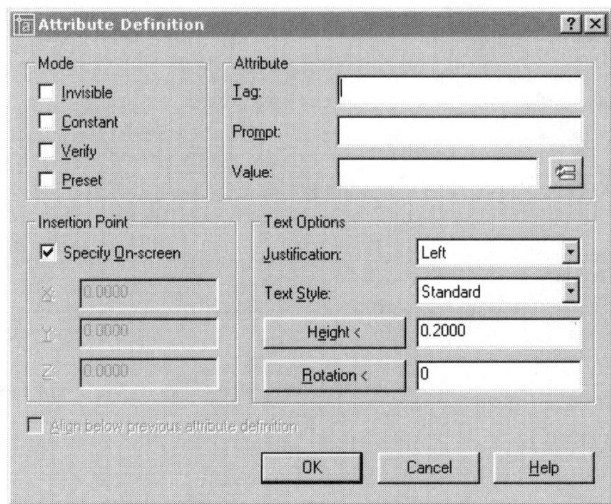

After clicking OK, AutoCAD prompts:

Specify start point: *(Pick a point to locate the attribute text.)*

DIALOG BOX OPTIONS

Mode options

☐ **Invisible** makes the attribute text invisible.

☐ **Constant** uses constant values for the attributes.

☐ **Verify** verifies the text after input.

☐ **Preset** presets the variable attribute text.

Attribute options

Tag identifies the attribute.

Prompt prompts the user for input.

Value sets the default value for the attribute.

Insert Field displays the Field dialog box; select a field, and then click **OK**. See **Field** command (*new to AutoCAD 2005*).

Insertion Point options

Pick point picks insertion point with cursor.

X specifies the x coordinate insertion point.

Y specifies the y coordinate insertion point.

Z specifies the z coordinate insertion point.

Text options

Justification sets the text justification.

Text style selects a text style.

Height specifies the height.

Rotation sets the rotation angle.

Additional option

Align below previous attribute definition places the text automatically below the previous attribute.

-ATTDEF Command

Command: -attdef

Current attribute modes: Invisible=N Constant=N Verify=N Preset=N

Enter an option to change [Invisible/Constant/Verify/Preset] <done>: *(Enter an option.)*

Enter attribute tag: *(Enter text, and then press* ENTER.*)*

Enter attribute prompt: *(Enter text, and then press* ENTER.*)*

Enter default attribute value: *(Enter text, and then press* ENTER.*)*

Specify start point of text or [Justify/Style]: *(Pick a point, or enter an option.)*

Specify height <0.200>: *(Enter a value.)*

Specify rotation angle of text <0>: *(Enter a value.)*

COMMAND LINE OPTIONS

Attribute mode selects the mode(s) for the attribute:

- **I** toggles visibility of attribute text in drawing (short for Invisible).
- **C** toggles fixed or variable value of attribute (short for Constant).
- **V** toggles confirmation prompt during input (short for Verify).
- **P** toggles automatic insertion of default values (short for Preset).

Start point indicates the start point of the attribute text.

Justify selects the justification mode for the attribute text.

Style selects the text style for the attribute text.

Height specifies the height of the attribute text; not displayed if the style specifies a height other than 0.

Rotation angle specifies the angle of the attribute text.

RELATED COMMANDS

AttDisp controls the visibility of attributes.

EAttEdit edits the values of attributes.

EAttExt extracts attributes to disk.

AttRedef redefines an attribute or block.

Block creates blocks with attributes.

Insert inserts a block and prompts for attribute values.

RELATED SYSTEM VARIABLES

AFlags holds the default value of modes in bit form:

AFlags	Meaning
0	No attribute mode selected.
1	Invisible.
2	Constant.
4	Verify.
8	Preset.

TIPS

- Constant attributes cannot be edited.
- Attribute tags cannot be null (have no value); attribute values may be null.
- You can enter any characters for the attribute tag, except a space or an exclamation mark. All characters are converted to uppercase.
- When you press ENTER at 'Attribute Prompt,' AutoCAD uses the attribute *tag* as the prompt.
- When you press ENTER at the 'Starting point:' prompt, **AttDef** automatically places the next attribute below the previous one.

*Block with attribute **value** (left) and attribute **tags** (right).*

- 'Attribute Prompt' and 'Default Attribute Value' are not displayed when constant mode is turned on. Instead, AutoCAD prompts 'Attribute Value.'

'AttDisp

V. 2.0 Controls the display of all attributes in the drawing (*short for ATTribute DISPlay*).

Command	Alias	Ctrl+	F-key	Alt+	Menu Bar	Tablet
attdisp	...	...	...	**VLA**	**View**	**L1**
					↳Display	
					↳Attribute Display	

Command: attdisp
Enter attribute visibility setting [Normal/ON/OFF] <Normal>: *(Enter an option.)*
Regenerating drawing.

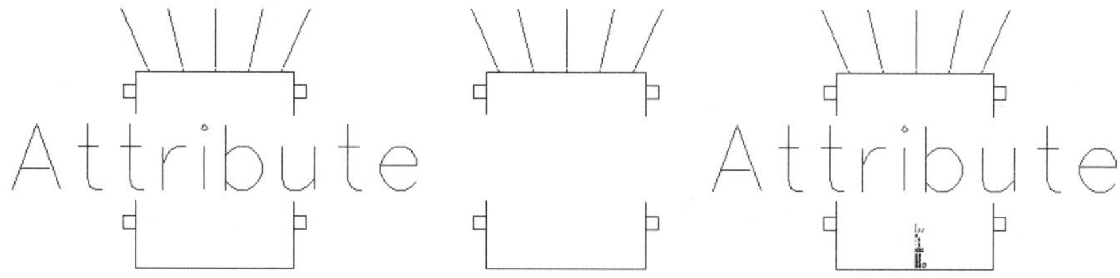

Attribute display **Normal** *(left),* **Off** *(center), and* **On** *(right).*

COMMAND LINE OPTIONS

Normal displays attributes according to **AttDef** setting.

ON displays all attributes, regardless of **AttDef** setting.

OFF displays no attributes, regardless of **AttDef** setting.

RELATED COMMAND

AttDef defines new attributes, including their default visibility.

RELATED SYSTEM VARIABLE

AttMode holds the current setting of **AttDisp**:

AttMode	Meaning
0	Off: no attributes are displayed.
1	Normal: invisible attributes are not displayed.
2	On: all attributes are displayed.

TIPS

- When **RegenAuto** is off, use **Regen** after **AttDisp** to see a change to attribute display.
- When you define invisible attributes, use **AttDisp** to view them.
- Use **AttDisp** to turn off the display of attributes, which increases display speed and reduces drawing clutter.

AttEdit

V. 2.0 Edits attributes in drawings (*short for ATTribute EDIT*).

Commands	Aliases	Ctrl+	F-key	Alt+	Menu Bar	Tablet
attedit	**ate** **atte** **ddatte**	...	...	**MOAS**	**Modify** ↳**Object** ↳**Attribute** ↳**Single**	**Y20**
-attedit	**-ate**			**MOAG**	**Modify** ↳**Object** ↳**Attribute** ↳**Global**	

Command: attedit
Select block reference: *(Pick a block.)*

Displays dialog box:

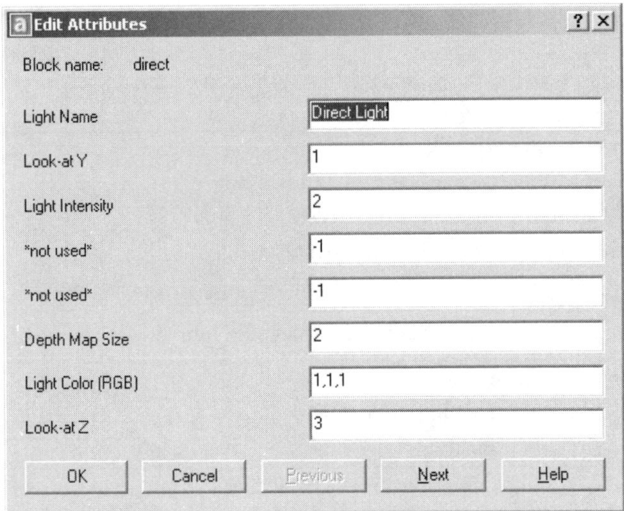

DIALOG BOX OPTIONS

Block Name names the selected block.

Attribute-specific prompts allow you to change attribute values.

Buttons

OK accepts the changes and closes the dialog box.

Cancel discards the changes and closes the dialog box.

Previous displays the previous list of attributes, if any.

Next displays the next list of attributes, if any.

-ATTEDIT Command

Command: -attedit

One-at-time attribute editing options

Edit attributes one at a time? [Yes/No] <Y>: *(Enter* **Y***.)*

Enter block name specification <*>: *(Press* **ENTER** *to edit all.)*

Enter attribute tag specification <*>: *(Press* **ENTER** *to edit all.)*

Enter attribute value specification <*>: *(Press* **ENTER** *to edit all.)*

Select Attributes: *(Select one or more attributes.)*

Select Attributes: *(Press* **ENTER***.)*

Enter an option [Value/Position/Height/Angle/Style/Layer/Color/Next] <N>: *(Enter an option, and then press* **ENTER***.)*

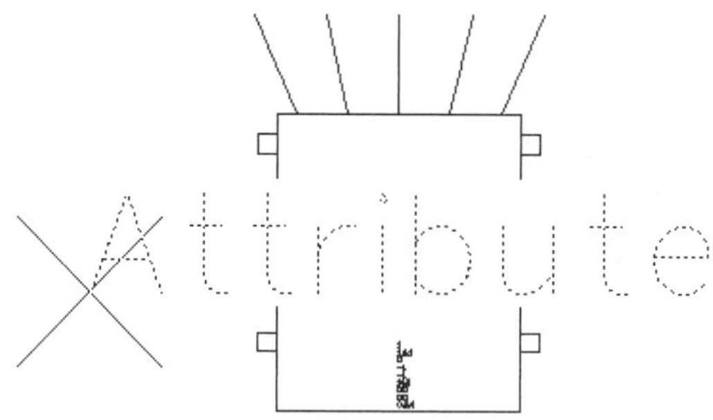

During single attribute editing, ***AttEdit*** *marks the current attribute with an 'X.'*

Global attribute editing options

Edit attributes one at a time? [Yes/No] <Y>: *(Enter* **N***.)*

Performing global editing of attribute values.

Edit only attributes visible on screen? [Yes/No] <Y>: *(Press* **ENTER***.)*

Enter block name specification <*>: *(Press* **ENTER***.)*

Enter attribute tag specification <*>: *(Press* **ENTER***.)*

Enter attribute value specification <*>: *(Press* **ENTER***.)*

Select Attributes: *(Select one or more attributes.)*

Select Attributes: *(Press* **ENTER***.)*

Enter string to change: *(Enter existing string, and then press* **ENTER***.)*

Enter new string: *(Enter new string, and then press* **ENTER***.)*

COMMAND LINE OPTIONS

Value changes or replaces the value of the attribute.

Position moves the text insertion point of the attribute.

Height changes the attribute text height.

Angle changes the attribute text angle.

Style changes the text style of the attribute text.

Layer moves the attribute to a different layer.

Color changes the color of the attribute text.

Next edits the next attribute.

RELATED SYSTEM VARIABLE

AttDia toggles use of **AttEdit** during the **Insert** command.

RELATED COMMANDS

AttDef defines an attribute's original value and parameter.

AttDisp toggles an attribute's visibility.

AttRedef redefines attributes and blocks.

EAttEdit edits all aspects of attributes.

Explode reduces an attribute to its tag.

TIPS

- Constant attributes cannot be edited with **AttEdit**.
- The DdEdit command also displays this dialog box (*new to AutoCAD 2005*).
- You can only edit attributes parallel to the current UCS.
- Unlike other text input to AutoCAD, attribute values are case-sensitive.
- To edit null attribute values, use **-AttEdit**'s global edit option, and enter **** (backslash) at the 'Enter attribute value specification' prompt.
- The wildcard characters **?** and ***** are interpreted literally at the 'Enter string to change' and 'Enter new string' prompts.
- To edit the different parts of an attribute, use the following commands:

Command	Edit Attribute
Attedit	Selects non-constant attribute *values* in one block.
-AttEdit	Selects attribute *values* and *properties* (such as position, height, and style) in one block or in all attributes.

- When selecting attributes for global editing, you may pick the attributes, or use the following selection modes: **Window**, **Last**, **Crossing**, **BOX**, **Fence**, **WPolygon**, and **CPolygon**.
- **AttEdit** does not trim leading and trailing spaces from attribute values. Be sure to avoid entering them to reduce unexpected results.
- You may use wildcards in the block name, tag, and value specifications:

#	matches any single numeric character.
@	matches any single alphabetic character.
.	matches any single non-alphabetic character.
*	matches any string? matches any single character.
~	matches anything but the following pattern.
[ ]	matches any single character enclosed.
[~]	matches any single character not enclosed.
[-]	matches any single character in the enclosed range.
'	treats the next character as a non-wild-card character.

AttExt

V. 2.0 Extracts attribute data from drawings to files on disk (*short for ATTribute EXTract*).

Commands	Alias	Ctrl+	F-key	Alt+	Menu Bar	Tablet
attext	**ddattext**	...	...	**FE**	**File**	...
				↳DXX	**↳Export**	
					↳DXX Extract	
-attext						

Command: attext

Displays dialog box:

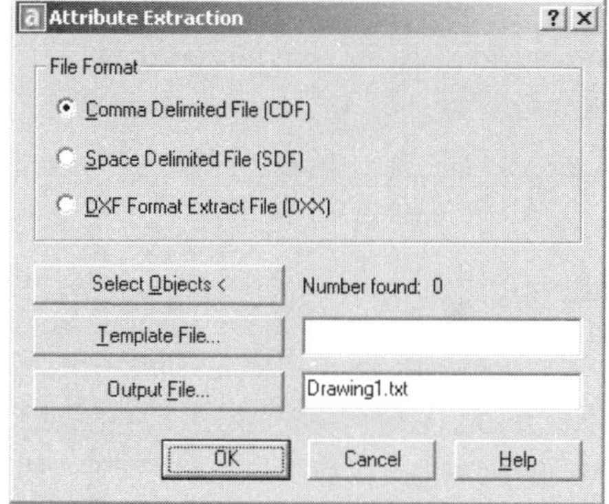

DIALOG BOX OPTIONS

File Format options

⦿**Comma Delimited File (CDF)** creates a CDF text file, where commas separate fields.

○**Space Delimited File (SDF)** creates an SDF text file, where spaces separate fields.

○**DXF Format Extract File (DXX)** creates an ASCII DXF-format file.

Additional options

Select Objects returns to the graphics screen to select attributes for export.

Template File specifies the name of the TXT template file for CDF and SDF files.

Output File specifies the name of the attribute output file, *.txt* for CDF and SDF formats, or *.dxx* for DXF format.

-ATTEXT Command

Command: -attext

Enter extraction type or enable object selection [Cdf/Sdf/Dxf/Objects] <C>: *(Enter an option.)*

Displays the ***Select Template File*** *dialog box; select the template file.*

Displays the ***Create Extract File*** *dialog box.*

COMMAND LINE OPTIONS

Cdf outputs attributes in comma-delimited format.

Sdf outputs attributes in space-delimited format.

Dxf outputs attributes in DXF format.

Objects selects objects from which to extract attributes.

RELATED COMMANDS

AttDef defines attributes.

EAttExt provides a smoother interface for extracting attributes.

RELATED FILES

.txt required extension for template file; extension for CDF and SDF files.

.dxx extension for DXF extraction files.

TIPS

- **CDF** is short for "Comma Delimited File"; it has one record for each block reference; a comma separates each field; single quotation marks delimit text strings.
- **SDF** is short for "Space Delimited File"; it has one record for each block reference; fields have fixed width padded with spaces; string delimiters are not used.
- **DXF** is short for "Drawing Interchange File"; it contains only block reference, attribute, and end-of-sequence DXF objects; no template file is required.
- CDF files use the following conventions:

  Specified field widths are the maximum width.
  Positive number fields have a leading blank.
  Character fields are enclosed in **' '** (single quotation marks).
  Trailing blanks are deleted.
  Null strings are **''** (two single quotation marks).
  Uses spaces; do not uses tabs.
  Use the C:DELIM and C:QUOTE records to change the field and string delimiters to another character.

- To output the attributes to the printer, specify:

Logical Filename	Meaning
CON	Displays on text screen.
PRN *or* **LPT1**	Prints to parallel port 1.
LPT2 *or* **LPT3**	Prints to parallel ports 2 or 3.

- Before you can specify the SDF or CDF option, you must create a template file.

'AttRedef

Rel.13 Redefines blocks and attributes (*short for ATTribute REDEFinition*).

Command	Alias	Ctrl+	F-key	Alt+	Menu Bar	Tablet
attredef	...	...	...	...	...	...

Command: redefine
Name of Block you wish to redefine: *(Enter name of block.)*
Select objects for new Block...
Select objects: *(Select one or more objects.)*
Select objects: *(Press ENTER.)*
Insertion base point of new block: *(Pick a point.)*

COMMAND LINE OPTIONS

Name of Block you wish to redefine specifies the name of the block to be redefined.

Select objects selects objects for the new block.

Insertion base point of new block picks the new insertion point.

RELATED COMMANDS

AttDef defines an attribute's original value and parameter.

AttDisp toggles an attribute's visibility.

EAttEdit edits the attribute's values.

Explode reduces an attribute to its tag.

TIPS

- Existing attributes retain their values.
- Existing attributes not included in the new block are erased.
- New attributes added to an existing block take on default values.

AttSync

2002 Updates blocks with new attribute definitions (*short for ATTribute SYNChronization*).

Command	Alias	Ctrl+	F-key	Alt+	Menu Bar	Tablet
attsync	...	...	...	...	...	...

Command: attsync
Enter an option [?/Name/Select] <Select>: *(Specify an option.)*
Select a block: *(Pick a block.)*
ATTSYNC block name? [Yes/No] <Yes>: *(Enter* **Y** *or* **N***.)*

COMMAND LINE OPTIONS

? lists the names of all blocks in the drawing.

Name enters the name of the block.

Select selects a single block with the cursor.

RELATED COMMANDS

AttDef defines an attribute.

BattMan edits the attributes in a block definition.

EAttEdit edits the attributes in block references.

TIPS

- This command is used together with other attributed-related commands, in the following order:
  1. **AttDef** and **Block** define attributes, and attach them to blocks.
  2. **Insert** inserts the block, and gives values to the attributes.
  3. **EAttEdit** changes the attributes in selected blocks, adding, deleting, or modifying the attribute definitions.
  4. **BAttMan** changes the attributes in the orignal blocks.
  5. **AttSync** updates the attributes to the new definition (**BAttMan** also performs this task).
- This command does not operate if the drawing lacks blocks with attributes. AutoCAD complains, "This drawing contains no attributed blocks."

Audit

Rel.11 Examines drawing files for structural errors.

Command	Alias	Ctrl+	F-key	Alt+	Menu Bar	Tablet
audit	...	...	...	**FUA**	**File** ⇘**Drawing Utilities** ⇘**Audit**	...

Command: audit

Fix any errors detected? [Yes/No] <N>: *(Type* **Y** *or* **N**.*)*

Sample output

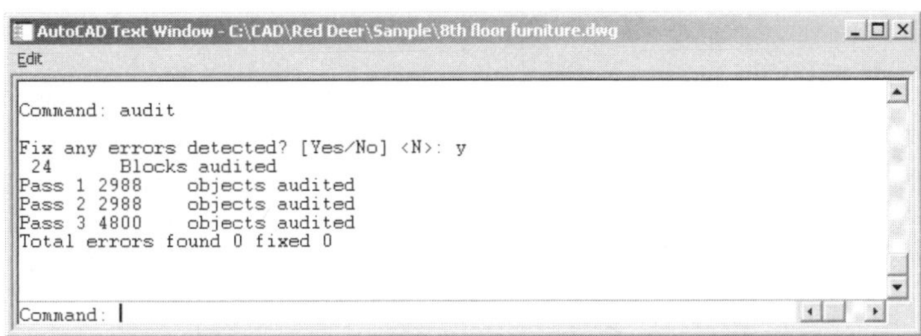

COMMAND LINE OPTIONS

N reports errors found; does not fix errors.

Y reports and fixes errors found in the drawing file.

RELATED COMMANDS

Save saves recovered drawings to disk.

Recover recovers damaged drawing files.

RELATED SYSTEM VARIABLE

AuditCtl create *.adt* audit log files when set to 1.

RELATED FILE

***.adt** is the audit log file, which records the auditing process.

TIPS

- The **Audit** command is a diagnostic tool for validating and repairing the contents of *.dwg* files.
- Objects with errors are placed in the **Previous** selection set. Use an editing command, such as **Copy**, to view the objects.
- If **Audit** cannot fix a drawing file, use the **Recover** command.

Background

Rel.14 Places a solid color, linear gradient, raster image, or the current view in the background of renderings, rendered viewports, and rendered 3dOrbit views.

Command	Alias	Ctrl+	F-key	Alt+	Menu Bar	Tablet
background	...	...	...	**VEB**	**View** ↳**Render** ↳**Background**	**Q2**

Command: background

Displays dialog box:

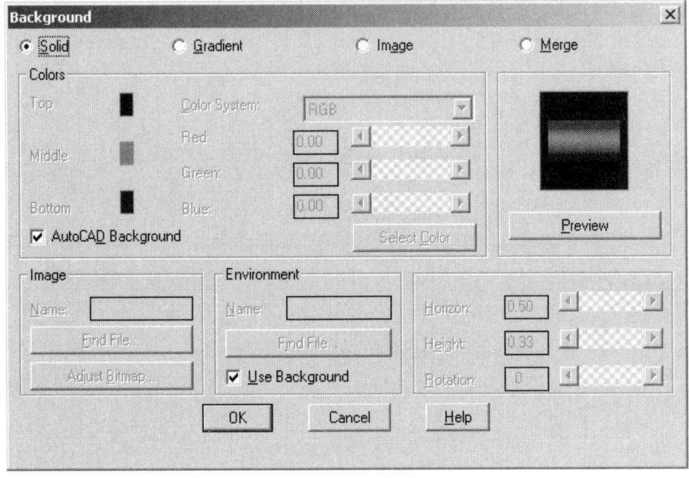

DIALOG BOX OPTIONS

Environment options

Name specifies the name of a raster file, which allows reflection and refraction effects on objects: mirror effect with the Photo Real renderer; or raytracing with the Photo Raytrace renderer.

Find File displays the Raytraced Environment Image dialog box; select an image file, and then click **Open**.

☑ **Use Background** specifies that objects reflect the background, whether a color, gradient, image, or merged image.

Solid view

Colors specifies a color from the RGB (red, green, blue) or HLS (hue, lightness, saturation) slider bars.

Select Color displays the Windows Color dialog box.

☑ **AutoCAD Background** sets the current AutoCAD background color.

Gradient view

Top specifies the top color for two- and three-color gradients.

Middle specifies the middle color for three-color gradients.

Bottom specifies the bottom color for two- and three-color gradients.

Horizon determines the center of the gradient as a percent of the viewport's height.

Height determines the start of the second color of a three-color gradient; automatically set to 0 for two-color gradients.

Rotation rotates the angle of the gradient.

Image view

Image Name specifies the name of the raster file to use as the background image.

Find File displays the file dialog box; allows selection of a *.bmp* (Windows bitmap), *.gif* (GIF), *.jpg* (JPEG), *.pcx* (PC Paintbrush), *.tga* (Targa), or *.tif* (TIFF) file.

Adjust Bitmap adjusts the position of a raster image; displays the Adjust Background Bitmap Placement dialog box.

Merge view

None. Displays the AutoCAD drawing as the background image.

Adjust Background Bitmap Placement dialog box

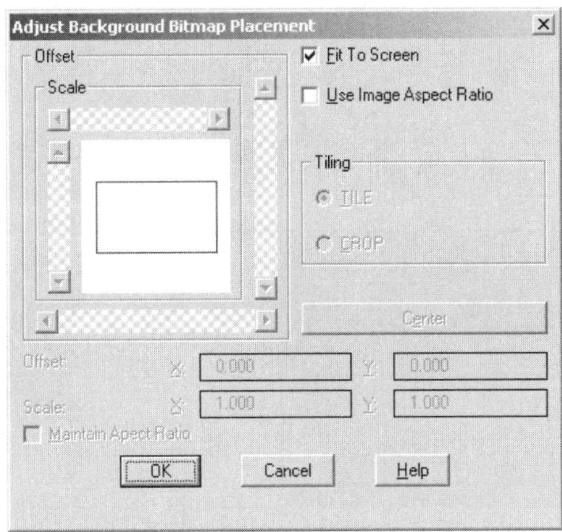

Offset uses the slider bars to position the image in the viewport.

☑ **Fit to Screen** stretches the image to fit the viewport.

☑ **Use Image Aspect Ratio** ensures the image is not distorted.

Tiling adjusts the size of the image when the image does not fit the viewport:

- ⊙ **Tile** repeats the image.
- ○ **Crop** cuts off the edges from the image to make it smaller.

Center centers the image in the viewport.

X,Y Offset changes the position of the image.

X,Y scale changes the size of the image.

RELATED COMMANDS

Fog creates fog-like effects.

ImageAttach loads raster images into drawings.

Render renders 3D objects in drawings.

Replay displays raster images in the current viewport.

3dOrbit displays background when in flat or Gouraud rendering mode.

TIPS

- Gradient backgrounds are useful for simulating a sunset (cyan-pink-orange) or underwater views (green-blue-black).
- To create 2-color gradients, set **Height** to 0.
- Image backgrounds are useful for placing the 3D rendered model in its environment, such as a rendered house on its building site.
- AutoCAD includes some background images in the *\acad 2005\textures* folder.
- The four types of background:

*Top: **Solid** option displays a uniform color (left); **Gradient** option displays a 2- or 3-color linear gradient (right).*

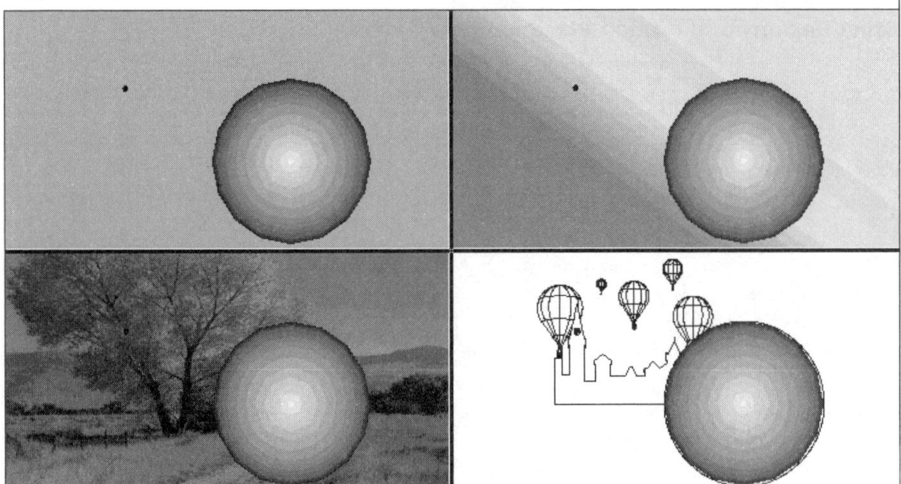

*Bottom: **Image** option displays a raster image (left); **Merge** option displays the current AutoCAD viewport (right).*

- As of AutoCAD 2005, this command also affects rendered viewports and rendered 3dOrbit views.

'Base

V. 1.0 Changes the insertion point of the drawing, located by default at (0,0,0).

Command	Alias	Ctrl+	F-key	Alt+	Menu Bar	Tablet
base	...	...	...	DKB	Draw ↳Block ↳Base	...

Command: base
Enter base point <0.0,0.0,0.0>: *(Pick a point.)*

COMMAND LINE OPTION

Enter base point specifies the x, y, z coordinates of the new insertion point.

RELATED COMMANDS

Block specifies the insertion point of new blocks.

Insert inserts another drawing into the current drawing.

Xref references other drawings.

RELATED SYSTEM VARIABLE

InsBase contains the current setting of the drawing insertion point.

TIPS

- Use this command to shift the insertion point of the current drawing.
- This command does not affect the current drawing. Instead, it comes into effect when you insert it or xref it into another drawing.

BAttMan

2002 Edits all aspects of attributes in blocks; works with one block at a time (*short for Block ATTribute MANager*).

Command	Alias	Ctrl+	F-key	Alt+	Menu Bar	Tablet
battman	...	...	...	**MOAB**	**Modify** ↳**Object** ↳**Attribute** ↳**Block Attribute Manager**	...

Command: battman

When the drawing contains no blocks with attributes, displays error message, "This drawing contains no attributed blocks."

When drawing contains at least one block with attributes, displays dialog box:

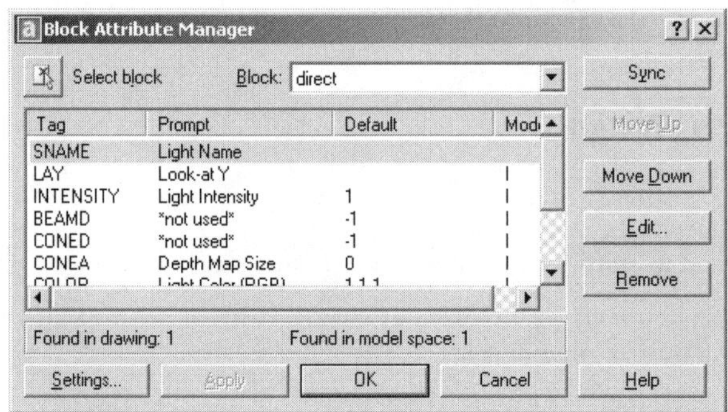

DIALOG BOX OPTIONS

Select block hides the dialog box, and then prompts, 'Select a block:'.

Block lists the names of blocks in the drawing, and displays the name of the selected block.

Sync changes the attributes in block insertions to match the changes made here.

Move Up moves the attribute tag up the list; constant attributes cannot be moved.

Move Down moves the attribute tag down the list.

Edit displays the Edit Attribute dialog box; see the **EAttEdit** command.

Remove removes the attribute tag and related data from the block; it does not operate when the block contains a single attribute.

Settings displays the Settings dialog box.

Apply applies the changes to the block definition.

Settings dialog box

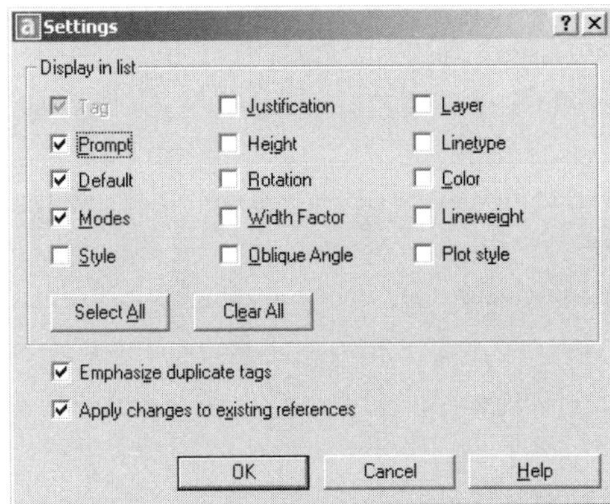

Display In List options

☑ **Tag** toggles (turns on and off) the display of the column of tags.

☑ **Prompt** toggles display of the column of attribute prompts.

☑ **Default** toggles the display of the attribute's default value.

☑ **Modes** toggles the display of the attribute's modes: invisible, constant, verify, and/or preset.

☐ **Style** toggles the display of the attribute's text style name.

☐ **Justification** toggles the display of the attribute's text justification.

☐ **Height** toggles the display of the attribute's text height.

☐ **Rotation** toggles the display of the attribute's text rotation angle.

☐ **Width Factor** toggles the display of the attribute's text width factor.

☐ **Oblique Angle** toggles the display of the attribute's text obliquing angle (slant).

☐ **Layer** toggles the display of the attribute's layer.

☐ **Linetype** toggles the display of the attribute's linetype.

☐ **Color** toggles the display of the attribute's color.

☐ **Lineweight** toggles the display of the attribute's lineweight.

☐ **Plot style** toggles the display of the attribute's plot style name (available only when plot styles are turned on).

Select All selects all display options.

Clear All clears all display options, except tag name.

Additional options

Emphasize duplicate tags:

☑ Highlights duplicate attribute tags in red.

☐ Does not highlight duplicate tags.

Apply changes to existing references:

☑ applies changes to all block instances that reference this definition in the drawing.

☐ applies the new attribute definitions only to newly-inserted blocks.

RELATED COMMANDS

AttDef defines attributes.

Block binds attributes to a symbol.

Insert inserts a block, and then allows you to specify the attribute data.

TIPS

- Use this command to edit and remove attribute definitions, as well as to change the order in which attributes appear.
- The **Sync** option does not change the values you assigned to attributes.
- When an attribute has a mode of Constant, it cannot be moved up or down the list.
- Turning on all the options displays a lot of data. To see all the data columns, you can stretch the dialog box.

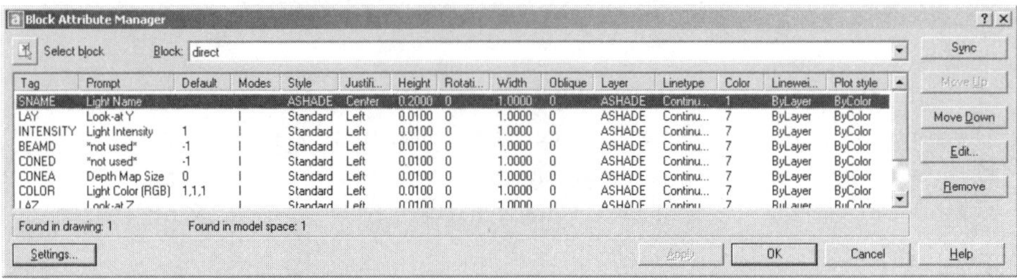

Tag	Prompt	Default	Modes	Style	Justifi...	Height	Rotati...	Width	Oblique	Layer	Linetype	Color	Linewei...	Plot style
SNAME	Light Name			ASHADE	Center	0.2000	0	1.0000	0	ASHADE	Continu...	1	ByLayer	ByColor
LAY	Look-at Y		I	Standard	Left	0.0100	0	1.0000	0	ASHADE	Continu...	7	ByLayer	ByColor
INTENSITY	Light Intensity	1	I	Standard	Left	0.0100	0	1.0000	0	ASHADE	Continu...	7	ByLayer	ByColor
BEAMD	"not used"	-1	I	Standard	Left	0.0100	0	1.0000	0	ASHADE	Continu...	7	ByLayer	ByColor
CONED	"not used"	-1	I	Standard	Left	0.0100	0	1.0000	0	ASHADE	Continu...	7	ByLayer	ByColor
CONEA	Depth Map Size	0	I	Standard	Left	0.0100	0	1.0000	0	ASHADE	Continu...	7	ByLayer	ByColor
COLOR	Light Color (RGB)	1,1,1	I	Standard	Left	0.0100	0	1.0000	0	ASHADE	Continu...	7	ByLayer	ByColor
LAZ	Look-at Z		I	Standard	Left	0.0100	0	1.0000	0	ASHADE	Continu...	7	ByLayer	ByColor

With the cursor, grab the edge of the dialog box to make it larger and smaller.

- An attribute definition cannot be changed to Constant via the Edit Attribute dialog box.
- The **Remove** option does not work when the block contains a single attribute.

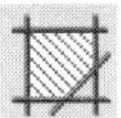

BHatch

Rel.12 Automatically applies associative hatch pattern objects within boundaries (*short for Boundary HATCH*).

Commands	Aliases	Ctrl+	F-key	Alt+	Menu Bar	Tablet
bhatch	**bh**	**...**	**...**	**DH**	**Draw**	**P9**
	h				**↳Hatch**	
-bhatch						

Command: bhatch

Displays dialog box:

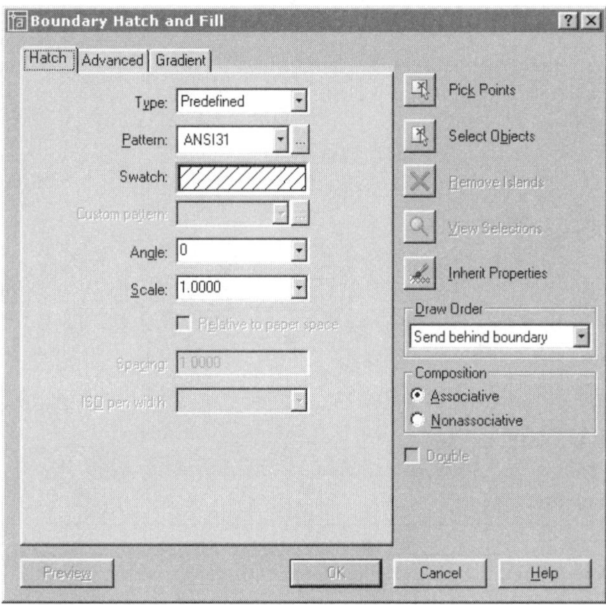

DIALOG BOX OPTIONS

Pick Points dismisses the dialog box. AutoCAD prompts, 'Select internal point:'. Pick a point inside an area; AutoCAD detects automatically the boundary, which will be filled with the hatch pattern.

Select Objects dismisses the dialog box. AutoCAD prompts, 'Select objects:'. Pick one or more objects, which will be filled with the hatch pattern.

Remove Islands removes *islands* (internal areas) from the hatch pattern selection set.

View Selections views hatch pattern selection set.

Inherit Properties sets the hatch pattern parameters from an existing hatch pattern.

Draw Order *(new to AutoCAD 2005):*

- **Do not assign** places hatch normally.
- **Send to back** places hatch behind all other overlapping objects in the drawings.
- **Bring to front** places hatch in front of all other overlapping objects.
- **Send behind boundary** places hatch behind its boundary.
- **Bring in front of boundary** places hatch in front of the boundary.

Composition determines the associativity of the hatch pattern:

⊙ **Associative** automatically updates the hatch when boundary or properties are modified; pattern is created as a hatch object.

○ **Nonassociative** means the hatch cannot be updated; pattern is created as a block.

Double:

☑ User-defined hatch is applied a second time at 90 degrees to the first pattern.

☐ User-defined hatch is applied once.

Preview displays a preview of the hatch pattern.

Hatch tab

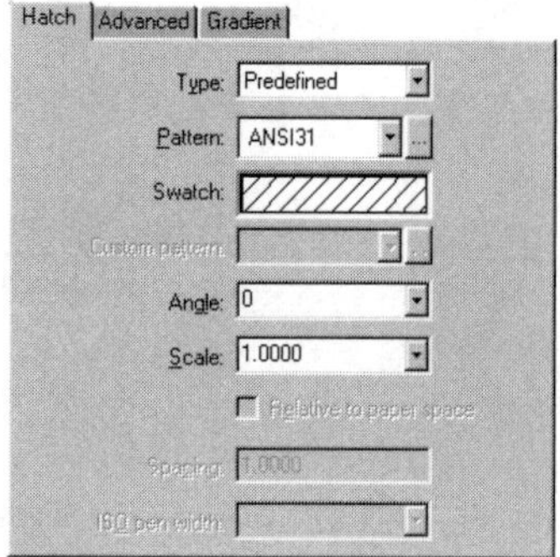

Type selects the pattern type:

Type	Meaning
Predefined	Hatches predefined by AutoCAD stored in the *acad.pat* and *acadiso.pat* files.
User Defined	Parallel-line hatches with spacing defined by you.
Custom	Hatches defined by *.pat* files added to AutoCAD's search path.

Pattern selects hatch pattern.

... displays Hatch Pattern Palette dialog box showing sample pattern types.

Swatch displays a non-scaled preview of the hatch pattern; click to display the Hatch Pattern Palette dialog box.

Custom Pattern lists the custom patterns, if any are available.

Angle specifies the hatch pattern rotation; default = 0 degrees.

Scale specifies the hatch pattern scale; default = 1.0.

Relative to Paper Space specifies the scale of the hatch pattern relative to paper space units; available only in layout mode.

Spacing specifies the spacing between the lines of a user-defined hatch pattern.

ISO Pen Width scales pattern according to pen width.

Advanced tab

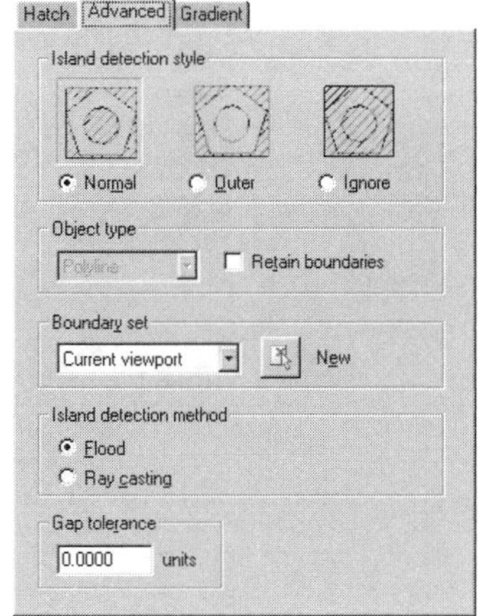

Island Detection Style

⊙ **Normal** turns hatching off and on each time it crosses a boundary; text is not hatched.

○ **Outer** hatches only the outermost areas; text is not hatched.

○ **Ignore** hatches everything within the boundary; text is hatched.

Object Type constructs the boundary from polyline or region objects.

Retain Boundaries:

☑ Boundary polylines or regions (created during the boundary hatcing process) are kept after **BHatch** finishes.

☐ Boundaries are discarded after hatching.

Boundary Set defines how objects are analyzed for defining boundaries; not available when Select Objects is used to define the boundary (default = current viewport).

New creates new boundary sets; AutoCAD dismisses the dialog box, and then prompts you to select objects.

Island Detection Method:

⊙ **Flood** includes islands as boundary objects.

○ **Ray Casting** runs an imaginary line from the pick point to the nearest object, and then traces the boundary in a counterclockwise direction; it excludes islands.

Gap Tolerance *(new to AutoCAD 2005)* specifies the maximum gap between boundary objects. (Objects don't need to touch to form a valid hatching boundary.) Range is 0 to 5000 units.

Gradient tab

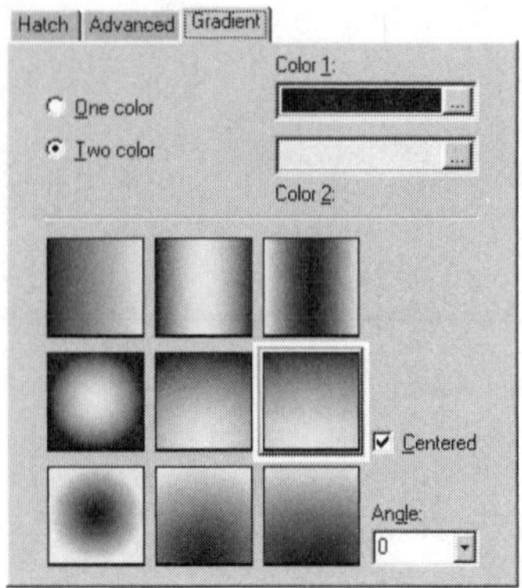

One Color options

⊙**One color** produces color-shade gradients.

Shade - Tint slider varies the second color between white (shade) and black (tint).

Two Color options

O**Two color** produces two-color gradients.

Color 1 selects the first color.

Color 2 selects the second color.

... displays the Select Color dialog box.

Centered:

☑ Centers the gradient in the hatch area.

☐ Moves gradient is moved up and to the left, simulating a natural light source on the object.

Angle rotates the gradient.

Hatch Pattern Palette dialog box

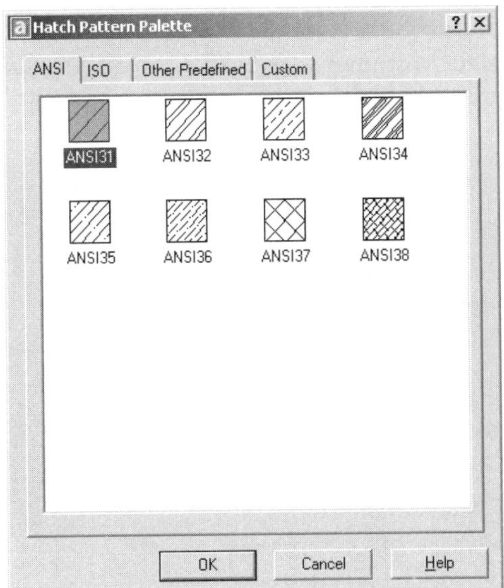

ANSI Patterns:

ANSI31 ANSI32 ANSI33 ANSI34

ANSI35 ANSI36 ANSI37 ANSI38

ISO Patterns:

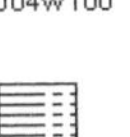

ISO02W100 ISO03W100 ISO04W100 ISO05W100

ISO06W100 ISO07W100 ISO08W100 ISO09W100

ISO10W100 ISO11W100 ISO12W100 ISO13W100

ISO14W100 ISO15W100

Other Predefined Patterns:

SOLID ANGLE AR-B816 AR-B816C

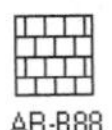

AR-B88 AR-BRELM AR-BRSTD AR-CONC

AR-HBONE AR-PARQ1 AR-RROOF AR-RSHKE

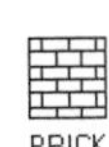

AR-SAND BOX BRASS BRICK

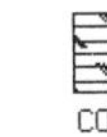

BRSTONE CLAY CORK CROSS

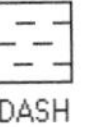

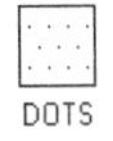

DASH DOLMIT DOTS EARTH

ESCHER FLEX GRASS GRATE

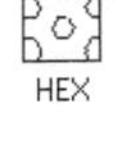

HEX HONEY HOUND INSUL

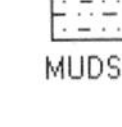
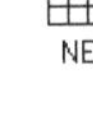

LINE MUDST NET NET3

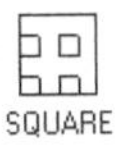

PLAST PLASTI SACNCR SQUARE

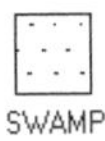

STARS STEEL SWAMP TRANS

TRIANG ZIGZAG

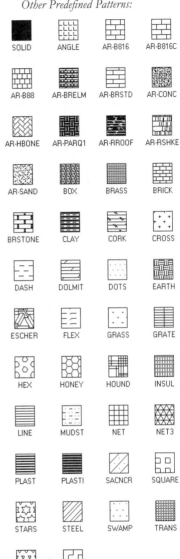

-BHATCH Command

Command: -bhatch

Current hatch pattern: ANSI31

Specify internal point or [Properties/Select/Remove islands/Advanced/DRaw order]: *(Pick a point, or specify an option.)*

COMMAND LINE OPTIONS

Internal Point option

Pick a point creates a boundary inside the area of your pick point.

Property options

Enter a pattern name or [?/Solid/User defined] <ANSI31>: *(Enter a name, or select an option.)*

Specify a scale for the pattern <1.0000>: *(Enter a scale factor.)*

Specify an angle for the pattern <0>: *(Enter an angle.)*

Enter a pattern name allows you to enter the name of the hatch pattern.

? lists the names of available hatch patterns.

Solid floods the area with a solid fill in the current color.

User defined creates a simple, user-defined hatch pattern.

Specify a scale specifies the hatch pattern angle (default = 0 degrees).

Specify an angle specifies the hatch pattern scale (default: = 1.0).

Select option

Select objects: *(Select one or more objects.)*

Select objects: *(Press* **ENTER**.*)*

Select objects selects one or more objects to fill with hatch pattern.

Remove islands options

Select island to remove: *(Select one object.)*

<Select island to remove>/Undo: *(Press* **ENTER**, *or type* **U**.*)*

Select island to remove selects the island to remove, which is not filled with the hatch pattern.

Undo adds the removed island.

Advanced options

Enter an option [Boundary set/Retain boundary/Island detection/Style/ Associativity/Gap tolerance]: *(Specify an option.)*

Boundary set defines the objects analyzed when a boundary is defined by a specified pick point.

Retain boundary:

- **On** the boundary created during the hatching process is kept after **-BHatch** finishes.
- **Off** the boundary is discarded.

Island detection:

- **On** objects within the outermost boundary are used as boundary objects.
- **Off** all objects within the outermost boundary are filled.

Style selects the hatching style: ignore, outer, or normal.

Associativity:

- **On** hatch pattern is associative.
- **Off** hatch pattern is not associative.

Gap tolerance specifies the largest gap permitted in the boundary (*new to AutoCAD 2005*):

Specify a boundary gap tolerance value <0>: *(Enter a value between 0 and 5000.)*

Draw Order options *(new to AutoCAD 2005)*

Enter draw order [do Not assign/send to Back/bring to Front/send beHind boundary/bring in front of bounDary] <send beHind boundary>: *(Specify an option.)*

do Not assign places hatch normally.

send to Back places hatch behind all other overlapping objects in the drawings.

bring to Front places hatch in front of all other overlapping objects.

send beHind boundary places hatch behind its boundary.

bring in front of bounDary places hatch in front of boundary.

RELATED COMMANDS

AdCenter places hatch patterns from other drawings.

Boundary traces polylines automatically around closed boundaries.

Convert converts Release 13 (and earlier) hatch patterns into Release 14-2005 format.

Hatch places nonassociative hatch patterns.

HatchEdit edits hatch patterns.

PsFill floods closed polylines with PostScript fill patterns.

ToolPalette stores and places selected hatch patterns and fill colors.

RELATED SYSTEM VARIABLES

DelObj toggles whether boundary is erased after hatch is placed.

GfAng specifies the angle of the gradient fill; ranges from 0 to 360 degrees.

GfClr1 specifies the first gradient fill color in RGB format, such as "RGB 000, 128, 255."

GfClr2 specifies the second gradient fill color in RGB format.

GfClrLum specifies the luminescence of one-color gradient fills, from 0.0 (black) to 1.0 (white).

GfClrState specifies whether the gradient fill is one-color or two-color.

GfName specifies the gradient fill pattern:

GfName	Meaning
1	Linear.
2	Cylindrical.
3	Inverted cylindrical.
4	Spherical.
5	Inverted spherical.
6	Hemispherical.
7	Inverted hemispherical.
8	Curved.
9	Inverted curved.

GfShift specifies whether the gradient fill is centered or is shifted to the upper-left.

HpAng specifies the current hatch pattern angle (default = 0).

HpBound specifies the hatch boundary object: polyline or region.

HpDouble specifies single or double hatching.

HpDrawOrder *(new to AutoCAD 2005)* controls the display order of the hatch pattern relative to other overlapping objects.

HpGalTol *(new to AutoCAD 2005)* reports the current gap tolerance.

HpName specifies the current hatch pattern name (up to 31 characters long):

HpName	Meaning
ANSI31	Default pattern name.
""	No current pattern name.
"."	Eliminate current pattern name.

HpScale specifies the current hatch pattern scale factor (default = 1).

HpSpace specifies the current hatch pattern spacing factor (default = 1).

OsnapHatch (*new to AutoCAD 2005*) determines whether object snap snaps to hatch patterns.

PickStyle controls the selection of hatch patterns:

PickStyle	Meaning
0	Neither groups nor hatches selected.
1	Groups selected (default).
2	Associative hatches selected.
3	Both selected.

SnapBase specifies the starting coordinates of hatch pattern (default = 0,0).

RELATED FILES

acad.pat contains the ANSI and other hatch pattern definitions.

acadiso.pat contains the ISO hatch pattern definitions.

TIPS

- The **BHatch** command first generates a boundary, and then hatches the inside area.
- Use the **Boundary** command to create just the boundary.
- **BHatch** stores hatching parameters in the pattern's extended object data.
- Bringing the pattern to the front (through the Draw Order option) makes it easier to edit the hatch.
- The ANSI31 patter is defined as 45° lines. To keep the angle at 45°, the Angle should be specified as 0.
- Set the **OsnapHatch** system variable to 0 to turn off object snapping of hatch and fill patterns.
- Hatches, solid fills, and gradient fills can be dragged into the Tool Palette window.
- The **Trim** command trims hatches (*new to AutoCAD 2005*).

'Blipmode

V. 2.1 Turns the display of pick-point markers, known as "blips," on and off.

Command	Alias	Ctrl+	F-key	Alt+	Menu Bar	Tablet
blipmode	...	...	...	...	...	...

Command: blipmode
Enter mode [ON/OFF] <OFF>: on

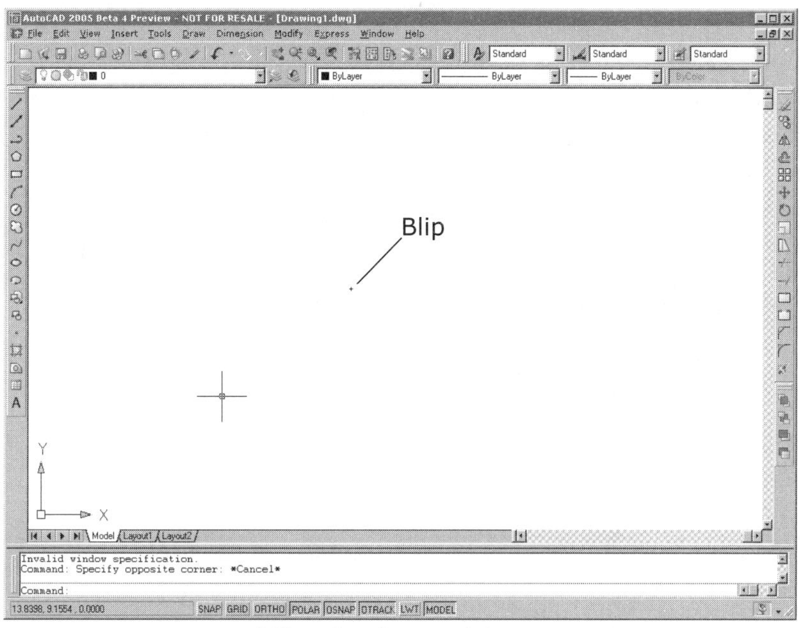

A blipmark at the center of the screen.

COMMAND LINE OPTIONS

ON turns on the display of pick-point markers.

OFF turns off the display of pick-point markers.

RELATED COMMANDS

Options allows blipmode toggling via a dialog box.

Redraw cleans blips off the screen.

RELATED SYSTEM VARIABLE

Blipmode contains the current blipmode setting.

TIPS

- You cannot change the size of the blipmark.
- Blipmarks are erased by any command that redraws the view, such as **Redraw**, **Regen**, **Zoom**, and **Vports**.

Block

V. 1.0 Defines a group of objects as a single named object; creates symbols.

Commands	Aliases	Ctrl+	F-key	Alt+	Menu Bar	Tablet
block	**b**	**...**	**...**	**DKM**	**Draw**	**N9**
	bmake				**↳Block**	
					↳Make	
-block	**-b**					

Command: block

Displays dialog box:

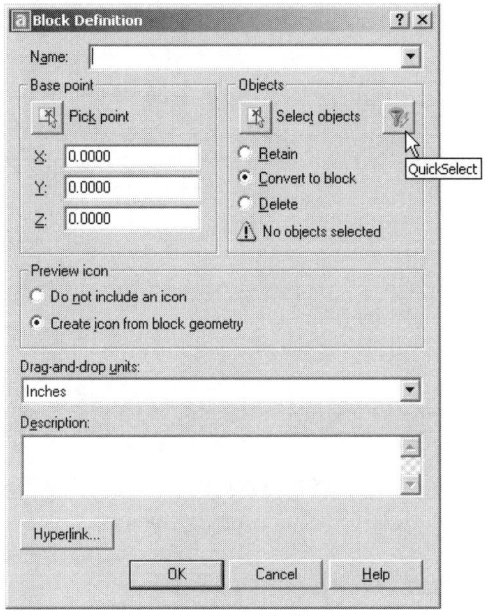

DIALOG BOX OPTIONS

Name names the block (*maximum = 255 characters).*

Base point options

Pick point dismisses the dialog box; AutoCAD prompts, 'Specify insertion base point:'. Pick a point that specifies the block's insertion point, usually the lower-left corner.

X, **Y**, and **Z** specify the x, y, z coordinates of the insertion point.

Objects options

Select objects dismisses the dialog box; AutoCAD prompts, 'Select objects:'. Select one or more objects that make up the block:

- ○ **Retain** leaves objects in place after the block is created.
- ⊙ **Convert to Block** erases objects making up the block, and replaces them with the block.
- ○ **Delete** erases the objects making up the block; the block is stored in drawing.

Quick Select displays the Quick Select dialog box; see the **QSelect** command.

Preview Icon options

⊙**Do not include an icon** does not create an icon.

○**Create icon from block geometry** creates a preview image of the block.

Insert units selects the units for the block when dragged from the DesignCenter.

Description allows you to enter a description of the block.

Hyperlink displays the Insert Hyperlink dialog box; see the **Hyperlink** command.

-BLOCK Command

Command: -block

Enter block name or [?]: *(Enter a name, or type **?**.)*

Specify insertion base point: *(Pick a point.)*

Select objects: *(Select one or more objects.)*

Select objects: *(Press **ENTER**.)*

COMMAND LINE OPTIONS

Block name allows you to enter the name of the block.

? lists the names of blocks stored in the drawing.

Insertion base point specifies the x, y coordinates of the block's insertion point.

Select objects selects the objects and attributes that make up the block.

RELATED COMMANDS

Explode reduces a block to its original objects.

Insert adds a block or another drawing to the current drawing.

Oops returns objects to the screen after creating the block.

WBlock writes a block to a file on disk as a drawing.

XRef displays another drawing in the current drawing.

RELATED SYSTEM VARIABLES

InsName default block name.

InsUnits drawing units for blocks dragged from the DesignCenter:

InsUnits	Meaning	InsUnits	Meaning
0	Unitless.	**11**	Angstroms.
1	Inches.	**12**	Nanometers.
2	Feet.	**13**	Microns.
3	Miles.	**14**	Decimeters.
4	Millimeters.	**15**	Decameters.
5	Centimeters.	**16**	Hectometers.
6	Meters.	**17**	Gigameters.
7	Kilometers.	**18**	Astronomical Units.
8	Microinches.	**19**	Light Years.
9	Mils.	**20**	Parsecs.
10	Yards.		

RELATED FILES

All *.dwg* drawing files can be inserted as blocks.

TIPS

- Blocks consist of these parts:

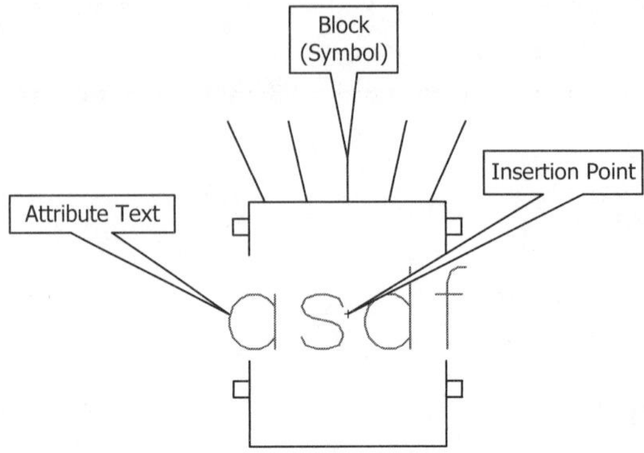

- The names of blocks have up to 255 alphanumeric characters, including $, -, and _.
- Use the **INSertion** object snap to select the insertion points of blocks.
- When blocks are created on layers other than layer 0, they are inserted on that layer.
- Blocks created on layer 0 are inserted on the current layer.
- AutoCAD has five types of blocks:

Block	Meaning
User block	Named blocks created by users.
Nested block	Blocks inside other blocks.
Unnamed block	Blocks created by AutoCAD.
Xref	Externally-referenced drawings.
Dependent block	Blocks in externally-referenced drawings.

- AutoCAD creates unnamed blocks, also called "anonymous blocks":

Name	Meaning
An*	Groups.
Dn*	Associative dimensions.
Un*	Created by AutoLISP or ObjectARx apps.
Xn*	Hatch patterns.

- AutoCAD automatically purges unreferenced anonymous blocks when drawings are first loaded.
- You cannot place anonymous blocks with **Insert**.

BlockIcon

2000 Creates preview images for all blocks in drawings created with AutoCAD Release 14 or earlier.

Command	Alias	Ctrl+	F-key	Alt+	Menu Bar	Tablet
blockicon	...	...	...	**FUU**	**File** ↳**Drawing Utilities** ↳**Update Block Icons**	...

Command: blockicon

Enter block names <*>: *(Enter a name, a wildcard pattern, or press* **ENTER** *for all names.)*

n **blocks updated.**

COMMAND LINE OPTION

Enter block names specifies the blocks for which to create icons; press ENTER to add icons to all blocks in the drawing.

RELATED COMMANDS

Block creates new blocks and their icons.

AdCenter displays the icons created by this command:

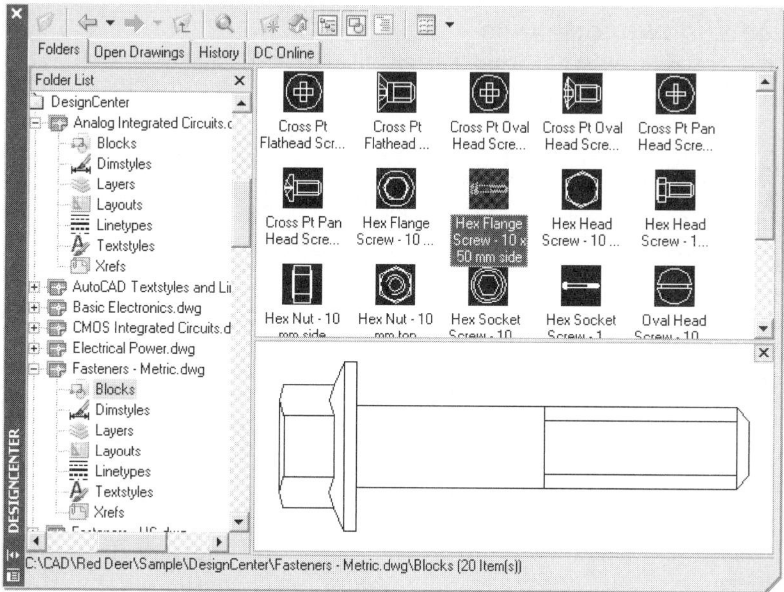

BmpOut

Rel.13 Exports the current viewport as a raster image in BMP bitmap format.

Command	Alias	Ctrl+	F-key	Alt+	Menu Bar	Tablet
bmpout	...	...	...	**FE** **↳BMP**	**File** **↳Export** **↳BMP**	...

Command: bmpout

Displays Create BMP File dialog box. Enter a file name, and then click ***Save****.*

Select objects or <all objects and viewports>: *(Press* **ENTER** *to select all objects, or select individual objects.)*

DIALOG BOX OPTION

Save saves drawings as BMP format raster files.

RELATED COMMANDS

JpgOut exports objects and viewports in JPEG format.

PngOut exports objects and viewports in PNG format.

TifOut exports objects and viewports in TIFF format.

WmfOut exports selected objects in WMF format.

RELATED WINDOWS COMMANDS

PRT SCR saves screen to the Clipboard.

ALT+PRT SCR saves the topmost window to the Clipboard.

TIPS

- The *.bmp* extension is short for "bitmap," a raster file standard for Windows.
- This command creates uncompressed *.bmp* files.

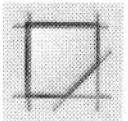

Boundary

Rel.12 Creates boundaries as polylines or 2D regions.

Command	Aliases	Ctrl+	F-key	Alt+	Menu Bar	Tablet
boundary	**bo**	...	...	**DB**	**Draw**	**Q9**
	bpoly				↳**Boundary**	
-boundary	**-bo**					

Command: boundary

Displays dialog box:

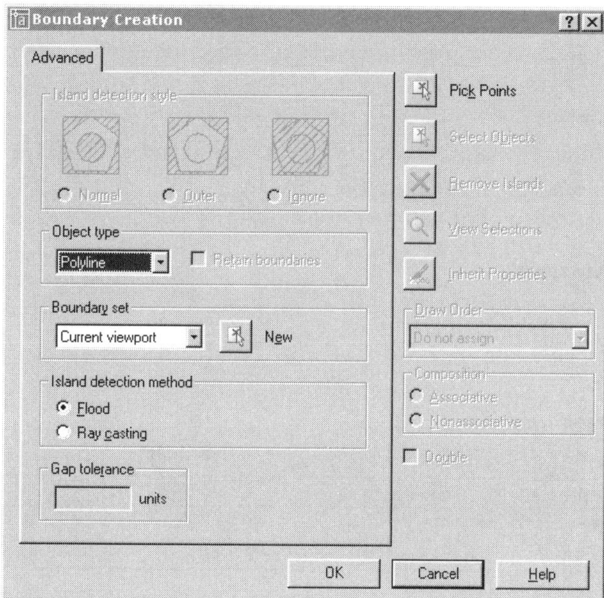

DIALOG BOX OPTIONS

Object Type options

- **Polyline** object.
- **Region** object.

Boundary Set options

Boundary Set defines the objects analyzed for defining boundaries; not available when Select Objects is used to define the boundary (default = current viewport).

New creates new boundary sets.

Island Detection options

⊙ **Flood** includes islands as boundary objects.

○ **Ray Casting** runs an imaginary line from the pick point to the nearest object, and then traces the boundary in a counterclockwise direction; excludes islands.

Button

Pick Points picks points inside of closed areas.

-BOUNDARY Command

Command: -boundary

Specify internal point or [Advanced options]: *(Pick a point, or type* **A***.)*

COMMAND LINE OPTIONS

Specify internal point creates a boundary based on the point you pick.

Advanced options options

Enter an option [Boundary set/Island detection/Object type]: *(Enter an option.)*

Boundary set defines the objects **-Boundary** analyzes when defining a boundary from a specified pick point. It chooses from a new set of objects, or all objects visible in the current viewport.

Island detection:

- **On** uses objects within the outermost boundary as boundary objects.
- **Off** fills objects within the outermost boundary.

Object type specifies polyline or region as the boundary object.

RELATED COMMANDS

PLine draws a polyline.

PEdit edits a polyline.

Region creates a 2D region from a collection of objects.

RELATED SYSTEM VARIABLE

HpBound specifies the default object used to create boundary:

HpBound	Meaning
0	Draw as region.
1	Draw as polyline (default).

TIPS

- Although the **Boundary Creation** dialog box looks similar to the **BHatch** command's **Advanced Options** dialog box, be aware of differences indicated by the grayed out sections.
- Use this command to measure irregular areas:
  1. Apply the **Boundary** command to an irregular area.
  2. Use the **Area** command to find the area and perimeter of the boundary.
- Use **Boundary** together with the **Offset** command to help create *poching*, areas partially covered by hatching.

Box

Rel.13 Draws a 3D box as a solid model.

Command	Alias	Ctrl+	F-key	Alt+	Menu Bar	Tablet
box	...	...	...	DIB	Draw ↳Solids ↳Box	J7

Command: box

Specify corner of box or [CEnter] <0,0,0>: *(Pick point 1, or type the* **CE** *option.)*

Specify corner or [Cube/Length]: *(Pick point 2, or enter an option.)*

Specify height: *(Pick point 3.)*

Second point: *(Pick point 4.)*

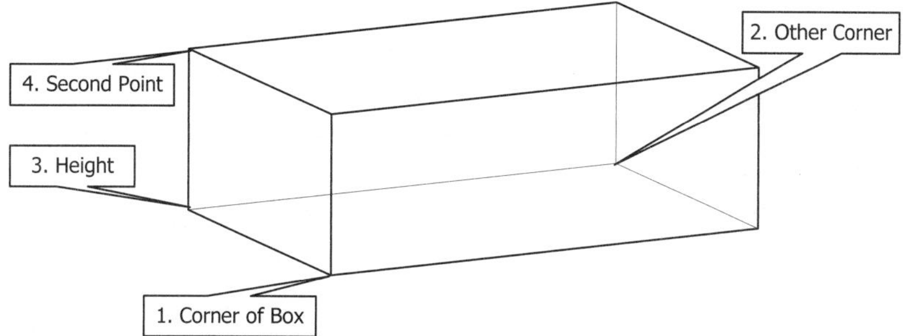

COMMAND LINE OPTIONS

Corner of box specifies one corner for the base of box.

Center draws the box about a center point.

Cube draws a cube box — all sides are the same length.

Length specifies the length, width, and height.

Height specifies the height of the box.

RELATED COMMANDS

Ai_Box draws a 3D wireframe (surface) box.

Cone draws a 3D solid cone.

Cylinder draws a 3D solid tube.

Sphere draws a 3D solid ball.

Torus draws a 3D solid donut.

Wedge draws a 3D solid wedge.

RELATED SYSTEM VARIABLE

DispSilh displays 3D objects as silhouettes after hidden-line removal and shading.

Break

V. 1.4 Removes portions of objects.

Command	Alias	Ctrl+	F-key	Alt+	Menu Bar	Tablet
break	br	...	...	MK	Modify ↳Break	W17

Command: break
Select object: *(Select one object — point 1.)*
Specify second break point or [First point]: *(Pick point 2, or type* **F**.*)*

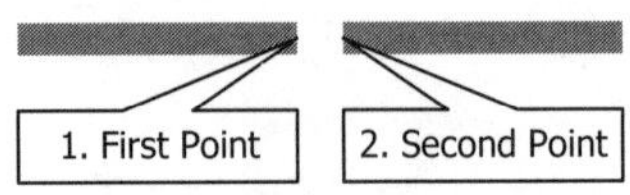

Breaking a line at two points.

COMMAND LINE OPTIONS

Select object selects one object to break; the pick point becomes the first break point, unless the **F** option is used at the next prompt.

@ uses the first break point's coordinates for the second break point.

 *First Point options*

Enter first point: *(Pick a point.)*
Enter second point: *(Pick a point.)*

First point specifies the first break point.

Second point specifies the second break point.

RELATED COMMANDS

Change changes the length of lines.

PEdit removes and relocates vertices of polylines.

Trim shortens the lengths of open objects.

TIPS

- Use this command to convert circles into arcs; pick the break point counter clockwise.
- This command can erase a portion of an object (as shown in the figure above) or remove the end of an open object.
- The second point does not need to be on the object; AutoCAD breaks the object at the point nearest to the pick point.
- The **Break** command works on the following objects: lines, arcs, circles, polylines, ellipses, rays, xlines, and splines, as well as objects made of polylines, such as donuts and polygons.

Browser / Browser2

Rel.14 Prompts for a Web address, and then launches the Web browser.

Commands	Alias	Ctrl+	F-key	Alt+	Menu Bar	Tablet
browser	...	...	...	**HD**	**Help** ↳**Autodesk On The Web**	**Y8**
browser2						

Command: browser

Enter Web location (URL) <http://www.autodesk.com>: *(Enter a Web address.)*

Launches browser:

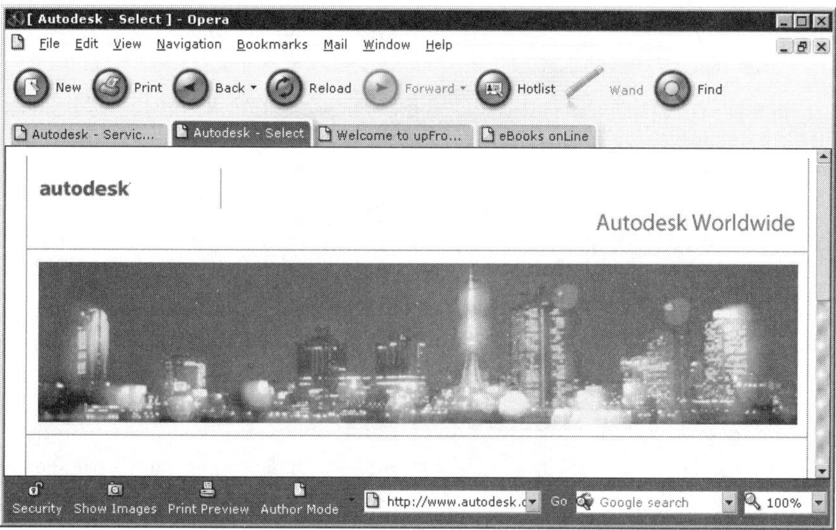

COMMAND LINE OPTION

Enter Web location specifies the URL; see the **AttachURL** command for information about URLs.

RELATED SYSTEM VARIABLE

InetLocation contains the name of the default URL.

TIPS

- *URL* is short for "uniform (or universal) resource locator," the universal file-naming system used on the Internet; also called a link or hyperlink.
- An example of a URL is http://www.autodeskpress.com, the Autodesk Press Web site.
- Many file dialog boxes also give you access to the Web browser; see the **Open** command.
- The undocumented **Browser2** command opens two URLs.

'Cal

Rel.12 Command-line algebraic and vector geometry calculator (*short for CALculator*).

Command	Alias	Ctrl+	F-key	Alt+	Menu Bar	Tablet
cal	...	...	...	...	...	...

Command: cal
>>Expression: *(Enter an expression, and then press* **ENTER**.*)*

COMMAND LINE OPTIONS

( )	Grouping of expressions.
[ ]	Vector expressions.
+	Addition.
-	Subtraction.
*****	Multiplication.
/	Division.
^	Exponentiation.
&	Vector product of vectors.
sin	Sine.
cos	Cosine.
tang	Tangent.
asin	Arc sine.
acos	Arc cosine.
atan	Arc tangent.
ln	Natural logarithm.
log	Logarithm.
exp	Natural exponent.
exp10	Exponent.
sqr	Square.
sqrt	Square root.
abs	Absolute value.
round	Round off.
trunc	Truncate.
cvunit	Converts units using *acad.unt*.
w2u	WCS to UCS conversion.
u2w	UCS to WCS conversion.
r2d	Radians-to-degrees conversion.
d2r	Degrees-to-radians conversion.
pi	The value PI (3.14159).
xyof	x,y coordinates of a point.
xzof	x,z coordinates of a point.
yzof	y,z coordinates of a point.
xof	x coordinate of a point.
yof	y coordinate of a point.
zof	z coordinate of a point.
rxof	Real x coordinate of a point.
ryof	Real y coordinate of a point.
rzof	Real z coordinate of a point.
cur	x, y, z coordinates of a picked point.
rad	Radius of object.
pld	Point on line, distance from.
plt	Point on line, using parameter *t*.
rot	Rotated point through angle about origin.
ill	Intersection of two lines.
ilp	Intersection of line and plane.
dist	Distance between two points.
dpl	Distance between point and line.
dpp	Distance between point and plane.
ang	Angle between lines.
nor	Unit vector normal.
ESC	Exits Cal mode.

RELATED COMMANDS

All.

RELATED SYSTEM VARIABLES

UserI1 — **UserI5** holds user-definable variables, which can be used to store integers.

UserR1 — **UserR5** holds user-definable variables, which can be used to store real numbers.

TIPS

- To use **Cal**, type an expression at the >> prompt. For example, to find the area of a circle (pi*r^2) with radius of 1.2 units:

  Expression >> pi*(1.2^2)
  4.52389

- **Cal** understands the following prefixes:

  ***** Scalar product of vectors.

  & Vector product of vectors.

- And the following suffixes:

  r Radian (degrees is the default).

  g Grad.

  ' Feet (unitless distance is the default).

  " Inches.

- Because **'Cal** is a transparent command, it can be used to perform a calculation in the middle of another command, and then return the value to that command. For example, to set the offset distance to the radius of a circle:

  Command: offset
  Specify offset distance or [Through] <Through>: 'cal
  >> Expression: rad
  >> Select circle, arc or polyline segment for RAD function: *(Pick circle.)*
  2.0
  Select object to offset or <exit>: *(And so on.)*

Camera

2000 Sets the camera and target coordinates to create 3D viewpoints.

Command	Alias	Ctrl+	F-key	Alt+	Menu Bar	Tablet
camera	...	...	...	...	...	...

Command: camera
Specify new camera position <7.6,4.5,21.1>: *(Pick a point.)*
Specify new camera target <7.6,4.5,0.0>: *(Pick a point.)*
Regenerating model.

COMMAND LINE OPTIONS

Specify new camera position specifies the x, y, z coordinates for the "look from" point.

Specify new camera target specifies the x, y, z coordinates for the "look at" point.

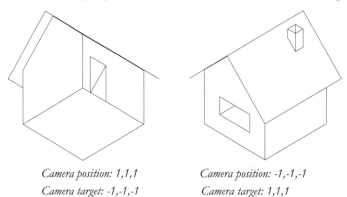

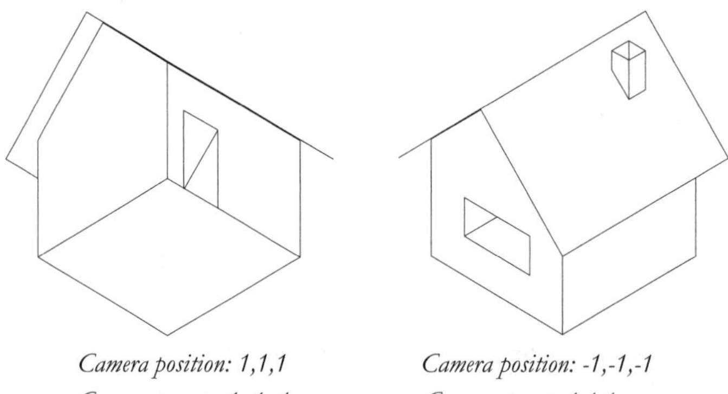

Camera position: 1,1,1
Camera target: -1,-1,-1

Camera position: -1,-1,-1
Camera target: 1,1,1

RELATED COMMANDS

DView has the CAmera option, which interactively sets a new 3D viewpoint.

VPoint sets a 3D viewpoint through x, y, z coordinates or angles.

3dOrbit sets a new 3D viewpoint interactively.

TIPS

- You may find the **3dOrbit** command easier to use than this command.
- Following the **Camera** command, you may need to use **Zoom Extents** to see the model.
- Use **Hide** or **SadeMode** to check whether you are looking at the model from above or below.

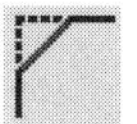

Chamfer

V. 2.1 Bevels the intersection of two lines, all vertices of 2D polylines, and the faces of 3D solid models.

Command	Alias	Ctrl+	F-key	Alt+	Menu Bar	Tablet
chamfer	cha	...	...	MC	Modify ⇨Chamfer	W18

Command: chamfer

(TRIM mode) Current chamfer Dist1 = 0.0, Dist2 = 0.0

Select first line or [Polyline/Distance/Angle/Trim/Method/mUltiple]: *(Pick an object, or enter an option.)*

Select second line: *(Pick an object.)*

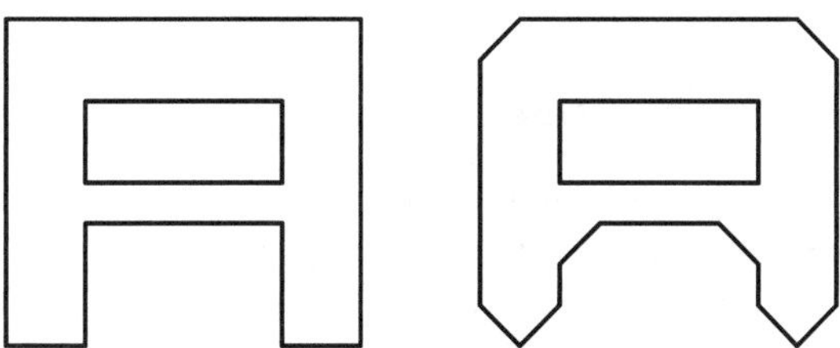

At left: original object drawn with polylines. At right: chamfered with ***Polyline*** *option.*

COMMAND LINE OPTIONS

Select first line selects the first line, arc, face, or edge.

Select second line selects the second line, arc, face, or edge.

mUltiple allows more than one pair of lines to be chamfered.

Polyline option

Select 2D polyline: *(Pick a polyline.)*

n **lines were chamfered**

Select 2D polyline chamfers *all* segments of a 2D polyline; if the polyline is not closed with the Close option, the first and last segments are not chamfered.

Distance options

Specify first chamfer distance <0.5000>: *(Enter a value.)*

Specify second chamfer distance <0.5000>: *(Enter a value.)*

First distance specifies the chamfering distance along the line picked first.

Second distance specifies the chamfering distance along the line picked second.

Angle options

Specify chamfer length on the first line <1.0000>: *(Enter a distance.)*

Specify chamfer angle from the first line <0>: *(Enter an angle.)*

Chamfer length specifies the chamfering distance along the line picked first.

Chamfer angle specifies the chamfering angle by an angle from the line picked first.

Trim options

Enter Trim mode option [Trim/No trim] <Trim>: *(Type* **T** *or* **N***.)*

Trim trims lines, edges, and faces are after chamfer.

No trim does not trim lines, edges, and faces after chamfer.

Intersecting lines before (at left) and after a no-trim chamfer (at right).

Method option

Enter trim method [Distance/Angle] <Angle>: *(Type* **D** *or* **A***.)*

Distance determines chamfer by two specified distances.

Angle determines chamfer by the specified angle and distance.

Chamfering 3D Solids — **Edge Mode** options

Command: chamfer

(TRIM mode) Current chamfer Dist1 = 0.0, Dist2 = 0.0

Polyline/.../<Select first line>: *(Pick a 3D solid model.)*

Select base surface: *(Pick surface edge on the solid model — see 1, below.)*

Next/<OK>: *(Enter* **N** *or select the next surface, or* **OK** *to end selection.)*

Enter base surface distance <0.0>: *(Enter a value.)*

Enter other surface distance <0.0>: *(Enter a value.)*

Loop/<Select edge>: *(Select an edge — see 2, below.)*

Loop/<Select edge>: *(Press* **ENTER***.)*

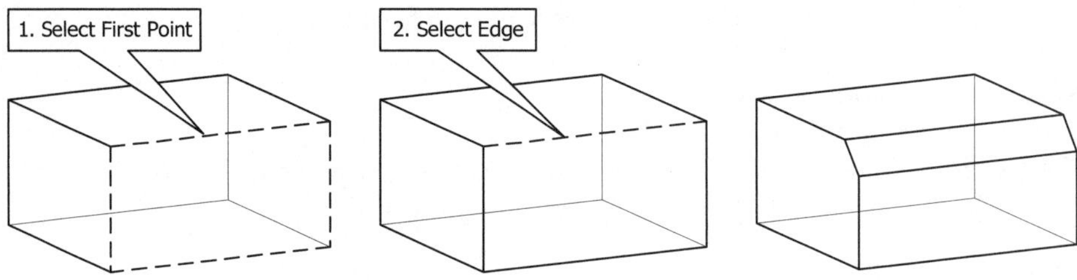

Chamfering 3D Solids — **Loop Mode** options

Command: chamfer

(TRIM mode) Current chamfer Dist1 = 0.0, Dist2 = 0.0

Polyline/.../<Select first line>: *(Pick a 3D solid model.)*

Select base surface: *(Pick a surface edge — see 1, below.)*

Next/<OK>: *(Enter* **N** *or select the next surface, or* **OK** *to end selection.)*

Enter base surface distance <0.0>: *(Enter a value.)*

Enter other surface distance <0.0>: *(Enter a value.)*

Loop/<Select edge>: *(Enter* **L***.)*

Edge/<Select edge loop>: *(Pick a surface edge — see 2, below.)*
Edge/<Select edge loop>: *(Press* **ENTER**.*)*

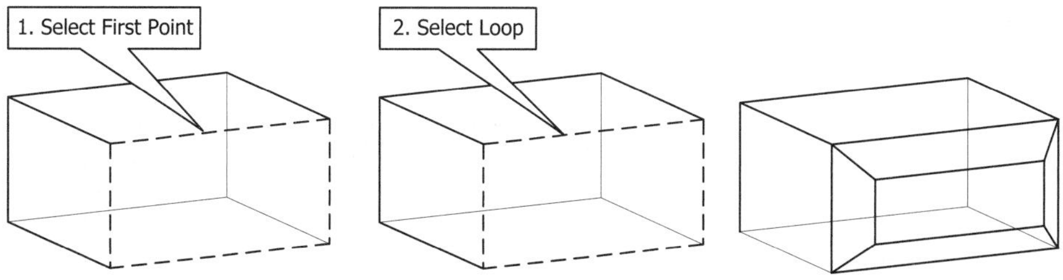

COMMAND LINE OPTIONS

Select first line selects the 3D solid.

Next selects the adjacent face, or press ENTER to accept face.

Enter base surface distance specifies the first chamfer distance (default = 0.5).

Enter other surface distance specifies the second chamfer distance (default = 0.5).

Loop selects all edges of the face.

Select edge selects a single edge of the face.

RELATED COMMANDS

Fillet rounds the intersection with a radius.

SolidEdit edits the faces and edges of solids.

RELATED SYSTEM VARIABLES

ChamferA is the first chamfer distance (default = 0.5).

ChamferB is the second chamfer distance (default = 0.5).

ChamferC is the length of chamfer (default = 1).

ChamferD is the chamfer angle (default = 0).

ChamMode is the toggles chamfer measurement:

ChamMode	Meaning
0	Chamfer by two distances (default).
1	Chamfer by distance and angle.

TrimMode determines whether lines/edges are trimmed after chamfer:

TrimMode	Meaning
0	Do not trim selected edges.
1	Trim selected edges (default).

TIPS

- When **TrimMode** is set to 1 and when lines do not intersect, **Chamfer** extends or trims the lines to intersect before chamfering.
- When the two objects are not on the same layer, **Chamfer** places the chamfer line on the current layer; the chamfer line takes on the layer, or current color and linetype.

Change

V. 1.0 Modifies the color, elevation, layer, linetype, linetype scale, lineweight, plot style, and thickness of most objects, and certain other properties of lines, circles, blocks, text, and attributes.

Command	Alias	Ctrl+	F-key	Alt+	Menu Bar	Tablet
change	-ch	...	...	...	...	...

Command: change
Select objects: *(Pick one or more objects.)*
Select objects: *(Press* **ENTER***.)*
Specify change point or [Properties]: *(Pick a point, an object, or type* **P***.)*
Enter property to change [Color/Elev/LAyer/LType/ltScale/LWeight/ Thickness/Plotstyle]: *(Enter an option.)*

COMMAND LINE OPTIONS

Specify change point selects an object to change:

(pick a line) indicates the new length of a line.

(pick a circle) indicates the new radius of a circle.

(pick a block) indicates the new insertion point or rotation angle of a block.

(pick text) indicates the new location of text.

(pick an attribute) indicates an attribute's new text insertion point, text style, height, rotation angle, text, tag, prompt, or default value.

ENTER changes the insertion point, style, height, rotation angle, and text of a text string.

Properties options

Color changes the color of the objects.

Elev changes the elevation of the objects.

LAyer moves the object to a different layers.

LType changes the linetype of the objects.

ltScale changes the scale of the linetypes.

LWeight changes the lineweight of the objects.

Thickness changes the thickness of any object, except blocks and 3D solids.

Plotstyle changes the plot style of the objects; available only when plot styles are turned on.

RELATED COMMANDS

AttRedef changes a block or attribute.

ChProp contains the properties portion of the **Change** command.

Color changes the current color setting.

Elev changes the working elevation and thickness.

LtScale changes the linetype scale.

Properties changes most aspects of all objects.

PlotStyle sets the plot style.

RELATED SYSTEM VARIABLES

CeColor contains the current color setting.

CeLType contains the current linetype setting.

CelWeight contains the current lineweight.

CircleRad contains the current circle radius.

CLayer contains the name of the current layer.

CPlotstyle contains the name of the current plot style.

Elevation contains the current elevation setting.

LtScale contains the current linetype scale.

TextSize contains the current height of text.

TextStyle contains the current text style.

Thickness contains the current thickness setting.

TIPS

- The **Change** command cannot change the size of donuts, the radius or length of arcs, the length of polylines, or the justification of text.
- Use this command to change the endpoints of a group of lines to a common vertex:

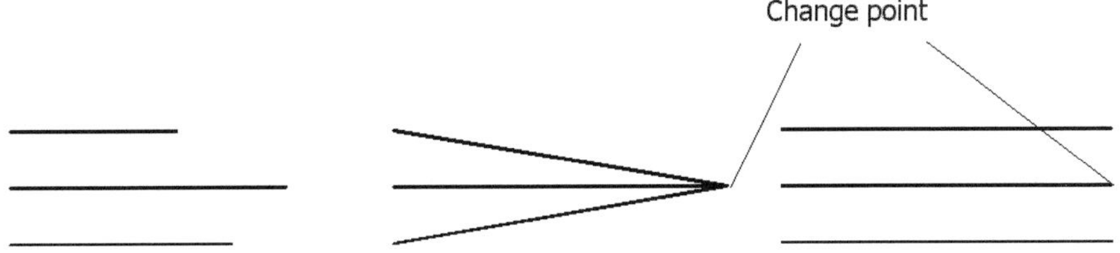

Change *with ortho mode turned off (center) and ortho mode turned on (at left).*

- Turn on ortho mode to extend or trim a group of lines, without needing a cutting edge (as do the **Extend** and **Trim** commands).
- The **PlotStyle** option is not displayed when plot styles are not turned on.

'CheckStandards

2001 Checks the drawing for adherence to standards previously specified by the **Standards** command.

Command	Alias	Ctrl+	F-key	Alt+	Menu Bar	Tablet
checkstandards	**chk**	...	...	**TSK**	**Tools** ↳**CAD Standards** ↳**Check**	...

Command: checkstandards

When the ***Standards*** *command has not set up standards for the drawing, AutoCAD displays this error message:*

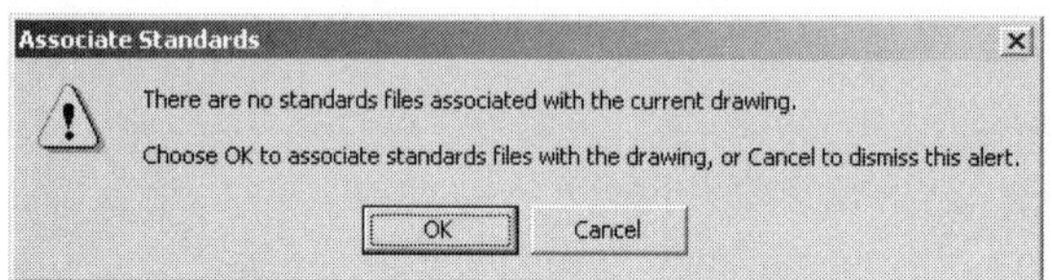

When standards have been set up for the drawing, this command checks the drawing against the CAD standards, and then displays this dialog box:

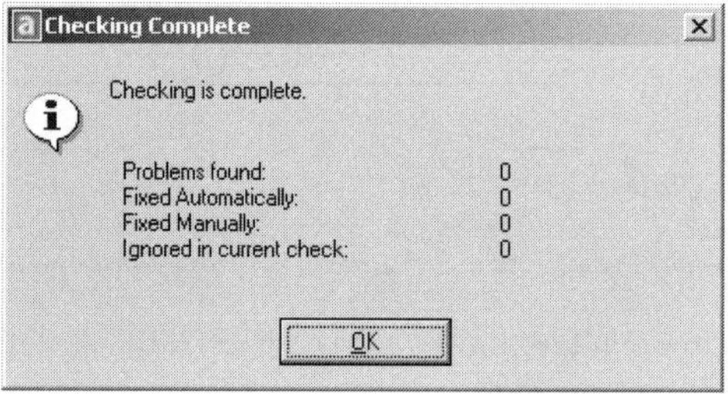

Click ***OK****.*

To change settings in the ***Configure Standards*** *dialog box, click* ***Settings****.*

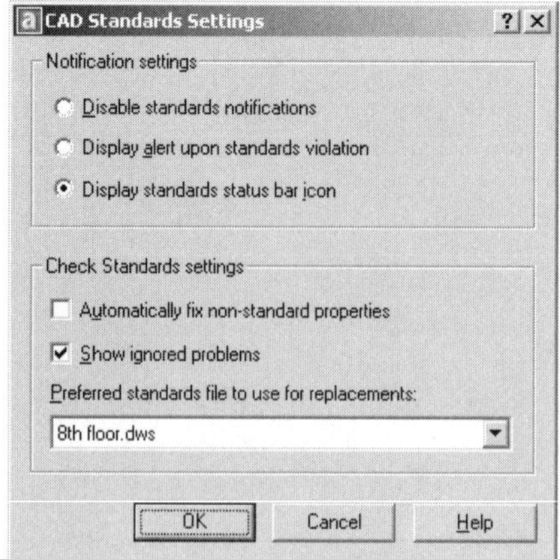

DIALOG BOX OPTIONS

Check Standards Settings dialog box

Notification settings:

○ Disable standards notifications.

○ Display alert upon standards violation.

⊙ Display standards status bar icon.

Automatically fix non-standard properties:

☑ Fixes properties not matching the CAD standard automatically.

☐ Allows you to step manually through the properties not matching the CAD standard.

Show ignored problems:

☑ Displays problems marked as ignored.

☐ Does not display ignored problems.

Preferred standards file to use for replacements selects the default *.dws* file.

Check Standards dialog box

When an object does not match the standards, AutoCAD displays this dialog box:

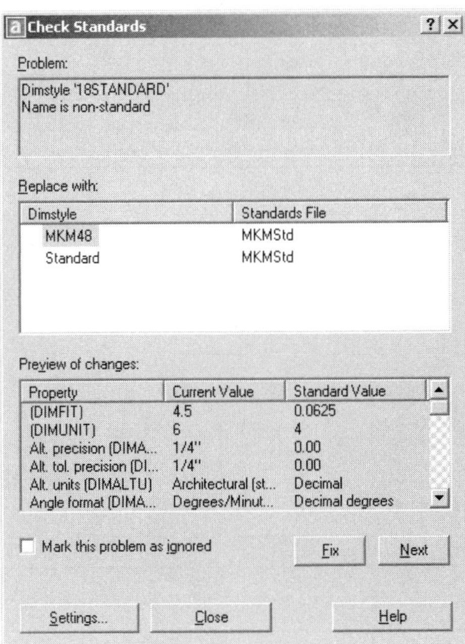

Problem describes the property in the drawing that does not match the standard; this dialog box displays one problem at a time.

Replace With lists linetypes, text styles, etc. found in the *.dws* standards file.

Mark this problem as ignored:

☑ Ignores nonstandard properties, and marks them with the user's login name; some errors are always ignored by AutoCAD, such as settings for layer 0 and DefPoints.

☐ Does not ignore nonstandard properties.

Fix replaces the nonstandard property with the selected standard; the color check mark icon means a fix is available.

Next displays the next nonstandard property.

Settings displays the Check Standards Settings dialog box.

Close closes the dialog box; displays this warning dialog box if standards are not fully checked:

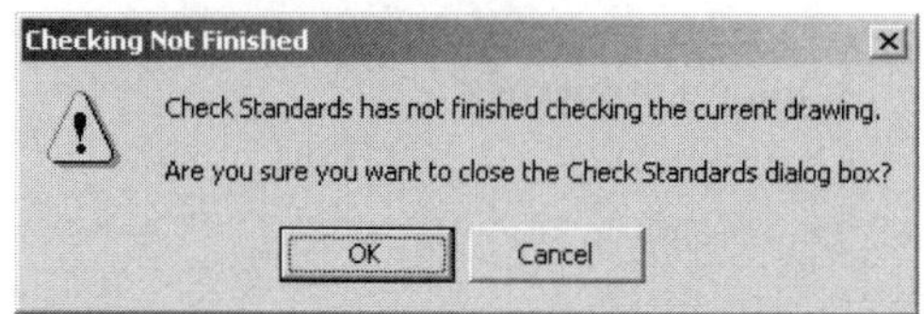

RELATED COMMANDS

Standards selects *.dws* standards files.

LayTrans translates layers between drawings.

RELATED SYSTEM VARIABLES

StandardsViolation determines whether alerts are displayed when a CAD standard is violated in the current drawing:

StandardsViolation	Meaning
0	Notification turned off.
1	Alert displayed when a violation occurs.
2	Icon displayed in status bar tray.

RELATED PROGRAM

DwgCheckStandards.Exe is the external Batch Standards Checker program that checks one or more drawings at a time. It checks drawings without needing AutoCAD. See Appendix B.

RELATED FILES

***.dws** drawing standards file; stored in DWG format.

***.chs** standard check file; stored in XML format.

TIPS

- While this dialog box is open, you can use the following shortcut keys:

Shortcut	Meaning
F4	Fix problem.
F5	Next problem.

- AutoCAD includes sample drawing standards files in the *support* folder.

ChProp

Rel.10 Modifies the color, layer, linetype, linetype scale, lineweight, plot style, and thickness of most objects.

Command	Alias	Ctrl+	F-key	Alt+	Menu Bar	Tablet
chprop	...	...	...	...	...	...

Command: chprop
Select objects: *(Select one or more objects.)*
Select objects: *(Press* **ENTER***.)*
Enter property to change [Color/LAyer/LType/ltScale/LWeight/Thickness/Plotstyle]: *(Enter an option.)*

COMMAND LINE OPTIONS

Color changes the color of the object.

LAyer moves the object to a different layer.

LType changes the linetype of the object.

ltScale changes the linetype scale.

LWeight changes the lineweight of the object.

Thickness changes the thickness of all objects, except blocks.

Plotstyle changes the plot style of the object; available only when plot styles are turned on.

RELATED COMMANDS

Change changes lines, circles, blocks, text and attributes.

Color changes the current color setting.

Elev changes the working elevation and thickness.

LtScale changes the linetype scale.

LWeight sets the lineweight options.

Properties changes most aspects of all objects.

PlotStyle sets the plot style.

RELATED SYSTEM VARIABLES

CeColor specifies the current color setting.

CeLType specifies the current linetype setting.

CelWeight specifies the current lineweight.

CLayer contains the name of the current layer.

CPlotstyle contains the name of the current plot style.

LtScale contains the current linetype scale.

Thickness specifies the current thickness setting.

TIPS

- Use the **Change** command to change the elevation of an object.
- The **Plotstyle** option is not displayed when plot styles are not turned on.

Circle

V. 1.0 Draws 2D circles by five different methods.

Command	Alias	Ctrl+	F-key	Alt+	Menu Bar	Tablet
circle	c	...	...	DC	Draw ↳Circle	J9

Command: circle
Specify center point for circle or [3P/2P/Ttr (tan tan radius)]: *(Pick a point, or enter an option.)*
Specify radius of circle or [Diameter]: *(Pick a point, or type* **D***.)*

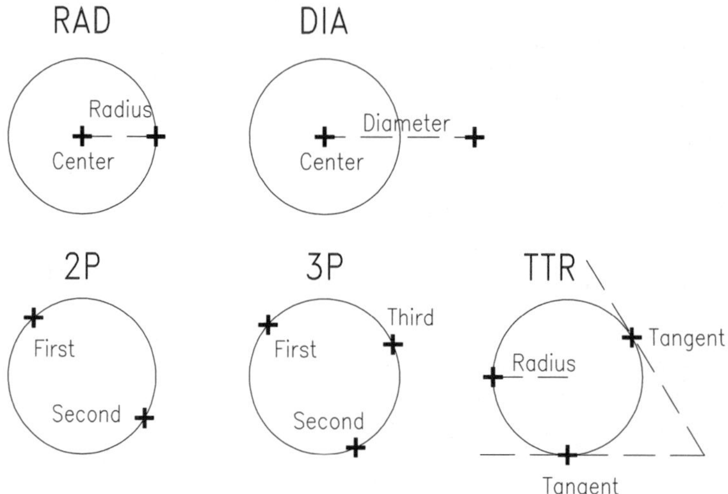

COMMAND LINE OPTIONS

Center and Radius or Diameter circle options
Specify center point for circle or [3P/2P/Ttr (tan tan radius)]: *(Pick a point, or enter an option.)*
Specify radius of circle or [Diameter] <0.5>: *(Pick a point, or type* **D***.)*
Specify diameter of circle <0.5>: *(Enter a value.)*

Center point indicates the circle's center point.

Radius indicates the circle's radius.

Diameter indicates the circle's diameter.

3P (three-point) circle options
Specify first point on circle: *(Pick a point.)*
Specify second point on circle: *(Pick a point.)*
Specify third point on circle: *(Pick a point.)*

First point indicates the first point on the circle.

Second point indicates the second point on the circle.

Third point indicates the third point on the circle.

2P (two-point) circle options

Specify first end point of circle's diameter: *(Pick a point.)*

Specify second end point of circle's diameter: *(Pick a point.)*

First end point indicates the first point on the circle.

Second end point indicates the second point on the circle.

TTR (tangent-tangent-radius) circle options

Specify point on object for first tangent of circle: *(Pick a point.)*

Specify point on object for second tangent of circle: *(Pick a point.)*

Specify radius of circle <0.5>: *(Enter a radius.)*

First tangent indicates the first point of tangency.

Second tangent indicates the second point of tangency.

Radius indicates the first point of radius.

Tan Tan Tan *(menu-only option)*:

From the **Draw** menu, select **Circle | Tan,Tan,Tan**. AutoCAD prompts:

Command: _circle Specify center point for circle or [3P/2P/Ttr (tan tan radius)]: _3p Specify first point on circle: _tan to *(Pick an object.)*

Specify second point on circle: _tan to *(Pick an object.)*

Specify third point on circle: _tan to *(Pick an object.)*

AutoCAD draws a circle tangent to the three points, if possible.

RELATED TOOLBAR ICONS

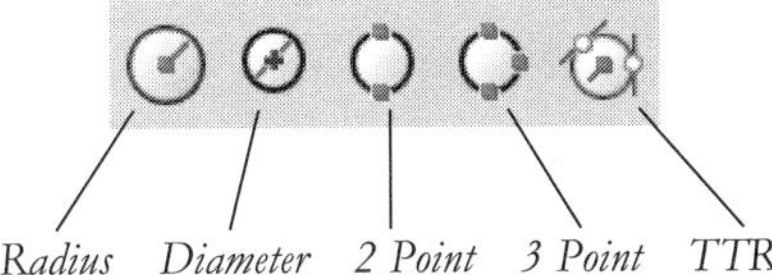

RELATED COMMANDS

Ai_CircTan draws a circle tangent to three objects.

Arc draws an arc.

Donut draws a solid-filled circle or donut.

Ellipse draws an elliptical circle or arc.

Sphere draws a 3D solid ball.

ViewRes controls the visual roundness of circles.

RELATED SYSTEM VARIABLE

CircleRad specifies the default circle radius.

TIPS

- The 3P (three-point) circle defines three points on the circle's circumference.
- When drawing a TTR (tangent, tangent, radius) circle, AutoCAD draws the circle with tangent points closest to the pick points; note that more than one circle placement is possible.
- Selecting **Draw | Circle | Tan, Tan, Radius** from the menu bar automatically turns on the TANgent object snap.
- Giving a circle thickness turns it into an open cylinder.

CleanScreenOn / CleanScreenOff

2004 Maximizes the drawing area.

Command	Alias	Ctrl+	F-key	Alt+	Menu Bar	Tablet
cleanscreenon	...	0	...	VC	**View** ↳**Clean Screen**	...
cleanscreenoff	...	0	...	VC	**View** ↳**Clean Screen**	...

Command: cleanscreenon

AutoCAD turns off the title bar, toolbars, and window edges, and maximizes the AutoCAD window to the full size of your computer screen:

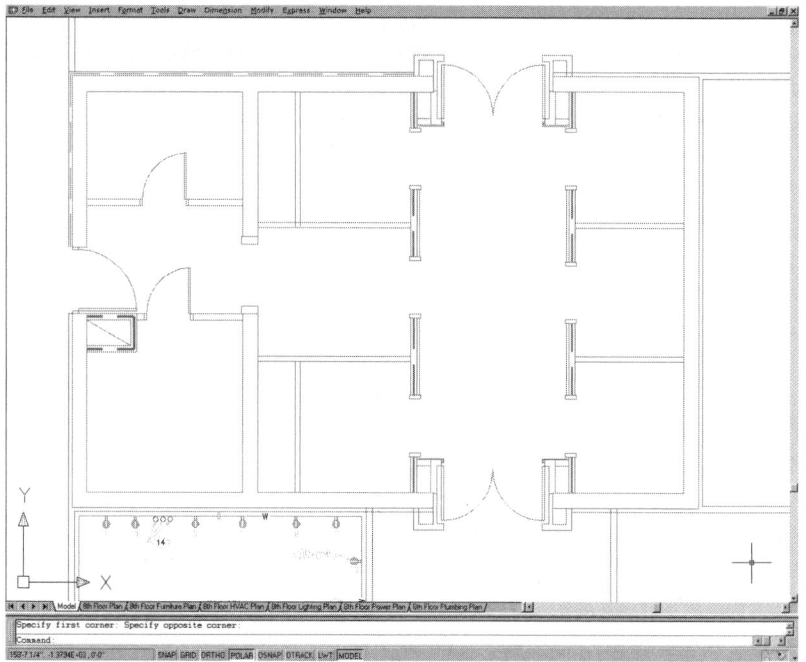

Command: cleanscreenoff

AutoCAD returns to normal.

COMMAND LINE OPTIONS

None.

TIPS

- For an even larger drawing area, turn off the scroll bars and layout tabs through the **Options | Display** dialog box, and drag the command prompt area into a window.
- To toggle this command, you can press CTRL+**0** (zero).

Close / CloseAll

2000 Closes the current drawing, or all drawings; does not exit AutoCAD.

Command	Alias	Ctrl+	F-key	Alt+	Menu Bar	Tablet
close	...	**F4**	...	**FC**	**File** **↳Close**	...
closeall	...	...	...	**WL**	**Window** **↳Close All**	...

Command: close

Displays dialog box if drawing not saved since last change:

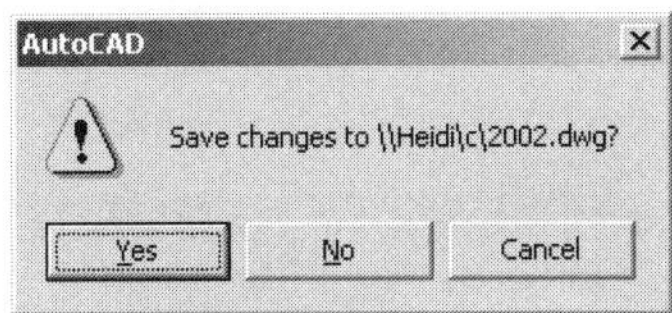

DIALOG BOX OPTIONS

Yes displays the Save Drawing As dialog box, or saves the drawing if it has been previously saved.

No exits the drawing without saving it.

Cancel returns to the drawing.

RELATED COMMANDS

Quit exits AutoCAD.

Open opens additional drawings, each in its own window.

CloseAll closes all drawings, displaying the same warning dialog box as illustrated above.

TIP

- As an alternative to this command, you can click the **x** button on the title bar:

'Color

V. 2.5 Sets the new working color.

Commands	Aliases	Ctrl+	F-key	Alt+	Menu Bar	Tablet
color	**ddcolor**	...	...	**OC**	**Format**	**U4**
	col				**↳Color**	
	colour					
-color						

Command: color

Displays dialog box:

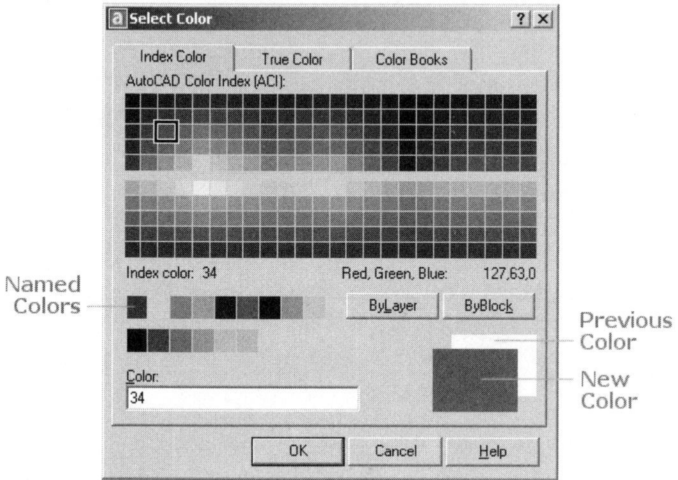

DIALOG BOX OPTIONS

Index Color tab

AutoCAD Color Index (ACI) selects one of AutoCAD's colors (short for "AutoCAD Color Index").

Bylayer sets color to BYLAYER (color 256); color 257 is ByEntity (*new to AutoCAD 2005*).

Byblock sets color to BYBLOCK.

Color sets the color by number or name.

True Color tab

Hue selects the hue (color), ranging from 0 (red) to 360 (violet).

Saturation selects the saturation (intensity of color), ranging from 0 (gray) to 100 (color).

Luminance selects the luminance (brightness of color), ranging from 0 (white) to 100 (black).

Color specifies the color number as hue, saturation, luminance.

Color model selects the type of color model:

- **HSL** is hue, saturation, luminance.
- **RGB** is red, green, blue.

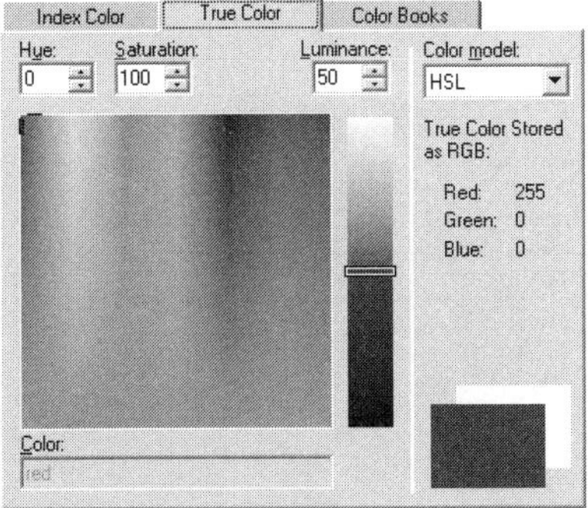

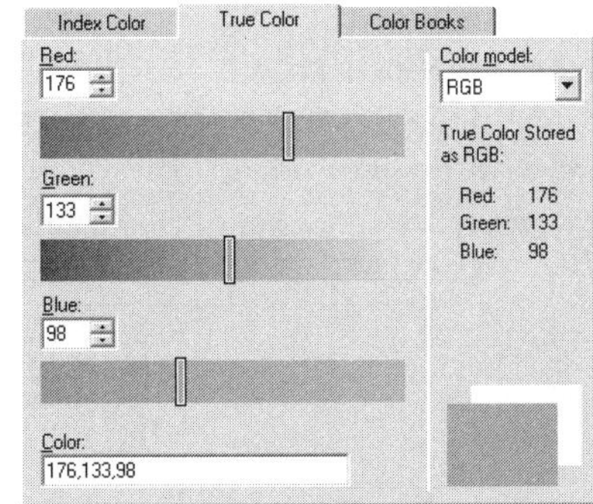

Red selects the range of red from 0 to 255.

Green selects the range of green from 0 to 255.

Blue selects the range of blue from 0 to 255.

Color specifies the color number as red, green, blue.

Color Book tab

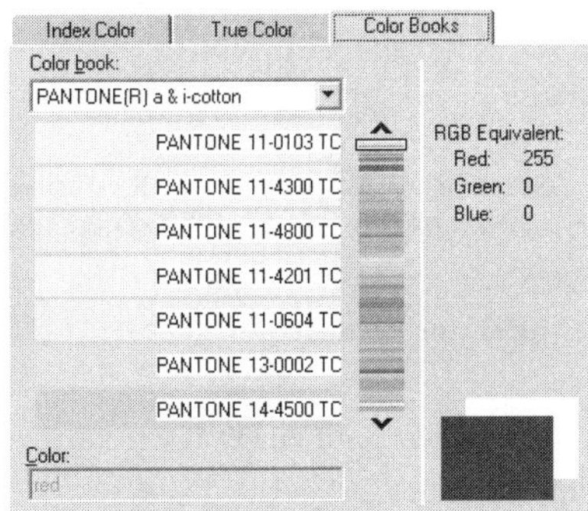

Color book selects a predefined collection of colors.

Color specifies the name of the Pantone, RAL, and DIC colors.

Historical notes: The **Pantone Color System** was designed in 1963 to specify color for graphic arts, textiles, and plastics, based on the assumption that colors are seen differently by different individuals. Designers typically work with Pantone's fan-format book of standardized colors <www.pantone.com> .

The **RAL** color system was designed in 1927 to standardize colors by limiting the number of color gradations, at first to just 30, but now to over 1,600. RAL (Reichs Ausschuß für Lieferbedingungen – German for "Imperial Committee for Supply Conditions") is administered by the German Institute for Quality Assurance and Labeling <www.ral.de>.

The **DIC Color Guide** is the Japanese standard for colors, developed by Dainippon Ink and Chemicals <www.dic.co.jp/eng/index.html> (*new to AutoCAD 2005*).

-COLOR Command

Command: -color

Enter default object color [Truecolor/COlorbook] <BYLAYER>: *(Enter a color number or name, or enter an option.)*

COMMAND LINE OPTIONS

BYLAYER sets the working color to the color of the current layer.

BYBLOCK sets the working color of inserted blocks.

color number sets working color using number (1 through 255), name, or abbreviation:

Color Number	Color Name	Abbreviation	Comments
1	Red	R	
2	Yellow	Y	
3	Green	G	
4	Cyan	C	
5	Blue	B	
6	Magenta	M	
7	White	W	Or black.
8 - 249			Additional colors.
250 - 255			Shades of gray.

Truecolor options

Red, Green, Blue: *(Specify color values separated by commas.)*

Red, Green, Blue specifies color by red, green, and blue in the range from 0 to 255:

Red, Green, Blue: 255,128,0

COlorbook options

Enter Color Book name: *(Specify a name, such as* ***Pantone****.)*

Enter color name: *(Enter a color name, such as* ***11-0103TC****.)*

Enter Color Book name specifies the name of a color book.

Enter color name specifies the name of the color.

RELATED COMMANDS

ChProp changes the color via the command line in fewer keystrokes.

Properties changes the color of objects via a dialog box.

RELATED SYSTEM VARIABLES

CeColor is the current object color setting.

TIPS

- 'BYLAYER' means that objects take on the color assigned to that layer.When BYLAYER objects on layer 0 are part of a block definition, they take on the properties of whichever layer the block was inserted or the layer to which the block was moved.
- 'BYBLOCK' objects in a block definition take on the color assigned to the block. This may be the color in effect at the time the block was inserted, or the color to which the block was changed.
- When more than one method is used to assign colors objects in a block, results may be confusing.
- White objects display as black when the background color is white.
- Color "0" cannot be specified; AutoCAD uses it internally as the background color.
- The *colorwh.dwg* drawing has 255-color and 16.7-million color wheels.

Compile

Rel.12 Compiles *.shp* shape and font files and *.pfb* font files into *.shx* files.

Command	Alias	Ctrl+	F-key	Alt+	Menu Bar	Tablet
compile	...	...	...	...	...	...

Command: compile

Displays Select Shape or Font File dialog box. Select a .shp or .pfb file, and then click ***Open****.*

DIALOG BOX OPTIONS

Open opens the *.shp* or *.pfb* file for compiling.

Cancel closes the dialog box without loading the file.

RELATED COMMANDS

Load loads compiled *.shx* shape files into the current drawing.

Style loads *.shx* and *.ttf* font files into the current drawing.

RELATED SYSTEM VARIABLE

ShpName is the current *.shp* file name.

RELATED FILES

.shp are AutoCAD font and shape source files.

.shx are AutoCAD compiled font and shape files.

.pfb are PostScript Type B font files.

TIPS

- As of AutoCAD Release 12, **Style** converts *.shp* font files on-the-fly; it is only necessary to use the **Compile** command to obtain an *.shx* font file.
- As of Release 14, AutoCAD no longer directly supports PostScript font files. Instead, use the **Compile** command to convert *.pfb* files to *.shx* format.
- TrueType fonts are not compiled.

Cone

Rel.11 Draws 3D solid cones with circular or elliptical bases.

Command	Alias	Ctrl+	F-key	Alt+	Menu Bar	Tablet
cone	...	...	...	DIO	Draw ↳Solids ↳Cone	M7

Command: cone

Cones with a circular base:

Current wire frame density: ISOLINES=4
Specify center point for base of cone or [Elliptical] <0,0,0>: *(Pick center point 1.)*
Specify radius for base of cone or [Diameter]: *(Specify radius 2, or type* **D***.)*
Specify height of cone or [Apex]: *(Specify height 3, or type* **A***.)*

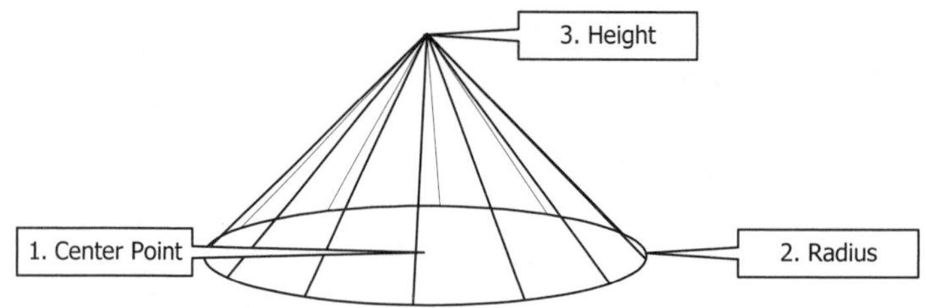

Cones with an elliptical base and an apex:

Elliptical base options
Specify center point for base of cone or [Elliptical] <0,0,0>: *(Type* **E***.)*
Specify axis endpoint of ellipse for base of cone or [Center]: *(Pick point 1, or type* **C***.)*
Specify second axis endpoint of ellipse for base of cone: *(Pick point 2.)*
Specify length of other axis for base of cone: *(Pick point 3.)*
Specify height of cone or [Apex]: *(Specify height 4, or type* **A***.)*

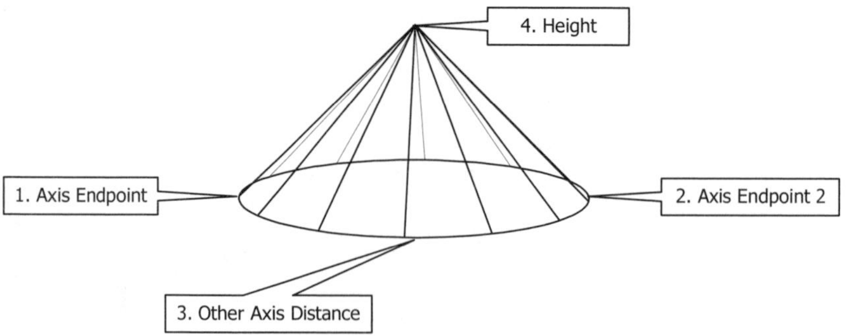

COMMAND LINE OPTIONS

Elliptical draws a cone with an elliptical base.

Center point specifies the center point of the cone's base.

Axis endpoint picks the endpoint of the axis defining the elliptical base.

Center of ellipse picks the center of the elliptical base.

Height specifies the height of the cone.

Apex determines the height and orientation of the cone's tip.

RELATED COMMANDS

Ai_Cone draws a 3D wireframe cone.

Box draws a 3D solid box.

Cylinder draws a 3D solid tube.

Sphere draws a 3D solid ball.

Torus draws a 3D solid donut.

Wedge draws a 3D solid wedge.

RELATED SYSTEM VARIABLES

DispSilh toggles display of silhouettes lines.

IsoLines specifies the number of isolines on solid surfaces:

IsoLines	Meaning
0	Minimum; no isolines.
4	Default.
12 *or* **16**	A reasonable value.
2047	Maximum value.

TIPS

- You define the elliptical base in two ways: by the length of the major and minor axes, or by the center point and two radii.
- To draw a cone at an angle, use the **Apex** option.
- Silhouette lines are displayed if **IsoLines** is set to 0. No need to hide or shade to see them.

Changed Command

The **Config** command now displays the Options dialog box; see the **Options** command.

Convert

Rel.14 Converts 2D polylines and associative hatches (created in R13 and earlier) to an optimized "lightweight" format.

Command	Alias	Ctrl+	F-key	Alt+	Menu Bar	Tablet
convert	...	...	...	...	...	...

Command: convert
Enter type of objects to convert [Hatch/Polyline/All] <All>: *(Enter an option.)*
Enter object selection preference [Select/All] <All>: *(Type* **S** *or* **A**.*)*

COMMAND LINE OPTIONS

Hatch converts associative hatch patterns from anonymous blocks to hatch objects; displays warning dialog box:

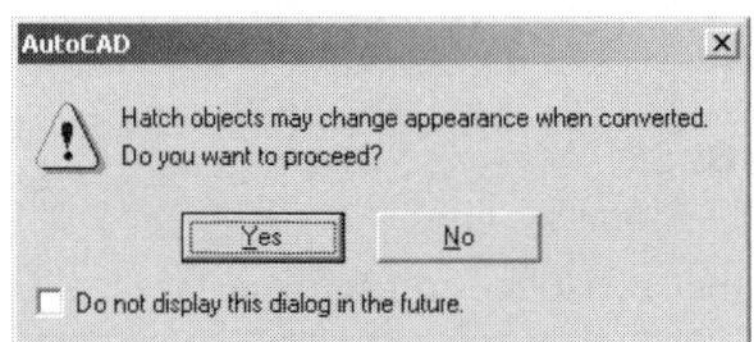

Yes converts hatch patterns to hatch objects.

No does not convert hatch patterns.

Polyline converts 2D polylines to Lwpolyline objects.

All converts all polylines and hatch patterns.

Select selects the hatch patterns and 2D polylines to convert.

RELATED COMMANDS

BHatch creates associative hatch patterns.

PLine draws 2D polylines.

RELATED SYSTEM VARIABLE

PLineType determines whether pre-Release 14 polylines are converted in AutoCAD.

PlineType	Meaning
0	Not converted; **PLine** creates old-format polylines.
1	Not converted; **PLine** creates lwpolylines.
2	Converted; **PLine** creates lwpolylines (default).

TIPS

- When a Release 13 or earlier drawing is opened in AutoCAD, it automatically converts most (not all) 2D polylines to lwpolylines; hatch patterns are not automatically updated.
- Hatch patterns are automatically updated the first time the **HatchEdit** command is applied, or when their boundaries are changed.
- Polylines are not converted when they contain curve fit segments, splined segments, extended object data in their vertices, or 3D polylines.
- **PLineType** affect the following commands: **Boundary** (polylines), **Donut**, **Ellipse** (**PEllipse** = 1), **PEdit** (when converting lines and arcs), **Polygon**, and **Sketch** (**SkPoly** = 1).

ConvertCTB

2001 Converts a plot style file from CTB color-dependent format to STB named format (*short for CONVERT Color TaBle*).

Command	Alias	Ctrl+	F-key	Alt+	Menu Bar	Tablet
convertctb	...	...	...	...	...	...

Command: convertctb

*Displays **Select File** dialog box.*

*1. Select a .ctb file, and then click **Open**. AutoCAD displays the Create File dialog box.*

*2. Specify the name of an .stb file, and then click **Save**.*

When AutoCAD has completed the conversion, it displays dialog box:

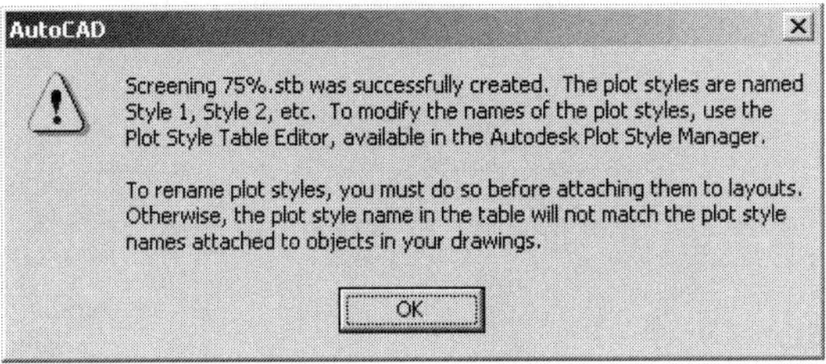

DIALOG BOX OPTIONS

Open opens the *.ctb* file.

Save saves as an *.stb* file.

RELATED COMMANDS

ConvertPStyles converts a drawing between color-dependent and named plot styles.

PlotStyle sets the plot style for a drawing.

RELATED FILES

***.ctb** are color-dependent plot style table files.

***.stb** are named plot style tables.

TIPS

- "Color-dependent plot styles" are used by older versions of AutoCAD, where the color of the object controls the pen selection.
- "Named plot styles" is the alternative introduced with AutoCAD 2000, which allows plotter-specific information to be assigned to layers and objects.

ConvertPStyles

2001 Converts a drawing between color-dependent and named plot styles (*short for Convert Plot STYLES*).

Command	Alias	Ctrl+	F-key	Alt+	Menu Bar	Tablet
convertpstyles	...	...	...	...	...	...

Command: convertpstyles

Displays warning dialog box:

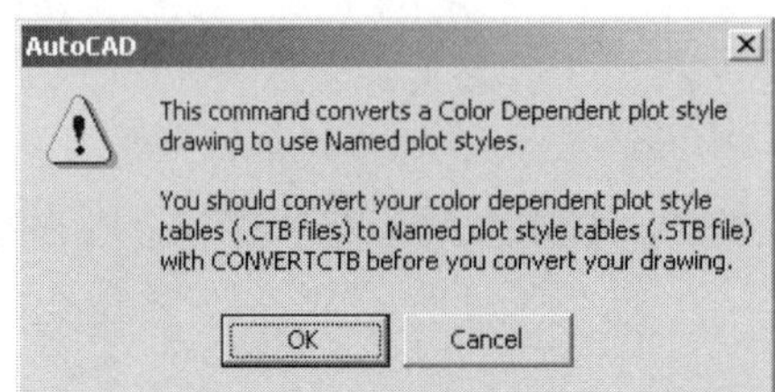

When AutoCAD has completed the conversion, it displays the message, "Drawing converted from Named plotstyle mode to Color Dependent mode."

DIALOG BOX OPTIONS

OK proceeds with the conversion.

Cancel prevents the conversion.

RELATED COMMANDS

ConvertCTB converts a plot style file from CTB color-dependent format to STB named format.

PlotStyle sets the plot style for a drawing.

TIPS

- "Color-dependent plot styles" was used by older versions of AutoCAD, where the color of the object controlled the pen selection.
- "Named plot styles" is the alternative introduced with AutoCAD 2000, which allows plotter-specific information to be assigned to layers and objects.

Copy

V. 1.0 Creates one or more copies of an object.

Command	Alias	Ctrl+	F-key	Alt+	Menu Bar	Tablet
copy	co	...	...	MY	Modify	V15
	cp				↳Copy	

Command: copy
Select objects: *(Select one or more objects — point 1.)*
Select objects: *(Press* ENTER.*)*
Specify base point: *(Pick point 2.)*
Specify second point of displacement or <use first point as displacement>: *(Pick point 3.)*
Specify second point of displacement: *(Press* ENTER.*)*

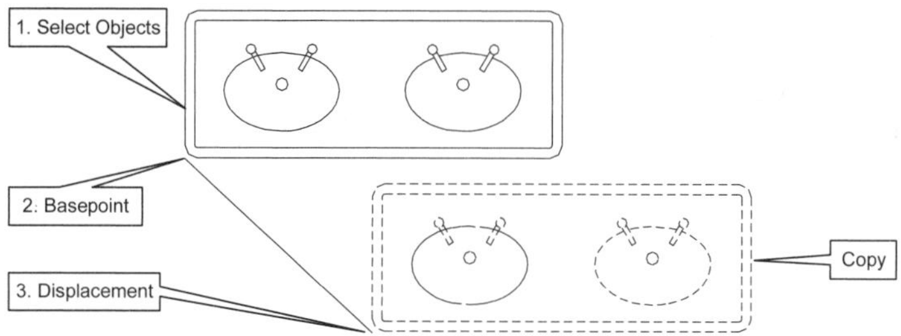

COMMAND LINE OPTIONS

Base point indicates the starting point.

Second point indicates the point to which to copy.

Displacement uses the base point as the displacement when you press ENTER.

RELATED COMMANDS

Array draws a rectangular or polar array of objects.

MInsert places an array of blocks.

Move moves an object to a new location.

Offset draws parallel lines, polylines, circles, and arcs.

TIPS

- The multiple option is the default, as of AutoCAD 2005; press ESC to end this command.
- Inserting a block multiple times is more efficient than placing multiple copies.
- Turn on ortho mode to copy objects in a precise horizontal and vertical direction; turn snap mode on to copy objects in precise increments; use object snap modes to copy objects precisely from one geometric feature to another.
- To copy an object by a known displacement, enter 0,0 as the 'Base point.' Then enter the known distance as the 'Second point.' Or, enter the known displacement as the first point, and press ENTER for the second point.

CopyBase

2000 Copies selected objects to the Clipboard with a specified base point (*short for COPY with BASEpoint*).

Command	Alias	Ctrl+	F-key	Alt+	Menu Bar	Tablet
copybase	...	Shift+C	...	EB	Edit ↳Copy with Base Point	...

Command: copybase
Specify base point: *(Pick a point.)*
Select objects: *(Select one or more objects.)*
Select objects: *(Press* **ENTER***.)*

COMMAND LINE OPTIONS

Specify base point specifies the base point.

Select objects selects the objects to copy to the Clipboard.

RELATED COMMANDS

CopyClip copies selected objects to the Clipboard with a base point equal to the lower-left extents of the selected objects.

PasteBlock pastes objects from the Clipboard into drawings as blocks.

PasteClip pastes objects from the Clipboard into drawings.

TIPS

- When **PasteBlock** pastes objects previously selected with the **CopyBase** command, AutoCAD prompts you 'Specify insertion point:', and then pastes the objects as a block with a name similar to A$C7E1B27BE.
- When specifying the **All** option at the 'Select objects:' prompt, **CopyBase** selects only objects visible in the current viewport.
- As of AutoCAD 2004, you can use the CTRL+SHIFT+C shortcut for this command.

CopyClip

Rel.12 Copies selected objects from the drawing to the Clipboard (*short for COPY to CLIPboard*).

Command	Alias	Ctrl+	F-key	Alt+	Menu Bar	Tablet
copyclip	...	C	...	EC	Edit ↳Copy	T14

Command: copyclip
Select objects: *(Select one or more objects.)*
Select objects: *(Press* **ENTER**.*)*

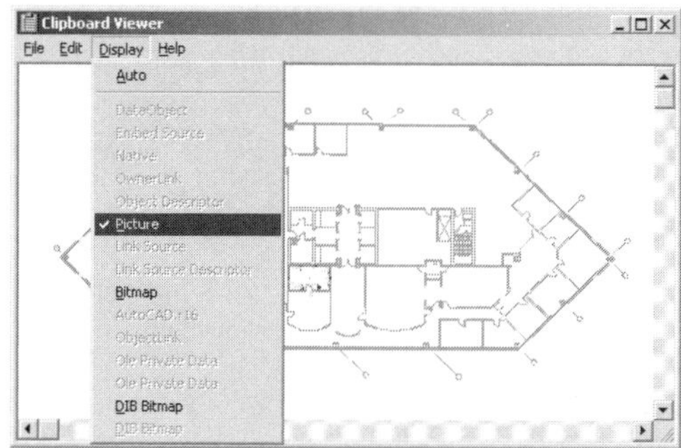

The Clipboard Viewer shows an object copied from AutoCAD to the Clipboard.

COMMAND LINE OPTION

Select objects selects the objects to copy to Clipboard.

RELATED COMMANDS

CopyBase copies objects to the Clipboard with a specified base point.

CopyHist copies Text window text to the Clipboard.

CopyLink copies the current viewport to the Clipboard.

CutClip cuts selected objects to the Clipboard.

PasteBlock pastes objects from the Clipboard into the drawing as a block.

PasteClip pastes objects from the Clipboard into the drawing.

RELATED WINDOWS COMMANDS

PRT SCR copies the entire screen to the Clipboard

ALT+PRT SCR copies the topmost window to the Clipboard.

TIPS

- Contrary to the AutoCAD *Command Reference*, text objects are *not* copied to the Clipboard in text format; instead, text is copied as an AutoCAD object.
- When the **All** option is specified at the 'Select objects' prompt, **CopyClip** selects only objects visible in the current viewport.

CopyHist

Rel.13 Copies all of the Text window text to the Clipboard (*short for COPY HISTory*).

Command	Alias	Ctrl+	F-key	Alt+	Menu Bar	Tablet
copyhist	...	...	...	**EH**	**Edit** ↳**Copy History**	...

Command: copyhist

AutoCAD copies all text in the history window to the Clipboard.

COMMAND LINE OPTIONS

None.

RELATED COMMAND

CopyClip copies selected text from the drawing to the Clipboard.

RELATED WINDOWS COMMAND

ALT+PRT SCR copies the Text window to the Clipboard in graphics format.

TIPS

- To copy a selected portion of Text window text to the Clipboard, highlight the text first, then select **Edit | Copy** from the Text window's menu bar.

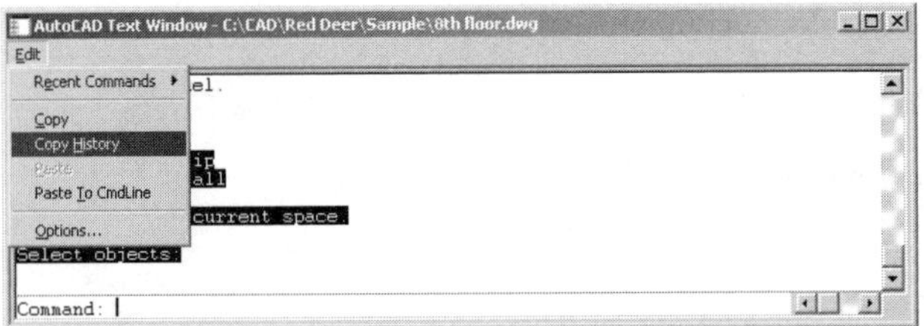

- To paste text to the command line, select **Edit | Paste to Cmdline** from the menu bar. However, this only works when the Clipboard contains text — not graphics.
- As an alternative, you can right-click in the **Text** window to bring up the cursor menu.

Quick Start Tutorial

Getting the Right Clipboard Result

You can use the Clipboard to display AutoCAD drawings in other Windows applications, such as word processing, desktop publishing, and paint programs. AutoCAD has two primary commands for copying objects from the drawing to the Clipboard: **CopyLink** and **CopyClip**. The result of using these commands, however, may surprise you, since each command produces a different result, depending on whether AutoCAD is in model or layout mode.

- **CopyClip** (**Edit | Copy**) copies selected objects from the drawing, as follows:

Command: copyclip
Select objects: all
Select objects: *(Press* **ENTER***.)*

Warning! When you select **All** objects, AutoCAD selects only those objects visible in the current viewport.

- **CopyLink** (**Edit | Copy Link**) copies everything visible in the current viewport. Use this command to capture a layout view with *all* viewports.

The two commands take on different meanings, as described below.

Model Mode

In *model mode*, AutoCAD copies only objects visible in the current viewport. If the drawing contains more than one viewport, you must select the correct viewport before copying objects to the Clipboard. **CopyClip** and **CopyLink** have the same effect:

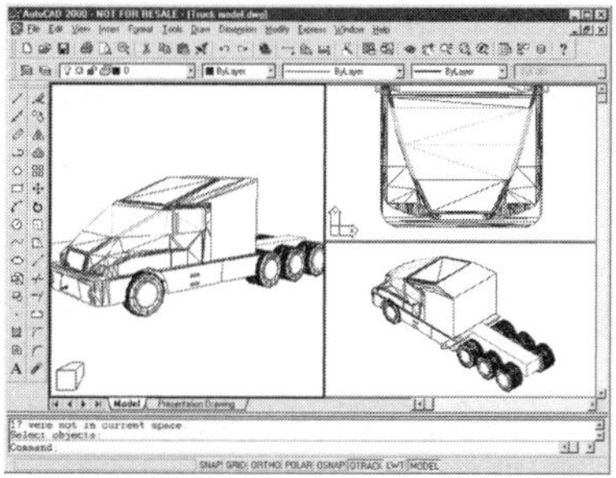

AutoCAD viewports in model mode.

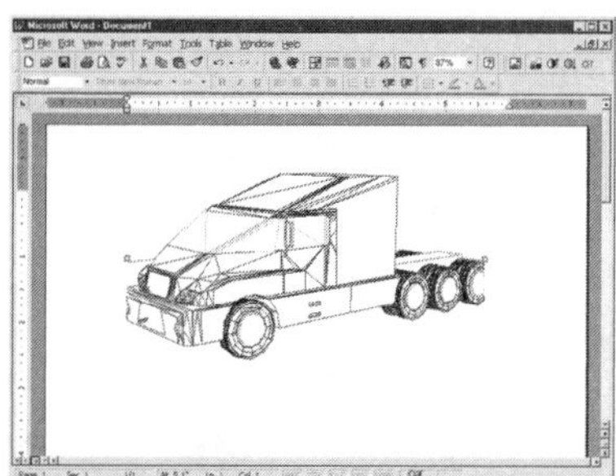

Drawing pasted in Word after either ***CopyClip*** *or* ***CopyLink****.*

Layout Mode

In layout mode's **PAPER** space, **CopyClip** copies objects drawn only in paper space — *nothing* drawn in model space is copied to the Clipboard, even if it is visible.

CopyLink copies *all* visible objects, whether drawn in paper space or model space. Notice that this command also copies the margin lines and gray background.

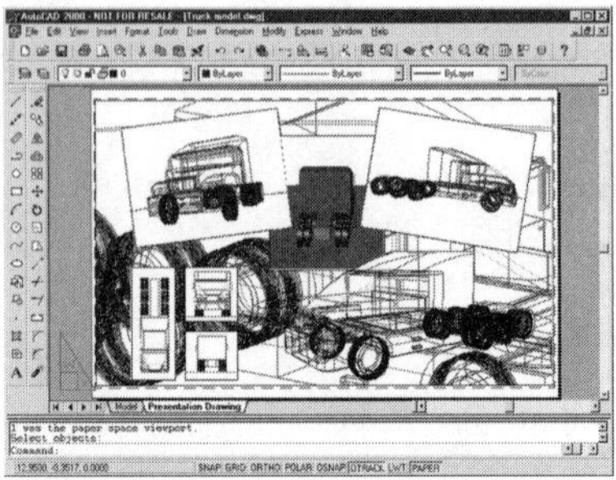

AutoCAD in layout mode's PAPER space.

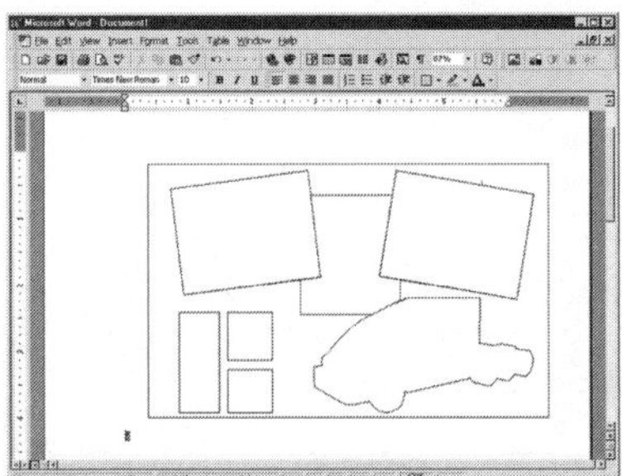

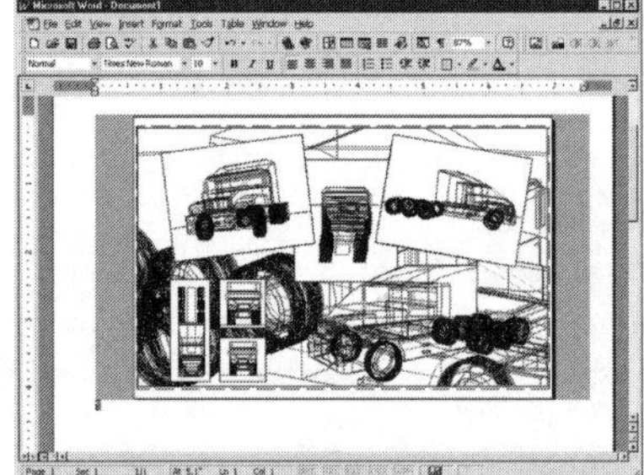

*Result in Word after using AutoCAD's **ClipClip** (left) and **CopyLink** (right).*

In layout mode's **MODEL** space, AutoCAD only copies objects drawn in model space that are visible in the selected viewport. **CopyClip** and **CopyLink** produce the same result when pasted in Word and other Windows applications.

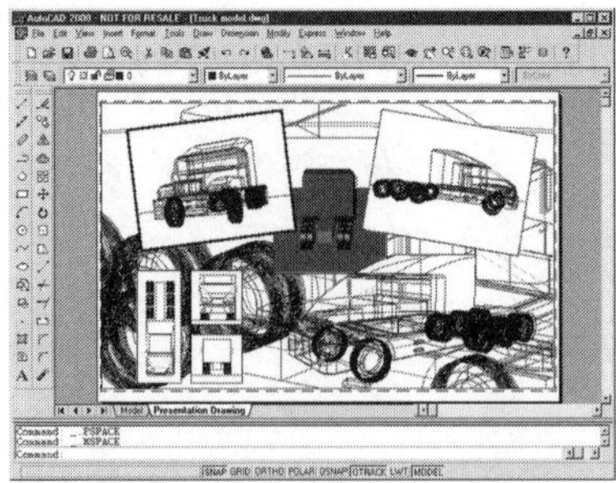

A viewport in layout mode's MODEL space.

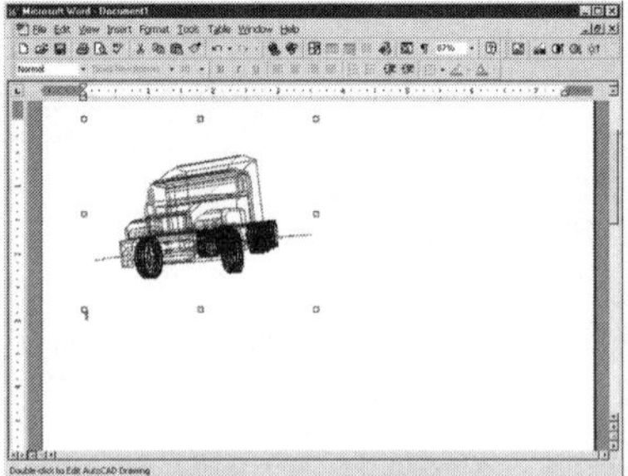

*Drawing pasted in Word after either **CopyClip** or **CopyLink**.*

CopyLink

Rel.13 Copies the current viewport to the Clipboard; optionally allows you to link the drawing back to AutoCAD.

Command	Alias	Ctrl+	F-key	Alt+	Menu Bar	Tablet
copylink	...	...		EL	Edit ↳Copy Link	...

Command: copylink

COMMAND LINE OPTIONS

None.

RELATED COMMANDS

CopyClip copies selected objects to the Clipboard.

CopyEmbed copies selected objects to the Clipboard.

CopyHist copies Text window text to the Clipboard.

CutClip cuts selected objects to the Clipboard.

PasteClip pastes objects from the Clipboard into the drawing.

RELATED WINDOWS COMMANDS

PRT SCR copies the entire screen to the Clipboard

ALT+PRT SCR copies the topmost window to the Clipboard.

TIPS

- In the other application, use the **Edit | Paste Special** commands to paste the AutoCAD image into the document; to link the drawing back to AutoCAD, select the **Paste Link** option.
- AutoCAD does not let you link a drawing to itself.
- This command copies everything in the current viewport (if in model space) or the entire drawing (if in paper space).
- **CopyEmbed** is identical to **CopyLink**, except that **CopyEmbed** prompts you to select objects.

Customize

2000i Creates and changes toolbars, toolbar macros and icons, and keyboard shortcuts.

Command	Aliases	Ctrl+	F-key	Alt+	Menu Bar	Tablet
customize	**toolbar**	...	...	**TC**	**Tools**	**T13**
	to				**↳Customize**	
	tbconfig					
+customize						

Command: customize

Displays dialog box:

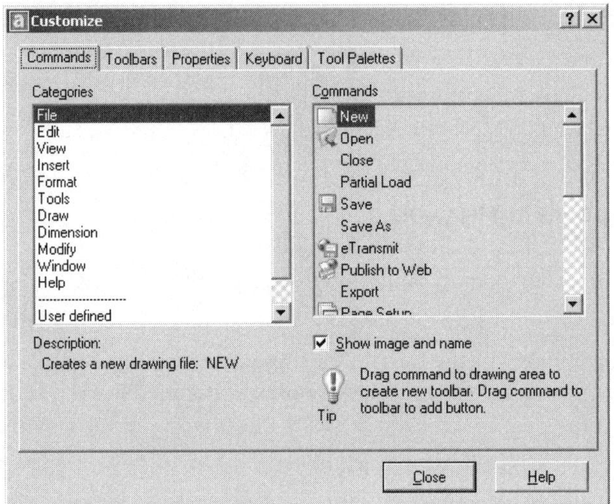

DIALOG BOX OPTIONS

Close closes the dialog box, and saves the changes.

Commands tab

Categories lists names of primary menu items.

Commands lists command names and related icons, if any; drag a command name out of the dialog box to start creating a new toolbar.

Show image and name

☑ Lists all command names and related icons.

☐Llists icons only; commands without an icon are shown as a gray square.

Toolbars tab

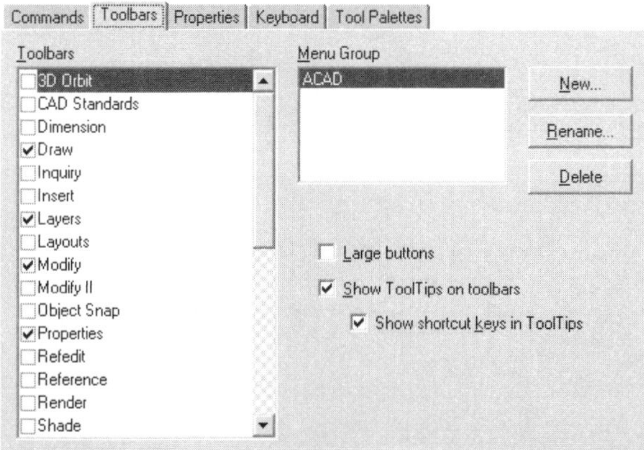

Toolbars lists the names of defined toolbars, and toggles their visibility.

Menu Group lists the names of menu groups loaded in AutoCAD.

New creates a new toolbar; displays New Toolbar dialog box, which prompts you to select the menu group to which the new toolbar should be added.

Rename changes the name of the selected toolbar; displays the Rename Toolbar dialog box.

Delete deletes the selected toolbar; displays a warning dialog box: "Are you sure you want to delete the toolbar from the menu group?"

Large buttons

- ☑ Toolbar buttons and icons are displayed at a larger size for better visibility on very-high resolution screens, or for visually-impaired users (40 pixels wide).
- ☐ Toolbar buttons and icons displayed at standard size (24 pixels wide).

Normal buttons (left) and large buttons (right).

Show tooltips on toolbars

- ☑ Displays a small yellow tag with the command name.
- ☐ Does not display tooltips.

Show shortcut keys in Tooltips

- ☑ Displays shortcut keystrokes, such as CTRL+N, in the tooltip.
- ☐ Does not display shortcut keystrokes.

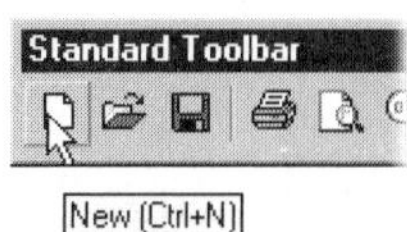

Tooltip with shortcut key CTRL+N.

Button Properties tab

When you first click the ***Button Properties*** *tab, AutoCAD prompts:*

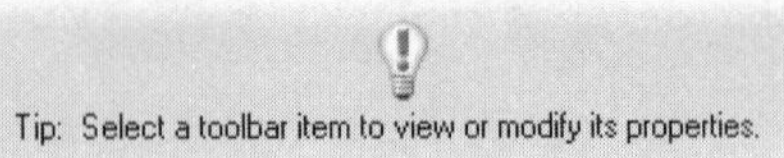

Tip: Select a toolbar item to view or modify its properties.

Select any toolbar button; AutoCAD displays the ***Button Properties*** *tab:*

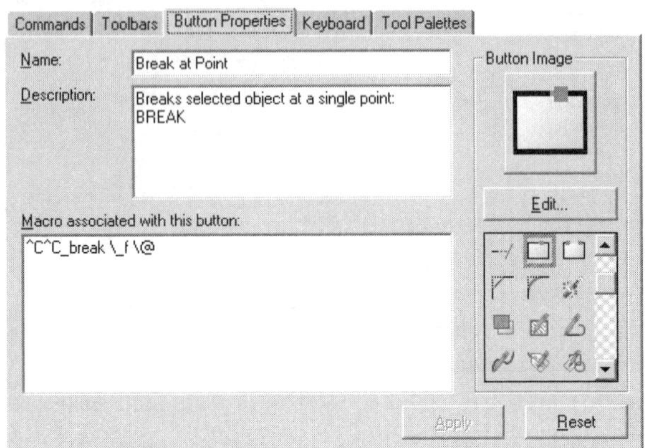

Name specifies the wording for the tooltip.

Description specifies the help text displayed on the status line.

Macro associated with this button specifies the macro executed when the button is clicked.

Edit edits the icon; displays the Button Editor dialog box.

Apply applies the changes to the button.

Reset changes the parameters back.

Keyboard tab

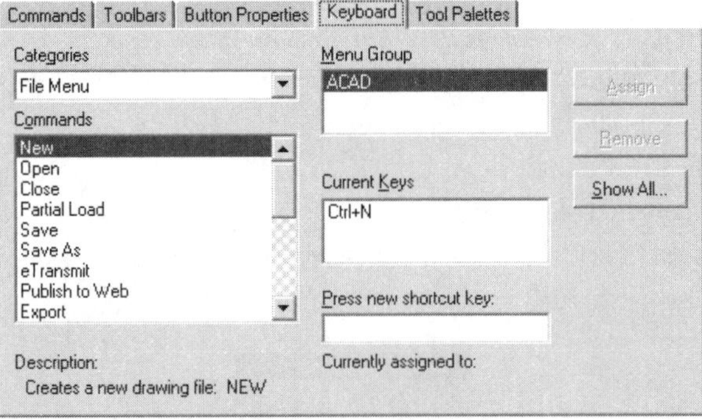

Categories allows you to select the name of a toolbar, menu item, or all AutoCAD commands.

Commands lists the commands, as specified by the Categories droplist.

Menu Group lists the names of menu groups loaded in AutoCAD.

Current Keys lists the shortcut key(s) assigned to the command; this option appears not to work with preassigned keys, such as CTRL+N.

Press new shortcut key selects the shortcut key: hold down CTRL or CTRL+SHIFT, plus a letter or number key between A and Z, and 1 and 0.

Assign assigns the shortcut keystroke to the command.

Remove removes the shortcut keystroke from the command.

Show All lists all assigned shortcut keys; displays the Shortcut Keys dialog box.

Tool Palettes tab

Redesigned in AutoCAD 2005.

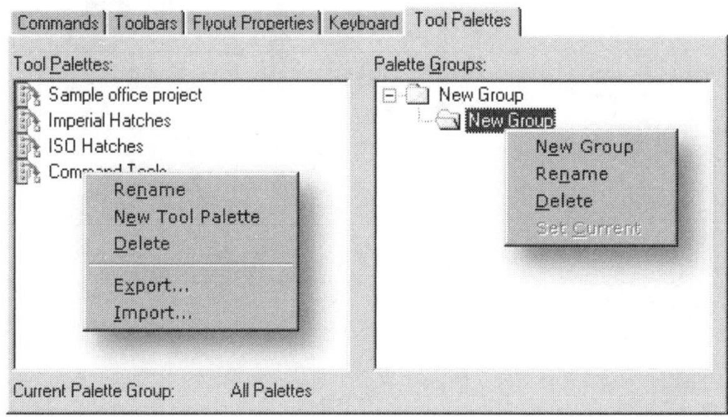

Tool Palettes shortcut menu

Rename renames the selected palette.

New Tool Palette creates a new, blank palette.

Delete removes the selected palette.

Import imports palettes (*.xtp* files).

Export exports the selected palette in an XML-like *.xtp* file.

Palette Groups shortcut menu

New Group creates new groups and subgroups.

Rename renames the selected group.

Delete removes the selected group.

Set Current selects the group to make current.

Shortcut Keys dialog box

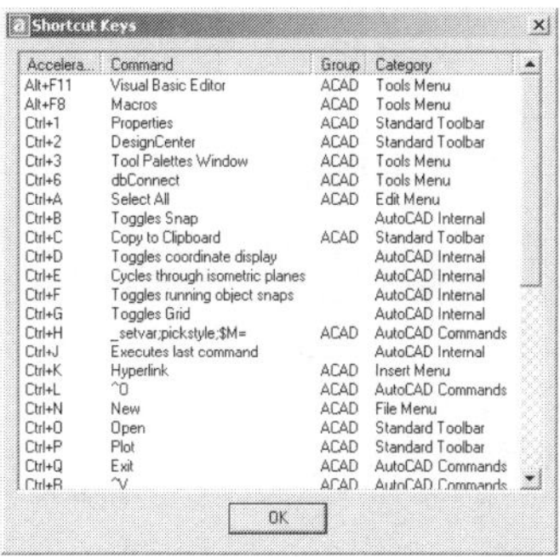

Shortcut Keys

Accelera...	Command	Group	Category
Alt+F11	Visual Basic Editor	ACAD	Tools Menu
Alt+F8	Macros	ACAD	Tools Menu
Ctrl+1	Properties	ACAD	Standard Toolbar
Ctrl+2	DesignCenter	ACAD	Standard Toolbar
Ctrl+3	Tool Palettes Window	ACAD	Tools Menu
Ctrl+6	dbConnect	ACAD	Tools Menu
Ctrl+A	Select All	ACAD	Edit Menu
Ctrl+B	Toggles Snap		AutoCAD Internal
Ctrl+C	Copy to Clipboard	ACAD	Standard Toolbar
Ctrl+D	Toggles coordinate display		AutoCAD Internal
Ctrl+E	Cycles through isometric planes		AutoCAD Internal
Ctrl+F	Toggles running object snaps		AutoCAD Internal
Ctrl+G	Toggles Grid		AutoCAD Internal
Ctrl+H	_setvar;pickstyle;$M=	ACAD	AutoCAD Commands
Ctrl+J	Executes last command		AutoCAD Internal
Ctrl+K	Hyperlink	ACAD	Insert Menu
Ctrl+L	^O	ACAD	AutoCAD Commands
Ctrl+N	New	ACAD	File Menu
Ctrl+O	Open	ACAD	Standard Toolbar
Ctrl+P	Plot	ACAD	Standard Toolbar
Ctrl+Q	Exit	ACAD	AutoCAD Commands
Ctrl+R	^V	ACAD	AutoCAD Commands

OK

OK closes the dialog box.

Button Editor dialog box

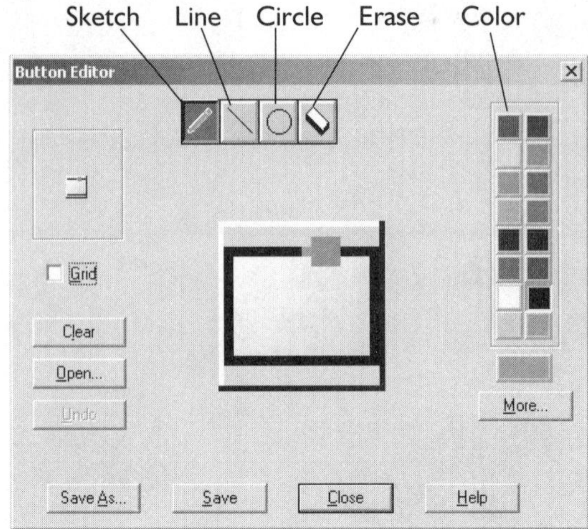

Grid:

☑ Displays a grid in the icon drawing area.

☐ Hides the grid.

Clear erases the icon.

Open opens a *.bmp* (bitmap) file, which can be used as an icon.

Undo undoes the last operation.

Save As saves the icon as a *.bmp* file.

Save saves the changes to the icon in AutoCAD's internal storage.

Close closes the dialog box.

More displays the Select Color dialog box.

RELATED COMMANDS

BmpOut exports selected objects in the current view to a *.bmp* file.

MenuLoad loads partial menus into AutoCAD.

-Toolbar opens and closes toolbars via the command line.

RELATED SYSTEM VARIABLE

TbCustomize toggles the ability to customize toolbars *(new to AutoCAD 2005).*

TIP

- Elements of a toolbar:

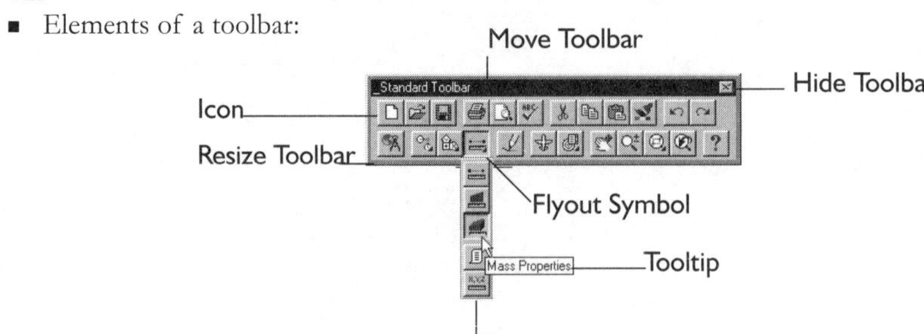

CutClip

Rel.12 Cuts the selected objects from the drawing to the Clipboard (*short for CUT to CLIPboard*).

Command	Alias	Ctrl+	F-key	Alt+	Menu Bar	Tablet
cutclip	...	**X**	...	**ET**	**Edit** **↳Cut**	**T13**

Command: cutclip
Select objects: *(Select one or more objects.)*
Select objects: *(Press* **ENTER**.*)*

COMMAND LINE OPTION

Select objects selects the objects to cut to Clipboard.

RELATED COMMANDS

BmpOut exports selected objects in the current view to a *.bmp* file.

CopyClip copies selected objects to the Clipboard.

CopyHist copies the Text window text to the Clipboard.

CopyLink copies the current viewport to the Clipboard.

PasteClip pastes objects from the Clipboard into the drawing.

WmfOut exports selected objects to *.wmf* files.

RELATED WINDOWS COMMANDS

PRT SCR copies the entire screen to the Clipboard

ALT+PRT SCR copies the topmost window to the Clipboard.

TIPS

- When the **All** option is specified at the 'Select objects:' prompt, **CutClip** selects objects visible in the current viewport only.
- In the other application, use the **Edit | Paste** or **Edit | Paste Special** commands to paste the AutoCAD image into the document; the **Paste Special** command lets you specify the pasted format.
- You can use the **Undo** command to return the "cut" objects to the drawing.

Cylinder

Rel.12 Draws a 3D solid cylinder with a circular or elliptical cross section.

Command	Alias	Ctrl+	F-key	Alt+	Menu Bar	Tablet
cylinder	...	...	...	DIC	Draw ↳Solids ↳Cylinder	L7

Command: cylinder
Current wire frame density: ISOLINES=4
Specify center point for base of cylinder or [Elliptical] <0,0,0>: *(Pick center point 1, or type* **E**.*)*
Specify radius for base of cylinder or [Diameter]: *(Specify radius 2, or type* **D**.*)*
Specify height of cylinder or [Center of other end]: *(Specify height 3, or type* **C**.*)*

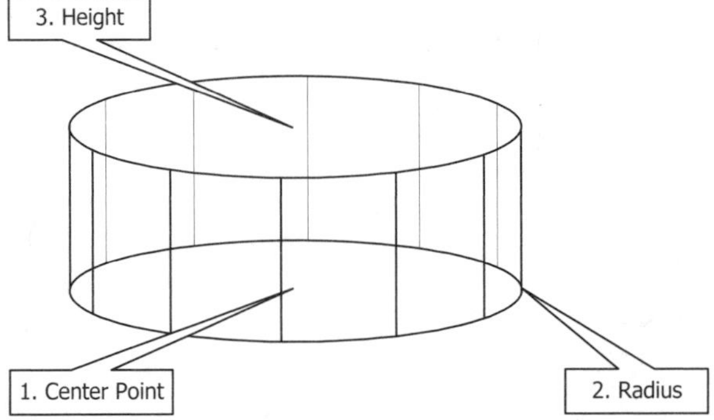

COMMAND OPTIONS

Center point specifies the center point of the cylinder's base.

Radius specifies the cylinder's radius.

Diameter specifies the cylinder's diameter.

Height specifies the cylinder's height.

Center of other end specifies the z orientation of the cylinder.

Elliptical options
Specify axis endpoint of ellipse for base of cylinder or [Center]: *(Pick point, or type* **C**.*)*
Specify second axis endpoint of ellipse for base of cylinder: *(Pick point.)*
Specify length of other axis for base of cylinder: *(Specify length.)*
Specify height of cylinder or [Center of other end]: *(Specify height, or type* **C**.*)*

Axis endpoint specifies one end of the elliptical axis.

Center specifies the center point of the elliptical base.

Second axis endpoint specifies the other end of the axis.

Length of other axis specifies the length of the other elliptical axis.

RELATED COMMANDS

Box draws a 3D solid box.

Cone draws a 3D solid cone.

Elevation turns a circle into a wireframe cylinder.

Extrude creates a cylinder with an arbitrary cross-section and sloped walls.

Sphere draws a 3D solid ball.

Torus draws a 3D solid donut.

Wedge draws a 3D solid wedge.

RELATED SYSTEM VARIABLES

DispSilh displays 3D objects as silhouettes after hidden-line removal and shading.

IsoLines determines the number of isolines on solid surfaces:

IsoLines	Meaning
0	Minimum (no isolines).
4	Default value.
12 *or* **16**	Reasonable values.
2047	Maximum value.

TIPS

- The **Ellipse** option draws a cylinder with an elliptical cross-section.
- Use the **Intersect** command to remove solid portions from the cylinder; use the **Union** command to add solid portions to the cylinder.
- Silhouette lines are displayed if **IsoLines** is set to 0. No need to hide or shade to see them.

DbConnect / DbClose

2000 Opens and closes the dbConnect Manager window to connect objects with rows in external database tables (*short for Data Base CONNECTion*).

Command	Alias	Ctrl+	F-key	Alt+	Menu Bar	Tablet
dbconnect	**dbc**	**6**	...	**TD**	**Tools ↳dbConnect**	**W12**
dbcclose	...	**6**	...	**TD**	**Tools ↳dbConnect**	...

Command: dbconnect

Displays window. (If a red x appears, the database is disconnected from the drawing.)

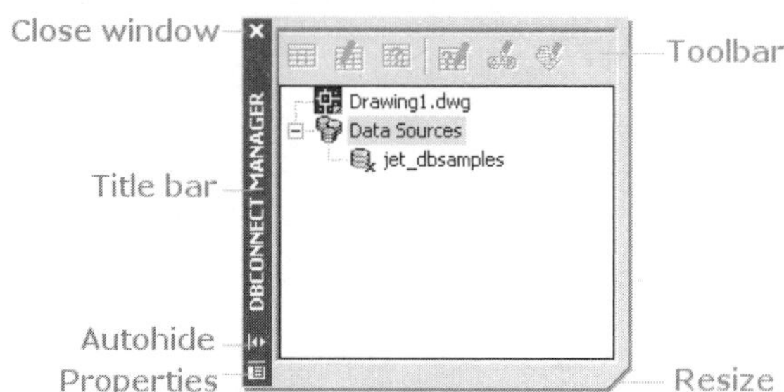

To connect the drawing with the database, right-click the database icon, and then select ***Connect****.*

The icons have the following meaning:

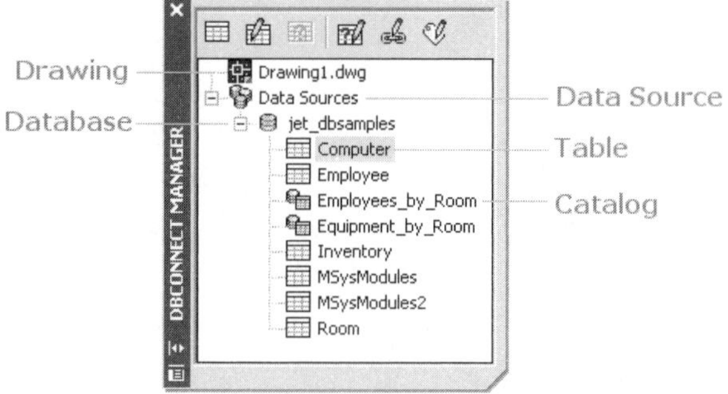

TOOLBAR OPTIONS

View Table opens an external database table in *read-only* mode; select a table, link template, or label template to make this button available.

Edit Table opens an external database table in *edit* mode; select a table, link template, or label template to make this button available.

Execute Query executes a query; select a previously-defined query to make this button available.

New Query displays the New Query dialog box when a table or link template is selected; displays the Query Editor when a query is selected.

New Link Template displays the New Link Template dialog box when a table is selected; displays the Link Template dialog box when a link template is selected; not available for link templates with links already defined in a drawing.

New Label Template displays the New Label Template dialog box when a table or link template is selected; displays the Label Template dialog box when a label template is selected.

View Table and **Edit Table** windows

*The **View Table** and **Edit Table** windows are identical, with the exception that **View Table** is read-only; hence all text in columns is grayed-out.*

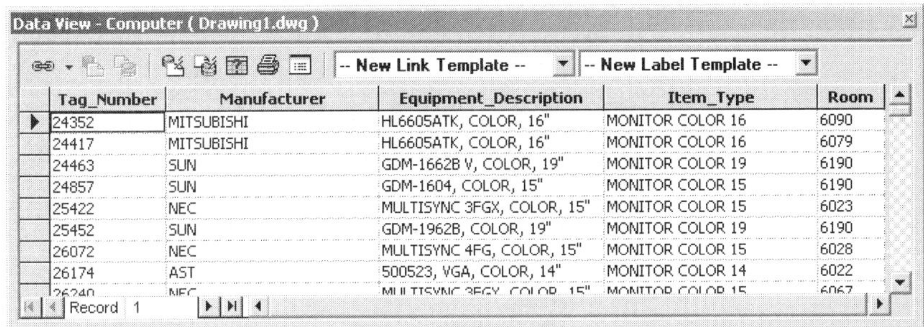

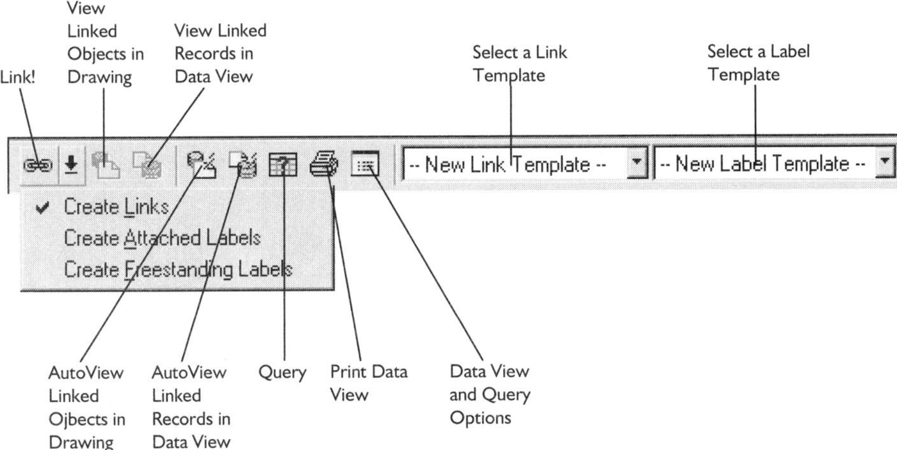

Link! creates a link or a label; click the drop list to select an option:

- **Create Links** turns on link creation mode.
- **Create Attached Labels** turns on the attached label creation mode.
- **Create Freestanding Labels** turns on freestanding label creation mode.

View Link Objects in Drawing highlights objects linked to selected records.

View Linked Records in Data View highlights records that are linked to selected objects in the drawing.

AutoView Linked Objects in Drawing automatically highlights objects that are linked to selected records.

AutoView Linked Records in Data View highlights records that are linked to selected objects in the drawing.

Query displays the New Query dialog box.

Print Data View prints the data in this window.

Data View and Query Options displays the Data View and Query Options dialog box.

Select a Link Template lists the names of previously-defined link templates.

Select a Label Template lists the names of previously-defined label templates.

New Query dialog box

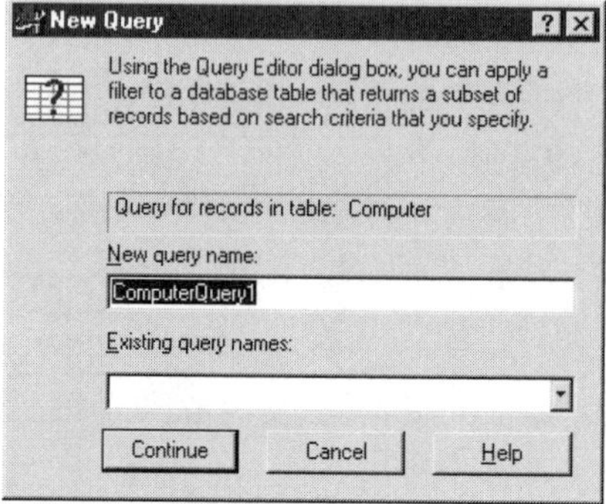

New query name specifies the name of the query.

Existing query names uses an existing query.

Continue displays the Query Editor dialog box.

Query Editor dialog box

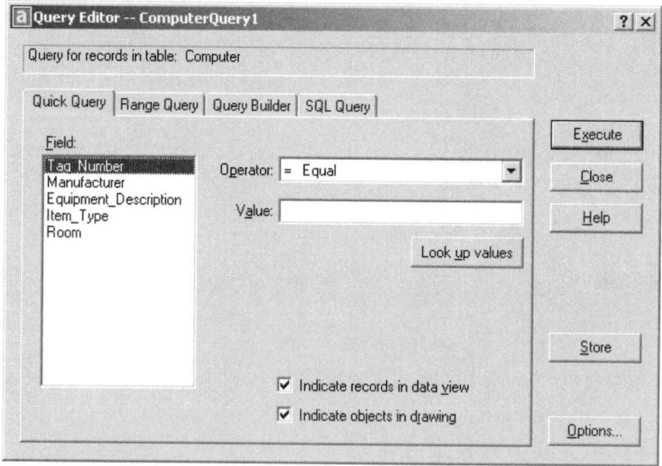

Execute executes the query.

Close closes the dialog box.

Store saves the settings.

Options displays the Data View and Query Options dialog box.

Quick Query tab

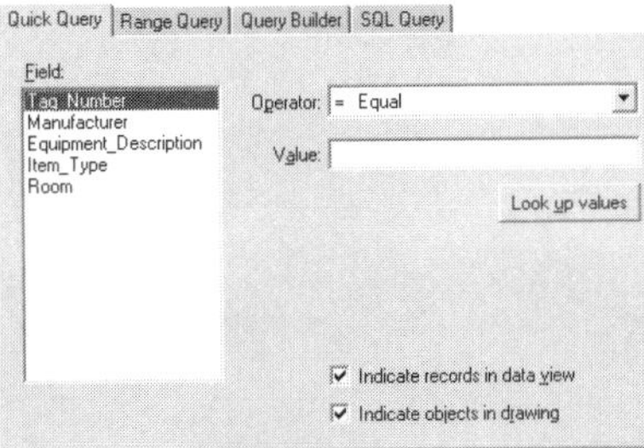

Field selects a field name.

Operator specifies a conditional operator:

Operator	Meaning
=	**Equal** — Exactly match the value (default).
<>	**Not equal** — Does not match the value.
>	**Greater than** — Greater than the value.
<	**Less than** — Less than the value.
>=	**Greater than or equal** — Greater than or equal to the value.
<=	**Less than or equal** — Less than or equal to the value.
Like	Contains the value; use the % wild-card character (equivalent to * in DOS).
In	Matches two values separated by a comma.
Is null	Does not have a value; used for locating records that are missing data.
Is not null	Has a value; used for excluding records that are missing data.

Value specifies the value for which to search.

Look up values displays a list of existing values:

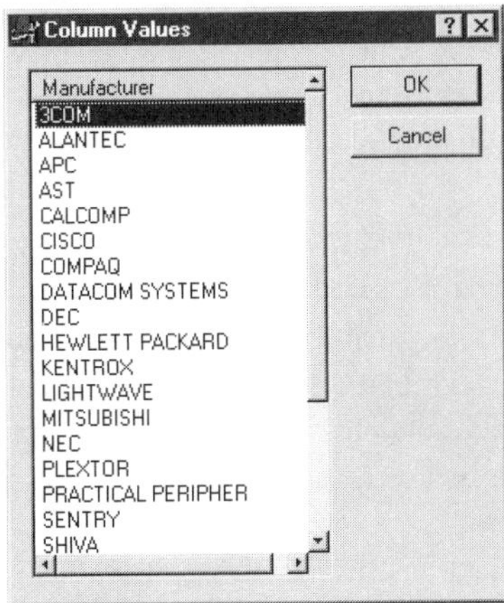

Range Query tab

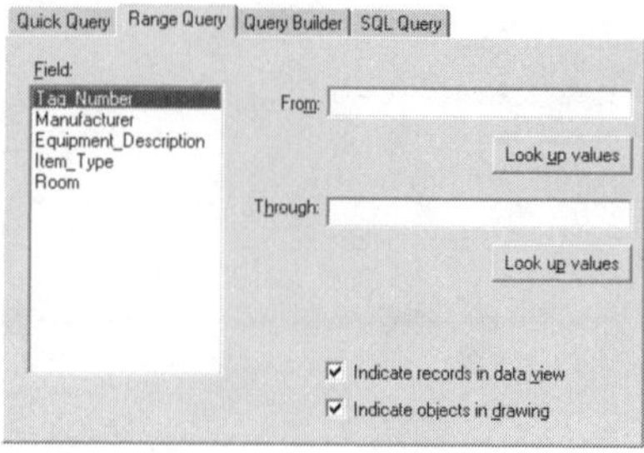

Field lists the names of fields in the current table.

From specifies the first value of the range.

Look Up Values displays the **Column Values** dialog box.

Through specifies the second value of the range.

☑ **Indicate Records in Data View** highlights records that match the search criteria in the Data View window.

☑ **Indicate Objects in Drawing** highlights objects that match the search criteria in drawings.

Query Builder tab

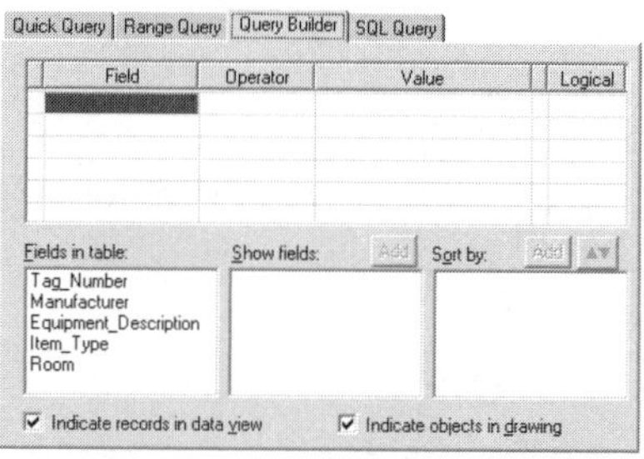

(specifies a opening parenthesis, which groups search criteria with parentheses; up to four sets can be nested. **)** specifies a closing parenthesis.

Field specifies a field name; double-click the cell to display the list of fields in the current table.

Operator specifies a logical operator; double-click to display the list of operators.

Value specifies a value for the query; click **...** to display a list of current values.

Logical specifies an And or Or operator; click once to add And; click again to change to Or.

Fields in table displays the fields in the current table; when no fields are selected, the query displays all fields from the table; double-click a field to add it to the list.

Show fields specifies the fields displayed by the Data View window; drag the field out of the list to remove it.

Add adds a field from the Fields in Table list to the Show Fields list.

Sort By specifies the sort order: the first field is the primary sort; to change the sort order, drag the field to another location in the list; press DELETE to remove a field from the list.

Add adds a field from the Fields in Table list to the Sort By list (default = ascending).

▲▼ reverses sort order.

☑ **Indicate Records in Data View** highlights records that match the search criteria in the Data View window.

☑ **Indicate Objects in Drawing** highlights objects that match the search criteria in drawings.

SQL Query tab

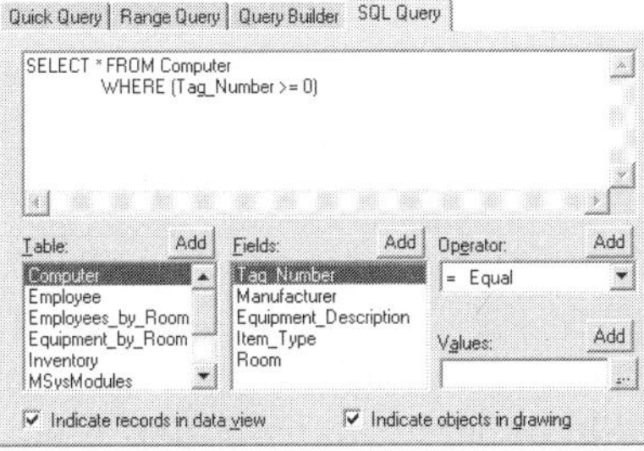

Table lists the names of all database tables available in the current data source.

Add adds the selected table to the SQL text editor.

Fields displays a list of field names in the selected database table.

Add adds the selected field to the SQL text editor.

Operator specifies the logical operator, which is added to the query (default = Equal).

Add adds the selected operator to the SQL text editor.

Values specifies a value for the selected field.

Add adds the value to the SQL text editor.

... lists available values for the field.

☑ **Indicate Records in Data View** highlights records that match the search criteria in the Data View window.

☑ **Indicate Objects in Drawing** highlights objects that match the search criteria in drawings.

Data View and Query Options dialog box

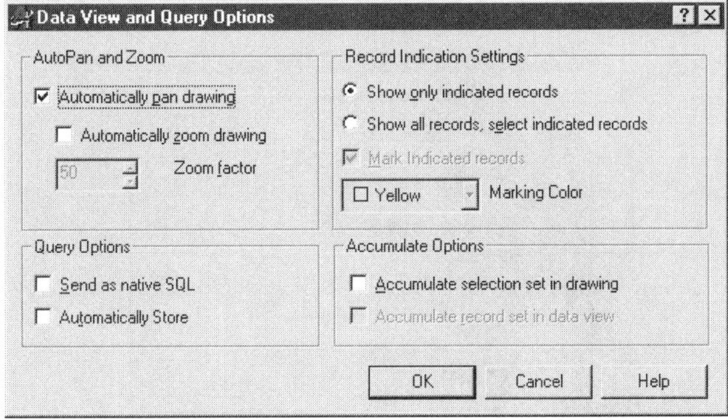

AutoPan and Zoom options

☑ **Automatically Pan Drawing** causes AutoCAD to pan the drawing automatically to display associated objects.

☐ **Automatically Zoom Drawing** causes AutoCAD to zoom the drawing automatically to display associated objects.

Zoom Factor specifies the zoom factor as a percentage of the viewport area:

Zoom Factor	Meaning
20	Minimum.
50	Default.
90	Maximum.

Query Options options

Send as Native SQL makes queries to database tables in:

☑ The format of the source table.

☐ SQL 92 format.

☐ **Automatically Store** automatically stores queries when they are executed (default = off).

Record Indication Settings:

⊙ **Show Only Indicated Records** displays the records associated with the current AutoCAD selections in the Data View window (default).

○ **Show All Records, Select Indicated Records** displays all records in the current database table.

☑ **Mark Indicated Records** colors linked records to differentiate them from unlinked records.

Marking Color specifies the marking color (default = yellow).

Accumulate Options options

Accumulate Selection Set in Drawing

☑ Adds objects to the selection set as data view records are added.

☐ Replaces the selection set each time data view records are selected.

Accumulate Record Set in Data View

☑ Adds records to the selection set as drawing objects are selected.

☐ Replaces the selection set each time drawing objects are selected.

New Link Template dialog box

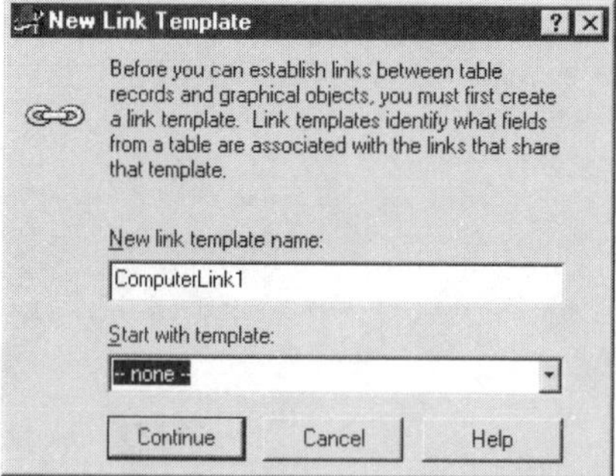

New link template name specifies the name of the link template.

Start with template reuses an existing template.

Continue displays the Link Template dialog box.

Link Template dialog box

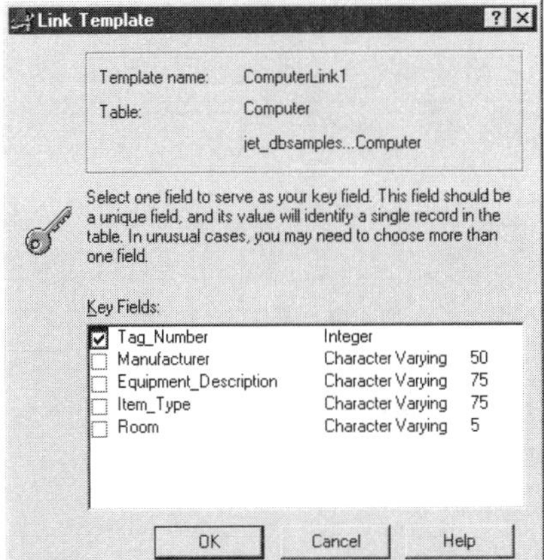

Key Fields selects one field name; you may select more than one field name, but AutoCAD warns you that too many key fields may slow performance.

New Label Template dialog box

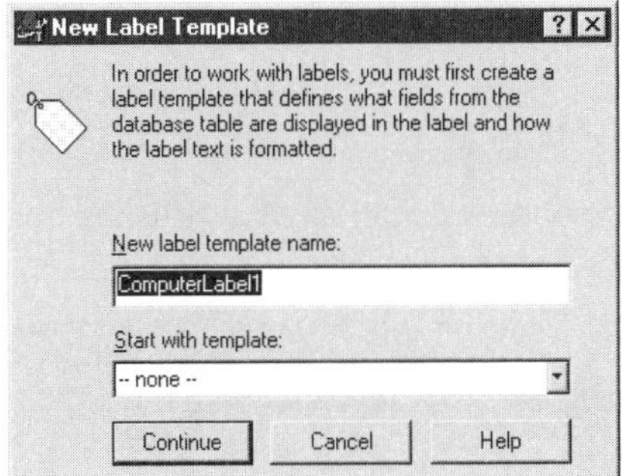

New label template name specifies the name of the label template.

Start with template reuses an existing template.

Continue displays the Label Template dialog box.

Label Template dialog box

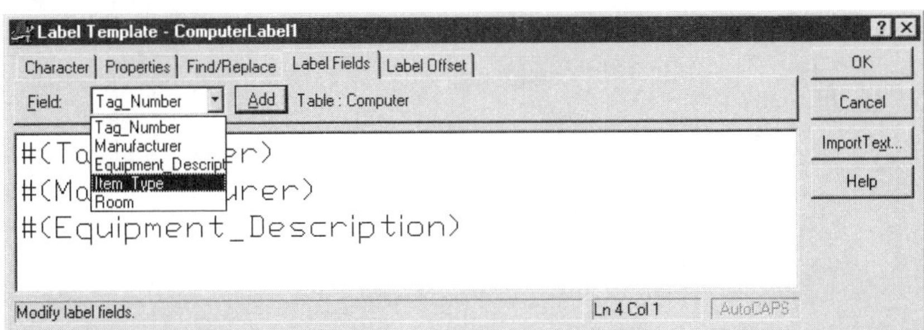

Field specifies the names of the fields.

Add adds the field to the label.

Label Offset displays the Label Offset tab:

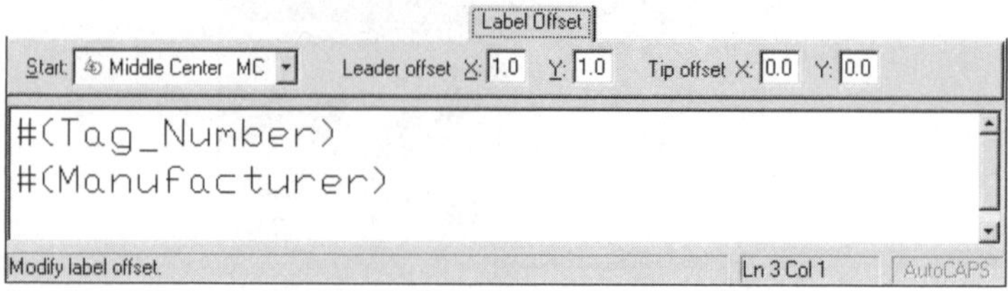

Start specifies the justification of the label starting point.

Leader offset specifies the x,y distance between the label's starting point and the leader line.

Tip offset specifies the offset distance to the leader tip or label text.

See ***MText*** *command for more information.*

RELATED COMMANDS

AttDef creates an attribute definition, akin to an internal database.

EAttExt exports attributes.

TIPS

- Query searches are case sensitive: "Computer" is not the same as "computer."
- OLE DB v2.0 must be installed before you use the **dbConnect Manager**.
- Leaders must have a length; to get rid of a leader, use a freestanding label.
- SQL is short for "structured query language," a standard method of querying databases.
- The properties of a link template can only be edited if it contains no links, and if the drawing is fully loaded (cannot be partially loaded).
- Before you can edit a record with an SQL Server table, you must define a *primary key*.

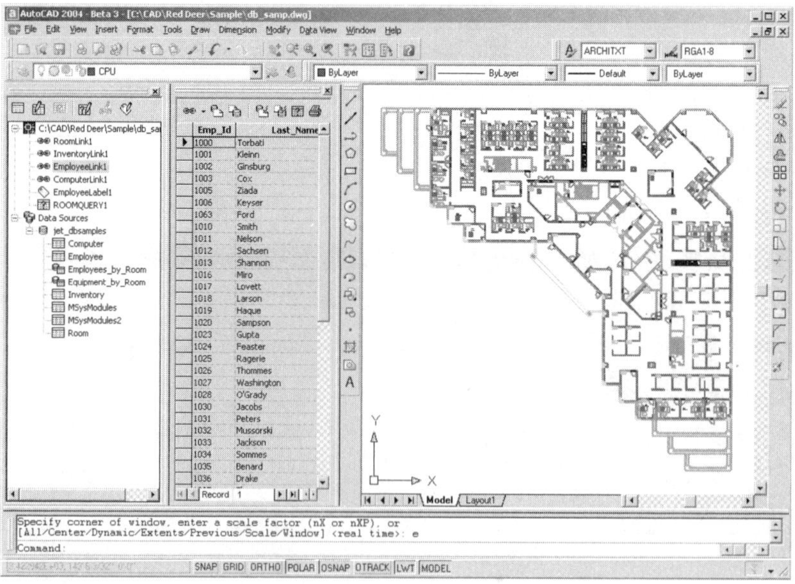

Constructing Your First Query

Step 1

From the **Tools** menu, select **dbConnect**. AutoCAD adds the **dbConnect** item to the menu.

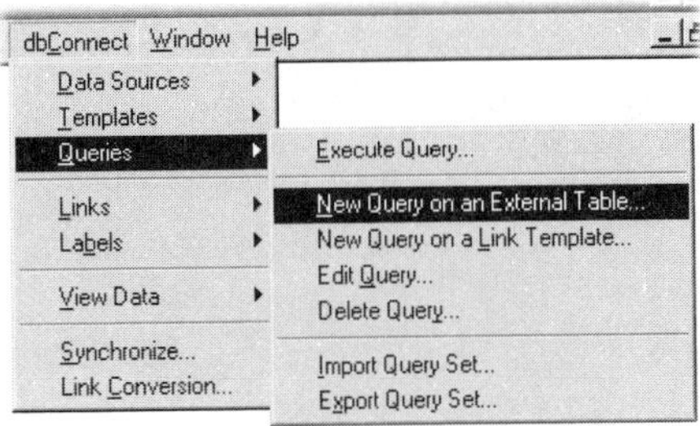

Step 2

From the **dbConnect** menu, select **Queries | New Query on an External Table**. AutoCAD displays the Select Data Object dialog box.

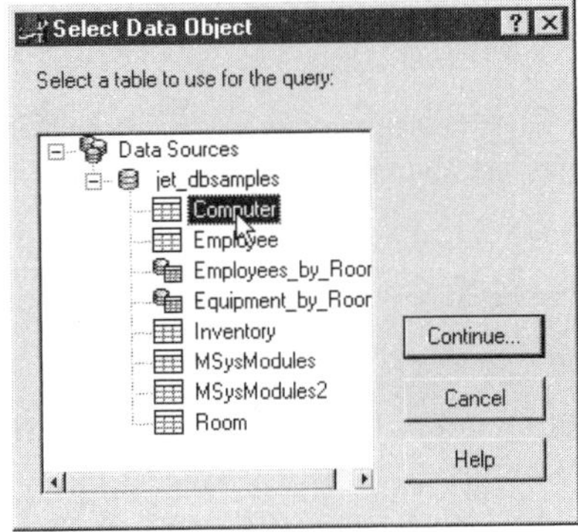

Step 3

Select a table, and then click **Continue**. AutoCAD displays the New Query dialog box.

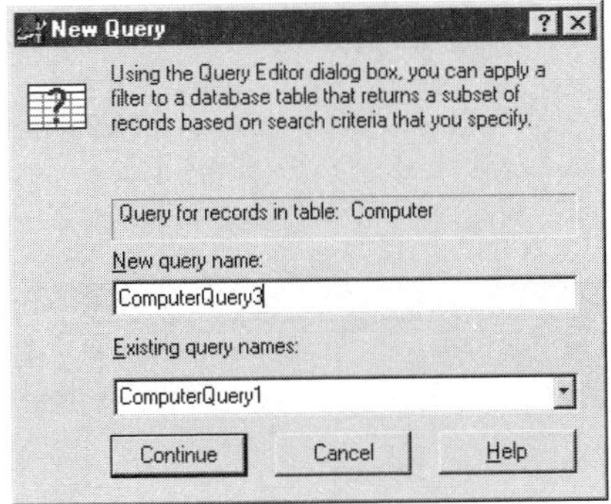

Step 4

Enter a name for the query in **New query name** text field, or select an existing name.

Step 5

Click **Continue**. AutoCAD opens the Query Editor dialog box.

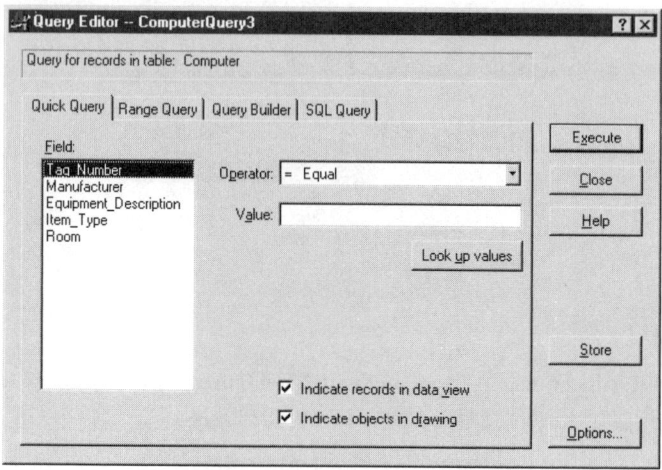

Step 6

Ensure the **Quick Query** tab is showing. Also, make sure that **Indicate records in data view** and **Indicate objects in drawing** are both checked.

Step 7

Select a field name from the **Field** list by highlighting it.

Step 8

Select an operator from the **Operator** list. For example, to match a value, select **= Equal**.

Step 9

Enter a value in the **Value** field, or click **Look up values**, and then select a value from the list of values already in the database.

Step 10

Click **Store** to save the query for reuse in the future.

Step 11

Click **Execute** to run the query. AutoCAD closes the dialog box, and then displays the records that match your selection in the **Data View** window.

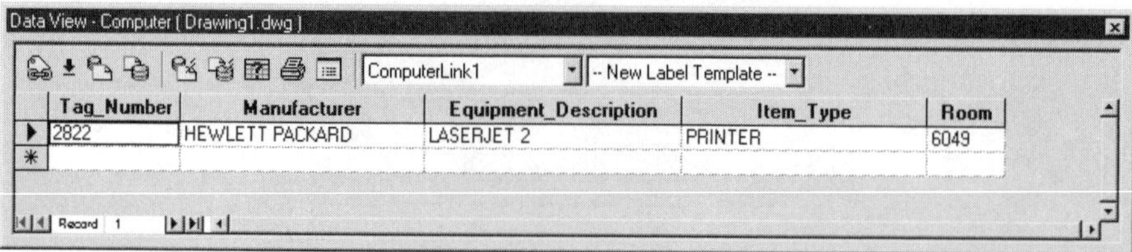

DblClkEdit

2000i Determines whether objects are edited by double-clicking the mouse button (*short for DouBLe CLicK EDITing*).

Command	Alias	Ctrl+	F-key	Alt+	Menu Bar	Tablet
dblclkedit	...	...	...	...	...	...

Command: dblclkedit
Enter double-click editing mode [ON/OFF] <ON>: *(Enter* **ON** *or* **OFF***.)*

COMMAND LINE OPTIONS

ON enables double-click editing.

OFF disables double-click editing.

RELATED COMMANDS

[Right-click] displays a shortcut menu with editing commands.

Properties displays the Properties dialog box.

TIPS

- When this command is turned on, double-clicking objects is the equivalent of entering a related editing command:

Object	Dialog Box	Equivalent Command
Attribute	Edit Attribute	**DdEdit** or **EAttEdit**
Block	Reference Edit	**RefEdit**
Hatch	Hatch Edit	**HatchEdit**
Leader	Multiline Text Editor	**MtEdit**
Mline	Multiline Edit Tools	**MlEdit**
Mtext	Multiline Text Editor	**MtEdit**
Table Cell	Multiline Text Editor	**TablEdit**
Text	Edit Text	**DdEdit**
Viewport	*none*	**VpMax**
Xref	Reference Edit	**RefEdit**

- In other cases, double-clicking objects displays the **Properties** window, and highlights the object with grips

DbList

V. 1.0 Lists information on all objects in the drawing (*short for Data Base LISTing*).

Command	Alias	Ctrl+	F-key	Alt+	Menu Bar	Tablet
dblist	...	...	...	...	...	...

Command: dblist

Sample listing:

```
AutoCAD Text Window - C:\CAD\Red Deer\Sample\db_samp.dwg
Edit
Command: dblist

                    LWPOLYLINE  Layer: "E-B-MULL"
                              Space: Model space
                     Handle = 3A
               Closed
     Constant width      0'-0"
                 area    11.25 square in. (0.0781 square ft.)
            perimeter    1'-2 1/2"

             at point  X=233'-4 33/64"  Y=270'-10 27/64"  Z=     0'-0"
             at point  X=233'-0 63/64"  Y=270'-6 7/8"  Z=     0'-0"
             at point  X=233'-2 9/16"  Y=270'-5 19/64"  Z=     0'-0"
Press ENTER to continue:
```

COMMAND LINE OPTIONS

ENTER continues display after pause.

ESC cancels database listing.

RELATED COMMANDS

Area lists the area and perimeter of objects.

Dist lists the 3D distance and angle between two points.

Id lists the 3D coordinates of a point.

List lists information about selected objects in the drawing.

TIP

- This command is typically used for debugging purposes, and is of little use for most users.

Removed Commands

DdAttDef was removed from AutoCAD 2000; it was replaced by **AttDef**.

DdAttE was removed from AutoCAD 2000; it was replaced by **AttEdit**.

DdAttExt was removed from AutoCAD 2000; it was replaced by **AttExt**.

DdChProp was removed from AutoCAD 2000; it was replaced by **Properties**.

DdColor was removed from AutoCAD 2000; it was replaced by **Color**.

DdEdit

Rel.11 Edits a single-line text, multiline text blocks, attribute values, and geometric tolerances *(short for Dynamic Dialog EDITor)*.

Command	Alias	Ctrl+	F-key	Alt+	Menu Bar	Tablet
ddedit	**ed**	...	...	**MOTE**	**Modify** ↳**Object** ↳**Text** ↳**Edit**	**Y21**

Command: ddedit

Select an annotation object or [Undo]: *(Select a text object, or type* **U**.*)*

Select single-line text placed by the **Text** *command, and the Edit Text dialog box is displayed*

Select paragraph text placed by the **MText** *command, and the Text Formatting bar is displayed.*

Select an attribute definition (not part of a block definition), and the Edit Attribute Definition dialog box is displayed.

Select a block with attributes, and the Enhanced Attribute Editor dialog box is displayed.

Select geometric tolerances, and the Geometric Tolerance dialog box is displayed.

Select an annotation object or [Undo]: *(Press* ESC *to exit dialog box.)*

COMMAND LINE OPTIONS

Undo undoes editing.

ESC ends the command.

DIALOG BOX OPTIONS

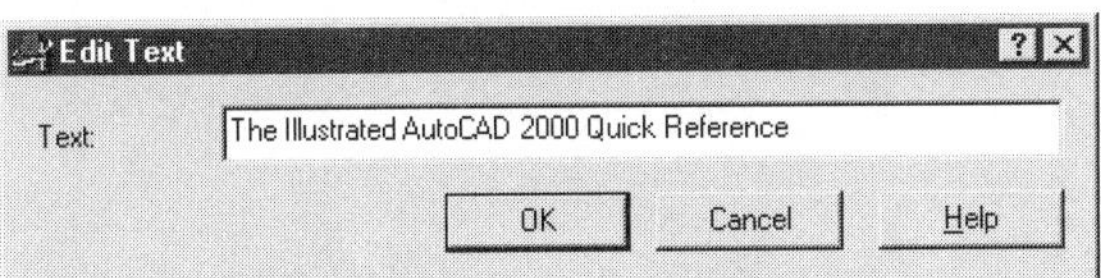

Text displays the selected text; editing options:

Keystroke	Meaning
Left or **Up**	Moves cursor one character left.
Right or **Down**	Moves cursor one character right.
HOME	Moves cursor to beginning of line.
END	Moves cursor to end of line.
DELETE	Erases character to right of cursor.
BACKSPACE	Erases character to left of cursor.

OK accepts editing changes, and dismisses dialog box.

Cancel ignores editing changes for the current annotation object, and dismisses dialog box.

RELATED COMMANDS

EAttEdit edits all text attributes connected with a block.

MtEdit edits paragraph text.

Properties edits all text *properties*, including the text itself.

RELATED SYSTEM VARIABLE

MTextEd specifies the name of the text editor used for editing multiline text.

TIPS

- **DdEdit** automatically repeats; press ESC to cancel the command.
- Between AutoCAD 2000 and 2004, this command did edit attribute text.
- As an alternative to this command, double-click text to bring up the appropriate editor.
- Text placed with the **Field**, **Leader**, and **QLeader** commands is mtext; double-click the text to edit.
- Text can also be edited through the Properties window.

Removed Commands

DdGrips was removed from AutoCAD 2000; it was replaced by the **Selection** tab of the **Options** command.

DDim was removed from AutoCAD 2000; it was replaced by **DimStyle**.

DdInsert was removed from AutoCAD 2000; it was replaced by **Insert**.

DdModify was removed from AutoCAD 2000; it was replaced by **Properties**.

'DdPtype

Rel.12 Sets the style and size of points (*short for Dynamic Dialog Point TYPE*).

Command	Alias	Ctrl+	F-key	Alt+	Menu Bar	Tablet
ddptype	...	...	...	**OP**	**Format** **↳Point Style**	**U1**

Command: ddptype

Displays dialog box:

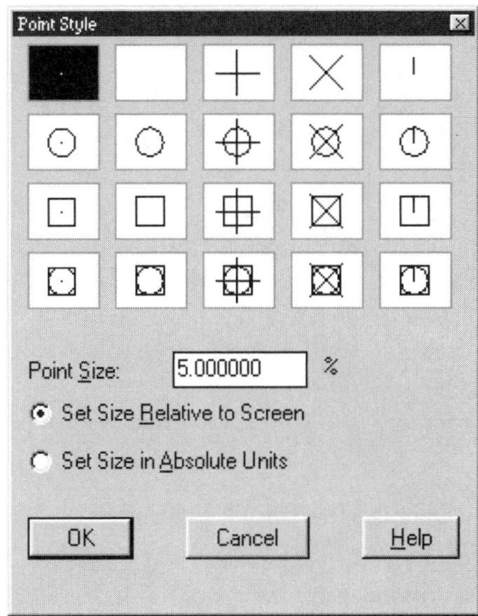

DIALOG BOX OPTIONS

Point size sets the size in percentage or pixels.

⊙**Set Size Relative to Screen** sets the size as a percentage of the total viewport height.

○**Set Size in Absolute Units** sets the size in drawing units.

RELATED COMMANDS

Divide draws points along an object at equally-divided lengths.

Point draws points.

Measure draws points a measured distance along an object.

Regen displays the new point format with a regeneration.

RELATED SYSTEM VARIABLES

PdSize contains the size of the point:

PdSize	Meaning
0	Point is 5% of viewport height (default).
positive	Absolute size in drawing units.
negative	Percentage of the viewport size.

PdMode determines the look of a point:

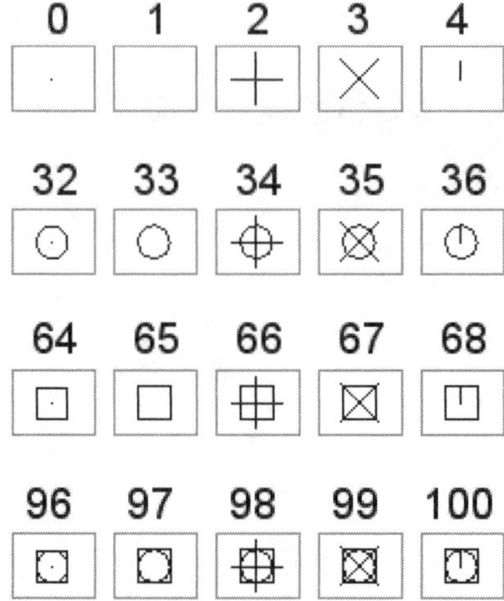

TIPS

- Points often cannot be seen in the drawing. To make them visible, change their mode and size.
- The two system variables affect all points in a drawing.

Removed Commands

DdEModes was removed from AutoCAD Release 14; it was replaced by the **Object Properties** toolbar.

DdLModes was removed from AutoCAD 2000; is was replaced by **Layer**.

DdLtype was removed from AutoCAD 2000; it was replaced by **Linetype**.

DdRename was removed from AutoCAD 2000; it was replaced by **Rename**.

DdRModes was removed from AutoCAD 2000; it was replaced by **DSettings**.

DdSelect was removed from AutoCAD 2000; it was replaced by the **Selection** tab of the **Options** command.

DdUcs and **DdUcsP** were removed from AutoCAD 2000; they were replaced by **UcsMan**.

DdUnits was removed from AutoCAD 2000; it was replaced by **Units**.

DdView was removed from AutoCAD 2000; it was replaced by **View**.

DdVPoint

Rel.12 Changes the 3D viewpoint through a dialog box (*short for Dynamic Dialog ViewPOINT*).

Command	Alias	Ctrl+	F-key	Alt+	Menu Bar	Tablet
ddvpoint	**vp**	...	...	**V3I**	**View** **↳3D Views** **↳Viewpoint Presets**	**N5**

Command: ddvpoint

Displays dialog box:

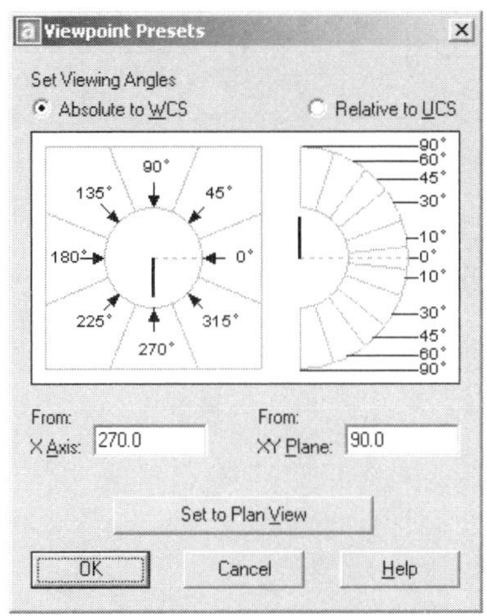

DIALOG BOX OPTIONS

Set Viewing Angles options

⊙ **Absolute to WCS** sets the view direction relative to the WCS.

○ **Relative to UCS** sets the view direction relative to the current UCS.

From X Axis measures the view angle from the x axis.

From XY Plane measures the view angle from the x,y plane.

Set to Plan View changes the view to plan view in the specified UCS.

RELATED COMMANDS

VPoint adjusts the viewpoint from the command line.

3dOrbit changes the 3D viewpoint interactively.

RELATED SYSTEM VARIABLE

WorldView determines whether viewpoint coordinates are in WCS or UCS.

TIPS

- After changing the viewpoint, AutoCAD performs an automatic **Zoom** extents.
- In the image tile shown below, the black arm indicates the new angle.
- In the image tile, the red arm indicates the current angle.
- To select an angle with your mouse:

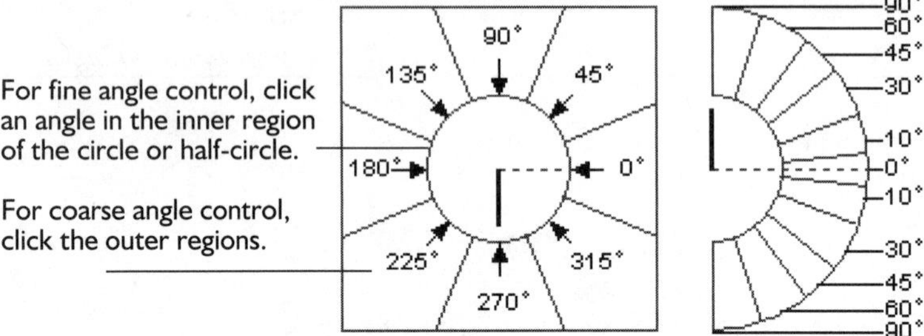

For fine angle control, click an angle in the inner region of the circle or half-circle.
For coarse angle control, click the outer regions.
90°
135°
45°
180°
0°
225°
315°
270°
90°
60°
45°
30°
10°
0°
10°
30°
45°
60°
90°

- WCS is short for "world coordinate system."
- UCS is short for "user-defined coordinate system."
- As an alternative, use the **3dOrbit** command to set the 3D viewpoint.

'Delay

v. 1.4 Delays the next command, in milliseconds.

Command	Alias	Ctrl+	F-key	Alt+	Menu Bar	Tablet
delay	...	...	...	...	...	...

Command: delay
Delay time in milliseconds: *(Enter the number of milliseconds.)*

COMMAND LINE OPTION

Delay time specifies the number of milliseconds by which to delay the next command.

RELATED COMMAND

Script initiates scripts.

TIPS

- Use **Delay** to slow down the execution of a script file.
- The maximum delay is 32767, just over 32 seconds.

DetachURL

Rel.14 Removes URLs from objects and areas.

Command	Alias	Ctrl+	F-key	Alt+	Menu Bar	Tablet
detachurl	...	...	...	...	...	...

Command: detachurl
Select objects: *(Select one or more objects.)*
Select objects: *(Press* **ENTER**.*)*

COMMAND LINE OPTION

Select objects selects the objects from which to remove URL(s).

RELATED COMMANDS

AttachUrl attaches a hyperlink to an object or an area.

Hyperlink attaches and removes hyperlinks via a dialog box.

SelectUrl selects all objects with attached hyperlinks.

TIPS

- When you select a hyperlinked area to detach, AutoCAD reports:

```
1. hyperlink ()
Remove, deleting the Area.
1 hyperlink deleted...
```

- When you select an object with no hyperlink attached, AutoCAD reports nothing.
- A URL (short for "uniform resource locator") is the universal file naming convention of the Internet; also called a link or hyperlink.

Dim

V. 1.2 Changes the prompt from 'Command' to 'Dim', allowing access to AutoCAD's original dimensioning mode (*short for DIMensions*).

Command	Alias	Ctrl+	F-key	Alt+	Menu Bar	Tablet
dim	...	...	...	...	...	...

Command: dim
Dim: *(Enter a dimension command from the list below.)*

COMMAND LINE OPTIONS

Aliases for the dimension commands are shown in UPPERCASE letters, such as "al" for ALigned.

ALigned draws linear dimensions aligned with objects (*first introduced with AutoCAD version 2.0*); replaced by **DimAligned**.

ANgular draws angular dimensions that measure angles (*ver. 2.0*); replaced by **DimAngular**.

Baseline continues dimensions from basepoints (*ver. 1.2*); replaced by **QDim** and **DimBaseline**.

CEnter draws '+' center marks on circles and arcs centers (*ver. 2.0*); replaced by **DimCenter**.

COntinue continues dimensions from previous dimensions' extension lines (*ver. 1.2*); replaced by the **QDim** and **DimContinue** commands.

Diameter draws diameter dimensions on circles, arcs, and polyarcs (*ver. 2.0*); replaced by the **QDim** and **DimDiameter** commands.

Exit returns to 'Command' prompt from 'Dim' prompt (*ver. 1.2*).

HOMetext returns dimension text to its original position (*ver. 2.6*); replaced by the **DimEdit** command's **Home** option.

HORizontal draws horizontal dimensions (*ver. 1.2*); replaced by **DimLinear**.

LEAder draws leaders (*ver. 2.0*); replaced by the **Leader** and **QLeader** commands.

Newtext edits text in associative dimensions (*ver. 2.6*); replaced by the **DimEdit** command's **New** option.

OBlique changes the angle of extension lines in associative dimensions (*rel. 11*); replaced by the **DimEdit** command's **Oblique** option.

ORdinate draws x- and y-ordinate dimensions (*Rel. 11*); replaced by **QDim** and **DimOrdinate**.

OVerride overrides current dimension variables (*Rel. 11*); replaced by **DimOverride**.

RAdius draws radial dimensions on circles, arcs, and polyline arcs (*ver. 2.0*); replaced by the **QDim** and **DimRadius** commands.

REDraw redraws the current viewport (same as **'Redraw**; *ver. 2.0*).

REStore restores dimensions to the current dimension style (*Rel. 11*); replaced by the **-DimStyle** command's **Restore** option.

ROtated draws linear dimensions at any angle (*ver. 2.0*); replaced by **DimLinear**.

SAve saves the current settings of dimension styles (*Rel. 11*); replaced by the **-DimStyle** command's **Save** option.

STAtus lists the current settings of dimension variables (*ver. 2.0*); replaced by the **-DimStyle** command's **Status** option.

STYle defines styles for dimensions (*ver. 2.5*); replaced by **DimStyle**.

TEdit changes locations and orientation of text in associative dimensions (*Rel. 11*); replaced by **DimTEdit**.

TRotate changes the rotation angle of text in associative dimensions (*Rel. 11*); replaced by the **DimTEdit** command's **Rotate** option.

Undo undoes the last dimension action (*ver. 2.0*); replaced by the **Undo** command.

UPdate updates selected associative dimensions to the current dimvar setting (*ver. 2.6*); replaced by the **-DimStyle** command's **Apply** option.

VAriables lists the values of variables associated with dimension styles, *not* dimvars (*Rel. 11*); replaced by the **-DimStyle** command's **Variables** option.

VErtical draws vertical linear dimensions (*ver. 1.2*); replaced by **DimLinear**.

RELATED DIM VARIABLES

Dim*xxx* specifies system variables for dimensions; see the **DimStyle** command.

DimAso determines whether dimensions are drawn associatively.

DimScale determines the dimension scale.

TIPS

- The 'Dim' prompt dimension commands are included for compatibility with AutoCAD Release 12 and earlier.
- Only transparent commands and dimension commands work at the 'Dim' prompt. To use other commands, you must exit the 'Dim' prompt with the **Exit** command, returning to the 'Command' prompt.
- The *defpoint* (short for "definition point") is used by earlier versions of AutoCAD to locate the extension lines. Defpoints appear as small dots on the DefPoints layer. When stretching dimensions, make sure you include the defpoints; otherwise the dimensions will not be updated automatically.
- As of AutoCAD 2002, defpoints are not used when **DimAssoc**=2 (the default); dimensions are attached directly to objects instead.

- Most dimensions consist of four basic components, as shown below:

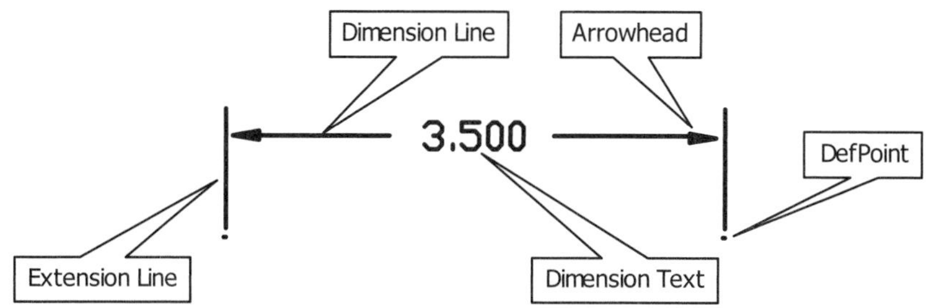

- As of AutoCAD 2005, dimension text can have a colored background, which is set globally through the **DimStyle** command, or overriden locally with the **Properties** command.
- All the components of an *associative dimension* are treated as a single object; components of a nonassociative dimension are treated as individual objects.
- For additional control over dimensions, see the following undocumented commands described earlier in this book:

  AiDimPrec
  AiDimStyle
  Ai_Dim_TextAbove
  Ai_Dim_TextCenter
  Ai_Dim_TextHome
  AiDimTextMove
- When editing dimension text, AutoCAD recognizes **<>** as *metacharacters* representing the dimension text of the measured value. For example, if AutoCAD measures a dimension as 25.4, then **<>mm** means 25.4mm.

Dim1

V. 2.5 Displays the 'Dim' prompt for a single dimensioning command, and then returns to the 'Command' prompt (*short for DIMension once*).

Command	Alias	Ctrl+	F-key	Alt+	Menu Bar	Tablet
dim1	...	...	...	...	...	...

Command: dim1
Dim: *(Enter a dimension command.)*

COMMAND LINE OPTION

Accepts all "original" dimension commands; see the ***Dim*** *command for the complete list.*

RELATED COMMANDS

DimStyle displays the dialog box for setting dimension variables.

Dim switches to AutoCAD's original dimensioning mode and remains there.

RELATED DIM VARIABLES

DimAso determines whether dimensions are drawn associatively.

DimTxt determines the height of text.

DimScale determines the dimension scale.

TIP

- Use **Dim1** when you require just a single, original dimension command.

DimAligned

Rel 13 Draws linear dimensions aligned with objects.

Command	Aliases	Ctrl+	F-key	Alt+	Menu Bar	Tablet
dimaligned	**dal**	...	...	**NG**	**Dimension**	**W4**
	dimali				**↳Aligned**	

Command: dimaligned

Specify first extension line origin or <select object>: *(Pick a point, or press* **ENTER** *to enable object selection, as shown by the following prompt.)*

Select object to dimension: *(Select an object.)*

Specify dimension line location or [Mtext/Text/Angle]: *(Pick a point, or enter an option.)*

Dimension text = *nnn*

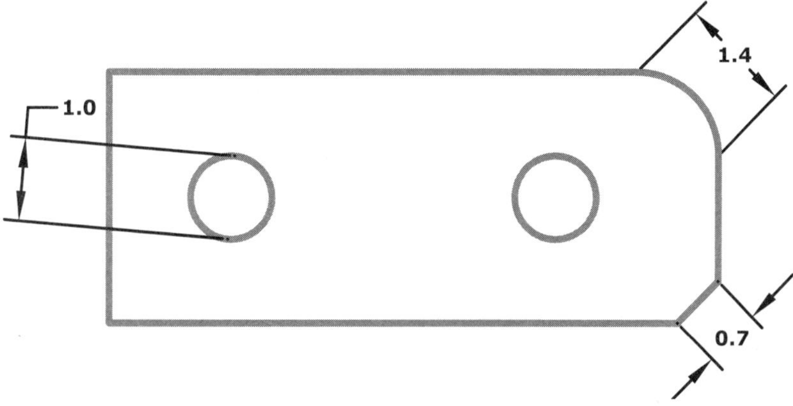

COMMAND LINE OPTIONS

Specify first extension line origin picks a point for the origin of the first extension line.

Specify second extension line origin picks a point for the origin of the second extension line.

Select object selects an object to dimension, after pressing **ENTER**.

Select object to dimension picks a line, circle, arc, polyline, or explodable object; individual segments of polylines are dimensioned.

Specify dimension line location picks a point from which to locate the dimension line and text.

Mtext changes the wording of the dimension text.

Text changes the position of the dimension text.

Angle changes the angle of the dimension text.

RELATED DIM COMMAND

DimRotated draws rotated dimension lines with a perpendicular extension lines.

Quick Start Tutorial

Editing Dimensions with Grips

Dimensions may be edited directly without first invoking an editing command. Click the dimension once to display grips (small blue squares). Click a grip to edit, as illustrated below. Double-click the dimension to display the Properties window; see the **Properties** command.

Linear Dimensions

Grips change the location of the dimension line, text, and extension lines.

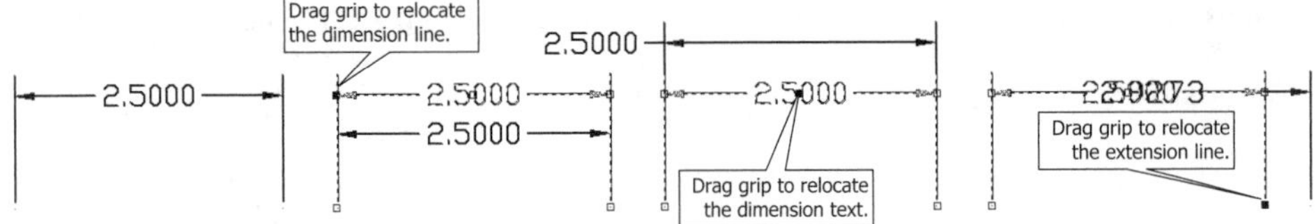

Aligned Dimensions

Grips change the location of the dimension line, text, and angle of the entire dimension.

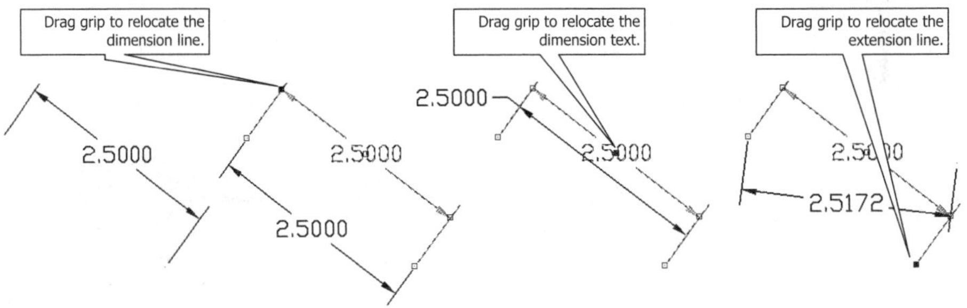

Angular Dimensions

Grips change the location of the dimension line, text, arrowhead location, and extension lines.

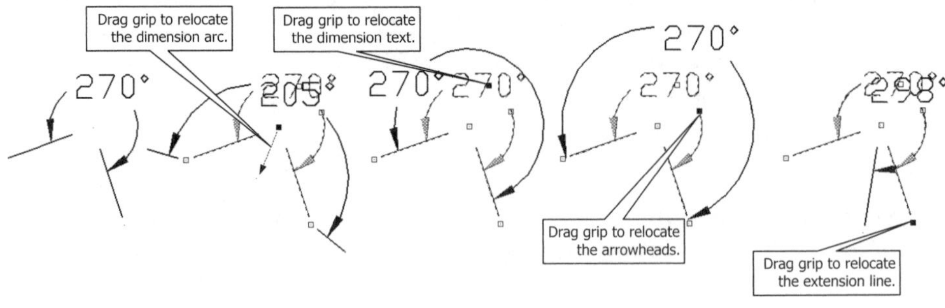

Diameter and Radius Dimensions

Grips change the location of the dimension line, text, and center mark.

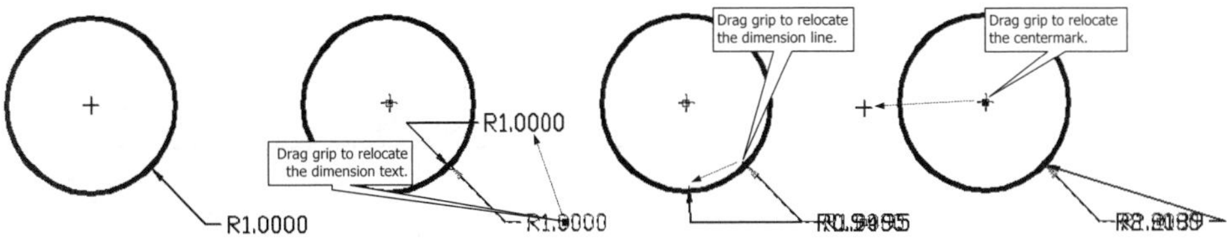

Ordinate Dimensions

Grips change the location of the ordinate line, text, and endpoint.

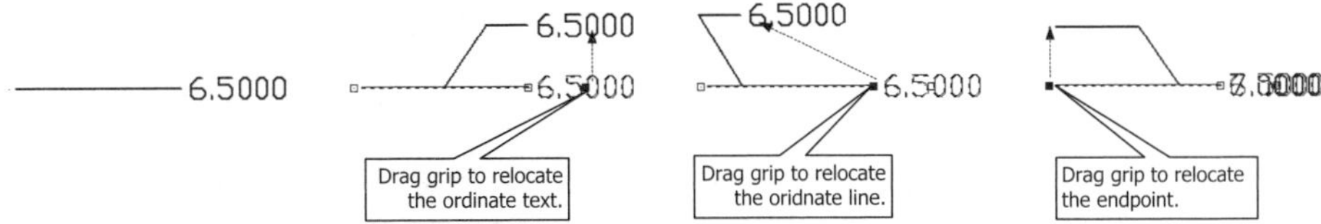

Leader Dimensions

Grips allow you to change the location of the dimension line, text, and extension lines.

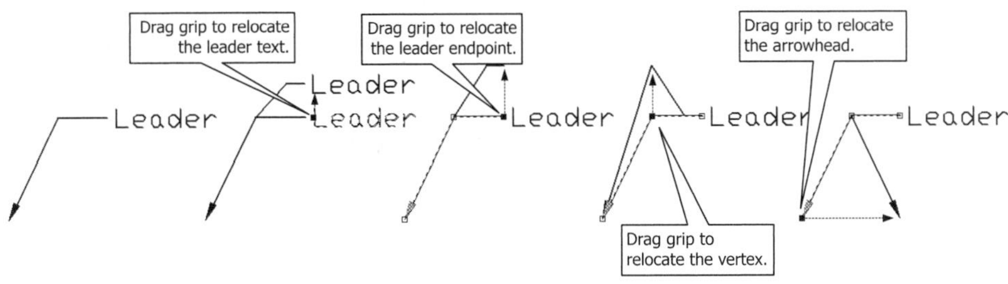

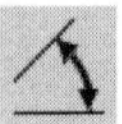

DimAngular

Rel 13 Draws dimensions that measure angles.

Command	Aliases	Ctrl+	F-key	Alt+	Menu Bar	Tablet
dimangular	dan	...	...	NA	Dimension	X3
	dimang				↳Angular	

Command: dimangular
Select arc, circle, line, or <specify vertex>: *(Select an object, or pick a vertex.)*
Specify second angle endpoint: *(Pick a point.)*
Specify dimension arc line location or [Mtext/Text/Angle]: *(Pick a point, or enter an option.)*
Dimension text = *nnn*

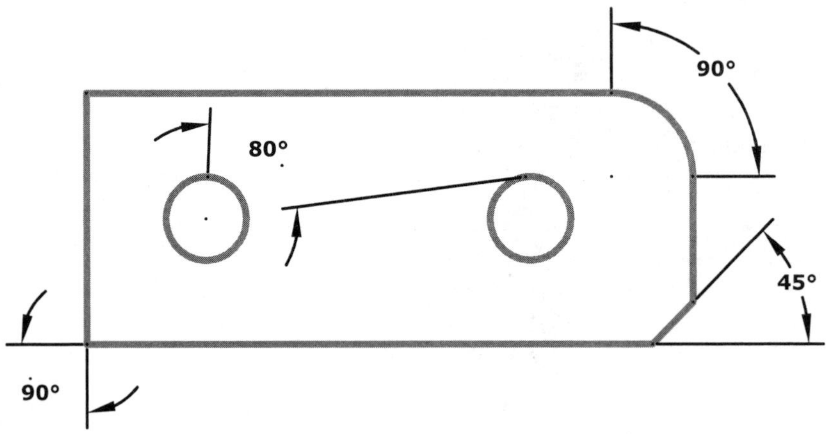

COMMAND LINE OPTIONS

Select arc measures the angle of the arc.

Circle prompts you to pick two points on the circle.

Line prompts you to pick two lines.

Specify vertex prompts you to pick points to make an angle.

Specify dimension arc/line location specifies the location of the angular dimension.

Mtext changes the wording of the dimension text.

Text changes the position of the text.

Angle changes the angle of the dimension text.

RELATED DIM COMMANDS

DimCenter places a center mark at the center of an arc or circle.

DimRadius dimensions the radius of an arc or circle.

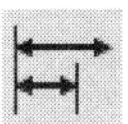

DimBaseline

Rel 13 Draws linear dimensions based on previous starting points.

Command	Aliases	Ctrl+	F-key	Alt+	Menu Bar	Tablet
dimbaseline	**dba**	...	...	**NB**	**Dimension**	...
	dimbase				**↳Baseline**	

Command: dimbaseline

Specify a second extension line origin or [Undo/Select] <Select>: *(Pick a point, or enter an option.)*

Dimension text = *nnn*

Specify a second extension line origin or [Undo/Select] <Select>: *(Pick a point, enter an option, or press* ESC *to exit command.)*

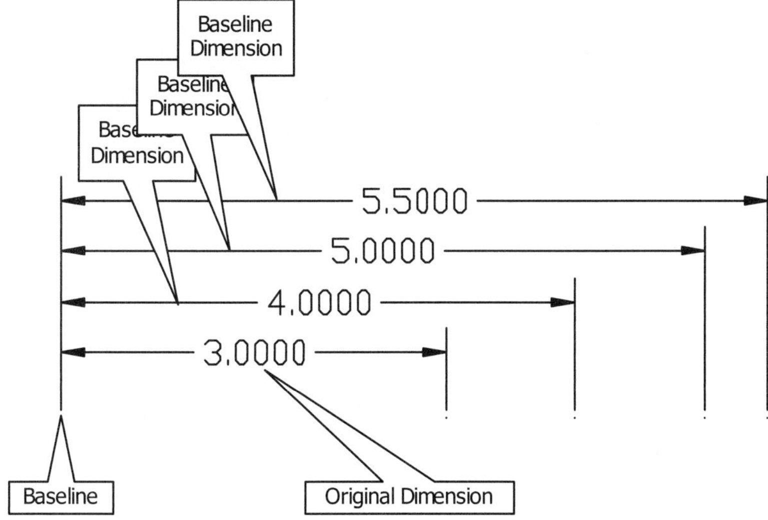

COMMAND LINE OPTIONS

Specify a second extension line origin positions the extension line of the next baseline dimension.

Select prompts you to select the base dimension.

Undo undoes the previous baseline dimension.

ESC exits the command.

RELATED DIM COMMANDS

Continue continues linear dimensioning from the last extension point.

QDim creates continuous or baseline dimensions quickly.

RELATED DIM VARIABLES

DimDli specifies the distance between baseline dimension lines.

DimSe1 suppresses the first extension line.

DimSe2 suppresses the second extension line.

DimCenter

Rel 13 Draws center marks and lines on arcs and circles.

Command	Alias	Ctrl+	F-key	Alt+	Menu Bar	Tablet
dimcenter	**dce**	...	...	**NM**	**Dimension** ↳**Center Mark**	**X2**

Command: dimcenter
Select arc or circle: *(Select an arc, circle, or explodable object.)*

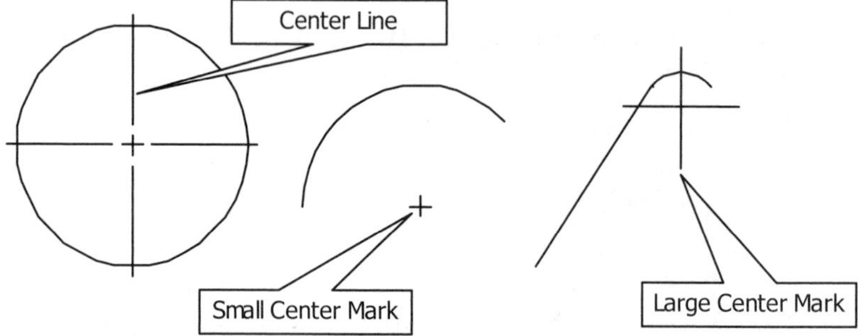

COMMAND LINE OPTION

Select arc or circle places the center mark at the center of the selected arc, circle, or polyarc.

RELATED DIM COMMANDS

DimAngular dimensions arcs and circles.

DimDiameter dimensions arcs and circles by diameter value.

DimRadius dimensions arcs and circles by radius value.

RELATED DIM VARIABLE

DimCen specifies the size and type of the center mark:

DimCen	Meaning
negative value	Draws center marks and lines.
0	Does not draw center marks or center lines.
positive value	Draws center marks.
0.09	Default value.

TIP

- Changing the center mark size for a dimension style does not update existing center lines and marks.

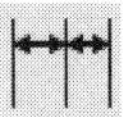

DimContinue

Rel 13 Continues dimension from the second extension line of previous dimensions.

Command	Aliases	Ctrl+	F-key	Alt+	Menu Bar	Tablet
dimcontinue	**dco**	...	...	**NC**	**Dimension**	...
	dimcont				**↳Continue**	

Command: dimcontinue

Specify a second extension line origin or [Undo/Select] <Select>: *(Pick a point, enter an option, or press* **ENTER** *to select a base dimension.)*

Dimension text = *nnn*

Specify a second extension line origin or [Undo/Select] <Select>: *(Pick a point, enter an option, or press* **ESC** *to exit command.)*

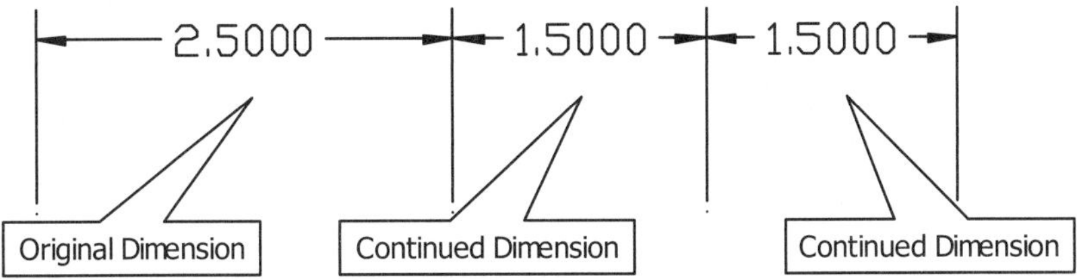

COMMAND LINE OPTIONS

Specify a second extension line origin positions the extension line of the next continued dimension.

Select prompts you to select the originating dimension.

Undo undoes the previous continued dimension.

ESC exits the command.

RELATED DIM COMMANDS

DimBaseline continues dimensioning from the first extension point.

QDim creates continuous or baseline dimensions quickly.

RELATED DIM VARIABLES

DimDli sets the distance between continuous dimension lines.

DimSe1 suppresses the first extension line.

DimSe2 suppresses the second extension line.

DimDiameter

Rel 13 Draws diameter dimensions on arcs, circles, and polyline arcs.

Command	Aliases	Ctrl+	F-key	Alt+	Menu Bar	Tablet
dimdiameter	ddi dimdia	...	...	ND	Dimension ↳Diameter	X4

Command: diameter
Select arc or circle: (Select *an arc, circle, or explodable object.)*
Dimension text = *nnn*
Specify dimension line location or [Mtext/Text/Angle]: *(Pick a point, or enter an option.)*

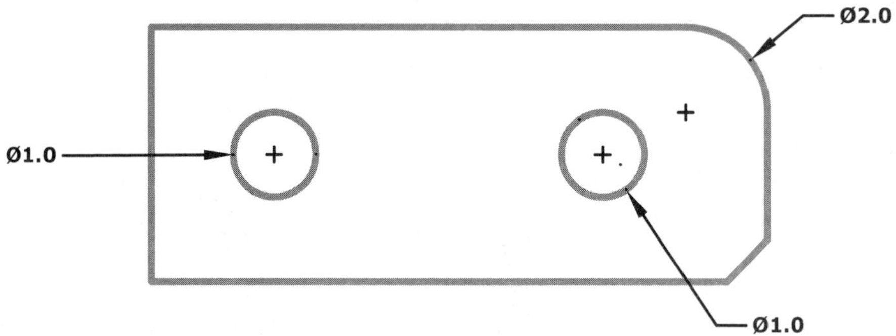

COMMAND LINE OPTION

Select arc or circle selects an arc, circle, or polyarc.

Specify dimension line location specifies the location of the angular dimension.

Mtext changes the wording of the dimension text.

Text changes the position of the dimension text.

Angle changes the angle of the dimension text.

RELATED DIM COMMANDS

DimCenter marks the center point of arcs and circles.

DimRadius draws the radius dimension of arcs and circles.

TIP

- To include the diameter symbol, use the **%%d** code or the Unicode **\U+2205** .

DimDisassociate

2002 Converts associative dimensions to non-associative (*the AutoCAD command that's most difficult to spell correctly*).

Command	Alias	Ctrl+	F-key	Alt+	Menu Bar	Tablet
dimdisassociate	...	...	...	...	...	...

Command: dimdisassociate
Select dimensions to disassociate...
Select objects: *(Select one or more dimensions, or enter* **All** *to select all dimensions.)*
Select objects: *(Press* **ENTER** *to end object selection.)*
nn **disassociated.**

COMMAND LINE OPTION

Select objects selects dimensions to convert to non-associative type.

RELATED DIM COMMANDS

DimReassociate converts dimensions from non-associative to associative.

DimRegen makes all dimensions associative automatically.

RELATED SYSTEM VARIABLE

DimAssoc determines whether newly-created dimensions are associative:

DimAssoc	Meaning
0	Dimension is created "exploded," so that all parts (such as dimension lines, arrowheads) are individual, ungrouped objects.
1	Dimension is created as a single object, but is not associative.
2	dimension is created as a single object, and is associative (default).

TIPS

- As you select objects, this command ignores non-dimensions, dimensions on locked layers, and those not in the current space (model or paper).
- The command displays a report of filtered and disassociated dimensions.
- The effects of this command can be reversed with the **U** and the **DimReassociate** commands.

DimEdit

Rel 13 Applies editing changes to dimension text.

Command	Aliases	Ctrl+	F-key	Alt+	Menu Bar	Tablet
dimedit	**ded**	...	...	**NQ**	**Dimension**	**Y1**
	dimed				**↳Oblique**	

Command: dimedit

Enter type of dimension editing [Home/New/Rotate/Oblique] <Home>: *(Enter an option, or press* **ENTER** *for Home.)*

Select objects: *(Select one or more objects.)*

Select objects: *(Press* **ENTER***.)*

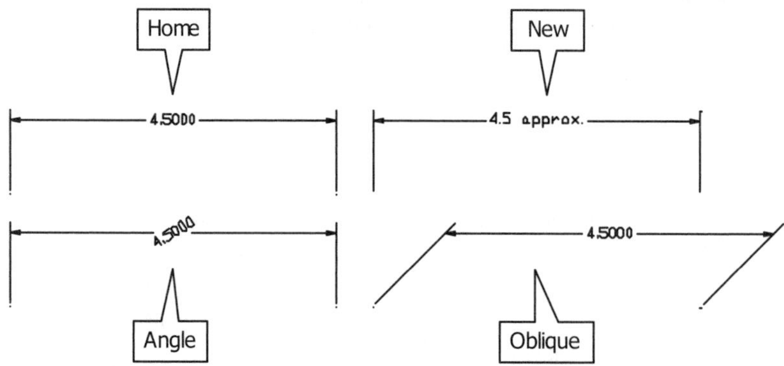

COMMAND LINE OPTIONS

Angle rotates the dimension text.

Home returns the dimension text to its original position.

Oblique rotates the extension lines.

New allows editing of the dimension text.

RELATED DIM COMMANDS

All.

RELATED DIM VARIABLES

Most.

TIPS

- When you enter dimension text with the **DimEdit** command's **New** option, AutoCAD recognizes **<>** as *metacharacters* representing existing text.
- Use the **Oblique** option to angle dimension lines by 30 0r 150 degrees, suitable for isometric drawings; use the **Style** command to oblique text by the same angle. See the **Isometric** command.
- **DimEdit** operates differently when used from the command line than from the menu bar: the **Oblique** option is the default when selected from the menu bar.

DimHorizontal

2004 Draws horizontal dimensions (*undocumented command*).

Command	Aliases	Ctrl+	F-key	Alt+	Menu Bar	Tablet
dimhorizontal	...	...	...	...	...	...

Command: dimhorizontal

Specify first extension line origin or <select object>: *(Pick a point, or press* **ENTER** *to select one object.)*

Specify second extension line origin: *(Pick a point.)*

Specify dimension line location or [Mtext/Text/Angle]: *(Pick a point, or select an option.)*

Dimension text = *nnn*

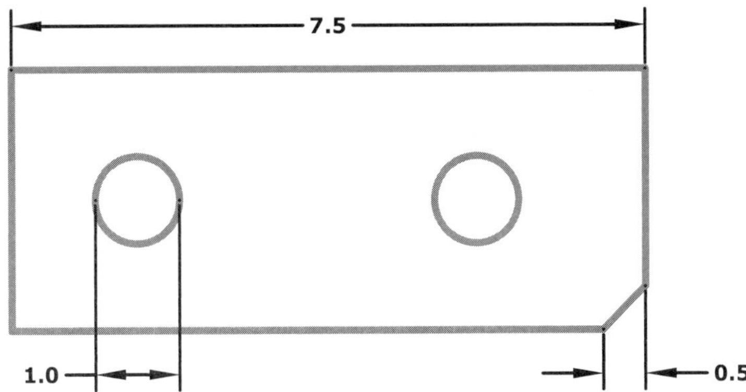

COMMAND LINE OPTIONS

Specify first extension line origin specifies the location of the first extension line's origin.

Select object dimensions a line, arc, or circle automatically.

Specify second extension line location specifies the location of the second extension line.

Specify dimension line location specifies the location of the dimension line.

Mtext displays the Text Formatting bar, which allows you to modify the dimension text; see the **MText** command.

Text prompts you to replace the dimension text on the command line: 'Enter dimension text <*nnn*>'.

Angle changes the angle of the dimension text: 'Specify angle of dimension text:'.

RELATED DIM COMMANDS

All.

RELATED DIM VARIABLES

Most.

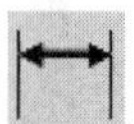

DimLinear

Rel 13 Draws horizontal dimensions.

Command	Aliases	Ctrl+	F-key	Alt+	Menu Bar	Tablet
dimlinear	**dli**	...	...	**NL**	**Dimension**	**W5**
	dimlin				**↳Linear**	

Command: dimlinear
Specify first extension line origin or <select object>: *(Pick a point, or press* ENTER *to select an object.)*
Specify second extension line origin: *(Pick a point.)*
Specify dimension line location or [Mtext/Text/Angle/Horizontal/Vertical/Rotated]: *(Pick a point, or enter an option.)*
Dimension text = *nnn*

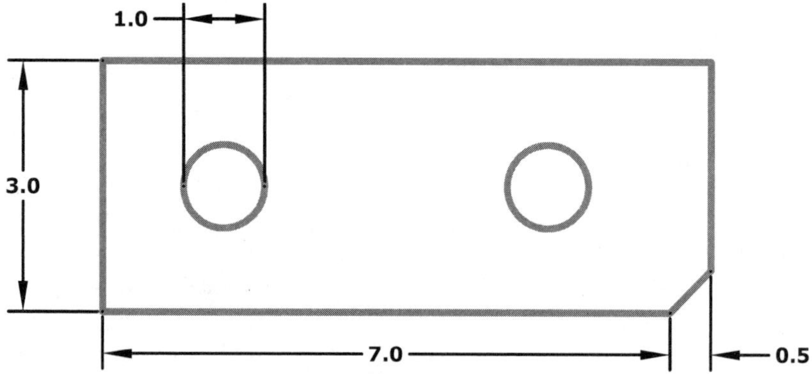

COMMAND LINE OPTIONS

Specify first extension line origin specifies of the origin of the first extension line.

Select object dimensions a line, arc, or circle automatically.

Specify second extension line location specifies the location of the second extension line.

Specify dimension line location specifies the location of the dimension line.

Mtext displays the Text Formatting bar, which allows you to modify the dimension text; see the **MText** command.

Text option
Enter dimension text <*nnn*>: *(Enter dimension text, or press* **ENTER** *to accept default value.)*

Enter dimension text prompts you to replace the dimension text on the command line.

Angle option
Specify angle of dimension text: *(Enter an angle.)*

Specify angle changes the angle of dimension text.

Horizontal option
Specify dimension line location or [Mtext/Text/Angle]: *(Pick a point or enter an option.)*

Horizontal option forces dimension to be horizontal.

Vertical option
Specify dimension line location or [Mtext/Text/Angle]: *(Pick a point or enter an option.)*

Vertical option forces dimension to be vertical.

Rotated option
Specify angle of dimension line <0>: *(Enter an angle.)*

Rotated option rotates the dimension.

Select Object options

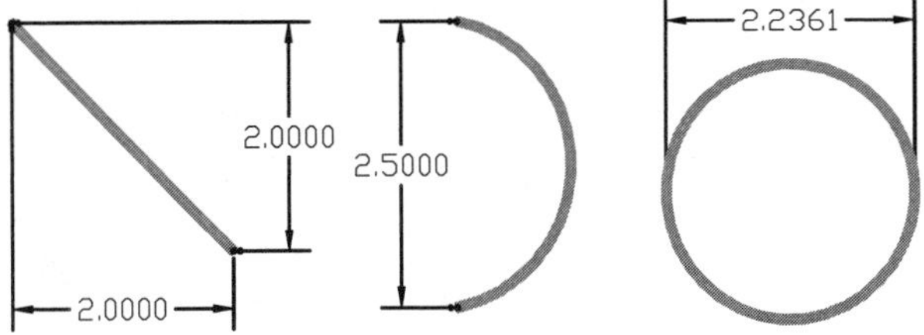

Specify first extension line origin or <select object>: *(Press* ENTER*.)*
Select object to dimension: *(Select one object.)*
Specify dimension line location or
[Mtext/Text/Angle/Vertical/Rotated]: *(Pick a point, or enter an option.)*
Dimension text = *nnn*

RELATED DIM COMMANDS

DimAligned draws linear dimensions aligned with objects.

QDim dimensions objects quickly.

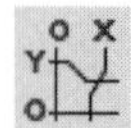

DimOrdinate

Rel 13 Draws x- and y-ordinate dimensions.

Command	Aliases	Ctrl+	F-key	Alt+	Menu Bar	Tablet
dimordinate	**dor**	...	...	**NO**	**Dimension**	**W3**
	dimord				**↳Ordinate**	

Command: dimordinate
Specify feature location: *(Pick a point.)*
Specify leader endpoint or [Xdatum/Ydatum/Mtext/Text/Angle]: *(Pick a point, or enter an option.)*
Dimension text = *nnn*

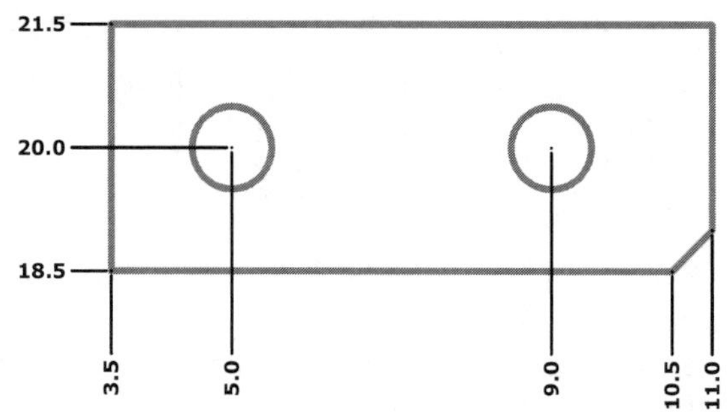

COMMAND LINE OPTIONS

Xdatum forces x ordinate dimension.

Ydatum forces y ordinate dimension.

Mtext displays the Text Formatting toolbar for modifying dimension text; see **MText** command.

Text prompts you to replace dimension text on the command line: 'Enter dimension text <*nnn*>'.

Angle changes the angle of dimension text: 'Specify angle of dimension text:'.

RELATED DIM COMMANDS

Leader draws leader dimensions.

Tolerance draws geometric tolerances.

RELATED TOOLBAR ICONS

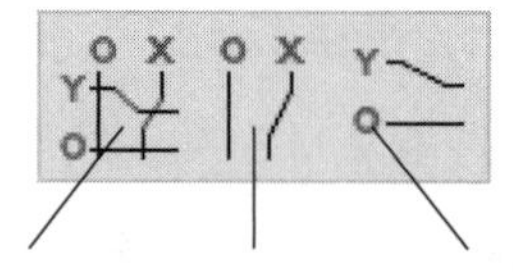

DimOrdinate XDatum YDatum

TIPS

- AutoCAD misuses the term "ordinate," which by definition is the distance from the x-axis only.
- The 0,0 point is determined by the current UCS. Use object snap to define a UCS origin at 0,0.
- Define the x=0 and y=0 ordinate dimensions, then use **DimBaseline** for the others.

DimOverride

Rel 13 Overrides the current dimension variables.

Command	Aliases	Ctrl+	F-key	Alt+	Menu Bar	Tablet
dimoverride	dov	...	...	NV	Dimension	Y4
	dimover				↳Override	

Command: dimoverride
Enter dimension variable name to override or [Clear overrides]: *(Enter the name of a dimension variable, or type* **C***.)*
Enter new value for dimension variable <*nnn*>: *(Enter a new value.)*
Select objects: *(Select one or more dimensions.)*
Select objects: *(Press* **ENTER***.)*

COMMAND LINE OPTIONS

Dimension variable to override requires you to enter the name of the dimension variable.

Clear removes the override.

New Value specifies the new value of the dimvar.

Select objects selects the dimension objects to which the change applies.

RELATED DIM COMMAND

DimStyle creates and modifies dimension styles.

RELATED DIM VARIABLES

All dimension variables.

DimRadius

Rel 13 Draws radial dimensions on circles, arcs, and polyline arcs.

Command	Aliases	Ctrl+	F-key	Alt+	Menu Bar	Tablet
dimradius	dra dimrad	...	...	NR	Dimension ↳Radius	X5

Command: dimradius
Select arc or circle: *(Select an arc, circle, or explodable object.)*
Dimension text = *nnn*
Specify dimension line location or [Mtext/Text/Angle]: *(Pick a point, or enter an option.)*

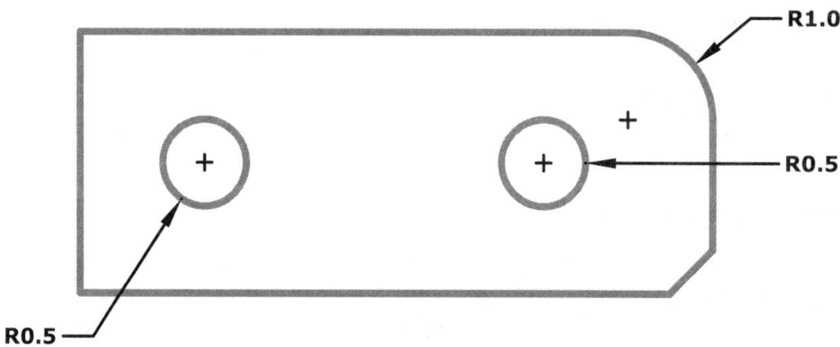

COMMAND LINE OPTIONS

Select arc or circle selects the arc, circle, or polyarc to dimension.

Specify dimension line location specifies the location of the angular dimension.

Mtext displays the Text Formatting bar, which allows you to modify the dimension text; see the **MText** command.

Text prompts you to replace the dimension text at the command line, 'Enter dimension text <*nnn*>'.

Angle changes the angle of dimension text, and prompts, 'Specify angle of dimension text:'.

RELATED DIM COMMANDS

DimCenter draws center marks on arcs and circles.

DimDiameter draws diameter dimensions on arcs and circles.

RELATED DIM VARIABLE

DimCen determines the size of the center mark.

DimReassociate

2002 Associates dimensions with objects.

Command	Alias	Ctrl+	F-key	Alt+	Menu Bar	Tablet
dimreassociate	...	...	...	**NN**	**Dimension** ↳**Reassociate Dimensions**	**X4**

Command: dimreassociate

Select dimensions to reassociate...

Select objects: *(Select one or more dimensions, or enter* **All** *to select all dimensions.)*

nn **found.** *nn* **were on a locked layer.** *nn* **were not in current space.**

Select objects: *(Press* **ENTER** *to end object selection.)*

The prompts that follow depend on the type of each dimension selected.

Specify first extension line origin or [Select object]: *(Pick a point, or select an object.)*

Select arc or circle <next>: *(Press* **ENTER** *to end object selection.)*

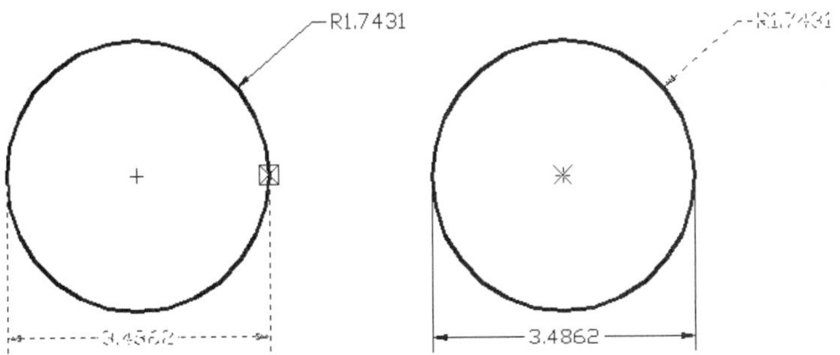

At left: the boxed X indicates that the linear dimension is associated with the circle;
At right: the X indicates the radius dimension is not associated.

COMMAND LINE OPTIONS

Select objects selects the dimensions to reassociate.

Specify first extension line origin picks a point to which to associate the dimension.

Select arc or circle picks a circle or arc to which to associate the dimension.

Next goes to the next circle or arc.

RELATED DIM COMMANDS

DimDisassociate converts dimensions from associative to non-associative.

DimRegen makes all dimensions associative automatically.

TIPS

- As you select objects, this command ignores non-dimensions, dimensions on locked layers, and those not in the current space (model or paper).
- AutoCAD displays a boxed **X** to indicate the object to which the dimension is associated; an unboxed **X** indicates an unassociated dimension. The markers disappear when a wheelmouse performs a zoom or pan.

DimRegen

2002 Updates the locations of associative dimensions.

Command	Alias	Ctrl+	F-key	Alt+	Menu Bar	Tablet
dimregen	...	...	...	...	...	...

Command: dimregen

COMMAND LINE OPTIONS

None.

RELATED DIM COMMANDS

DimDisassociate converts dimensions from associative to non-associative.

DimReassociate converts dimensions from non-associative to associative.

TIPS

- This command is meant for use after three conditions:

  The drawing has been edited by a version of AutoCAD prior to 2004.

  The drawing contains dimensions associated with an external reference, and the xref has been edited.

  The drawing is in layout mode, model space is active, and a wheelmouse has been used to pan or zoom.

DimRotated

2004 Draw rotated dimensions (*undocumented command*).

Command	Alias	Ctrl+	F-key	Alt+	Menu Bar	Tablet
dimrotated	...	...	...	...	...	...

Command: dimrotated
Specify angle of dimension line <0>: *(Enter a rotation angle.)*
Specify first extension line origin or <select object>: *(Pick a point, or press* **ENTER** *to select one object.)*
Specify second extension line origin: *(Pick a point.)*
Specify dimension line location or [Mtext/Text/Angle]: *(Pick a point, or select an option.)*
Dimension text = *nnn*

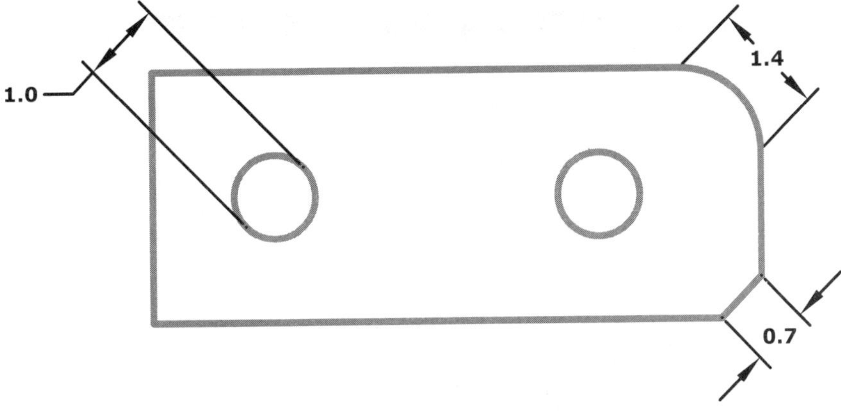

Rotated dimensions placed at 45 degrees.

COMMAND LINE OPTIONS

Specify angle of dimension line specifies the angle at which the dimension line is rotated.

Specify first extension line origin specifies the origin of the first extension line.

Select object dimensions a line, arc, or circle automatically.

Specify second extension line location specifies the location of the second extension line.

Specify dimension line location specifies the location of the dimension line.

Mtext displays the Text Formatting bar, which allows you to modify the dimension text; see the **MText** command.

Text prompts you to replace the dimension text on the command line: 'Enter dimension text <*nnn*>'.

Angle changes the angle of dimension text: 'Specify angle of dimension text:'.

RELATED DIM COMMANDS

All.

DimStyle

Rel 13 Creates and edits dimstyles (*short for DIMension STYLE*).

Commands	Aliases	Ctrl+	F-key	Alt+	Menu Bar	Tablet
dimstyle	d dst	...	...	NS	Dimension ↳Style	Y5
	dimsty ddim			OD	Format ↳Dimension Style	
-dimstyle	...	...	...	NU	Dimension ↳Update	

Command: dimstyle

Displays dialog box:

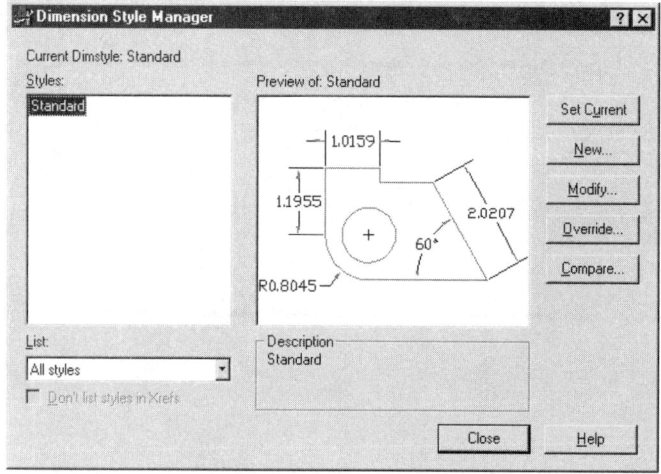

DIALOG BOX OPTIONS

Styles lists the names of dimension styles in the drawing.

List modifies the style names listed under Styles:

- **All styles** lists all dimenion style names stored in the current drawing (default).
- **Styles in use** list only those dimstyles used by dimensions in the drawing.

Don't list styles in Xrefs

☑ Does not list dimension styles found in externally-referenced drawings.

☐ Lists xref dimension styles under Styles (default).

Buttons

Set Current sets the selected style as the current dimension style.

New creates a new dimension style via Create New Dimension Style dialog box.

Modify modifies an existing dimension style; displays the Modify Dimension Style dialog box.

Override allows temporary changes to a dimension style; displays the Override Dimension Style dialog box.

Compare lists the differences between dimension variables of two styles; displays the Compare Dimension Styles dialog box.

CURSOR MENU OPTIONS

*Right-click a dimension style name under **Styles:***

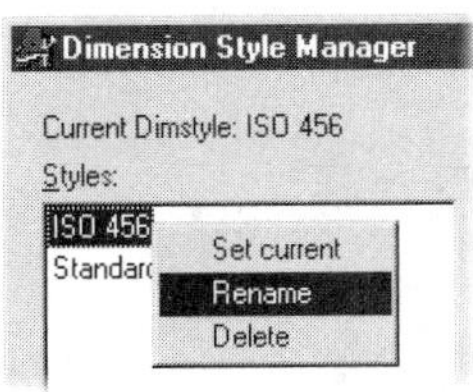

Set Current sets the selected dimension style as the current style.

Rename renames dimension styles.

Delete erases selected dimension styles from the drawing; you cannot erase the Standard style, or styles that are in use.

Create New Dimension Style dialog box

When creating a new dimension style, typically you start by making changes to an existing style.

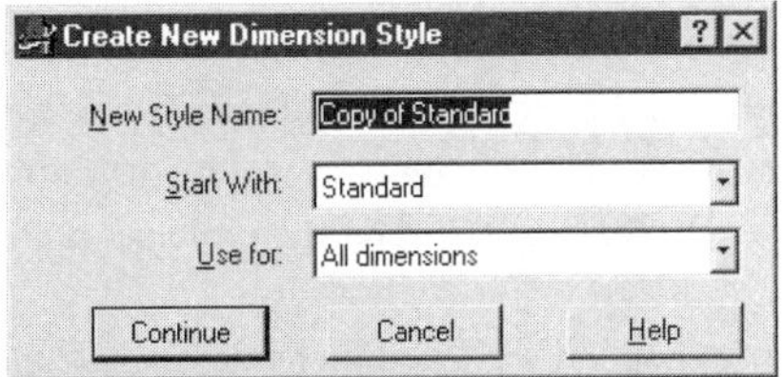

New Style Name specifies the name of the new dimension style.

Start With lists the names of the current dimension style(s), which are used as the template for the new dimension style.

Use for creates a substyle that applies to a specific type of dimension type: linear, angular, radius, diameter, ordinate, leaders and tolerances.

Continue continues to the next dialog box, New Dimension Style.

Cancel dismisses this dialog box, and returns to the Dimension Style Manager dialog box.

New Dimension Style dialog box

OK records the changes made to dimension properties, and returns to the Dimension Manager dialog box.

Cancel cancels the changes, and returns to the Dimension Manager dialog box.

Lines and Arrows tab

Sets the format of dimension lines, extension lines, arrowheads, and center marks.

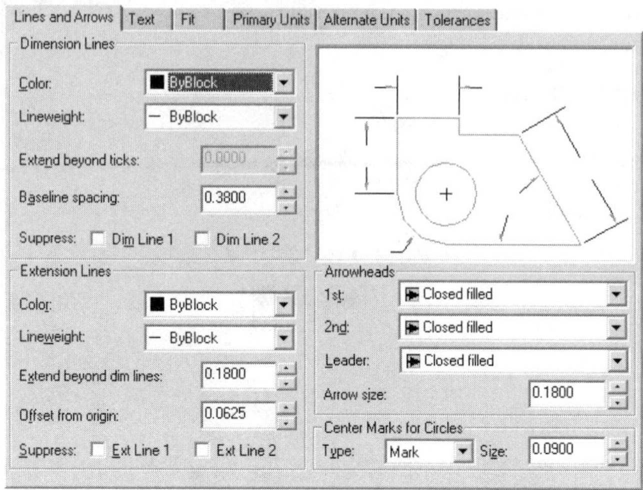

Dimension Lines options

Color specifies the color of the dimension line; select Other to display the Select Color dialog box (stored in dimension variable DimClrD; default = ByBlock).

Lineweight specifies the lineweight of the dimension line (DimLwD; default = ByBlock).

Extend beyond ticks specifies the distance the dimension line extends beyond the extension line; used with oblique, architectural, tick, integral, and no arrowheads (DimDlE; default = 0).

Baseline spacing specifies the spacing between the dimension lines of a baseline dimension (DimDlI; default = 0.38).

☐ **Suppress** suppresses the first and second dimension lines when outside the extension lines (DimSd1 and DimSd2).

Extension Lines options

Color specifies the color of the extension line; select Other to display the Select Color dialog box (stored in dimension variable DimClrE ; default=ByBlock).

Lineweight specifies the lineweight of the extension line (DimLwE; default=ByBlock).

Extend beyond dim lines specifies the distance the extension line extends beyond the dimension line; used with oblique, architectural, tick, integral, and no arrowheads (DimExe; default = 0.18).

Offset from origin specifies the distance from the origin point to the start of the extension lines (DimExO; default = 0.0625).

☐ **Suppress** suppresses the first and second extension lines (DimSe1 and DimSde1).

Arrowheads options

1st specifies the name of the arrowhead to use for the first end of the dimension line (DimBlk1; default = closed filled).

*To use a custom arrowhead, select **User Arrow** to display the Select Custom Arrow Block dialog box:*

- Closed filled
- Closed blank
- Closed
- Dot
- Architectural tick
- Oblique
- Open
- Origin indicator
- Origin indicator 2
- Right angle
- Open 30
- Dot small
- Dot blank
- Dot small blank
- Box
- Box filled
- Datum triangle
- Datum triangle filled
- Integral
- None

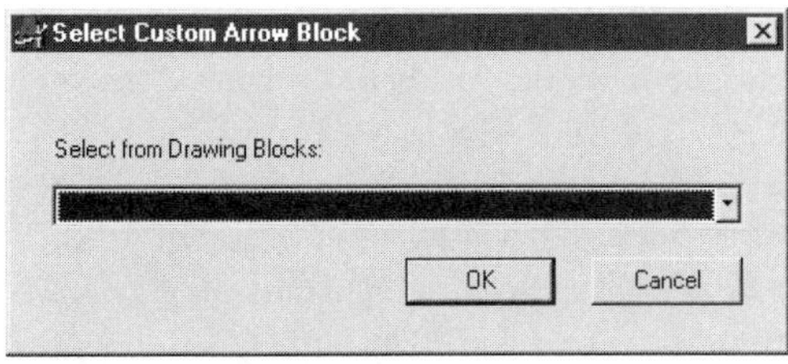

2nd specifies the arrowhead for the second dimension line; select User Arrow to display the Select Custom Arrow Block dialog box (DimBlk2; default = closed filled).

Leader specifies the arrowhead for the leader; select User Arrow to display the Select Custom Arrow Block dialog box (DimLdrBlk; default = closed filled).

Arrow Size specifies the size of arrowheads (DimASz; default = 0.18).

Center Marks for Circles options

Type specifies the type of center mark (DimCen; default = 0.09):

Type	Meaning
Mark	Places a center mark (DimCen > 0).
Line	Places a center mark and center lines (DimCen < 0).
None	Places no center mark or center line (DimCen=0).

Size specifies the size of the center mark or center line (DimCen; default = 0.09).

Text tab

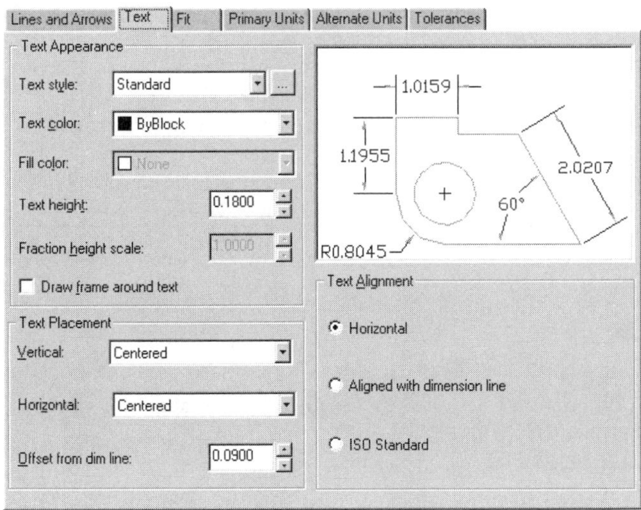

Text Appearance options

Text style specifies the text style name for dimension text (DimTxSty; default = Standard).

... displays the Text Style dialog box; see the Style command.

Text color specifies the color of the dimension line; select Other to display the Select Color dialog box (DimClrT; default = ByBlock).

Fill color specifies the color of the background behind the text (*new to AutoCAD 2005*).

Text height specifies the height of the dimension text, when the height defined by the text style is 0 (DimTxt; default = 0.18).

Fraction height scale scales fraction text height relative to dimension text; AutoCAD multiples this value by the text height (DimTFac; default = 1.0).

☐ **Draw frame around text** draws a rectangle around dimension text; when on, dimension variable **DimGap** is set to a negative value (DimGap).

Text Placement options

⦿ **Vertical** specifies the vertical justification of dimension text relative to the dimension line (DimTad; default = 1):

Vertical	Meaning
Centered	Centers dimension text in the dimension line (DimTad = 0).
Above	Places text above the dimension line (DimTad = 1).
Outside	Places text on the side of the dimension line farthest from the first defining point (DimTad = 2).
JIS	Places text in conformity with JIS DimTad= 3).

○ **Horizontal** specifies the horizontal justification of dimension text along the dimension and extension lines (DimJust; default = 0):

Horizontal	Meaning
Centered	Centers dimension text along the dimension line between the extension lines (DimJust = 0).
1st Extension Line	Left-justifies the text with the first extension line (DimJust = 1).
2nd Extension Line	Right-justifies the text with the second extension line (DimJust = 2).
Over 1st Extension Line	Places the text over the first extension line (DimJust = 3).
Over 2nd Extension Line	Places the text over the second extension line (DimJust = 4).

○ **Offset from dimension line** specifies the text gap, the distance between dimension text and the dimension line (DimGap; default = 0.09).

Text Alignment options

Horizontal forces dimension text to be always horizontal (DimTih = on; DimToh = on).

Aligned with dimension line forces dimension text to be aligned with the dimension line (DimTih = off; DimToh = off).

ISO Standard forces text to be aligned with the dimension line when inside the extension lines; forces text to be horizontal when outside the extension lines (DimTih = off; DimToh = on).

Fit tab

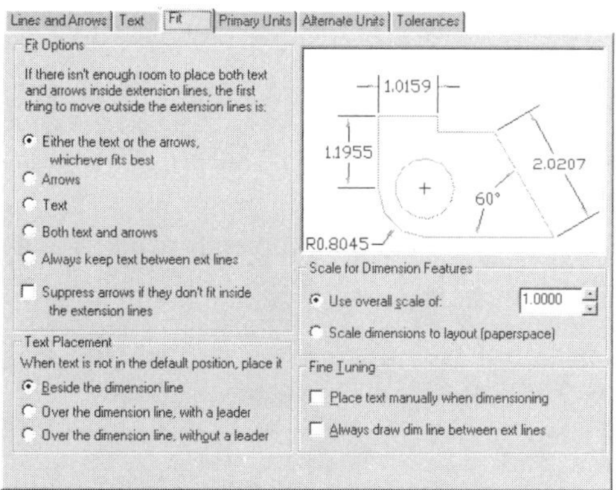

Fit Options options

If there isn't enough room to place both text and arrows inside extension lines, the first thing to move outside the extension lines is:

⊙ **Either the text or the arrows, whichever fits best** places dimension text and arrowheads between the extension lines when space is available; when space is not available for both, the text or the arrowheads are placed outside the extension lines, whichever fits best; if there is room for neither, both are placed outside the extension lines (DimAtFit = 3; default).

○ **Arrows** places arrowheads between the extension lines when there is not enough room for arrowheads and dimension text (DimAtFit = 2).

○ **Text** places text between extension lines when there is not enough room for arrowheads and dimension text (DimAtFit = 1).

○ **Both text and arrows** places both outside the extension lines when there is not enough room for dimension text and arrowheads (DimAtFit = 0).

○ **Always keep text between ext lines** forces text between the extension lines (DimTix; default = off).

☐ **Suppress arrows if they don't fit inside the extension lines** suppresses arrowheads when there is not enough room between the extension lines.

Text Placement:

⊙ **Beside the dimension line** places dimension text beside the dimension line (DimTMove =0; default).

○ **Over the dimension line, with a leader** draws a leader when dimension text is moved away from the dimension line (DimTMove = 1).

○ **Over the dimension line, without a leader:** does not draw a leader when dimension text is moved away from the dimension line (DimTMove = 2).

Scale for Dimension Features

⊙ **Use overall scale of** specifies the scale factor for all dimensions in the drawing; affects text and arrowhead sizes, distances, and spacings (DimScale; default = 1.0).

○ **Scale dimension to layout (paper space)** determines the scale factor of dimensions in layout mode; based on the scale factor between the current model space viewport and the layout (DimScale = 0; default = off).

Fine Tuning options

☐ **Place text manually when dimensioning** places text at the position picked at the 'Dimension line location' prompt (DimUpt).

☐ **Always draw dim line between ext lines** forces the dimension line between the exten sion lines (DimTofl).

Primary Units tab

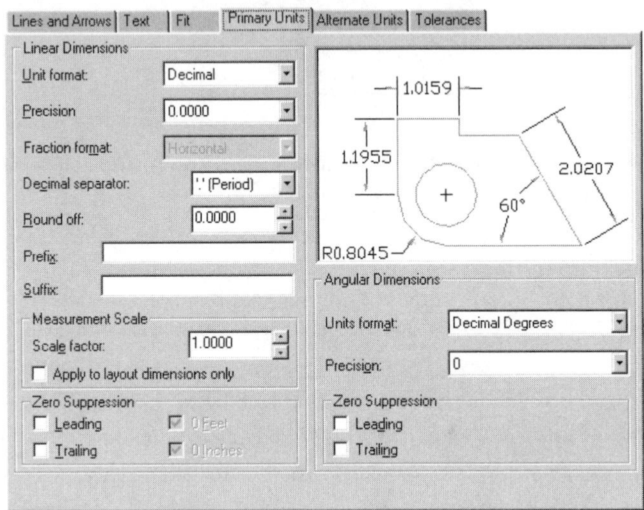

Linear Dimensions options

Unit Format specifies the linear units format; does not apply to angular dimensions (DimLUnit; default = Decimal):

DimLUnit	Meaning
1	Scientific.
2	Decimal (default).
3	Engineering.
4	Architectural.
5	Fractional.
6	Windows desktop setting.

Precision specifies the number of decimal places (or fractional accuracy) for linear dimensions (DimDec; default = 4).

Fraction Format specifies the stacking format of fractions (DimFrac; default = 0):

DimFrac	Meaning
0	Horizontal stacked: $\frac{1}{2}$ (default).
1	Diagonal stacked: ½
2	Not stacked: 1/2.

Decimal Separator specifies the separator for decimal formats (DimDSep; default = .).

Round Off specifies the format for rounding dimension values; does not apply to angular dimensions (DimRnd; default = 0).

Prefix specifies a prefix for dimension text (DimPost; default = nothing); you can use the following control codes to show special characters:

Control Code	Meaning
%%nnn	Character specified by ASCII number *nnn*.
%%o	Turns on and off overscoring.
%%u	Turns on and off underscoring.
%%d	Degrees symbol (°).
%%p	Plus/minus symbol (±).
%%c	Diameter symbol (Ø).
%%%	Percentage sign (%).

Suffix specifies a suffix for dimension text (DimPost; default = nothing); you can use the control codes listed above to show special characters.

Measurement Scale options

Scale factor specifies a scale factor for linear measurements, except for angular dimensions (DimLFac; default = 1.0); use this, for example, to change dimension values from imperial to metric.

☐ **Apply to layout dimensions only** specifies that the scale factor is applied only to dimensions created in layout mode or paper space (stored as a negative value in DimLFac; default = off).

Zero Suppression options:

☐ **Leading** suppresses leading zeros in all decimal dimensions (DimZin = 4).

☐ **Trailing** suppresses trailing zeros in all decimal dimensions (DimZin = 8).

☑ **0 Feet** suppresses zero feet of feet-and-inches dimensions (DimZin = 0).

☑ **0 Inches** suppresses zero inches of feet-and-inches dimensions (DimZin = 2).

DimZin	Meaning
0	Suppresses zero feet and precisely zero inches (default).
1	Includes zero feet and precisely zero inches.
2	Includes zero feet and suppresses zero inches.
3	Includes zero inches and suppresses zero feet.
4	Suppresses leading zeros in decimal dimensions.
8	Suppresses trailing zeros in decimal dimensions.
12	Suppresses leading and trailing zeros.

Angular Dimensions options

Units Format specifies the format of angular dimensions (DimAUnit; default = 0):

DimAUnit	Meaning
0	Decimal degrees (default).
1	Degrees/minutes/seconds.
2	Grads.
3	Radians.

Precision specifies the precision of angular dimensions (DimADec; default = 0).

Zero Suppression options

Same as for linear dimensions (DimAZin; default = 0).

Alternate Units tab

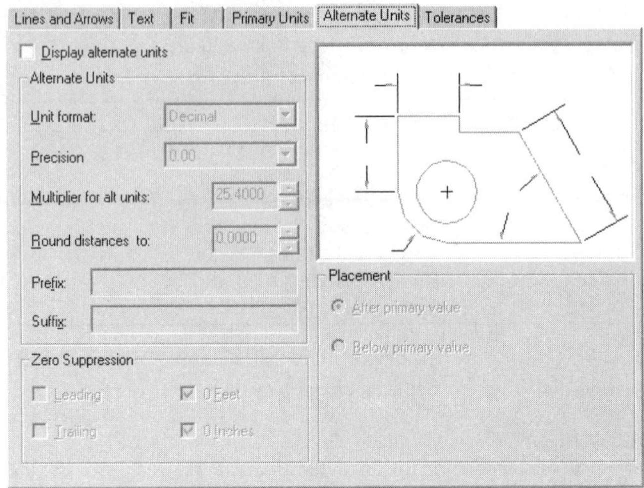

☐ **Display alternate units** adds alternate units to dimension text (DimAlt).

Alternate Units options

Unit Format specifies the alternate units formats (DimAltU; default = Decimal).

Precision specifies the number of decimal places or fractional accuracy (DimAltD; default=2).

Multiplier for alt units specifies the conversion factor between primary and alternate units (DimAltF; default = 25.4).

Round distances to specifies the format for rounding dimension values; does not apply to angular dimensions (DimAltRnd; default = 0.0000).

Prefix specifies a prefix for dimension text (DimAPost; default = nothing); you can use control codes to show special characters.

Suffix specifies a suffix for dimension text (DimAPost; default = nothing).

Zero Suppression options

☐ **Leading** suppresses leading zeros in all decimal dimensions (DimAltZ = 4).

☐ **Trailing** suppresses trailing zeros in all decimal dimensions (DimAltZ = 8).

☑ **0 Feet** suppresses zero feet of feet-and-inches dimensions (DimAltZ = 0).

☑ **0 Inches** suppresses zero inches of feet-and-inches dimensions (DimAltZ = 2).

Placement:

⊙ **After primary units** places alternate units behind the primary units (DimAPost).

○ **Below primary units** places alternate units below the primary units.

Tolerances tab

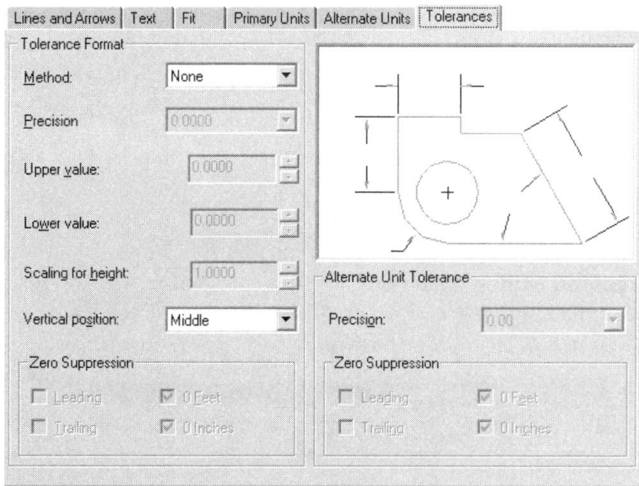

Tolerance Format options

Method specifies the tolerance format:

- **None** does not display tolerances (DimTol = 0; default).
- **Symmetrical** places ± after the dimension (DimTol=0; DimLim=0).
- **Deviation** places + and – symbols (DimTol = 1; DimLim = 1).
- **Limits** places maximum over minimum value (DimTol=0; DimLim=1):

  Maximum value = dimension value + upper value.

  Minimum value = dimension value - lower value.
- **Basic** boxes the dimension text (DimGap=negative value).

Precision specifies the number of decimal places for tolerance values (DimTDec; default=4).

Upper value specifies the upper tolerance value (DimTp; default = 0).

Lower value specifies the lower tolerance value (DimTm; default = 0).

Scaling for height specifies the scale factor for tolerance text height (DimTFac; default = 1.0).

Vertical position specifies the vertical text position for symmetrical and deviation tolerances (DimTolJ; default = 1):

Vertical	Meaning
Top	Aligns the tolerance text with the top of the dimension text (DimTolJ = 2).
Middle	Aligns the tolerance text with the middle of the dimension text (DimTolJ = 1).
Bottom	Aligns the tolerance text with the bottom of the dimension text (DimTolJ = 0).

Zero Suppression options

☐ **Leading** suppresses leading zeros in all decimal dimensions (DimTZin = 4).

☐ **Trailing** suppresses trailing zeros in all decimal dimensions (DimTZin = 8).

☑ **0 Feet** suppresses zero feet of feet-and-inches dimensions (DimTZin = 0).

☑ **0 Inches** suppresses zero inches of feet-and-inches dimensions (DimTZin = 2).

Alternate Unit Tolerance options

Precision specifies the precision — the number of decimal places — of tolerance text (DimAltTd; default = 2).

Zero Suppression *options:* the same as for tolerance format; stored in **DimAltTz**.

Modify Dimension Style dialog box

This dialog box is identical to the New Dimension Style dialog box.

Override Dimension Style dialog box

This dialog box is identical to the New Dimension Style dialog box.

Compare Dimension Styles dialog box

The list is blank when AutoCAD finds no differences. When ***With*** *is set to* ***<none>*** *or the same style as* ***Compare****, AutoCAD displays all dimension variables.*

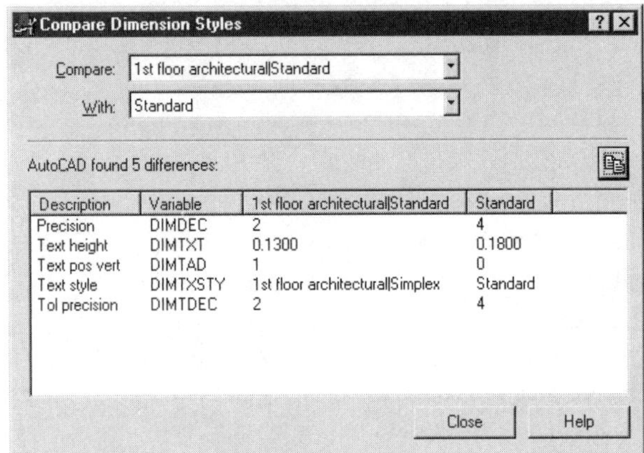

Compare displays the name of one dimension style.

With displays the name of the second dimension style.

Copy to Clipboard copies the style comparison text to the Clipboard, which can be pasted in another Windows application.

Description describes the dimension variable.

Variable names the dimension variable.

Close closes the dialog box.

-DIMSTYLE Command

Command: -dimstyle
Current dimension style: Standard
Enter a dimension style option
[Save/Restore/STatus/Variables/Apply/?] <Restore>: *(Enter an option.)*
Current dimension style: Standard
Enter a dimension style name, [?] or <select dimension>: *(Enter an option.)*
Select dimension: *(Select a dimension in the drawing.)*
Current dimension style: Standard

COMMAND LINE OPTIONS

Save saves current *dimvar* (dimension variable) settings as a named *dimstyle* (dimension style).

Restore retrieves dimvar settings from a named dimstyle.

STatus lists dimvars and current settings.

Variables lists dimvars and their current settings.

Apply updates selected dimension objects with current dimstyle settings.

? lists names of dimstyles stored in drawing.

INPUT OPTIONS

~dimvar (*tilde prefix*) lists the differences between current and selected dimstyle.

ENTER lists the dimvar settings for the selected dimension object.

RELATED DIM COMMANDS

DDim changes dimvar settings.

DimScale determines the scale of dimension text.

RELATED DIM VARIABLES

All.

DimStyle contains the name of the current dimstyle.

TIPS

- At the 'Dim' prompt, the **Style** command sets the text style for the dimension text and does *not* select a dimension style.
- Dimstyles cannot be stored to disk, except in a drawing.
- Read dimstyles from other drawings with the **XBind Dimstyle** command.

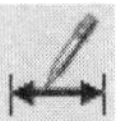

DimTEdit

Rel 13 Dynamically changes the location and orientation of text in dimensions.

Command	Alias	Ctrl+	F-key	Alt+	Menu Bar	Tablet
dimtedit	...	...	...	NX	Dimension ↳Align Text	Y2

Command: dimtedit
Select dimension: *(Select a dimension in the drawing.)*
Specify new location for dimension text or [Left/Right/Center/Home/Angle]: *(Pick a point, or enter an option.)*

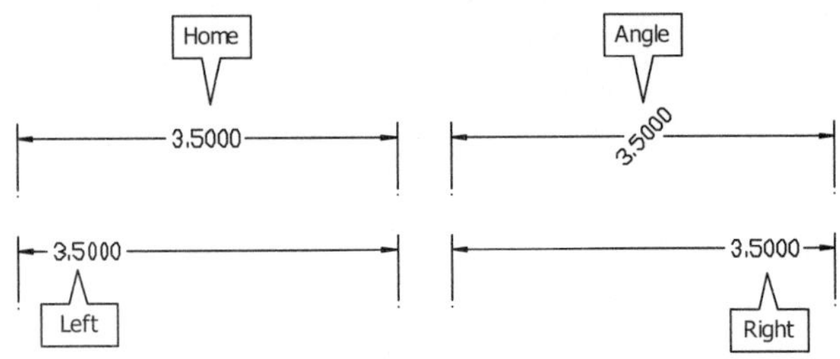

COMMAND LINE OPTIONS

Select dimension selects the dimension to edit.

Angle rotates the dimension text.

Center centers the text on the dimension line.

Home returns the dimension text to original position.

Left moves the dimension text to the left.

Right moves the dimension text to the right.

RELATED DIM VARIABLES

DimSho specifies whether dimension text is updated dynamically while dragged.

DimTih specifies whether dimension text is drawn horizontally or aligned with the dimension line.

DimToh specifies whether dimension text is forced inside the dimension lines.

RELATED TOOLBAR ICONS

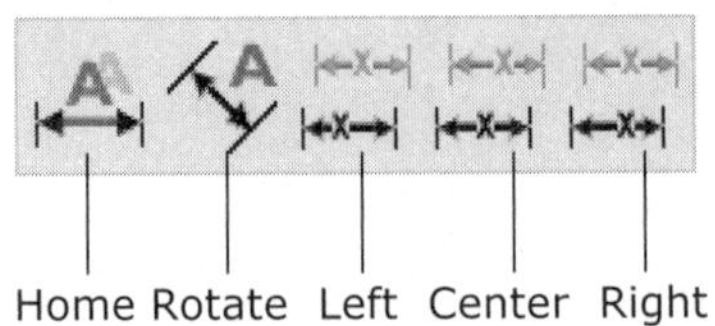

Home Rotate Left Center Right

TIPS

- This command works only with dimensions created created with **DimAssoc** = 1 or 2; use the **DdEdit** command to edit text in non-associative dimensions.
- An angle of 0 returns dimension text to its default orientation.

DimVertical

2004 Draw vertical dimensions (*undocumented command*).

Command	Alias	Ctrl+	F-key	Alt+	Menu Bar	Tablet
dimvertical	...	...	...	...	...	...

Command: dimvertical
Specify first extension line origin or <select object>: *(Pick a point, or press* **ENTER** *to select one object.)*
Specify second extension line origin: *(Pick a point.)*
Specify dimension line location or [Mtext/Text/Angle]: *(Pick a point, or select an option.)*
Dimension text = *nnn*

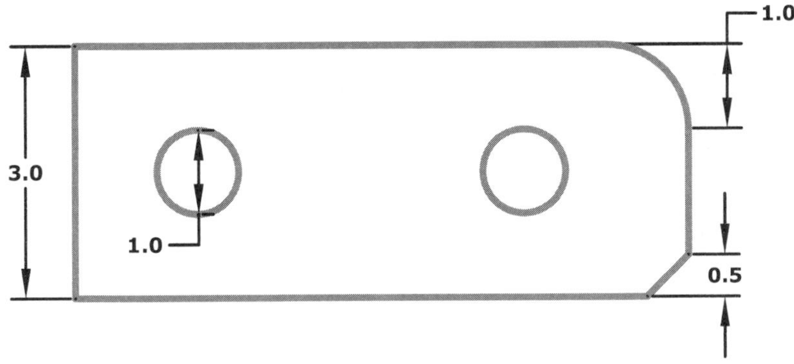

COMMAND LINE OPTIONS

Specify first extension line origin specifies the origin of the first extension line.

Select object dimensions a line, arc, circle, or explodable object automatically.

Specify second extension line location specifies the location of the second extension line.

Specify dimension line location specifies the location of the dimension line.

Mtext displays the Multiline Text Editor dialog box, which allows you to modify the dimension text; see the **MText** command.

Text prompts you to replace the dimension text on the command line: 'Enter dimension text <*nnn*>'.

Angle changes the angle of dimension text: 'Specify angle of dimension text:'.

RELATED DIM COMMANDS

All.

RELATED DIM VARIABLES

Most.

'Dist

V. 1.0 Lists the 3D distances and angles between two points (*short for DISTance*).

Command	Alias	Ctrl+	F-key	Alt+	Menu Bar	Tablet
dist	di	...	...	TYD	Tools ↳Inquiry ↳Distance	T8

Command: dist
Specify first point: *(Pick a point.)*
Specify second point: *(Pick another point.)*

Sample result:

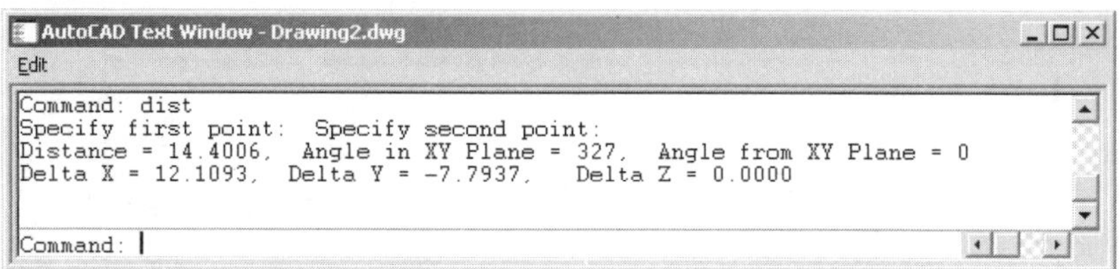

```
AutoCAD Text Window - Drawing2.dwg
Edit
Command: dist
Specify first point:  Specify second point:
Distance = 14.4006,  Angle in XY Plane = 327,  Angle from XY Plane = 0
Delta X = 12.1093,  Delta Y = -7.7937,   Delta Z = 0.0000

Command:
```

COMMAND LINE OPTIONS

Specify first point determines the start point of distance measurement.

Specify second point determines the end point.

RELATED COMMANDS

Area calculates the area and perimeter of objects.

Id lists the 3D coordinates of points.

Length reports the length of open objects.

List lists information about selected objects.

MassProp reports on 2D regions and 3D solid models.

RELATED SYSTEM VARIABLE

Distance specifies the last calculated distance.

TIPS

- Use object snaps to measure precisely the distance between two geometric features.
- When the z-coordinate is left out, **Dist** uses the current elevation for z.

Divide

V. 2.5 Places points or blocks at equal distances along objects.

Command	Alias	Ctrl+	F-key	Alt+	Menu Bar	Tablet
divide	div	...	...	DOD	Draw ↳Point ↳Divide	V13

Command: divide
Select object to divide: *(Select one object.)*
Enter the number of segments or [Block]: *(Enter a number, or enter **B**.)*

Polyline (left) divided by ten points (right).

COMMAND LINE OPTIONS

Select object to divide selects a single open or closed object.

Enter the number of segments specifies the number of segments; must be a number between 2 and 32767.

Block options
Enter name of block to insert: *(Enter the name of a block.)*
Align block with object? [Yes/No] <Y>: *(Enter **Y** or **N**.)*
Enter the number of segments: *(Enter a number.)*

Enter name of block to insert specifies the block to insert (found in the current drawing).

Align block with object aligns the block's x axis with the object.

RELATED COMMANDS

Block creates the block to use with the **Divide** command.

Insert places a single block in the drawing.

MInsert places an array of blocks in the drawing.

Measure places points or blocks at measured distances.

RELATED SYSTEM VARIABLES

PdMode sets the style of the point drawn.

PdSize sets the size of the point, in pixels.

TIPS

- The first dividing point on a closed polyline is its initial vertex; on circles, the first dividing point is in the 0-degree direction from the center.
- The points or blocks are placed in the **Previous** selection set, so that you can select them with the next 'Select Objects' prompt.
- Objects are unchanged by this command.

Donut

V. 2.5 Draws solid-filled circles as a wide polyline consisting of a pair of arcs.

Commands	Alias	Ctrl+	F-key	Alt+	Menu Bar	Tablet
donut	do	...	...	DD	Draw ↳Donut	K9
doughnut						

Command: donut
Specify inside diameter of donut <0.5000>: *(Enter a value.)*
Specify outside diameter of donut <1.0000>: *(Enter a value.)*
Specify center of donut or <exit>: *(Pick a point.)*
Specify center of donut or <exit>: *(Press* **ENTER** *to exit command.)*

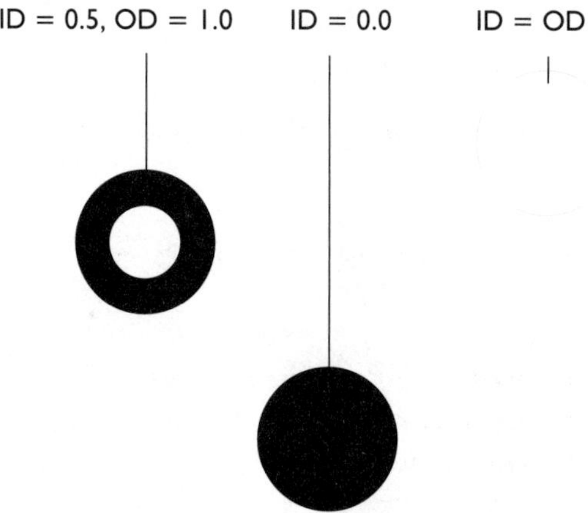

Default donut (left), solid donut (center), and polyline circle (right).

COMMAND LINE OPTIONS

Inside diameter indicates the inner diameter by entering a number or picking two points.

Outside diameter indicates the outer diameter.

Center of donut indicates the donut's center point by specifying coordinates, or picking points.

Exit exits the command.

RELATED COMMAND

Circle draws circles.

RELATED SYSTEM VARIABLES

DonutId specifies the default donut internal diameter.

DonutOd specifies the default donut outside diameter.

Fill toggles the filling of the donut, as well as wide polylines, hatch, and 2D solids.

TIPS

- This command repeats itself until cancelled.
- Donuts are made of two polyline arcs.

'Dragmode

V. 2.0 Controls the display of objects during dragging operations.

Command	Alias	Ctrl+	F-key	Alt+	Menu Bar	Tablet
dragmode	...	...	...	...	...	...

Command: dragmode
Enter new value [ON/OFF/Auto] <Auto>: *(Enter an option.)*

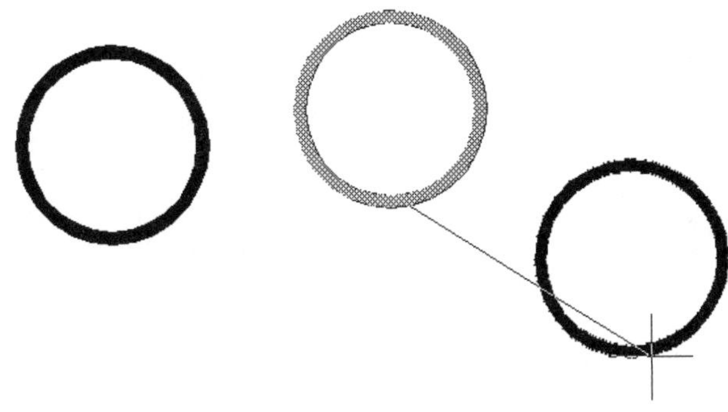

Highlight image (center) and drag image (right).

COMMAND LINE OPTIONS

ON enables dragging display only with the **Drag** option.

OFF turns off all dragging displays.

Auto allows AutoCAD to determine when to display drag image.

COMMAND MODIFIER

Drag displays drag images when DragMode = on.

RELATED SYSTEM VARIABLE

DragMode is the current drag setting:

DragMode	Meaning
0	No drag image.
1	On if required.
2	Automatic.

TIP

- Turn off **DragMode** and **Highlight** in very large drawings to help speed up editing.

DrawOrder

Rel. 14 Controls the display of overlapping objects.

Command	Alias	Ctrl+	F-key	Alt+	Menu Bar	Tablet
draworder	**dr**	...	...	**TO**	**Tools** **↳Display Order**	**T9**

Command: draworder
Select objects: *(Select one or more objects.)*
Select objects: *(Press* **ENTER**.*)*
Enter object ordering option
[Above objects/Under objects/Front/Back] <Back>: *(Enter an option.)*

Text ***under*** *solid (left) and text* ***above*** *solid (right).*

COMMAND LINE OPTIONS

Select objects selects the objects to be moved.

Above object forces selected objects to appear above the reference object.

Under object forces selected objects to appear below the reference object.

Front forces selected objects to the top of the display order.

Back forces selected objects to the bottom of the display order.

RELATED COMMAND

TextToFront *(new to AutoCAD 2005)* brings text and/or dimenions on top of overlappping objects.

RELATED SYSTEM VARIABLES

DrawOrderCtl *(new to AutoCAD 2005)* controls draw order:

DrawOrderCtl	Meaning
0	Turns off draw order.
1	Objects displayed with draw order.
2	New objects, created from objects with draw order, take on draw order of object selected first.
3	Combines options 1 and 2 (default).

HpDrawOrder *(new to AutoCAD 2005)* specifies display order of hatch patterns and fills; see the **BHatch** command.

TIPS

- When you pick more than one object for reordering, AutoCAD maintains the relative display order of the selected objects. The order in which you select objects has no effect on display order.
- When **DrawOrderCtl** is set to 3, editing operations may take longer.
- *Draw order inheritance* means that new objects created from objects with draw order are assigned the display order of the object selected first.

'DSettings

2000 Controls the most-common settings for drafting operations (*short for Drafting SETTINGS*).

Commands	Aliases	Ctrl+	F-key	Alt+	Menu Bar	Tablet
dsettings	**ds**	**...**	**...**	**TF**	**Tools**	**W10**
	se				**↳Drafting Settings**	
	ddrmodes					
+dsettings						

Command: dsettings

Displays dialog box:

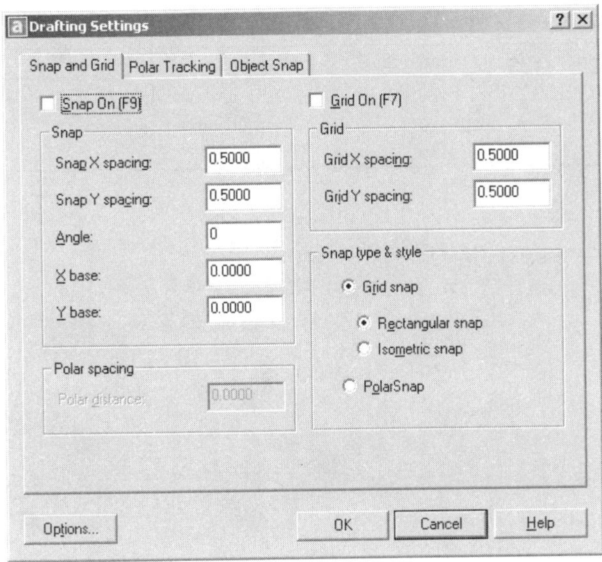

DIALOG BOX OPTIONS

Options displays the Options dialog box; see the **Options** command.

OK saves the changes to settings, and closes the dialog box.

Cancel discards the changes to settings, and closes the dialog box.

Snap and Grid tab

☐ **Snap On (F9)** turns on and off snap mode (the setting is stored in system variable SnapMode; default = off).

☐ **Grid On (F7)** turns on and off the grid display (GridMode; default = off).

Snap options

Snap X spacing specifies the snap spacing in the x direction (SnapUnit; default = 0.5).

Snap Y spacing specifies the snap spacing in the y direction (SnapUnit; default = 0.5).

Angle specifies the snap rotation angle (SnapAng; default = 0).

X base specifies the x coordinate for the snap origin (SnapBase; default = 0).

Y base specifies the y coordinate for the snap origin (SnapBase; default = 0).

Polar Spacing options

Polar distance specifies the snap distance, when Snap type & style is set to Polar snap; when 0, the polar snap distance is set to the value of Snap X spacing (PolarDist; default = 13).

Grid options

Grid X spacing specifies the spacing of grid dots in the x direction; when 0, the grid spacing is set to the value of Snap X spacing (GridUnit; default = 0.5).

Grid Y spacing specifies the spacing of grid dots in the y direction; when 0, the grid spacing is set to the value of Snap Y spacing (GridUnit; default = 0.5).

Snap Type & Style options

⊙**Grid snap** specifies non-polar snap (SnapType; default = on).

⊙**Rectangular snap** specifies rectangular snap (SnapStyl; default = on).

○**Isometric snap** specifies isometric snap mode (SnapStyl; default = off).

○**Polar snap** specifies polar snap (SnapType; default = on).

SnapType	SnapStyl	Meaning
0 (off)	**0** (off)	Rectangular snap.
0 (off)	**1** (on)	Isometric snap.
1 (on)	**0** (off)	Polar snap.

Polar Tracking tab

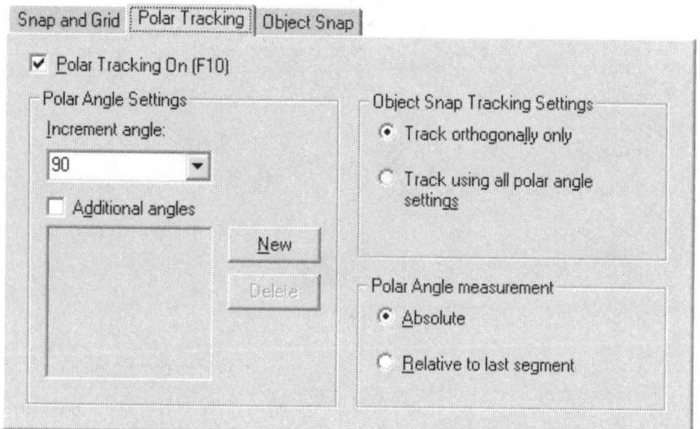

☑ **Polar Tracking On (F10)** turns on and off polar tracking (**AutoSnap**; default = off).

Polar Angle Settings options

Increment angle specifies the increment angle displayed by the polar tracking alignment path; select a preset angle — 90, 60, 45, 30, 22.5, 18, 15, 10, or 5 degrees — or enter a value (PolarAng).

☑ **Additional angles** allows you to set additional polar tracking angles (PolarMode; default=off).

New adds up to ten polar tracking alignment angles (PolarAddAng; default = 0;15;23;45).

Delete deletes added angles.

Object Snap Tracking Settings options

⊙**Track orthogonally only** displays orthogonal tracking paths when object snap tracking is on (PolarMode).

○**Track using all polar angle settings** tracks cursor along polar angle tracking path when object snap tracking is turned on (PolarMode).

Polar Angle Measurement options

⦿ **Absolute** forces polar tracking angle along the current user coordinate system (UCS).

○ **Relative to Last Segment** forces polar tracking angles on the last-created object.

Object Snap tab

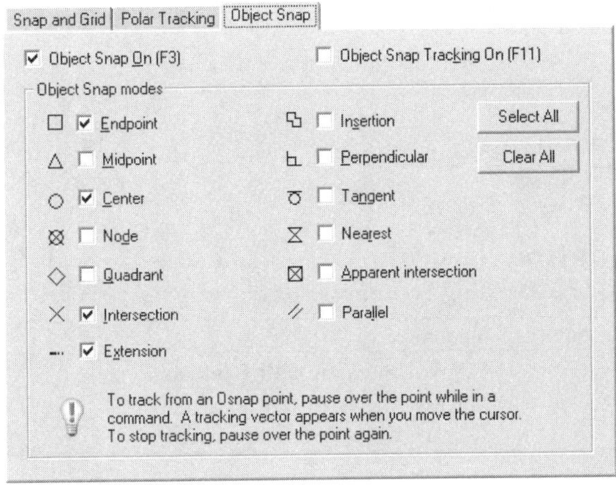

☐ **Object Snap On (F3)** turns on and off running object snaps (OsMode).

☐ **Object Snap Tracking On (F11)** toggles object snap tracking (AutoSnap).

Object Snap Modes options

☑ **ENDpoint** snaps to the nearest endpoint of a line, multiline, polyline segment, ray, arc, and elliptical arc; and to the nearest corner of a trace, solid, and 3D face.

☐ **MIDpoint** snaps to the midpoint of a line, multiline, polyline segment, solid, spline, xline, arc, ellipse, and elliptical arc.

☑ **CENter** snaps to the center of an arc, circle, ellipse, and elliptical arc.

☐ **NODe** snaps to a point.

☐ **QUAdrant** snaps to a quadrant point (90 degrees) of an arc, circle, ellipse, and elliptical arc.

☑ **INTersection** snaps to the intersection of a line, multiline, polyline, ray, spline, xline, arc, circle, ellipse, and elliptical arc; edges of regions; it does not snap to the edges and corners of 3D solids.

☑ **EXTension** displays an extension line from the endpoint of objects; snaps to the point where two objects would intersect if they were infinitely extended; does not work with the edges and corners of 3D solids; automatically turns on intersection mode. (Do not turn on apparent intersection at the same time as extended intersection.)

☐ **INSertion** snaps to the insertion point of text, block, attribute, or shape.

☐ **PERpendicular** snaps to the perpendicular of a line, multiline, polyline, ray, solid, spline, xline, arc, circle, ellipse, and elliptical arc; snaps from a line, arc, circle, polyline, ray, xline, multiline, and 3D solid edge; in this case *deferred perpendicular mode* is automatically turned on.

☐ **TANgent** snaps to the tangent of an arc, circle, ellipse, or elliptical arc; deferred tangent snap mode is automatically turned on when more than one tangent snap is required.

☐ **NEArest** snaps to the nearest point on a line, multiline, point, polyline, spline, xline, arc, circle, ellipse, and elliptical arc.

☐ **APParent intersection** snaps to the apparent intersection of two objects that do not actually intersect but appear to intersect in 3D space; works with a line, multiline, polyline, ray, spline, xline, arc, circle, ellipse, and elliptical arc; does not work with edges and corners of 3D solids.

☐ **PARallel** snaps to a parallel point when AutoCAD prompts for a second point.

Clear All turns off all object snap modes.

Select All turns on all object snap modes.

+DSETTINGS Command

Command: +dsettings

Tab Index <0>: *(Enter a digit.)*

COMMAND LINE OPTION

Tab Index displays the **Drafting Settings** dialog box with the associated tab:

Tab Index	Meaning
0	Displays the **Snap and Grid** tab (default).
1	Displays the **Polar Tracking** tab.
2	Displays the **Object Snap** tab.

RELATED COMMANDS

Grid sets the grid spacing and toggles visibility.

Isoplane selects the working isometric plane.

Ortho toggles orthographic mode.

Snap sets the snap spacing and isometric mode.

RELATED SYSTEM VARIABLES

AutoSnap controls AutoSnap, polar tracking, and object snap tracking.

GridMode indicates the current grid visibility:

GridMode	Meaning
0	Off (default).
1	On.

GridUnit indicates the current grid spacing (default = 0.0).

PolarAddAng specifies user-defined polar angles, separated by semicolons.

PolarAng specifies the increments of the polar angle.

PolarDist specifies the polar snap distance.

SnapAng specifies the current snap rotation angle (default = 0).

SnapBase sets the base point of the snap rotation angle (default = 0,0).

OsMode holds the current object snap modes:

OsMode	Meaning	OsMode	Meaning
0	NONe	**128**	PERpendicular
1	ENDpoint	**256**	TANgent
2	MIDpoint	**512**	NEArest
4	CENter	**1024**	QUIck
8	NODe	**2048**	APParent Intersection
16	QUAdrant	**4096**	EXTension
32	INTersection		
64	INSertion		
8192	PARallel		

PolarMode holds the settings for polar and object snap tracking:

PolarMode	Meaning
0	Measures polar angles based on current UCS (absolute); tracks orthogonally; doesn't use additional polar tracking angles; and acquires object tracking points automatically.
1	Measures polar angles from selected objects (relative).
2	Uses polar tracking settings in object snap tracking.
4	Uses additional polar tracking angles (via **PolarAng**).
8	Acquires object snap tracking points when SHIFT is pressed.

SnapIsoPair specifies the current isoplane.

SnapMode sets the current snap mode setting.

SnapStyl specifies the snap style setting.

SnapUnit sets the current snap spacing (default = 1,1).

TIPS

- Use snap to set the cursor movement increment.
- Use the grid as a visual display to help you better gauge distances.
- Use these CTRL and function keys to change modes during commands:

Mode	Ctrl Key	Function Key
Grid	CTRL+G	F7
Isoplane	CTRL+E	F5
Object Snap	CTRL+F	F3
Object Snap Tracking	CTRL+W	F11
Ortho	CTRL+L	F8
Polar Tracking	CTRL+U	F10
Snap	CTRL+B	F9

- Use object snaps to draw precisely to geometric features.
- New to AutoCAD 2005 is **m2p**, a running object snap, which snaps to the midpoint of two picked points.

DsViewer

Rel.13 Displays the bird's-eye view window; provides real-time pan and zoom (*short for DiSplay VIEWer*).

Command	Alias	Ctrl+	F-key	Alt+	Menu Bar	Tablet
dsviewer	av	...	...	VW	View ↳Aerial View	K2

Command: dsviewer

*Displays the **Aerial View** window:*

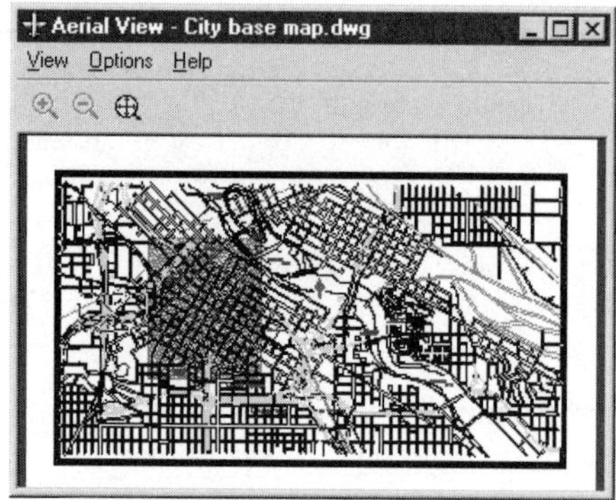

MENU BAR OPTIONS

View menu

Zoom In increases centered zoom by a factor of 2.

Zoom Out decreases centered zoom by a factor of 2.

Global displays entire drawing in Aerial View window.

Options menu

Auto Viewport updates the **Aerial View** automatically with the current viewport.

Dynamic Update updates the **Aerial View** automatically with editing changes in the current viewport.

Realtime Zoom updates the drawing in real time as you zoom in the **Aerial View** window.

TOOLBAR ICONS

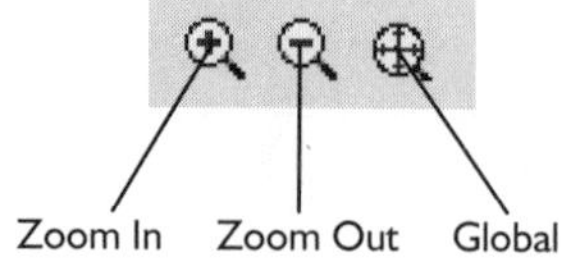

RELATED COMMANDS

Pan moves the drawing view.

View creates and displays named views.

Zoom makes the view larger or smaller.

TIPS

- The purposes of the **Aerial View** are to let you see the entire drawing at all times, and to zoom and pan without entering the **Zoom** and **Pan** commands, or selecting items from the menu.
- The parts of the **Aerial View** window:

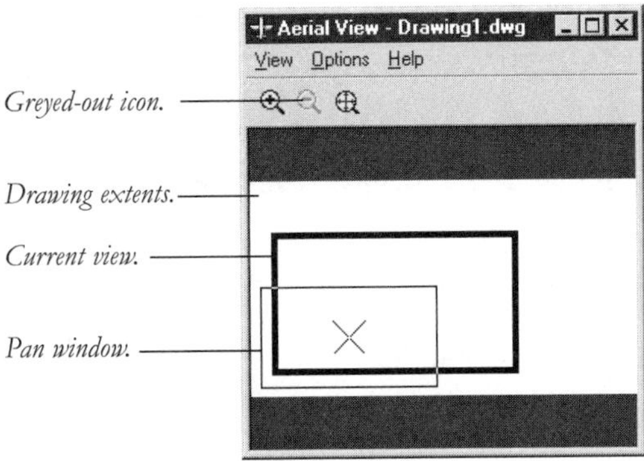

- *Warning!* When in paper space, the **Aerial View** window shows only paper space objects.

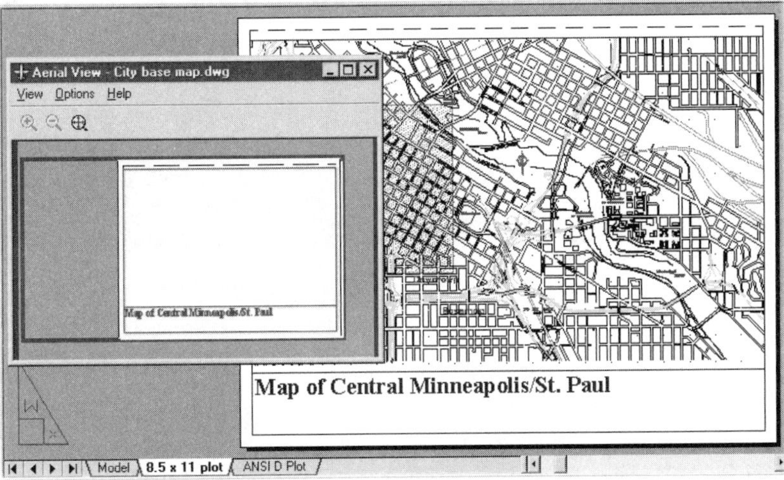

- To switch quickly between **Pan** (default) and **Zoom** modes, click on the **Aerial View** window.

Removed Command

DText was removed from AutoCAD 2000; it was combined with the **Text** command.

DView

Rel.10 Dynamically zooms and pans 3D drawings; turns on perspective mode; superceded by the **3dOrbit** command *(short for Dynamic VIEW)*.

Command	Alias	Ctrl+	F-key	Alt+	Menu Bar	Tablet
dview	dv	...	...	...	...	...

Command: dview

Select objects or <use DVIEWBLOCK>: *(Select objects, or press* **ENTER.**)

Enter option [CAmera/TArget/Distance/POints/PAn/Zoom/TWist/CLip/Hide/Off/Undo]: *(Enter an option.)*

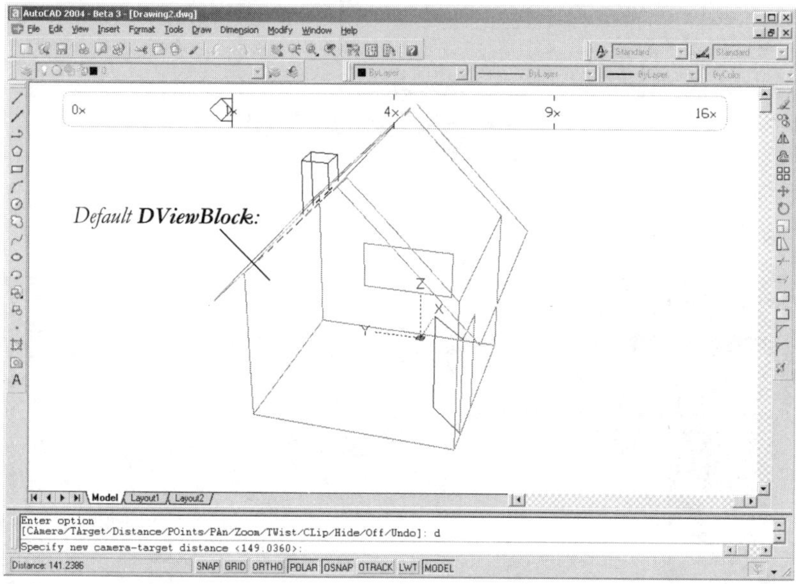

COMMAND LINE OPTIONS

CAmera indicates the camera angle relative to the target.

Toggle switches between input angles.

TArget indicates the target angle relative to the camera.

Distance indicates the camera-to-target distance; turns on perspective mode.

POints indicates both the camera and target points.

PAn pans the view dynamically.

Zoom zooms the view dynamically.

TWist rotates the camera.

CLip options

Enter clipping option [Back/Front/Off] <Off>:

Back clip options

Specify distance from target or [ON/OFF] <0.0>:

ON turns on the back clipping plane.

OFF turns off the back clipping plane.

Distance from target indicates the location of the back clipping plane.

Front clip options

Specify distance from target or [set to Eye(camera)/ON/OFF] <1.0>:

Eye positions the front clipping plane at the camera.

Distance from target indicates the location of the front clipping plane.

Off turns off view clipping.

Hide removes hidden lines.

Off turns off the perspective view.

Undo undoes the most recent **DView** action.

eXit exits **DView**.

RELATED COMMANDS

Hide removes hidden lines from non-perspective views.

Pan pans non-perspective views.

VPoint selects non-perspective viewpoints of 3D drawings.

Zoom zooms non-perspective views.

3dOrbit creates 3D views interactively, in parallel or perspective mode.

RELATED SYSTEM VARIABLES

BackZ specifies the back clipping plane offset.

FrontZ specifies the front clipping plane offset.

LensLength specifies the perspective view lens length, in millimeters.

Target specifies the UCS 3D coordinates of target point.

ViewCtr specifies the 2D coordinates of current view center.

ViewDir specifies the WCS 3D coordinates of camera offset from target.

ViewMode specifies the perspective and clipping settings.

ViewSize specifies the height of view.

ViewTwist specifies the rotation angle of current view.

RELATED SYSTEM BLOCK

DViewBlock is the alternate viewing object displayed during **DView**.

TIPS

- The view direction is from the camera to target.
- Press ENTER at the 'Select objects' prompt to display the DViewBlock house. You can replace the house block with your own by redefining the DViewBlock block.
- To view a 3D drawing in one-point perspective, use the **Zoom** option.
- Menus and transparent zoom and pan are not available during **DView**. Once the view is in perspective mode, you cannot use **Sketch**, **Zoom**, and **Pan**.

Removed Commands

DwfOut was removed from AutoCAD 2004; it was made part of **Plot**.

DwfOutD was removed from AutoCAD 2000; it was combined with **DwfOut**.

Two- and 3-point Perspectives

In two-point perspective, the camera and the target are at the same height. Vertical lines remain vertical. In three-point perspective, the camera and target are at different heights.

Step 1: Two-Point Perspective

1. Start **DView** and select all objects:

   Command: dview
   Select objects: all
   Select objects: *(Press* **ENTER***.)*

Step 2: Place Camera and Target

1. The **POints** option combines the **TArget** and **CAmera** options into one step:

   CAmera/TArget/Distance/POints/.../Undo/<eXit>: po

2. Use the **.xy** filter to pick the target point on the floorplan:

   Enter target point <0.4997, 0.4999, 0.4997>: .xy
   of *(Pick target point.)*

3. Enter a number for your eye height, such as 5'10" or 180cm:

   (need Z): *(Enter height.)*

4. Use the **.xy** filter to pick the camera point:

   Enter camera point <0.4997, 0.4999, 1.4997>: .xy
   of *(Pick camera point.)*

5. Type the same z coordinate for the camera height:

   (need Z): *(Enter same height as in #3, above.)*

Step 3: Turn On Perspective Mode

1. The **Distance** option turns on perspective mode:

   CAmera/TArget/Distance/POints/.../Undo/<eXit>: d

2. In perspective mode, the UCS icon becomes a perspective icon. Use the slider bar to set the distance while in **Distance** mode:

   New camera/target distance <1.0943>: *(Move slider bar.)*

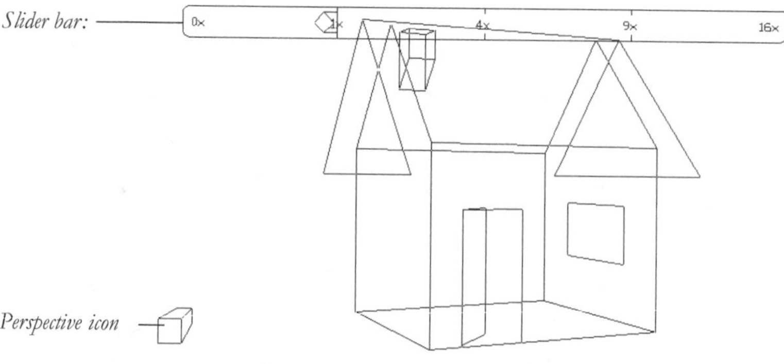

Step 4: Three-Point Perspective

In three-point perspective, the target and camera heights differ. Most commonly, the camera is higher than the target, so that you look down on the 3D scene.

1. Follow the earlier steps, but change the camera and target heights, as follows:

   Command: dview
   Select objects: all
   1 found Select objects: *(Press* **ENTER**.*)*
   CAmera/TArget/Distance/POints/.../Undo/<eXit>: po

2. For target height, enter the height of an object you are looking at, such as a window or table:

   Enter target point <0.4997, 0.4999, 0.4997>: .xy
   of *(Pick target point.)*
   (need Z): *(Enter a height.)*

3. For the camera height, enter your eye height or a larger number for a bird's-eye view:

   Enter camera point <0.4997, 0.4999, 1.4997>: .xy
   of *(Pick camera point.)*
   (need Z): *(Enter a height greater than in #2, above.)*
   CAmera/TArget/Distance/POints/.../Undo/<eXit>: d
   New camera/target distance <1.0943>: *(Adjust distance.)*

4. Use the **Hide** option to create a hidden-line view:

   CAmera/TArget/Distance/POints/.../Hide/Off/Undo/<eXit>: h

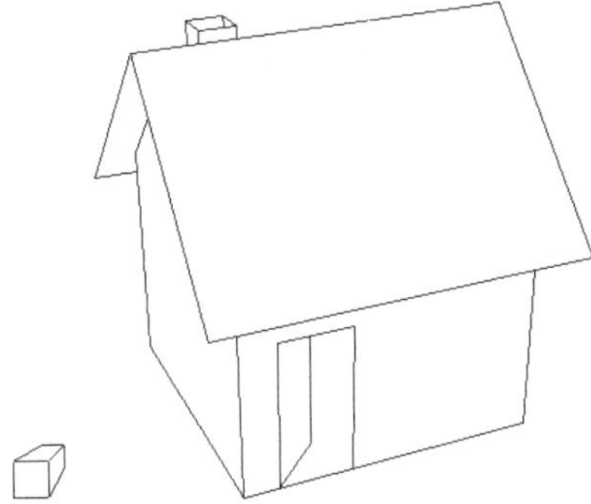

Hidden-line view in three-point perspective mode.

Step 5: Exit Dview

1. Exit **DView**:

   CAmera/TArget/Distance/POints/.../Undo/<eXit>: *(Press* **ENTER**.*)*

   The view remains in perspective mode. While in perspective mode, the **Zoom**, **Pan**, and **DsViewer** commands and scroll bars do not work.

2. To exit perspective mode, use the **Plan** command.

DwgProps

2000 Records and reports information about drawings (*short for DraWinG PROPertieS*).

Command	Alias	Ctrl+	F-key	Alt+	Menu Bar	Tablet
dwgprops	...	...	...	**FI**	**File ⇘Drawing Properties**	...

Command: dwgprops

Displays tabbed dialog box; see below.

DIALOG BOX OPTIONS

OK records the changes, and exits the dialog box.

Cancel discards the changes, and exits the dialog box.

General tab

Displays information about the drawing obtained from the operating system:

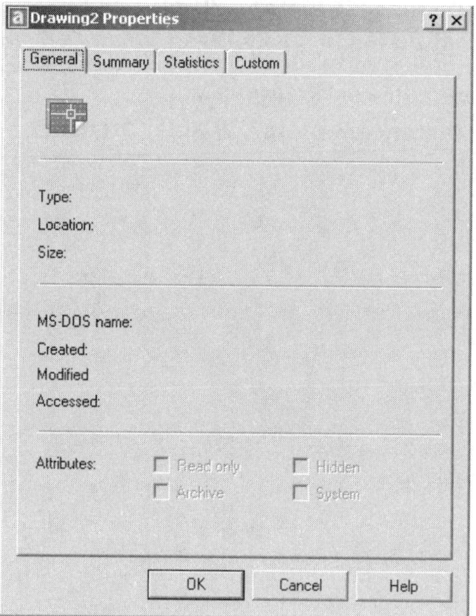

File Type indicates the type of file.

Location indicates the location of the file.

Size indicates the size of the file.

MS-DOS Name indicates MS-DOS file name truncated to eight characters with three-letter extension.

Created indicates the date and time the file was first saved.

Modified indicates the date and time the file was last saved.

Accessed indicates the date and time the file was last opened.

Attributes options

Read-Only indicates the file cannot edited or erased.

Archive indicates the file has been changed since it was last backed up.

Hidden indicates the file cannot be seen in file listings.

System indicates the file is a system file; DWG drawing files never have this attribute turned on.

Summary tab

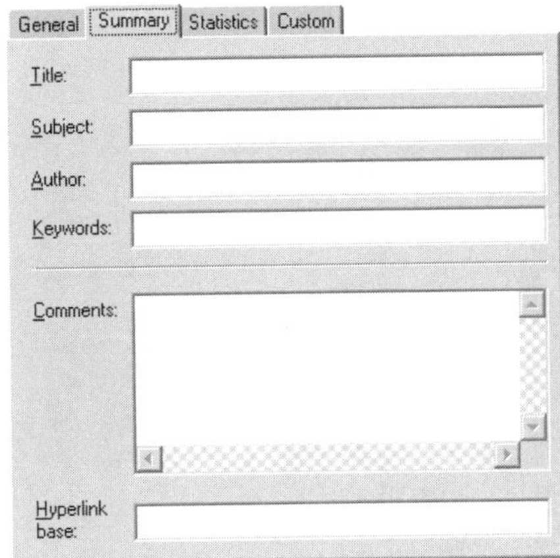

Title specifies a title for this drawing; is usually different from the file name.

Subject specifies a subject for this drawing.

Author specifies the name of the drafter of this drawing.

Keywords specifies keywords used by the operating system's Find or Serach commands to locate drawings.

Comments contains comments on this drawing.

Hyperlink Base specifies the base address for relative links in the drawing, such as http://www .upfrontezine.com; may be an operating system path name, such as *c:\autocad 2005*, or a network drive name. Stored in system variable HyperlinkBase.

Statistics tab

Displays information about the drawing obtained from the drawing:

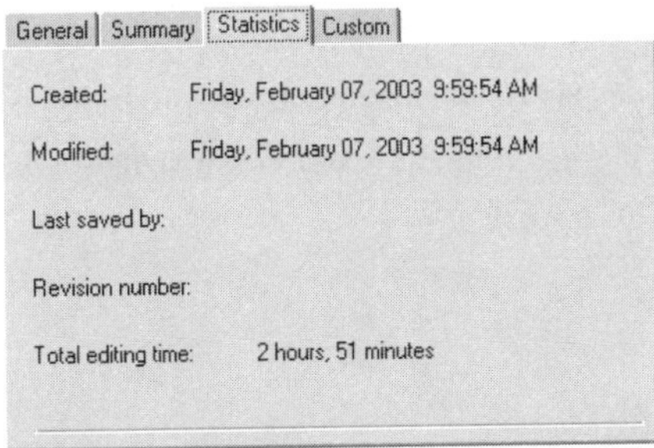

Created indicates the date and time the drawing was first opened, as stored in system variable TdCreate.

Modified indicates the date and time the drawing was last opened or modified, as stored in system variable TdUpdate.

Last saved by indicates who last accessed this drawing file, as stored in the LoginName system variable.

Revision number indicates the revision number; usually blank.

Total editing time indicates the total amount of time that the drawing has been open, as stored in system variable TdInDwg.

Custom tab

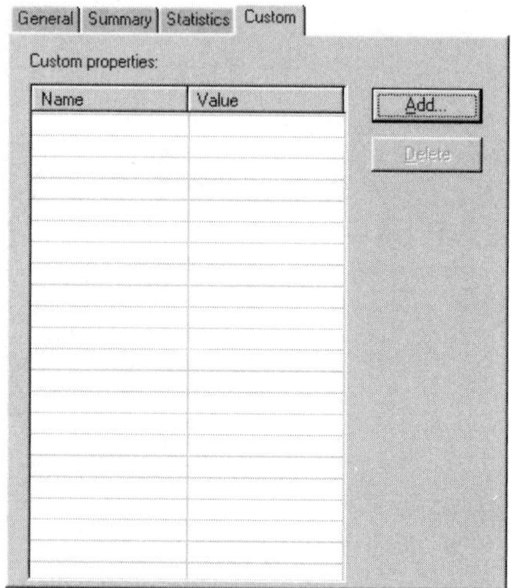

Custom Properties options

Add displays Add Custom Property dialog box; enter a name and a value *(new to AutoCAD 2005)*. To edit values, click them under **Value**.

Delete removes the custom property *(new to AutoCAD 2005)*.

RELATED COMMANDS

Properties lists information about objects in the drawing.

Status lists information about the drawing.

TIPS

- Use the Custom Properties tab to include project data in drawings.
- Use the **Find** button in DesignCenter to search for drawings containing values in the Custom Properties tab.

DxbIn

<u>**Ver. 2.1**</u> Imports *.dxb* files into drawings (*short for Drawing eXchange Binary INput*).

Command	Alias	Ctrl+	F-key	Alt+	Menu Bar	Tablet
dxbin	...	...	...	**IE**	**Insert** **↳Drawing Exchange Binary**	...

Command: dxbin

*Displays **Select DXB File** dialog box. Select a .dxb file, and then click **Open**.*

DIALOG BOX OPTION

Open opens the *.dxb* file, and inserts it in the drawing.

RELATED COMMANDS

DxfIn reads DXF-format files.

Plot writes DXB-format files when configured for an ADI plotter.

TIPS

- To produce *.dxb* files, configure AutoCAD with the ADI plotter driver: after starting the **PlotterManager** command's Add-a-Plotter wizard, select "AutoCAD DXB File" as the manufacturer.
- This command was created for an early Autodesk software product called CAD\camera, which converted raster scans into the DXB vector format.

DxfIn

V. 2.0 Imports *.dxf* files into drawings (*short for Drawing interchange Format INput; undocumented command*).

Command	Alias	Ctrl+	F-key	Alt+	Menu Bar	Tablet
dxfin	...	...	...	FO ↳DXF	File ↳Open ↳DXF	...

Command: dxfin

Displays Select File dialog box. Select a .dxf file, and then click ***Open.***

DIALOG BOX OPTION

Open opens the *.dxf* file.

RELATED COMMANDS

DxbIn reads DXB-format files.

DxfOut writes DXF-format files.

TIPS

- Alternatively, you can use the **Open** command to open *.dxf* files.
- The *.dxf* file comes in two styles: *complete* and *partial*:

  Complete *.dxf* files contain all data required to reproduce a complete drawing file.

  Partial *.dxf* files must be imported into existing drawings.
- To load a complete *.dxf* file, AutoCAD requires the current drawing to be empty. Partial *.dxf* files can be imported or inserted into any drawing, empty or not.
- When you try to import a complete *.dxf* file but the drawing is not new, some versions of AutoCAD complain, "DXFIN requires a new drawing." To create an empty drawing, use **New** with the **Start from Scratch** option.
- If you need to import the complete *.dxf* file into a non-empty drawing, use the **Insert** command, and insert the *.dxf* file with the **Explode** option turned on.
- Autodesk documents the DXF format at usa.autodesk.com/adsk/servlet/item?siteID=123112&id=752569

DxfOut

V. 2.0 Writes *.dxf* files for parts of or the entire AutoCAD drawing to exchange with other programs (*short for Drawing interchange Format OUTput; undocumented command*).

Command	Alias	Ctrl+	F-key	Alt+	Menu Bar	Tablet
dxfout	...	...	...	FA ↳DXF	File ↳Save As ↳DXF	...

Command: dxfout

Displays Save Drawing As dialog box. Enter a file name, and then click ***Save***.

DIALOG BOX OPTIONS

Save saves the drawing as a *.dxf* file.

Files of type creates a *.dxf* file compatible with these versions of AutoCAD:

- AutoCAD 2004 (compatible with 2005).
- AutoCAD 2000 (compatible with 2000, 2000i, and 2002, and AutoCAD LT).
- AutoCAD Release 12 and AutoCAD LT Release 2.

Tools | Options dialog box

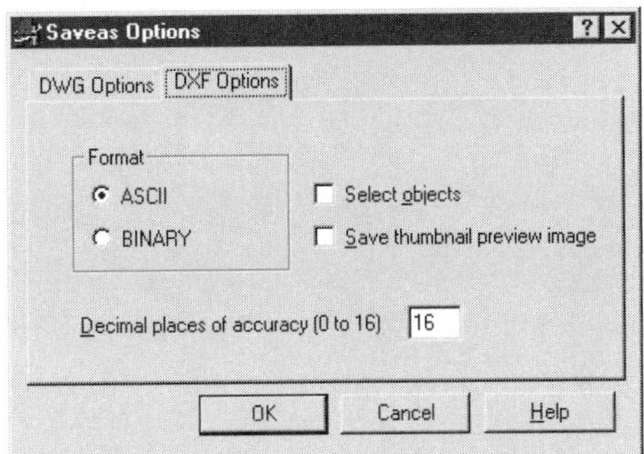

Format options

⊙**ASCII** creates a file in text format, which can be read by humans, and imported by most applications.

○**Binary** creates a binary file with a smaller file size, but it cannot be read by all applications.

Additional options

☐ **Select objects** selects objects to export, instead of the entire drawing.

☐ **Save thumbnail preview image** includes a preview image in the *.dxf* file.

Decimal places of accuracy (0 to 16) specifies the decimal places of accuracy.

RELATED COMMANDS

DxfIn reads DXF-format files.

AcisOut saves solid model objects in the drawing as ACIS-compatible SAT format.

SaveAs writes drawings in DWG format.

TIPS

- Use the ASCII DXF format to exchange drawings with other CAD and graphics programs.
- Some applications, such as those for CNC (computer numerically controlled) machines, require 4 decimal places.
- Binary *.dxf* files are much smaller and are created much faster than ASCII binary files; few applications, however, read binary *.dxf* files.
- The AutoCAD Release 12 dialect of DXF is the most compatible with other applications.
- Drawing files created by AutoCAD 2000 are compatible with AutoCAD 2000i and 2002.
- *Caution!* When saving as an earlier release in DXF format, AutoCAD 2005 erases or converts some objects into simpler objects.

EAttEdit

2002 Edits attribute values and properties in a selected block *(short for Enhanced ATTribute EDITor).*

Command	Alias	Ctrl+	Key	Alt+	Menu Bar	Tablet
eattedit	...	...	...	**MOAS**	**Modify ⇘Object ⇘Attribute ⇘Single**	...

Command: eattedit

If drawing contains no blocks with attributes, AutoCAD complains, "This drawing contains no attributed blocks."

When attributes exist, the command continues:

Select a block: *(Select a single block.)*

Displays dialog box:

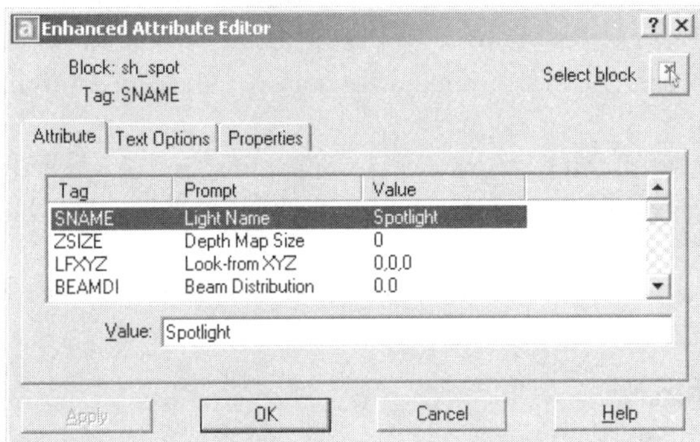

DIALOG BOX OPTIONS

Select block selects another block for attribute editing.

Apply applies the changes to the attributes.

Attribute tab

Value modifies the value of the selected attribute; neither the tag nor the prompt can be modified by this command.

Text Options tab

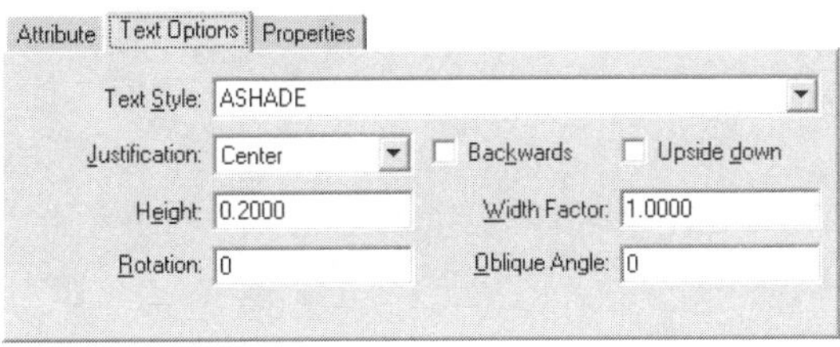

Text Style selects a text style name from the list; text styles are defined by the **Style** command; default is Standard.

Justification selects a justification mode from the list; default is left justification.

Height specifies the text height; can be changed only when height is set to 0.0 in the text style.

Rotation specifies the rotation angle of the attribute text; default = 0 degrees.

☐ **Backwards** displays the text backwards.

☐ **Upside Down** displays the text upside-down.

Width Factor specifies the relative width of characters; default = 1.

Oblique Angle specifies the slant of characters; default = 0 degrees.

Properties tab

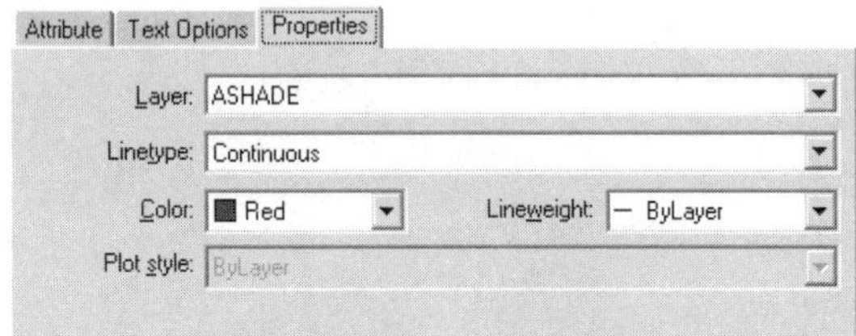

Layer selects a layer name from the list; layers are defined by the **Layer** command.

Linetype selects a linetype name from the list; linetypes are loaded into the drawing with the **Linetype** command.

Color selects a color from the list; to select from the full 255-color spectrum, select Other.

Lineweight selects a lineweight from the list; to turn on the display of lineweights, click LWT on the status bar.

Plot style selects a plot style name from the list; available only if plot styles are enabled in the drawing.

RELATED COMMANDS

AttDef creates attribute definitions.

Block attaches attributes to objects.

BAttMan manages attributes.

EAttExt extracts attributes to a file.

TIPS

- This command edits only attribute values and their properties; to edit all aspects of an attribute, use the **BAttMan** command.
- If you select a block with no attributes, AutoCAD complains, 'The selected block has no editable attributes.' When you select an object that isn't a block, AutoCAD complains, 'Error selecting entity.'
- Blocks on locked layers cannot be edited.

EAttExt

2002 Extracts attribute data to file via a step-by-step procedure *(short for Enhanced ATTribute EXTraction).*

Command	Alias	Ctrl+	Key	Alt+	Menu Bar	Tablet
eattext	...	...	...	...	...	...

Command: eattext

Displays dialog box.

DIALOG BOX OPTIONS

Back goes back to the previous step.

Next proceeds to the next step.

Cancel exits the dialog box.

Help displays helpful information.

Select Drawing page

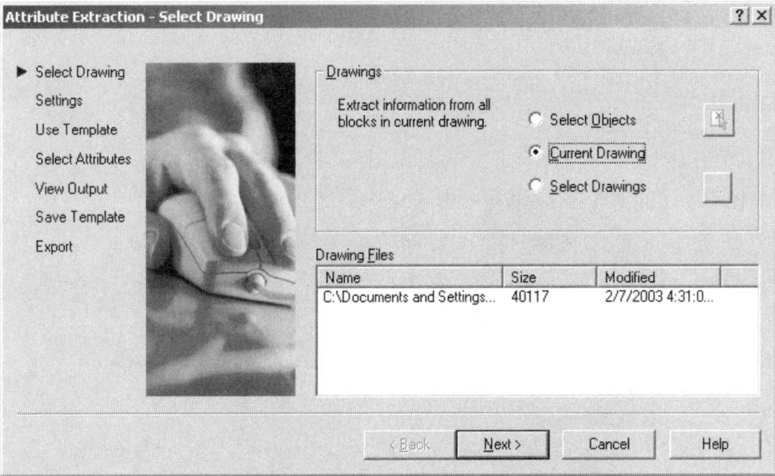

Drawings specifies the location of the blocks containing attributes:

- ○ **Select objects** selects the specific blocks in the current drawing.
- ⊙ **Current drawing** selects all blocks containing attributes in the current drawing.
- ○ **Select drawings** selects drawings located on your computer or network.

Drawing Files lists the names of drawing files from which attributes will be extracted.

Settings page

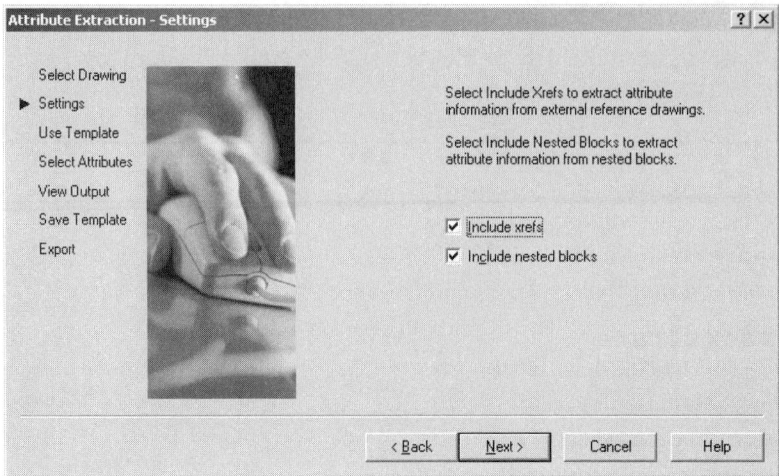

☑ **Include xrefs** specifies that attributes in blocks stored in externally-referenced drawings should also be extracted.

☑ **Include nested blocks** specifies that nested blocks (blocks within blocks) should also be searched for attribute data.

Use Template page

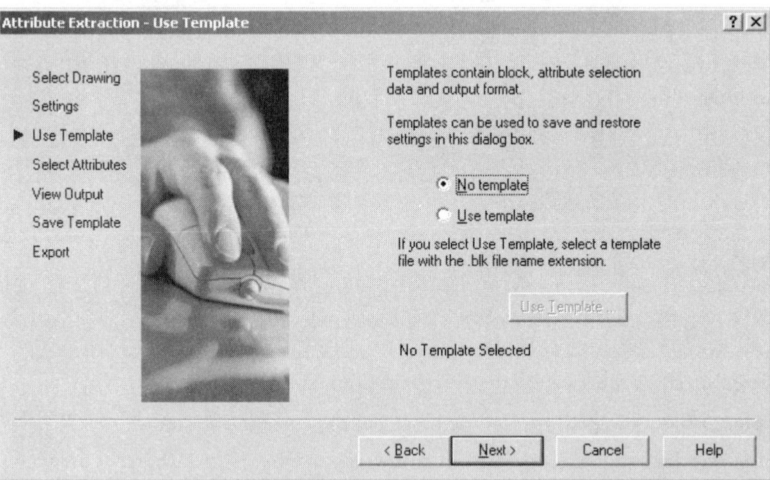

⊙ **No template** means you will be creating a template to specify the data output format.

○ **Use template** means that AutoCAD uses the existing output format, previously stored in a *.blk* file.

Use Template displays the Open dialog box; select a Block Template (*.blk*) file. AutoCAD does not include any *.blk* files; you can create one later during the Save Template step.

Select Attributes page

List of Blocks Found in Drawing(s)   *List of Attributes for each Block*

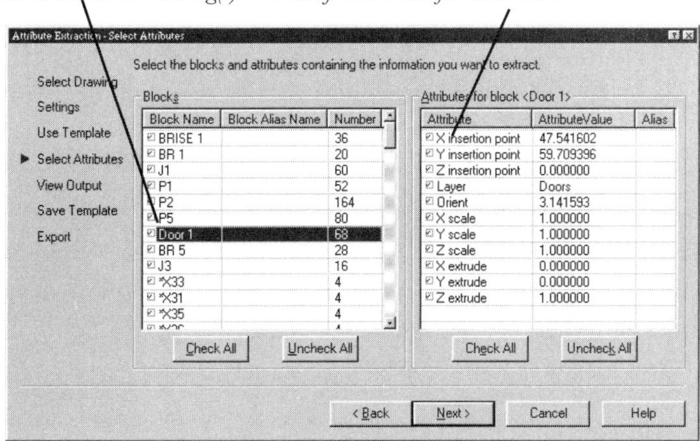

☑ *or* ☐ Includes or excludes blocks and attributes by checking or unchecking boxes in left or right columns, respectively.

Check All selects all blocks or attributes.

Uncheck All unselects all blocks or attributes.

Alias allows you to enter an alias for each block and attribute; an *alias* is an alternative name.

Block Information	Attribute Information
Block Name	Attribute.
Block Alias	Attribute Value.
Number	Alias.

View Output page

Note: This page can take a long time to complete on slow computers with drawings containing a large number of blocks.

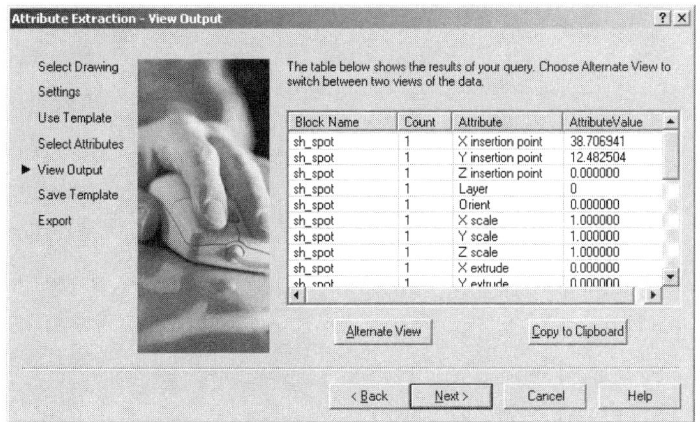

Alternate view 1: Block Name, Count, and attribute data.

Block Name	C...	X insertion ...	Y in...	Z i...	Layer	Orient
BRISE 1	4	35.316713	24.1...	0.0...	Interior ...	0.0015
BR 1	4	62.408729	21.7...	0.0...	0	0.0000
J1	4	67.008472	26.7...	0.0...	0	3.9269
J1	4	40.322257	22.4...	0.0...	0	2.3561
BRISE 1	4	48.593551	10.8...	0.0...	Interior ...	0.0015
BRISE 1	4	45.938184	13.5...	0.0...	Interior ...	0.0015
P1	4	46.223056	16.7...	0.0...	0	2.3561
P2	52	0.000000	0.00...	0.0...	Interior ...	0.0000
J1	4	50.928832	11.8...	0.0...	0	2.3561
BRISE 1	4	37.972081	21.4...	0.0...	Interior ...	0.0015

Alternate view 2: Block Name, Count, and block data.

Block Name lists the name of each block; block names prefixed with *X are hatch patterns created by AutoCAD. Click the header to sort block names in alphabetical order.

Count specifies the number of times the block appears.

Alternate View displays the data in two formats:

- Information about each block.
- Information about each attribute.

Alternate View 1	Alternate View 2
Block Name.	Block Name.
Count.	Count.
Attribute.	X, Y, Z Insertion.
Attribute Value.	Layer.
	Orient.
	X, Y, Z scale.
	X, Y, Z Extrusion.
	User Defined.

Copy to Clipboard copies the displayed data to the Windows Clipboard in tab-delimited format (each field is separated by a tab), which can be pasted into a spreadsheet or document with the **Edit | Paste** (CTRL+v) command.

A field:

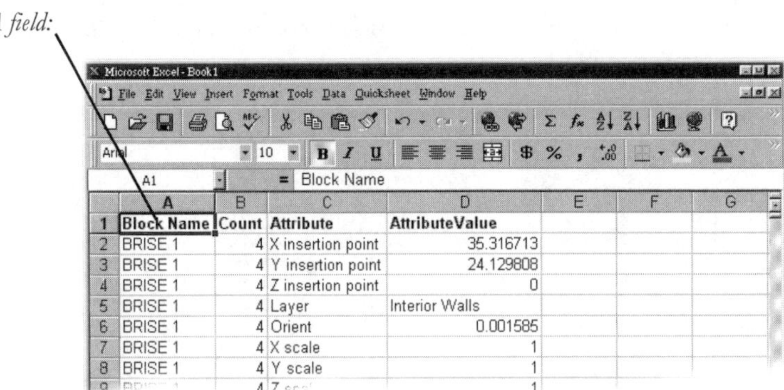

	A	B	C	D
1	Block Name	Count	Attribute	AttributeValue
2	BRISE 1	4	X insertion point	35.316713
3	BRISE 1	4	Y insertion point	24.129808
4	BRISE 1	4	Z insertion point	0
5	BRISE 1	4	Layer	Interior Walls
6	BRISE 1	4	Orient	0.001585
7	BRISE 1	4	X scale	1
8	BRISE 1	4	Y scale	1

Save Template page

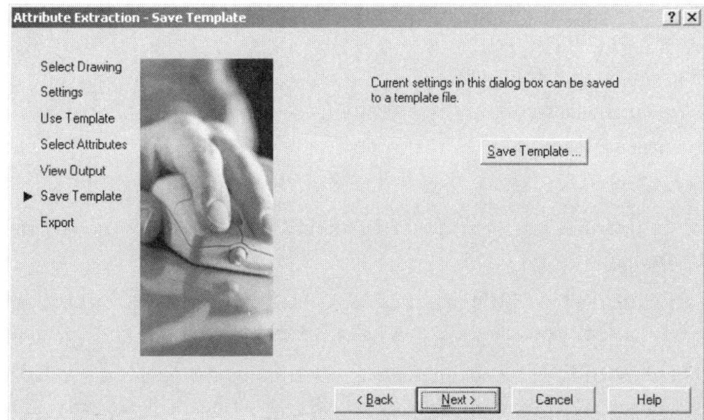

Save Template saves the block and attribute selections (made in the Select Attributes step) to a *.blk* (Block template) file; this *.blk* file can be reused during the Use Template step.

Export page

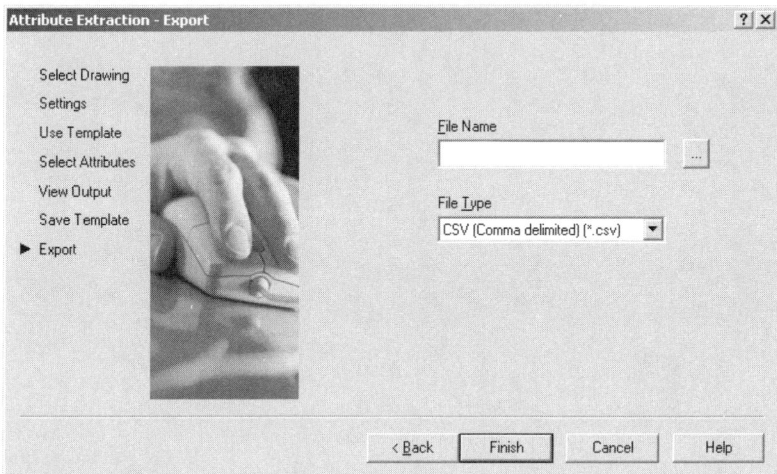

File Name specifies the name of the file that holds the extracted attribute data.

File Type selects the format of the data:

- **CSV** (comma separated values) separates fields with a comma.
- **TXT** (tab delimited values) separates fields with a tab.
- **XLS** (Excel spreadsheet format) saves the data in a proprietary format.

Finish completes the attribute data extraction process.

RELATED COMMANDS

AttExt exports to DXF format, as well as to comma- and tab-delimited formats; it is an older method of attribute extraction.

AttDef creates attribute definitions.

Block attaches attributes to objects.

BAttMan manages attributes.

TIPS

- This command does not export attribute data in DXF format; if you require this format, use the **AttExt** command.
- When the attribute data has been exported, you may open the file in another program, such as WordPerfect (word processing), Lotus 1-2-3 (spreadsheet), or Access (database).
- A drawing (and its xrefs) can contain many attributes. For example, the *1st Floor.dwg* sample drawing contains nearly a thousand blocks, which take up nearly 8,000 rows in a spreadsheet.
- You can use this command to create a crude BOM (bill of material). During the **View Output** step, click the **Copy to Clipboard** button, and paste the data in the AutoCAD drawing.

Edge

Rel.12 Toggles the visibility of 3D faces.

Command	Alias	Ctrl+	F-key	Alt+	Menu Bar	Tablet
edge	...	...	...	**DFE**	**Draw** ↳**Surfaces** ↳**Edge**	...

Command: edge
Specify edge of 3dface to toggle visibility or [Display]: *(Pick an edge, or type* **D**.*)*
Enter selection method for display of hidden edges [Select/All] <All>: *(Type* **S** *or* **A**.*)*
Select objects: *(Select one or more objects.)*
Specify edge of 3dface to toggle visibility or [Display]: *(Press* ESC *to end the command.)*

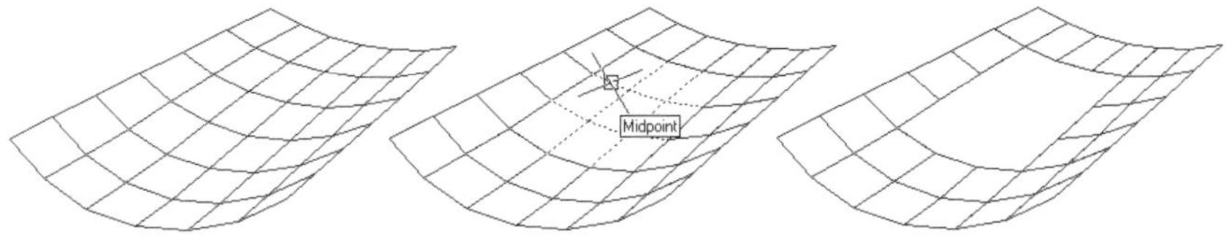

3d faces (left), edges selected with **Edge** *(center), and invisible edges (right).*

COMMAND LINE OPTIONS

Specify edge selects edge to make invisible.

Display options

Select highlights invisible edges.

All selects all hidden edges, and regenerates them.

RELATED COMMAND

3dFace creates 3D faces.

RELATED SYSTEM VARIABLE

SplFrame toggles visibility of 3D face edges.

TIPS

- Make edges invisible to improve the appearance of 3D objects.
- **Edge** applies only to objects made of 3D faces; it does not work with polyface meshes or solid models.
- Use the **Explode** command to convert meshed objects into 3D faces.
- Re-execute **Edge** to display an edge that has been made invisible.
- The command repeats until you press ENTER or ESC at the 'Specify edge of 3dface to toggle visibility or [Display]:' prompt.

EdgeSurf

Rel.10 Draws 3D polygon meshes as Coons surfaces/patches between four boundaries (*short for EDGE-defined SURFace*).

Command	Alias	Ctrl+	F-key	Alt+	Menu Bar	Tablet
edgesurf	...	...	...	**DFD**	**Draw** ↳**Surfaces** ↳**Edge Surface**	**R8**

Command: edgesurf

Current wire frame density: SURFTAB1=6 SURFTAB2=6

Select object 1 for surface edge: *(Pick an object.)*

Select object 2 for surface edge: *(Pick an object.)*

Select object 3 for surface edge: *(Pick an object.)*

Select object 4 for surface edge: *(Pick an object.)*

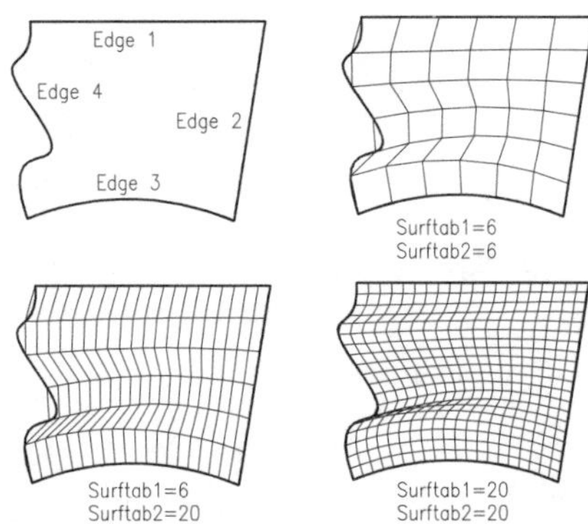

COMMAND LINE OPTION

Select object picks an edge.

RELATED COMMANDS

3dMesh creates 3D meshes by specifying every vertex.

3dFace creates 3D meshes of irregular vertices.

PEdit edits meshes created by **EdgeSurf**.

TabSurf creates tabulated 3D surfaces.

RuleSurf creates ruled 3D surfaces.

RevSurf creates 3D surfaces of revolution.

RELATED SYSTEM VARIABLES

SurfTab1 is the current m-density of meshing; the maximum mesh density is 32767.

SurfTab2 is the current n-density of meshing; maximum is 32767.

TIP

- The four boundary edges can be made from lines, arcs, and open 2D and 3D polylines; the edges must meet at their endpoints.

'Elev

V. 2.1 Sets elevation and thickness for creating extruded 3D objects (*short for ELEVation*).

Command	Alias	Ctrl+	F-key	Alt+	Menu Bar	Tablet
elev	...	...	...	...	...	...

Command: elev
Specify new default elevation <0.0000>: *(Enter a value for elevation.)*
Specify new default thickness <0.0000>: *(Enter a value for thickness.)*

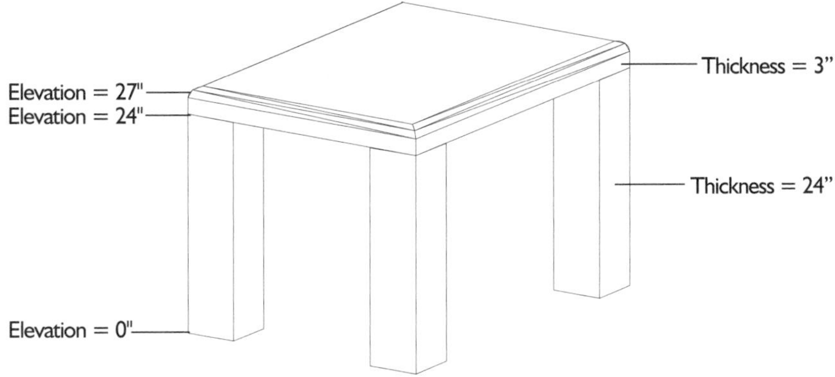

COMMAND LINE OPTIONS

Elevation changes the base elevation from z = 0.

Thickness extrudes new 2D objects in the z-direction.

RELATED COMMANDS

Change changes the thickness and z coordinate of objects.

Move moves objects, even in the z direction.

Properties changes the thickness of objects.

RELATED SYSTEM VARIABLES

Elevation stores the current elevation setting.

Thickness stores the current thickness setting.

TIPS

- The current value of elevation is used whenever the z coordinate is not supplied.
- Thickness is measured up from the current elevation in the positive z-direction.

 Ellipse

V. 2.5 Draws ellipses by four different methods, and draws elliptical arcs and isometric circles.

Command	Alias	Ctrl+	F-key	Alt+	Menu Bar	Tablet
ellipse	**el**	...	...	**DE**	**Draw** **↳Ellipse**	**M9**

Command: ellipse
Specify axis endpoint of ellipse or [Arc/Center]: *(Pick a point, or enter an option.)*
Specify other endpoint of axis: *(Pick a point.)*
Specify distance to other axis or [Rotation]: *(Pick a point, or type* **R***.)*

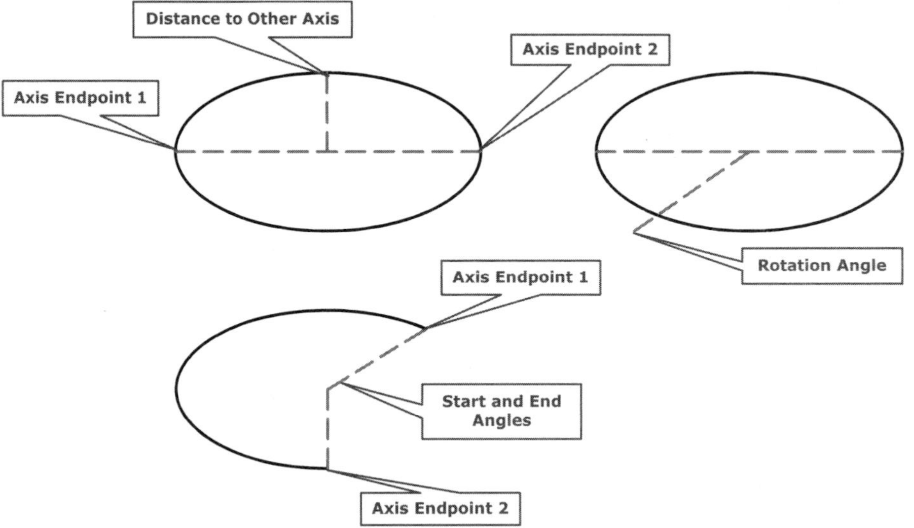

COMMAND LINE OPTIONS

Specify axis endpoint of ellipse indicates the first endpoint of the major axis.

Specify other endpoint of axis indicates the second endpoint of the major axis.

Specify distance to other axis indicates the half-distance of the minor axis.

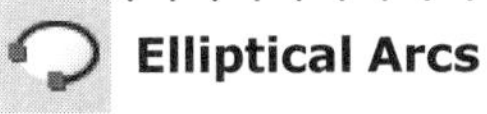

Elliptical Arcs

Command: ellipse

Specify axis endpoint of elliptical arc or [Center]: *(Pick a point, or type* **C**.*)*

Specify other endpoint of axis: *(Pick a point.)*

Specify distance to other axis or [Rotation]: *(Pick a point, or type* **R**.*)*

Specify start angle or [Parameter]: *(Enter an angle, or type* **P**.*)*

Specify end angle or [Parameter/Included angle]: *(Enter an angle, or enter an option.)*

COMMAND LINE OPTIONS

Specify start angle indicates the starting angle of the elliptical arc.

Specify end angle indicates the ending angle of the elliptical arc.

Parameter indicates the starting angle of the elliptical arc; draws the arc with this formula:

p(u)=**c**+(**a***cos(u))+(**b***sin(u))

Parameter	Meaning
a	Major axis.
b	Minor axis.
c	Center of ellipse.

Included angle indicates an angle measured relative to the start angle, rather than 0 degrees.

Center option

Specify center of ellipse: *(Pick a point.)*

Specify center of ellipse indicates the center point of the ellipse.

Rotation option

Specify rotation around major axis: *(Enter an angle.)*

Specify rotation around major axis indicates a rotation angle around the major axis:

Rotation	Meaning
0 degrees	Minimum rotation: creates a round ellipse, like a circle.
89.4 degrees	Maximum rotation: creates a very thin ellipse.

Isometric Circles

This option appears only when system variable ***SnapStyl*** *is set to* ***1*** *(isometric snap mode).*

Command: ellipse

Specify axis endpoint of ellipse or [Arc/Center/Isocircle]: *(Type* **I**.*)*

Specify center of isocircle: *(Pick a point.)*

Specify radius of isocircle or [Diameter]: *(Pick a point, or type* **D**.*)*

COMMAND LINE OPTIONS

Specify center indicates the center point of the isocircle.

Specify radius indicates the radius of the isocircle.

Diameter indicates the diameter of the isocircle.

RELATED COMMANDS

IsoPlane sets the current isometric plane.

PEdit edits ellipses (when drawn with a polyline).

Snap controls the setting of isometric mode.

RELATED SYSTEM VARIABLES

PEllipse determines how the ellipse is drawn:

PEllipse	Meaning
0	Draws ellipse with the ellipse object (default).
1	Draws ellipse as a series of polyline arcs.

SnapIsoPair sets the current isometric plane:

SnapIsoPair	Meaning
0	Left (default).
1	Top.
2	Right.

SnapStyl specifies regular or isometric drawing mode:

SnapStyl	Meaning
0	Standard (default).
1	Isometric.

TIPS

- Previous to AutoCAD Release 13, **Ellipse** constructed the ellipse as a series of short polyline arcs. The **PEllipse** system variable controls how ellipses are drawn. When 0, true ellipses are drawn; when 1, a polyline approximation of an ellipse is drawn.
- When **PEllipse** = 1, the **Arc** option is not available.
- Use ellipses to draw circles in isometric mode. When **Snap** is set to isometric mode, **Ellipse**'s **Isocircle** option projects a circle into the working isometric drawing plane. Use CTRL+E to toggle isoplanes.
- The **Isocircle** option only appears in the option prompt when **Snap** is set to isometric mode.

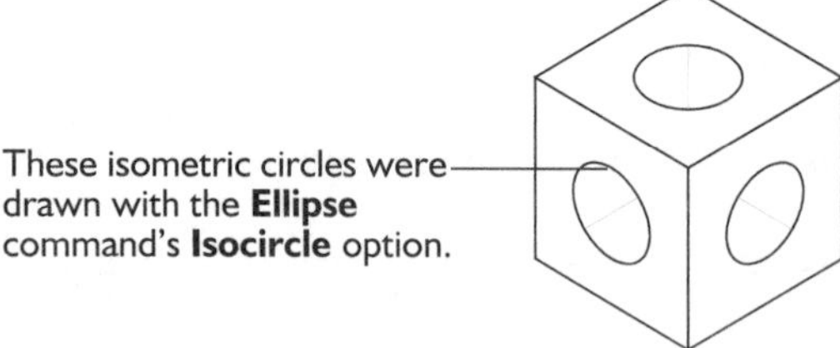

These isometric circles were drawn with the **Ellipse** command's **Isocircle** option.

- See the **Isoplane** command for a tutorial on creating isometric objects.

Removed Commands

End was removed from AutoCAD Release 14; it was replaced by **Quit**.

EndToday was removed from AutoCAD 2004.

Erase

V. 1.0 Erases objects from drawings.

Command	Alias	Ctrl+	Key	Alt+	Menu Bar	Tablet
erase	e	...	Del	ME	Modify ↳Erase	V14
					Edit ↳Clear	U14

Command: erase
Select objects: *(Select one or more objects.)*
Select objects: *(Press* **ENTER** *to end object selection.)*

COMMAND LINE OPTION

Select objects selects the objects to erase.

RELATED COMMANDS

Break erases a portion of a line, circle, arc, or polyline.

Oops returns the most-recently erased objects to the drawing.

Trim cuts off the end of a line, arc, and other objects.

Undo returns the erased objects to the drawing.

TIPS

- The **Erase L** command erases the last-drawn item visible in the current viewport.
- **Oops** brings back the most-recently erased objects; use **U** to bring back other erased objects.
- Objects on locked and frozen layers cannot be erased. To erase them, change the layers to unlocked and thawed.
- *Warning!* The **Erase All** command erases all objects in the current space (model or layout), except on locked, frozen, and/or off layers.

 eTransmit

2000i Transmits drawings and related files as email messages *(short for Electronic TRANSMITtal).*

Commands	Alias	Ctrl+	Key	Alt+	Menu Bar	Tablet
etransmit	...	...	...	**FT**	**File** ↳**eTransmit**	...
-etransmit						

Command: etransmit

If the drawing has not been saved, displays error dialog box:

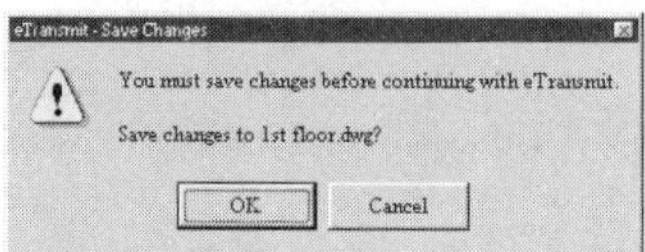

OK saves drawing, and then displays the Create Transmittal dialog box.

Cancel cancels the **eTransmit** command.

DIALOG BOX OPTIONS

Dialog box was redesigned in AutoCAD 2005.

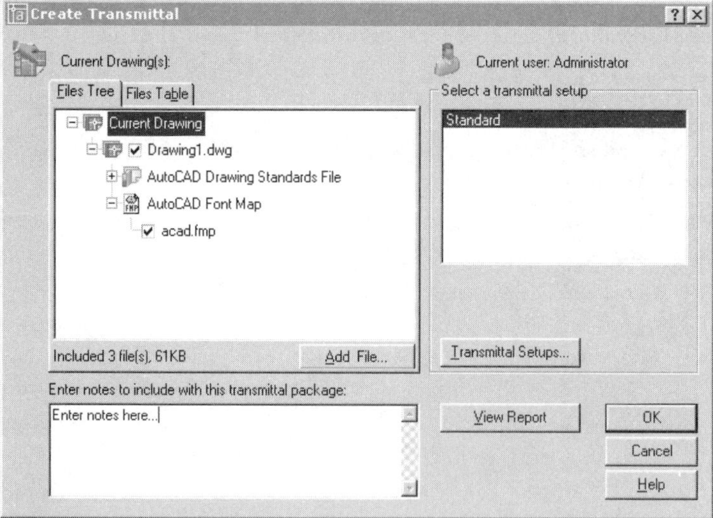

Files Tree tab

Sheets tab displays sheets included with transmittal.

Files Tree tab displays names of drawing and support files, grouped by category.

Files Table tab displays file names in alphabetical order.

Select a Transmittal Setup lists pre-defined setups.

Transmital Setups displays the Transmittal Setups dialog box.

Enter notes to include with this transmital package provides space for entering notes.

View Report displays the View Transmittal Report dialog box.

Files Tree tab

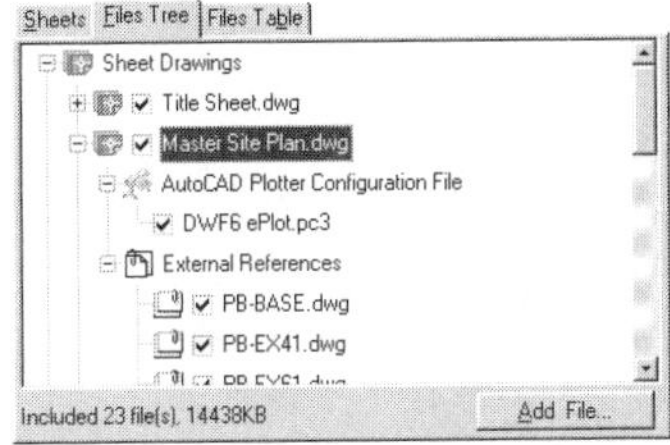

☑ File included in transmittal.

☐ File excluded from transmittal.

Add File displays the Add File to Transmittal dialog box, to add files to the transmittal.

Files Table tab

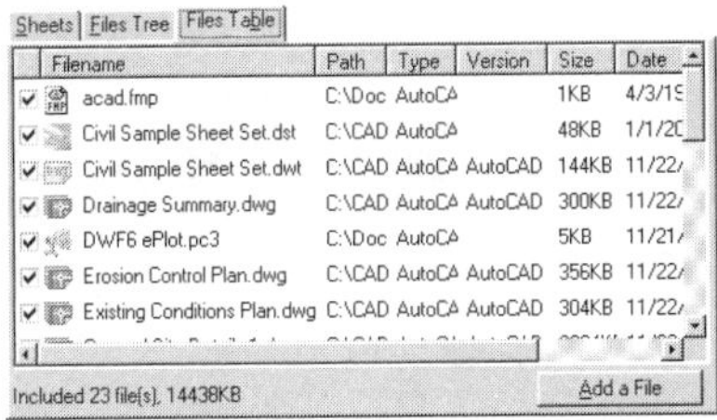

Transmittal Setup dialog box

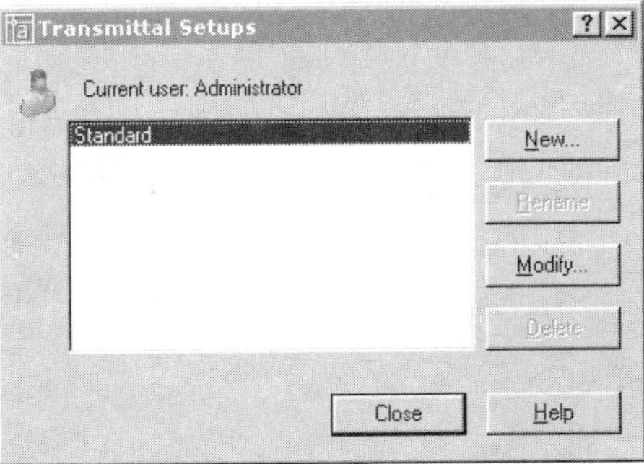

New creates new setups; displays New Transmittal Setup dialog box.

Rename renames setups, except for "Standard."

Modify changes setups; displays Modify Transmittal Setup dialog box.

Delete removes setups, except for "Standard."

New Transmittal Setup dialog box

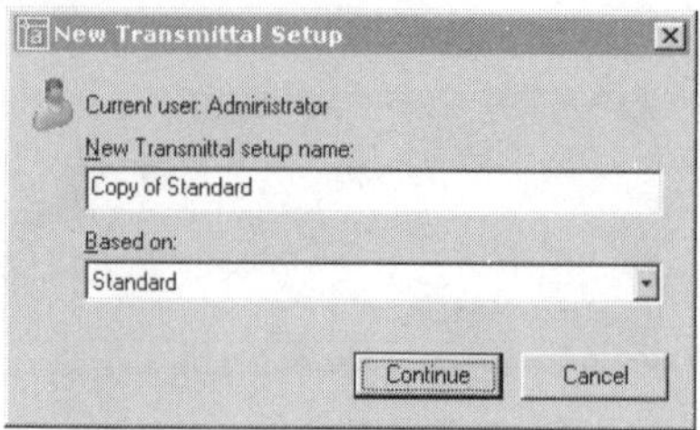

New Transmittal Setup Name specifies the name of the new setup.

Based on selects the setup to copy.

Continue displays the Modify Transmittal Setup dialog box.

Modify Transmittal Setup dialog box

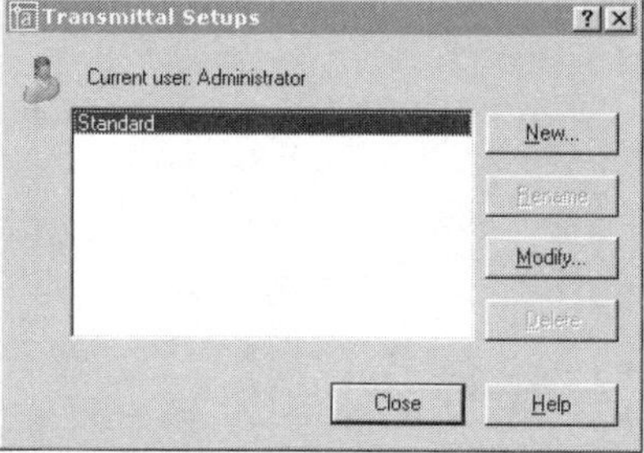

New creates new setups; displays New Transmittal Setup dialog box.

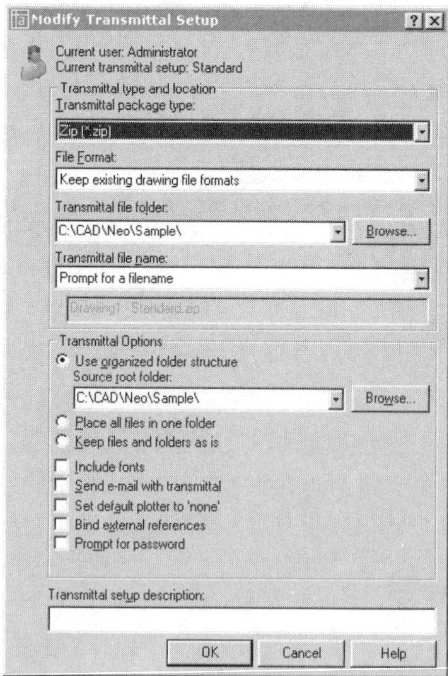

Transmittal Type and Location options

Transmittal Package Type options

Folder (set of files) transmits uncompressed files in new and existing folders. Best option for transmitting files on CDs or by FTP.

Self-Extracting Executable (*.exe) transmits files in a compressed, self-extracting file. Uncompress by double-clicking the file.

Zip (*.zip) (*default*) transmits files as a compressed ZIP file. Uncompress the file using the PkZip or WinZip programs. Best option when sending by email.

File Format options

Keep existing drawing file formats transmits files in their native format, including AutoCAD 2005.

AutoCAD 2004/LT 2004 Drawing Format transmits files in AutoCAD 2004 format.

AutoCAD 2000/LT 2000 Drawing Format transmits files in AutoCAD 2000 format. *Caution!* Those objects and properties not found in AutoCAD 2000 may be changed or lost.

Transmittal File Folder specifies the location in which to collect the files. When no location is specified, the transmittal is created in the same folder as the *.dst* file.

Transmittal File Name options

Prompt for a File Name prompts the user for the file name.

Overwrite if Necessary uses the specified file name, and overwrites the existing file of the same name.

Increment File Name if Necessary uses the specified file name, and appends a digit to avoid overwriting the existing file of the same name.

Transmittal options

⊙ **Use Organized Folder Structure** preserves the folder structure in the transmittal, but makes the changes listed below; allows you to specify the name of the folder tree.

○ **Place All Files in One Folder locates** all files in a single folder.

○ **Keep Files and Folders As Is** preserves the folder structure in the transmittal.

☐ **Include Fonts** includes all *.ttf* and *.shx* font files in the transmittal.

☐ **Send email with transmittal** opens the computer's default email software.

☐ **Set Default Plotter to 'None'** resets the plotter to None for all drawings in the transmittal.

☐ **Bind External References** merges all xrefs into the drawing *(new to AutoCAD 2005)*.

☐ **Prompt for Password** displays a dialog box for specifying a password; not available when the Folder archive type is selected.

Transmittal setup description allows you to describe the setup.

-ETRANSMIT command

Command: -etransmit

Enter an option [Create transmittal package/Report only/CUrrent setup/CHoose setup/Sheet set] <Report only>: *(Enter an option.)*

COMMAND LINE OPTIONS

Command line options were changed in AutoCAD 2005.

Create Transmittal Package creates transmittal packages from the current drawing and all support files; uses settings in the current transmittal setup.

Report Only displays the Save Report File As dialog box; enter a file name, and then click Save. Saves the report in plain ASCII text format; does not create the transmittal package.

Current Setup lists the name of the current transmittal setup.

Choose Setup selects the transmittal setup.

Sheet Set specifies the sheet set and transmittal setup to use for the transmittal package; available only when a sheet set is open in the drawing.

RELATED COMMANDS

Archive creates archive sets of drawings, support files, and sheet sets.

Publish creates a *.dwf* file with drawing sheets.

PublishToWeb saves the drawing as an HTML file.

TIPS

- The **eTransmit** command is an expanded version of the **Pack'n Go** command, a bonus command first included with AutoCAD Release 14.
- The **Password** button is *not* available when you select the **Folder** option.
- Including a TrueType font (*.ttf*) is touchy, because sending a copy of the font might infringe on its copyright. All *.shx* and *.ttf* files included with AutoCAD may be transmitted. For this reason, the **Include Fonts** option is turned off by default. In addition, not including fonts saves some file space; recall that smaller files take less time to transmit via the Internet.
- "Self-extracting executable" means that the files are compressed into a single file with the *.exe* extension. The email recipient double-clicks the file to extract (uncompress) the files. The benefit is that the recipient does not need to have a copy of PkUnzip or WinZip on their computer; the drawbacks are that a virus could hide in the *.exe* file, and some e-mail programs disallow receipt of attachments with *.exe* extensions..
- The benefit to converting the drawing to AutoCAD 2000 or Release 14 format is that clients with older versions of the software can read the file; the drawback is that R2000-specific objects are either erased or modified to a simpler format. Two problems that can occur include: (1) lineweights are no longer displayed; and (2) database links and freestanding labels are converted to Release 14 links and displayable attributes. (Lineweights are restored when the drawing is opened again in AutoCAD 2005.)
- If your computer cannot compress the files (or if your recipient cannot uncompress them), you are probably lacking the software needed to uncompress the *.zip* file. (Do not confuse *.zip* files with Iomega's ZIP disk drive; the two having nothing in common, except the name.) You can obtain the PkUnzip and WinZIP utilities as freeware or shareware.
- Files occasionally become corrupted when sent by email; the transmittal may need to be resent. If you continue to have problems, you may need to change a setting in your email software. Try changing the attachment encoding method from BinHex or Uuencode to MIME.

Explode

V. 2.5 Explodes polylines, blocks, associative dimensions, hatches, multilines, 3D solids, regions, bodies, and meshes into their constituent objects.

Command	Alias	Ctrl+	F-key	Alt+	Menu Bar	Tablet
explode	x	...	...	MX	Modify ↳Explode	Y22

Command: explode
Select objects: *(Select one or more objects.)*
Select objects: *(Press* **ENTER** *to end object selection.)*

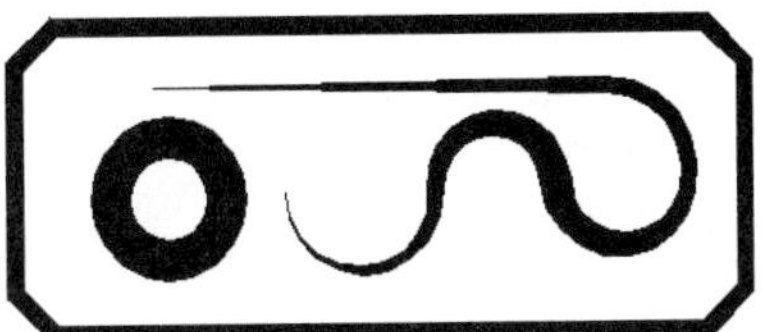

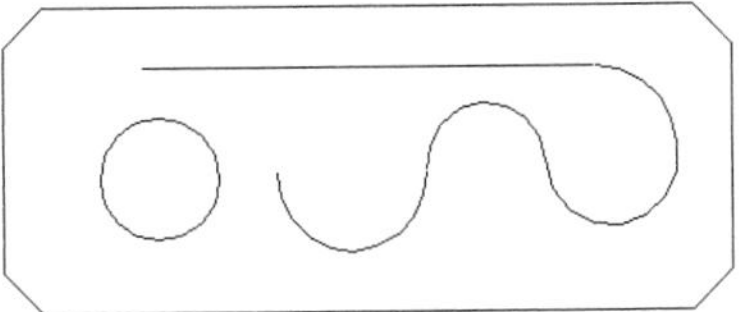

Polylines (left) exploded into lines and arcs (right).

COMMAND LINE OPTION

Select objects selects the objects to explode.

RELATED COMMANDS

Block recreates a block after an explode.

PEdit converts a line into a polyline.

Region converts 2D objects into a region.

Undo reverses the effects of explode.

Xplode provides control over the explosion process.

RELATED SYSTEM VARIABLE

ExplMode toggles whether non-uniformly scaled blocks can be exploded:

ExplMode	Meaning
0	Does not explode (compatible with AutoCAD Release 12 and earlier).
1	Does explode (default).

TIPS

- As of Release 13, AutoCAD can explode blocks inserted with unequal scale factors, mirrored blocks, and blocks created by the **MInsert** command.
- You cannot explode xrefs and dependent blocks (blocks from xref drawings), or single-line text.
- Parts making up exploded blocks and associative dimensions of BYBLOCK color and linetype are displayed in color White (or Black when the background color is white) and linetype Continuous.

- The **Explode** command alters objects, as follows:

Objects	Exploded Into
Arcs in non-uniformly scaled blocks	Elliptical arc.
Associative dimensions	Lines, solids, and text.
Blocks	Constituent parts.
Circles in non-uniformly scaled blocks	Ellipse.
Mtext and field text	Text
Multilines	Lines.
Polygon meshes	3D faces.
Polyface meshes	3D faces, lines, and points.
Tables	Lines.
2D polylines	Lines and arcs; width and tangency information lost.
3D polylines	Lines.
3D solids	Regions and bodies.
Regions	Lines, arcs, ellipses, and splines.
Bodies	Single bodies, regions, and curves.

- Resulting objects become the previous selection set.

Export

Rel.13 Saves drawings in formats other than DWG and DXF.

Command	Alias	Ctrl+	F-key	Alt+	Menu Bar	Tablet
export	...	...	...	**FE**	**File** ↳**Export**	**W24**

Command: export

Displays dialog box:

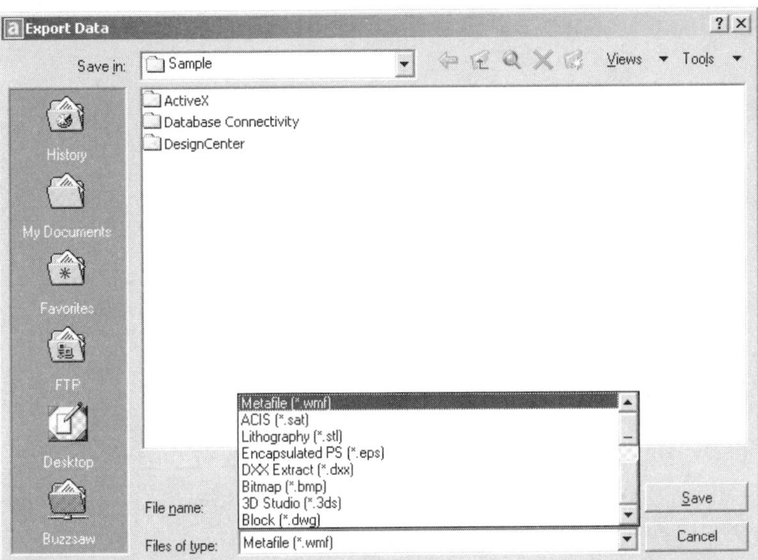

DIALOG BOX OPTIONS

Save in selects the folder (subdirectory) and drive into which to export the file.

Back returns to the previous folder (ALT+1).

Up One Level moves up one level in the folder structure (ALT+2).

Search the Web displays a simple Web browser that accesses the Autodesk Web site (ALT+3).

Delete erases the selected file(s) or folder (DEL).

Create New Folder creates a new folder (ALT+5).

Views displays files and folders in a list or with details.

Tools lists several additional commands, including the Options dialog box — available only with encapsulated PostScript; see the **PsOut** command.

File name specifies the name of the file, or accepts the default.

Save as type selects the file format in which to save the drawing.

Save saves the drawing.

Cancel dismisses the dialog box, and returns to AutoCAD.

RELATED COMMANDS

AttExt exports attribute data in the drawing in CDF, SDF, or DXF formats.

CopyClip exports the drawing to the Clipboard.

CopyHist exports text from the text screen to the Clipboard.

Import imports several vector and raster formats.

LogFileOn saves the command line text as ASCII text in the *acad.log* file.

MassProp exports the mass property data as ASCII text in an *.mpr* file.

MSlide exports the current viewport as an *.sld* slide file.

Plot exports the drawing in many vector and raster formats.

SaveAs saves the drawing in AutoCAD's DWG format.

SaveImg exports the rendering in TIFF, Targa, or BMP formats.

TIPS

- The **Export** command exports the current drawing in the following formats:

Extension	Meaning
3DS	3D Studio file: **3dsOut** command.
BMP	Device-independent bitmap file: **BmpOut** command.
DWG	AutoCAD drawing file : **WBlock** command.
DWF	Drawing Web format file: **DwfOut** command.
DXF	AutoCAD drawing interchange file: **DxfOut** command.
DXX	Attribute extract DXF file: **AttExt** command.
EPS	Encapsulated PostScript file: **PsOut** command.
SAT	ACIS solid object file: **AcisOut** command.
STL	Solid object stereo-lithography file: **StlOut** command.
WMF	Windows metafile: **WmfOut** command.

- This command acts as a "shell": it launches other AutoCAD commands that perform the actual export function, as noted above.
- When drawings are exported in formats other than *.dwg* or *.dxf*, information is lost, such as layers and attributes.

Removed (and Restored) Command

The **ExpressTools** (also known as "bonus CAD tools" in earlier versions of AutoCAD) were removed from AutoCAD 2002, but returned in AutoCAD 2004.

Extend

V. 2.5 Extends the length of lines, rays, open polylines, arcs, and elliptical arcs to boundary objects.

Command	Alias	Ctrl+	F-key	Alt+	Menu Bar	Tablet
extend	ex	...	...	MD	Modify ↳Extend	W16

Command: extend
Current settings: Projection=UCS Edge=None
Select boundary edges ...
Select objects: *(Select one or more objects.)*
Select objects: *(Press* ENTER.*)*
Select object to extend or shift-select to trim or [Project/Edge/Undo]: *(Select an object, or enter an option.)*
Select object to extend or shift-select to trim or [Project/Edge/Undo]: *(Press* ENTER *to end the command.)*

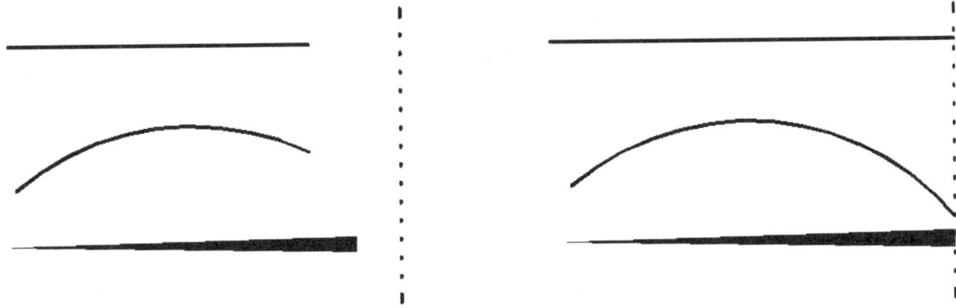

Line, arc, and variable-width polyline before (left) and after (right) extended to dotted line.

COMMAND LINE OPTIONS

Select objects selects the objects to be used for the extension boundary.

Select objects to extend selects the objects that will be extended.

Shift-select to trim trims objects when you hold down the SHIFT key.

Undo undoes the most recent extend operation.

Project options
Enter a projection option [None/Ucs/View] <Ucs>: *(Enter an option.)*

None extends objects to boundary (Release 12-compatible).

Ucs extends objects in the x,y-plane of the current UCS.

View extends objects in the current view plane.

Edge options
Enter an implied edge extension mode [Extend/No extend] <No extend>: *(Enter an option.)*

Extend extends to implied boundary.

No extend extends only to actual boundary (Release 12-compatible).

RELATED COMMANDS

Change changes the length of lines.

Lengthen changes the length of open objects.

SolidEdit extends the face of a solid object.

Stretch stretches objects wider and narrower.

Trim reduces the length of lines, polylines and arcs.

RELATED SYSTEM VARIABLES

EdgeMode toggles boundary mode for the **Extend** and **Trim** commands:

EdgeMode	Meaning
0	Use actual edges; Release 12 compatible (default).
1	Use implied edge.

ProjMode toggles projection mode for the **Extend** and **Trim** commands:

ProjMode	Meaning
0	None; Release 12 compatible.
1	Current UCS (default).
2	Current view plane.

TIPS

- The following objects (even when inside blocks) can be used as boundaries:

2D polyline	Line
3D polyline	Ray
Arc	Region
Circle	Spline
Ellipse	Text
Floating viewport	Xline
Hatch	

- When a wide polyline is the edge, **Extend** extends to the polyline's centerline.
- Pick the object a second time to extend it to a second boundary line.
- Circles and other closed objects are valid edges: the object is extended in the direction nearest to the pick point.
- Extending a variable-width polyline widens it proportionately; extending a splined polyline adds a vertex.

Extrude

Rel.11 Creates 3D solids by extruding 2D objects, optionally with tapered sides.

Command	Alias	Ctrl+	F-key	Alt+	Menu Bar	Tablet
extrude	**ext**	...	...	**DIX**	**Draw** ↳**Solids** ↳**Extrude**	**P7**

Command: extrude
Select objects: *(Select one or more objects.)*
Select objects: *(Press* **ENTER**.*)*
Specify height of extrusion or [Path]: *(Enter an value, or type* **P**.*)*
Specify angle of taper for extrusion <0>: *(Enter an angle, or press* **ENTER**.*)*

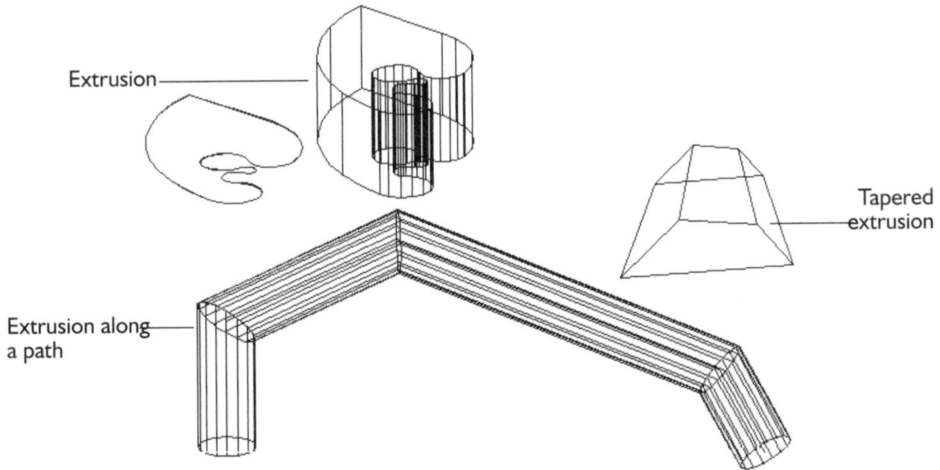

COMMAND LINE OPTIONS

Select objects selects the 2D objects to extrude.

Specify height of extrusion specifies the extrusion height.

Specify angle of taper for extrusion specifies the taper angle; ranges from -90 to +90 degrees.

Path option
Select extrusion path or [Taper angle]: *(Select an object, or type* **T**.*)*

Select extrusion path specifies the path for the extrusion.

RELATED COMMANDS

Revolve creates a 3D solid by revolving a 2D object.

Elev gives thickness to non-solid objects to extrude them.

TIPS

- This command extrudes the following objects:

  Circles
  Ellipses
  Donuts
  Closed polylines
  Polygons
  Closed splines
  Regions
  3D faces

- This command can use the following objects as extrusion paths:

  Lines
  Polylines
  Arcs
  Elliptical arcs
  Circles
  Ellipses
  Splines

- You *cannot* extrude polylines with less than 3, or more than 500, vertices; similarly, you cannot extrude crossing or self-intersecting polylines.
- Objects within a block cannot be extruded; use the **Explode** command first.
- The taper angle must be between 0 (default) and 90 degrees.
- Positive angles taper in from the base; negative angles taper out.
- This command does not work when a taper angle is less than -90 degrees or more than +90 degrees.
- **Extrude** also does not work if the combination of angle and height makes the object's extrusion walls intersect.

Field

2005 Places automatically updatable field text in drawings.

Command	Alias	Ctrl+	F-key	Alt+	Menu Bar	Tablet
Field	...	...	...	**DA**	**Insert** ↳**Field**	...

Command: field

Displays dialog box, whose content varies depending on the field name selected:

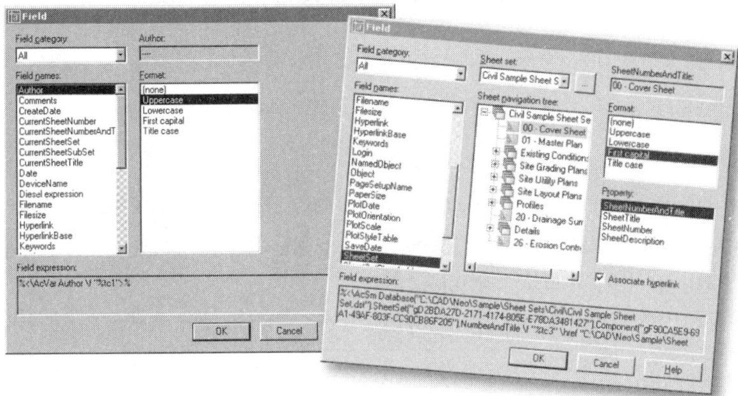

Select a field name and its options.

Click ***OK****. AutoCAD places the field as mtext:*

MTEXT Current text style: "Standard" Text height: 0.2000

Specify start point or [Height/Justify]: *(Pick a point, or enter an option.)*

If the field has no value, AutoCAD inserts dashes as placeholders.

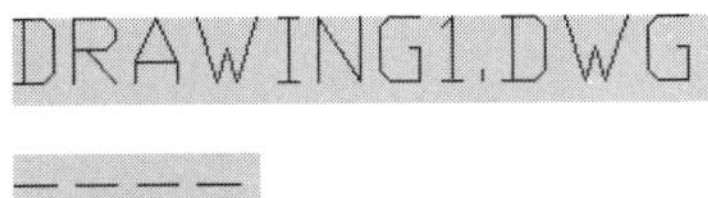

DIALOG BOX OPTIONS

Field category lists groups of fields: All, Date & Time, Document, Linked, Objects, Other, Plot, and Sheetset.

Field names lists that names of fields that can be placed in drawings.

Format lists optional formats available.

Field Expression illustrates the field code to be inserted in the drawing; cannot be edited directly.

Additional options appear, depending on the field name selected.

RELATED COMMANDS

UpdateField forces selected fields to update their values.

Find finds field text in drawings.

MText places text in drawings.

AttDef defines attribute text.

RELATED SYSTEM VARIABLES

FieldDisplay toggles the gray background to fields (default = 1, on).

FieldEval determines when fields are updated (default = 31, all on):

FieldEval	Meaning
0	Fields are not updated (static).
1	Fields are updated when drawing is opened.
2	Fields are updated when drawing is saved.
4	Field are updated when plotted.
8	Fields are updated with the **eTransmit** command.
16	Fields are updated when drawing is regenerated.

TIPS

- To edit fields, follow these steps:
  1. Double-click a field.
  2. In the mtext editor, right-click the field text.
  3. From the shortcut menu, select **Edit Field.**
- Follow the above steps to convert fields to mtext, but select **Convert Field to Text**.
- The gray background helps identify field text in drawings.
- AutoCAD supports the following field names:

Field Category	Field Names	Comment
Date & Time	Create Date	*Date drawing created.*
	Date	*Current date and time.*
	Plot Date	
	Save Date	*Date last saved.*
Document	Author	
	Comments	
	Filename	
	Filesize	
	HyperlinkBase	
	Keywords	
	Subject	
	Title	
Linked	Hyperlink	
Objects	NamedObject	*Blocks, layers, and so on.*
	Object	*Object selected from drawing.*
Other	Diesel Expression	
	System Variable	
Plot	DeviceName	
	Login	
	PageSetup	
	PaperSize	
	PlotDate	
	PlotOrientation	
	PlotScale	
	PlotStyleTable	
Sheetset	CurrentSheetNumber	
	CurrentSheetNumberandTitle	
	CurrentSheetSet	
	CurrentSheetSubset	
	CurrentSheetTitle	
	SheetSet	
	SheetSetPlaceholder	
	SheetView	

FileOpen

Rel.12 Opens drawing files without dialog boxes (*undocumented command*).

Command	Alias	Ctrl+	F-key	Alt+	Menu Bar	Tablet
fileopen	...	...	...	...	...	...

Command: fileopen
Enter name of drawing to open <filename.dwg>: *(Enter a file name.)*

COMMAND LINE OPTIONS

Enter name of drawing specifies the name of the *.dwg* file to open.

RELATED COMMANDS

Close closes the current drawing.

Open opens multiple drawing files.

RELATED SYSTEM VARIABLES

DbMod detects whether the drawing was changed since being opened.

SDI allows only one drawing to be open in AutoCAD at a time.

TIPS

- Use this command in menu and toolbar macros to open a drawing file when you don't want to display a dialog box.
- This command can only be used in SDI (single drawing interface) mode; if you use this command when two or more drawings are open, AutoCAD complains, "The SDI variable cannot be reset unless there is only one drawing open. Cannot run FILEOPEN if SDI mode cannot be established."
- Use **QSave** before using **FileOpen**; otherwise, **FileOpen** displays the following dialog box:

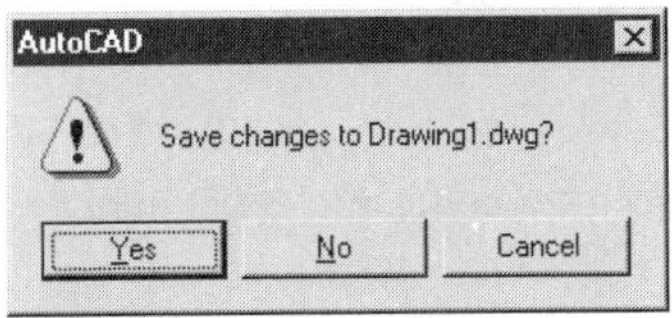

When you click **Yes**, AutoCAD closes the current drawing, and displays the **Select Drawing File** dialog box.

Removed Command

The **Files** command was removed from Release 14. In its place, use Windows Explorer.

'Fill

V. 1.4 Toggles whether hatches and wide objects — traces, multilines, solids, and polylines — are displayed and plotted with fills or as outlines.

Command	Alias	Ctrl+	F-key	Alt+	Menu Bar	Tablet
fill	...	...	...	...	...	...

Command: fill
Enter mode [ON/OFF] <ON>: *(Type* **ON** *or* **OFF***.)*

Fill on (left) and off (right) with a wide polyline, donut, and 2D solid.

COMMAND LINE OPTIONS

ON turns on fill after the next regeneration.

OFF turns off fill after the next regeneration.

RELATED SYSTEM VARIABLE

FillMode holds the current setting of fill status:

FillMode	Meaning
0	Fill mode is off.
1	Fill mode is on (default).

RELATED COMMAND

Regen changes the display to reflect the current fill or no-fill status.

TIPS

- The state of fill (or no fill) does not come into effect until the next regeneration:

  Command: regen
  Regenerating model.

- Traces, solids, and polylines are only filled in plan view, regardless of the setting of **Fill**.
- Since filled objects take longer to regenerate, redraw, and plot, consider leaving fill off during editing and plotting. During plotting, use a wide pen for filled areas.
- **Fill** affects objects derived from polylines, including donuts, polygons, rectangles, and ellipses, when created with **PEllipse** = 1.
- **Fill** does *not* affect TrueType fonts, which have their own system variable — **TextFill** — that toggles their fill-no fill status.
- **Fill** does not toggle in rendered mode.

Fillet

V. 1.4 Joins intersecting lines, polylines, arcs, circles, and 3D solids with a radius.

Command	Alias	Ctrl+	F-key	Alt+	Menu Bar	Tablet
fillet	**f**	...	...	**MF**	**Modify** ↳**Fillet**	**W19**

Command: fillet

Current settings: Mode = TRIM, Radius = 0.0

Select first object or [Polyline/Radius/Trim/mUltiple]: *(Select an object, or enter an option.)*

Select second object: *(Select another object.)*

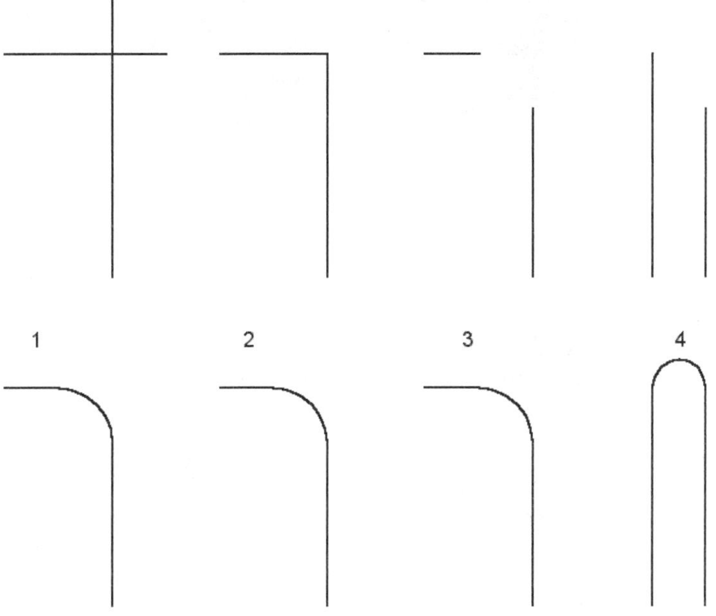

*Lines before (top) and after (bottom) applying the **Fillet** command:*
1. Crossing lines. 2. Touching lines. 3. Non-intersecting lines. 4. Parallel lines.

COMMAND LINE OPTIONS

Select first object selects the first object to be filleted.

Select second object selects the second object to be filleted.

mUltiple prompts you to select additional pairs of objects to fillet.

Polyline option

Select 2D polyline: *(Pick a polyline.)*

Select 2D polyline fillets all vertices of a 2D polyline; 3D polylines cannot be filleted.

Radius option

Specify fillet radius <0.0>: *(Enter a value.)*

Specify fillet radius specifies the filleting radius.

Trim option

Enter Trim mode option [Trim/No trim] <Trim>: *(Type* **T** *or* **N***.)*

Trim trims objects when filleted.

No trim does not trim objects.

Filleting 3D Solids

Select an edge or [Chain/Radius]: *(Pick an edge, or enter an option.)*

Select an edge or [Chain/Radius]: *(Press* **ENTER***.)*

n **edge(s) selected for fillet.**

Box before (left) and after (right) applying ***Fillet*** *and* ***Render*** *to a 3D solid.*

Edge option

Select an edge or [Chain/Radius]: *(Pick an edge.)*

Select edge selects a single edge.

Chain option

Select an edge or [Chain/Radius]: *(Type* **C***.)*

Select an edge chain or [Edge/Radius]: *(Pick an edge.)*

Select an edge chain selects all tangential edges.

Radius option

Select an edge or [Chain/Radius]: *(Type* **R***.)*

Enter fillet radius <1.0000>: *(Enter a value.)*

Enter fillet radius specifies the fillet radius.

RELATED COMMANDS

Chamfer bevels intersecting lines or polyline vertices.

SolidEdit edits 3D solid models.

RELATED SYSTEM VARIABLES

FilletRad specifies the current filleting radius.

TrimMode toggles whether objects are trimmed.

TIPS

- Pick the end of the object you want filleted; the other end will remain untouched.
- The lines, arcs, or circles need not touch.
- As a faster substitute for the **Extend** and **Trim** commands, use the **Fillet** command with the radius of zero.
- If the lines to be filleted are on two different layers, the fillet is drawn on the current layer.

- The fillet radius must be smaller than the length of the lines. For example, if the lines to be filleted are 1.0m long, the fillet radius must be less than 1.0m.
- Use the **Close** option of the **PLine** command to ensure a polyline is filleted at all vertices.

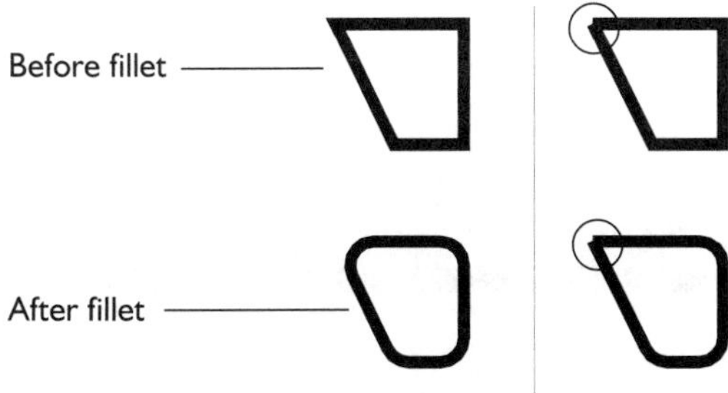

*Polyline closed with **Close** option (left) and without (right).*

- Filleting polyline segments from different polylines joins them into a single polyline.
- Filleting a line or an arc with a polyline joins it to the polyline.
- Filleting a pair of circles does not trim them.
- As of AutoCAD Release 13, the **Fillet** command fillets a pair of parallel lines; the radius of the fillet is automatically determined as half the distance between the lines.
- The **mUltiple** option was added to AutoCAD 2004.

'Filter

Rel.12 Creates filter lists that are applied to create selection sets.

Command	Alias	Ctrl+	F-key	Alt+	Menu Bar	Tablet
filter	fi	...	...	...	...	...

Command: filter

Displays dialog box:

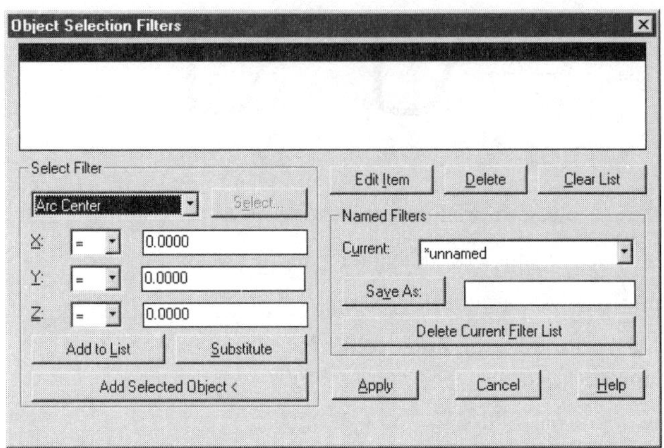

After you click ***Apply****, AutoCAD continues in the command area:*

Applying filter to selection.

Select objects: *(Select one or more objects.)*

Select objects: *(Press* **ENTER** *to end object selection.)*

Exiting filtered selection.

AutoCAD highlights filtered objects with grips.

DIALOG BOX OPTIONS

Select Filter options

Select displays all items of the specified type in the drawing.

X, Y, Z specifies the object's coordinates.

Add to List adds the current select-filter option to the filter list.

Substitute replaces a highlighted filter with the selected filter.

Add Selected Object selects the object to be added from the drawing.

Edit Item edits the highlighted filter item.

Delete deletes the highlighted filter item.

Clear list clears the entire filter list.

Named Filter options

Current selects the named filter from the list.

Save As saves the filter list with a name and the *.nfl* extension.

Delete Current Filter List deletes the named filter.

Apply closes the dialog box, and applies the filter operation.

COMMAND LINE OPTION

Select options selects the objects to be filtered; use the **All** option to select all non-frozen objects in the drawing.

RELATED COMMANDS

Any AutoCAD command with a 'Select objects' prompt.

QSelect creates a selection set quickly, via a dialog box.

Select creates a selection set via the command line.

RELATED FILE

***.nfl** is a named filter list.

TIPS

- The selection set created by **Filter** is accessed via the **P** (previous) selection option.
- Alternatively, **'Filter** is used transparently at the 'Select objects:' prompt.
- **Filter** cannot find objects when the color and linetype are set to BYLAYER.
- **Filter** uses the following grouping operators:

  **Begin OR _with_ **End OR
  **Begin AND _with_ **End AND
  **Begin XOR _with_ **End XOR
  **Begin NOT _with_ **End NOT

- **Filter** uses the following relational operators:

Operator	Meaning
<	Less than.
<=	Less than or equal to.
=	Equal to.
!=	Not equal to.
>	Greater than.
>=	Greater than or equal to.
*	All values.

- Save selection sets by name to an *.nfl* (short for *named filter*) file on disk for use in other drawings or editing sessions.

Filtering Selection Sets

In this tutorial, you erase all construction lines (xlines) from a drawing:

Step 1

Start the **Erase** command, and then invoke the **Filter** command transparently:

Command: erase
Select objects: 'filter

AutoCAD displays the Object Selection Filters dialog box:

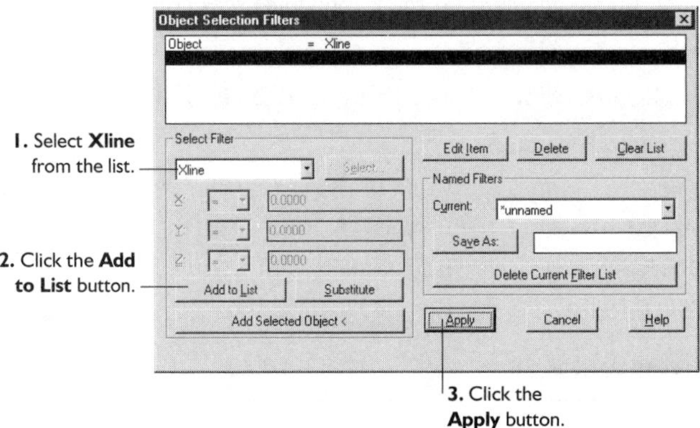

Step 2

In the **Select Filter** section:

Select **Xline** from the drop list.

Click the **Add to List** button.

Click **Apply**. Notice that the **Filter** command continues by displaying prompts on the command line, 'Applying filter to selection.'

To specify that **Filter** should search the entire drawing, enter the **All** option:

Select objects: all
n **found** *n* **were filtered out.**

Step 3

Exit the **Filter** command by pressing ENTER:

Select objects: *(Press* ENTER *to end object selection.)*
Exiting filtered selection. <Selection set: 7>
n **found**

AutoCAD resumes the **Erase** command. Press ENTER to end it:

Select objects: *(Press* ENTER *to the end the command.)*

AutoCAD uses the selection set created by the **Filter** command to erase the xlines.

Find

2000 Finds, and optionally replaces, text in drawings.

Command	Alias	Ctrl+	F-key	Alt+	Menu Bar	Tablet
find	...	...	...	**EF**	**Edit** ↳**Find**	**X10**

Command: find

Displays dialog box:

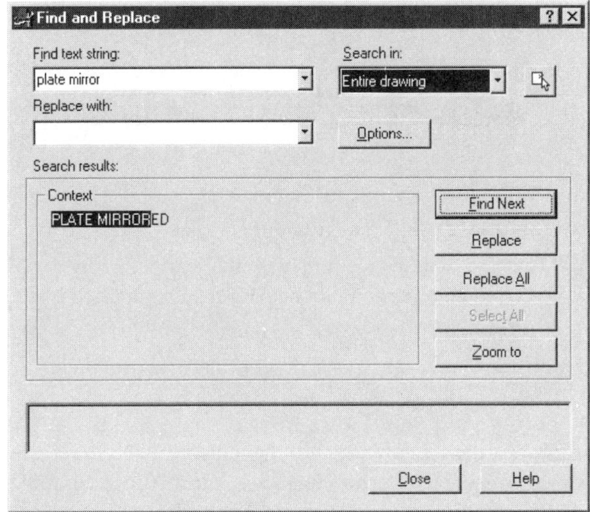

DIALOG BOX OPTIONS

Find text string specifies the text to find; enter text or click the down arrow to select one of the most recent lines of text searched for.

Replace with specifies the text to be replaced (only if replacing found text); enter text or click the down arrow to select one of the most recent lines of text searched for.

Search in:

- **Current Selection** searches the current selection set for text; click the **Select objects** button to create a selection set, if necessary.
- **Entire Drawing** searches the entire drawing.

Select objects allows you to select objects in the drawing; press ENTER to return to dialog box.

Options displays the Find and Replace Options dialog box.

Search Results options

Context displays the found text in its context.

Find Next finds the text entered in the **Find text string** field.

Replace replaces a single found text with the text entered in the **Replace with field**.

Replace All replaces all instances of the found text.

Select All selects all objects containing the text entered in the **Find text string** field; AutoCAD displays the message "AutoCAD found and selected *n* objects that contain "...".”

Zoom to zooms in to the area of the current drawing containing the found text.

Find and Replace Options dialog box

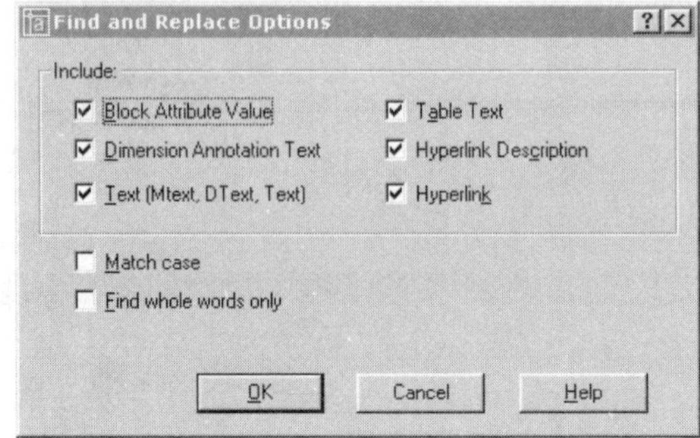

Include options

☑ **Block Attribute Value** finds text in attributes.

☑ **Dimension Annotation Text** finds text in dimensions.

☑ **Text (Mtext, DText, Text)** finds text in paragraph text placed by the **MText** command, single-line text placed by the **Text** command, and field text placed by the **Field** command.

☑ **Table Text** finds text in tables *(new to AutoCAD 2005)*.

☑ **Hyperlink Description** finds text in the description of a hyperlink.

☑ **Hyperlink** finds text in a hyperlink.

☐ **Match case** finds text that exactly matches the uppercase and lowercase pattern; for example, when searching for "Quick Reference," AutoCAD would find "Quick Reference" but not "quick reference."

☐ **Find whole words only** finds text that exactly matches whole words; for example, when searching for "Quick Reference," AutoCAD would find "Quick Reference" but not "Quickly Reference."

RELATED COMMANDS

AttDef creates attribute text.

DdEdit edits text.

Dim*xxx* creates dimension text.

Field places updatable text.

Hyperlink creates hyperlinks and hyperlink descriptions.

MText creates paragraph text.

Properties edits selected text.

QSelect finds text objects.

Text creates single-line text.

TIPS

- To find database links, use the **dbConnect** command.
- The **QSelect** command places text in a selection set.

Fog

Rel.14 Creates fog-like effects in renderings to add visual distance cues that suggest the apparent distance of objects from cameras.

Command	Alias	Ctrl+	F-key	Alt+	Menu Bar	Tablet
fog	...	...	...	**VEF**	**View** ↳**Render** ↳**Fog**	**P2**

Command: fog

Displays dialog box:

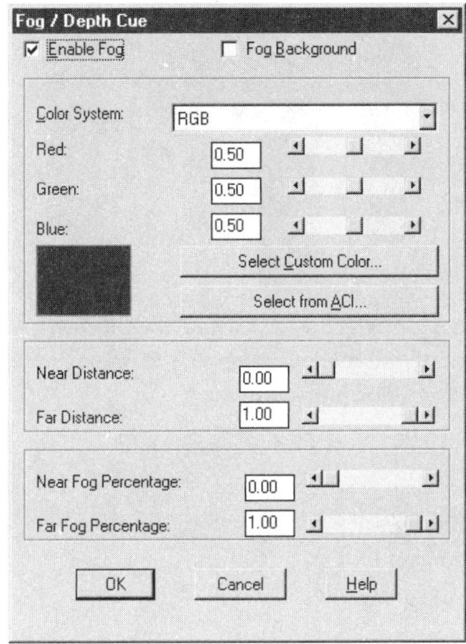

DIALOG BOX OPTIONS

☑ **Enable Fog** turns fog effect on; allows turning fog off without affecting parameters.

☑ **Fog Background** applies fog effect to background; see the **Background** command.

Color System selects the color of the fog by either RGB (red, green blue) or HLS (hue, lightness, saturation) methods.

Near Distance defines where the fog effect begins; value is the percentage distance between camera and the back clipping plane.

Far Distance defines where the fog effect ends.

Near Fog specifies the percentage of fog effect at the near distance; ranges between 0 and 100%.

Far Fog specifies the percentage of fog effect at the far distance; ranges between 0 and 100%.

RELATED COMMAND

Render creates renderings with optional fog effect.

TIPS

- Apply the **Fog** command, and then use the **Render** command to see the effect.
- The fog can be any color:

Color	RGB	HLS	Effect
White	1,1,1	0,1,0	Fog.
Black	0,0,0	0,0,0	Distance.
Green	0,1,0	0.33,0.5,0	Alien mist.

- The effect of using White as the fog color:

- It can be tricky getting the fog effect to work. Use the **3dOrbit** command to set the back clipping plane at the back of the model, or where you want the fog to have its full effect. Then use **Fog** to set up the following parameters:

  Near distance: 0.70

  Far distance: 1.00

  Near fog percentage: 0.00

  Far fog percentage: 1.00

Removed Command

GifIn was removed from Release 14. In its place, use **Image**.

'GotoUrl

2000 Goes to hyperlinks contained by objects.

Command	Alias	Ctrl+	F-key	Alt+	Menu Bar	Tablet
gotourl	...	...	...	...	...	...

Command: gotourl
Select objects: *(Select one or more objects.)*
Select objects: *(Press* ENTER *to end object selection.)*
browser Enter Web location (URL) <http://www.autodesk.com>: http://www.upfrontezine.com
AutoCAD launches your computer's default Web browser, and attempts to access the URL.

COMMAND LINE OPTION

Select objects selects one or more objects containing a hyperlink.

RELATED COMMANDS.

Browser launches the Web browser.

Hyperlink attaches, edits, and removes hyperlinks from objects.

TIPS

- This command is meant for use by macros and menus.
- AutoCAD uses this command for the shortcut menu's **Hyperlink | Open** option.

'GraphScr

V. 2.1 Puts the graphics window in front of the text window.

Command	Alias	Ctrl+	F-key	Alt+	Menu Bar	Tablet
graphscr	...	...	F2	...	...	...

Command: graphscr

AutoCAD displays the drawing window:

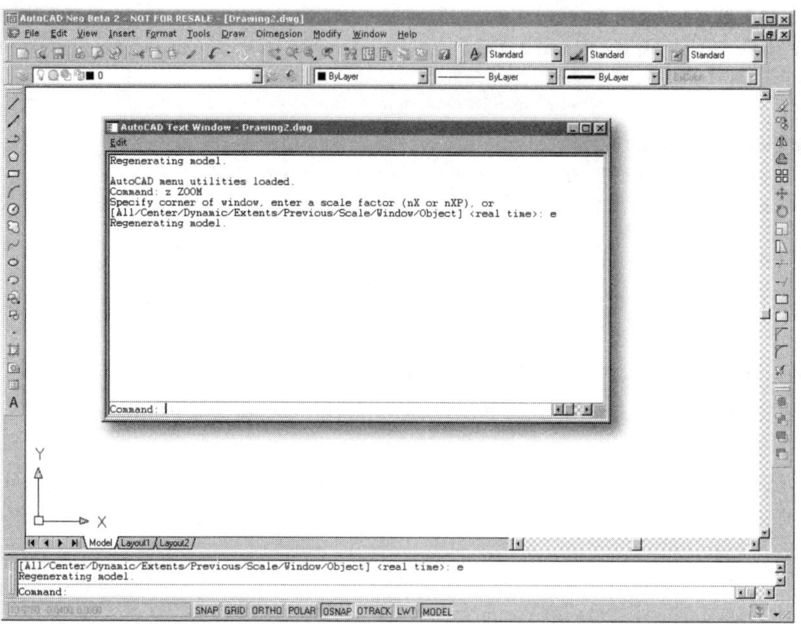

Text window (on top) and graphics windows (underneath).

COMMAND LINE OPTIONS

None.

RELATED COMMANDS

CopyHist copies text from the Text window to the Clipboard.

TextScr switches from the graphics window to the Text window.

RELATED SYSTEM VARIABLE

ScreenMode indicates whether the current screen is in text or graphics mode:

ScreenMode	Meaning
0	Text window.
1	Graphics window (default).
2	Dual screen displaying both text and graphics.

TIP

- The Text window appears frozen when a dialog box is active. Click the dialog box's **OK** or **Cancel** buttons to regain access to the Text window.

'Grid

<u>**V. 1.0**</u> Displays a grid of reference dots within the current drawing limits.

Command	Alias	Ctrl+	F-key	Alt+	Status Bar	Tablet
grid	...	G	F7	...	GRID	...

Command: grid

Specify grid spacing(X) or [ON/OFF/Snap/Aspect] <0.5000>: *(Enter a value, or enter an option.)*

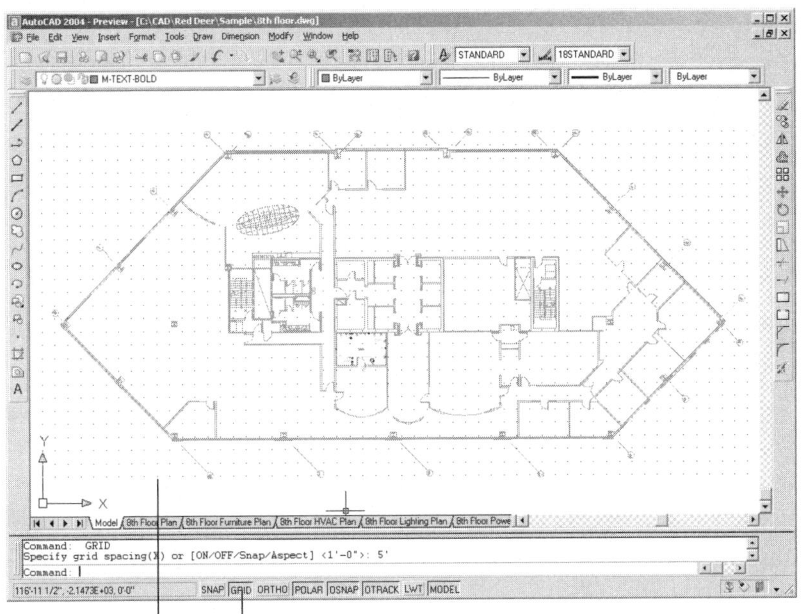

Grid dots. Click ***GRID*** *to toggle the grid display; right-click for menu.*

COMMAND LINE OPTIONS

Specify grid spacing(X) sets the x and y direction spacing; an *X* following the value sets the grid spacing to a multiple of the current snap setting.

ON turns on grid markings.

OFF turns off grid markings.

Snap makes the grid spacing the same as the snap spacing.

Aspect options

Specify the horizontal spacing(X) <0.0000>: *(Enter a value.)*

Specify the vertical spacing(Y) <0.0000>: *(Enter a value.)*

Horizontal sets the spacing in the x-direction.

Vertical sets the spacing in the y-direction.

RELATED COMMANDS

Options sets the grid via a dialog box.

Limits sets the limits of the grid in WCS.

Snap sets the snap spacing.

RELATED SYSTEM VARIABLES

GridMode is the current grid visibility:

GridMode	Meaning
0	Grid is off (default).
1	Grid is on.

GridUnit specifies the current grid x,y spacing.

LimMin specifies the x,y coordinates of the lower-left corner of the grid display.

LimMax specifies the x,y coordinates of the upper-right corner of the grid display.

SnapStyl displays a normal or isometric grid:

SnapStyl	Meaning
0	Normal (default).
1	Isometric grid.

TIPS

- The grid is most useful when set to the snap spacing, or to a multiple of the snap spacing.
- When the grid spacing is set to 0, it matches the snap spacing.
- You can set a different grid spacing in each viewport, and a different grid spacing in the x and y directions.
- Rotate the grid with the **Snap** command's **Rotate** option.
- The **Snap** command's **Isometric** option displays an isometric grid.
- If a very dense grid spacing is selected, the grid will take a long time to display; press ESC to cancel the display.
- AutoCAD will not display a grid that is too dense, and returns the message, "Grid too dense to display."
- Grid markings are not plotted; to create a plotted grid, use the **Array** command to place an array of points.

Group

Rel.13 Creates named selection sets of objects.

Commands	Aliases	Ctrl+	F-key	Alt+	Menu Bar	Tablet
group	g	Shift+A	...	...	...	X8
-group	-g					

Command: group

Displays dialog box:

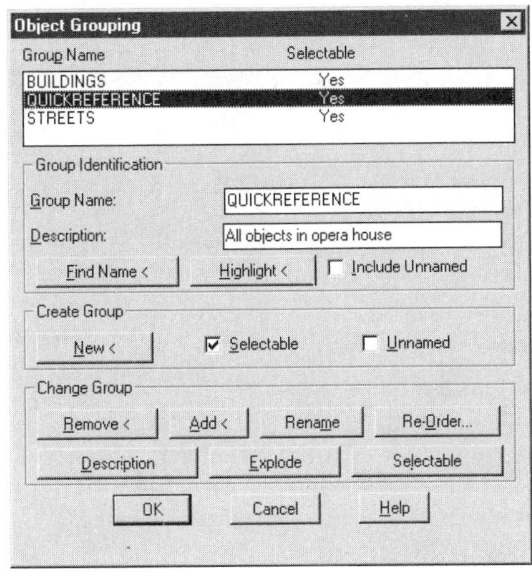

DIALOG BOX OPTIONS

Group Name lists the names of groups in the drawing.

Group Identification options

Group Name displays the name of the current group.

Description describes the group; may be up to 64 characters long.

Find Name lists the name(s) of group(s) that a selected object belongs to.

Highlight highlights the objects included in the current group.

☐ **Include Unnamed** lists unnamed groups in the dialog box.

Create Group options

New selects objects for the new group.

☑ **Selectable** toggles selectability: picking one object picks the entire group.

☐ **Unnamed** creates an unnamed group; AutoCAD gives the name ***A***n*, where *n* is a number that increases with each group.

Change Group options

Remove removes objects from the current group.

Add adds objects to the current group.

Rename renames the group.

Re-order changes the order of objects in the group; displays Order Group dialog box.

Description changes the description of the group.

Explode removes the group description; does not erase group members.

Selectable toggles selectability.

Order Group dialog box

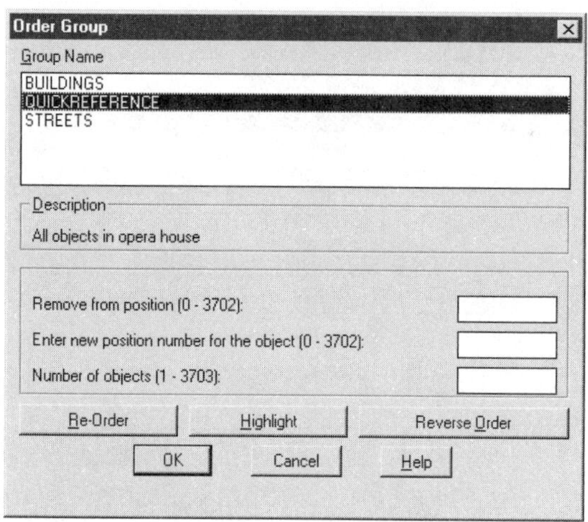

Group Name lists the names of groups in the current drawing.

Description describes the selected group.

Remove from position (0 - *n*) selects the object to move.

Replace at position (0 - *n*) moves the group name to a new position.

Number of objects (1 - *n*) lists the number of objects to reorder.

Re-Order applies the order changes.

Highlight highlights the objects in the current group.

Reverse Order reverses the order of the groups.

-GROUP Command

Command: -group

Enter a group option

[?/Order/Add/Remove/Explode/REName/Selectable/Create] <Create>:

(Enter an option.)

COMMAND LINE OPTIONS

? lists the names and descriptions of currently-defined groups.

Order changes the order of objects within the group.

Add adds objects to the group.

Remove removes objects from the group.

Explode removes the group definition from the drawing.

REName renames the group.

Selectable toggles whether the group is selectable.

Create creates a newly-named group from the objects selected.

RELATED COMMANDS

Block creates named symbols from a group of objects.

Select creates selection sets.

RELATED SYSTEM VARIABLE

PickStyle toggles whether groups are selected by the usual selection process:

PickStyle	Meaning
0	Groups and associative hatches are not selected.
1	Groups are included in selection sets (default).
2	Associate hatches are included in selection sets.
3	Both are selected.

TIPS

- You can toggle groups on and off with the CTRL+SHIFT+A shortcut keystroke.
- Consider a group as a named selection set; unlike a regular selection set, a group is not "lost" when the next group is created.
- Group descriptions can be up to 64 characters long.
- Anonymous groups are unnamed; AutoCAD refers to them as ***A***n.

Hatch

V. 1.4 Draws non-associative crosshatch patterns within closed boundaries.

Command	Alias	Ctrl+	F-key	Alt+	Menu Bar	Tablet
hatch	-h	...	...	...	...	...

Command: hatch

Enter a pattern name [?/Solid/User defined] <ANSI31>: *(Enter a pattern name, or enter an option.)*

Specify a scale for the pattern <1.0000>: *(Enter a scale factor.)*

Specify an angle for the pattern <0>: *(Enter a rotation angle.)*

Select objects: *(Press* ENTER*, or select one or more objects.)*

Select objects: *(Press* ENTER *to end object selection.)*

COMMAND LINE OPTIONS

Enter a pattern name specifies the valid name of a hatch pattern; include one of the following optional, undocumented style parameters, such as:

Enter a pattern name [?/Solid/User defined]: ansi31,o

Style	Meaning
N	Hatches alternate boundaries (Normal).
O	Hatches only outermost boundary (Outermost).
I	Hatches everything within outermost boundary (Ignore).

Specify a scale specifies the hatch pattern scale.

Specify an angle specifies the hatch pattern angle.

Select objects selects the objects that make up the hatch pattern boundary.

Pattern Name options

? lists the hatch pattern names: 'Enter pattern(s) to list <*>:'.

Solid specifies a solid fill drawn with the current color.

User Defined options

Specify angle for crosshatch lines <0>: *(Enter an angle.)*

Specify spacing between the lines <1.0000>: *(Enter a spacing distance.)*

Double hatch area? [Yes/No] <N>: *(Type* **Y** *or* **N***.)*

Select objects to define hatch boundary or <direct hatch>,

Select objects: *(Select one or more objects.)*

Select objects: *(Press* ENTER *to end object selection.)*

Specify angle for crosshatch lines specifies the hatching angle.

Specify spacing between the lines specifies the distance between lines.

Double hatch area? determines whether a second set of hatch lines is drawn at 90 degrees to the first.

Select objects to define hatch boundary applies the hatch within objects.

Direct hatch specifies points that bound the hatch area.

Direct Hatch options

Press ENTER *at the first 'Select objects' prompt.*

Retain polyline boundary? [Yes/No] <N>: *(Enter* **Y** *or* **N***.)*

Specify start point: *(Pick a point.)*
Specify next point or [Arc/Close/Length/Undo]: *(Pick a point, or enter option.)*
Specify next point or [Arc/Close/Length/Undo]: *(Pick a point, or enter option.)*
Specify next point or [Arc/Close/Length/Undo]: *(Enter* **C** *to close.)*
Specify start point for new boundary or <apply hatch>: *(Press* **ENTER** *to apply.)*

Retain polyline boundary?

Yes leaves boundary in place after the hatch is complete.

No erases boundary after the hatch is complete.

Specify start point begins drawing the hatch boundary.

Specify next point draws straight lines.

Arc draws arc hatch boundaries; prompts 'Enter an arc boundary option [Angle/ CEnter/ CLose/Direction/Line/Radius/Second pt/Undo/Endpoint of arc] <Endpoint>"; see the **Arc** command.

Close closes the hatch boundary.

Length continues the boundary by a specified distance; prompts 'Specify length of line.'

Undo undoes the last-drawn segment.

RELATED COMMANDS

BHatch automatically places associative hatching.

Boundary automatically creates polyline or region boundary.

Explode reduces hatch pattern to its constituent lines.

Snap changes the hatch pattern's origin.

RELATED SYSTEM VARIABLES

HpAng specifies the current hatch pattern angle.

HpDouble specifies doubled hatch pattern.

HpName specifies the current hatch pattern name.

HpScale specifies the current hatch pattern scale.

HpSpace specifies the current hatch pattern spacing.

SnapBase controls the origin of the hatch pattern.

SnapAng controls the angle of the hatch pattern.

RELATED FILES

acad.pat holds hatch definition file for ANSI patterns.

acadiso.pat holds hatch definition file for ISO patterns.

TIPS

- The **Hatch** command draws non-associative hatch patterns; the pattern remains in place when its boundary is edited.
- For complex hatch areas, you may find it easier to outline the area with a polyline — using object snap — or with the **Boundary** or **BHatch** commands.
- To create the hatch as lines, precede the pattern name with * (asterisk).
- This command fills areas with solid color, but does not apply gradient fills; use the **BHatch** or **-BHatch** commands instead.
- For a listing of hatch patterns included with AutoCAD, see the **BHatch** command.

HatchEdit

Rel.13 Edits associative hatch objects.

Commands	Alias	Ctrl+	F-key	Alt+	Menu Bar	Tablet
hatchedit	**he**	...	...	...	**Modify** **↳Hatch**	**Y16**
-hatchedit						

Command: hatchedit
Select associative hatch object: *(Select one hatch object.)*

Displays dialog box:

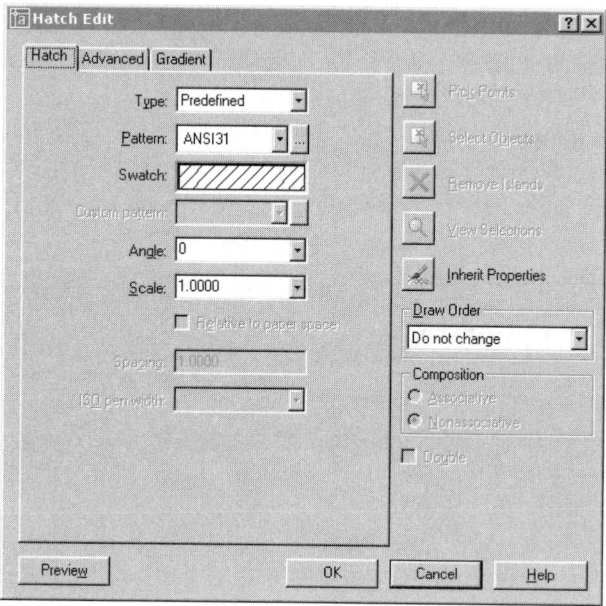

DIALOG BOX OPTIONS

Most options are grayed-out because they are unavailable.

Preview dismisses the dialog box temporarily, so that you can see the effect of the editing changes on the hatch pattern.

Inherit Properties allows you to pick another hatch pattern whose properties will apply to the selected hatch pattern.

Draw Order *(new to AutoCAD 2005):*

- **Do not assign** places hatch normally.
- **Send to back** places hatch behind all other overlapping objects in the drawings.
- **Bring to front** places hatch in front of all other overlapping objects.
- **Send behind boundary** places hatch behind its boundary.
- **Bring in front of boundary** places hatch in front of boundary.

Hatch tab

Pattern Type selects the source of the hatch pattern:

Source	Meaning
Predefined	*acad.pat* (default).
User-defined	Create hatch pattern on the fly.
Custom	Select hatch from another *.pat* file.

Pattern selects a pattern name from the drop list.

... selects a pattern from a tabbed dialog box.

Custom Pattern names the custom hatch pattern.

Scale specifies the hatch pattern scale.

Angle specifies the hatch pattern angle.

Spacing specifies the spacing between pattern lines.

Advanced tab

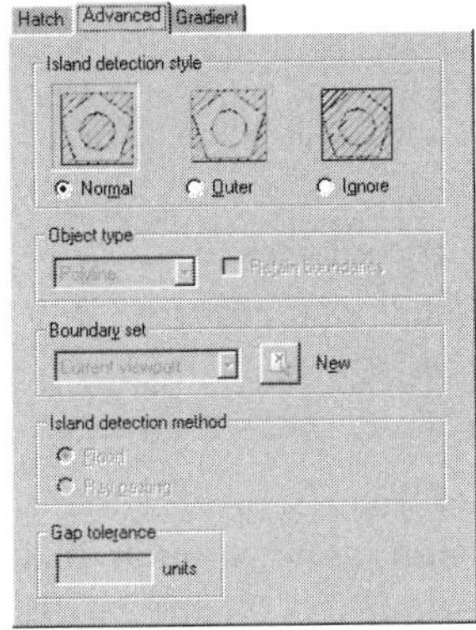

Island Detection Style options

⊙ **Normal** hatches the alternate boundaries (default).

○ **Outer** hatches only the outermost boundary.

○ **Ignore** hatches everything within the outermost boundary.

Gradient tab

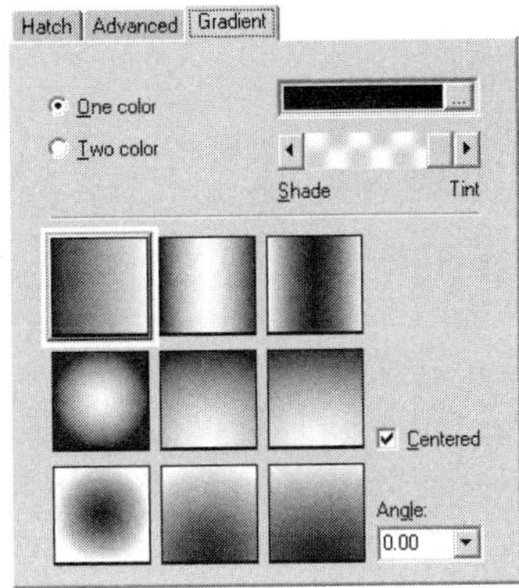

See ***BHatch*** *command for options.*

-HATCHEDIT Command

Command: -hatchedit

Select associative hatch object: *(Select one object.)*

Enter hatch option [Disassociate/Style/Properties/DRaw order] <Properties>: *(Enter an option.)*

COMMAND LINE OPTIONS

Select selects a single associative hatch pattern.

Disassociate removes associativity from hatch pattern.

Style options

Enter hatching style [Ignore/Outer/Normal] <Normal>: *(Enter an option.)*

Normal hatches alternate boundaries (default).

Outer hatches only outermost boundary.

Ignore hatches everything within outermost boundary.

Properties options

*See **Hatch** command.*

Pattern name specifies the name of a valid hatch pattern.

? lists the available hatch pattern names.

Solid replaces the hatch pattern with solid fill.

User defined creates a new on-the-fly hatch pattern.

Scale changes the hatch pattern scale.

Angle changes the hatch pattern angle.

Draw Order options *(new to AutoCAD 2005)*

Enter draw order [do Not assign/send to Back/bring to Front/send beHind boundary/bring in front of bounDary] <send beHind boundary>: *(Specify an option.)*

do Not assign places hatch normally.

send to Back places hatch behind all other overlapping objects in the drawings.

bring to Front places hatch in front of all other overlapping objects.

send beHind boundary places hatch behind its boundary.

bring in front of bounDary places hatch in front of boundary.

RELATED COMMANDS

BHatch applies associative hatch patterns.

Explode explodes a hatch pattern block into lines.

Properties changes properties of hatch patterns.

RELATED SYSTEM VARIABLES

DelObj toggles whether boundary is erased after hatch is placed.

GfAng specifies the angle of the gradient fill; ranges from 0 to 360 degrees.

GfClr1 specifies the first gradient fill color in RGB format, such as "RGB 000, 128, 255."

GfClr2 specifies the second gradient fill color in RGB format.

GfClrLum specifies the luminescence of a one-color gradient fill; ranges from 0.0 (black) to 1.0 (white).

GfClrState specifies whether the gradient fill is one-color or two-color.

GfName specifies the gradient fill pattern.

GfShift specifies whether the gradient fill is centered or is shifted to the upper-left.

HpAng specifies the current hatch pattern angle (default = 0).

HpBound specifies the hatch boundary object.

HpDouble specifies single or double hatching.

HpDrawOrder *(new to AutoCAD 2005)* controls the display order of the hatch pattern relative to other overlapping objects.

HpGalTol *(new to AutoCAD 2005)* reports the current gap tolerance.

HpName specifies the current hatch pattern name (up to 31 characters long).

HpScale specifies the current hatch pattern scale factor (default = 1).

HpSpace specifies the current hatch pattern spacing factor (default = 1).

PickStyle controls the selection of hatch patterns.

SnapBase specifies the starting coordinates of hatch pattern (default = 0,0).

TIPS

- **HatchEdit** works with associative and non-associative hatch objects.
- AutoCAD cannot change a non-associative hatch to associative, or vice versa; you can use the **Explode** command to reduce the hatch to lines.
- As an alternative to entering this command, double-click hatch patterns to display the **Hatch Edit** dialog box.
- To select a solid fill, click the outer edge of the hatch pattern, or use a crossing window selection on top of the solid fill.
- Even though the Hatch Edit dialog box looks identical to the BHatch dialog box, many options are not available in the Hatch Edit dialog box.

'Help

V. 1.0 Lists information for using AutoCAD's commands.

Command	Alias	Ctrl+	F-key	Alt+	Menu Bar	Tablet
help	**'?**	**...**	**F1**	**HH**	**Help** **↳Help**	**Y7**

Command: help

Displays window:

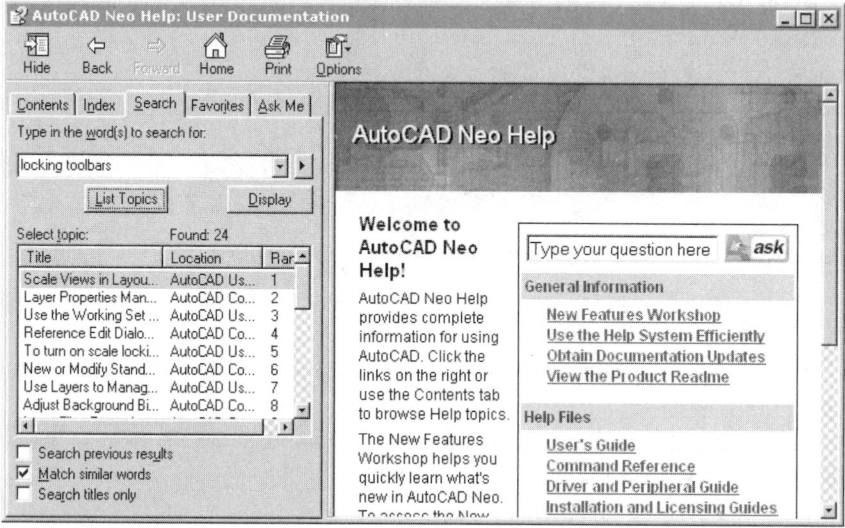

RELATED COMMANDS

All.

RELATED FILE

acad.chm is AutoCAD's primary help file.

TIP

- Because **'Help**, **'?**, and **F1** are transparent commands, you can use them during another command to get help with the command's options.

Hide

V. 2.1 Removes hidden lines from 3D drawings.

Command	Alias	Ctrl+	F-key	Alt+	Menu Bar	Tablet
hide	hi	...	...	VH	View ↳Hide	M2

Command: hide
Regenerating drawing.
Removing hidden lines: *nn*

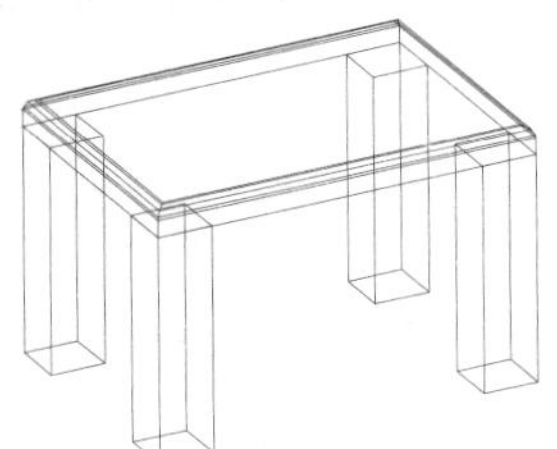

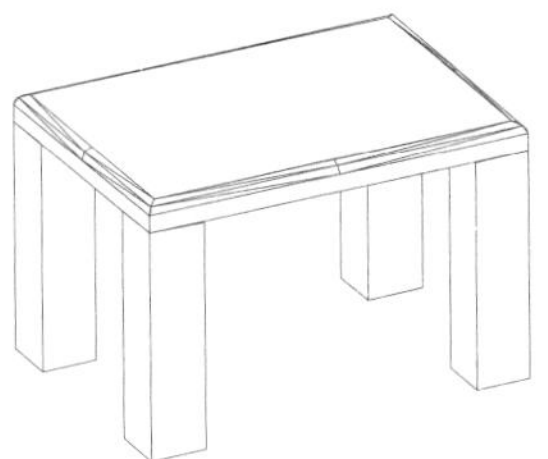

3D object without (left) and with (right) hidden lines removed.

COMMAND LINE OPTIONS

None.

RELATED COMMANDS

HLSettings specifies settings for lines displayed by the **Hide** command.

MView removes hidden lines during plots of paper space drawings.

Plot removes hidden lines during plots of 3D drawings.

Regen returns the view to wireframe.

Render performs realistic renderings of 3D models.

ShadeMode performs quick renderings and quick hides of 3D models.

VPoint selects the 3D viewpoint.

3dOrbit removes hidden lines of perspective in 3D views.

RELATED SYSTEM VARIABLES

HaloGap specifies the size of gap where a hidden object intersects with a visible object.

HidePrecision determines the calculated accuracy of hidden-line removal.

HideText determines whether text is involved in the hidden-line calculations.

ObscuredColor specifies the color of hidden lines.

ObscuredLType specifies the linetype of hidden lines.

IntersectionDisplay determines whether polylines are displayed at the intersection of faces.

TIPS

- This command considers the following objects as opaque: circles, 2D solids, traces, wide 2D polylines, 3D faces, polygon meshes, any objects with thickness, as well as regions and 3D solids.
- For a faster hide, use the **ShadeMode** command's **Hide** option.
- Use **MSlide** (or **SaveImg**) to save the hidden-line view as *.sld* (or *.tif*, *.tga*, or *.gif*) files. View the saved images with **VSlide** (or **Replay**).
- Freezing layers speeds up the hide process, because **Hide** ignores objects on frozen layers.
- **Hide** does consider the visibility of text and attributes, depending on the setting of the **HideText** system variable.
- The figure below shows (left to right) a 3D model, normal hidden line removal, and hidden lines turned on (see **HlSettings** command):

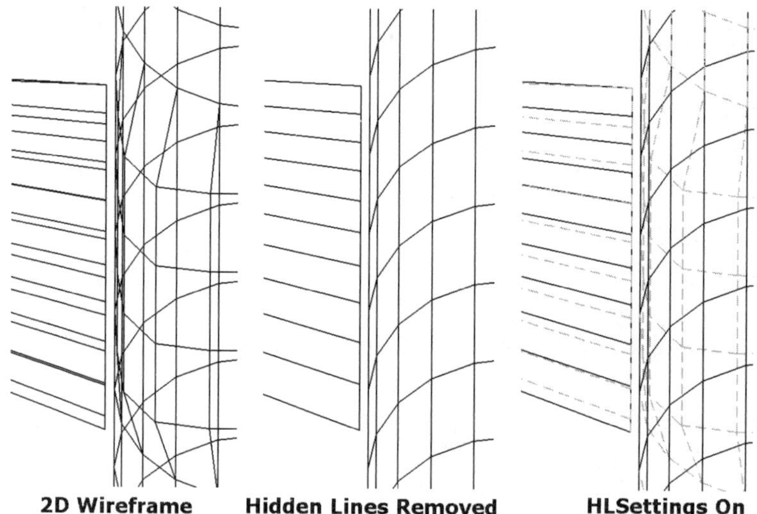

2D Wireframe **Hidden Lines Removed** **HLSettings On**

- To create a hidden-line view when plotting in paper space, select the **HidePlot** option of the **MView** command.
- Set **DispSilh** = 1 to show silhouette lines and hides tesselations; set to 0 to hide silhouette lines and show tesselations.
- **FacetRes** controls the smoothness of hidden, shaded, and rendered objects. A better value than the default of 0.5 is 2.

Removed Command

HpConfig was removed from AutoCAD 2000; it was replaced by the **Plot** command.

HLSettings

2004 Controls the display of hidden (obscured) lines.

Command	Alias	Ctrl+	F-key	Alt+	Menu Bar	Tablet
hlsettings	...	...	...	...	...	...

Command: hlsettings

Displays dialog box:

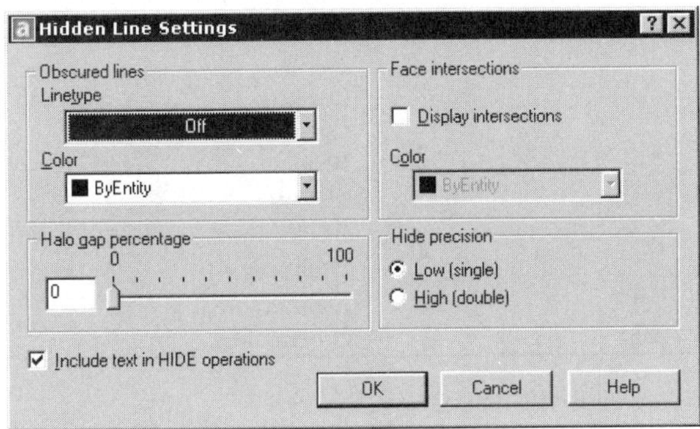

DIALOG BOX OPTIONS

Obscured Lines options

Linetype selects linetype by which to display hidden lines; select **Off** to display no hidden lines.

Color selects the color in which to display hidden lines.

Face Intersection options

Display intersections:

☑ Displays polylines at the intersection of faces.

☐ Does not display face intersections.

Color selects the color of the face intersections.

Additional options

Halo Gap Percentage specifies the distance to shorten a *haloed* line.

Hide Precision options:

⊙ **Low (single)** calculates the display of hidden lines with single-precision; uses less memory.

○ **High (double)** calculates the display of hidden lines with double-precision; more accurate.

Include text in HIDE operation:

☑ Text is hidden by objects, and hides other objects.

☐ Text is ignored: it is not hidden, and does not hide other objects.

RELATED COMMANDS

Hide performs hidden-line removal.

ShadeMode performs quick hides of 3D models.

RELATED SYSTEM VARIABLES

HaloGap specifies the size of gap where a hidden object intersects with a visible object.

HidePrecision determines the calculated accuracy of hidden-line removal.

HideText determines whether text is involved in the hidden-line calculations.

IntersectionColor specifies the color of the polylines displayed at the intersection of faces.

IntersectionDisplay determines whether polylines are displayed at the intersection of faces.

ObscuredColor specifies the color of hidden lines:

ObscuredColor	Meaning
0	Hidden lines are not displayed.
1	Hidden lines displayed in red.
2	Yellow.
3	Green.
4	Cyan.
5	Blue.
6	Magenta.
7	Black (white).
8	Dark gray.
9	Light gray.
10 - 255	Other colors.

ObscuredLtype specifies the linetype of hidden lines:

ObscuredLtype	Meaning
0	Hidden lines are not displayed.
1	Solid.
2	Dashed (traditional Hidden).
3	Dotted.
4	Short Dash.
5	Medium Dash.
6	Long Dash.
7	Double Short Dash.
8	Double Medium Dash.
9	Double Long Dash.
10	Medium Long Dash.
11	Sparse Dash.

TIPS

- AutoCAD sometimes calls hidden lines "obscured lines."
- A *haloed* line is a hidden line that is shortened at the point where it intersects another object. The value of the gap is the percentage of the current unit value; it is independent of zoom level.
- An *intersection polyline* is a polyline displayed at the intersection of 3D faces.
- This command affects the display of the **Hide** and **ShadEdge** commands; it does not work with the **3dOrbit** command's **Hide** option.
- To disable the display of hidden-lines, set the **Linetype** to off (ObscuredLtype = 0).

Hyperlink

2000 Attaches hyperlinks (URLs) to objects in drawings.

Commands	Alias	Ctrl+	F-key	Alt+	Menu Bar	Tablet
hyperlink	...	k	...	IH	**Insert ↳Hyperlink**	...
-hyperlink						

Command: hyperlink
Select objects: *(Select one or more objects.)*
Select objects: *(Press* ENTER *to end object selection.)*

Displays dialog box:

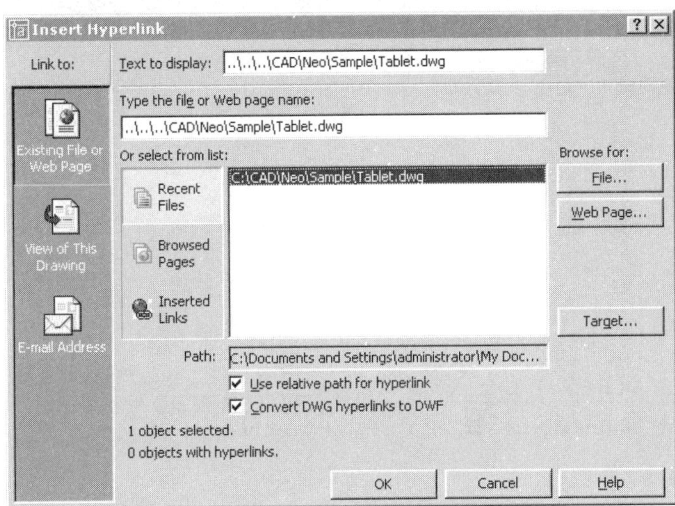

COMMAND LINE OPTION

Select object selects the objects to which to attach the hyperlink.

DIALOG BOX OPTIONS

Existing File or Web Page page

Text to display describes the hyperlink; the description is displayed by the tooltip; when blank, the tooltip displays the URL.

Type the file or Web page name specifies the hyperlink (URL) to associate with the selected objects; the hyperlink may be any file on your computer, on any computer you can access on your local network, or on the Internet.

Or select from list:

- **Recent files** lists drawings recently opened in AutoCAD.
- **Browsed pages** lists pages recently viewed with Web browsers, and other software.
- **Inserted links** lists URLs recently entered in Web browsers.

File opens the Browse the Internet - Select Hyperlink dialog box.

Web Page starts up a simple Web browser.

Target specifies a location in the file, such as a target in an HTML file, a named view in AutoCAD, or a page in a spreadsheet document. Displays Select Place in Document dialog box.

Path displays the full path and filename to the hyperlink; only the file name appears when **Use relative path for hyperlink** is checked.

☑ **Use relative path for hyperlink** toggles use of the path for relative hyperlinks in the drawing; when "" (null), the drawing paths stored in **AcadPrefix** are used.

☑ **Convert DWG hyperlinks to DWF** preserves hyperlinks when drawings are exported in DWF format (*new to AutoCAD 2005*).

Remove link removes the hyperlink from the object; this button appears only if you select an object that already has a hyperlink.

View of This Drawing page

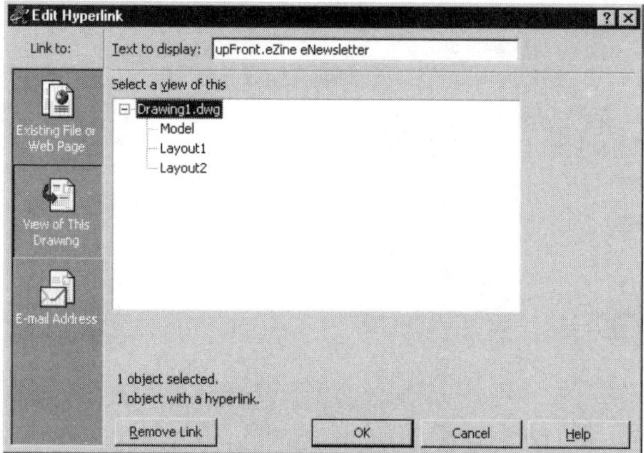

Select a view of this selects a layout, named view, or named plot.

Email Address page

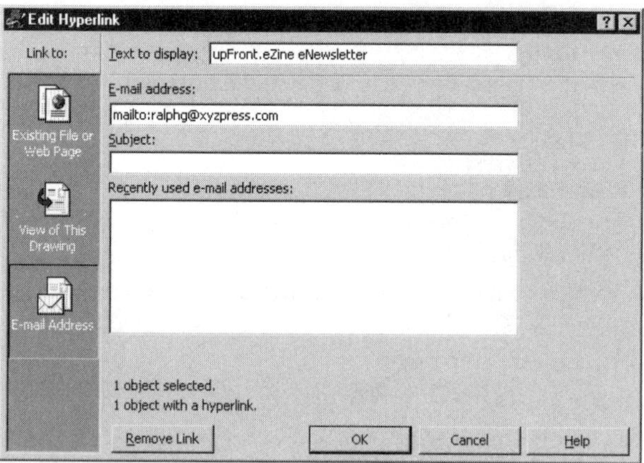

Email address specifies the email address; the *mailto:* prefix is added automatically.

Subject specifies the text that will be added to the Subject line.

Recently used e-mail addresses lists the email addresses recently entered.

Select Place in Document dialog box

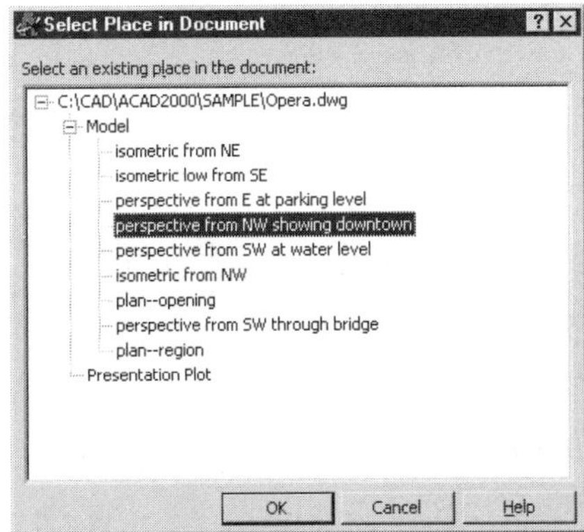

Select an existing place in the document selects a named location.

-HYPERLINK Command

Command: -hyperlink
Enter an option [Remove/Insert] <Insert>: *(Enter an option.)*
Enter hyperlink insert option [Area/Object] <Object>: *(Enter an option.)*
Select objects: *(Select one or more objects.)*
Select objects: *(Press* **ENTER** *to end object selection.)*
Enter hyperlink <current drawing>: *(Enter hyperlink address.)*
Enter named location <none>: *(Optional: Enter a bookmark.)*
Enter description <none>: *(Optional: Enter a description.)*

COMMAND LINE OPTIONS

Remove removes a hyperlink from selected objects or areas.

Insert adds a hyperlink to selected objects or areas.

Select objects selects the object to which the hyperlink will be added.

Enter hyperlink specifies the filename or hyperlink address.

Enter named location (*optional*) specifies a location within the file or hyperlink.

Enter description (*optional*) specifies a description of the hyperlink.

RELATED COMMANDS

HyperlinkOptions toggles the display of the hyperlink cursor, shortcut menu, and tooltip.

HyperlinkOpen opens hyperlinks (URL) via the command line.

HyperlinkBack returns to the previous URL.

HyperlinkFwd moves forward to the next URL; works only when the HyperlinkBack command was used, otherwise AutoCAD complains, "** No hyperlink to navigate to **".

HyperlinkStop stops the display of the current hyperlink.

GoToUrl displays specific Web pages.

PasteAsHyperlink pastes hyperlinks to selected objects.

RELATED SYSTEM VARIABLE

HyperlinkBase specifies the path for relative hyperlinks in the drawing; when "" (null), the drawing paths stored in **AcadPrefix** are used.

TIPS

- If the drawing has never been saved, AutoCAD is unable to determine the default *relative folder*. For this reason, the **Hyperlink** command prompts you to save the drawing.
- By using hyperlinks, you can create a *project document* consisting of drawings, contacts (word processing documents), project timelines, cost estimates (spreadsheets pages), and architectural renderings. To do so, create a "title page" of an AutoCAD drawing with hyperlinks to the other documents.
- To edit a hyperlink with **Hyperlink**, select object(s), make editing changes, and click **OK**.
- To remove a hyperlink with **Hyperlink**, select the object, and then click **Remove Link**.
- An object can have just one hyperlink attached to it; more than one object, however, can share the same hyperlink.
- The alternate term for "hyperlink" is *URL*, which is short for "uniform resource locator," the universal file naming system used by the Internet.

'HyperlinkOpen/Back/Fwd/Stop

2000 Controls the display of hyperlinked pages (*undocumented commands*).

Commands	Alias	Ctrl+	F-key	Alt+	Menu Bar	Tablet
hyperlinkopen	...	...	...	...	...	...
hyperlinkback						
hyperlinkfwd						
hyperlinkstop						

Command: hyperlinkopen
Enter hyperlink <current drawing>: *(Enter a hyperlink option.)*
Enter named location <none>: *(Optional: Enter a bookmark.)*

Displays the specified Web page or file, if possible.

Command: hyperlinkback

Returns to the previous hyperlinked page.

Command: hyperlinkstop

Stops displaying the Web page.

Command: hyperlinkfwd

*Hyperlinks to the next page; can be used only after the **HyperLinkBack** command.*

COMMAND LINE OPTIONS

Enter hyperlink enters a URL (uniform resource locator) or a filename.

Enter named location enters a named view or other valid target.

RELATED COMMANDS

Hyperlink attaches a hyperlink to objects.

HyperlinkOptions specifies the options for hyperlinks.

GoToUrl displays a specific Web page.

RELATED TOOLBAR ICONS

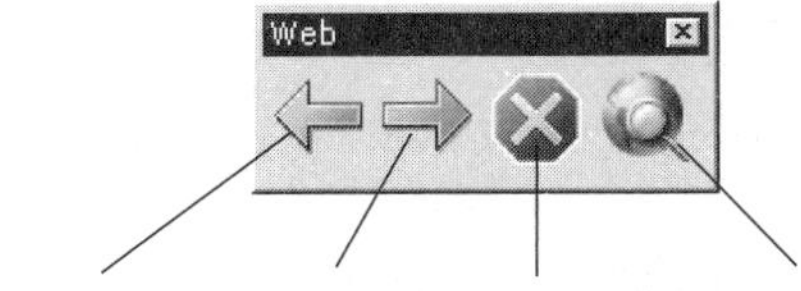

Go Back *Go Forward* *Stop Navigation* *Browse the Web*

Go Back goes to the previous hyperlink; executes the **HyperlinkBack** command.

Go Forward goes to the next hyperlink; executes the **HyperlinkFwd** command.

Stop Navigation stops loading the current hyperlink file; executes the **HyperlinkStop** command.

Browse the Web displays the Web browser; executes the **Browser** command.

'HyperlinkOptions

2000 Toggles the display of the hyperlink cursor, shortcut menu, and tooltip (*the longest command name in AutoCAD*).

Command	Alias	Ctrl+	F-key	Alt+	Menu Bar	Tablet
hyperlinkoptions	...	...	...	...	...	

Command: hyperlinkoptions
Display hyperlink cursor tooltop and shortcut menu? [Yes/No] <Yes>: *(Type* **Y** *or* **N***.)*

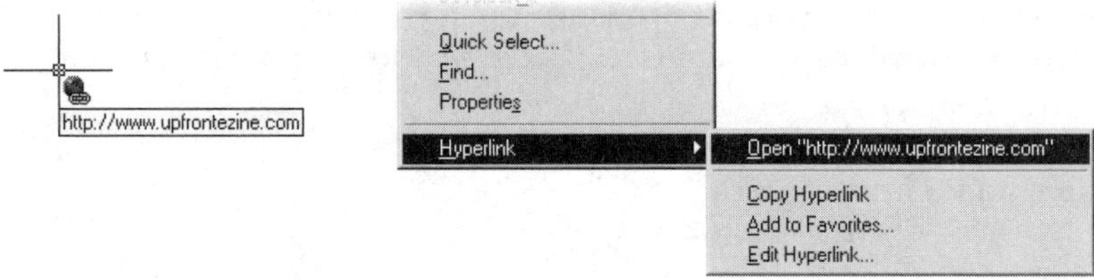

Hyperlink cursor and tooltip (left); hyperlink shortcut menu (right).

COMMAND LINE OPTIONS

Display hyperlink cursor and shortcut menu toggles the display of the hyperlink cursor and shortcut menu.

Display hyperlink tooltip toggles the display of the hyperlink tooltip.

CURSOR MENU OPTIONS

Select an object containing a hyperlink; then right-click to display the cursor menu.

***Hyperlink** options:*

Open "url" launches the appropriate applications and loads the file referenced by the URL.

Copy Hyperlink copies hyperlink data to the Clipboard; use the **PasteAsHyperlink** command to paste the hyperlink to selected objects.

Add to Favorites adds the hyperlink to a favorites list.

Edit Hyperlink displays the Edit Hyperlink dialog box; see the **Hyperlink** command.

RELATED COMMANDS

Hyperlink attaches a hyperlink to objects.

Options determines options for most other aspects of AutoCAD.

RELATED SYSTEM VARIABLE

HyperlinkBase specifies the path for relative hyperlinks in the drawing; when "" (null), the drawing paths stored in **AcadPrefix** are used.

TIP

- Answering "n" to this command makes hyperlinks unavailable.

'Id

V. 1.0 Identifies the 3D coordinates of specified points (*short for IDentify*).

Command	Alias	Ctrl+	F-key	Alt+	Menu Bar	Tablet
id	...	...	...	TYI	Tools ↳Inquiry ↳Id Point	U9

Command: id
Specify point: *(Pick a point.)*

Sample output:

X = 1278.0018 Y = 1541.5993 Z = 0.0000

COMMAND LINE OPTION

Specify point picks a point.

RELATED COMMANDS

List lists information about picked objects.

Point draws points.

RELATED SYSTEM VARIABLE

LastPoint contains the 3D coordinates of the last picked point.

TIPS

- The **Id** command stores the picked point in the **LastPoint** system variable. Access that value by entering @ at the next prompt for a point value.
- Invoke the **Id** command to set the value of the **LastPoint** system variable, which can be used as relative coordinates in another command.
- If a 2D point is specified, the z-coordinate displayed by **Id** is the current elevation setting; otherwise, the z-coordinate is that of the specified point.
- When you use **Id** with an object snap, then the z-coordinate is the object-snapped value.

Image

Rel.14 Controls the attachment of raster images.

Commands	Aliases	Ctrl+	F-key	Alt+	Menu Bar	Tablet
image	**im**	...	...	**IM**	**Insert ⇘Image Manager**	**T3**
-image	**-im**					

Command: image

Displays dialog box:

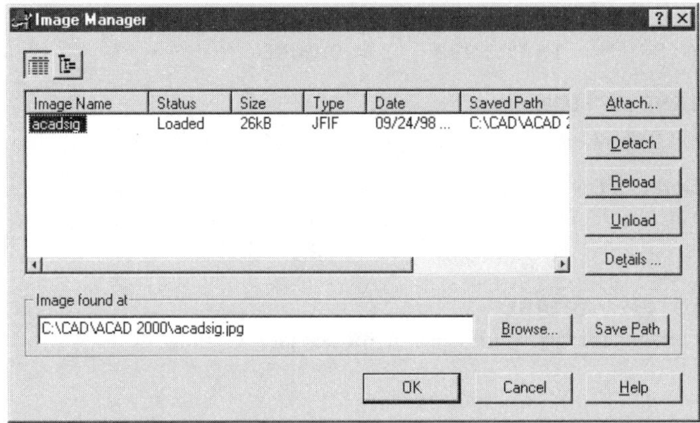

DIALOG BOX OPTIONS

Attach displays the Attach Image File dialog box; see the **ImageAttach** command.

Detach erases the image from the drawing.

Reload reloads the image file into the drawing.

Unload removes the image from memory without erasing the image.

Save Path describes the drive and folder location of the file.

Browse searches for the image file; displays the Attach Image File dialog box.

Details describes the technical details of the images; displays dialog box.

Image File Details dialog box

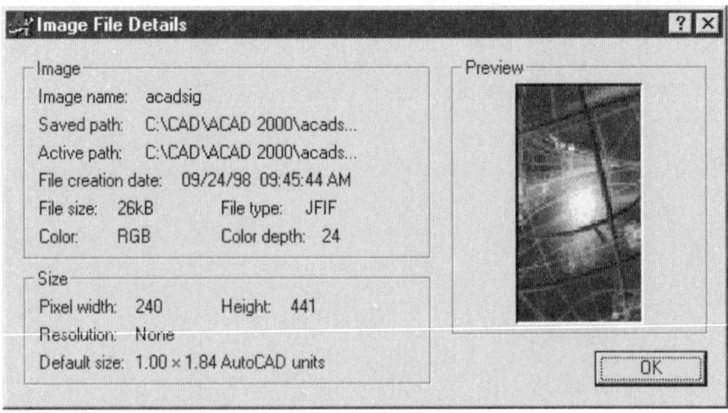

-IMAGE Command

Command: -image

Enter image option [?/Detach/Path/Reload/Unload/Attach] <Attach>: *(Enter an option.)*

COMMAND LINE OPTIONS

? lists currently-attached image files.

Detach erases images from drawings.

Path lists the names of images in the drawing.

Reload reloads image files into drawings.

Unload removes images from memory without erasing them.

Attach displays the Attach Image File dialog box; see the **ImageAttach** command.

RELATED COMMANDS

ImageAdjust controls the brightness, contrast, and fading of images.

ImageAttach attaches images to the current drawing.

ImageClip creates clipping boundaries around images.

ImageFrame toggles the display of image frames.

ImageQuality toggles the display between draft and high-quality mode.

Transparency changes the transparency of images.

Xref attaches *.dwg* drawings as externally-referenced files.

RELATED SYSTEM VARIABLE

ImageHlt toggles whether the entire image is highlighted.

TIPS

- The **Image** command handles raster images of these color depths:

Depth	Colors
Bitonal	Black and white (monochrome).
8-bit gray	256 shades of gray.
8-bit color	256 colors.
24-bit color	16.7 million colors.

- AutoCAD can display one or more images in any viewport.
- There is no theoretical limit to the number and size of images.

ImageAdjust

Rel.14 Controls brightness, contrast, and fade of attached raster images.

Commands	Alias	Ctrl+	F-key	Alt+	Menu Bar	Tablet
imageadjust	**iad**	...	...	**MOIA**	**Modify** ↳**Object** ↳**Image** ↳**Adjust**	**X20**
-imageadjust						

Command: imageadjust
Select image(s): *(Select one or more image objects.)*
Select image(s): *(Press* **ENTER** *to end object selection.)*

Displays dialog box:

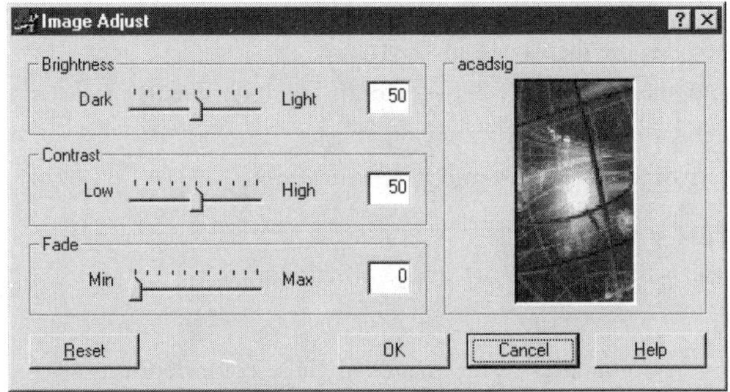

DIALOG BOX OPTIONS

Brightness options

Dark reduces the brightness of the image when values are closer to 0.

Light increases the brightness of the image when values are closer to 100.

Contrast options

Low reduces the image contrast when values are closer to 0.

High increases the image contrast when values are closer to 100.

Fade options

Min reduces the image fade when values are closer to 0.

Max increases the image fade when values are closer to 100.

Reset resets the image to its original parameters; default values are:

Parameter	Original Setting
Brightness	50
Contrast	50
Fade	0

-IMAGEADJUST Command

Command: -imageadjust

Select image(s): *(Select one or more image objects.)*

Select image(s): *(Press* **ENTER** *to end object selection.)*

Enter image option [Contrast/Fade/Brightness] <Brightness>: *(Enter an option.)*

COMMAND LINE OPTIONS

Contrast option

Enter contrast value (0-100) <50>: *(Enter a value.)*

Enter contract value adjusts the contrast between 0% contrast and 100% contrast (*default = 50*).

Fade option

Enter fade value (0-100) <0>: *(Enter a value.)*

Enter fade option adjusts the fading between 0% faded and 100% faded (*default = 0*).

Brightness option

Enter brightness value (0-100) <50>: *(Enter a value.)*

Enter brightness value adjusts the brightness between 0% bright and 100% bright (*default = 50*).

RELATED COMMANDS

Image controls the loading of raster image files in drawings.

ImageAttach attaches images to the current drawing.

ImageClip creates clipping boundaries on images.

ImageFrame toggles display of image frames.

ImageQuality toggles display between draft and high-quality mode.

Transparency changes the transparency of images.

TIPS

- **Brightness** ranges from 0 (*left*) to 50 (*center*) and 100 (*right*).

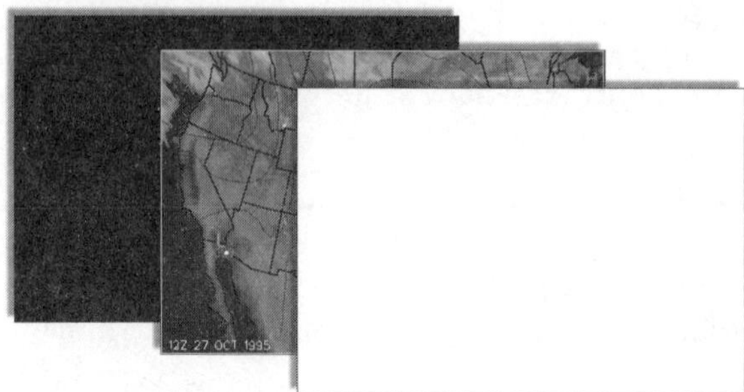

- **Contrast** ranges from 0 (*left*) to 50 (*center*) and 100 (*right*).

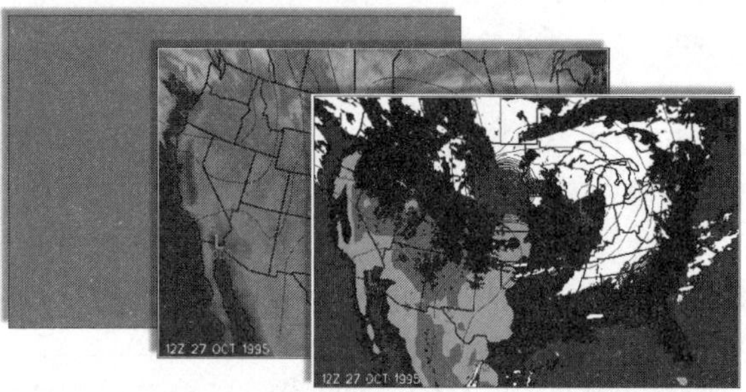

- **Fade** ranges from 0 (*left*) to 50 (*center*) and 100 (*right*).

ImageAttach

Rel.14 Selects raster files to attach to drawings.

Command	Alias	Ctrl+	F-key	Alt+	Menu Bar	Tablet
imageattach	**iat**	...	...	**II**	**Insert** **↳Raster Image**	...

Command: imageattach

Displays file dialog box:

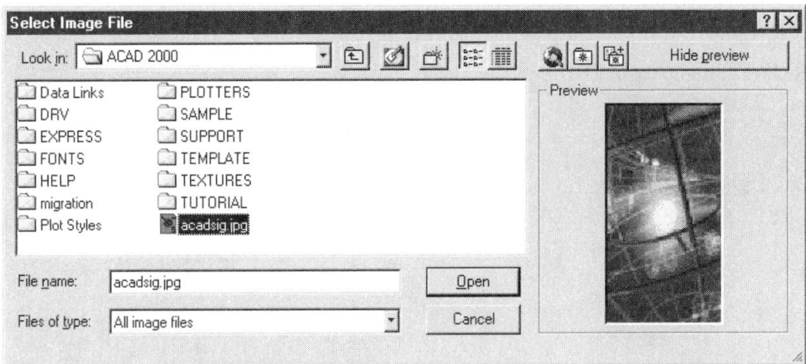

Select an image file, and then click ***Open.***

Displays dialog box:

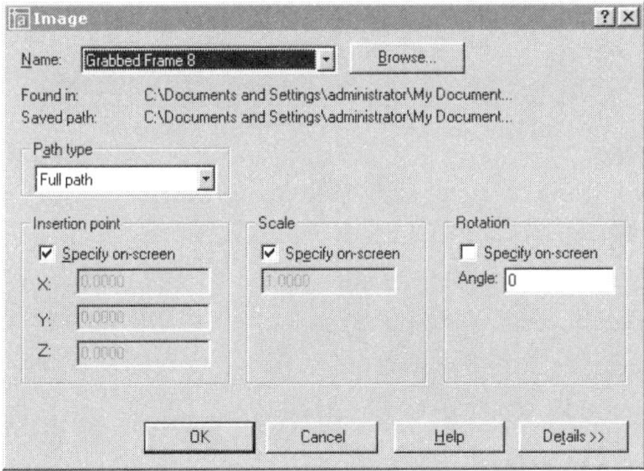

DIALOG BOX OPTIONS

Name selects names from a list of previously-attached image names.

Browse selects files; displays Select Image File dialog box.

Path Type options *(new to AutoCAD 2005)*

Full Path saves the path to the image file.

Relative Path saves the relative to the current drawing file.

No Path does not save the path.

Insertion Point options

☑ **Specify On-Screen** specifies the insertion point of the image in the drawing, after the dialog box is dismissed.

X, Y, Z specifies the x, y, z coordinates of the lower-left corner of the image.

Scale options

☑ **Specify on-screen** specifies the scale of the image (relative to the lower-left corner) in the drawing after the dialog box is dismissed.

Scale specifies the scale of the image; a positive value enlarges the image, while negative values reduce the image.

Rotation options

☐ **Specify on-screen** specifies the rotation angle of the image about the lower-left corner in the drawing, after the dialog box is dismissed.

Angle specifies the angle to rotate the image; positive angles rotate the image counterclockwise.

Button

Details expands the dialog box to display information about the image:

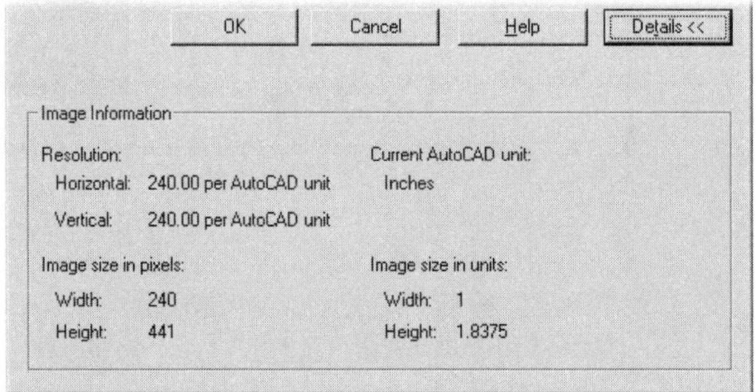

RELATED COMMANDS

Image controls the loading of raster image files in drawings.

ImageAttach attaches images to the current drawing.

ImageClip creates clipping boundaries on images.

ImageFrame toggles the display of image frames.

ImageQuality adjusts the quality of images.

Transparency changes the transparency of images.

RELATED SYSTEM VARIABLE

InsUnits specifies the drawing units for the inserted image.

TIPS

- For a command-line version of the **ImageAttach** command, use the **-Image** command's **Attach** option.
- This dialog box no longer selects units from the **Current AutoCAD Unit** list box; as of AutoCAD 2000, use the **InsUnits** system variable.

ImageClip

Rel.14 Clips raster images.

Command	Alias	Ctrl+	F-key	Alt+	Menu Bar	Tablet
imageclip	icl	...	...	MCI	Modify ↳Clip ↳Image	X22

Command: imageclip
Select image to clip: *(Select one image object.)*
Enter image clipping option [ON/OFF/Delete/New boundary] <New>: *(Enter an option.)*

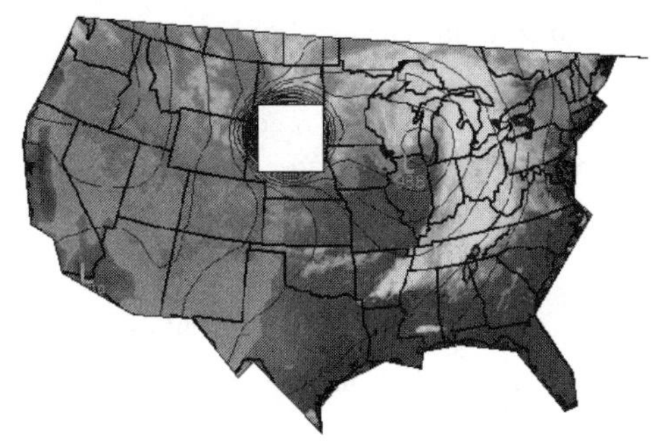

COMMAND LINE OPTIONS

Select image selects one image to clip.

ON turns on a previous clipping boundary.

OFF turns off the clipping boundary.

Delete erases the clipping boundary.

New Boundary options
Enter clipping type [Polygonal/Rectangular] <Rectangular>: *(Enter an option.)*

Polygonal creates a polygonal clipping path.

Rectangular creates a rectangular clipping boundary.

Polygonal options
Specify first point: *(Pick a point.)*
Specify next point or [Undo]: *(Pick a point, or type* **U***.)*
Specify next point or [Undo]: *(Pick a point, or type* **U***.)*
Specify next point or [Close/Undo]: *(Pick a point, or enter an option.)*
Specify next point or [Close/Undo]: *(Enter* **C** *to close the polygon.)*

Specify first point specifies the start of the first segment of the polygonal clipping path.

Specify next point specifies the next vertex.

Undo undoes the last vertex.

Close closes the polygon clipping path.

Rectangular options

Specify first corner point: *(Pick a point.)*

Specify opposite corner point: *(Pick a point.)*

Specify first corner point specifies one corner of the rectangular clip.

Specify opposite corner point specifies the second corner.

When you select an image with a clipped boundary, AutoCAD prompts:

Delete old boundary? [No/Yes] <Yes>: *(Type **N** or **Y**.)*

Delete old boundary?

Yes removes the previously-applied clipping path.

No exits the command.

RELATED COMMANDS

Image controls the loading of raster image files in the drawing.

ImageAdjust controls the brightness, contrast, and fading of the image.

ImageAttach attaches an image in the current drawing.

ImageFrame toggles the display of the image's frame.

ImageQuality toggles the display between draft and high-quality mode.

Transparency changes the transparency of the image.

XrefClip clips *.dwg* drawing files attached as external-reference files.

TIPS

- You can use object snap modes on the image's frame, but not on the image itself.
- To clip a hole in the image, create the hole, then double back on the same path:

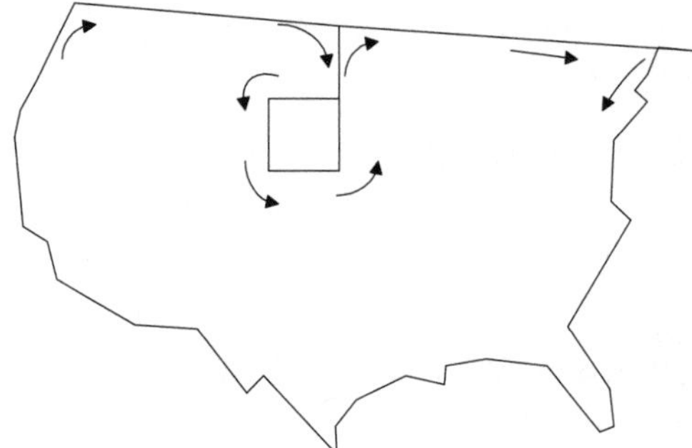

- For a rounded clipping path, apply the **PEdit** command.

ImageFrame

Rel.14 Toggles the display of the frame around raster images.

Command	Alias	Ctrl+	F-key	Alt+	Menu Bar	Tablet
imageframe	...	...	...	MOIF	**Modify** ↳**Object** ↳**Image** ↳**Frame**	...

Command: imageframe
Enter image frame setting [ON/OFF] <ON>: *(Type* **ON** *or* **OFF**.*)*

Image frame turned on (left) and turned off (right).

COMMAND LINE OPTIONS

ON turns on the display of all image frames (default).

OFF turns off the display of all image frames.

RELATED SYSTEM VARIABLE

ImageHlt toggles whether the entire image is highlighted:

ImageHlt	Meaning
0	Highlights image frame.
1	Highlights entire image (default).

RELATED COMMANDS

Image controls the loading of raster image files in the drawing.

ImageAdjust controls the brightness, contrast, and fading of the image.

ImageAttach attaches an image in the current drawing.

ImageClip creates a clipping boundary on the image.

ImageQuality toggles the display between draft and high-quality mode.

Transparency changes the transparency of the image.

TIPS

- *Warning!* When **ImageFrame** is turned off, you cannot select the image.
- Frames are turned on (or off) in all viewports.
- Image frames are plotted, if turned on.

ImageQuality

Rel.14 Toggles the quality of raster images.

Command	Alias	Ctrl+	F-key	Alt+	Menu Bar	Tablet
imagequality	...	...	...	MOIQ	Modify ↳Object ↳Image ↳Quality	...

Command: imagequality

Enter image quality setting [High/Draft] <High>: *(Type* **H** *or* **D***.)*

COMMAND LINE OPTIONS

High displays images at a higher quality.

Draft displays images at a lower quality.

RELATED COMMANDS

Image controls the loading of raster image files in the drawing.

ImageAdjust controls the brightness, contrast, and fading of the image.

ImageAttach attaches an image in the current drawing.

ImageClip creates a clipping boundary on the image.

ImageFrame toggles the display of the image's frame.

Transparency changes the transparency of the image.

TIPS

- High quality displays the image more slowly; draft quality displays the image more quickly.
- I find that draft quality looks better (crisper) than high quality (blurred), but others disagree.
- This command affects the display only; AutoCAD always plots images in high quality.

Import

Rel.13 Imports vector and raster files into drawings.

Command	Alias	Ctrl+	F-key	Alt+	Menu Bar	Tablet
import	imp	...	...	...	...	T2

Command: import

Displays dialog box:

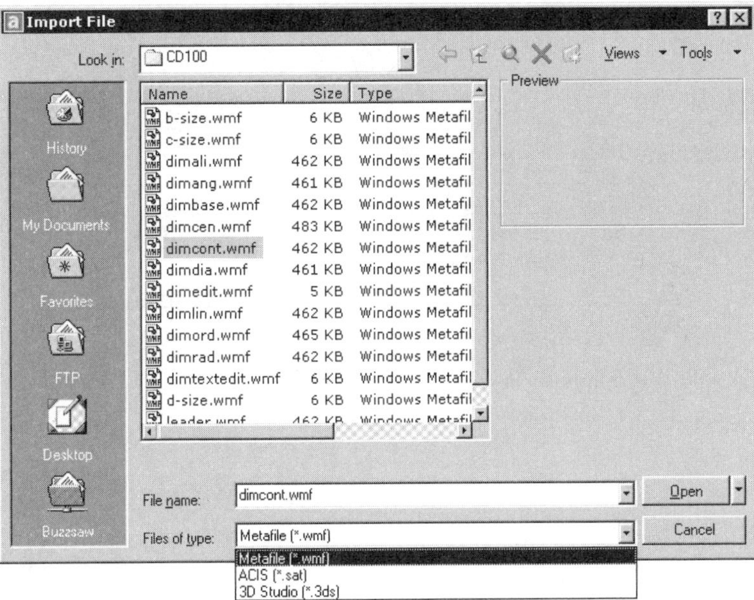

DIALOG BOX OPTIONS

Look in selects the folder (*subdirectory*) and drive from which to import the file.

File name specifies the name of the file, or accepts the default.

File of type selects the file format in which to import the file.

Open imports the file.

Cancel dismisses the dialog box, and returns to AutoCAD.

RELATED COMMANDS

AppLoad loads AutoLISP, VBA, and ObjectARX routines.

DxbIn imports a DXB file.

Export exports the drawing in several vector and raster formats.

Load imports SHX shape objects.

Insert places another drawing in the current drawing as a block.

InsertObj places an OLE object in the drawing via the Clipboard.

LsLib imports landscape objects.

MatLib imports rendering material definitions.

MenuLoad loads menu files into AutoCAD.

Open opens AutoCAD (any version) *.dwg* and *.dxf* files.

PasteClip pastes objects from the Clipboard.

PasteSpec pastes or links object from the Clipboard.

Replay displays renderings in TIFF, Targa, or GIF formats.

VSlide displays *.sld* slide files.

XBind imports named objects from another *.dwg* file.

XRef displays another *.dwg* file in the current drawing.

TIPS

- The **Import** command acts as a "shell" command; it launches other AutoCAD commands that perform the actual import function. Other options may be available with the actual command, such as insertion point and scale.

Format	Meaning
Metafile	Windows metafile WMF; executes **WmfIn**.
ACIS	ASCII SAT; executes **AcisIn**.
3D Studio	3D Studio 3DS format, executes **3dsIn**.

- To import *.dxf* files, use the **Open** command.
- AutoCAD no longer imports PostScript and EPS files.

Removed Commands

INetCfg was removed from AutoCAD 2000; it was replaced by Windows' Internet configuration.

INetHelp was removed from AutoCAD 2000; it was replaced by AutoCAD's standard online help.

Insert

V. 1.0 Inserts previously-defined blocks into drawings.

Commands	Aliases	Ctrl+	F-key	Alt+	Menu Bar	Tablet
insert	**i**	...	...	**IB**	**Insert**	**T5**
	inserturl				**↳Block**	
-insert	**-i**					

Command: insert

Displays dialog box

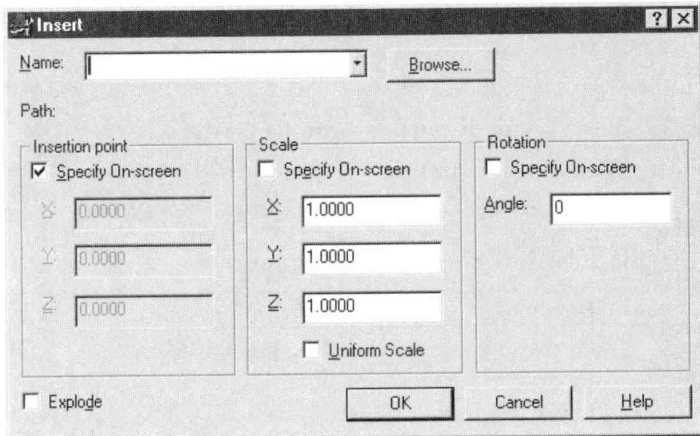

DIALOG BOX OPTIONS

Name selects the name from a list of previously-inserted blocks.

Browse displays the Select Drawing File dialog box; select a block in either file format:

- **Drawing (*.dwg)** AutoCAD drawing file.
- **DXF (*.dxf)** drawing interchange file.

Insertion point options

☑ **Specify On-Screen** specifies the insertion point in the drawing, after closing dialog box.

X, Y, Z specifies the x, y, z coordinates of the lower-left corner of the block.

Scale options

☑ **Specify on-screen** specifies the scale of the block (relative to the lower-left corner) in the drawing, after the dialog box is dismissed.

X, Y, Z specifies the x,y,z scale of the block; positive values enlarge the block, while negative values reduce the block.

☐ **Uniform Scale** forces the y and z scale factors to be the same as the x scale factor.

Rotation options

☐ **Specify on-screen** specifies the rotation angle of the block (about the lower-left corner) in the drawing after the dialog box is dismissed.

Angle specifies the angle to rotate the block; positive angles rotate the block counterclockwise.

Additional option

☐ **Explode** explodes the block upon insertion.

-INSERT Command

Enter block name or [?]: *(Enter name, or* **?***.)*

Specify insertion point or [Scale/X/Y/Z/Rotate/PScale/PX/PY/PZ/PRotate]: *(Pick a point, or enter an option.)*

Enter X scale factor, specify opposite corner, or [Corner/XYZ] <1>: *(Enter a scale factor, or enter an option.)*

Enter Y scale factor <use X scale factor>: *(Enter a scale factor, or press* **ENTER***.)*

Specify rotation angle <0>: *(Enter a rotation angle.)*

COMMAND LINE OPTIONS

Block name specifies the name of the block to be inserted.

? lists the names of blocks stored in the drawing.

Specify insertion point specifies the lower-left corner of the block's insertion point.

P supplies a predefined block name, scale, and rotation values.

X scale factor indicates the x scale factor.

Corner indicates the x and y scale factors by pointing on the screen.

XYZ displays the x, y, and z scale submenu.

INPUT OPTIONS

In response to the 'Block Name' prompt, you can enter:

Option	Meaning
~	Display a dialog box of drawings stored on disk: **`Block name:`** `~`
*	Insert block exploded: **`Block name:`** `*filename`
=	Redefine existing block with a new block: **`Block name:`** `oldname=newname`

In response to the 'Insertion point' prompt, you can enter:

Option	Meaning
Scale	Specify x, y, and z-scale factors.
PScale	Preset x, y, and z-scale factors.
XScale	Specify x scale factor.
PxScale	Preset x scale factor.
YScale	Specify y scale factor.
PyScale	Preset y scale factor.
ZScale	Specify z scale factor.
PzScale	Preset z scale factor.
Rotate	Specify rotation angle.
PRotate	Preset rotation angle.

RELATED COMMANDS

Block creates a block out of a group of objects.

Explode reduces inserted blocks to their constituent objects.

MInsert inserts a block as a blocked rectangular array.

Rename renames blocks.

WBlock writes blocks to disk.

XRef displays drawings stored on disk in the drawing.

RELATED SYSTEM VARIABLES

ExplMode toggles whether non-uniformly scaled blocks can be exploded:

ExplMode	Meaning
0	Cannot explode; AutoCAD Release 12 compatible.
1	Can be exploded (default).

InsBase specifies the name of the most-recently inserted block.

InsUnits specifies the drawing units for the inserted block.

TIPS

- You can insert any other AutoCAD drawing into the current drawing.
- A *preset* scale factor or rotation means the dragged image is shown at that scale, but you can enter a new scale when inserting.
- Drawings are normally inserted as a block; prefix the filename with an * (*asterisk*) to insert the drawing as separate objects.
- Redefine all blocks of the same name in the current drawing by adding the = (*equal*) suffix after its name at the 'Block name' prompt.
- Insert a mirrored block by supplying a negative x- or y-scale factor, such as:

  X scale factor: -1

- AutoCAD converts a negative z scale factor into its absolute value, which always makes it positive.
- As of AutoCAD Release 13, you can explode a mirrored block and a block inserted with different scale factors when the system variable **ExplMode** is turned on.
- The **Insert** command no longer imports drawings in XML format as of AutoCAD 2004.
- When inserting block with attached xrefs, the xrefs are retained (*new to AutoCAD 2005*).

InsertObj

Rel.13 Places OLE objects as a linked or embedded objects (*short for INSERT OBJect*).

Command	Alias	Ctrl+	F-key	Alt+	Menu Bar	Tablet
insertobj	io	...	...	IO	Insert ↳OLE Object	T1

Command: insertobj

Displays dialog box:

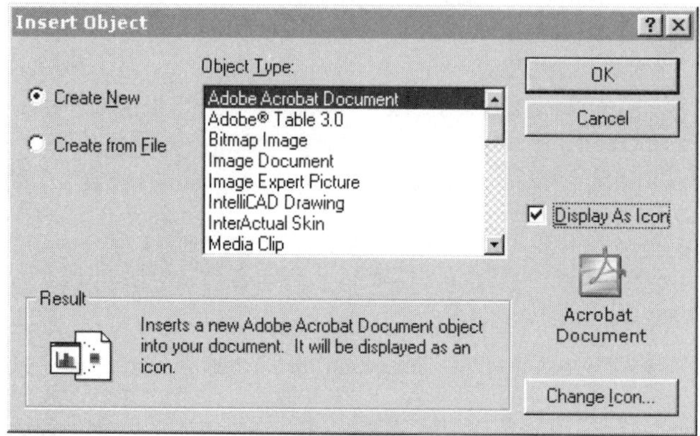

DIALOG BOX OPTIONS

⊙**Create New** creates new objects in other applications, and then embeds them in the drawing.

○**Create from File** selects a file to embed or link in the current drawing.

Object Type selects an object type from the list; the related application automatically launches if you select the Create New option.

☑ **Display As Icon** displays the object as an icon, rather than as itself.

Change Icon selects another icon.

RELATED COMMANDS

OleLinks controls the OLE links.

PasteSpec places an object from the Clipboard in the drawing as a linked object.

RELATED SYSTEM VARIABLES

MsOleScale determines the scale of OLE objects placed in model space (*new to AutoCAD 2005*).

OleHide toggles the display of OLE objects.

OleQuality determines the plot quality of OLE objects.

OleStartup specifies whether the source apps of embedded OLE objects load for plotting.

RELATED WINDOWS COMMANDS

Edit | Copy copies an object to the Clipboard into another Windows application.

File | Update updates an OLE object from another application.

Removed Command

InsertUrl was removed from AutoCAD 2000; it was replaced by the **Insert** command's **Browse | Search the Web** option.

Interfere

Rel.11 Determines the interference of two or more 3D solid objects; optionally creates a 3D solid body of the volumes in common.

Command	Alias	Ctrl+	F-key	Alt+	Menu Bar	Tablet
interfere	inf	...	...	DII	Draw ↳Solids ↳Interference	...

Command: interfere

Select first set of solids:

Select objects: *(Select one or more solid objects.)*

Select objects: *(Press* **ENTER** *to end object selection.)*

Select second set of solids:

Select objects: *(Select one or more solid objects.)*

Select objects: *(Press* **ENTER** *to end object selection.)*

Comparing 1 solid against 1 solid.

Interfering solids (first set) : 1

(second set) : 1

Interfering pairs : 1

Create interference solids? [Yes/No] <N>: *(Enter* **Y** *or* **N***.)*

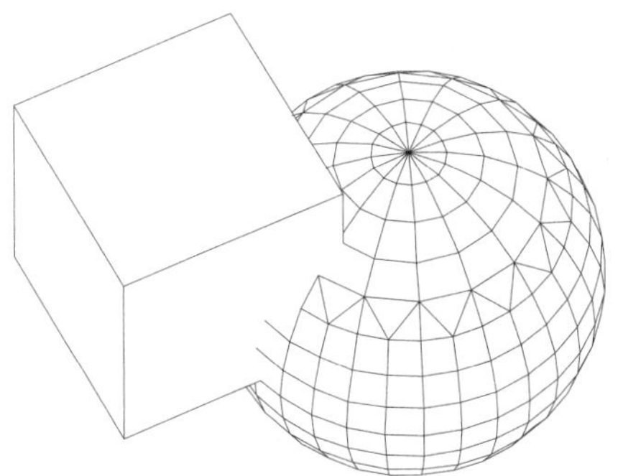

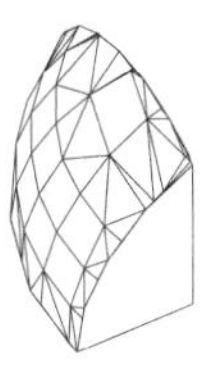

A pair of interfering solids (left) and the interference (right).

COMMAND LINE OPTIONS

Select objects checks all solids in a single selection set for interference with one another.

Create interference solids creates a solid representing the volume of interference.

RELATED COMMANDS

Intersect creates a new volume from the intersection of two volumes.

Section creates a 2D region from a 3D solid.

Slice slices a 3D solid with a plane.

TIP

- When three or more solids interfere, AutoCAD asks "Highlight pairs of interfering solids?"

Intersect

Rel.11 Creates 3D solids of 2D regions through the Boolean intersection of two or more solids or regions.

Command	Alias	Ctrl+	F-key	Alt+	Menu Bar	Tablet
intersect	**in**	...	...	**MNI**	**Modify** **↳Solids Editing** **↳Intersect**	**X17**

Command: intersect
Select objects: *(Select one or more solid objects.)*
Select objects: *(Press* **ENTER** *to end object selection.)*

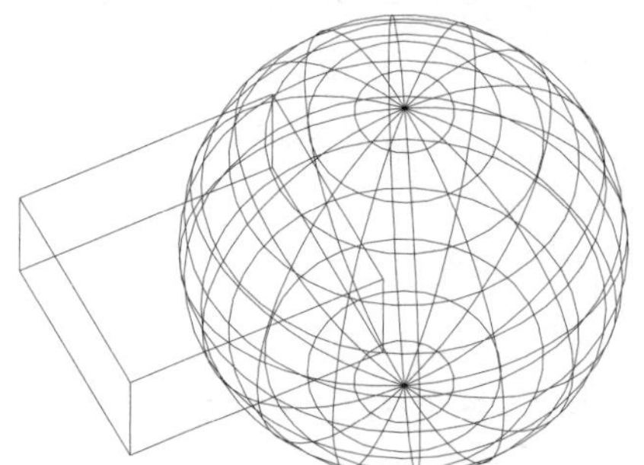
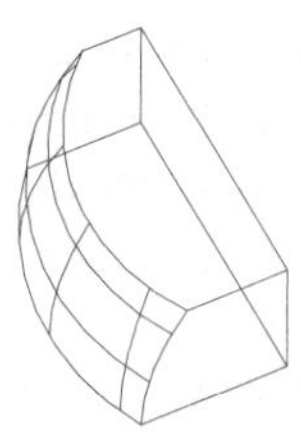

Two intersecting solids (left) and the resulting intersection (right).

COMMAND LINE OPTION

Select objects selects two or more objects to intersect.

RELATED COMMANDS

Interfere creates a new volume from the interference of two or more volumes.

Subtract subtracts one 3D solid from another.

Union joins 3D solids into a single body.

TIPS

- You can use this command on 2D regions and 3D solids.
- The **Interference** and **Intersect** command may seem similar. Here is the difference between the two:

  Intersect *erases* all of the 3D solid parts that do not intersect.

  Interfere optionally *creates a new object* from the intersection; it does not erase the original objects.

'Isoplane

V. 2.0 Changes the crosshair orientation and grid pattern among the three isometric drawing planes.

Command	Alias	Ctrl+	F-key	Alt+	Menu Bar	Tablet
isoplane	...	E	F5	...	...	...

Command: isoplane
Enter isometric plane setting [Left/Top/Right] <Top>: *(Enter an option, or press* ENTER.*)*

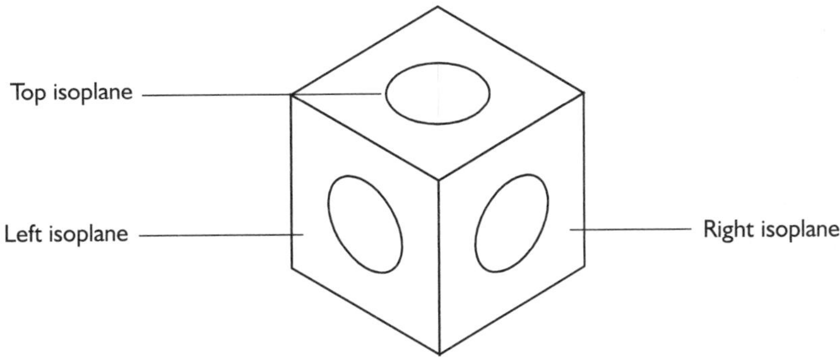

COMMAND LINE OPTIONS

Left switches to the left isometric plane.

Top switches to the top isometric plane.

Right switches to the right isometric plane.

ENTER switches to the next isometric plane in the following order: left, top, right.

RELATED COMMANDS

Options displays a dialog box for setting isometric mode and planes.

Ellipse draws isocircles.

Snap turns on isometric drawing mode.

RELATED SYSTEM VARIABLE

SnapIsoPair contains the current isometric plane.

GridMode toggles grid visibility.

GridUnit specifies the current grid x,y spacing.

LimMin holds the x,y coordinates of the lower-left corner of the grid display.

LimMax holds the x,y coordinates of the upper-right corner of the grid display.

SnapStyl displays a normal or isometric grid:

SnapStyl	Meaning
0	Normal (default).
1	Isometric grid.

Quick Start Tutorial

Creating Isometric Dimensions

AutoCAD's dimensions must be modified for isometric drawings, so that the dimension text looks "correct" in isometric mode. This involves two steps — (1) creating isometric text styles and (2) changing dimension variables — repeated three times, once for each isoplane.

Step 1

How to create isometric text styles:

1. From the menu bar, select **Format | Text Style**.
2. When the **Text Style** dialog box appears, click **New**.
3. Enter **isotop** for the name of the new text style, which is used for text in the top isoplane.
4. Click **OK**.
5. When the **Text Style** dialog box reappears, select *Simplex.Shx* from **Font Name**.
6. Change the **Oblique Angle** to **-30**.
7. Click **Apply**.
8. Create text styles for the other two isoplanes:

Style Name	Font Name	Oblique Angle
IsoTop	*simplex.shx*	-30
IsoRight	*simplex.shx*	30
IsoLeft	*simpelx.shx*	30

Enter these values into the **Text Style** dialog box, click **Apply**, and then **Close**.

Step 2

How to create isometric dimension styles:

1. Create the dimension styles for the three isoplanes by selecting **Format | Dimension Styles** from the menu bar. These dimension variables must be changed:
2. Create a new dimension style:

   Click **New**.

   Enter **IsoLeft** in the **New Style Name** field.

   Click **Continue**.

3. Force dimension text to align with the dimension line:

   Select the **Text** tab.

   Select **Aligned with dimension line** in the **Text Alignment** section.

4. Specify text style for dimension text:

   Select **ISOLEFT** from the **Text Style** list box.

   Click **OK**.

5. One of the three needed dimension styles has been created. Create dimstyles for all isoplanes using these parameters:

Dimstyle Name	Text Style
Isotop	IsoTop
Isoright	IsoRight
Isoleft	IsoLeft

6. Click **Close** to exit the **Dimension Style Manager** dialog box.

Step 3

To place linear dimensions in an isometric drawing, you must use the **DimAligned** command, because it aligns the dimension along the isometric axes: place all dimensions in one isoplane, and then switch to the next isoplane.

1. Press **F5** to switch to the appropriate isoplane, such as **Top**.
2. Use the **DimStyle** command to select the associated dimension style, such as **IsoTop**.
3. Place the dimension with the **DimAligned** command; it is helpful to use **INTersection** object snaps.
4. Use the **DimEdit** command's **Oblique** option to skew the dimension by 30 or -30 degrees, as follows:

IsoPlane	DimStyle	Oblique Angle
Top	IsoTop	30
Left	IsoLeft	30
Right	IsoRight	-30

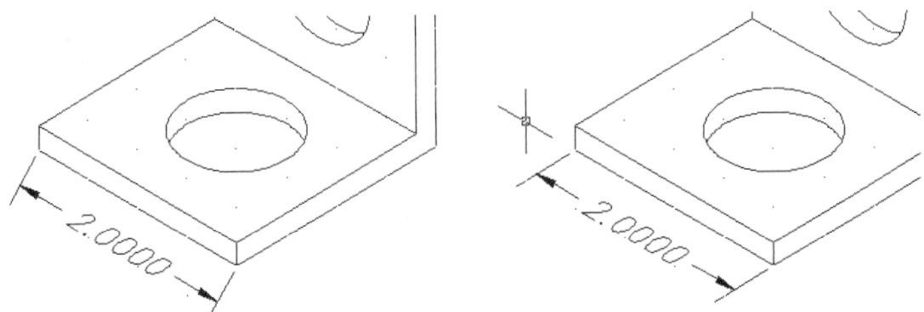

*Aligned dimension text before (left) and after (right) applying **DimEdit's Oblique** option.*

5. To place a leader, use the **Standard** dimstyle and **Standard** text style.

JpgOut

2004 Exports drawings in JPEG format.

Command	Alias	Ctrl+	F-key	Alt+	Menu Bar	Tablet
jpgout	...	...	...	...	...	...

Command: jpgout

*Displays Create Raster File dialog box. Specify a filename, and then click **Save**.*

Select objects or <all objects and viewports>: *(Select objects, or press **ENTER** to select all objects and viewports.)*

COMMAND LINE OPTIONS

Select objects selects specific objects.

All objects and viewports selects all objects and all viewports, whether in model space or in layout mode.

RELATED COMMANDS

BmpOut exports drawings in BMP (bitmap) format.

Image places raster images in the drawing.

PngOut exports drawings in PNG (portable network graphics) format.

TifOut exports drawings in TIFF (tagged image file format) format.

TIPS

- The rendering effects of the **ShadEdge** command are preserved, but not of the **Render** command.
- The drawback to saving drawings in JPEG format is that the image is less clear than in other formats; the advantage is that JPEG files are highly compressed.

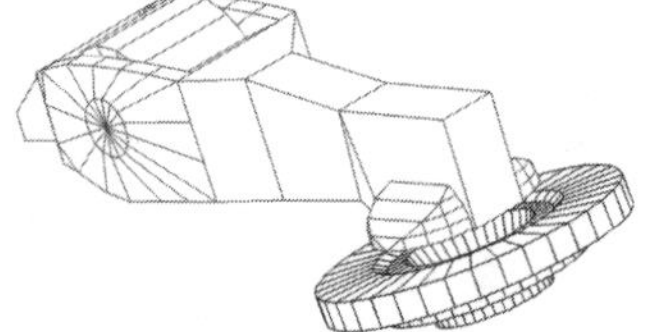

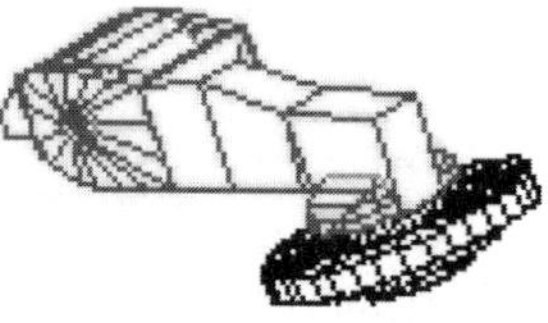

Zoomed-in original image in AutoCAD (left), and enlarged JPEG image with artifacts (right).

- This command provides no options for specifying the level of compression.
- JPEG files are often used by digital cameras and Web pages.
- JPEG is short for "joint photographic expert group."

JustifyText

2001 Changes the justification of text.

Command	Alias	Ctrl+	F-key	Alt+	Menu Bar	Tablet
justifytext	...	...	...	**MOTJ**	**Modify** **↳Object** **↳Text** **↳Justify**	...

Command: justifytext
Select objects: *(Select one or more text objects.)*
Select objects: *(Press* **ENTER** *to end object selection.)*
Enter a justification option
[Left/Align/Fit/Center/Middle/Right/TL/TC/TR/ML/MC/MR/BL/BC/BR]
<Left>: *(Enter an option, or press* **ENTER***.)*

COMMAND LINE OPTIONS

Select objects selects one or more text objects in the drawing.

Align aligns the text between two points with adjusted text height.

Fit fits the text between two points with fixed text height.

Center centers the text along the baseline.

Middle centers the text horizontally and vertically.

Right right-justifies the text.

TL justifies to top-left.

TC justifies to top-center.

TR justifies to top-right.

ML justifies to middle-left.

MC justifies to middle-center.

MR justifies to middle-right.

BL justifies to bottom-left.

BC justifies to bottom-center.

BR justifies to bottom-right.

RELATED COMMANDS

Text places text in the drawing.

DdEdit edits text.

ScaleText changes the size of text.

Style defines text styles.

Properties changes the justification, but moves text.

TIPS

- This command works with text, mtext, leader text, and attribute text.
- When the justification is changed, the text does not move.

'Layer

V. 1.0 Controls the creation, status, and visibility of layers.

Commands	Aliases	Ctrl+	F-key	Alt+	Menu Bar	Tablet
layer	la ddlmodes	...	...	OL	Format ↳Layer	U5
-layer	-la					

Command: layer

Displays dialog box (new user interface with AutoCAD 2005):

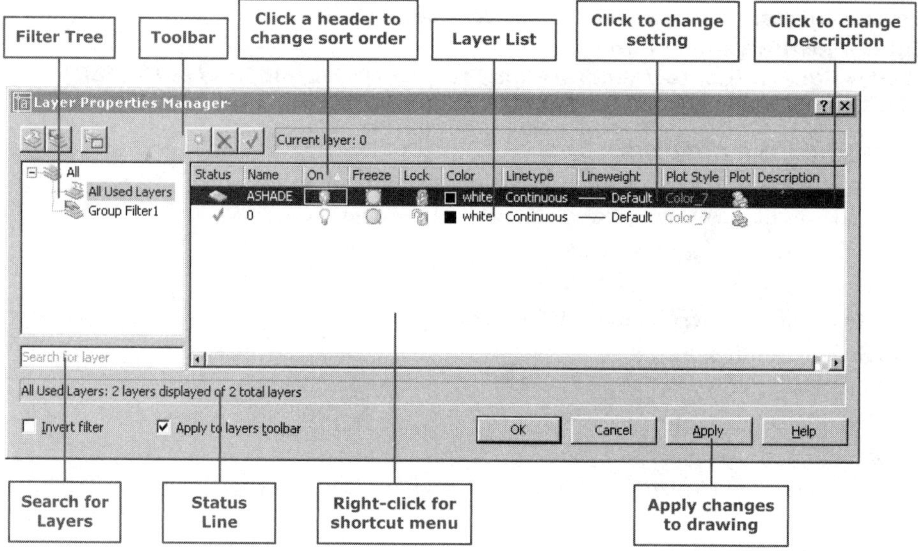

DIALOG BOX OPTIONS

Click a header to sort alphabetically (A-Z); click a second time for reverse-alphabetical sort (Z-A).

Status reports the layer status: current layer, empty layer, layer in use, or layer filter *(new to AutoCAD 2005).*

Name lists the names of layers in the current drawing.

On toggles layers between on and off.

Freeze in all VP toggles layers between thawed and frozen in all viewports.

Lock toggles layers between unlocked and locked.

Color specifies the color for objects on layers.

Linetype specifies the linetype for objects on layers.

Lineweight specifies the lineweight for objects on layers.

Plot Style specifies the plot style for objects on layers.

Plot toggles layers between plot and no-plot.

Current VP Freeze specifies whether the layer is frozen in the current viewport (available only in layout mode).

NewVP Freeze specifies whether the layer is frozen in the new viewports (available only in layout mode).

Description provides space for you to describe the layer *(new to AutoCAD 2005)*.

☐ **Invert filter** inverts the display of layer names; for example, when **Show all used layers** is selected, the **Invert filter** option causes all layers with no content to be displayed; alternatively, press ALT+I.

☑ **Apply to layers toolbar** applies the filter to the layer names displayed by the Layers toolbar; alternatively, press ALT+T.

Apply applies the changes to the layers without exiting the dialog box.

OK exits the dialog box, and applies changes to the layers.

Cancel exits the dialog box, and leaves layers unchanged.

DIALOG BOX TOOLBAR

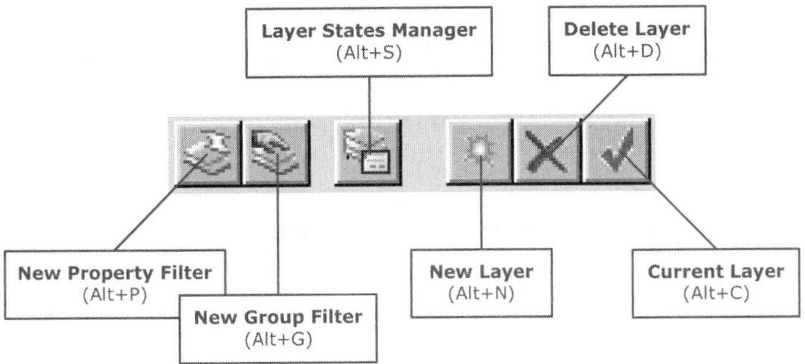

New Property Filter displays New Filter Properties dialog box.

New Group Filter adds an item to the group filter list, initially named "Group Filter 1."

Layer States Manager displays Layer States Manager dialog box.

New Layer adds layers to the drawing, initially named "Layer1."

Delete Layer purges the selected layer; some layers cannot be deleted, as described by the warning dialog box:

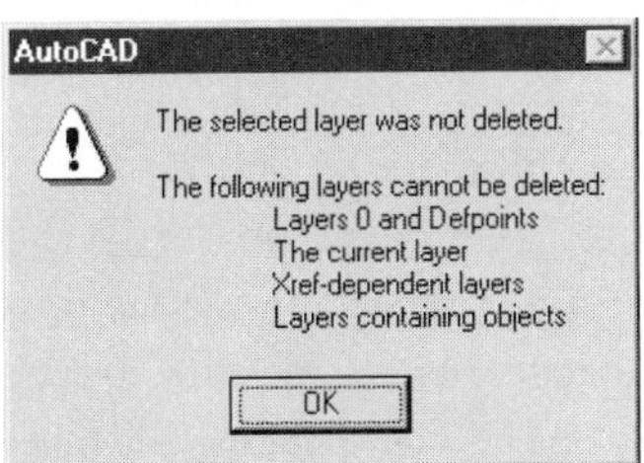

Current Layer sets the selected layer as the current layer.

RIGHT-CLICK MENUS

As of AutoCAD 2005, most layer options are found in right-click menus.

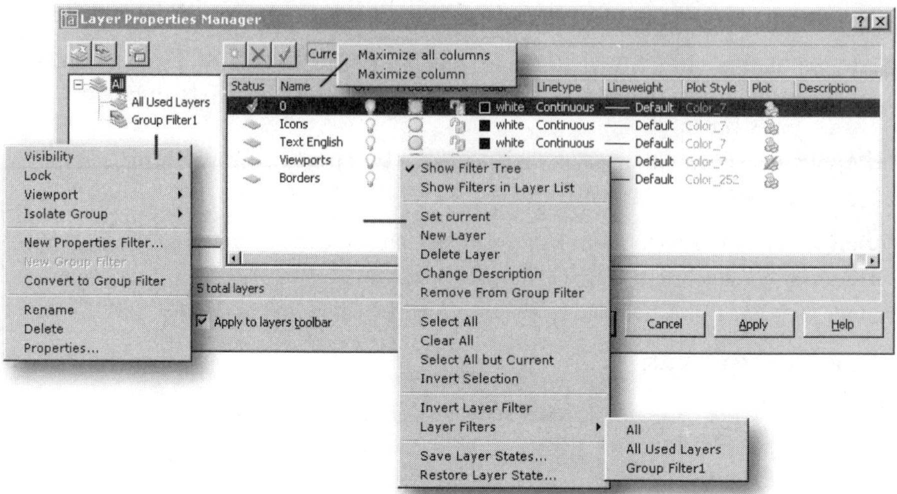

Column Headers right-click menu

Maximize Column changes the selected column width to display all column content.

Maximize All Columns changes the width of all columns to display content of all columns. To display compete headers, you must resize the column by dragging the header separators. You can resize the dialog box.

Filter Tree right-click menu

Depending on where you right-click in the filter tree area, some menu options may be grayed-out, indicating they are unavailable.

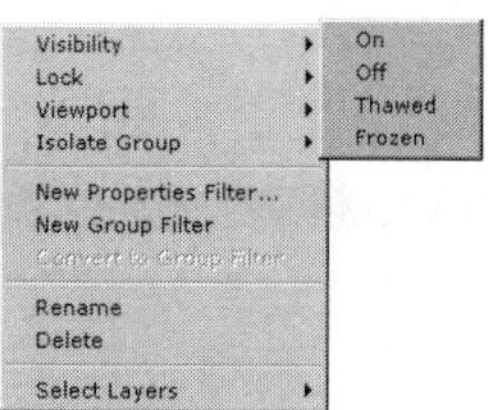

Visibility toggles the visibility of all layers in the selected filter or group:

On displays, plots, and regenerates objects; includes objects during hidden-line removal.

Off does not display or plot objects; does not include objects during hidden-line removal, and the drawing is not regenerated when the layer is turned on.

Thawed reverses the action of the Frozen option.

Frozen does not display or plot objects; includes objects during hidden-line removal, and the drawing is regenerated when the layer is thawed.

Lock determines whether objects can be edited:

Lock prevents objects from being edited; all other operations that don't involve editing are permitted, such as object snaps.

Unlock allows objects to be edited.

Viewport specifies whether layers are frozen in layout mode (unavailable in model space):

Freeze applies Current VP Freeze to all layers (in the filter or group) of the current viewport.

Thaw turns off Current VP Freeze for layers in the filter or group.

Isolate Group turns off all layers not part of the filter or group; only layers that are part of the filter or group are visible in the drawing (*new to AutoCAD 2005*). In model space, this option applies to all layers; in layout mode, this option applies selectively, depending on the suboption selected:

All Viewports freezes all layers not in the filter or group.

Active Viewport Only freezes all layers (not in the filter or group) in the current viewport only.

New Properties Filter displays the Layer Filter Properties dialog box.

New Group Filter creates a new layer group filter named "Group Filter 1," which you can rename to something more meaningful. To add layers to the group:

1. Hold down the CTRL key, and then select layer names in the layer list (right-hand pane).
2. Drag the layers onto the group filter name (left-hand pane).

As an alternative, follow these steps:

1. Right-click the group filter name.
2. From the shortcut menu, select **Select Layers | Add**.
3. In the drawing, click on objects whose layer you want to add to the group.
4. Press ENTER when done selecting objects.

Convert to Group Filter converts the selected filter to a group; the name does not change, but the icon changes to indicate a group filter.

Rename changes the name of the filter or group. As an alternative, click the name twice, slowly, and then enter a new name.

Delete erases the filter or group; does not erase the layers in the filter or group. The All, All Used Layers, and Xref filters cannot be erased.

Properties displays the Layer Filter Properties dialog box (available only when a filter is selected).

Select Layers allows you to add and remove layers by selecting objects in the drawing (available only when a group is selected):

Add allows you to select objects in the drawing; press ENTER to return to the layer dialog box.

Replace removes existing layers from the group, and adds the newly-selected layers.

Layer List right-click menu

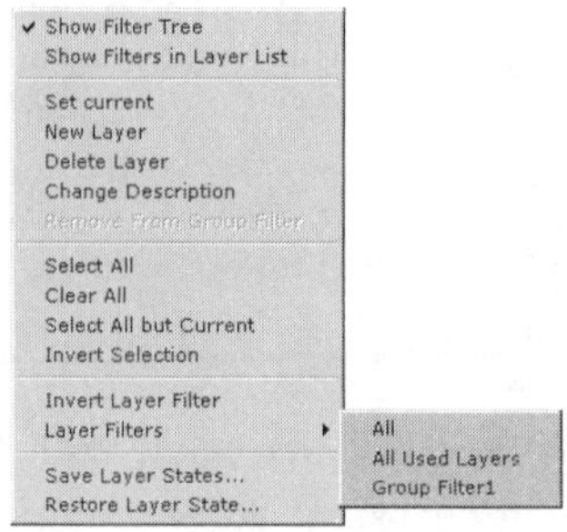

Show Filter Tree toggles filter tree view (left-hand pane).

Show Filters in Layer List toggles the display of filters in the layer list view (right-hand pane); when off (no check mark), only layers are shown.

Set Current sets the selected layer as the current layer. As alternatives, press ALT+C or click the Set Current button on the dialog box's toolbar.

New Layer creates a new layer, names it "Layer1," and gives it the properties of the currently-selected layer. As alternatives, press ALT+N or click the New Layer button on the dialog box's toolbar.

Delete Layer erase the selected layers from the drawing; layers 0, Defpoints, the current layer, xref-dependent layers, and those containing objects cannot be erased. As alternatives, press ALT+D or click the Delete Layer button on the dialog box's toolbar.

Change Description adds and changes layer description text. As an alternative, click the description twice, and then edit the text.

Remove from Group Filter removes the selected layers from the select group.

Select All selects all layers; alternatively, press CTRL+A.

Clear All unselects all layers; alternatively, click a single layer.

Select All but Current selects all layers, except the current layer.

Invert Selection unselects selected layers, and selects all other layers.

Invert Layer Filter displays all layers not in the selected filter; alternatively, press ALT+I.

Layer Filters displays a submenu listing the names of filters; alternatively, turn on the filter tree view (left-hand pane). Select a filter name to apply it to the layer list.

Save Layer States displays the New Layer State to Save dialog box.

Restore Layer State displays the Layer States Manager dialog box; alternatively, press ALT+S.

Layer Filters Properties dialog box

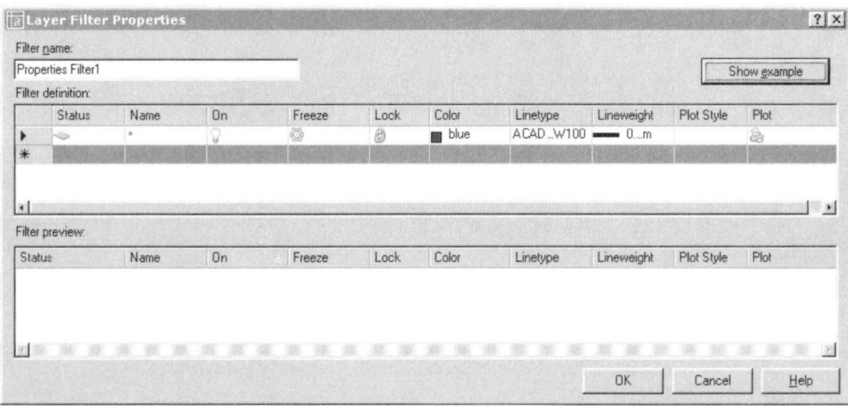

Filter name names the filter.

Show Example displays the filter result in the Filter Preview area (lower half).

Filter definition defines the parameters of the filter:

- **Status** specifies the status of the layer.
- **Layer name** specifies the names of layers to filter; see wildcard metacharacters below.
- **On** selects the layers that are on, off, or both; leave blank for both.
- **Freeze** selects the layers that are frozen, thawed, or both.
- **Lock** selects the layers that are locked, unlocked, or both.
- **Colors** selects the layers of a specific color.
- **Lineweight** selects the layers with a specific lineweight.
- **Linetype** selects the layers with a specific linetype.
- **Plot style** selects the layers with a specific plot style.
- **Plot** selects the layers that plot, do not plot, or both.
- **Current VP Freeze** selects the layers that are frozen, thawed, or both, in the current viewport.
- **New VP Freeze** selects the layers that are frozen, thawed, or both, in the new viewport.

Right-click menus

Right-click filter headers:

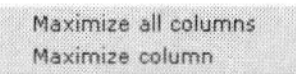

Maximize Column changes the selected column width to display all column content.

Maximize All Columns changes the width of all columns to display content of all columns.

Right-click filter definition row:

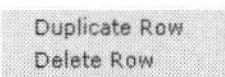

Duplicate Row makes a duplicate of the filter definition.

Delete Row erases the row.

New Layer State to Save dialog box

To access this dialog box, right-click in the Layer List, and select ***Save Layer States*** *from the shortcut menu.*

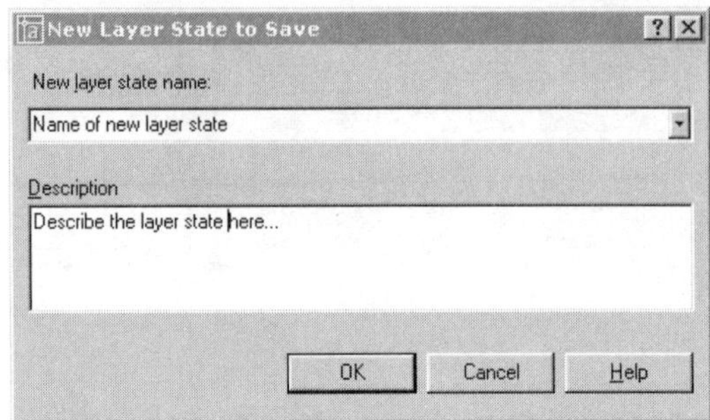

New layer state name specifies the name of the layer state.

Description provides an optional description of the layer state.

OK displays the Layer States Manager dialog box (described below).

Layer States Manager dialog box

To access this dialog box, right-click in the Layer List, and select ***Restore Layer States*** *from the shortcut menu. Alternatively, click the Layer States Manager button, or press* ALT+S.

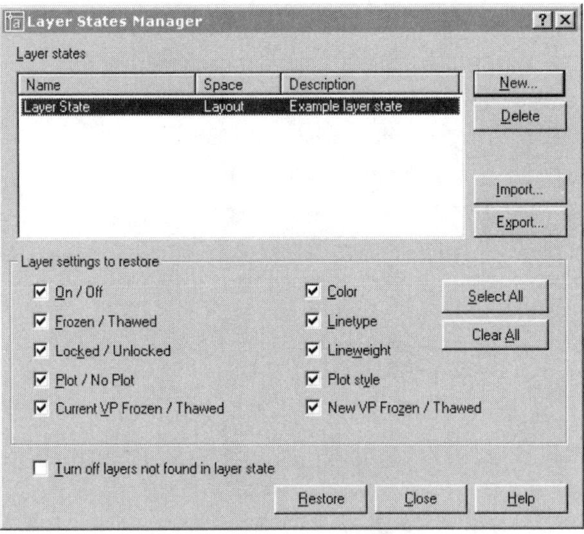

To rename a layer state, click its name twice, and then edit the name.

To edit the description, click the description twice, and then edit the text.

To sort layer states, click the headers – Name, Space, and Description.

Layer states lists the names of layer states that can be restored.

New displays the New Layer State to Save dialog box.

Delete deletes the selected layer state.

Import imports *.las* (layer state) files; layer states can be shared among drawings.

Export exports the selected layer state to *.las* files.

Layer settings to restore selects the settings to restore.

Select All selects all settings.

Clear All unselects all settings (turns them off).

☐ **Turns off layers not found in layer state** turns off layers for which settings were not saved.

Restore closes the dialog box, and restores the layer state.

-LAYER Command

Command: -layer

Current layer: "0"

Enter an option [?/Make/Set/New/ON/OFF/Color/Ltype/LWeight/Plot/PStyle/Freeze/Thaw/LOck/Unlock/stAte]: *(Enter an option.)*

COMMAND LINE OPTIONS

Groups and filters cannot be created by this command.

Color indicates the color for all objects drawn on the layer.

Freeze disables the display of the layer.

LOck locks the layer.

Ltype indicates the linetype for all objects drawn on the layer.

LWeight specifies the lineweight.

Make creates a new layer, and makes it the current layer.

New creates a new layer.

OFF turns off the layer.

ON turns on the layer.

Plot determines whether the layer is plotted.

PStyle specifies the plot style (available only when plot styles are attached to the drawing).

Set makes the layer the current layer.

stAte sets and saves layer states.

Thaw unfreezes the layer.

Unlock unlocks the layer.

? lists the names of layers in the drawing.

stAte options

Enter an option [?/Save/Restore/Edit/Name/Delete/Import/EXport]: *(Enter an option).*

? lists the names of layer states in the drawing.

Save saves the layer state and properties by name; properties include on, frozen, lock, plot, newvpfreeze, color, linetype, lineweight, and plot style.

Restore restores a named state.

Edit changes the settings of named state.

Name renames named states.

Delete erases named states from the drawing.

Import opens layer state *.las* files.

Export saves a selected named state to a *.las* file.

RELATED COMMANDS

LayerP returns to the previous layer.

Change moves objects to different layers via the command line.

Properties moves objects to different layers via a dialog box.

LayTrans translates layer names.

Purge removes unused layers from drawings.

Rename renames layers.

View controls visibility of layers with named views.

VpLayer controls the visibility of layers in paper space viewports.

RELATED SYSTEM VARIABLE

CLayer contains the name of the current layer.

RELATED FILE

.*las are layer state files, which use the DXF format.

TIPS

- A *frozen* layer cannot be seen or edited.
- A *locked* layer can be seen, but not edited.
- Layer **Defpoints** is a non-plotting layer.
- To create more than one new layer at a time, use commas to separate layer names in **-Layer**.
- For new layers to take on properties of an existing layer, select the layer before clicking **New**.
- Wildcard metacharacters used in creating filters:

Char	Meaning
*	Matches any one or more characters.
?	Matches any single character.
@	Matches any alphabetic character (A - Z).
#	Matches any numeric digit (0 - 9)
.	Matches any non-alphanumeric character (!@#$%^&*, etc.)
~	Matches anything but the pattern of characters.
[ ]	Matches any one enclosed character.
[~]	Matches any character not enclosed.
[-]	Matches a range of characters.
`	Allows literal use of metacharacters.

- If layer names appear to be missing, they have been filtered from the list.

LayerP

2002 Undoes changes made to layer settings *(short for LAYER Previous).*

Command	Alias	Ctrl+	F-key	Alt+	Menu Bar	Tablet
layerp	...	...	...	...	...	...

Command: layerp
Restored previous layer status

COMMAND LINE OPTIONS

None.

RELATED COMMANDS

LayerPMode toggles layer-previous mode on and off.

Layer creates and sets layers and modes.

RELATED SYSTEM VARIABLE

CLayer specifies the name of the current layer.

TIPS

- This command acts like an "undo" command for changes made to layers only, such as changes to the layer's color or lineweight.
- **LayerPMode** command must be turned on for the **LayerP** command to work.
- The **LayerP** command does not undo the renaming, deleting, purging, or creating of layers.

LayerPMode

<u>**2002**</u> Turns on and off layer-previous mode *(short for LAYER Previous MODE).*

Command	Alias	Ctrl+	F-key	Alt+	Menu Bar	Tablet
layerpmode	...	...	...	...	...	...

Command: layerpmode
Enter LAYERP mode [ON/OFF] <ON>: *(Type* **ON** *or* **OFF***.)*

COMMAND LINE OPTIONS

On turns on layer previous mode.

Off turns off layer previous mode.

RELATED COMMANDS

LayerP returns the drawing to the previous layer.

Layer creates and sets layers and modes.

TIP

- When layer previous mode is on, AutoCAD tracks changes to layers.

Layout

2000 Creates and deletes paper space layouts on the command line.

Commands	Alias	Ctrl+	F-key	Alt+	Menu Bar	Tablet
layout	**lo**	...	...	**IL**	**Insert** ↳**Layout**	...
-layout						

Command: layout *or* -layout

Enter layout option [Copy/Delete/New/Template/Rename/SAveas/Set/?] <set>: *(Enter an option.)*

COMMAND LINE OPTIONS

Copy copies a layout to create a new layout.

Delete deletes a layout; the Model tab cannot be deleted.

New creates a new layout tab, automatically generating the name for the layout (default = first unused tab in the form Layout*n*), which you may override.

Template displays the Select File dialog box, which allows you to select a *.dwg* drawing or *.dwt* template file to use as a template for a new layout. If the file has layouts, it displays the Insert Layout(s) dialog box.

Rename renames a layout.

SAveas saves the layouts in a drawing template (*.dwt*) file. The last current layout is used as the default for the layout to save.

Set makes a layout current.

? lists the layouts in the drawing in a format similar to the following:

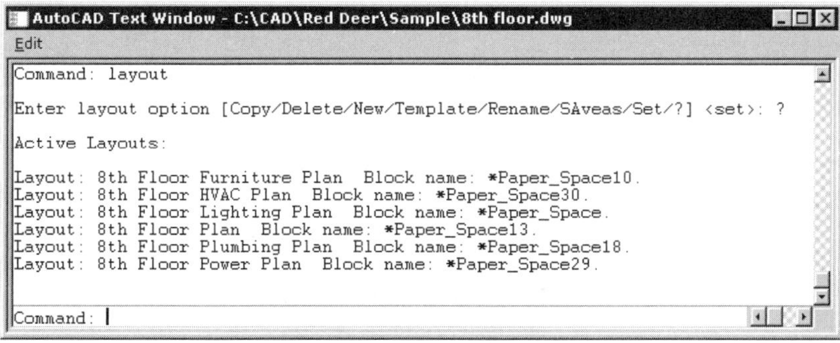

SHORTCUT MENU

Right-click any layout tab:

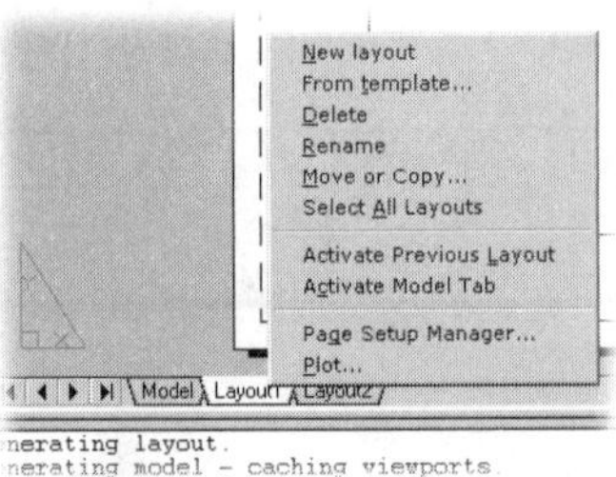

New layout creates a new layout with the default name of **Layout***n*.

From template displays the Select File and Insert Layout dialog boxes.

Delete deletes the selected layout; displays a warning dialog box:

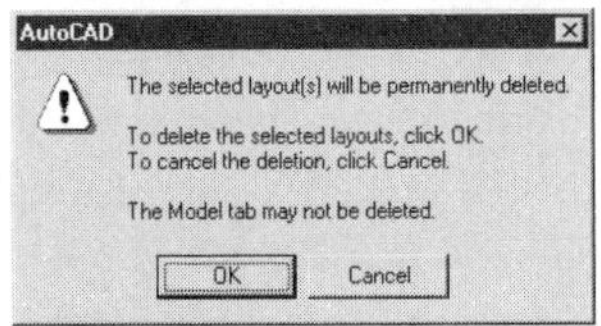

Rename Displays the Rename Layout dialog box:

Move or Copy displays the Move or Copy dialog box.

Select All Layouts selects all layouts.

Activate Previous Layout returns to the previously-accessed layout, useful when a drawing has many layouts (*new to AutoCAD 2005*).

Activate Model Tab returns to the Model tab (*new to AutoCAD 2005*).

Page Setup displays the Page Setup dialog box; see the **PageSetup** command.

Plot displays the Plot dialog box; see the **Plot** command.

Insert Layout(s) dialog box

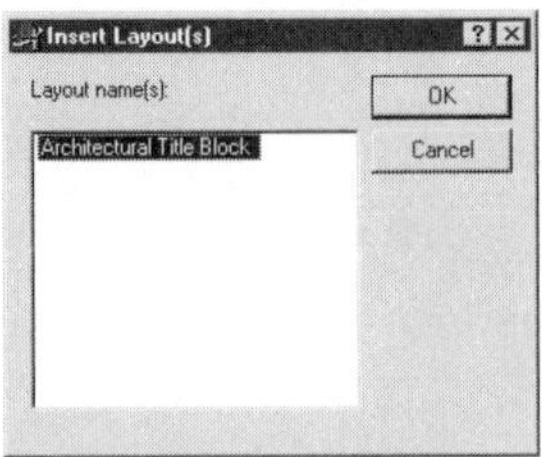

Layout names(s) lists the names of layouts found in the selected template or drawing; you may select more than one layout at a time by holding down the CTRL key.

OK adds the selected layouts to the current drawing.

Cancel dismisses the dialog box and cancels the command.

Move or Copy dialog box

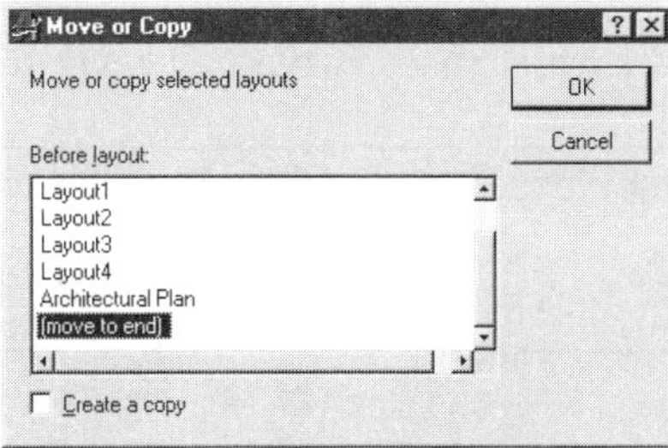

Before layout selects a layout to appear before the current layout.

Move to end moves the current layout to the end of layouts.

☐ **Create a copy** makes copies of selected layouts.

RELATED COMMAND

LayoutWizard creates and deletes paper space layouts via wizard.

TIPS

- "Layout" is the new name for paper space as of AutoCAD 2000.
- A layout name can be up to 255 characters long; the first 31 characters are displayed in the tab.
- To switch between layouts, click tabs located below the drawing:

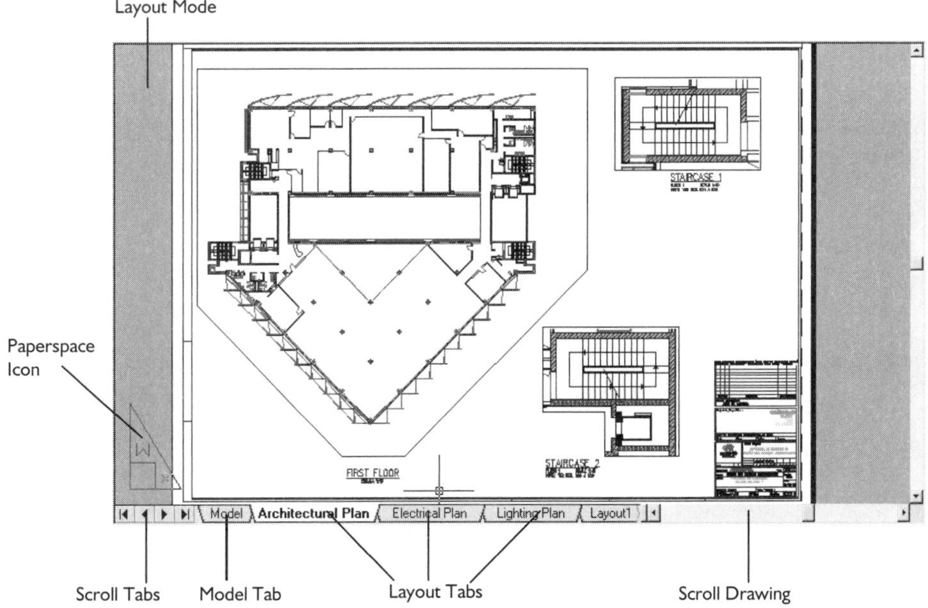

- The **Model** tab cannot be deleted, renamed, moved, or copied.
- A drawing can contain up to 255 layouts.

LayoutWizard

2000 Creates and deletes paper space layouts via a wizard.

Command	Alias	Ctrl+	F-key	Alt+	Menu Bar	Tablet
layoutwizard	...	...	...	**ILW**	**Insert** ↳**Layout** ↳**Layout Wizard**	...
				TZC	**Tool** ↳**Wizards** ↳**Create Layout**	

Command: layoutwizard

Displays dialog box:

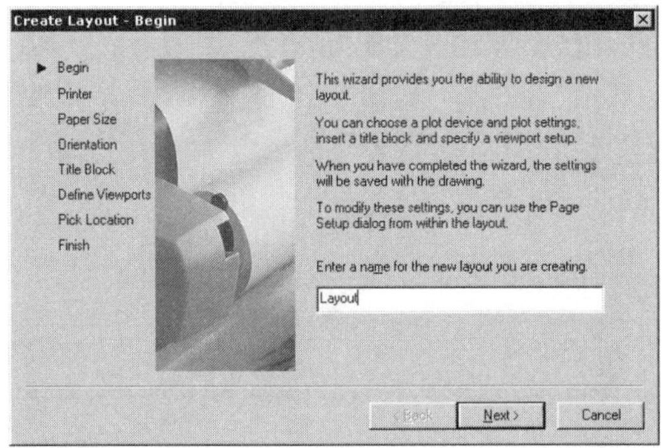

DIALOG BOX OPTIONS

Enter a name specifies the name for the layout.

Buttons

Back displays the previous dialog box.

Next displays the next dialog box.

Cancel cancels the command.

Begin page

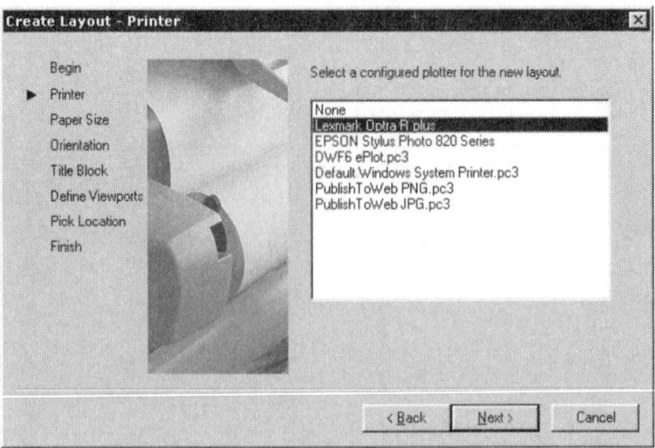

Select a configured plotter selects a printer or plotter to output the layout.

Paper Size page

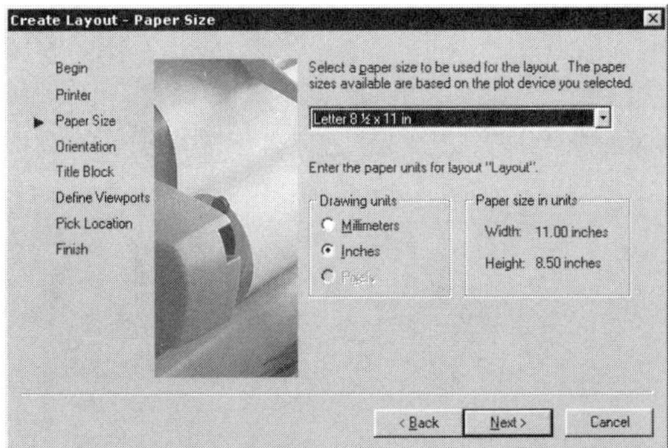

Select a paper size... selects a size of paper supported by the output device.

Enter the paper units

⊙ **Millimeters** measures paper size in metric units.

O **Inches** measures paper size in Imperial units.

O **Pixels** measures paper size in dots per inch.

Orientation page

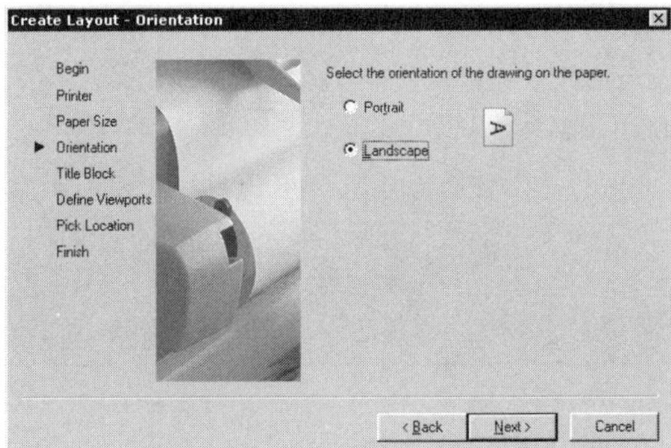

Select the orientation

- **Portrait** plots the drawing vertically.
- **Landscape** plots the drawing horizontally.

Title Block page

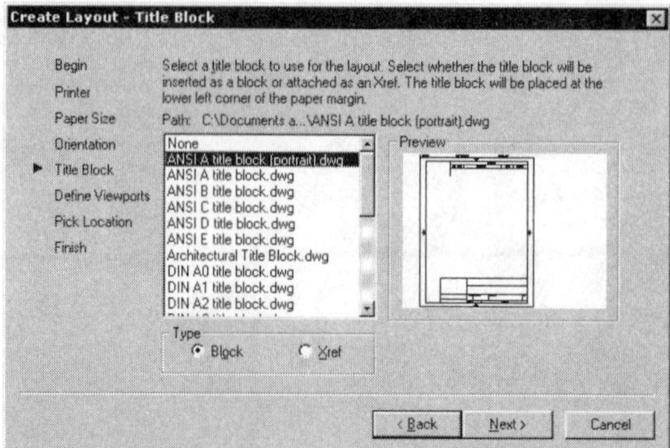

Select a title block... specifies a title border for the drawing as a block or an xref:

⊙**Block** inserts the title border drawing.

○**Xref** references the title border drawing.

Define Viewports page

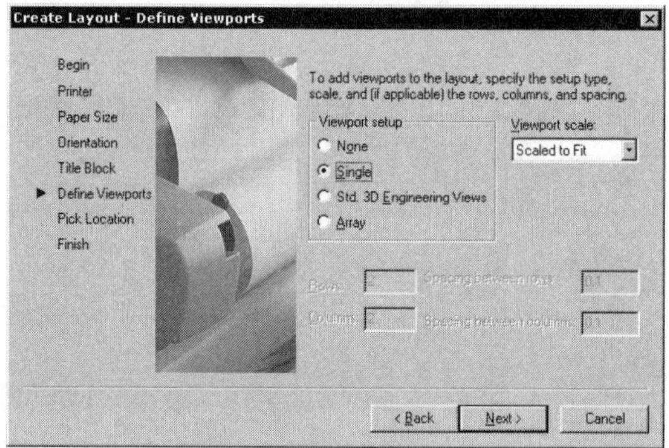

Viewport Setup

○ **None** creates no viewport.

⊙ **Single** creates a single paper space viewport.

○ **Std. 3D Engineering Views** creates top, front, side, and isometric views.

○ **Array** creates a rectangular array of viewports.

Viewport scale

- **Scaled to Fit** fits the model to the viewport.
- *mm*:*nn* specifies a scale factor, ranging from 1:1 to 1/128":1'0".

Rows specifies the number of rows for arrayed viewports.

Columns specifies the number of columns for arrayed viewports.

Spacing between rows specifies the vertical distance between viewports.

Spacing between columns specifies the horizontal distance between viewports.

Pick Location page

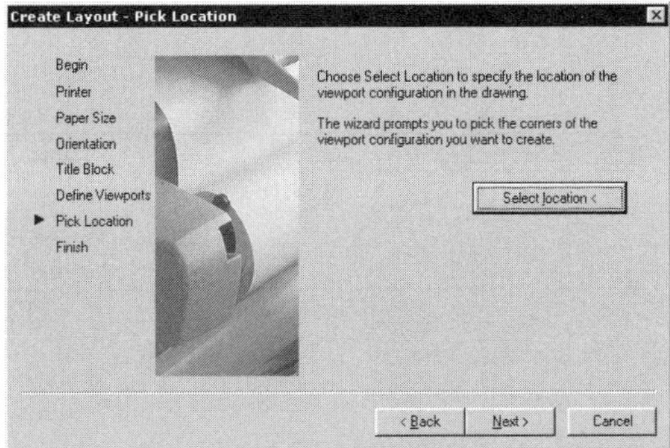

Select location specifies the corners of a rectangle holding the viewports. AutoCAD prompts:

Regenerating layout.
Specify first corner: *(Pick a point.)*
Specify opposite corner: *(Pick a point.)*
Regenerating model.

AutoCAD returns to the Create Layout dialog box.

Finish page

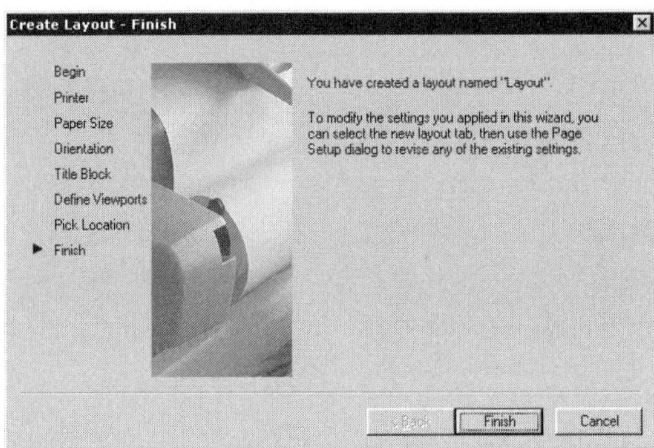

Finish exits the dialog box and creates the layout.

RELATED COMMANDS

Layout creates a layout on the command line.

RELATED SYSTEM VARIABLES

CTab contains the name of the current tab.

LayTrans

2002 Translates layer names *(short for LAYer TRANSlation).*

Command	Alias	Ctrl+	F-key	Alt+	Menu Bar	Tablet
laytrans	...	...	...	**TSL**	**Tools** ↳**CAD Standards** ↳**Layer Translator**	...

Command: laytrans

Displays dialog box:

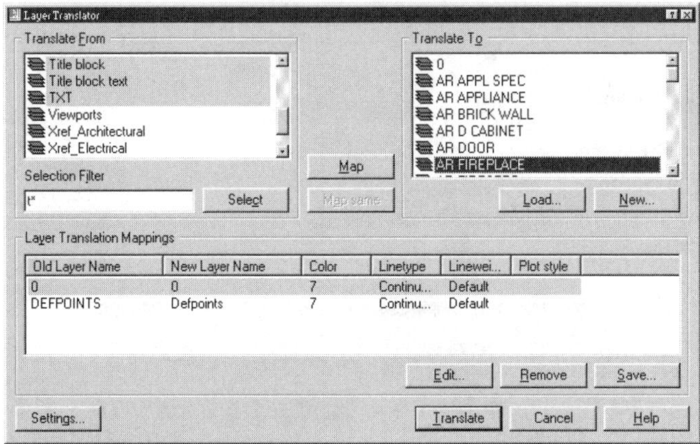

DIALOG BOX OPTIONS

Translate From options

Translate From lists the names of layers in the current drawing; icons indicate whether the layer is being used (being referenced):

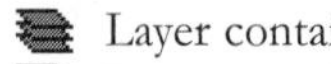

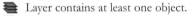

Layer contains at least one object.

Layer contains no objects, and can be purged.

Selection Filter specifies a subset of layer names; see Wildcard Metacharacters in the **Layer** command.

Select highlights the layer names that match the selection filter.

Map maps the selected layer(s) in the Translate From column to the selected layer in the Translate To column.

Map same maps layers automatically with the same name.

Translate To options

Translate To lists layer names in the drawing opened with the Load button.

Load accesses the layer names in another drawing via the Select Drawing File dialog box.

New creates a new layer via the New Layer dialog box.

Layer Translation Mappings options

Edit edits the linetype, color, lineweight, and plot style settings via the Edit Layer dialog box; identical to the New Layer dialog box.

Remove removes the selected layer from the list.

Save saves the matching table to a *.dws* (drawing standard) file.

Settings specifies translation options via the Settings dialog box.

Translate changes the names of layers, as specified by the Layer Translation Mappings list.

New Layer dialog box

Identical to the ***Edit Layer*** *dialog box.*

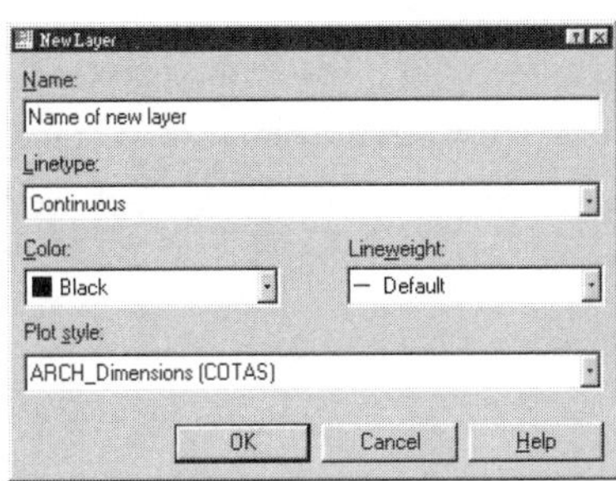

Name specifies the name of the layer, up to 255 characters long.

Linetype selects a linetype from those available in the drawing.

Color selects a color; select **Other** for the Select Color dialog box.

Lineweight selects a lineweight.

Plot style selects a plot style from those available in the drawing; this option is not available if plot styles have not been enabled in the drawing.

Settings dialog box

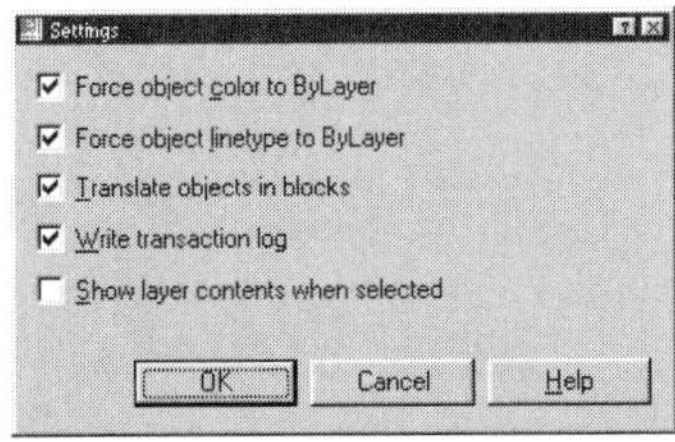

☑ **Force object color to Bylayer** forces every translated layer to take on color Bylayer.

☑ **Force object linetype to Bylayer** forces every translated layer to take on linetype Bylayer.

☑ **Translate objects in blocks** forces objects in blocks to take on new layer assignments.

☑ **Write transaction log** writes the results of the translation to a *.log* file, using the same filename as the drawing. When command is complete, AutoCAD reports:

`Writing transaction log to` *`filename`*`.log.`

☐ **Show layer contents when selected** lists the names of selected layers only in the Translate From list.

RELATED COMMANDS

Standards creates the standards for checking drawings.

CheckStandards checks the current drawing against a list of standards.

Layer creates and sets layers and modes.

RELATED FILES

.dws drawing standard file; saved in DWG format.

.log log file recording layer translation; saved in ASCII format.

TIPS

- You can purge unused layers (those prefixed by a white icon) within the Layer Translator dialog box:

  1. Right click any layer name in the **Translate From** list.

  2. Select **Purge Layers**. The layers are removed from the drawing.

- You can load layers from more than one drawing file; duplicate layer names are ignored.

Leader

Rel.13 Draws leader lines with one or more lines of text.

Command	Alias	Ctrl+	F-key	Alt+	Menu Bar	Tablet
leader	lead	...	...	...	...	R7

Command: leader

Specify leader start point: *(Pick a starting point, 1.)*

Specify next point: *(Pick the shoulder point, 2.)*

Specify next point or [Annotation/Format/Undo] <Annotation>: *(Pick another point, 3, or enter an option.)*

Specify next point or [Annotation/Format/Undo] <Annotation>: *(Press* **ENTER** *to specify the text. 4.)*

Enter first line of annotation text or <options>: *(Enter text, 5, or press* **ENTER** *for options.)*

Enter next line of annotation text: *(Press* **ENTER***.)*

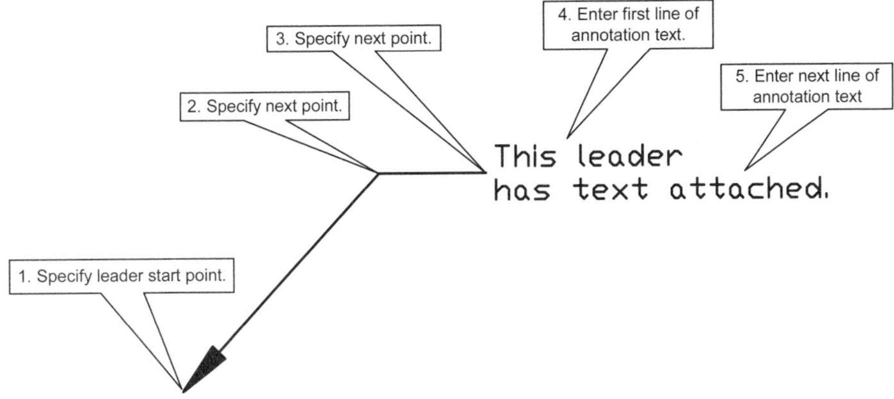

COMMAND LINE OPTIONS

Specify start point specifies the location of the arrowhead.

Specify next point positions the leader line's vertex.

Undo undoes the leader line to the previous vertex.

Format options

Enter leader format option [Spline/STraight/Arrow/None] <Exit>: *(Enter an option.)*

Spline draws the leader line as a NURBS (short for non uniform rational Bezier spline) curve.

STraight draws a straight leader line (default).

Arrow draws the leader with an arrowhead (default).

None draws the leader with no arrowhead.

Annotation options

Enter first line of annotation text or <options>: *(Press* **ENTER**.*)*

Enter an annotation option [Tolerance/Copy/Block/None/Mtext] <Mtext>: *(Enter an option.)*

Enter first line of annotation text specifies the leader text.

Tolerance places one or more tolerance symbols; see the **Tolerance** command.

Copy copies text from another part of the drawing.

Block places a block; see the **-Insert** command.

None specifies no annotation.

MText displays the Text Formatting toolbar; see the **MText** command.

RELATED DIM VARIABLES

DimAsz specifies the size of the arrowhead and the hookline.

DimBlk specifies the type of arrowhead.

DimClrd specifies the color of the leader line and the arrowhead.

DimGap specifies the gap between hookline and annotation (gap between box and text).

DimScale specifies the overall scale of the leader.

TIPS

- This command draws several types of leader:

- The text in a leader is an mtext (multiline text) object.
- Use the \P metacharacter to create line breaks in leader text.
- Autodesk recommends using the **QLeader** command, which has a dialog box for settings and more options.

Lengthen

Rel.13 Lengthens and shortens open objects by several methods.

Command	Alias	Ctrl+	F-key	Alt+	Menu Bar	Tablet
lengthen	**len**	...	...	**MG**	**Modify** **↳Lengthen**	**W14**

Command: lengthen
Select an object or [DElta/Percent/Total/DYnamic]: *(Select an open object.)*
Current length: ***n.nnnn***

COMMAND LINE OPTIONS

Select an object displays length and included angle; does not change the object.

DElta option
Enter delta length or [Angle] <0.0000>: *(Enter a value, or type* **A***.)*
Specify second point: *(Pick a point.)*
Select an object to change or [Undo]: *(Select an open object, or type* **U***.)*

Enter delta length changes the length by the specified amount.

Angle changes the angle by the specified value.

Undo undoes the most-recent lengthening operation.

Percent option
Enter percentage length <100.0000>: *(Enter a value.)*
Select an object to change or [Undo]: *(Select an open object, or type* **U***.)*

Enter percent length changes the length to a percentage of the current length.

Undo undoes the most-recent lengthening operation.

Total option
Specify total length or [Angle] <1.0000)>: *(Enter a value, or type* **A***.)*
Select an object to change or [Undo]: *(Select an open object, or type* **U***.)*

Specify total length changes the length by an absolute value.

Angle changes the angle by the specified value.

Undo undoes the most-recent lengthening operation.

DYnamic option
Select an object to change or [Undo]: *(Select an open object, or type* **U***.)*
Specify new end point: *(Pick a point.)*

Specify new end point dynamically changes the length by dragging.

Undo undoes the most-recent lengthening operation.

TIPS

- **Lengthen** command only works with open objects, such as lines, arcs, and polylines; it does not work with closed objects, such as circles, polygons, and regions.
- **DElta** option changes the length or angle using the following measurements: (1) the distance from the endpoint of the selected object to the pick point; or (2) the incremental length measured from the endpoint of the angle.

Light

Rel.12 Places several types of lights for use by **Render**.

Command	Alias	Ctrl+	F-key	Alt+	Menu Bar	Tablet
light	...	...	...	**VEL**	**View** ↳**Render** ↳**Light**	**O1**

Command: light

Displays dialog box:

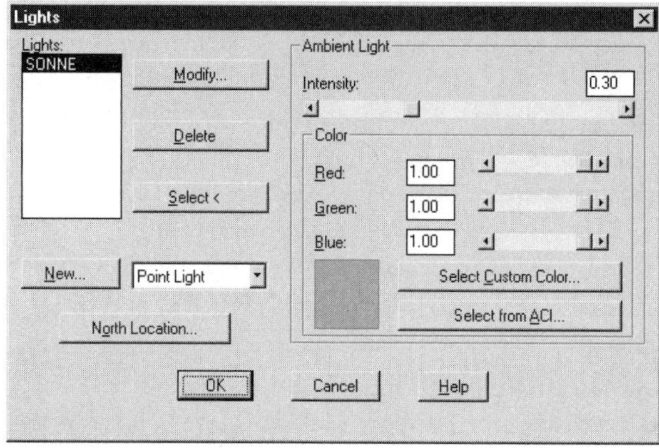

DIALOG BOX OPTIONS

Lights lists the currently-defined lights in the drawing.

Modify modifies an existing light in the drawing; displays Modify Light dialog box.

Delete deletes the selected light.

Select selects a light from the drawing.

New creates a new point, spot, or direct light; displays New Light dialog box.

North Location selects the direction for North; displays North Location dialog box.

Ambient Light option

Ambient Light Intensity adjusts the intensity of ambient light from 0 (dark) to 1.0 (bright).

Color options

Red adjusts the level of red from 0 (black) to 1.0 (full red).

Green adjusts the level of green from 0 (black) to 1.0 (full green).

Blue adjusts the level of blue from 0 (black) to 1.0 (full blue).

Select Custom Color displays Windows' Color dialog box.

Select from ACI displays AutoCAD's Select Color dialog box.

North Location dialog box

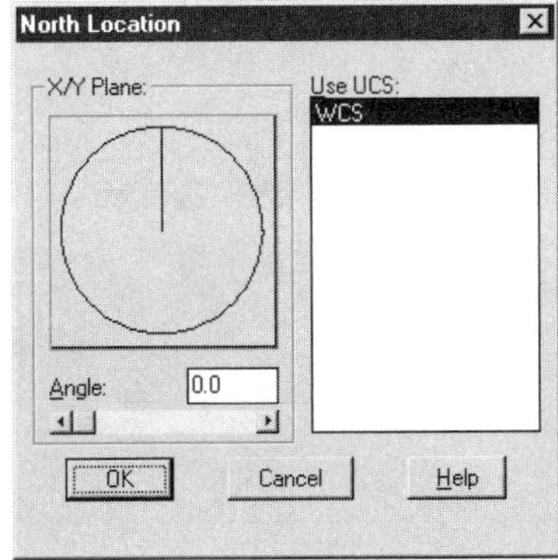

XY Plane options

Angle selects the angle from icon; enter a number, or drag the slider bar.

Use UCS selects a named UCS (user-defined coordinate system).

New Point Light dialog box

Point lights must be positioned; they radiate light in all directions.

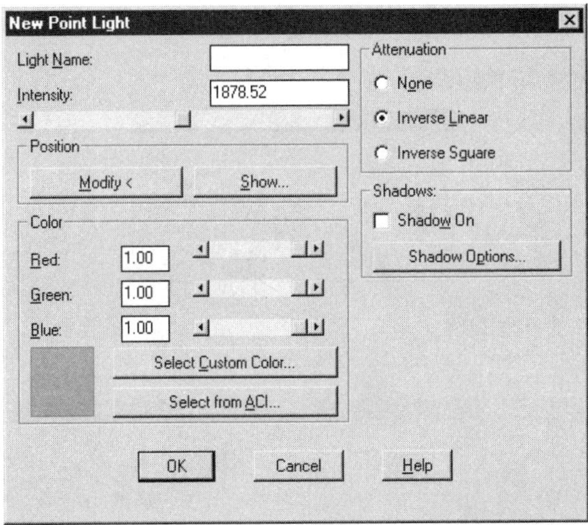

Light Name names the light; maximum = 8 characters, no spaces.

Intensity specifies the intensity of the light, from 0 (turned off) to 31.33.

Attenuation options

○ **None** specifies a light that does not diminish in intensity with distance.

⊙ **Inverse Linear** specifies a light whose intensity decreases with distance.

○ **Inverse Square** specifies a light whose intensity decreases with the square of the distance.

Position options

Modify changes the location of the light.

Show displays Show dialog box.

Shadows options

☐ **Shadows On** turns on shadow casting.

Shadow Options displays Shadow Options dialog box.

Shadow Options dialog box

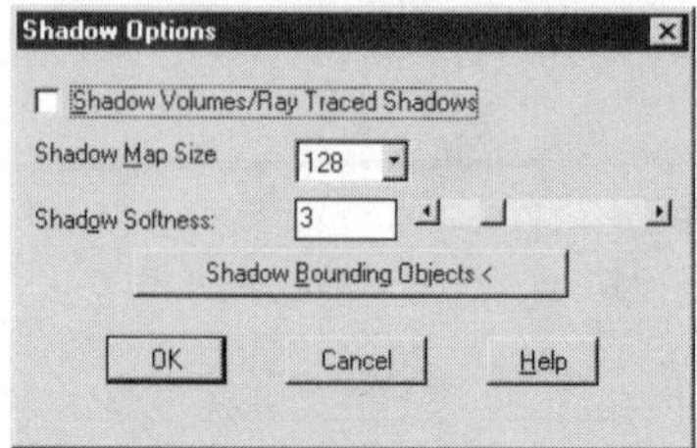

☐ **Shadow Volumes/Raytrace Shadows** creates volumetric shadows; raytracer creates ray-traced shadows; disables shadow map.

Shadow Map Size specifies the size of one side of the shadow map; ranges from 64 to 4096 pixels; larger values give more accurate shadows.

Shadow Softness specifies the number of pixels at the shadow's edge blended with the underlying image; ranges from 1 to 10.

Shadow Bounding Objects selects objects to clip the shadow maps.

Color selects a color for the light.

Show Light Position dialog box

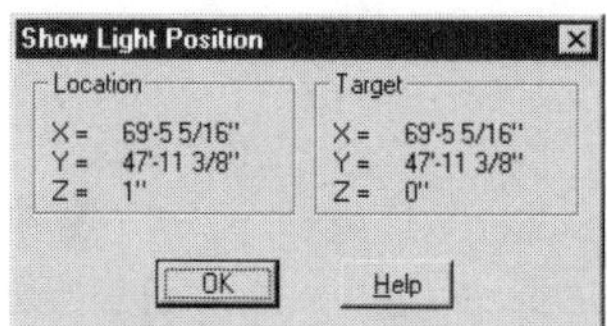

New Distant Light dialog box

Distant lights must be positioned; they radiate light in parallel rays in specified directions.

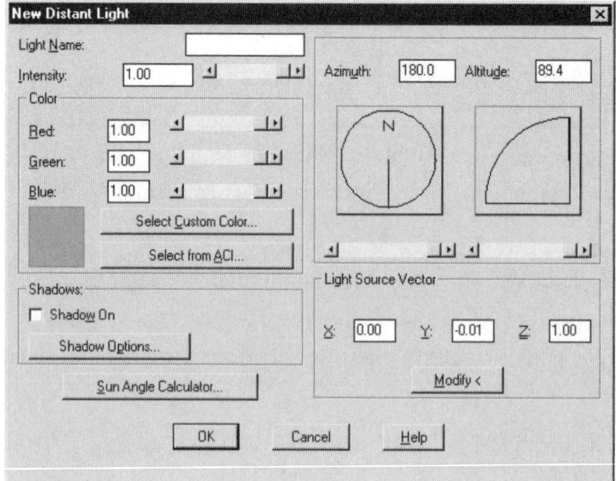

Name names the light; maximum = 8 characters.

Intensity specifies the intensity of the light; 0 is off.

Color specifies the color of the light.

Shadows creates shadows.

Azimuth sets light's position between -180 and 180 degrees.

Altitude sets an angle for the light between 0 and 90 degrees.

Light Source Vector options

X specifies the vector x coordinate; ranges from -1.0 to 1.0.

Y specifies the vector y coordinate; ranges from -1.0 to 1.0.

Z specifies the vector z coordinate; ranges from -1.0 to 1.0.

Modify changes the position of the light.

Sun Angle Calculator displays Sun Angle Calculator dialog box.

Sun Angle Calculator dialog box

This calculator eliminates the need to specify the azimuth, altitude, and light source vectors for a distant light.

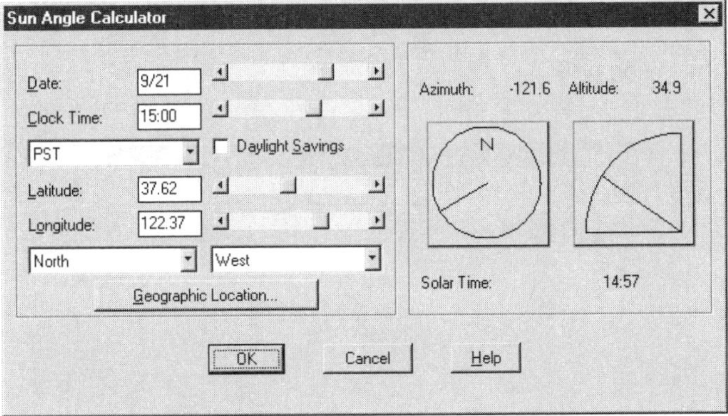

Date displays today's date or any date of the year.

Clock Time displays the current time or any time of day.

Latitude displays the latitude on earth.

Longitude displays the longitude.

Geographic Locator displays dialog box (see below).

Geographic Locator dialog box

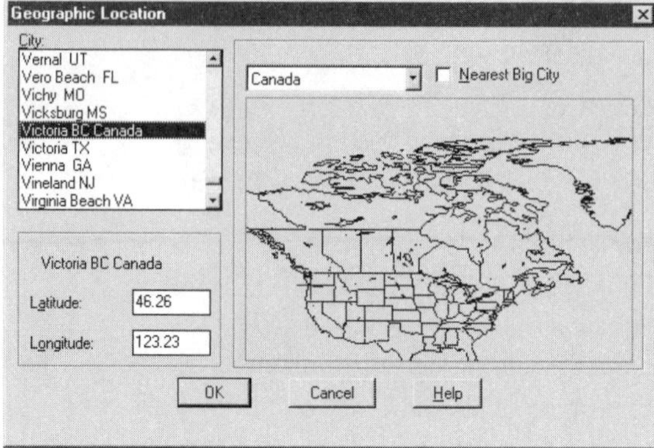

City selects the name of a city.

Latitude displays the latitude of the city.

Longitude displays the longitude of the city.

Nearest Big City selects a city from its list closest to your pick point.

New Spotlight dialog box

Spotlights radiate a cone of light, from the light to a spot centered on the target position.

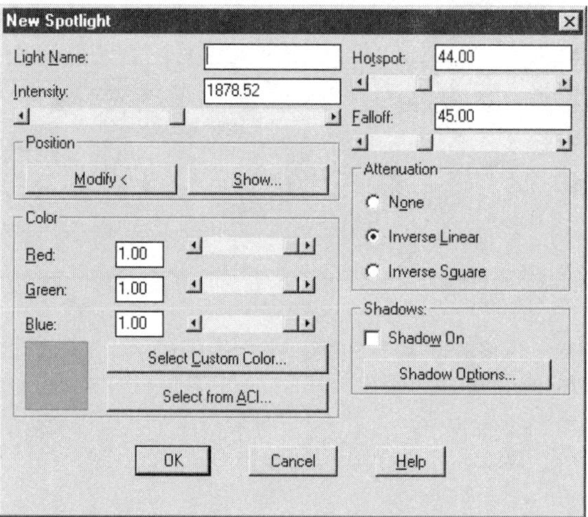

Light Name names the light; maximum = 8 characters, no spaces.

Intensity specifies the intensity of the light, from 0 (*turned off*) to 31.33.

Attenuation options

○**None** specifies that the light's intensity does not diminish with distance.

⊙**Inverse Linear** specifies that the light's intensity decreases with distance.

○**Inverse Square** specifies that the light's intensity decreases with the square of the distance.

Position options

Modify changes the location of the light.

Show displays the Show Light Position dialog box.

Shadows options

☐ **Shadows On** turns on shadow casting.

Shadow Options displays the Shadows Options dialog box.

Color specifies the color of the light.

RELATED COMMANDS

Render renders the drawing.

Scene specifies the lights and view to use in rendering.

RELATED FILES

In \autocad 2005\support folder:

direct.dwg is the direct light block.

overhead.dwg is the overhead drawing block.

sh_spot.dwg is the spotlight drawing block.

Direct (left), Overhead (center), and Spotlight (right).

TIPS

- This command works in model space only.
- When the drawing has no lights defined, AutoCAD assumes ambient light.
- Ambient light ensures every object in the scene has illumination; ambient light is an omnipresent light source.
- Set ambient light to 0 to turn off for night scenes.
- While it is not necessary to define any lights to use the **Render** command, a light must be included in a **Scene** definition for the **Render** command to make use of the light.
- In a spotlight, the light beam travels from the *light location* (light block) to the *light target.*
- Place one distant light to simulate the Sun; distant lights have parallel light beams with constant intensity.
- Place several point lights as light bulbs (*lamps*); a point light beams light in all directions, with inverse linear, inverse square, or constant intensity.
- Spotlights emit light in a cone.

DEFINITIONS

Constant light — attenuation is 0; default intensity is 1.0.

Inverse linear light — light strength decreases to ½-strength two units of distance away, and ¼-strength four units away; default intensity is ½ extents distance.

Inverse square light— light strength decreases to ¼-strength two units away, and $^1/_8$-strength four units away; default intensity is ½ the square of the extents distance.

Extents distance — distance from minimum lower-left coordinate to the maximum upper-right.

RGB color — three primary colors — red, green, blue — shaded from black to white.

HLS color — changes colors by hue (color), lightness, and saturation (less gray).

Hotspot — brightest cone of light; beam angle ranges from 0 to 160 degrees (default: 45 degrees).

Falloff — angle of the full light cone; field angle ranges 0 to 160 degrees (default: 45 degrees).

'Limits

V. 1.0 Defines the 2D limits in the WCS for the grid markings and the **Zoom All** command; optionally prevents specifying points outside of limits.

Command	Alias	Ctrl+	F-key	Alt+	Menu Bar	Tablet
limits	...	...	...	OA	Format ↳Drawing Limits	V2

Command: limits

In model space:

Reset Model space limits:

In paper space:

Reset Paper space limits:

In either model or paper space:

Specify lower left corner or [ON/OFF] <0.0000,0.0000>: *(Pick a point, or type* **ON** *or* **OFF***.)*

Specify upper right corner <12.0000,9.0000>: *(Pick a point.)*

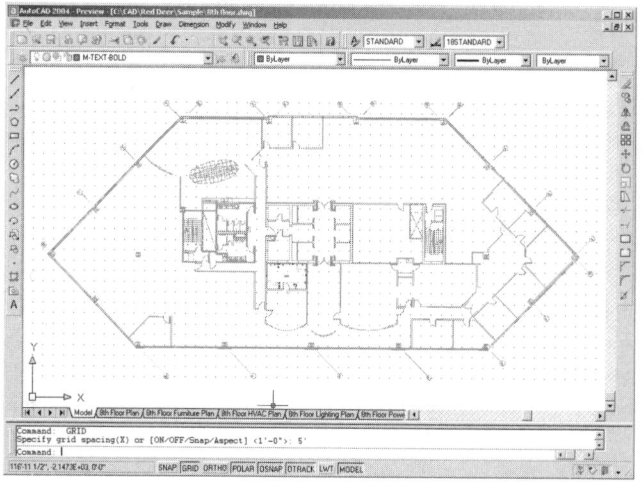

COMMAND LINE OPTIONS

OFF turns off limits checking.

ON turns on limits checking.

ENTER retains limits values.

RELATED COMMANDS

Grid displays grid dots, which are bounded by limits.

Zoom displays the drawing's extents or limits with the **All** option.

RELATED SYSTEM VARIABLES

LimCheck toggles the limit's drawing check.

LimMin specifies the lower-right 2D coordinates of current limits.

LimMax specifies the upper-left 2D coordinates of current limits.

Line

V. 1.0 Draws straight 2D and 3D lines.

Command	Alias	Ctrl+	F-key	Alt+	Menu Bar	Tablet
line	l	...	...	DL	Draw ↳Line	J10

Command: line
Specify first point: *(Pick a starting point.)*
Specify next point or [Undo]: *(Pick another point, or type* **U***.)*
Specify next point or [Undo]: *(Pick another point, or type* **U***.)*
Specify next point or [Close/Undo]: *(Pick another point, or enter an option.)*
Specify next point or [Close/Undo]: *(Press* **ENTER** *to end the command.)*

Single Segment Line:

Multi Segment Line:

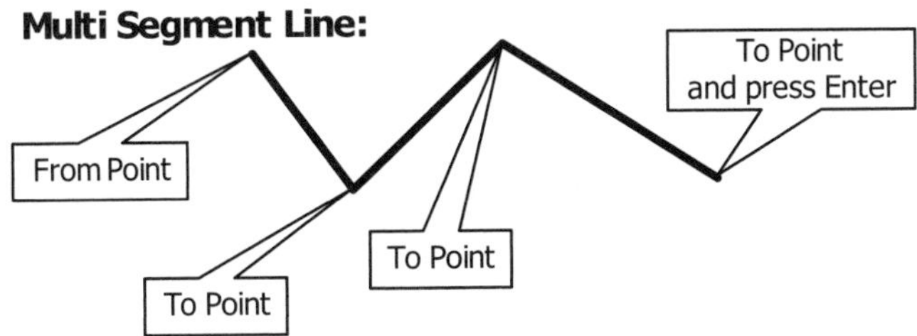

Closed, MultiSegment Line:

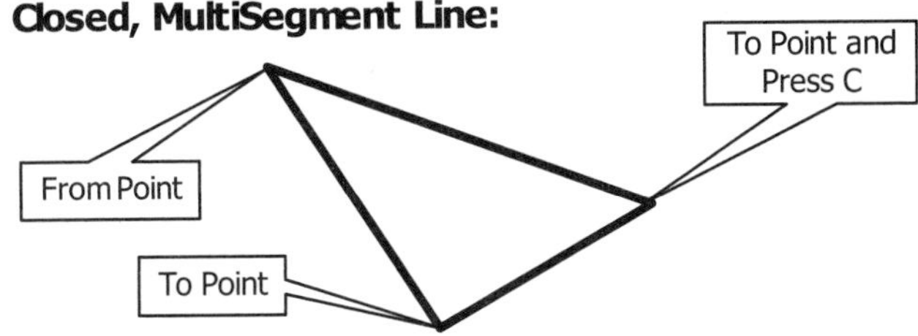

COMMAND LINE OPTIONS

Close closes the line from the current point to the starting point.

Undo undoes the last line segment drawn.

ENTER continues the line from the last endpoint at the 'From point' prompt; terminates the **Line** command at the 'To point' prompt.

RELATED COMMANDS

MLine draws up to 16 parallel lines.

PLine draws polylines and polyline arcs.

Trace draws lines with width.

Ray creates semi-infinite construction lines.

XLine creates infinite construction lines.

RELATED SYSTEM VARIABLES

Elevation specifies the distance above (or below) the x,y plane a line is drawn.

Lastpoint specifies the last-entered coordinate triple (x,y,z-coordinate).

Thickness determines the thickness of the line.

TIPS

- To draw 2D lines, enter x,y coordinate pairs; the z coordinate takes on the value of the **Elevation** system variable.
- To draw 3D lines, enter x,y,z coordinate triples.
- When system variable **Thickness** is not zero, the line has thickness, which makes it a plane perpendicular to the current UCS.

'Linetype

V. 2.0 Loads linetype definitions into the drawing, creates new linetypes, and sets the working linetype.

Commands	Aliases	Ctrl+	F-key	Alt+	Menu Bar	Tablet
linetype	**lt**	...	...	**ON**	**Format**	**U3**
	ltype				↳**Linetype**	
	ddltype					
-linetype	**-lt**					
	-ltype					

Command: linetype

Displays dialog box:

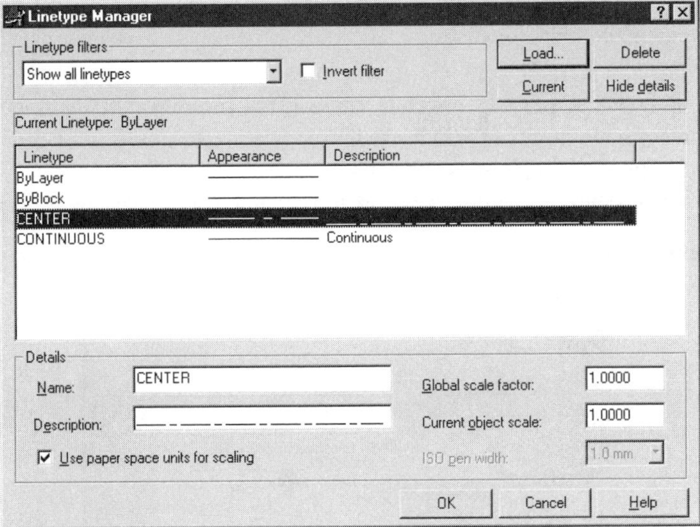

DIALOG BOX OPTIONS

Linetype filters displays the following groups of linetypes:

- **Show all linetypes** displays all linetypes defined in the current drawing.
- **Show all used linetypes** displays all linetypes being used.
- **All xref dependent linetypes** displays linetypes in externally-referenced drawings.

☐ **Invert filter** inverts the display of layer names; for example, when **Show all used linetypes** is selected, the **Invert filter** option displays all linetypes not used in the drawing.

Buttons

Loads displays the Load or Reload Linetypes dialog box.

Current sets the selected layer as the current layer.

Delete purges the selected linetypes; some linetypes cannot be deleted, as described by the warning dialog box:

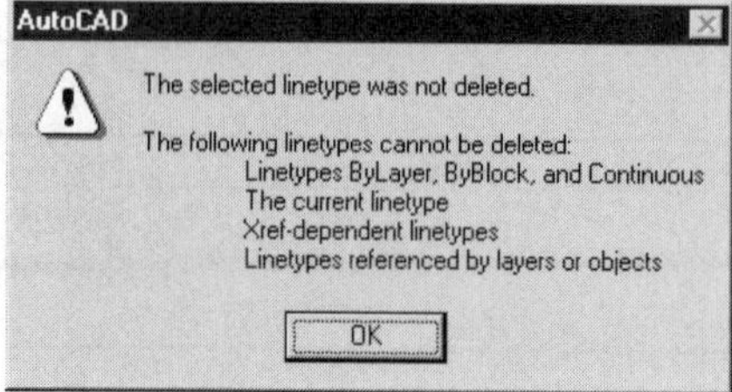

Show/Hide Details toggles the display of the Details portion of the Linetype Properties Manager dialog box.

Details options

Name names the selected linetype.

Description displays the description associated with the linetype.

☑ **Use paper space units for scaling** specifies that paper space linetype scaling is used, even in model space.

Global scale factor specifies the scale factor for all linetypes in the drawing.

Current object scale specifies the individual object scale factor for all subsequently-drawn linetypes, multiplied by the global scale factor.

ISO pen width applies standard scale factors to ISO (international standards) linetypes.

Load or Reload Linetypes dialog box

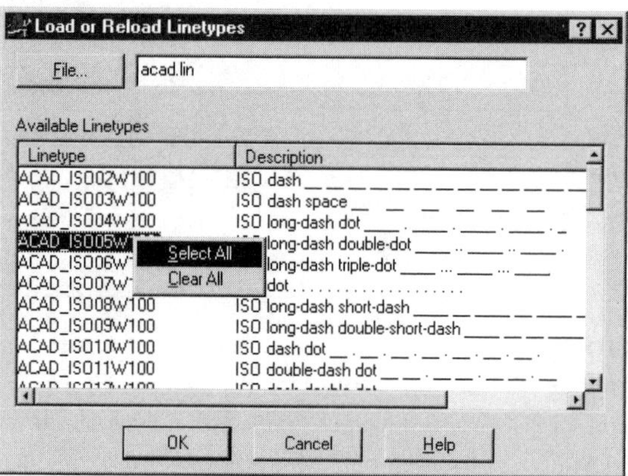

File names the *.lin* linetype definition file.

SHORTCUT MENU

Right-click any linetype name in the ***Linetype Manager*** *dialog box:*

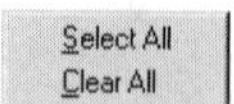

Select All selects all linetypes.

Clear All selects no linetypes.

-LINETYPE Command

Command: -linetype

Enter an option [?/Create/Load/Set]: *(Enter an option.)*

COMMAND LINE OPTIONS

Create creates new user-defined linetypes.

Load loads linetypes from linetype definition (*.lin*) files.

Set sets the working linetype.

? lists the linetypes loaded into the drawing.

RELATED COMMANDS

Change changes objects to a new linetype; changes linetype scale.

ChProp changes objects to a new linetype.

LtScale sets the scale of the linetype.

Rename changes the name of the linetype.

RELATED SYSTEM VARIABLES

CeLtype specifies the current linetype setting.

LtScale specifies the current linetype scale.

PsLtScale specifies the linetype scale relative to paper scale.

PlineGen controls how linetypes are generated for polylines.

TIPS

- The only linetypes defined initially in a new AutoCAD drawing are:

  Continuous draws unbroken lines.

  Bylayer specifies linetype by the layer setting.

  Byblock specifies linetypes by the block definition.

- Linetypes must be loaded from *.lin* definition files before being used in a drawing.
- When loading one or more linetypes, it is faster to load all linetypes, and then use the **Purge** command to remove the linetype definitions that have not been used in the drawing.
- As of AutoCAD Release 13, objects can have independent linetype scales.

RELATED FILE

- The following standard linetypes are in *\autocad 2005\support\acad.lin*:

ACAD_ISO02W100	ISO dash __ __ __ __ __ __ __ __ __ __ __ __ __
ACAD_ISO03W100	ISO dash space __ __ __ __ __ __
ACAD_ISO04W100	ISO long-dash dot ____ . ____ . ____ . ____ . _
ACAD_ISO05W100	ISO long-dash double-dot ____ .. ____ .. ____ .
ACAD_ISO06W100	ISO long-dash triple-dot ____ ... ____ ... ____
ACAD_ISO07W100	ISO dot . . . .
ACAD_ISO08W100	ISO long-dash short-dash ____ __ ____ __ ____ _
ACAD_ISO09W100	ISO long-dash double-short-dash ____ __ __ ____
ACAD_ISO10W100	ISO dash dot __ . __ . __ . __ . __ . __ .
ACAD_ISO11W100	ISO double-dash dot __ __ . __ __ . __ __ . __
ACAD_ISO12W100	ISO dash double-dot __ . . __ . . __ . . __ . .
ACAD_ISO13W100	ISO double-dash double-dot __ __ . . __ __ . .
ACAD_ISO14W100	ISO dash triple-dot __ . . . __ . . . __ . . .
ACAD_ISO15W100	ISO double-dash triple-dot __ __ . . . __ __ .
BATTING	Batting SSSSSSSSSSSSSSSSSSSSSSSSSSSSSSSSSSSSSSSSSS
BORDER	Border __ __ . __ __ . __ __ . __ __ . __ __ .
BORDER2	Border (.5x) __.__.__.__.__.__.__.__.__.__.
BORDERX2	Border (2x) ____ ____ . ____ ____ . __
CENTER	Center ____ _ ____ _ ____ _ ____ _ ____ _ ____
CENTER2	Center (.5x) ___ _ ___ _ ___ _ ___ _ ___ _ ___
CENTERX2	Center (2x) ________ __ ________ __ _____
DASHDOT	Dash dot __ . __ . __ . __ . __ . __ . __
DASHDOT2	Dash dot (.5x) _._._._._._._._._._._._._._.
DASHDOTX2	Dash dot (2x) ____ . ____ . ____ . ___
DASHED	Dashed __ __ __ __ __ __ __ __ __ __ __ __ __ _
DASHED2	Dashed (.5x) _ _ _ _ _ _ _ _ _ _ _ _ _ _ _ _
DASHEDX2	Dashed (2x) ____ ____ ____ ____ ____ ___
DIVIDE	Divide ____ . . ____ . . ____ . . ____ . . ____
DIVIDE2	Divide (.5x) __..__..__..__..__..__..__..__.._
DIVIDEX2	Divide (2x) ________ . . ________ . . _
DOT	Dot . . . .
DOT2	Dot (.5x) ....
DOTX2	Dot (2x) . . . .
FENCELINE1	Fenceline circle ----0-----0----0-----0----0---
FENCELINE2	Fenceline square ----[]-----[]----[]-----[]----
GAS_LINE	Gas line ----GAS----GAS----GAS----GAS----GAS---
HIDDEN	Hidden __ __ __ __ __ __ __ __ __ __ __ __ __ __ _
HIDDEN2	Hidden (.5x) _ _ _ _ _ _ _ _ _ _ _ _ _ _ _ _ _
HIDDENX2	Hidden (2x) ____ ____ ____ ____ ____ ____ ____
HOT_WATER_SUPPLY	Hot water supply ---- HW ---- HW ---- HW ----
PHANTOM	Phantom ______ __ __ ______ __ __ ______
PHANTOM2	Phantom (.5x) ___ _ _ ___ _ _ ___ _ _ ___ _ _
PHANTOMX2	Phantom (2x) ____________ ____ ____ _
TRACKS	Tracks -|-|-|-|-|-|-|-|-|-|-|-|-|-|-|-|-|-|-|-
ZIGZAG	Zig zag /\/

List

V. 1.0 Lists information about selected objects in the drawing.

Command	Alias	Ctrl+	F-key	Alt+	Menu Bar	Tablet
list	**li**	...	...	**TYL**	**Tools**	**U8**
	ls				↳**Inquiry**	
					↳**List**	

Command: list
Select objects: *(Select one or more objects.)*
Select objects: *(Press* **ENTER** *to end object selection.)*

Sample output:

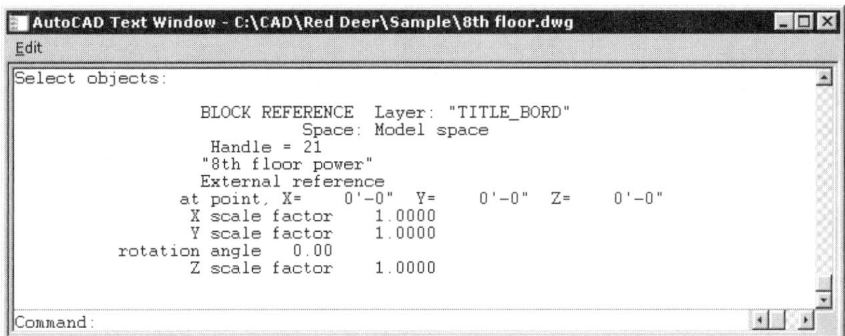

COMMAND LINE OPTIONS

ENTER continues the display.

ESC cancels the display.

F2 returns to graphics screen.

RELATED COMMANDS

Area calculates the area and perimeter of selected objects.

DbList lists information about *all* objects in the drawing.

MassProp calculates the properties of 2D regions and 3D solids.

TIPS

- The **List** command lists the following information only under certain conditions:

Information	Condition
Color	When not set BYLAYER.
Linetype	When not set BYLAYER.
Thickness	When not 0.
Elevation	When z coordinate is not 0.
Extrusion direction	When z axis differs from current UCS.

- Object handles are described by hexadecimal numbers.

Removed Command

ListURL was removed from AutoCAD 2000; it was replaced by **-Hyperlink**.

Load

V. 1.0 Loads SHX-format shape files into drawings.

Command	Alias	Ctrl+	F-key	Alt+	Menu Bar	Tablet
load	...	...	...	...	...	...

Command: load

Displays Load Shape File dialog box. Select an .shx file, and the click ***Open.***

COMMAND LINE Options

None.

RELATED AUTOCAD COMMAND

Shape inserts shapes into the current drawing.

RELATED FILES

***.shp** are source code for shape files.

***.shx** are compiled shape files.

In \autocad 2005\support folder:

gdt.shx and ***gdt.shp*** are geometric tolerance shapes used by the **Tolerance** command.

ltypeshp.shx and ***ltypeshp.shp*** are linetype shapes used by the **Linetype** command.

TIPS

- Shapes are more efficient than blocks, but are harder to create.
- The **Load** command cannot load *.shx* files meant for fonts. For example, AutoCAD complains, "gdt.shx is a normal text font file, not a shape file."
- Do not confuse this command with the AutoLISP **(load)** function, which loads *.lsp* files.

LogFileOn / LogFileOff

Rel.13 Turns on and off the command logging to *.log* files.

Command	Alias	Ctrl+	F-key	Alt+	Menu Bar	Tablet
logfileon	...	...	...	...	...	...
logfileoff	...	...	...	...	...	...

Command: logfileon

AutoCAD begins recording command-line text to the log file.

Command: logfileoff

AutoCAD stops recording command-line text, and closes the log file.

COMMAND LINE OPTIONS

None.

RELATED AUTOCAD COMMAND

CopyHist copies all command text from the Text window to the Clipboard.

RELATED SYSTEM VARIABLES

LogFileName specifies the name of the log file.

LogFilePath specifies the path for the log files for all drawings in a session.

LogFileMode toggles whether text window is written to log file:

LogFileMode	Meaning
0	Text not written to file (default).
1	Text written to file.

TIPS

- AutoCAD places a dashed line at the end of each log file session.
- If log file recording is left on, it resumes when AutoCAD is next loaded, which can result in very large log files.
- The default log file name is the same as the drawing name, and is stored in folder *C:\Documents and Settings\username\Local Settings\Application Data\Autodesk\AutoCAD 2005\R16.1\enu*. You can give the file a different folder and name with the **Options** command's **Files** tab, or with system variables **LogFileName** and **LogFilePath**.
- *Historical note:* In some early versions of AutoCAD, CTRL+Q meant "quick screen print," which output the current screen display to the printer. CTRL+Q reappeared in AutoCAD Release 14 to record command text to a file. As of AutoCAD 2004, the CTRL+Q shortcut quits AutoCAD, instead of toggling the log file — curious, given that there already is a keyboard shortcut, ALT+F4, that quits AutoCAD.

LsEdit

Rel.14 Edits the properties of landscape objects (*short for LandScape EDIT*).

Command	Alias	Ctrl+	F-key	Alt+	Menu Bar	Tablet
lsedit	...	...	...	VEE	View ↳Render ↳Landscape Edit	...

Command: lsedit

Select a landscape object: *(Select a single landscape object.)*

Displays dialog box:

DIALOG BOX OPTIONS

Height changes height of the object, by entering a new value or moving the slider bar.

Position moves the object to another position in the drawing.

Geometry options

⊙**Single Face** renders faster, but is less realistic.

○**Crossing Face** produces more realistic ray-traced shadows.

☑ **View Aligned** forces object always to face the camera.

RELATED COMMANDS

LsLib lets you add and remove raster images from the *render.lli* file.

LsNew places a landscape object in the drawing.

Render renders the landscape object.

TIP

- Landscape objects are rendered only when using the **Render** command's photoreal or photo ray trace options.

LsLib

Rel.14 Maintains libraries of landscape objects *(short for LandScape LIBrary).*

Command	Alias	Ctrl+	F-key	Alt+	Menu Bar	Tablet
lslib	...	...	...	**VEC**	**View** ↳**Render** ↳**Landscape Library**	...

Command: lslib

Select a landscape object: *(Select a single landscape object.)*

Displays dialog box:

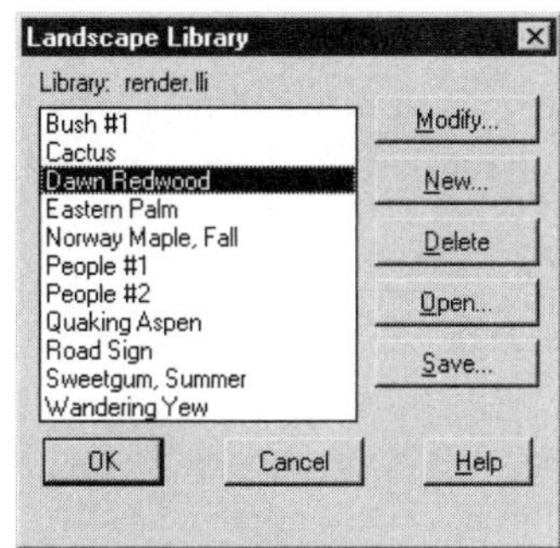

DIALOG BOX OPTIONS

Library indicates current *.lli* landscape library filename; selects a landscape object.

Buttons

Modify changes the properties of landscape objects; displays Landscape Library Edit dialog box.

New assigns default values to landscape objects; displays Landscape Library New dialog box.

Delete removes landscape objects from the library.

Open opens landscape library files; displays the Open Landscape Library dialog box.

Save saves landscape objects to *.lli* files; displays dialog box.

Landscape Library Edit dialog box

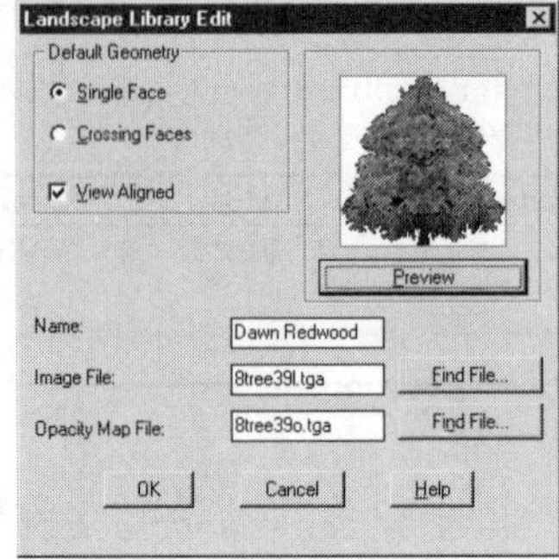

Default Geometry options

⦿ **Single Face** renders faster, but is less realistic.

○ **Crossing Face** produces more realistic ray-traced shadows.

☑ **View Aligned** forces the object always to face the camera.

Preview previews the landscape image.

Name names the landscape object.

Image File specifies the type of raster file, *.bmp*, *.png*, *.gif*, *.jpg*, *.pcx*, *.tga*, or *.tif*.

Opacity Map File names the raster file that provides opacity.

Find File finds the file; displays the Find Image File dialog box.

Landscape Library New dialog box

Options are identical to those found in the ***Landscape Library Edit*** *dialog box.*

RELATED COMMANDS

LsEdit edits the properties of a landscape object.

LsNew places a landscape object in the drawing.

MatLib provides a library of surface textures.

LsNew

Rel.14 Places landscape objects in drawings (*short for LandScape NEW*).

Command	Alias	Ctrl+	F-key	Alt+	Menu Bar	Tablet
lsnew	...	...	...	**VEN**	**View** ↳**Render** ↳**Landscape New**	...

Command: lsnew

Displays dialog box:

DIALOG BOX OPTIONS

Preview views the raster image.

Height changes the height of the object by entering a new value or moving the slider bar.

Position moves the object to another position in the drawing.

Geometry options

⦿ **Single Face** renders faster, but is less realistic.

○ **Crossing Face** produces more realistic ray-traced shadows.

☑ **View Aligned** forces the object always to face the camera.

RELATED COMMANDS

LsLib adds and removes raster images from the *render.lli* file.

LsEdit edits the properties of a landscape object.

Render renders the landscape object.

TIPS

- A *landscape object* is defined as a **Plant** object in the AutoCAD database.
- Turn on **View Aligned** when you want the landscape object — such as a tree — always to face the camera.

- Turn off **View Aligned** to fix the orientation of the landscape object, such as a store front.

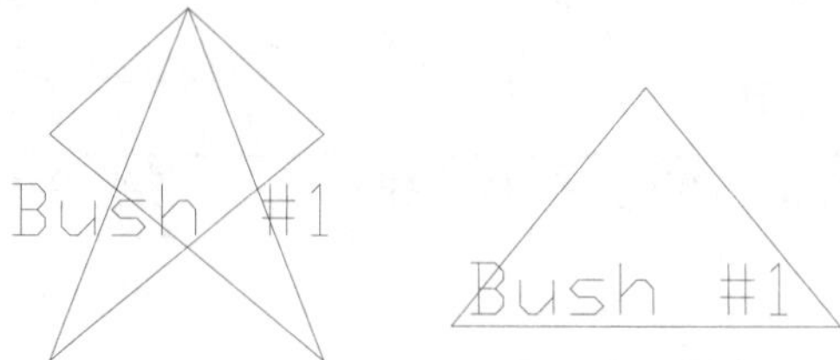

*A landscape object with **crossing faces** (left) and **single face** (right).*

- The grips at the base, top, and corners of landscape objects have special meaning:

Grip	Meaning
Top	Changes the object's height.
Bottom corner	Rotates (if not view aligned) and scales the object.
Base	Moves the object.

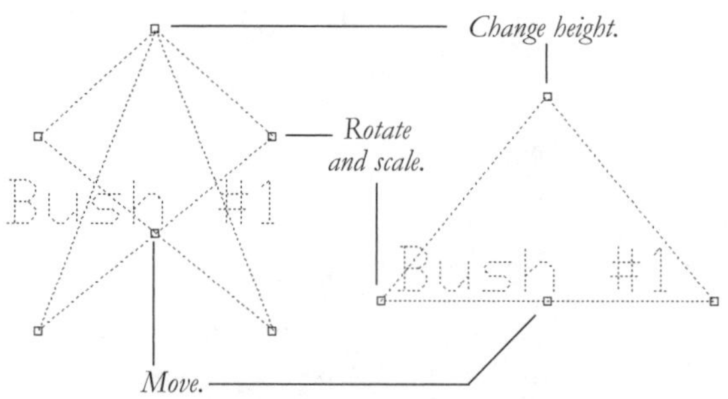

- An *opacity map* determines which part of a raster image is opaque and which is transparent. The opacity map should be a bi-color (*black and white*) image file.

- The landscape object does not appear until you use the **Render** command:

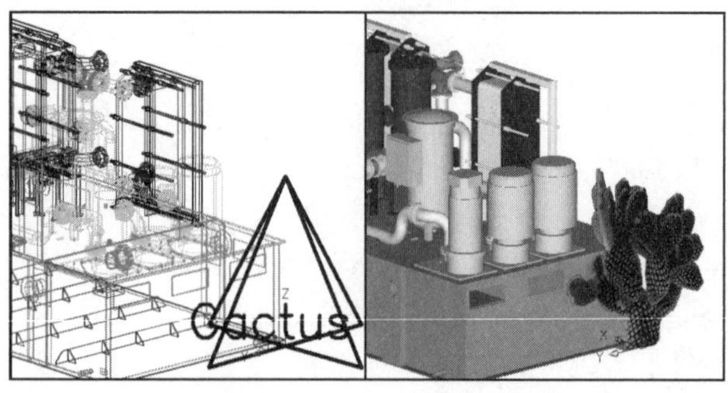

'LtScale

V. 2.0 Sets the global scale factor of linetypes (*short for Line Type SCALE*).

Command	Alias	Ctrl+	F-key	Alt+	Menu Bar	Tablet
ltscale	**lts**	...	...	...	...	...

Command: ltscale
Enter new linetype scale factor <1.0000>: *(Enter a scale factor.)*
Regenerating drawing.

1x dashed linetype *0.5x dashed linetype* *2x dashed linetype*

COMMAND LINE OPTION

Enter new linetype scale factor changes the global scale factor of all linetypes in drawings.

RELATED COMMANDS

ChProp changes the linetype scale of one or more objects.

Properties changes the linetype scale of objects.

Linetype loads, creates, and sets the working linetype.

RELATED SYSTEM VARIABLES

LtScale contains the global linetype scale factor.

CeLtScale specifies the current object linetype scale factor relative to the global scale.

PlineGen controls how linetypes are generated for polylines.

PsLtScale specifies that the linetype scale is relative to paper space.

TIPS

- If the linetype scale is too large, the linetype appears solid.
- If the linetype scale is too small, the linetype appears as a solid line that redraws very slowly.
- In addition to setting the scale with the **LtScale** command, the *acad.lin* file contains each linetype in three scales: normal, half-size, and double-size.
- You can change the linetype scaling of individual objects, which is then multiplied by the global scale factor specified by the **LtScale** command.

'LWeight

2000 Sets the current lineweight (*display width*) of objects.

Commands	Aliases	Ctrl+	Status Bar	Alt+	Menu Bar	Tablet
lweight	**lw**	**...**	**LWT**	**OW**	**Format**	**W14**
	lineweight				**↳Lineweight**	
-lweight						

Command: lweight

Displays dialog box:

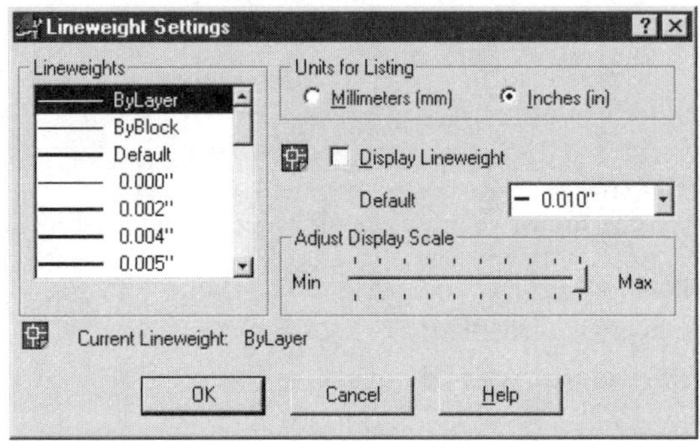

DIALOG BOX OPTIONS

Lineweights lists lineweight values.

Units for Listing specifies the units of lineweights:

⊙ **Millimeters (mm)** specifies lineweight values in millimeters.

○ **Inches (in)** specifies lineweight values in inches.

☐ **Display Lineweight** toggles the display of lineweights; when checked, lineweights are displayed.

Default specifies the default lineweight for layers (default = 0.01" or 0.25 mm).

Adjust Display Scale controls the scale of lineweights in the Model tab, which displays lineweights in pixels.

SHORTCUT MENU OPTIONS

Right-click ***LWT*** *on status bar to display shortcut menu:*

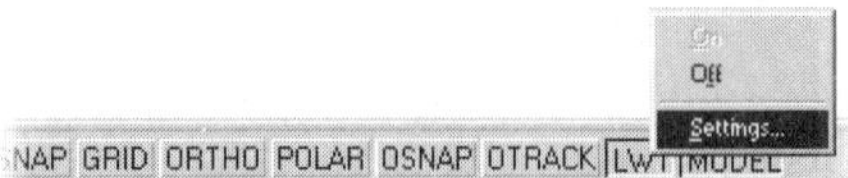

On turns on lineweight display.

Off turns off lineweight display.

Settings displays Lineweight Settings dialog box.

-LWEIGHT Command

Command: -lweight

Enter default lineweight for new objects or [?]: *(Enter a value, or type* **?***.)*

COMMAND LINE OPTIONS

Enter default lineweight specifies the current lineweight; valid values include Bylayer, Byblock, and Default.

? lists the valid values for lineweights:

```
ByLayer ByBlock Default
 0.000" 0.002"  0.004"  0.005" 0.006"  0.007"
 0.008" 0.010"  0.012"  0.014" 0.016"  0.020"
 0.021" 0.024"  0.028"  0.031" 0.035"  0.039"
 0.042" 0.047"  0.055"  0.062" 0.079"  0.083"
```

RELATED SYSTEM VARIABLES

LwDefault specifies the default linewidth; default = 0.01" or 0.25 mm.

LwDisplay toggles the display of lineweights in the drawing.

LwUnits determines whether the lineweight is measured in inches or millimeters.

TIPS

- To create custom lineweights for plotting, use the **Plot Style Table Editor**.
- A lineweight of 0 plots the lines at the thinnest width of which the plotter is capable, usually one pixel or one dot wide.

Replaced Command

MakePreview was removed from AutoCAD Release 14; it was replaced by the **RasterPreview** system variable, which controls the creation of previews when drawings are saved.

Markup / MarkupClose

2005 Opens and closee the Markup Set Manager window.

Command	Alias	Ctrl+	F-key	Alt+	Menu Bar	Tablet
markup	**msm**	**7**	...	...	**Tools** ↳**Markup Set Manager**	...
markupclose	...	**7**	...	...	**Tools** ↳**Markup Set Manager**	...

Command: markup

Displays the Markup Set Manager window.

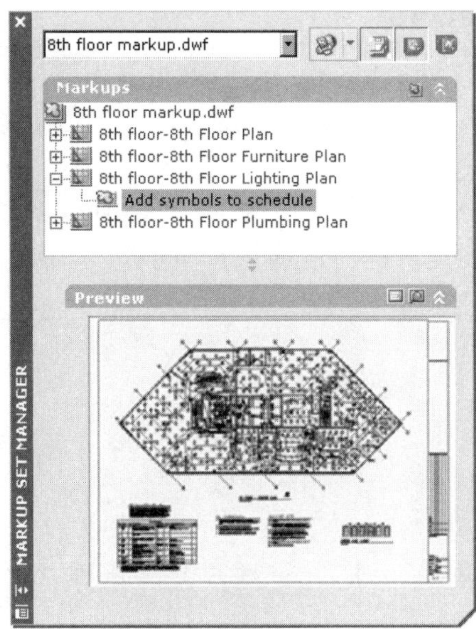

To open a markup set, select ***Open****. AutoCAD displays the Open Markup DWF dialog box. Select a .dwf file, and then click* ***Open****.*

If the file contains no markup data, AutoCAD complains, 'Filename.dwf does not contain any markup data. Would you like to open this DWF file in the viewer?' Click ***Yes*** *to open the file in Composer; click* ***No*** *to cancel.*

Command: markupclose

Closes the Markup Set Manger window.

WINDOW OPTIONS

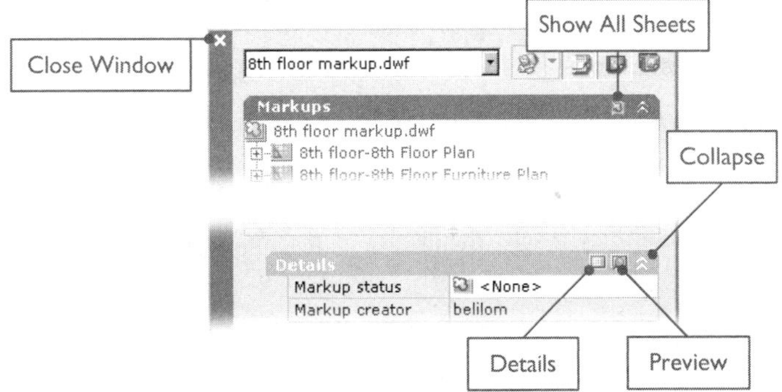

Close Window closes the Markup Set Manager window; reopen with CTRL+7.

Show All Sheets shows all sheets with markups.

Collapse collapses the preview/details area.

Details displays details about the markup sheet.

Preview displays a bitmap image of the sheet.

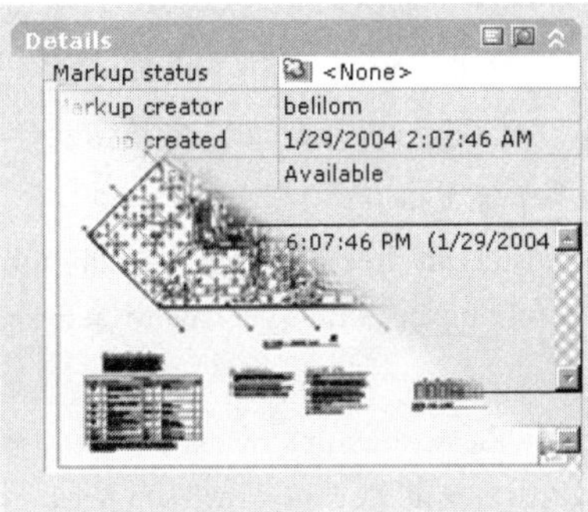

TOOLBAR

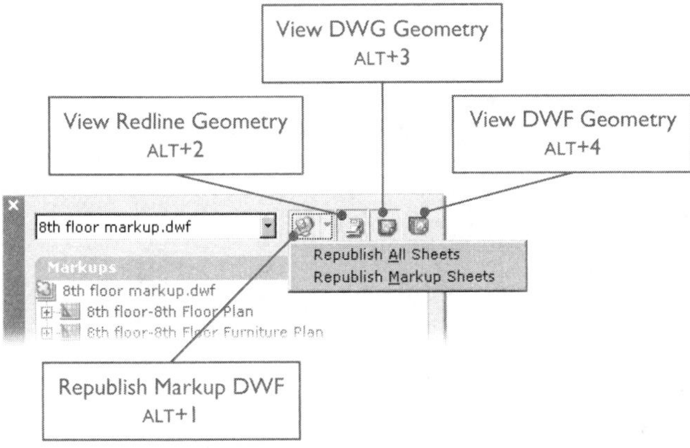

 **Republish Markup DWF** displays a menu (ALT+1):

Republish All Sheets republishes all sheets as a background job using **Publish**.

Republish Markup Sheets republishes only sheets with markups.

 **View Redline Geometry** toggles the display of markup objects (ALT+2).

 **View DWG Geometry** toggles the display of the drawing (ALT+3).

 **View DWF Geometry** toggles the display of the *.dwf* file (ALT+4).

SHORTCUT MENU

Right-click a markup file:

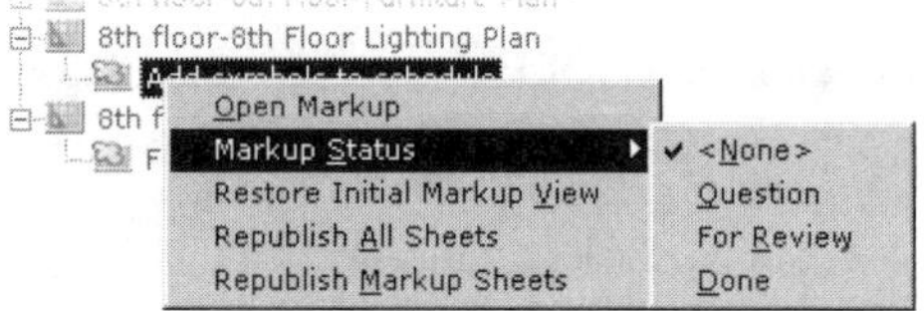

Open Markup opens the markup data in AutoCAD; alternatively, double-click the markup.

Markup Status displays a submenu for changing the status of a markup sheet:

 **None** indicates no change in status.

Question indicates markup has questions to be answered.

For Review indicates markup changes need to be reviewed.

Done indicates markup is done.

Republish All Sheets republishes all sheets as a background job using **Publish**.

Republish Markup Sheets republishes only sheets with markups.

RELATED COMMANDS

OpenDwfMarkup opens *.dwf* files containing markup data.

RmlIn imports *.rml* redline markup files created by Volo View.

RELATED SYSTEM VARIABLES

MsmState reports whether the Markup Set Manager window is open.

TIPS

- MSM is short for "markup set manager."
- This command reads only *.dwf* files marked-up with Composer; it does not read other *.dwf* files.

MassProp

Rel.11 Reports the mass properties of 3D solid models, bodies, and 2D regions (*short for MASS PROPerties*).

Command	Alias	Ctrl+	F-key	Alt+	Menu Bar	Tablet
massprop	...	...	...	**TYM**	**Tools** ↳**Inquiry** ↳**Region/Mass Properties**	**U7**

Command: massprop

Select objects: *(Select one or more regions and/or solid model objects.)*

Select objects: *(Press* **ENTER***.)*

Example output of a solid sphere:

```
AutoCAD Text Window - Drawing2.dwg
Edit

 ----------------   SOLIDS    ----------------

Mass:                    627.6327
Volume:                  627.6327
Bounding box:         X: -10.9140  --  -0.2913
                      Y: -34.3786  --  -23.7559
                      Z: -5.3114  --  5.3114
Centroid:             X: -5.6026
                      Y: -29.0673
                      Z: 0.0000
Moments of inertia:   X: 537373.7625
                      Y: 26783.2515
                      Z: 557074.6721
Press ENTER to continue:
Products of inertia: XY: 102211.6596
                     YZ: 0.0000
                     ZX: 0.0000
Radii of gyration:    X: 29.2607
                      Y: 6.5325
                      Z: 29.7923
Principal moments and X-Y-Z directions about centroid:
                      I: 7082.3418 along [1.0000 0.0000 0.0000]
                      J: 7082.3418 along [0.0000 1.0000 0.0000]
                      K: 7082.3418 along [0.0000 0.0000 1.0000]

Write analysis to a file? [Yes/No] <N>:
```

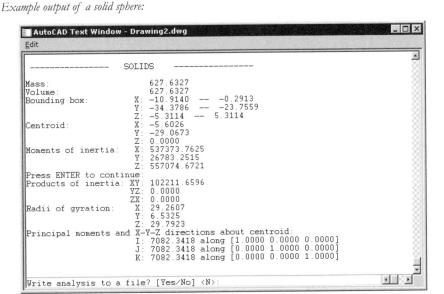

COMMAND LINE OPTIONS

Select objects selects the solid model objects — 2D regions, 3D solids, and bodies — to analyze.

Write analysis to a file:

Yes writes mass property reports to *.mpr* files.

No doesn't write reports to file.

RELATED COMMAND

Area calculates the area and perimeter of non-solid objects.

RELATED FILE

***.mpr** is the file to which **MassProp** writes its results (mass properties report).

TIPS

- This command can be used with 2D regions as well as 3D solids; it cannot be used with 3D surface models or 2D non-region objects.
- Mass properties are computed as if the selected regions were unioned, and as if the selected solids were unioned.
- As of Release 13, AutoCAD's solid modeling no longer allows you to apply a material density to a solid model. All solids and bodies have a density of 1.
- AutoCAD only analyzes regions coplanar (lying in the same plane) to the first region selected.

DEFINITIONS

Area

– total surface area of the selected 3D solids, bodies, or 2D regions.

Bounding Box

– the lower-right and upper-left coordinates of a rectangle enclosing 2D regions.

– the x,y,z coordinate triple of a 3D box enclosing 3D solids or bodies.

Centroid

– the x,y,z coordinates of the center of 2D regions.

– the center of mass of 3D solids and bodies.

Mass

– equal to the volume, because density = 1; not calculated for regions.

Moment of Inertia

– for 2D regions = **Area** * **Radius**^{2}

– for 3D bodies = **Mass** * **Radius**$^{2.}$

Perimeter

– total length of inside and outside loops of 2D regions (not calculated for 3D solids).

Product of Inertia

– for 2D regions = **Mass** * **Distance** (of centroid to y,z axis) * **Distance** (of centroid to x,z axis).

– for 3D bodies = **Mass** * **Distance** (of centroid to y,z axis) * **Distance** (of centroid to x,z axis)

Radius of Gyration

– for 2D regions and 3D solids = (**MomentOfInertia** / **Mass**)$^{1/2}$

Volume

– 3D space occupied by a 3D solid or body (not calculated for regions).

MatchCell

2005 Matches the properties of table cells.

Command	Alias	Ctrl+	F-key	Alt+	Menu Bar	Tablet
matchcell	...	...	...	...	...	...

Command: matchcell
Select source cell: *(Select a cell in a table.)*
Select destination cell: *(Select one or more cells.)*

COMMAND LINE OPTIONS

Select source cell gets property settings from the source cell.

Select destination cell passes property settings to the destination cells.

RELATED COMMANDS

Table creates new tables in drawings.

TableStyle defines table styles.

MatchProp copies properties between objects other than cells.

RELATED SYSTEM VARIABLE

CTable Style specifies the name of the current table style.

TIPS

- Use this command to copy formatting from one cell to another.
- Use the **MatchProp** command to copy properties from one table to another.

'MatchProp

Rel.14 Matches the properties between selected objects (*short for MATCH PROPerties*).

Command	Alias	Ctrl+	F-key	Alt+	Menu Bar	Tablet
matchprop	**ma**	...	...	**MM**	**Modify**	**Y14**
	painter				**⇨Match Properties**	

Command: matchprop

Select source object: *(Select a single object.)*

Current active settings: Color Layer Ltype Ltscale Lineweight Thickness PlotStyle Text Dim Hatch Polyline Viewport

Select destination object(s) or [Settings]: *(Pick one or more objects, or type* **S***.)*

Select destination object(s) or [Settings]: *(Press* **ENTER** *to exit command.)*

COMMAND LINE OPTIONS

Select source object gets property settings from the source object.

Select destination object(s) passes property settings to the destination objects.

Settings displays dialog box:

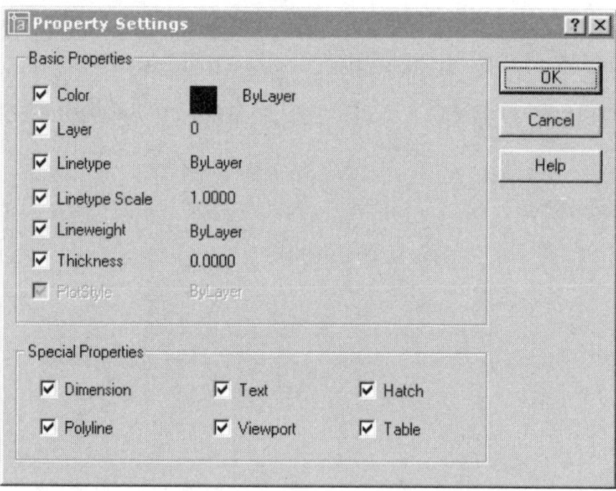

DIALOG BOX OPTIONS

Basic Properties options

☑ **Color** specifies the color for destination objects; not available when OLE objects are selected.

☑ **Layer** specifies the layer name for destination objects; not available when OLE objects are selected.

☑ **Linetype** specifies the linetype for destination objects; not available when attributes, hatch patterns, mtext, OLE objects, points, or viewports are selected.

☑ **Linetype Scale** specifies the linetype scale for destination objects; n attributes, hatch patterns, mtext, OLE objects, points, or viewports ar

☑ **Lineweight** specifies the lineweight for destination objects.

☑ **Thickness** specifies the thickness for destination objects; available only fo can have thickness: arcs, attributes, circles, lines, mtext, points, 2D polylines and traces.

☑ **Plot Style** specifies the plot style; not available when **PStylePolicy** = 1 (color- nt plot style mode) or when OLE objects are selected.

Special Properties options

☑ **Dimension** copies the dimension style of dimension, leader, and tolerance objects.

☑ **Text** copies the text style of text and mtext objects.

☑ **Hatch** copies the hatch pattern of hatched objects.

☑ **Polyline** copies the width and linetype generation of polylines; curve fit, elevation, and variable width properties are not copied.

☑ **Viewport** copies all properties of viewport objects, except clipping, UCS-per-viewport, and freeze-thaw settings.

☑ **Table** copies style of table objects (*new to AutoCAD 2005*).

RELATED COMMAND

Properties changes most aspects of one selected object.

RELATED SYSTEM VARIABLE

PStylePolicy determines whether the **PlotStyle** option is available.

TIPS

- In other Windows applications, this command is known as **Format Painter**.
- Use the **MatchCell** command to copy properties from one table cell to another.

MatLib

Rel.13 Imports and exports material-look definitions for use by the **RMat** command (*short for MATerial LIBrary*).

Command	Alias	Ctrl+	F-key	Alt+	Menu Bar	Tablet
matlib	...	...	...	**VEY**	**View** **↳Render** **↳Materials Library**	**Q1**

Command: matlib

Displays dialog box:

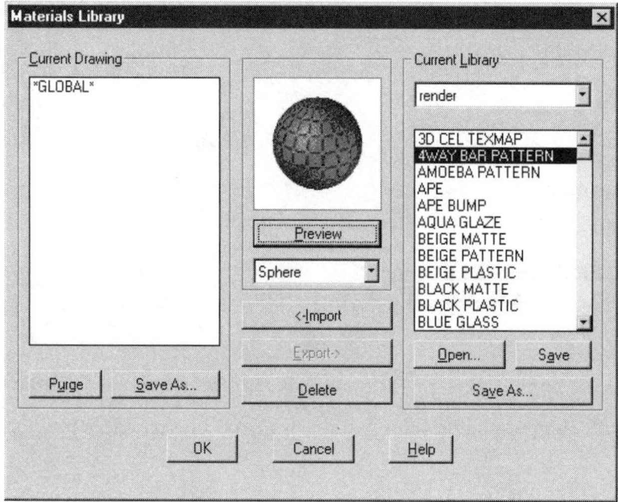

DIALOG BOX OPTIONS

Import brings the selected material definition into the drawing; when there is a conflict, displays Reconcile Imported Material Names dialog box.

Preview previews the selected material mapped to sphere and box objects.

Export adds material definitions to *.mli* library files; if there is a conflict, displays the Reconcile Exported Material Names dialog box, which is identical to Reconcile Imported Material Names dialog box.

Purge deletes unattached material definitions from the Materials list.

Save saves to *.mli* files.

Delete deletes selected material definitions from the Materials or Library lists.

Open loads material definitions from *.mli* files; displays file dialog box.

Reconcile Imported Material Names dialog box

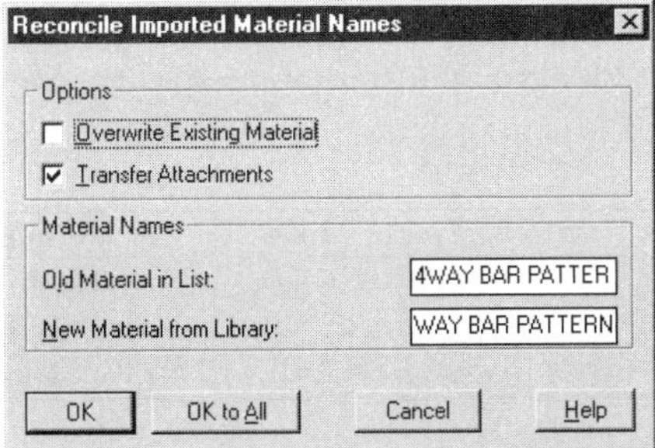

Options options

☐ **Overwrite Existing Material** overwrites existing material definition with selected material definition.

☑ **Transfer Attachments** keeps objects attached to material definition.

Material Names options

Old Material in List allows you to edit the name of the material.

New Material from Library allows you to edit the name of the material.

RELATED COMMAND

RMat attaches a material definition to objects, colors, and layers.

RELATED FILE

render.mli is the material library that contains the material definitions.

TIPS

- This command can be accessed from the **RMat** dialog box: click **Materials Library** button.
- To use materials in rendering, you must take these steps:
  1. Use **MatLib** to load material definitions.
  2. Use **RMat** to attach the definitions to objects, layers, or colors.
  3. In the Render dialog box, turn on the **Apply Materials** option.
- A *material* defines the look of a rendered object: coloring, reflection or shine, roughness, and ambient reflection.
- Materials only appear with **Render**; they do no appear with the **Shade** command.
- By default, a drawing contains a single material definition, called *GLOBAL*, with the default parameters for color, reflection, roughness, and ambience.

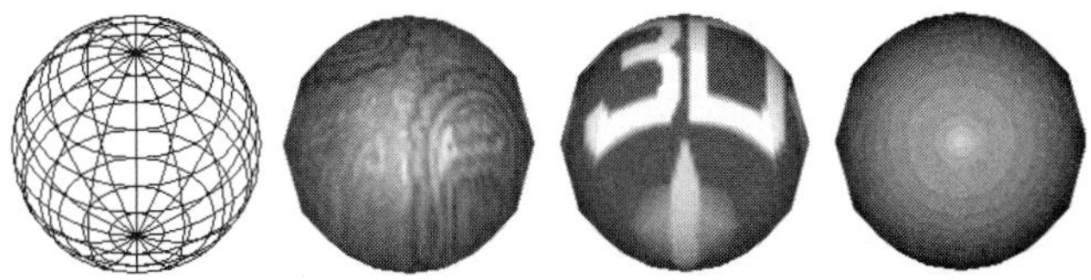

Materials applied to spheres.

- Materials do not define the density of 3D solids and bodies.

Measure

V. 2.5 Places points or blocks at constant intervals along lines, arcs, circles, and polylines.

Command	Alias	Ctrl+	F-key	Alt+	Menu Bar	Tablet
measure	**me**	...	...	**DOM**	**Draw** ↳**Point** ↳**Measure**	**V12**

Command: measure

Select object to measure: *(Pick a single object.)*

Specify length of segment or [Block]: *(Enter a value, or type **B**.)*

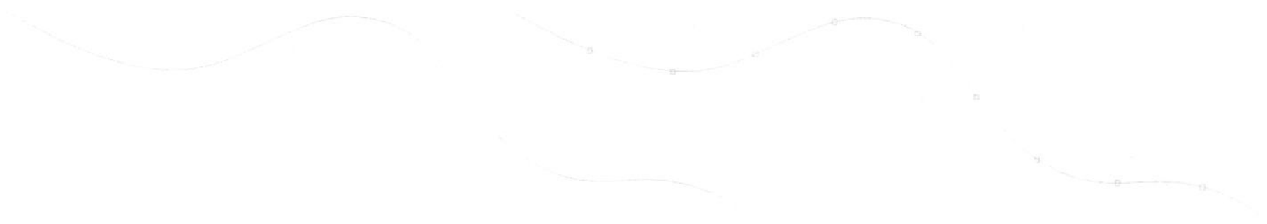

Polyline (left); measured with ten points (right).

COMMAND LINE OPTIONS

Select object selects a single object for measurement.

Specify length of segment indicates the distance between markers.

Block options

Enter name of block to insert: *(Enter name.)*

Align block with object? [Yes/No] <Y>: *(Enter **Y** or **N**.)*

Specify length of segment: *(Enter a value.)*

Enter name of block indicates the name of the block to use as a marker; the block must already exist in the drawing.

Align block with object? aligns the block's x axis with the object.

RELATED COMMANDS

Block creates blocks that can be used with the **Measure** command.

Divide segments objects.

RELATED SYSTEM VARIABLES

PdMode controls the shape of a point.

PdSize controls the size of a point.

TIPS

- You must define the block before it can be used with this command.
- The **Measure** command does not place points (or blocks) at the beginning and end of measured objects.

Removed Command

MeetNow was removed from AutoCAD 2004.

Menu

V. 1.0 Loads *.mnc*, *.mns*, and *.mnu* menu files.

Command	Alias	Ctrl+	F-key	Alt+	Menu Bar	Tablet
menu	...	...	...	...	...	...

Command: menu

Displays the Select Menu File dialog box.

Select a .mnc, .mns, or .mnu file, and then click ***Open****.*

COMMAND LINE OPTIONS

None.

RELATED COMMANDS

MenuLoad loads a partial menu file.

Tablet configures digitizing tablet for use with overlay menus.

RELATED SYSTEM VARIABLES

MenuName specifies the name of the currently-loaded menu file.

MenuCtl determines whether sidescreen menu pages switch in parallel with commands entered at the keyboard.

MenuEcho suppresses menu echoing.

ScreenBoxes specifies the number of menu lines displayed on the side menu.

RELATED FILES

***.mnc** compiled menu files; stored in binary format.

***.mnc** source menu files; stored in ASCII format.

***.mnu** menu template files; stored in ASCII format.

TIPS

- AutoCAD automatically compiles *.mns* and *.mns* files into *.mnc* files for faster loading.
- The *.mnu* file defines the function of the screen menu, menu bar, cursor menu, icon menus, digitizing tablet menus, pointing device buttons, toolbars, help strings, and the AUX: device.
- To access the menu source code, use the **Tools | Customize | Edit Custom Files | Current Menu** command. AutoCAD displays the current *.mns* file in Notepad.
- AutoCAD 2005 adds the *custom.mnu* file for customizing menus independent of *acad.mnu*.

MenuLoad /MenuUnload

Rel.13 Loads and unloads parts of menu files.

Command	Alias	Ctrl+	F-key	Alt+	Menu Bar	Tablet
menuload	...	...	...	**TC**	**Tools ↳Customize Menus**	**Y9**
menuunload	...	...	...	**TC**	**Tools ↳Customize Menus**	**Y9**

Command: menuload *or* menuunload

Both commands display the same tabbed dialog box:

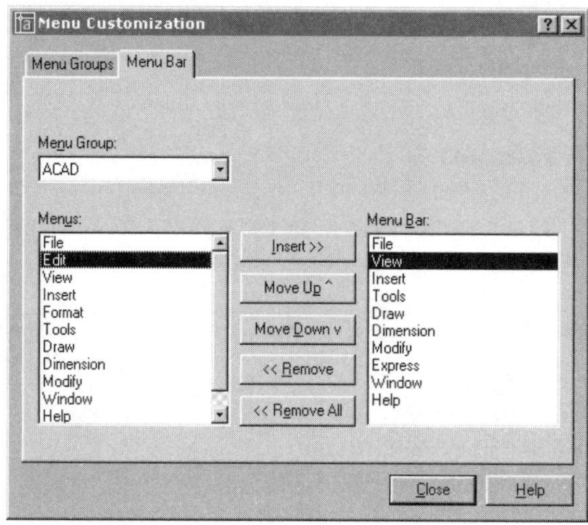

DIALOG BOX OPTIONS

Menu Bar tab

Menu Group selects a menu group or file.

Menus provides the names of the menu items in the selected menu group.

Menu Bar provides the names of menu items on the menu bar.

Insert >> inserts a menu item on the menu bar, immediately above the selected item.

Move Up ^ moves the menu item along the menu bar (*new to AutoCAD 2005*).

Move Down V moves the menu item along the menu bar (*new to AutoCAD 2005*).

< Remove removes a menu item from the menu bar.

<< Remove All removes all menu items from the menu bar.

Menu Groups tab

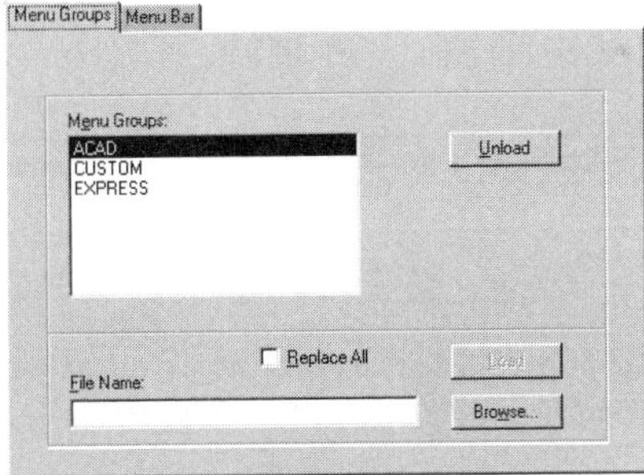

Menu Groups lists the names of loaded menu groups and files.

Unload unloads selected menu group.

☐ **Replace All** replaces all currently-loaded menus with the newly-loaded menu.

Load loads the selected menu group into AutoCAD.

File Name displays the name of the menu file.

Browse displays the Select Menu File dialog box.

Close closes the dialog box.

Help provides context-sensitive help.

RELATED COMMANDS

Menu loads full menu files.

MenuUnload unloads parts of menu files.

Tablet configures digitizing tablets for use with overlay menus.

RELATED SYSTEM VARIABLES

MenuName specifies the name of the currently-loaded menu file.

MenuEcho suppresses menu echoing.

RELATED FILES

.*mnc is the compiled menu file; stored in binary format.

.*mns is the source menu file; stored in ASCII format.

.*mnu is the menu template file; stored in ASCII format.

TIPS

- The **MenuLoad** command allows you to add *partial* menus to the menu bar, without replacing the entire menu structure. Use the **Menu** command to load complete menus, overwriting the existing menu.
- The **MenuUnload** command allows you to remove partial menu files, such as *db_con.mnu* and *accov.mns*.
- AutoCAD 2005 adds the *custom.mnu* file for customizing menus independent of *acad.mnu*.

MInsert

V. 2.5 Inserts an array of blocks as a single block (*short for Multiple INSERT*).

Command	Alias	Ctrl+	F-key	Alt+	Menu Bar	Tablet
minsert	...	...	...	...	...	...

Command: minsert
Enter block name or [?]: *(Enter a name, type* **?**, *or enter* **~** *to select a .dwg file.)*
Specify insertion point or [Scale/X/Y/Z/Rotate/PScale/PX/PY/PZ/PRotate]: *(Pick a point, or enter an option.)*
Enter X scale factor, specify opposite corner, or [Corner/XYZ] <1>: *(Enter a value, pick a point, or enter an option.)*
Enter Y scale factor <use X scale factor>: *(Enter a value, or press* **ENTER**.*)*
Specify rotation angle <0>: *(Enter a value, or press* **ENTER**.*)*
Enter number of rows (---) <1>: *(Enter a value.)*
Enter number of columns (|||) <1>: *(Enter a value.)*
Enter distance between rows or specify unit cell (---): *(Enter a value.)*
Specify distance between columns (|||): *(Enter a value.)*

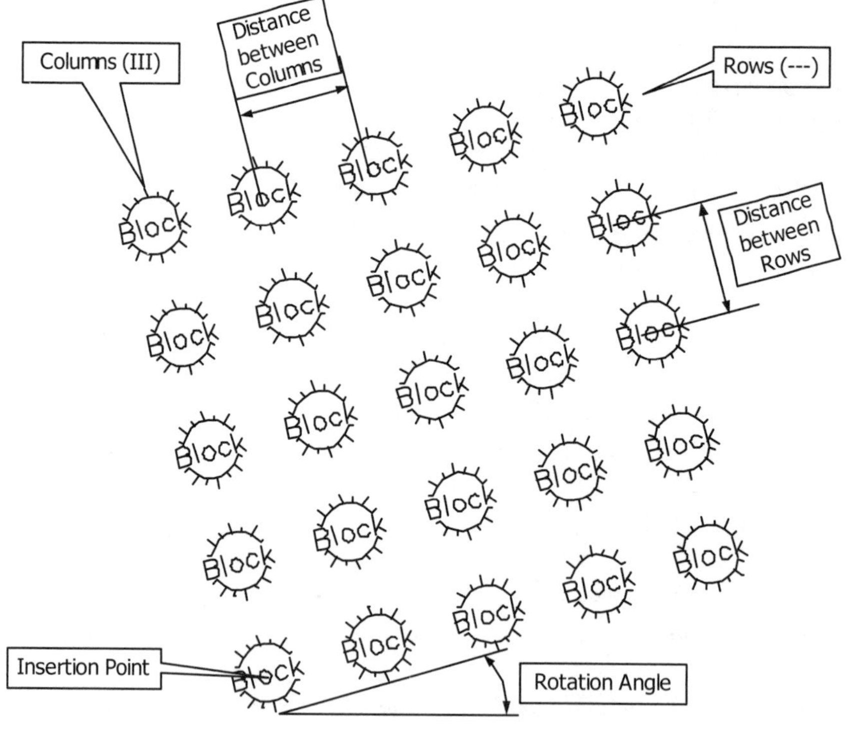

COMMAND LINE OPTIONS

Enter block name indicates the name of the block to be inserted; the block must already exist in the drawing.

? lists the names of blocks stored in the drawing.

Specify insertion point specifies the x,y coordinates of the first block.

P supplies predefined scale and rotation values.

X scale factor indicates the x scale factor.

Specify opposite corner specifies a second point that indicates the x,y-scale factor.

Corner indicates the x and y scale factors by picking two points on the screen.

XYZ specifies x, y, and z scaling.

Specify rotation angle specifies the angle for the array.

Number of rows specifies the number of horizontal rows.

Number of columns specifies the number of vertical columns.

Distance between rows specifies the distance between rows.

Specify unit cell shows the cell distance by picking two points on the screen.

Distance between columns specifies the distance between columns.

RELATED COMMANDS

3dArray creates 3D rectangular and polar arrays.

Array creates 2D rectangular and polar arrays.

Block creates a block.

TIPS

- The array placed by the **MInsert** command is a single block.
- You *cannot* explode blocks created by this command.
- You may redefine blocks created by the **MInsert** command.

Mirror

V. 2.0 Creates a mirror copy of a group of objects in 2D space.

Command	Alias	Ctrl+	F-key	Alt+	Menu Bar	Tablet
mirror	mi	...	...	MI	Modify ↳Mirror	V16

Command: mirror
Select objects: *(Pick one or more objects.)*
Select objects: *(Press* ENTER *to end object selection.)*
Specify first point of mirror line: *(Pick a point.)*
Specify second point of mirror line: *(Pick another point.)*
Delete source objects? [Yes/No] <N>: *(Type* **Y** *or* **N**.*)*

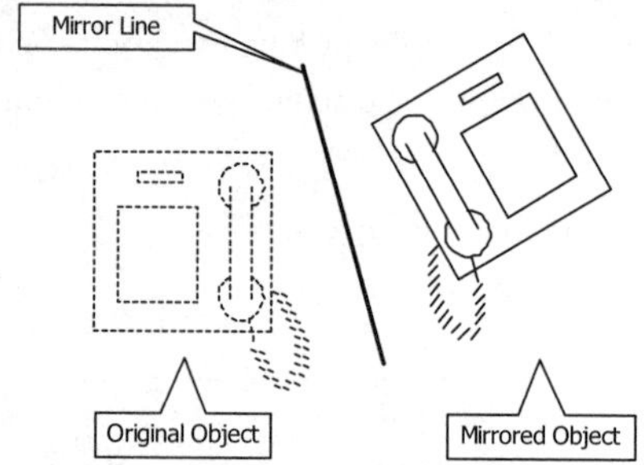

COMMAND LINE OPTIONS

Select objects selects the objects to mirror.

First point specifies the starting point of the mirror line.

Second point specifies the end point of the mirror line.

Delete source objects deletes selected objects.

RELATED COMMANDS

Copy creates non-mirrored copies of objects.

Mirror3d mirrors objects in 3D-space.

RELATED SYSTEM VARIABLE

MirrText determines whether text is mirrored by this command. (Prior to AutoCAD 2005, this variable defaulted to 1. With AutoCAD 2005, it defaults to 0.)

TIPS

- The **Mirror** command is excellent for cutting your drawing work in half for symmetrical objects. For double-symmetrical objects, use **Mirror** twice.
- Although you can mirror a viewport in paper space, this does not mirror the model space objects inside the viewport.
- Turn on **Ortho** mode to ensure that the mirror is perfectly horizontal or vertical.
- The mirror line becomes a mirror plane in 3D; it is perpendicular to the x,y plane of the UCS containing the mirror line.

Mirror3d

Rel.11 Mirrors objects about a plane in 3D space.

Command	Alias	Ctrl+	F-key	Alt+	Menu Bar	Tablet
mirror3d	...	...	...	**M3M**	**Modify** ↳**3D Operation** ↳**Mirror 3D**	**W21**

Command: mirror3d
Select objects: *(Pick one or more objects.)*
Select objects: *(Press* **ENTER** *to end object selection.)*
Specify first point of mirror plane (3 points) or
[Object/Last/Zaxis/View/XY/YZ/ZX/3points] <3points>: *(Pick a point, or enter an option.)*
Delete old objects? <N>: *(Type* **Y** *or* **N**.*)*

COMMAND LINE OPTIONS

Select objects selects the objects to be mirrored in space.

Specify first point specifies the first point of the mirror plane.

Object selects a circle, arc or 2D polyline segment as the mirror plane.

Last selects the last-picked mirror plane.

View specifies that the current view plane is the mirror plane.

XY specifies that the x,y plane is the mirror plane.

YZ specifies that the y,z plane is the mirror plane.

ZX specifies that the z,x plane is the mirror plane.

Zaxis defines the mirror plane by a point on the plane and the normal to the plane, i.e., the z axis.

3points defines three points on the mirror plane.

RELATED COMMANDS

Align translates and rotates objects in 2D planes and 3D space.

Mirror mirrors objects in 2D space.

Rotate3d rotates objects in 3D space.

RELATED SYSTEM VARIABLE

MirrText determines whether text is mirrored by the **Mirror** command:

MirrText	Meaning
0	Text is not mirrored about the horizontal axis (*default*).
1	Text is mirrored

MlEdit

Rel.13 Edits multilines (*short for MultiLine EDITor*).

Command	Alias	Ctrl+	F-key	Alt+	Menu Bar	Tablet
mledit	...	...	...	...	Modify ↳Multiline	Y19
-mledit						

Command: mledit

Displays dialog box:

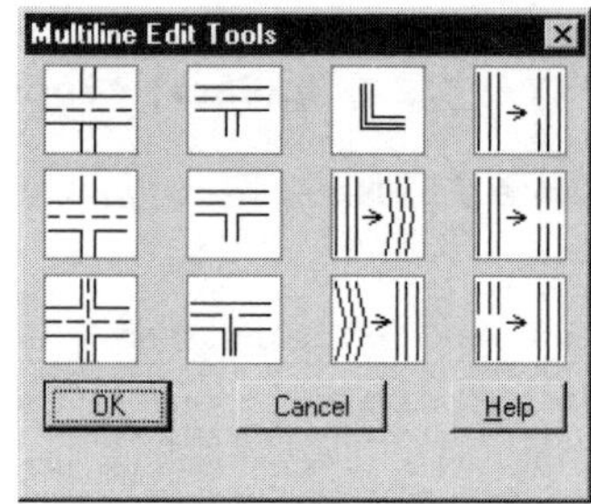

DIALOG BOX OPTIONS

Closed Cross closes the intersection of two multilines.

Open Cross opens the intersection of two multilines.

Merged Cross merges a pair of multilines: opens exterior lines; closes interior lines.

Closed Tee closes T-intersections.

Open Tee opens T-intersections.

Merged Tee merges T-intersection by opening exterior lines and closing interior lines.

Corner Joint creates corner joints with pairs of intersecting multilines.

Add Vertex adds vertcies (*joints*) to multiline segments.

Delete Vertex removes vertices from multiline segments.

Cut Single places gaps in a single line of multilines.

Cut All places gaps in all lines of multilines.

Weld All removes gaps from multilines.

-MLEDIT Command

Command: -mledit

Enter mline editing option [CC/OC/MC/CT/OT/MT/CJ/AV/DV/CS/CA/WA]: *(Enter an option.)*

COMMAND LINE OPTIONS

AV adds vertices.

DV deletes vertices.

CC closes crossings.

OC opens crossings.

MC merges crossings.

CT closes tees.

OT opens tees.

MT merges tees.

CJ creates corner joints.

CS cuts a single line.

CA cuts all lines.

WA welds all lines.

U undoes the most-recent multiline edit.

RELATED COMMANDS

MLine draws up to 16 parallel lines.

MlStyle defines the properties of a multiline.

RELATED SYSTEM VARIABLES

CMlJust specifies the current multiline justification mode:

CMlJust	Meaning
0	Top (default).
1	Middle.
2	Bottom.

CMlScale specifies the current multiline scale factor (default = 1.0).

CMlStyle specifies the current multiline style name (default = " ").

RELATED FILE

.mln is the multiline style definition file.

TIPS

- Use the **Cut All** option to open up a gap before placing door and window symbols in a multiline wall.
- Use the **Weld All** option to close up a gap after removing the door or window symbol in a multiline.
- Use the **Stretch** command to move a door or window symbol in a multiline wall.
- When you open a gap in a multiline, AutoCAD does not cap the sides of the gap. You may need to add the endcaps with the **Line** command.

MLine

Rel.13 Draws up to 16 parallel lines (*short for Multiple LINE*).

Command	Alias	Ctrl+	F-key	Alt+	Menu Bar	Tablet
mline	**ml**	...	...	**DM**	**Draw** ↳**Multiline**	**M10**

Command: mline

Current settings: Justification = Top, Scale = 1.00, Style = STANDARD

Specify start point or [Justification/Scale/STyle]: *(Pick a point, or enter an option.)*

Specify next point: *(Pick a point.)*

Specify next point or [Undo]: *(Pick a point, or type* **U***.)*

Specify next point or [Close/Undo]: *(Pick a point, or else enter an option.)*

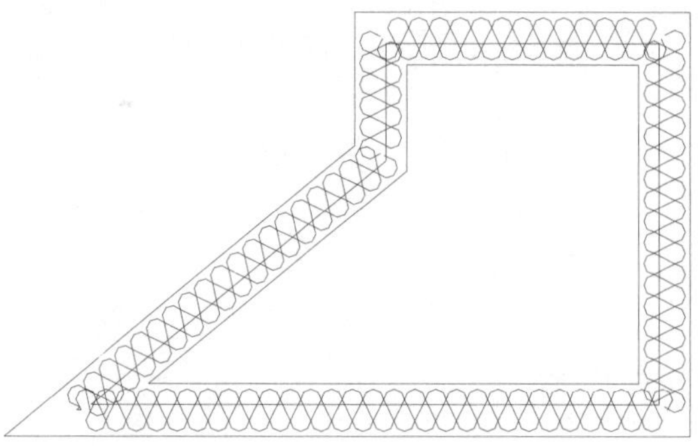

COMMAND LINE OPTIONS

Specify start point indicates the start of the multiline.

Specify next point indicates the next vertex.

Undo removes the most recently-added segment.

Close closes the multiline to its start point.

Justification options

Enter justification type [Top/Zero/Bottom] <top>: *(Enter an option.)*

Top draws the top line of the multiline at the cursor; remainder of multiline is "below" the cursor.

Zero draws the center (*zero offset point*) of the multiline at the cursor.

Bottom draws the bottom of the multiline at the cursor; remainder of the multiline is "above" the cursor.

Scale option

Enter mline scale <1.00>: *(Enter a value.)*

Enter mline scale specifies the scale of the width of the multiline; see Tips for examples.

STyle options

Enter mline style name or [?]: *(Enter style name, or type* **?***.)*

Enter mline style name specifies the name of the multiline style.

? lists the names of the multiline styles defined in drawing.

RELATED COMMANDS

MlEdit edits multilines.

MlProp defines the properties of a multiline.

RELATED SYSTEM VARIABLES

CMlJust specifies the current multiline justification:

CMlJust	Meaning
0	Top (default).
1	Middle.
2	Bottom.

CMlScale specifies the current multiline scale factor (default = 1.0).

CMlStyle specifies the current multiline style name (default = "").

RELATED FILE

***.mln** is the multiline style definition file.

TIPS

- Examples of scale factors:

Scale	Meaning
1.0	Default scale factor.
2.0	Draws multiline twice as wide.
0.5	Draws multiline half as wide.
-1.0	Flips multiline.
0	Collapses multiline to a single line.

- Multiline styles are stored in *.mln* files in DXF-like format.

MlStyle

Rel.13 Defines the characteristics of multilines (*short for MultiLine STYLE*).

Command	Alias	Ctrl+	F-key	Alt+	Menu Bar	Tablet
mlstyle	...	...	...	**OM**	**Format** **↳Multiline Style**	**V5**

Command: mlstyle

Displays dialog box:

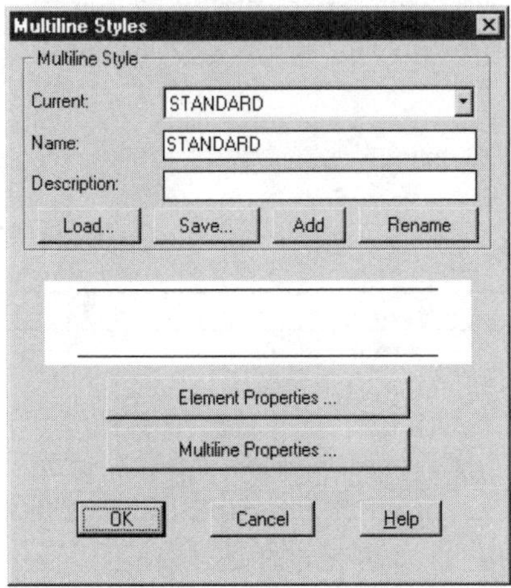

DIALOG BOX OPTIONS

Multiline Style options

Current lists the currently-loaded multiline style names (default = STANDARD).

Name names the new multiline style, or renames existing styles.

Description describes the multiline style, with up to 255 characters.

Load loads styles from the multiline library file *acad.mln* or other *.mln* files; displays the Load Multiline Styles dialog box.

Save saves a multiline style or renames a style; displays dialog box.

Add adds the multiline style from the Name box to the Current list.

Remove removes the multiline style from the Current list.

Additional options

Element Properties specifies properties of multiline elements; displays the Element Properties dialog box.

Multiline Properties specifies additional properties for multilines; displays the Multiline Properties dialog box.

Load Multiline Styles dialog box

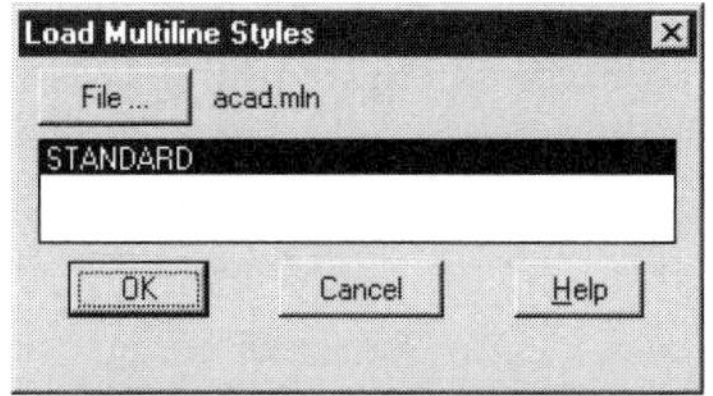

File selects an *.mln* multiline definition file.

Element Properties dialog box

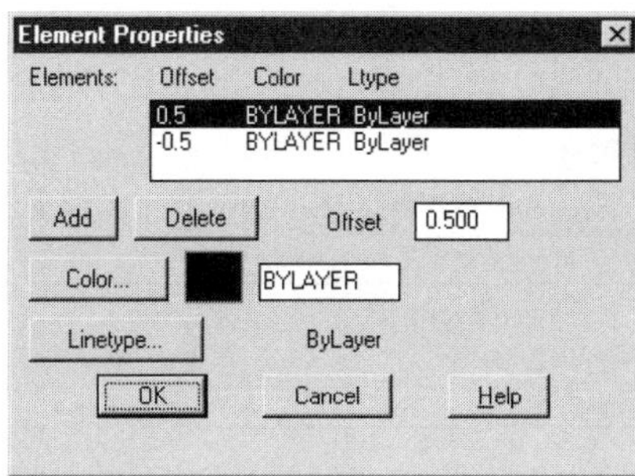

Add adds an element (line).

Delete deletes an element.

Offset specifies the distance from origin to element.

Color specifies the element color; displays Select Color dialog box.

Linetype specifies the element linetype; displays Select Linetype dialog box.

Multiline Properties dialog box

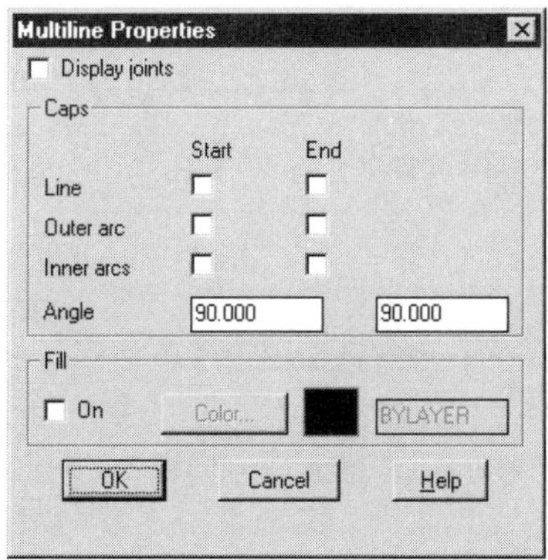

☐ **Display Joints** toggles the display of joints (miters) at vertices; affects all multiline segments.

Caps options

☐ **Line** draws a straight line start and/or end cap.

☐ **Outer Arc** draws an arc to cap the outermost pair of lines.

☐ **Inner Arcs** draws an arc to cap all inner pairs of lines.

Angle specifies the angle for straight line caps.

Fill options

☐ **On** specifies the fill color.

Color displays the Select Color dialog box.

RELATED COMMANDS

MlEdit edits multilines.

MLine draws up to 16 parallel lines.

RELATED SYSTEM VARIABLES

CMlJust specifies the current multiline justification:

CMlJust	Meaning
0	Top (default).
1	Middle.
2	Bottom.

CMlScale specifies the current multiline scale factor (default = 1.0).

CMlStyle specifies the current multiline style name (default = " ").

RELATED FILE

acad.mln is the multiline style definition file.

TIPS

- Use the **MlEdit** command to create (or close up) gaps to place door and window symbols in multiline walls.
- The multiline scale factor has the following effect on the look of a multiline:

Scale	Meaning
1.0	The default scale factor.
0.5	Draws multiline half as wide.
2.0	Draws multiline twice as wide, not twice as long.
-1.0	Flips multiline about its origin.
0.0	Collapses multiline to a single line.

- The *.mln* file describes multiline styles in a DXF-like format.
- You cannot change the element or multiline properties once the drawing contains a multiline using the style.

Model

2000 Switches to Model tab.

Command	Alias	Ctrl+	F-key	Alt+	Menu Bar	Tablet
model	...	...	...	...	...	...

Command: model

Switches to the model tab.

COMMAND LINE OPTIONS

None.

RELATED COMMANDS

Layout creates layouts.

MSpace switches to model space.

RELATED SYSTEM VARIABLE

Tilemode switches between model tab and layout tab.

TIPS

- This command automatically sets **TileMode** to 1.
- As an alternative to this command, you can select the **Model** tab:

Model Layout1 Layout2

- The **Model** tab replaces the **TILE** button on the status bar of AutoCAD Release 13 and 14.
- *Historical note:* The system variable is named "Tilemode," because model space can only display tiled viewports. (Paper space, or layout mode, can display overlapping viewports.) Turning off tiled-viewport mode meant AutoCAD was switching to paper space, where viewports no longer had to be tiled.

  Going back further, it was a graphic board manufacturer, Control Systems, that first figured out how to make AutoCAD display four tiled viewports at once. Autodesk added the feature to AutoCAD Release 10.

  All of which leads to a question I cannot answer: Why can't viewports be tiled in model space?

Move

V. 1.0 Moves one or more objects to a new location.

Command	Alias	Ctrl+	F-key	Alt+	Menu Bar	Tablet
move	**m**	...	...	**MV**	**Modify** ↳**Move**	**V19**

Command: move
Select objects: *(Select one or more objects.)*
Select objects: *(Press* **ENTER** *to end object selection.)*
Specify base point or displacement: *(Pick a point.)*
Specify second point of displacement or <use first point as displacement>:
(Pick a point, or press **ENTER***.)*

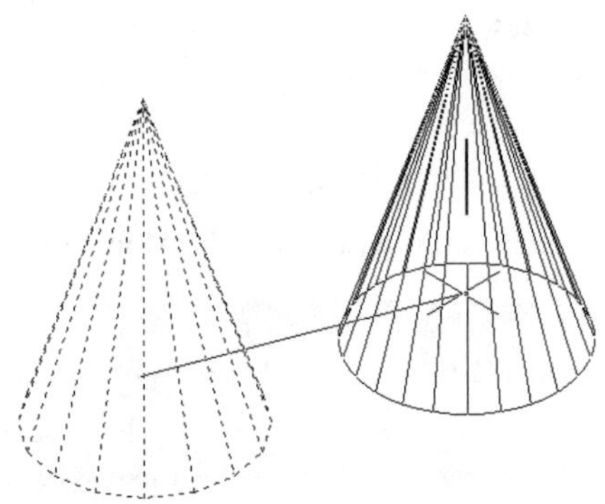

COMMAND LINE OPTIONS

Select objects selects the objects to copy.

Specify base point indicates the starting point for the move.

Displacement specifies relative x,y,z displacement when you press **ENTER** at the next prompt.

Specify second point of displacement indicates the distance to move.

RELATED COMMANDS

Copy copies the selected objects.

MlEdit moves the vertices of a multiline.

PEdit moves the vertices of a polyline.

TIP

- When you press **ENTER** at the 'Specify Second Point of Displacement' prompt, AutoCAD uses the first point as the "displacement."

  For example, when you enter **4,3** at the 'Specify base point' prompt and press **ENTER** at the second prompt, then AutoCAD moves the objects 4 units in the x direction and 3 units in the y.

MRedo

2004 Reverses the effect of the **Undo** command (*short for Multiple* REDO).

Command	Alias	Ctrl+	F-key	Alt+	Menu Bar	Tablet
mredo	...	...	...	...	...	...

Command: mredo
Enter number of actions or [All/Last]: *(Enter an option.)*

COMMAND LINE OPTIONS

Enter number of actions redoes the specified number of steps.

All redoes all commands undone.

Last redoes the last command.

RELATED COMMANDS

Redo redoes a single undo.

U undoes a single command.

Undo undoes one or more commands.

TIPS

- The **MRedo** button on the toolbar lists the redoable actions:

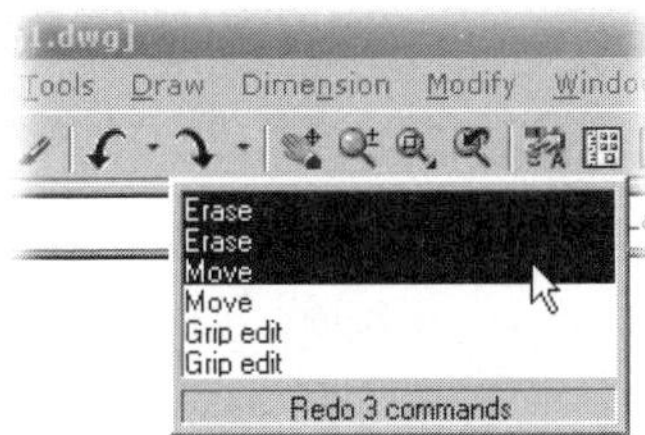

- This command allows you to undo several undoes, but does not allow you to skip over actions.

MSlide

Ver.2.0 Saves the current viewport as *.sld* slide files on disk (*short for Make SLIDE*).

Command	Alias	Ctrl+	F-key	Alt+	Menu Bar	Tablet
mslide	...	...	...	...	...	...

Command: mslide

Displays Create Slide File dialog box. Specify a file name, and then click ***Save****.*

COMMAND LINE OPTIONS

None.

RELATED COMMANDS

Save saves the current drawing as a DWG-format drawing file.

SaveImg saves the current view as a TIFF, Targa, or GIF-format raster file.

VSlide displays an SLD-format slide file in AutoCAD.

RELATED FILES

.sld files store slides created by this command.

.slb files store libraries of slides.

RELATED AUTODESK PROGRAM

SlideLib.exe compiles a group of slides into an SLB-format slide library file.

TIPS

- You view slides with the **VSlide** command.
- Slides were a predecessor to viewing raster and vector images inside AutoCAD.
- Slide files are used to create the images in palette dialog boxes.

MSpace

Rel.11 Switches the drawing from paper space to model space (*short for Model SPACE*).

Command	Alias	Ctrl+	F-key	Alt+	Menu Bar	Tablet
mspace	ms	...	...	...	...	L4

Command: mspace

*In model tab, AutoCAD complains, '** Command not allowed in Model Tab **'.*

If in model space in a layout tab, AutoCAD complains, 'Already in model space.'

In a layout tab (paper space), AutoCAD switches to model space in layout mode, and highlights a viewport:

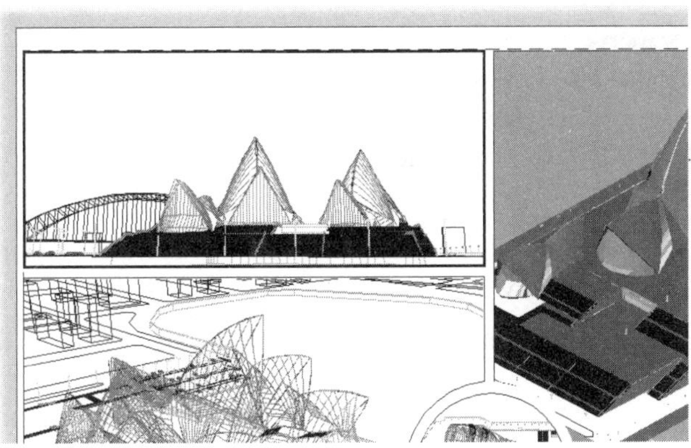

COMMAND LINE OPTIONS

None.

RELATED COMMANDS

PSpace switches from model space to paper space.

Model switches from layout mode to model mode.

Layout switches from model mode to layout mode.

RELATED SYSTEM VARIABLES

MaxActVp specifies the maximum number of viewports with visible objects; default=64.

TileMode specifies the current setting of tiled viewports.

TIPS

- To switch quickly between paper space and model space, click the **MODEL** or **PAPER** button on the status bar:

4.0679, 0.4813, 0.0000 SNAP GRID ORTHO POLAR OSNAP OTRACK LWT MODEL

4.0679, 0.4813, 0.0000 SNAP GRID ORTHO POLAR OSNAP OTRACK LWT PAPER

Click ***PAPER*** *to switch from* ***paper space*** *to* ***model space.***

- AutoCAD clears the selection set when moving between paper space and model space.

MTEdit

Rel.13 Edits mtext objects (*short for Multiline Text EDITor; undocumented command*).

Command	Alias	Ctrl+	F-key	Alt+	Menu Bar	Tablet
mtedit	...	...	...	...	...	...

Command: mtedit

Select an MTEXT object: *(Pick an mtext object.)*

*Displays Text Formatting toolbar; see **MText** command.*

COMMAND LINE OPTION

Select an MTEXT object selects one paragraph text object for editing.

RELATED COMMANDS

DdEdit displays the text editor appropriate for the text object.

Properties changes the properties of an mtext object.

MtProp specifies the properties of an mtext object.

RELATED SYSTEM VARIABLE

MTextEd specifies the name of the external text editor to place and edit multiline text.

TIPS

- This command displays the same dialog box as the **DdEdit** command when an mtext object is selected.
- You can also invoke the mtext editor by double-clicking mtext.

MText

Rel.13 Creates multiline, or paragraph, text objects that fit the width defined by the boundary box (*short for Multline TEXT*).

Command	Alias	Ctrl+	F-key	Alt+	Menu Bar	Tablet
mtext	**t**	...	...	**DXM**	**Draw**	**J8**
	mt				**↳Text**	
					↳Multiline Text	
-mtext	**-t**					

Command: mtext
Current text style: "Standard" Text height: 0.20
Specify first corner: *(Pick a point.)*
AutoCAD displays the mtext bounding box:

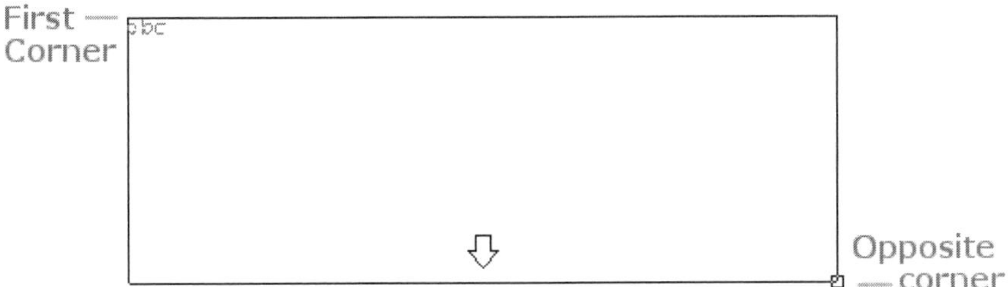

Specify opposite corner or [Height/Justify/Line spacing/Rotation/Style/Width]: *(Pick another point, or enter an option.)*

Displays toolbar.

TOOLBAR OPTIONS

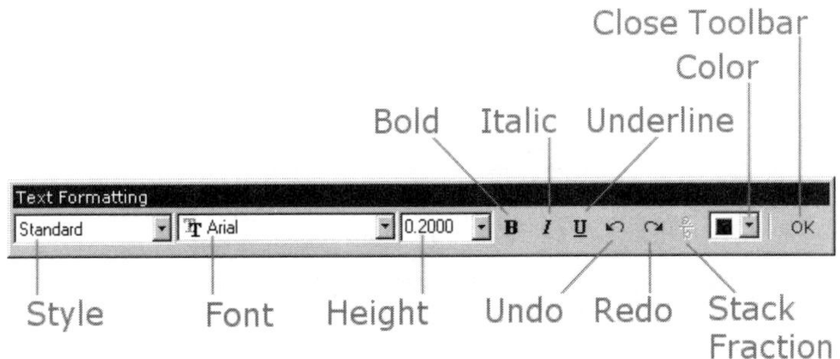

Style selects a predefined text style; see **Style** command.

Font selects a TrueType (*.ttf*) or AutoCAD (*.shx*) font name (default=TXT).

Height specifies the height of the text in units (default = 0.2 units for Imperial drawings, 2.5 units for metric drawings).

Bold boldfaces the text, if permitted by the font.

Italic *italicizes* the text, if permitted by the font.

Underline underlines the text.

Undo undoes the last action.

Redo undoes the last undo.

Stack Fraction stacks a pair of characters separated by slash.

Color selects color for text; choose **Other Color** to display Select Color dialog box.

OK closes the toolbar, and exits the **MText** command.

TAB BAR OPTIONS

Drag Indent Markers
Top: First line
Bottom: Paragraph

Click anywhere to create tabs

Tab: 1.3896

Indents and Tabs...
Set Mtext Width...

The illustrated AutoCAD
Reference

Right click for menu

Indents and Tabs displays Indents and Tabs dialog box.

Set MText Width displays Set MText Width dialog box.

Set Mtext Width dialog box

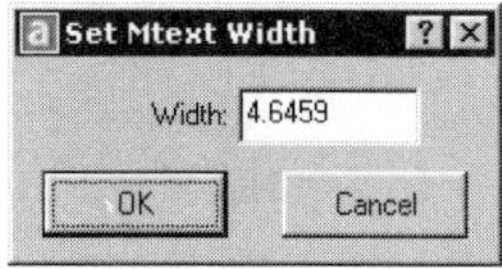

Width changes the width of the mtext bounding box; as an alternative, you can change the boundary box's size by dragging its right and bottom borders.

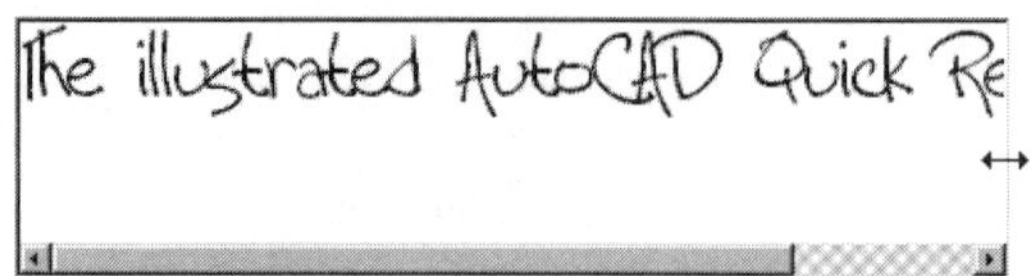

SHORTCUT MENU

Right-click text to display menu:

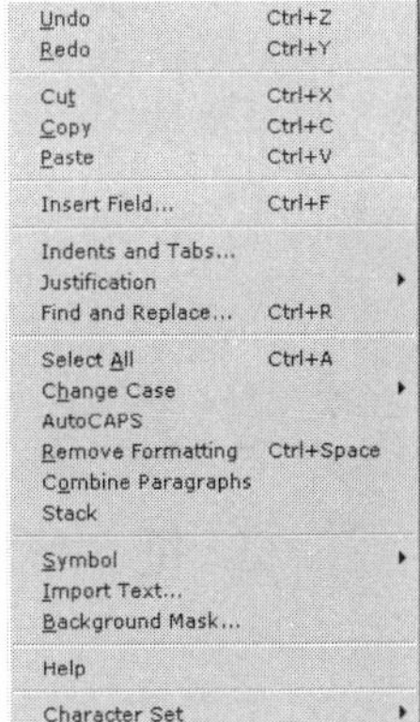

Undo (CTRL+Z) undoes the last action.

Redo (CTRL+Y) undoes the last undo.

Cut (CTRL+X) removes the selected text, and places it in the Clipboard.

Copy (CTRL+C) copies the selected text, and places it in the Clipboard.

Paste (CTRL+V) inserts text stored in the Clipboard.

Insert Field (CTRL+F) displays the Field dialog box; select a field, and then click **OK**. See the **Field** command *(new to AutoCAD 2005)*.

Indents and Tabs displays the Indents and Tabs dialog box.

Justification selects a justification mode:

Mode	Meaning
TL	Top left (default).
TC	Top center.
TR	Top right.
ML	Middle left.
MC	Middle center.
MR	Middle right.
BL	Bottom left.
BC	Bottom center.
BR	Bottom right.

Find and Replace (CTRL+R) displays the Find and Replace dialog box.

Select All (CTRL+A) selects all the text in the bounding box.

Change Case changes the case of the text:

- **UPPERCASE** (CTRL+SHIFT+U) changes selected characters to uppercase.
- **lowercase** (CTRL+SHIFT+L) changes selected characters to lowercase.

AutoCAPS places text as uppercase as it is typed.

Remove Formatting removes bold, italic, and underlined formatting from selected text.

Combine Paragraphs combines selected text into a single paragraph.

Stack stacks text on either side of the / , ^ , or # symbol.

Symbol inserts symbols in the text:

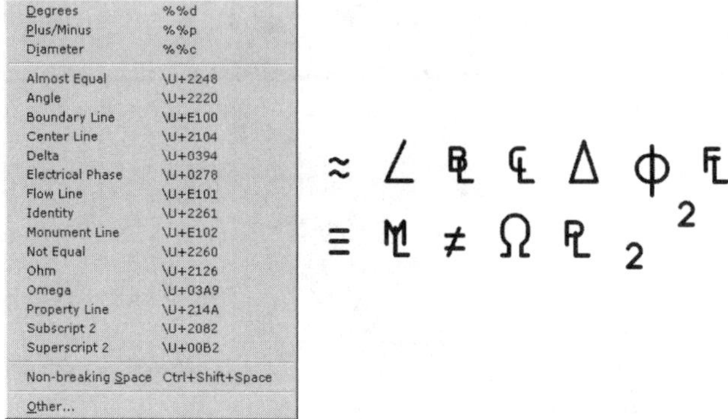

Click **Other** for **Character Map** dialog box.

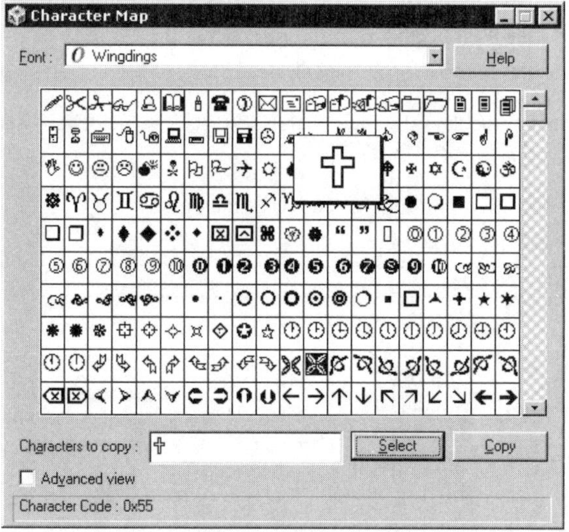

Import Text imports text from ASCII (*.txt*) and RTF format files; displays the Open dialog box. *Caution!* The maximum size of text file is limited to 32KB.

Background Mask displays the Background Mask dialog box (*new to AutoCAD 2005*).

Help displays online help, just like pressing **F1**.

Character Set selects alternate character sets, such as Western, Hebrew, and Thai.

Indents and Tabs dialog box

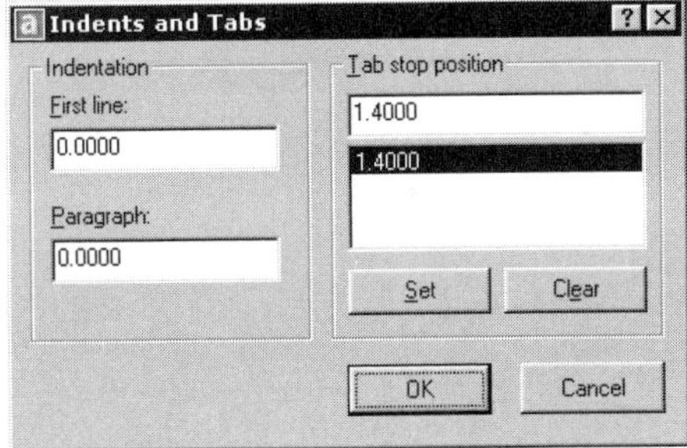

Indentation options

First line specifies the indent distance for the first line of a paragraph; enter a negative number to create a hanging indent.

Paragraph specifies the indent distance for the entire paragraph.

Tab Stop Position options

Set adds the tab position.

Clear removes the selected tab position.

Find and Replace dialog box

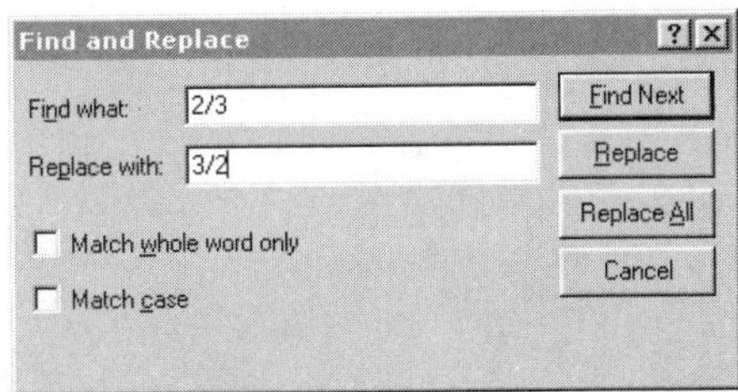

Find what specifies the text to search for. This dialog box searches only text within the bounding box; to search for text in the entire drawing, use the **Find** command.

Replace with specifies the replacement text; leave blank to search only.

Match whole word only

☑ Matches the entire word(s).

☐ Matches parts of the word(s).

Match case

☑ Matches the case of the words.

☐ Ignores the word case.

Find Next finds the next occurance of the word(s).

Replace replaces the found occurance.

Replace All replaces all occurances.

Cancel dismisses the dialog box.

Background Mask dialog box

New to AutoCAD 2005.

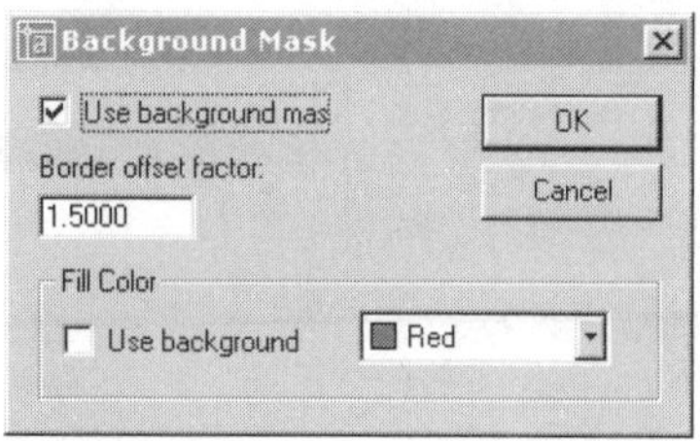

☑ **Use background mask** toggles the display of the background mask. Note that the background mask applies to the full width of the mtext block, rather than the width of text.

Border offset factor determines the distance that the "margin" extends beyond the text. The *factor* is based on the text height: 1.0 means there is no offset; 1.5 means the offset distance is 1.5 times the text height. Maximum value = 5.0; minimum value = 1.0.

Fill Color options

Use background:

☑ Uses the background color (usually white or black).

☐ Uses the specified color; for more colors, chose Select Color.

-MTEXT Command

Command: -mtext

Current text style: STANDARD. Text height: 0.2000

Specify first corner: *(Pick a point.)*

Specify opposite corner or [Height/Justify/Rotation/Style/Width]: *(Pick another point.)*

MText: *(Enter text.)*

MText: *(Press* **ENTER** *to end the command.)*

COMMAND LINE OPTIONS

Height specifies the height of UPPERCASE text (*default = 0.2 units*).

Justify specifies a justification mode.

Rotation specifies the rotation angle of the boundary box.

Style selects the text style for multiline text (default = STANDARD).

Width sets the width of the boundary box; a width of 0 eliminates the boundary box.

RELATED COMMANDS

Properties changes all aspects of mtext.

MtProp changes properties of multiline text.

MtEdit edits mtext.

PasteSpec pastes formatted text from the Clipboard into the drawing.

Style creates a named text style from a font file.

RELATED SYSTEM VARIABLE

MTextEd names the external text editor for placing and editing multiline text.

TIPS

- Use the **MTextEd** system variable to define a different text editor.
- The **Import Text** option is limited to ASCII (unformatted) and RTF (rich text format) text files no more than 32KB in size.
- To import Word documents, copy the text to the Clipboard, and then press CTRL+V in the MText editor. Most, but not all, formatting is retained.
- To import formatted text, copy text from the word processor to the Clipboard, and then use AutoCAD's **PasteSpec** command.
- To link text in the drawing with a word processor, use the **InsertObj** command. When the word processor updates, the linked text is updated in the drawing.
- The mtext editor displays the diameter symbol as "%%c" and nonbreaking spaces as hollow rectangles, but these are displayed correctly in drawings.
- Stacked text can be created on either side of the following symbols:

  Carat (^) stacks text as left-justified tolerance values.

  Forward slash (/) stacks text as center-justified fractional-style values; the slash is converted to a horizontal bar.

  Pound sign (#) stacks text with a tall diagonal bar.

  Use the stack tool a second time to unstack stacked text.

MtProp

Rel.13 Changes the properties of multiline text (*short for Multline Text PROPerties; undocumented command*).

Command	Alias	Ctrl+	F-key	Alt+	Menu Bar	Tablet
mtprop	...	...	...	...	...	...

Command: mtprop

Select an MText object: *(Pick an mtext object.)*

Displays the Text Formatting toolbar; see the ***MText*** *command.*

COMMAND LINE OPTIONS

See the ***MText*** *command.*

RELATED COMMANDS

DdEdit edits multiline text.

MText places multiline text.

Style creates a named text style from a font file.

TIP

- You can also invoke the mtext editor by double-clicking mtext.

Multiple

V. 2.5 Automatically repeats commands that do not repeat on their own.

Command	Alias	Ctrl+	F-key	Alt+	Menu Bar	Tablet
multiple	...	...	...	...	...	...

Command: multiple
Enter command name to repeat: *(Enter command name.)*

This command can also be used as a command modifier:
Command: multiple circle
3P/2P/TTR/<Center point>: *(Pick a point, or enter an option.)*
Diameter/<Radius>: *(Enter an option.)*
circle 3P3P/2P/TTR/<Center point>: *(Pick a point, or enter an option.)*
Diameter/<Radius>: *(Enter an option.)*
circle 3P3P/2P/TTR/<Center point>: *(Press* **ESC** *to end command.)*

COMMAND LINE OPTIONS

Enter command name to repeat specifies the name of the command to repeat.

ESC stops the command from automatically repeating itself.

COMMAND INPUT OPTIONS

SPACEBAR repeats the previous command.

CLICK repeats a command by clicking on any blank spot of the tablet menu.

RELATED COMMANDS

Redo undoes an undo.

U undoes the previous command; undoes one multiple command at a time.

RELATED COMMAND MODIFIERS

' *(apostrophe)* allows the use of some commands within another command.

. *(period)* forces the use of undefined commands.

- *(dash)* forces the display of prompts on the command line for some commands.

+ *(plus)* prompts for the tab number of tabbed dialog boxes.

_ *(underscore)* uses the English command in international versions of AutoCAD.

(*(open parenthesis)* executes AutoLISP functions on the command line.

$(*(dollar and parenthesis)* executes Diesel functions on the command line.

TIPS

- Use the **Multiple** command to repeat commands that do not repeat on their own. This command does not cause options to repeat.
- *Warning!* Multiple **U** will undo all edits in the drawing.

MView

Rel.11 Creates and manipulates overlapping viewports (*short for Make VIEWports*).

Command	Alias	Ctrl+	F-key	Alt+	Menu Bar	Tablet
mview	mv	R	...	...	...	M4

Command: mview

*In Model tab, AutoCAD complains, '** Command not allowed in Model Tab **'.*

In a layout tab, AutoCAD prompts:

Specify corner of viewport or
[ON/OFF/Fit/Shadeplot/Lock/Object/Polygonal/Restore/2/3/4]<Fit>: *(Pick a point, or enter an option.)*
Specify opposite corner: *(Pick a point.)*
Regenerating drawing.

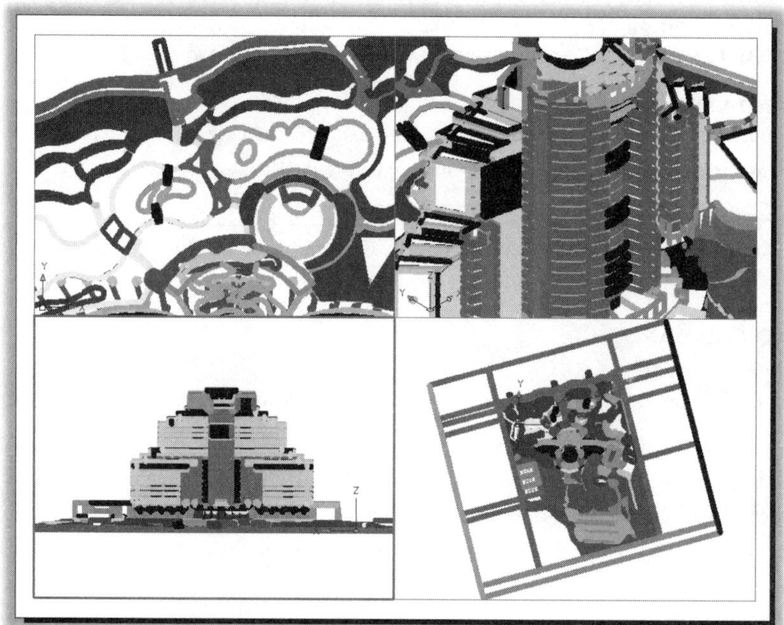

COMMAND LINE OPTIONS

Specify corner of viewport indicates the first point of a single viewport (default).

Fit creates a single viewport that fits the screen.

Shadeplot creates a hidden-line or shaded view during plotting and printing.

Lock locks the selected viewport.

Object converts a circle, closed polyline, ellipse, spline, or region into a viewport.

OFF turns off a viewport.

ON turns on a viewport.

Polygonal creates a multisided viewport of straight lines and arcs.

Restore restores a saved viewport configuration.

2 options
Enter viewport arrangement [Horizontal/Vertical] <Vertical>: *(Enter an option.)*
Specify first corner or [Fit] <Fit>: *(Pick a point, or enter an option.)*

Horizontal stacks two viewports.

Vertical places two viewports side-by-side (default).

3 options
[Horizontal/Vertical/Above/Below/Left/Right]<Right>: *(Enter an option.)*
Specify first corner or [Fit] <Fit>: *(Pick a point, or enter an option.)*

Horizontal stacks the three viewports.

Vertical places three side-by-side viewports.

Above places two viewports above the third.

Below places two viewports below the third.

Left places two viewports to the left of the third.

Right places two viewports to the right of the third (*default*).

4 options
Specify first corner or [Fit] <Fit>: *(Pick a point, or enter an option.)*

Fit creates four identical viewports that fit the viewport.

First Point indicates the area of the four viewports (default).

Shadeplot options
Shade plot? [As displayed/Wireframe/Hidden/Rendered] <As displayed>:
(Enter an option.)

As displayed plots the drawing as displayed.

Wireframe plots the drawing as a wireframe.

Hidden plots the drawing with hidden lines removed.

Rendered plots the drawing rendered.

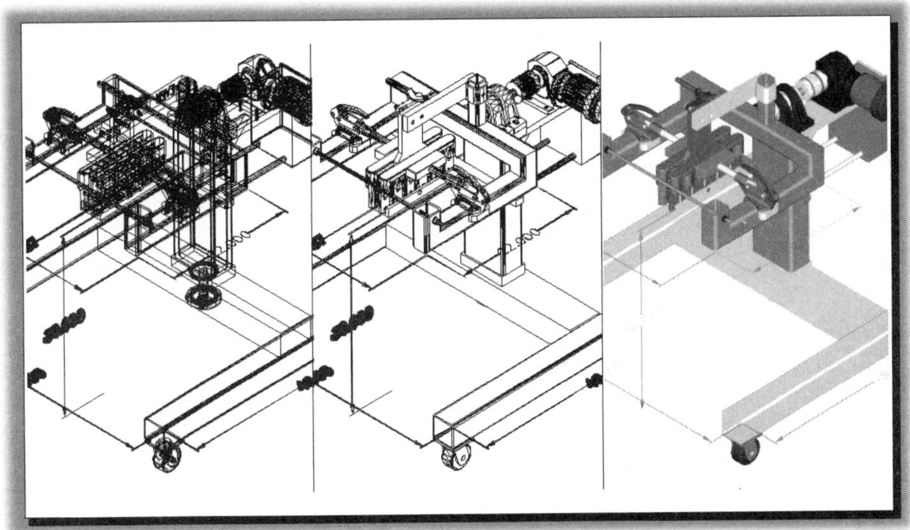

Wireframe Hidden-line Rendered

RELATED COMMANDS

Layout creates new layouts.

MSpace switches to model space.

PSpace switches to paper space before creating viewports.

RedrawAll redraws all viewports.

RegenAll regenerates all viewports.

VpLayer controls the visibility of layers in each viewport.

VPorts creates tiled viewports in model space.

Zoom zooms a viewport relative to paper space via the XP option.

RELATED SYSTEM VARIABLES

CvPort specifies the number of the current viewport.

MaxActVp controls the maximum number of visible viewports:

MaxActVP	Meaning
1	Minimum.
64	Default.
32767	Maximum.

TileMode controls the availability of overlapping viewports.

TIPS

- Although the system variable **MaxActVp** limits the number of simultaneously-visible viewports, the **Plot** command plots all viewports.
- **TileMode** must be set to 0 to switch to paper space and to use the **MSpace** command.
- **Snap**, **Grid**, **Hide**, **Shade**, and so on can be set separately in each viewport.
- Some of this command's options are also available from the Properties window: select a viewport, right-click, and then select **Properties** from shortcut menu.

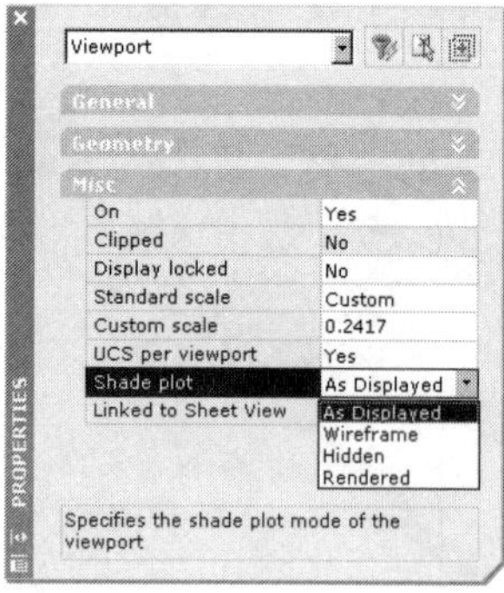

- Press CTRL+R to switch between viewports.

- The preset viewports created by the **MView** command have these shapes:

Fit option

Fit	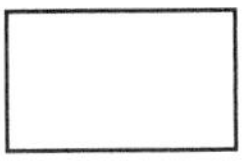	Creates a single viewport.

2 options

Horizontal	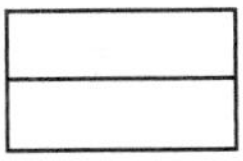	Creates one viewport over another viewport.
Vertical	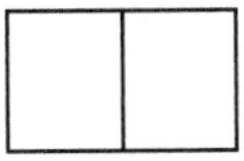	Creates one viewport beside another (default).

3 options

Horizontal	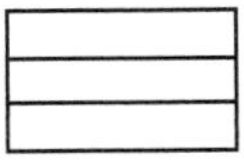	Creates three viewports over each other.
Vertical	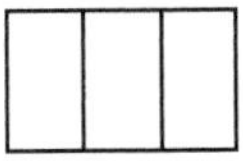	Creates three viewports side-by-side.
Above	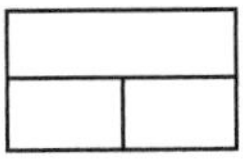	Creates one viewport over of two viewports.
Below	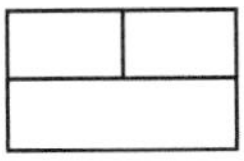	Creates one viewport below two viewports.
Left	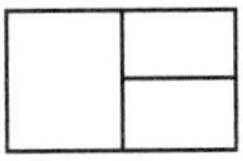	Creates one viewport left of two viewports.
Right	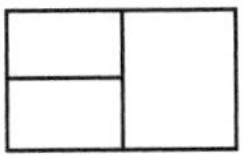	Creates one viewport right of two viewports (default).

4 option

4	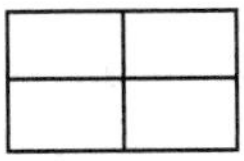	Splits the current viewport into four viewports.

MvSetup

Rel. 11 Quickly sets up a drawing, complete with a predrawn border. Optionally sets up multiple viewports, sets the scale, and aligns views in each viewport (*short for Model View SETUP*).

Command	Alias	Ctrl+	F-key	Alt+	Menu Bar	Tablet
mvsetup	**mvs**	...	...	...	...	...

Command: mvsetup

When in model space:

Enable paper space? [No/Yes] <Y>: *(Type* **Y** *or* **N**.*)*

Command prompts in model tab (not paper space):

Enter units type [Scientific/Decimal/Engineering/Architectural/Metric]: *(Enter an option.)*

Enter the scale factor: *(Specify a distance.)*

Enter the paper width: *(Specify a distance.)*

Enter the paper height: *(Specify a distance.)*

Command prompts in layout mode (paper space):

Enter an option [Align/Create/Scale viewports/Options/Title block/Undo]: *(Enter an option.)*

COMMAND LINE OPTIONS

Align options

Pans the view to align a base point with another viewport.

Enter an option [Angled/Horizontal/Vertical alignment/Rotate view/Undo]: *(Enter an option.)*

Angled specifies the distance and angle from a base point to a second point.

Horizontal aligns views horizontally with a base point in another viewport.

Vertical alignment aligns views vertically with a base point in another viewport.

Rotate view rotates the view about a base point.

Undo undoes the last action.

Create options

Enter option [Delete objects/Create viewports/Undo] <Create>: *(Enter an option.)*

Delete objects erases existing viewports.

Create viewports creates viewports in these configurations:

Layout	Meaning
0	No layout.
1	Single viewport.
2	Standard engineering layout.
3	Array viewports along x and y axes.

Undo undoes the last action.

Scale Viewports options
Select the viewports to scale...

Select objects: *(Pick a viewport.)*

Select objects: *(Press* **ENTER** *to end object selection.)*

Set the ratio of paper space units to model space units...

Enter the number of paper space units <1.0>: *(Enter a value.)*

Enter the number of model space units <1.0>: *(Enter a value.)*

Select objects selects one or more viewports.

Enter the number of paper space units scales the objects in the viewport with respect to drawing objects.

Enter the number of model space units scales the objects in the viewport with respect to drawing objects.

Options options
Enter an option [Layer/LImits/Units/Xref] <exit>: *(Enter an option.)*

Layer specifies the layer name for the title block.

Limits specifies whether to reset limits after title block insertion.

Units specifies inch or millimeter paper units.

Xref specifies whether title is inserted as a block or as an external reference.

Title Block options
Enter title block option [Delete objects/Origin/Undo/Insert] <Insert>: *(Enter an option.)*

Delete objects erases an existing title block from the drawing.

Origin relocates the origin.

Undo undoes the last action.

Insert displays the available title blocks.

RELATED SYSTEM VARIABLE

TileMode specifies the current setting of TileMode.

RELATED FILES

mvsetup.dfs is the MvSetup default settings file.

acadiso.dwg is a template drawing with ISO (international standards) defaults.

Plus all *.dwt* template drawings.

RELATED COMMANDS

LayoutWizard sets up the viewports via a "wizard."

TIPS

- When option **2 (Std. Engineering)** is selected at the **Create** option, the following views are created (counterclockwise from upper left):

  Top view.

  Isometric view.

  Front view.

  Right view.

- To create the title block, **MvSetup** searches the path specified by the **AcadPrefix** variable. If the appropriate drawing cannot be found, **MvSetup** creates the default border.

- **MvSetup** makes use the following predefined title blocks:

  0: None
  1: ISO A4 Size(mm)
  2: ISO A3 Size(mm)
  3: ISO A2 Size(mm)
  4: ISO A1 Size(mm)
  5: ISO A0 Size(mm)
  6: ANSI-V Size(in)
  7: ANSI-A Size(in)
  8: ANSI-B Size(in)
  9: ANSI-C Size(in)
  10: ANSI-D Size(in)
  11: ANSI-E Size(in)
  12: Arch/Engineering (24 x 36in)
  13: Generic D size Sheet (24 x 36in)

- The metric A0 size is similar to the imperial E-size, while the metric A4 size is similar to A-size.

- This command provides the following preset scales (scale factor shown in parentheses):

Architectural Scales	Scientific Scales	Decimal Scales	Engineering Scales	Metric Scales
(480) 1/40"=1'	**(4.0) 4 TIMES**	**(4.0) 4 TIMES**	**(120) 1"=10'**	**(5000) 1:5000**
(240) 1/20"=1'	**(2.0) 2 TIMES**	**(2.0) 2 TIMES**	**(240) 1"=20'**	**(2000) 1:2000**
(192) 1/16"=1'	**(1.0) FULL**	**(1.0) FULL**	**(360) 1"=30'**	**(1000) 1:1000**
(96) 1/8"=1'	**(0.5) HALF**	**(0.5) HALF**	**(480) 1"=40'**	**(500) 1:500**
(48) 1/4"=1'	**(0.25) QUARTER**	**(0.25) QUARTER**	**(600) 1"=50'**	**(200) 1:200**
(24) 1/2"=1'			**(720) 1"=60'**	**(100) 1:100**
(16) 3/4"=1'			**(960) 1"=80'**	**(75) 1:75**
(12) 1"=1'			**(1200) 1"=100'**	**(50) 1:50**
(4) 3"=1'				**(20) 1:20**
(2) 6"=1'				**(10) 1:10**
(1) FULL				**(5) 1:5**
				(1) FULL

- You can add your own title block with the **Add** option. Before doing so, create the title block as an AutoCAD drawing.

Quick Start Tutorial

Using MvSetup

MvSetup has many options, but does not present them in a logical fashion. To set up a drawing with **MvSetup**, follow these basic steps:

Step 1

Start the **MvSetup** command:

Command: mvsetup
Enter an option [Align/Create/Scale viewports/Options/Title block/Undo]: *(Type* **C***.)*

Step 2

Select options:

Enter an option [Layer/LImits/Units/Xref] <exit>: *(Type* **L***.)*

Decide on the layer for the title block with the **Layer** option. Specify the paper space units with the **Units** option.

Step 3

Place title block:

Enter title block option [Delete objects/Origin/Undo/Insert] <Insert>: *(Type* **I***.)*

Place the title block with the **Title block** option's **Insert** option.

Step 4

Create viewports:

Enter option [Delete objects/Create viewports/Undo] <Create>: *(Type* **C***.)*

Set up the viewports with the **Create** option's **Create viewports** option. For standard drawings, select option **#2, Std. Engineering**.

Step 5

Scale the viewports. Make the object the same size in all four viewports with the **Scale viewports** option. When you are prompted to 'Select objects', select the four *viewports*, not the objects in the viewports.

Step 6

Align the views in each viewport with the **Align** option.

- You can interrupt the **MvSetup** command at any time with the ESC key, and then resume the command to complete the setup.
- Save your work when done!

NetLoad

2005 Loads *.dll* files written with Microsoft's .Net programming interface.

Command	Alias	Ctrl+	F-key	Alt+	Menu Bar	Tablet
netload	...	...	...	...	...	...

Command: netload

Displays the Choose .Net Assembly dialog box.

Select a .dll file, and then click ***Open****.*

COMMAND LINE OPTIONS

None.

TIP

- Some parts of AutoCAD 2005 written in .Net include the new Layer dialog box and the migration utility.

New

Rel.12 Starts new drawings from template drawings, from scratch, or through step-by-step drawing setup "wizards."

Command	Alias	Ctrl+	F-key	Alt+	Menu Bar	Tablet
new	...	**N**	...	**FN**	**File** ↳**New**	**T24**

Command: new

AutoCAD displays one of three interfaces, depending on the settings of the ***FileDia*** *and* ***Startup*** *variables.*

FileDia	Startup	New
1	1	Displays Startup wizard.
1	0	Displays Select Template dialog box.
0	1 or 0	Prompts for *.dwt* file at command line.

DIALOG BOX OPTIONS

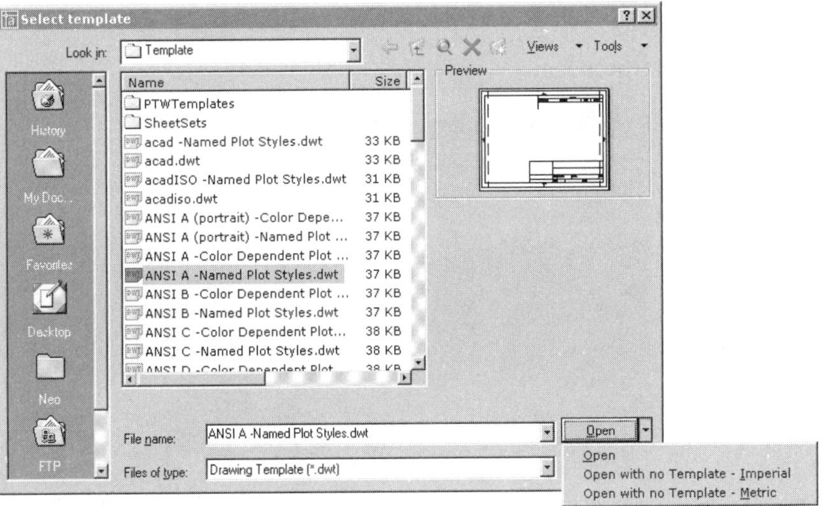

Open opens the selected template file.

Open with no template - Imperial opens the *acad.dwt* file.

Open with no template - Metric opens the *acadiso.dwg* file.

WIZARD OPTIONS

The Startup wizard is displayed when AutoCAD first starts.

The Create New Drawing wizard is similar, and is displayed when subsequent new drawings are opened.

Open a Drawing view

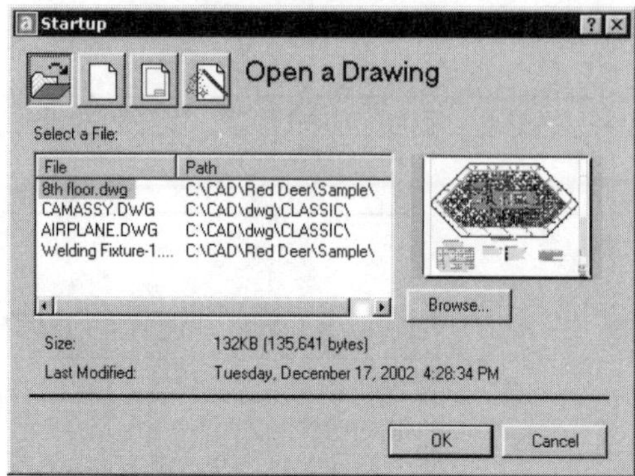

(This view is not found in the Create New Drawing wizard.)

Select a File selects one of the four drawings listed.

Browse displays the Select File dialog box; see the **Open** command.

Start from Scratch view

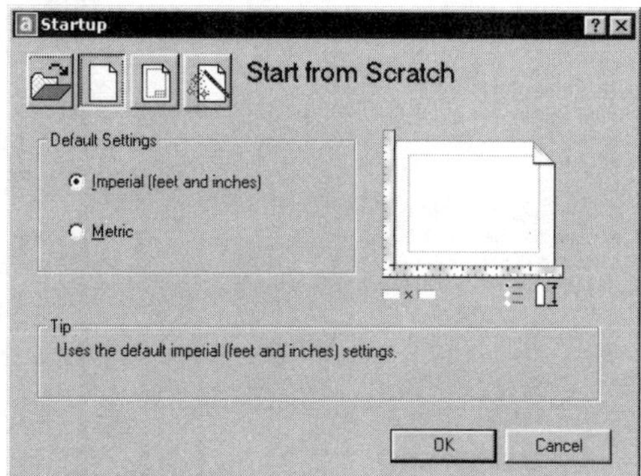

⊙**English** creates a new drawing based on the *acad.dwt* (English units) template file.

○**Metric** creates a new drawing based on the *acadiso.dwt* (metric units) template file.

Use a Template view

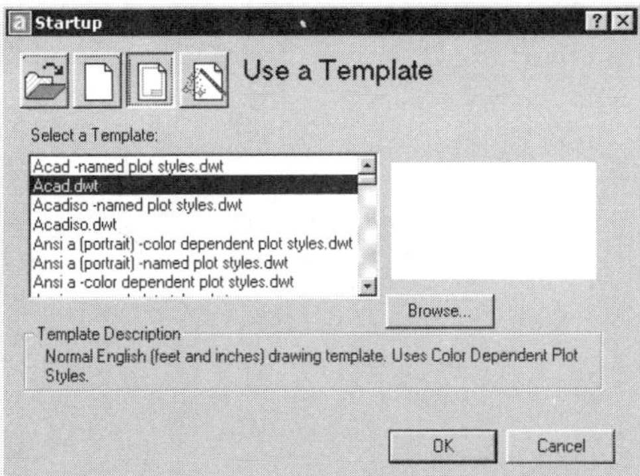

Select a Template creates a new drawing based on the selected *.dwt* template file.

Browse displays the Select a Template File dialog box.

Use a Wizard view

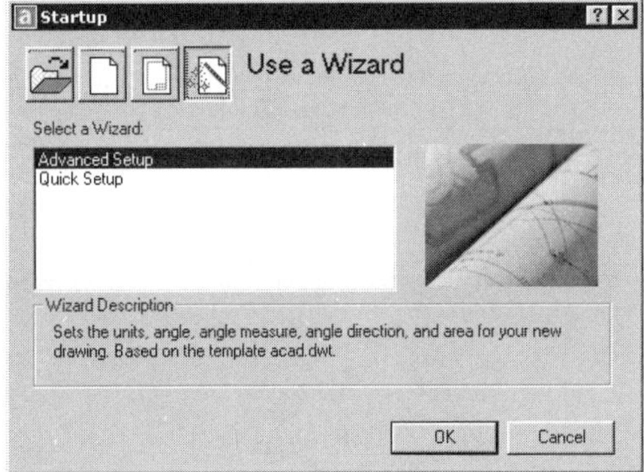

Select a Wizard

- **Advanced Setup** sets up a new drawing in several steps.
- **Quick Setup** sets up a new drawing in two steps.

Quick Setup wizard

Units page

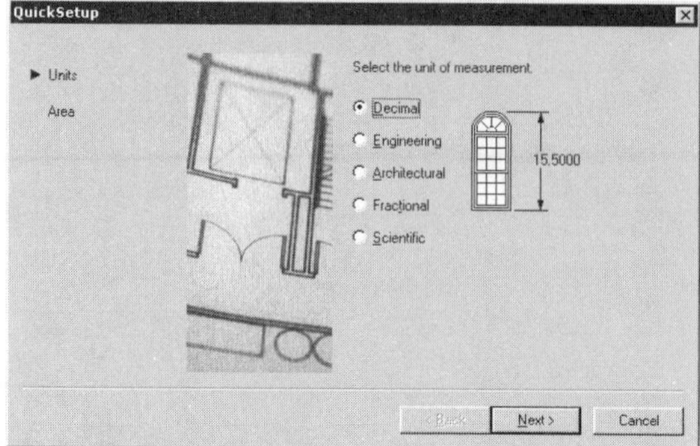

⊙ **Decimal** displays units in decimal (or "metric") notation (default): 123.5000.

○ **Engineering** displays units in feet and decimal inches: 10'-3.5000".

○ **Architectural** displays units in feet, inches, and fractional inches: 10' 3-1/2".

○ **Fractional** displays units in inches and fractions: 123 1/2.

○ **Scientific** displays units in scientific notation: 1.235E+02.

Buttons

Cancel cancels the wizard, and returns to the previous drawing.

Back moves back one step.

Next moves forward one step.

Area page

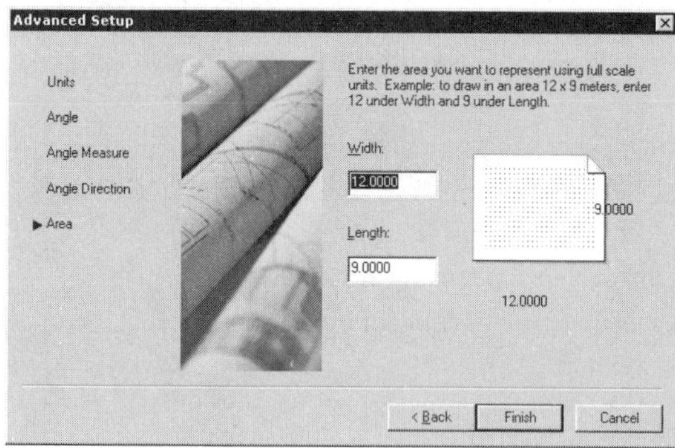

Width specifies the width of the drawing in real-world (not scaled) units; default = 12 units.

Length specifies the length or depth of the drawing in real-world units; default = 9 units.

Advanced Setup wizard

Units page

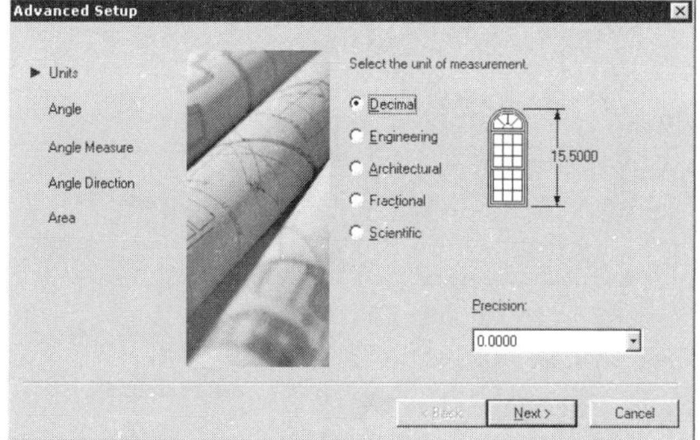

⊙ **Decimal** displays units in decimal (or "metric") notation (default): 123.5000.

O **Engineering** displays units in feet and decimal inches: 10'-3.5000".

O **Architectural** displays units in feet, inches, and fractional inches: 10' 3-1/2".

O **Fractional** displays units in inches and fractions: 123 1/2.

O **Scientific** displays units in scientific notation: 1.235E+02.

O **Precision** selects the precision of display up to 8 decimal places or 1/256.

Angle page

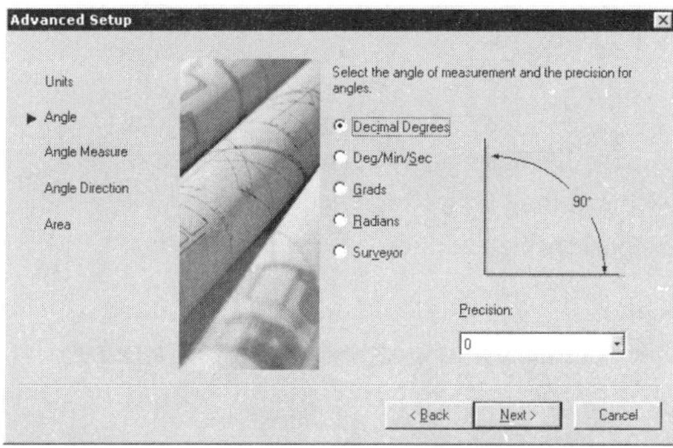

⊙ **Decimal Degrees** displays decimal degrees (default): 22.5000.

O **Deg/Min/Sec** displays degrees, minutes, and seconds: 22 30.

O **Grads** displays grads: 25g.

O **Radians** displays radians: 25r.

O **Surveyor** displays surveyor units: N 25d0'0" E.

Precision selects a precision ranging up to 8 decimal places.

Angle Measure page

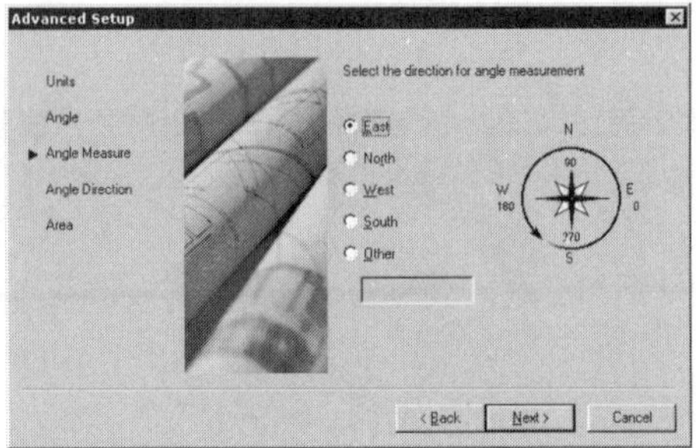

⊙ **East** specifies that zero degrees points East (default).

○ **North** specifies that zero degrees points North.

○ **West** specifies that zero degrees points West.

○ **South** specifies that zero degrees points South.

○ **Other** specifies any of the 360 degrees as zero degrees.

Angle Direction page

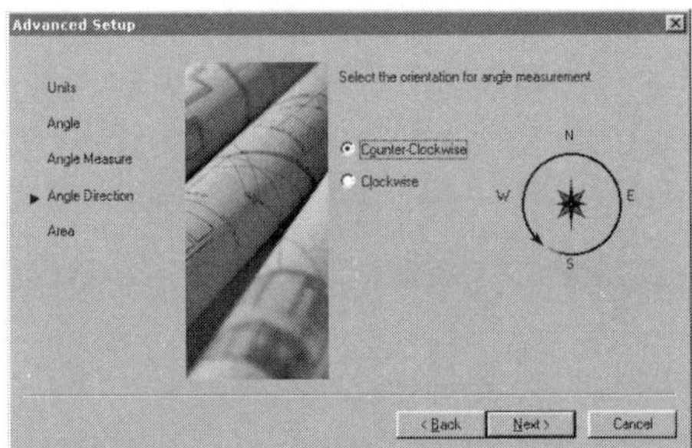

Counter-Clockwise measures positive angles counterclockwise from 0 degrees (default).

Clockwise measures positive angles clockwise from 0 degrees.

Area page

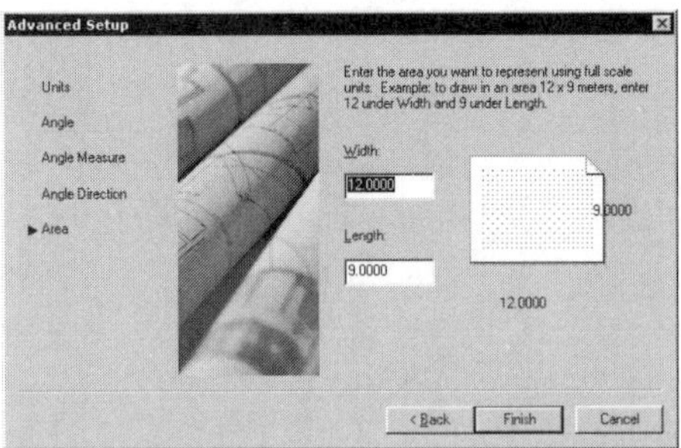

Width specifies the width of the drawing in real-world (not scaled) units; default = 12 units.

Length specifies the length or depth of the drawing in real-world units; default = 9 units.

Command Line Switches

*Switches used by **Target** field on **Shortcut** tab in the AutoCAD desktop icon's Properties dialog box:*

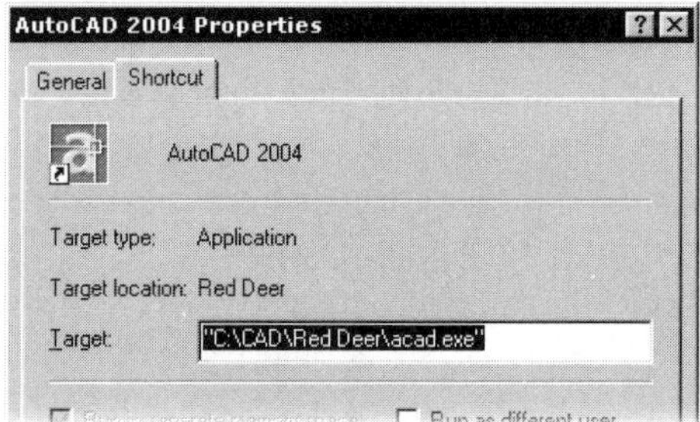

/b runs a script file after AutoCAD starts; uses the following format:

acad.exe "\acad 2005\drawing.dwg" /b "file name.scr"

/c specifies the path for alternative hardware configuration file; default = *acad2005.cfg*.

/layout specifies the layout to display *(new to AutoCAD 2005)*.

/nologo suppresses the display of the AutoCAD logo screen.

/nossm prevents Sheetset Manager window from loading *(new to AutoCAD 2005)*.

/p specifies a user-defined profile to customize AutoCAD's user interface.

/r restores the default pointing device.

/s specifies additional support folders; maximum is 15 folders, with each folder name separated by a semicolon.

/set specifies the *.dst* sheet set file to load *(new to AutoCAD 2005)*.

/t specifies the *.dwt* template drawing to use.

/v specifies the named view to display upon startup of AutoCAD.

Command Line Options

*When **FileDia** = 0, AutoCAD prompts you at the command line:*

Command: new

Enter template file name or [. (for none)] <*default .dwt file path name*>: *(Enter the path and name of a .dwt , .dwg, or .dws file.)*

Alternatively, enter the following options:

ENTER accepts the default template drawing file.

. (period) eliminates use of a template; AutoCAD uses either *acad.dwt* or *acadiso.dwt*, depending on the setting of the **MeasureInit** system variable.

~ (tilde) forces the display of the Select Template dialog box.

RELATED COMMANDS

QNew starts a new drawing based on a predetermined template file.

SaveAs saves the drawing in *.dwg* or *.dwt* formats; creates template files.

RELATED SYSTEM VARIABLES

DbMod indicates whether the drawing has changed since being loaded.

DwgPrefix indicates the path to the drawing.

DwgName indicates the name of the current drawing.

FileDia displays prompts at the 'Command:' prompt.

Startup determines whether the dialog box or the wizard is displayed.

MeasureInit determines whether the units are imperial or metric.

RELATED FILES

wizard.ini holds the names and descriptions of template files.

***.dwt** are template files stored in *.dwg* format.

TIPS

- Until you give the drawing a name, AutoCAD names it *drawing1.dwg*.
- The default template drawing is *acad.dwg*.
- Edit and save *.dwt* template drawings to change the defaults for new drawings.
- When you press CTRL+N, AutoCAD's behavior differs from Microsoft Office programs: Office programs display a new document that takes on the properties of the current document.
- Turning off **FileDia** (set to 0) is meant for use with macros and programs.
- The **Startup** system variable can be set in the Options dialog box: from the **Tools** menu, select **Options**; click the **System** tab (not the Open and Save tab!). In the General Options section, the Startup option offers two settings:

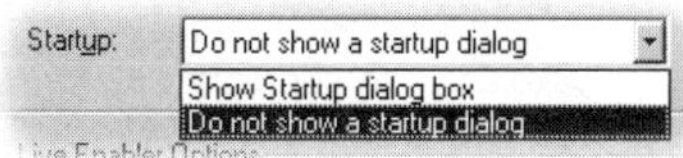

  Do not show a startup dialog displays the Select Template dialog box.

  Show Startup dialog displays the Startup and Create New Drawing wizards.

- The **Today** window was removed with AutoCAD 2004.

NewSheetset

2005 Runs the New Sheetset wizard.

Command	Alias	Ctrl+	F-key	Alt+	Menu Bar	Tablet
newsheetset	...	...	...	**TZS**	**Tools** ↳**Wizards** ↳**New Sheet Set**	...

Command: newsheetset

Displays the New Sheet Set wizard.

WIZARD OPTIONS

Begin page

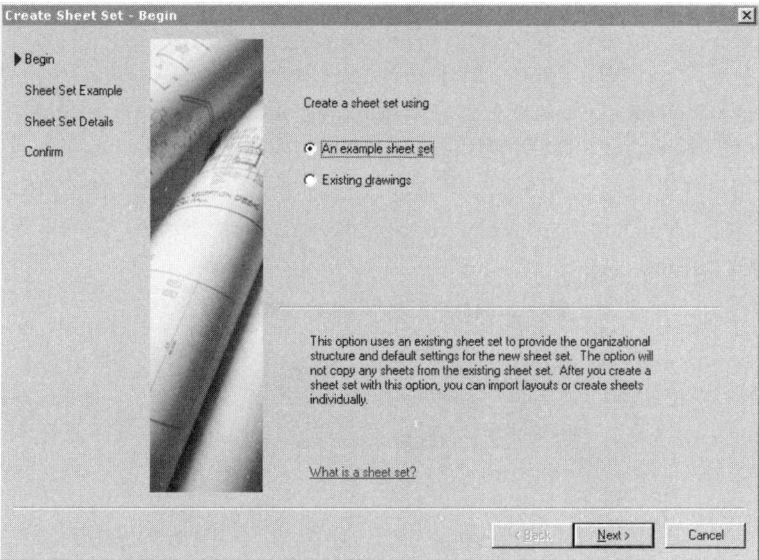

Create a sheet set using:

⊙**An example sheet set** creates a new sheet set based on "templates."

O**Existing drawings** selects one or more folders holding drawings, which are imported into the sheet set.

Click **Next** to continue.

Example Sheet Set options

Sheet Set Example page

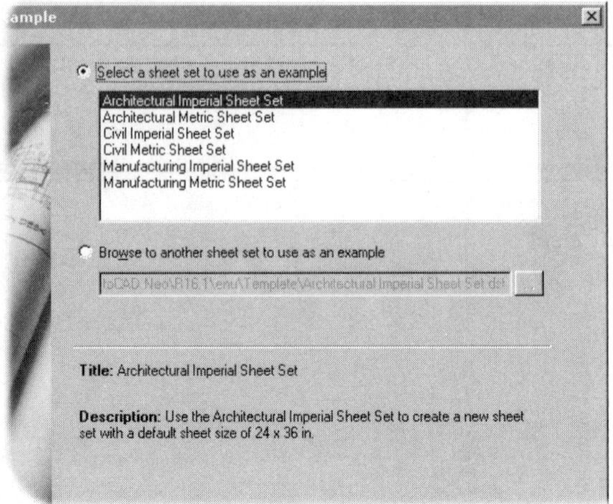

⊙**Select a sheet set to use as an example:** selects one of the sheet sets provided by Autodesk.

○**Browse to another sheet set to use as an example** opens a *.dst* sheet set data file.

... Opens the Browse for Sheet Set dialog box; select a *.dst* file, and then click **Open**.

Click **Next** to continue.

Sheet Set Details page

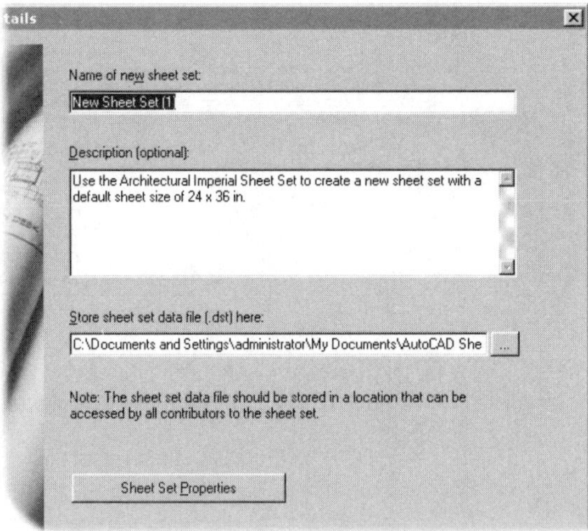

Name of new sheet set names the the sheet set.

Description describes the sheet seet.

Store sheet set data file here specifies the location of the *.dst* sheet set data file.

... displays the Browse for Sheet Set Folder dialog box; select a folder, and then click **Open**.

Sheet Set Propeties displays the Sheet Set Properties dialog box.

Click **Next** to continue.

Finishes with the Confirm page, illustrated below.

Existing Drawings options

Sheet Set Details page

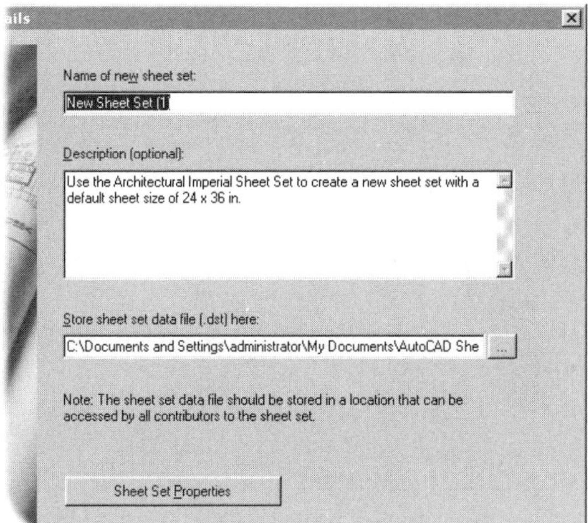

Name of new sheet set names the sheet set.

Description describes the sheet seet.

Store sheet set data file here specifies the location of the sheet set data file.

... displays the Browse for Sheet Set Folder dialog box; select a folder, and then click **Open**.

Sheet Set Propeties displays the Sheet Set Properties dialog box.

Click **Next** to continue.

Choose Layouts page

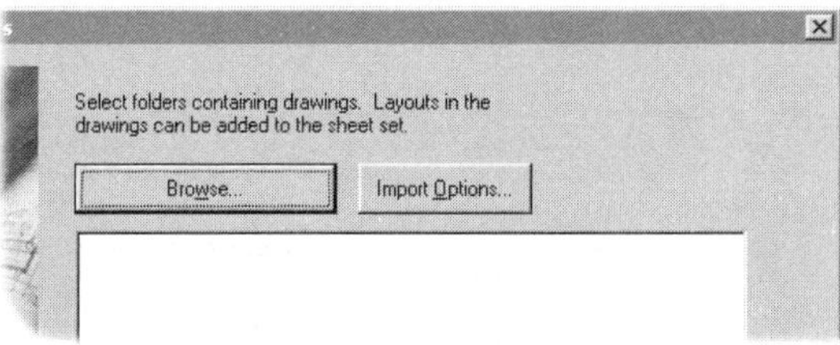

Browse displays Browse for Folder dialog box; select a folder, and then click **OK**.

Import Options displays the Import Options dialog box.

Click **Next** to continue.

Confirm page

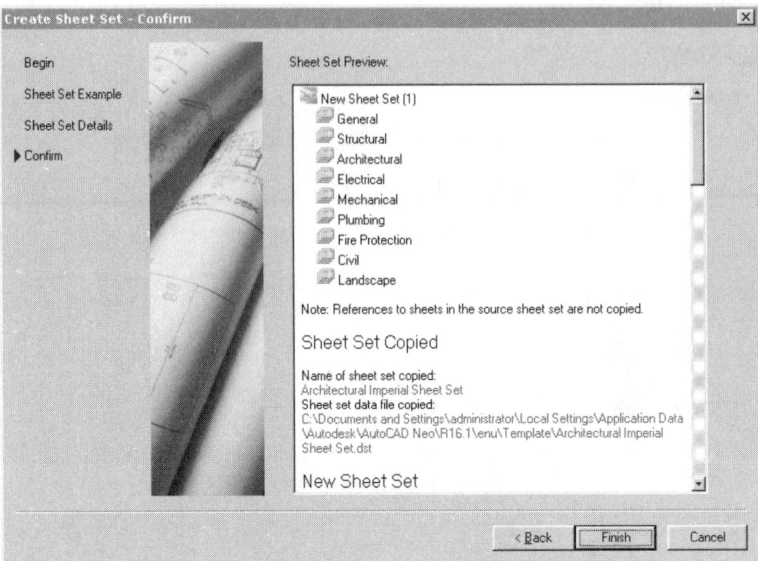

To make a copy of the settings, select all text, and then press **Ctrl+C**. (In a text editor, press **Ctrl+V** to paste the text in a document.)

Click **Back** to change and correct settings.

Click **Finish** to create the new sheet set.

Sheet Set Properties dialog box

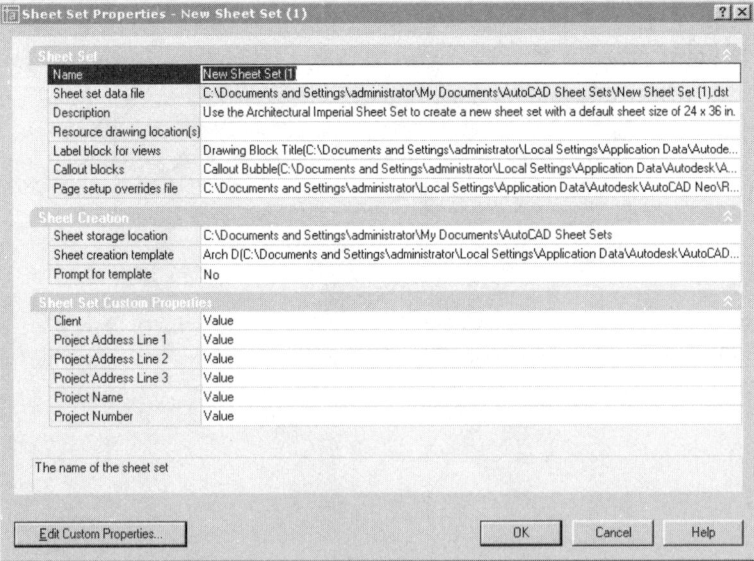

Edit Custom Properties displays the Custom Properties dialog box.

Custom Properties dialog box

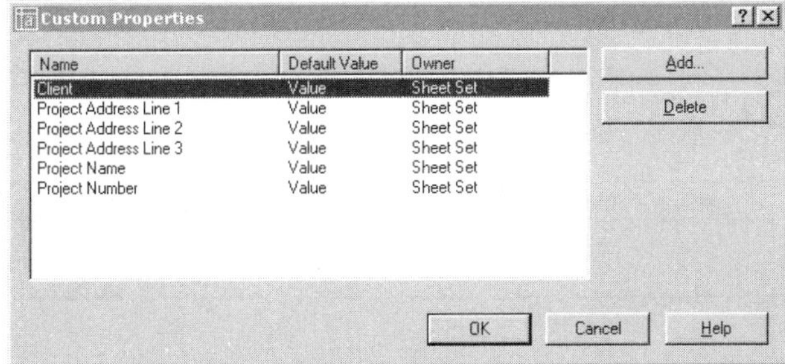

Add displays the Add Custom Property dialog box.

Delete removes the selected custom property without warning; click **Cancel** to undo erasure.

Add Custom Property dialog box

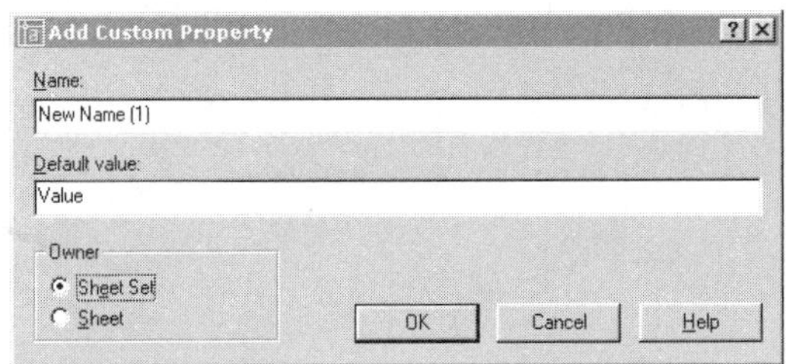

Name names the custom property.

Default value specifies the default value.

Owner:

⦿ **Sheet set** indicates that the custom property belongs to sheet set.

○ **Sheet** indicates that custom property belongs to sheet.

Import Options dialog box

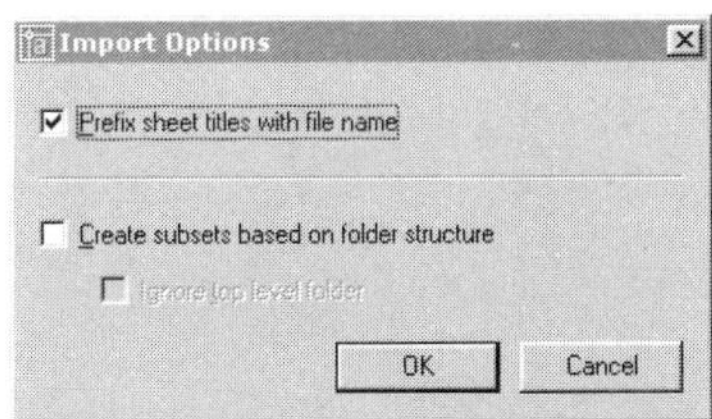

☑ **Prefix sheet titles with file name** tags sheet set names with file names.

☐ **Create subsheets based on folder structure** generates sheets based on folder names.

☐ **Ignore top level folder** does not include topmost folder in sheet name generation.

RELATED COMMANDS

OpenSheetset opens existing sheet sets.

Sheetset displays the Sheet Set Manager window.

Offset

V. 2.5 Draws parallel lines, arcs, circles and polylines; repeats automatically until cancelled.

Command	Alias	Ctrl+	F-key	Alt+	Menu Bar	Tablet
offset	**o**	**...**	**...**	**MS**	**Modify** **↳Offset**	**V17**

Command: offset
Specify offset distance or [Through] <Through>: *(Enter a number, or type* **T***.)*
Select object to offset or <exit>: *(Select an object.)*
Specify point on side to offset: *(Pick a point.)*
Select object to offset or <exit>: *(Select another object, or press* **ENTER** *to end the command.)*

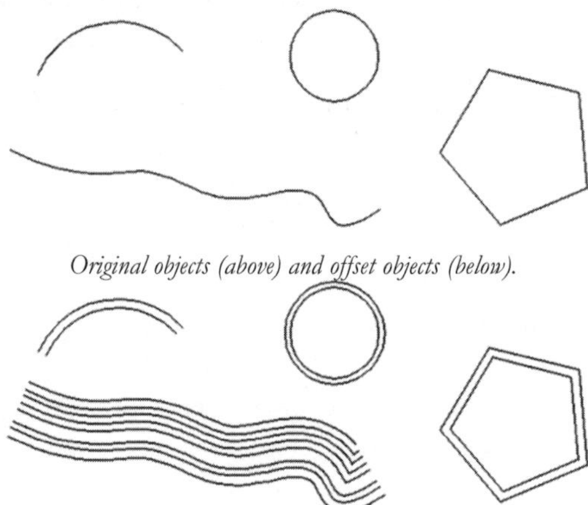

Original objects (above) and offset objects (below).

COMMAND LINE OPTIONS

Offset distance specifies the perpendicular distance to offset.

Through indicates the offset distance.

ESC exits the command.

Through options
Select object to offset or <exit>: *(Select an object.)*
Specify through point: *(Pick a point.)*

Select object to offset selects the object to be offset.

Specify through point specifies the offset distance.

RELATED COMMANDS

Copy creates one or more copies of a group of objects.

MLine draws up to 16 parallel lines.

RELATED SYSTEM VARIABLES

OffsetDist specifies the current offset distance.

OffsetGapType determines how to close gaps created by offset polylines.

OleConvert

2000 Converts OLE objects, if possible (*undocumented command*).

Command	Alias	Ctrl+	F-key	Alt+	Menu Bar	Tablet
oleconvert	...	...	...	...	...	...

Select an OLE object before entering the command.

Command: oleconvert

Displays dialog box:

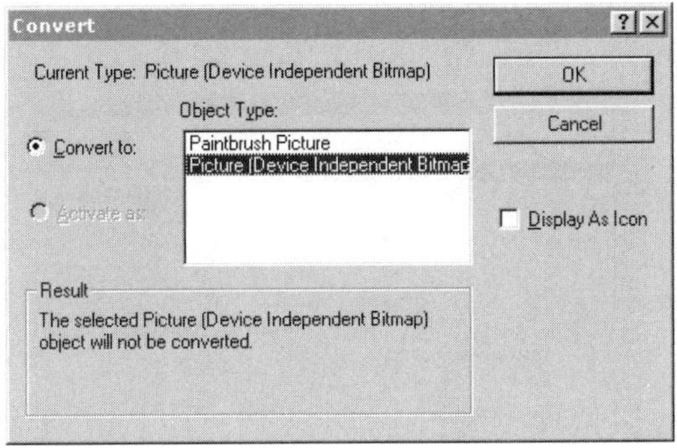

DIALOG BOX OPTIONS

Object Type selects the type of object to convert to; list of types varies, depending on the object.

⊙**Convert to** converts the OLE object to another type.

○ **Activate as** opens the OLE object with the source application.

☐ **Display as Icon** displays the OLE object as an icon.

RELATED COMMANDS

InsertObj places an OLE object in the drawing.

PasteSpec pastes objects from the Clipboard as linked objects in the drawing.

OleReset returns the OLE object to its original form.

RELATED SYSTEM VARIABLES

OleHide toggles the display of OLE objects.

OleQuality determines the plot quality of OLE objects.

OleStartup loads the source application for embedded OLE objects before plotting.

TIPS

- In many cases, this command is unable to convert OLE objects.
- You can also access this command by: (1) selecting OLE object, (2) right-clicking, and (3) selecting **OLE | Convert** from the shortcut menu.
- OLE is short for "object linking and embedding." It is technology invented by IBM for its OS/2 operating system. OLE allows documents and objects from other applications to exist in "foreign" documents — documents that would not otherwise accept unknown objects.

OleLinks

Rel.13 Changes, updates, and cancels OLE links between the drawing and other Windows applications (*short for Object Linking and Embedding LINKS*).

Command	Alias	Ctrl+	F-key	Alt+	Menu Bar	Tablet
olelinks	...	...	...	EO	Edit ↳OLE Links	...

Command: olelinks

When no OLE links are in the drawing, the command does nothing.

When at least one OLE object is in the drawing, displays dialog box:

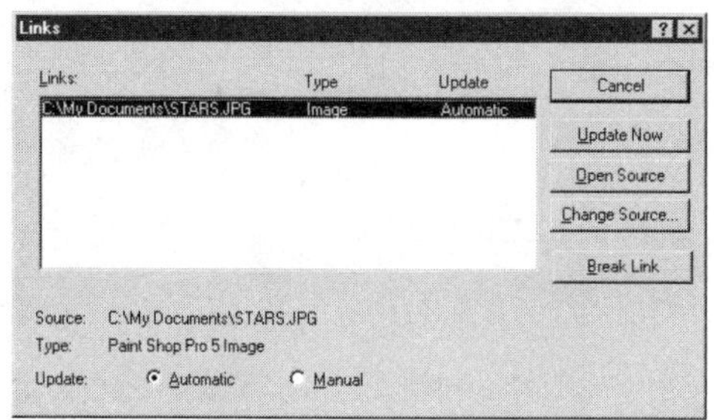

DIALOG BOX OPTIONS

Links displays a list of linked objects: source filename, type of file, and update mode — automatic or manual.

Update selects either automatic or manual updates.

Update Now updates selected links.

Open Source starts the source application program.

Break Link cancels the OLE link; keeps the object in place.

Change Source displays the Change Source dialog box.

RELATED COMMANDS

InsertObj places an OLE object in the drawing.

PasteSpec pastes objects from the Clipboard as linked objects in the drawing.

OleScale specifies the size, scale, and properties of selected OLE objects.

RELATED SYSTEM VARIABLES

MsOleScale specifies the scale of OLE objects in model space.

OleHide toggles the display of OLE objects.

OleQuality determines the plot quality of OLE objects.

OleStartup loads the source applications for embedded OLE objects before plotting.

RELATED WINDOWS COMMANDS

Edit | Copy copies objects from the source application to the Clipboard.

File | Update updates the linked object in the source application.

OleOpen / OleReset

2000 Opens OLE objects in their source application; resets OLE objects (*undocumented commands*).

Command	Alias	Ctrl+	F-key	Alt+	Menu Bar	Tablet
oleopen	...	...	...	...	...	...
olereset	...	...	...	...	...	...

Select an OLE object before entering the command.

Command: oleopen

Opens the OLE object in its source application. For example, selecting a text document opens it in the default word processor.

Command: olereset

Returns the OLE object to it original form.

RELATED COMMANDS

InsertObj places an OLE object in the drawing.

PasteSpec pastes objects from the Clipboard as linked objects in the drawing.

RELATED SYSTEM VARIABLES

MsOleScale specifies the scale of OLE objects in model space.

OleHide toggles the display of OLE objects.

OleQuality determines the plot quality of OLE objects.

OleStartup loads the source apps for embedded OLE objects before plotting.

TIPS

- You can also access these commands by: (1) selecting OLE object, (2) right-clicking, and (3) selecting **OLE | Open** from the shortcut menu (or selecting **OLE | Reset**).
- If you do not first select an OLE object, AutoCAD complains, 'Unable to find OLE object. Object must be selected before entering the command.'

OleScale

2000 Modifies the properties of OLE objects.

Command	Alias	Ctrl+	F-key	Alt+	Shortcut Menu	Tablet
olescale	...	...	...	...	...	...

Command: olescale

Displays dialog box:

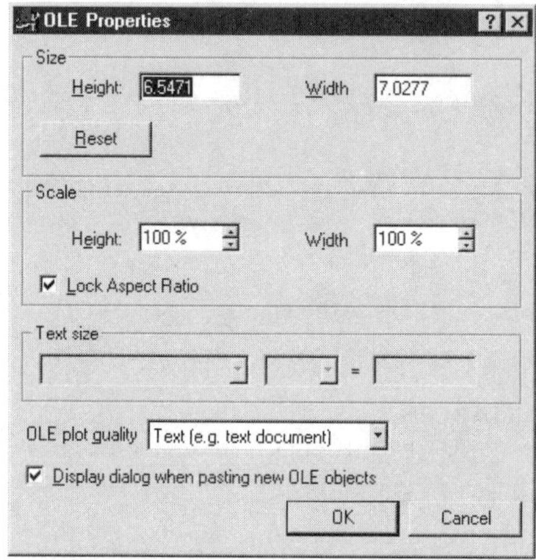

Alternatively, right-click a selected OLE object, and choose ***OLE | Text Size*** *from the shortcut menu. AutoCAD displays an abbreviated dialog box:*

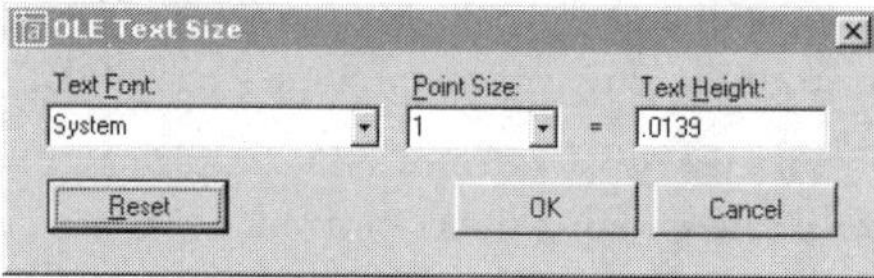

DIALOG BOX OPTIONS

Size options

Height changes the height of the OLE object; displays the current height.

Width changes the width of the OLE object; displays the current width.

Reset resets the OLE object to its original size when first inserted into the drawing.

Scale options

Height changes the height of the OLE object by a percentage of the original height.

Width changes the width of the OLE object by a percentage of the original width.

☑ **Lock Aspect Ratio** changes the Width size and ratio to match the Height, and vice versa.

Text Size options

Font displays the fonts used by the OLE object, if any.

Point Size displays the text height in point sizes; limited to the point sizes available for the selected font, if any (1 point = $^1/_{72}$ inch).

Text Height specifies the text height in drawing units.

OLE Plot Quality determines the quality of the pasted object when plotted:

Plot Quality	Meaning
Line Art	Text is plotted as text; no colors or shading are preserved; some graphical images are not plotted, while others are plotted as monochrome images (black and white only, no shades of gray).
Text	All text formatting is preserved; text is plotted as graphics, which plots less cleanly than the **Line Art** setting; graphics are plotted less cleanly and at reduced colors than **Graphics** and **Photograph** settings.
Graphics	Graphics are plotted at a reduced number of colors (fewer shades of gray or "posterization"); all text formatting is preserved; text is plotted as graphics, but more cleanly than with **Text** and **Photograph** settings.
Photograph	Graphics are plotted at reduced resolution and colors; all text formatting is preserved; text is plotted as graphics.
High Quality	Graphics are plotted at full resolution and colors.
Photograph	All text formatting preserved; text plotted more eanly than with **Text** and **Photograph** settings.

Display dialog when pasting new OLE object:

☑ Displays the OLE Properties dialog box automatically when inserting an OLE object.

☐ Does not display the dialog box.

RELATED SYTSTEM VARIABLES

MsOleScale specifies the scale of OLE objects in model space.

OleHide toggles the display of OLE objects in the drawing and in plots.

OleQuality specifies the quality of displaying and plotting of embedded OLE objects.

OleStartup loads the source application of an embedded OLE object for plots.

RELATED COMMANDS

MsOleScale sets the scale for text in OLE objects (*new to AutoCAD 2005*).

InsertObj inserts an OLE object into the drawing.

OleLinks modifies the link between the object and its source.

PasteSpec allows you to paste an object with a link.

TIPS

- Change **OleStartup** to 1 to load the OLE source application, which may help improve the plot quality of OLE objects.
- The **OLE Plot Quality** list box determines the quality of the pasted object when plotted. I recommend the **Line Art** setting for text, unless the text contains shading and other graphical effects.

Oops

V. 1.0 Restores the last-erased group of objects; restores objects removed by the **Block** and **-Block** commands.

Command	Alias	Ctrl+	F-key	Alt+	Menu Bar	Tablet
oops	...	...	...	...	...	...

Command: oops

COMMAND LINE OPTIONS

None.

RELATED COMMANDS

Block, Erase, WBlock: use **Oops** after these command to return erased objects.

U undoes the most recent command.

TIPS

- **Oops** restores only the most-recently erased object; use the **Undo** command to restore earlier objects.
- Use **Oops** to bring back objects after turning them into a block with the **Block** and **WBlock** commands.

Open

Rel.12 Loads one or more drawings into AutoCAD.

Command	Alias	Ctrl+	F-key	Alt+	Menu Bar	Tablet
open	openurl	O	...	FO	File ↳Open	T25

Command: open

Displays dialog box:

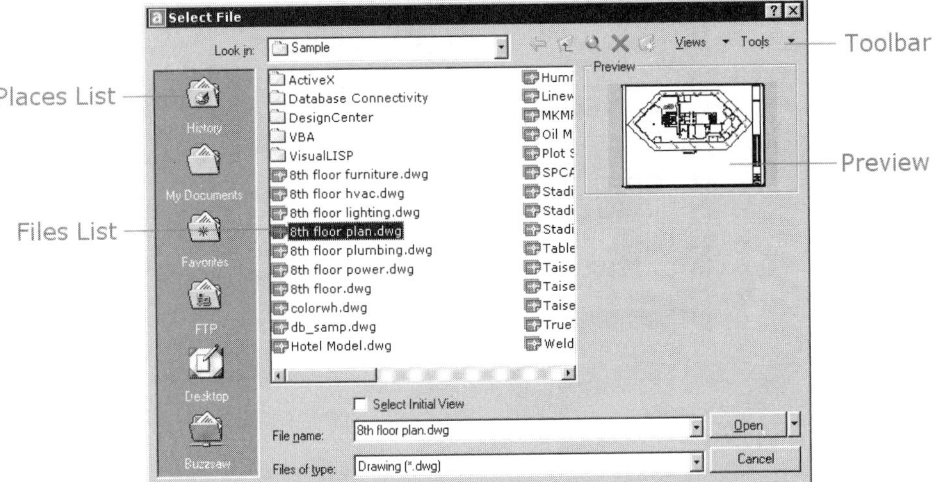

DIALOG BOX OPTIONS

Look in selects the network drive, hard drive, or folder (subdirectory).

Preview displays a preview image of AutoCAD drawings.

☐ **Select initial view** selects a named view from a dialog box, if the drawing has saved views. (After drawing is opened, AutoCAD displays the Select Initial View dialog box listing named views.)

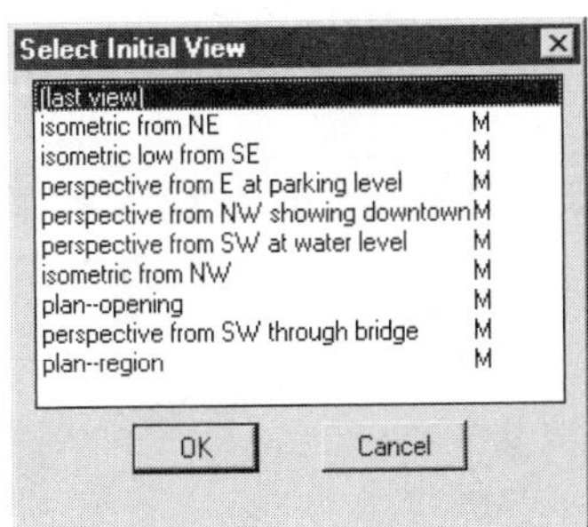

M indicates a view created in model space.

P indicates a view created in paper space (layout mode).

File name specifies the name of the drawing.

Files of type specifies the type of file:

- **Drawing (*.dwg)** AutoCAD drawing file.
- **Standard (*.dws)** drawing standards file; see the **Standards** command.
- **DXF (*.dxf)** drawing interchange file; see the **DxfIn** command.
- **Drawing Template File (*.dwt)** template drawing file; see the **New** command.

Open opens the selected drawing file(s); to open more than one drawing at a time:

- In the files list, hold down the SHIFT key to select a contiguous range of files:

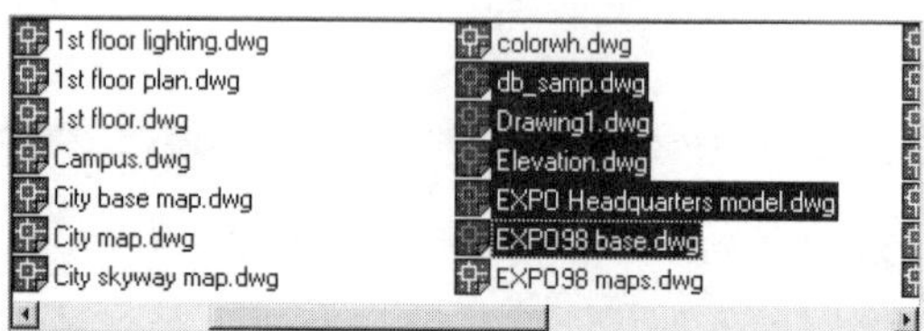

- Hold down the CTRL key to select two or more non-continuous files:

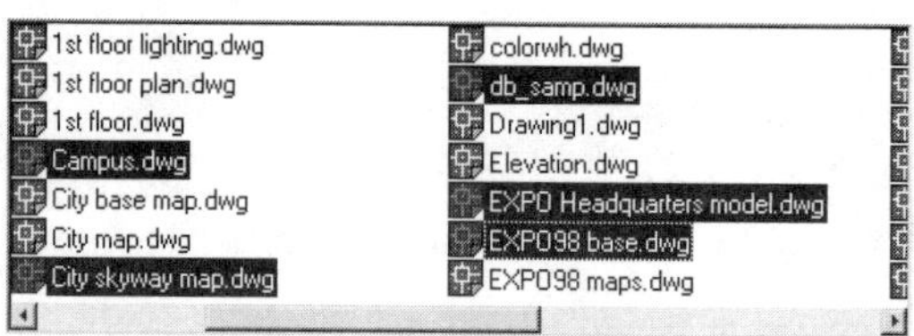

Cancel dismisses the dialog box without opening a file.

Open button options

Open opens the drawing.

Open as read-only loads the drawing, but you cannot save changes to the drawing except under another file name. AutoCAD displays "Read Only" on the title bar:

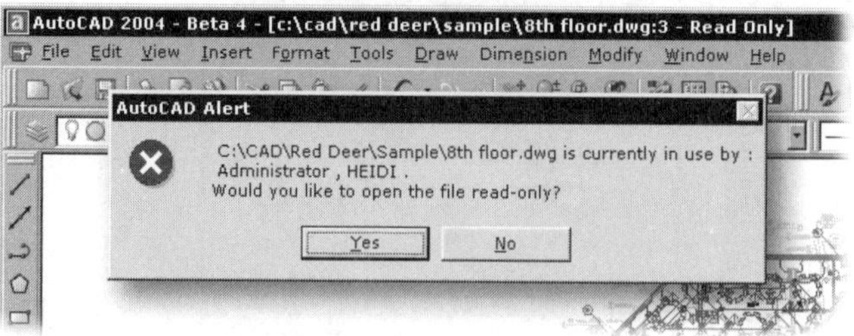

Partial Open loads selected layers or named views; displays the Partial Open dialog box; not available for *.dxf* and template files.

Partial Open Read-Only partially loads the drawing in read-only mode.

Partial Open dialog box

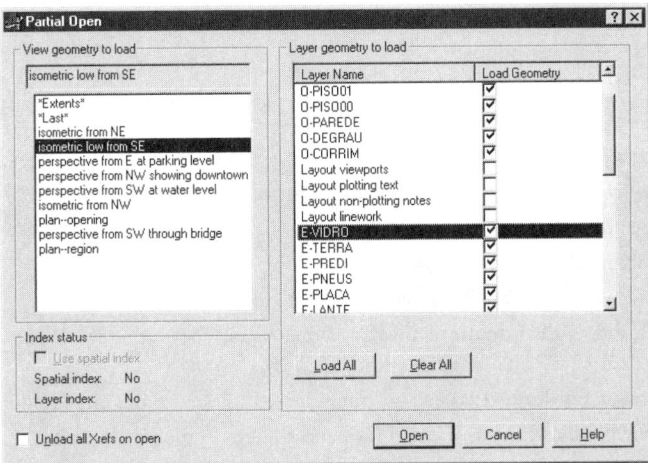

View geometry to load selects the model space views to load; paper space views are not available for partial loading.

Layer Geometry to Load options

Load Geometry selects the layers to load.

Load All loads all layers.

Clear All deselects all layer names.

Index Status options

Available only when the drawing was saved with spatial indices.

Use spatial index determines whether to use the spatial index for loading, if available.

Spatial Index indicates whether the drawing contains the spatial index.

Layer Index indicates whether the drawing contains the layer index.

Additional options

☐ **Unload all xrefs on open** loads externally-referenced drawings when opening the drawing.

Open opens the drawing, and then partially loads the geometry.

TOOLBAR ICONS

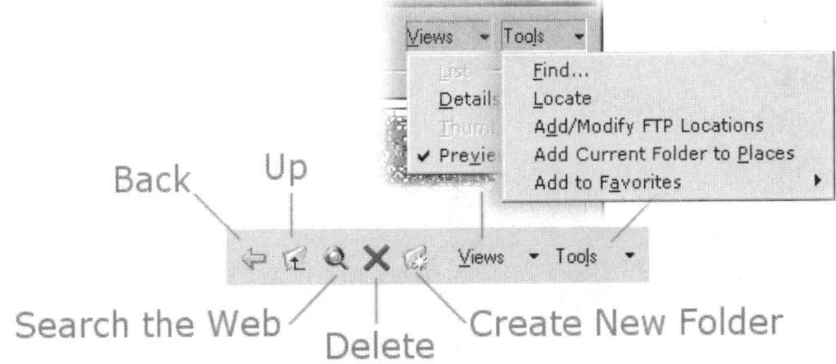

Back returns to the previous folder (keyboard shortcut ALT+1).

Up moves up one level to the next folder or drive (ALT+2).

Search the Web displays the Browse the Web window (ALT+3); see **Browser** command.

Delete removes the selected file(s); does not delete folders or drives (DEL).

Create New Folder creates new folders (ALT+5).

Views provides display options:

- **List** displays the file and folder names only.
- **Details** displays file and folder names, type, size, and date.
- **Thumbnails** displays thumbnail images of *.dwg* files.
- **Preview** toggles display of the preview window.

Tools provides file-oriented tools:

- **Find** displays the Find dialog box for searching files.
- **Locate** searches for the file along AutoCAD's search paths.
- **Add/Modify FTP Locations** displays a dialog box for storing the logon names and passwords for FTP (file transfer protocol) sites.
- **Add Current Folder to Places** adds the selected folders to the places sidebar.
- **Add to Favorites** adds the selected files and folders to the Favorites list.

PLACES LIST

History displays files opened by AutoCAD during the last four weeks.

My Documents displays files and folders in the *my documents* folder.

Favorites displays files and folders in the *favorites* folder.

FTP displays the **FTP Locations** list.

Desktop displays the contents of the *\desktop* folder.

Buzzsaw.com goes to the www.buzzsaw.com Web site.

The Point A and RedSpark items were removed from AutoCAD 2004.

SHORTCUT MENUS

Places List menu

Right-click icons in the Places List:

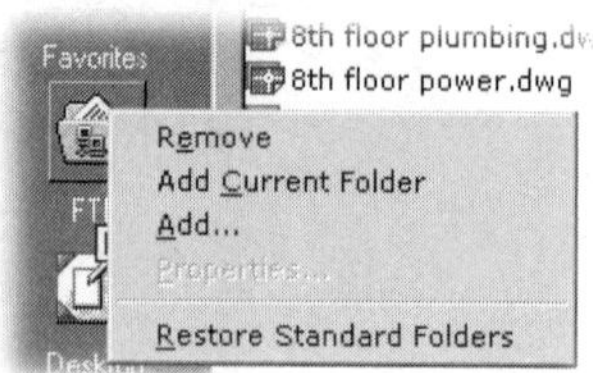

Remove removes a folder from the list.

Add Current Folder adds the selected folder to the list; you can also drag a folder from the file list into the places list.

Add displays the Add Places Item dialog box.

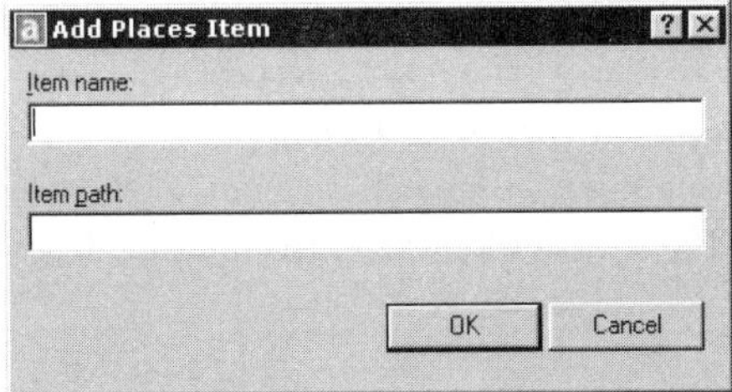

Properties displays the Places Item Properties dialog box, which is identical to the Add Places Item dialog box, and is available only for items you add.

Restore Standard Folders restores the folders shown above.

File List menu

Right-click the file list without selecting a file or folder:

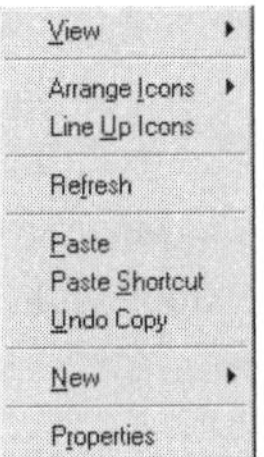

View switches between filename views: large icons, small icons, list, thumbnails, and details.

Arrange Icons arranges icons by name, type, size, and date.

Line Up Icons places icons in an orderly pattern.

Refresh updates the folder listing.

Paste pastes a file from the Clipboard.

Paste Shortcut pastes a file from the Clipboard as a shortcut.

Undo Copy undoes the copy-paste operation; available only after a copy or paste.

New creates a new folder (subdirectory) or shortcut.

Properties displays the Properties dialog box of the folder selected by the Look in list.

Right-click a file or folder name:

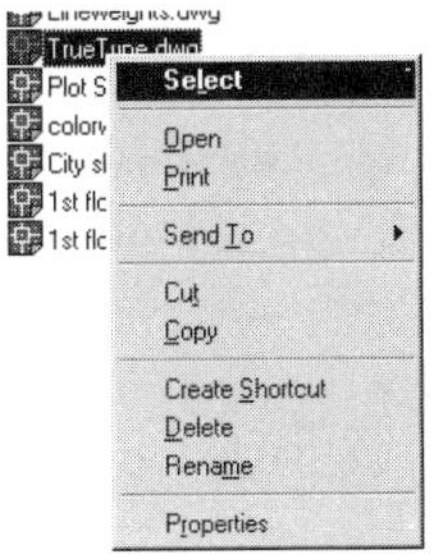

Select opens the drawing in AutoCAD.

Open also opens the drawing in AutoCAD.

Print does not work.

Send To copies the file to another drive; may not work with some software.

Cut cuts the file to the Clipboard.

Copy copies the file to the Clipboard.

Create Shortcut creates a shortcut icon for the selected file.

Delete erases the file; displays a warning dialog box.

Rename renames the file; displays a warning dialog box if you change the extension.

Properties displays the Properties dialog box in read-only mode; use the **DwgProps** command within AutoCAD to change the settings.

Command Line Options

When FileDia = 0:

Command: open

Enter name of drawing to open <.>: *(Enter name of drawing file.)*

Opening an AutoCAD 2004 format file

Enter the (optional) path and name of the *.dwg*, *.dxf*, *.dws* file. If you leave out the extension, AutoCAD assumes a *.dwg* file. Enter tilde (~) to display the Select File dialog box.

RELATED SYSTEM VARIABLES

TaskBar displays a button on the Windows taskbar for every open drawing (*new to 2005*).

DbMod indicates whether the drawing has been modified.

DwgCheck checks if the drawing was last edited by AutoCAD.

DwgName contains the drawing's filename.

DwgPrefix contains the drive and folder of the drawing.

DwgTitled indicates whether the drawing has a name other than *drawing1.dwg*.

FileDia toggles the interface between dialog boxes and the command-line.

FullOpen indicates whether the drawing is fully or partially opened.

RELATED COMMANDS

FileOpen opens drawings without the dialog box.

SaveAs saves drawings with new names.

PartiaLoad loads additional portions of partially-opened drawings.

TIPS

- When drawing filenames are dragged from Windows Explorer into open drawings, they are inserted as blocks; when dragged to AutoCAD's titlebar, they are opened as drawings.
- DXF and template files cannot be partially opened.
- After a drawing is partially opened, use **PartiaLoad** to load additional parts of the drawing.
- When a partially-opened drawing contains a bound xref, only the portion of the xref defined by the selected view is bound to the partially-open drawing.
- The **OpenUrl** alias is a holdover from an earlier release of AutoCAD that used the command to open drawings over the Internet.

Removed Command

OpenUrl was removed from AutoCAD 2000; it was replaced by **Open**'s FTP option.

OpenDwfMarkup

2005 Opens *.dwf* markup files in AutoCAD, and then loads the Markup Set Manager window.

Command	Alias	Ctrl+	F-key	Alt+	Menu Bar	Tablet
opendwfmarkup	...	...	...	**FK**	**File** ↳**Load Markup Set**	...

Command: opendwfmarkup

Displays the Open Markup DWF dialog box. Select a .dwf file, and then click ***Open****.*

If the .dwf file contains no markup data, AutoCAD complains, 'Filename.dwf does not contain any markup data. Would you like to open this DWF file in the viewer?' Click ***Yes*** *to open the file in Composer; click* ***No*** *to cancel.*

COMMAND OPTIONS

See ***Markup*** *command.*

RELATED COMMANDS

Markup displays the Markup Set Manager, and also opens *.dwf* files containing markup data.

MarkupClose closes the Markup Set Manager window.

RmlIn imports *.rml* redline markup files created by Volo View.

RELATED SYSTEM VARIABLES

MsmState reports whether the Markup Set Manager window is open.

FileDia determines whether this command displays a dialog box or command-line prompts.

TIP

- MSM is short for “markup set manager.” DWF is short for “drawing Web format.”

OpenSheetset

2005 Loads sheet sets into the current drawing.

Command	Alias	Ctrl+	F-key	Alt+	Menu Bar	Tablet
opensheetset	...	...	...	**FE**	**File** **↳Open Sheet Set**	...
-opensheetset						

Command: opensheetset

Displays Open Sheet Set dialog box. Select a .dst file, and then click ***Open****.*

AutoCAD displays the Sheet Set Manager. See ***Sheetset*** *command.*

-OPENSHEETSET Command

Command: -opensheetset

Enter name of sheet set to open: *(Enter .dst file name.)*

Enter the path and name of a *.dst* file, or enter tilde (~) to display the file dialog box.

RELATED COMMANDS

NewSheetset creates new sheet sets.

Sheetset opens the Sheet Set Manager window.

SheetsetClose closes the Sheet Set Manager window.

RELATED SYSTEM VARIABLES

SsFound records path and file name of the sheet set.

SsLocate toggles whether AutoCAD opens sheet sets associated with the drawing.

SsmAutoOpen toggles whether AutoCAD displays the Sheet Set Manager window with the drawing being opened.

SsmState reports whether the Sheet Set Manager is active.

FileDia determines whether this command displays a dialog box or command-line prompts.

TIPS

- The **Sheetset** command automatically opens the previously-opened sheet set file.
- SSM is short for "sheet set manager."

Options

2000 Sets system and user preferences.

Commands	Aliases	Ctrl+	F-key	Alt+	Menu Bar	Tablet
options	op	...	...	TN	Tools	Y10
	gr				↳Options	
	preferences					
	ddgrips					
	ddselect					
+options						

Command: options

Displays dialog box.

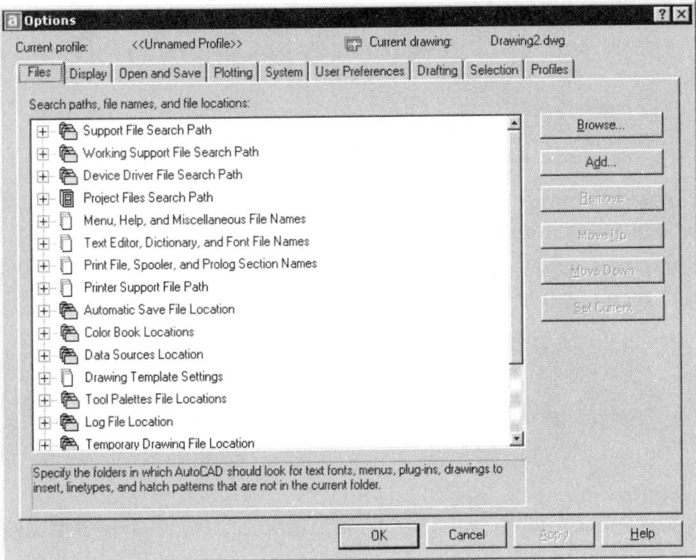

DIALOG BOX OPTIONS

OK applies the changes, and closes the dialog box.

Cancel cancels the changes, and closes the dialog box.

Apply applies the changes, and keeps the dialog box open.

Files tab

Search paths, file names, and file locations specifies the folders and support files used by AutoCAD.

Browse displays the Browse for Folder dialog box for selecting folders, and the Select a File dialog box for selecting files.

Add adds an item below the selected path or file name.

Remove removes the selected item without warning; click CANCEL to undo the removal.

Move Up moves the selected item above (or before) the preceding item; applies to search paths only.

Move Down moves the selected item below the following item; applies to search paths only.

Set Current makes current the selected project names and spelling dictionaries only.

Display tab

indicates the setting saved in a system variable.

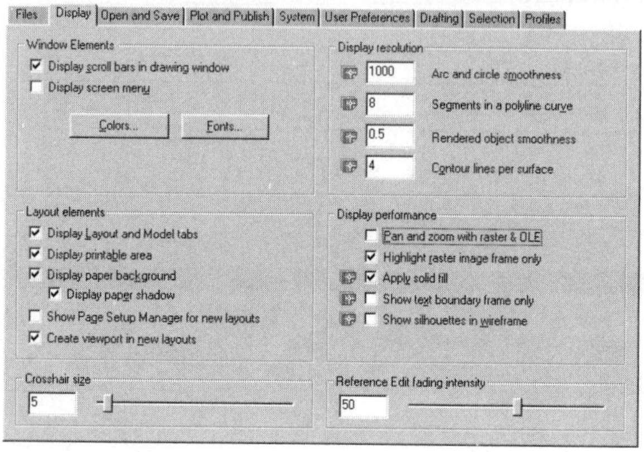

Window Elements options

☑ **Display scroll bars in drawing window** toggles the presence of the horizontal and vertical scroll bars.

☐ **Display screen menu** toggles the presence of the screen menu.

Colors displays the Color Options dialog box to select colors for the AutoCAD graphics and text windows.

Fonts displays the Command Line Window Font dialog box to select the font for text on the command line.

Layout Elements options

☑ **Display Layout and Model tabs** toggles the presence of the Model and Layout tabs.

☑ **Display printable area** toggles the display of dashed margin lines in layout modes.

☑ **Display paper background** toggles the presence of the page in layout modes.

☑ **Display paper shadow** toggles the presence of the drop shadow under the page in layout modes.

☐ **Show Page Setup Manager for new layouts** specifies whether the Page Setup dialog box is displayed when you create a new layout. Use this dialog box to set options related to paper and plot settings.

☑ **Create viewport in new layouts** toggles the automatic creation of a single viewport for new layouts.

Crosshair size specifies the size of the crosshair cursor; range is 1% to 100% of the viewport (stored in system variable CursorSize); default = 5%.

Display Resolution options

Arc and circle smoothness controls the displayed smoothness of circles, arcs, and other curves; range is 1 to 20000 (ViewRes); default = 100.

Segments in a polyline curve specifies the number of line segments used to display polyline curves; range is -32767 to 32767 (SplineSegs); default = 8.

Rendered object smoothness controls the displayed smoothness of shaded and rendered curveds; range is 0.01 to 10 (FaceTRes) default = 0.5.

Contour lines per surface specifies the number of contour lines on solid 3D objects; range is 0 to 2047 (IsoLines); default = 4.

Display Performance options

☐ **Pan and zoom with raster and OLE** toggles the display of raste and OLE images during realtime pan and zoom (RtDisplay).

☑ **Highlight raster image frame only** highlights only the frame, and not the entire raster image, when on (ImageHlt).

☑ **Apply solid fill** toggles the display of solid fills in multilines, traces, solids, solid fills, and wide polylines; this option does not come into effect until you click OK, and then use the Regen command (FillMode).

☐ **Show text boundary frame only** toggles the display of rectangles in place of text; this option does not come into effect until you click OK, and then use the Regen command (QTextMode).

☐ **Show silhouettes in wireframe** toggles the display of silhouette curves for 3D solid objects; when off, tesselation lines are drawn when hidden-line removal is applied to the 3D object (DispSilh).

Reference fading intensity specifies the amount of fading during in-place reference editing; range is 0% to 90% (XFadeCtl); default = 50%.

Open and Save tab

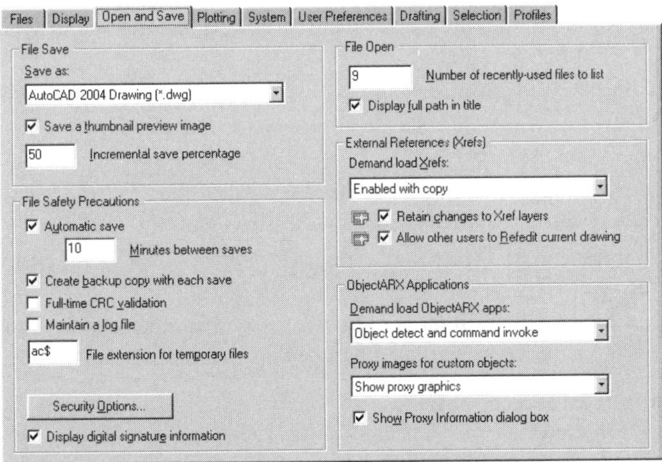

File Save options

Save as specifies the default file format used by the **Save** and **SaveAs** commands; default = "AutoCAD 2004 Drawing (*.dwg)."

Save a thumbnail preview image displays Thumbnail Preview Settings dialog box.

Incremental save percentage indicates the percentage of wasted space allowed in a drawing file before a full save is performed; range is 0% to 100% (ISavePercent); default = 50.

File Safety Precautions options

☑ **Automatic save** automatically saves the drawing at prescribed time intervals (SaveFile and SaveFilePath).

Minutes between saves specifies the duration between automatic saves (SaveTime); default = 10 minutes.

☑ **Create backup copy with each save** creates backup copies when drawings are saved; (ISavBak).

☐ **Full-time CRC validation** performs cyclic redundancy check (*CRC*) error-checking each time an object is read into the drawing.

☐ **Maintain a log file** saves the Text window text to a log file (LogFileMode); default = off.

File extension for temporary files specifies the filename extension for temporary files created by AutoCAD (NodeName); default = .ac$.

Security Options displays the Security Options dialog box; see the SecurityOptions command.

☑ **Display digital signature information** displays digital signature information when opening files with valid digital signatures (SigWarn).

File Open options

Number of recently-used files to list specifies the number of recently-opened filenames to list in the Files menu; minimum = 0; default and maximum = 9.

☑ **Display full path in title** displays the drawing file's path in AutoCAD's titlebar.

External References (Xrefs) options

Demand load Xrefs specifies the style of demand loading of externally-referenced drawings a.k.a. xrefs (XLoadCtl):

- **Disabled** turns off demand loading.
- **Enabled** turns on demand loading to improve performance, but the drawing cannot be edited by another user; default.
- **Enabled with copy** turns on demand loading; loads a copy of the drawing so that another user can edit the original.

☑ **Retain changes to Xref layers** saves changes to properties for xref-dependent layers (VisRetain).

☑ **Allow other users to Refedit current drawing** allows another user to edit the current drawing when referenced by another drawing (XEdit).

ObjectARX Applications options

Demand load ObjectARX apps demand-loads an ObjectARx application when the drawing contains proxy objects (DemandLoad):

- **Disable load on demand** turns off demand loading.
- **Custom object detect** demand-loads the application when the drawing contains proxy objects.

- **Command invoke** demand-loads the application when a command of the application is invoked.
- **Object detect and command invoke** demand-loads the application when the drawing contains proxy objects, or when one of the application's commands is invoked; default.

Proxy images for custom objects specifies how proxy objects are displayed:

- **Do not show proxy graphics** does not display proxy objects.
- **Show proxy graphics** displays proxy objects.
- **Show proxy bounding box** displays a rectangle instead of the proxy object.

☐ **Show Proxy Information dialog box** displays a warning dialog box when a drawing contains proxy objects (ProxyNotice).

Plot and Publish tab

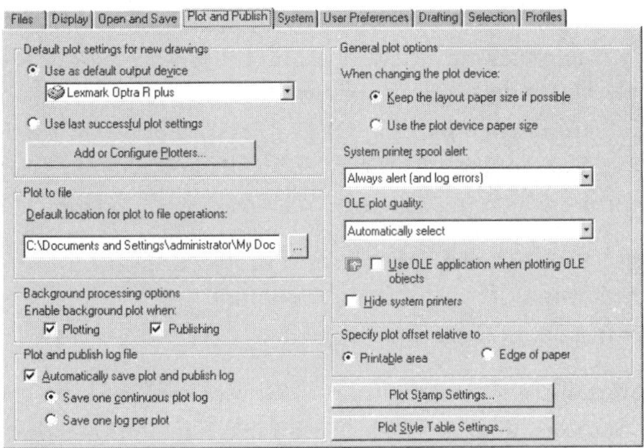

Default Plot Settings for New Drawings options

⊙ **Use as default output device** selects the default output device.

○ **Use last successful plot settings** reuses the plot settings from the last successful plot.

Add or configure plotters displays Autodesk Plotter Manager window; see PlotterManager command.

Plot to File options

Default location for plot to file operations specifies the folder in which to store plot files; default=*documents and settings\\username\\my documents.*

... displays a dialog box for selecting the folder.

Background processing options *(new to AutoCAD 2005)*

☑ **Plotting** executes the Plot command in the background (BackgroundPlot).

☑ **Publishing** executes the Publish command in the background (BackgroundPlot).

Plot and publish log file options *(new to AutoCAD 2005)*

☑ **Automatically save plot and publish log file:**

⊙ **Save one continuous plot log** saves all log data in a single file.

○ **Save one log per plot** saves log data in separate files.

General Plot Options options

When changing the plot device:

⊙ **Keep the layout paper size if possible** uses the paper size specified by the Page Setup dialog box's Layout Settings tab, provided the output device can handle the paper size (PaperUpdate); default = on.

○ **Use the plot device paper size** uses the paper size specified by the PC3 plotter configuration file (PaperUpdate); default = off.

System printer spool alert displays an alert when a spooled drawing has a conflict:

- **Always alert (and log errors)** displays alert, and logs the error message.
- **Alert first time only (and log errors)** displays the alert once, but logs all error messages.
- **Never alert (and log first error)** does not display an alert, but logs the first error message.
- **Never alert (do not log errors)** neither displays an alert, nor logs any error messages.

OLE plot quality determines the quality of OLE objects when plotted (OleQuality); default = Automatically Select; see the OleScale command.

☐ **Use OLE application when plotting OLE objects** launches the application that created the OLE object when plotting a drawing with an OLE object (OleStarup).

☐ **Hide system printers** hides the names of Windows system printers not specific to CAD.

Specify Plot Offset Relative To options *(new to AutoCAD 2005)*

⊙ **Printable area** measures offsets to plotter margins.

○ **Edge of paper** measures offsets to paper edges.

Plot Stamp Settings displays the Plot Stamp dialog box; see PlotStamp command.

Plot Style Table Settings displays the Plot Style Table Settings dialog box.

System tab

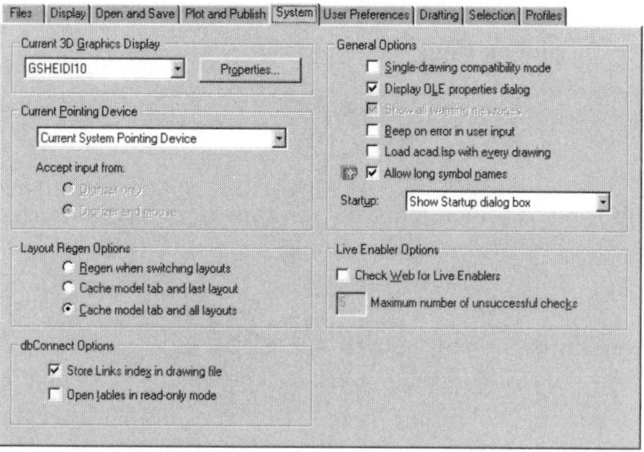

Current 3D Graphics Display options

Current 3D Graphics Display selects the 3D graphics display driver (default = GSHEIDI10).

Properties displays the 3D Graphics System Configuration dialog box.

Current Pointing Device options

Current Pointing Device selects the pointing device driver:

- **Current System Pointing Device** selects the pointing device used by Windows.
- **Wintab Compatible Digitizer** selects a Wintab-compatible digitizer driver.

Accept input from *(available only when a digitizing tablet is attached):*

○ **Digitizer only** reads input from the digitizer, and ignores the mouse.

⊙ **Digitizer and mouse** reads input from the digitizer and the mouse.

Layout Regen options

○ **Regen when switching layouts** regenerates the drawing each time layouts are switched.

○ **Cache model tab and last layout** saves the display list of the model tab and last layout accessed.

⊙ **Cache model tab and all layouts** saves the display list of the model tab and all layouts.

dbConnect Options options

☑ **Store links index in drawing file** stores the database index in the drawing file.

☐ **Open tables in read-only mode** opens database tables in read-only mode.

General Options options

☐ **Single-drawing compatibility mode** forces the Single-drawing Interface (SDI), which limits AutoCAD to opening a single drawing at a time; this may be required for compat ibility with some third-party applications (SDI).

☑ **Display OLE Properties dialog** displays the OLE Properties dialog box after an OLE object is inserted in the drawing; see the OleScale command.

☑ **Show all warning messages** displays all dialog boxes with the Don't Display This Warning Again option.

☐ **Beep on error in user input** beeps the computer when AutoCAD detects a user error.

☐ **Load acad.lsp with every drawing** loads the *acad.lsp* file with every drawing (AcadLspAsDoc).

☑ **Allow long symbol names** allows symbol names — layers, dimension styles, blocks, linetypes, text styles, layouts, UCS names, views, and viewport configurations — to be up to 255 characters long, and to include letters, numbers, blank spaces, and most punctua tion marks; when off, names are limited to 31 characters, and spaces may not be used (ExtNames).

Startup:

- **Show Startup dialog box** displays Startup dialog box when AutoCAD launches or Create New Drawing dialog box when AutoCAD is already running.
- **Do not show a startup dialog** displays the Select Template File dialog box.

Live Enabler Options options

☐ **Check Web for Live Enablers** checks whenever an Internet connection is present.

Maximum number of unsuccessful checks limits how often AutoCAD attempts to call home; default = 5.

User Preferences tab

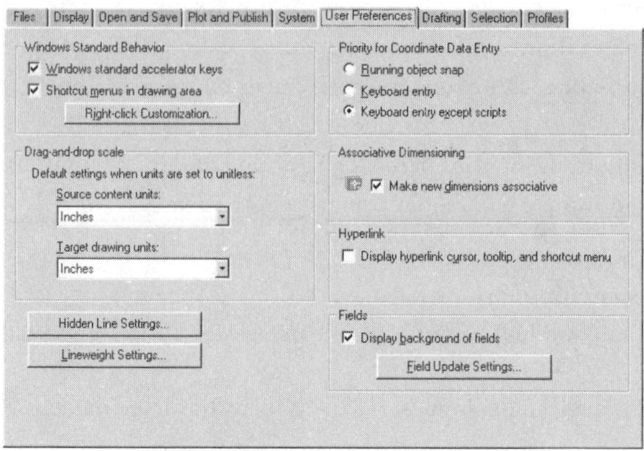

Windows Standard Behavior options

☑ **Windows standard accelerator keys** uses Windows' keyboard accelerators; when off, uses DOS-based AutoCAD keyboard accelerators. The differences

Accelerator	On	Off
CTRL+C	CopyClip	Cancel command.
CTRL+O	Open	Ortho toggle.
CTRL+V	Paste	Viewports switch.

☑ **Shortcut menus in drawing area:** toggles the effect of right-clicking in the drawing area: when on, displays a shortcut menu; when off, is equivalent to pressing the ENTER key (ShortCutMenu).

Right-click customization displays the Right-Click Customization dialog box (ShortCutMenu).

Drag-and-Drop Scale options *(formerly AutoCAD DesignCenter)*

Source content units specifies the default units when an object is inserted into the drawing from AutoCAD DesignCenter; Unspecified-Unitless means the object is not scaled when inserted (InsUnitsDefTarget); default = inches or mm.

Target drawing units specifies the default units when "insert units" are not specified by the InsUnits system variable (InsUnitsDefTarget); default = inches or mm.

Hidden Line Settings displays the Hidden Line Settings dialog box; see HlSettings command.

Lineweight Settings displays the Lineweight Settings dialog box; see Lineweight command.

Priority for Coordinate Data Entry options

○ **Running object snap** means that osnap overrides coordinates entered at the keyboard (OSnapCoord); default = off.

○ **Keyboard entry** means that coordinates entered at the keyboard override osnaps (OSnapCoord); default = off.

⊙ **Keyboard entry except scripts** means that coordinates entered at the keyboard override running object snaps, except when coordinates are provided by a script (OSnapCoord); default = on.

Associative Dimensioning options

☑ **Make new dimension associtive** means that dimensions are associated with objects; when off, dimensions are associated with defpoints (DimAssoc).

Hyperlink options

☑ **Display hyperlink cursor and shortcut menu** displays:

Hyperlink cursor, which looks like chain links and the planet earth, when the cursor passes over an object containing a hyperlink.

Hyperlink option on shortcut menu when right-clicking an object containing a hyperlink.

See the HyperlinkOptions command.

Field options *(new to AutoCAD 2005)*

☑ **Display background of fields** displays gray behind field text (FieldDisplay).

Field Update Settings displays the Field Update Settings dialog box; see UpdateField command.

Drafting tab

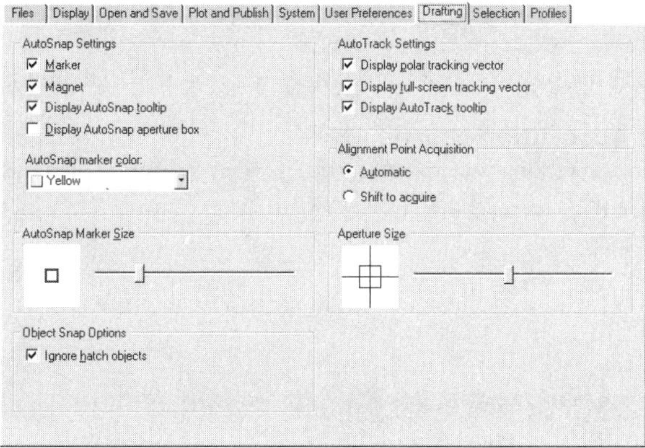

AutoSnap Settings options

☑ **Marker** displays the AutoSnap icon (AutoSnap).

☑ **Magnet** turns on the AutoSnap magnet (AutoSnap).

☑ **Display AutoSnap tooltip** displays the AutoSnap tooltip (AutoSnap).

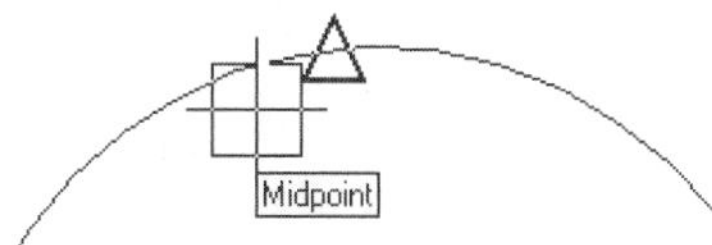

☐ **Display AutoSnap aperture box** displays the AutoSnap aperture box (ApBox).

AutoSnap marker color specifies the color of the AutoSnap icons; choose from seven colors; default = yellow.

AutoSnap marker size sets the size for the AutoSnap icon; range is 1 to 20 pixels.

Object Snap Options options *(new to AutoCAD 2005)*

☑ **Ignore hatch objects** prevents AutoCAD from snapping to hatch objects.

AutoTrack Settings options

☑ **Display polar tracking vector** displays the Polar Tracking vectors at specific angles (TrackPath).

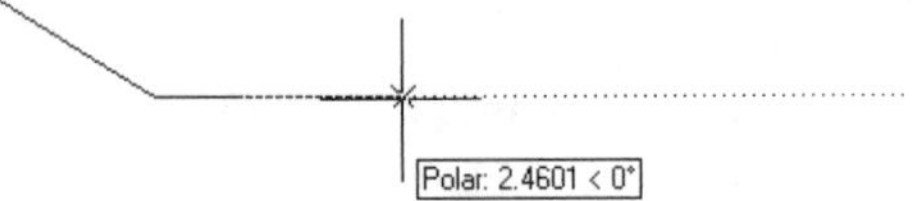

Polar tracking vector (dotted line) and AutoTrack tootip.

☑ **Display full-screen tracking vector** displays the tracking vectors (TrackPath).

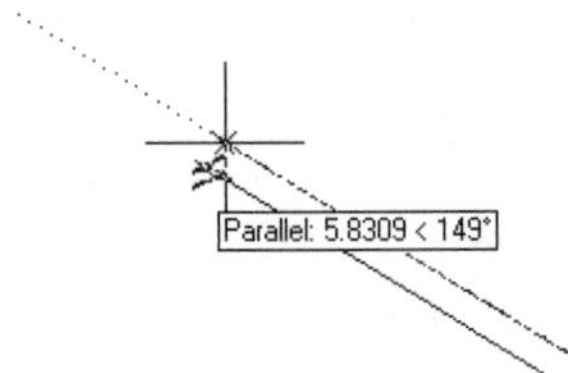

Full-screen tracking vector and AutoTrack tooltip.

☑ **Display AutoTrack tooltip** displays the AutoTrack tooltip (AutoSnap).

Alignment Point Acquisition options

⊙ **Automatic** displays tracking vectors automatically when aperture moves over an object snap.

○ **Shift to acquire** displays tracking vectors when pressing SHIFT and moving the aperture over an object snap.

Aperture Size sets the size for the aperture; range is 1 to 50 pixels (Aperture); default = 10.

Selection tab

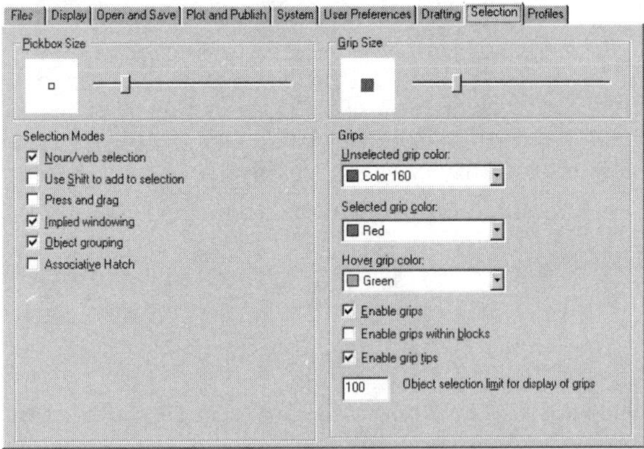

Pickbox Size specifies the size of the pickbox; range is 1 to 20 pixels (PickBox); default = 3.

Grip size specifies the size of grips; range is from 1 to 20 pixels (GripSize); default = 5.

Selection Modes options *(replaces DdSelect command)*

☑ **Noun/verb selection** allows you to select an object before executing an editing command (PickFirst).

☐ **Use Shift to add to selection** requires you to press SHIFT to add or remove objects from the selection set, a la Windows (PickAdd).

☐ **Press and drag** allows you to create the selection window by dragging (PickDrag).

☑ **Implied windowing** creates a selection window when you pick a point in the drawing that does not pick an object (PickAuto).

☑ **Object grouping** selects the entire group when an object in the group is selected (PickStyle).

☐ **Associative hatch** selects boundary objects, along with the associative hatch patterns (PickStyle).

Grips options *(replaces the DdGrips command)*

Unselected grip color specifies color of unselected (cold) grips (GripColor); default = blue.

Selected grip color specifies the color of selected (hot) grip (GripHot); default = red.

Hover grip color specifies the color of the grip when the cursor hovers over it; default = green.

☑ **Enable grips** displays grips on selected objects (Grips).

☐ **Enable grips within blocks** displays all grips for every object in the selected block; when off, a single grip at the block's insertion point is displayed (GripBlock).

☑ **Enable grip tips** toggles the display of grip tips on custom objects (GripTips).

Object selection limit for display of grips limits the number of selected objects that display grips (GripObjLimit); default = 100.

Profiles tab

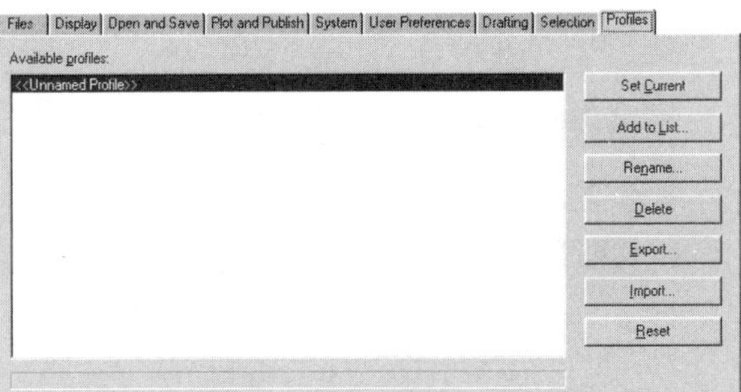

Available Profiles lists available profiles, which customize the AutoCAD user interface.

Set Current sets the selected profile as the current profile.

Add to List displays the Add Profile dialog box; allows you to enter a name and description for the new profile.

Rename displays the Change Profile dialog box; allows you to change the name and the description of the selected profile.

Delete erases the selected profile; the current profile cannot be erased.

Export exports profiles as *.arg* files.

Import imports *.arg* profiles into AutoCAD.

Reset resets the selected profile to AutoCAD's default settings.

Color Options dialog box

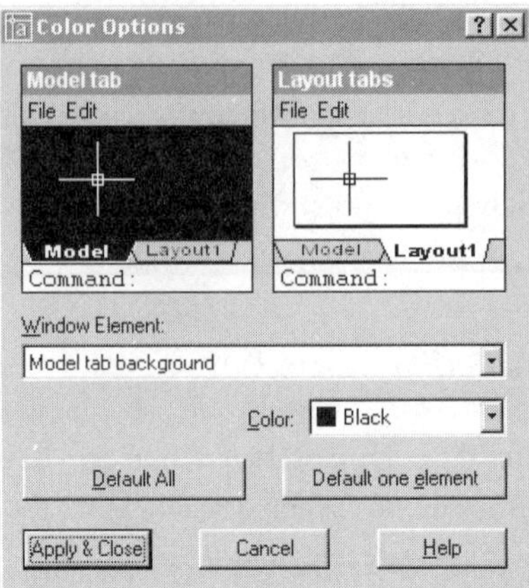

Window Element selects the element whose color is to be changed.

Color sets the color for the element.

Default All returns all colors to the original out-of-the-box setting.

Default One Element returns the selected element to its original color.

Command Line Window Font dialog box

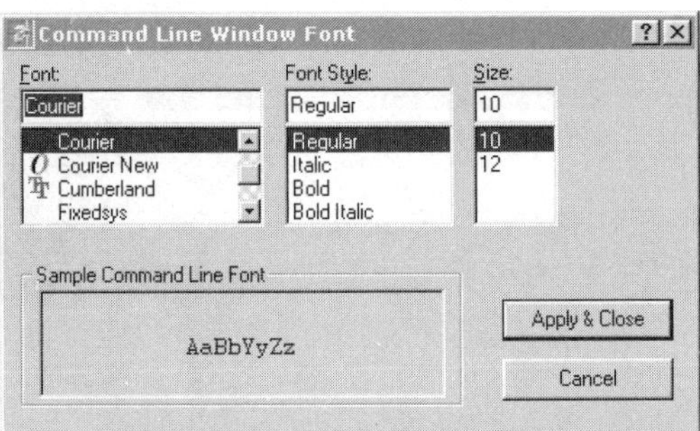

Font selects the font for the command line window.

Font Style changes the style.

Size specifies the size of the text.

Thumbnail Preview Settings dialog box *(new to AutoCAD 2005)*

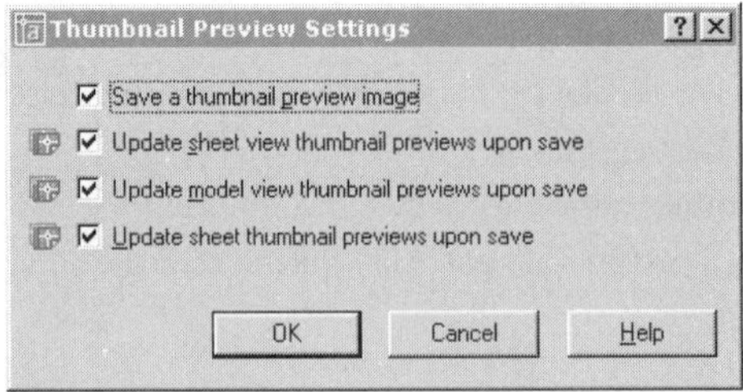

☑ **Save a Thumbnail Preview Image** save the current view of the drawing as a bitmap preview image the *.dwg* file (RasterPreview).

☑ **Update Sheet View Thumbnail Previews Upon Save** updates the sheet view preview images on the Sheet Set Manager's *View List* tab (UpdateThumbnail).

☑ **Update Sheet Thumbnail Previews Upon Save** updates the model space preview images on the Sheet Set Manager's *Resources* tab (UpdateThumbnail).

☑ **Update Model View Thumbnail Previews Upon Save** updates the sheet preview images on the Sheet Set Manager's *Sheet List* tab (UpdateThumbnail).

Plot Style Table Settings dialog box *(new to AutoCAD 2005)*

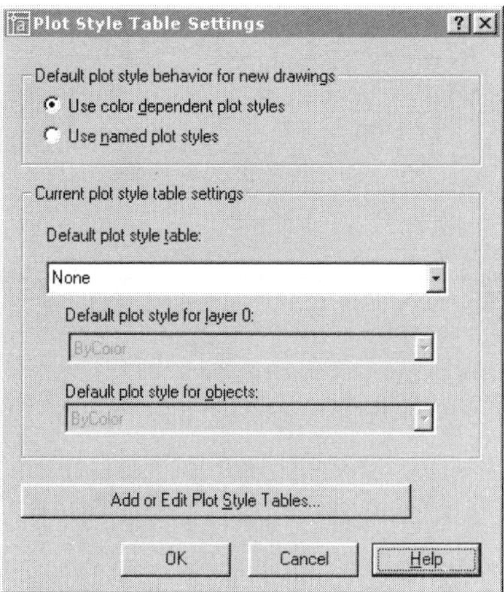

Default Plot Style Behavior for New Drawings (PStylePolicy)

⊙ **Use Color Dependent Plot Styles** uses color-dependent plot styles in new drawing; plotted colors are defined by a number between 1 to 255.

○ **Use Named Plot Styles** uses named plot styles in new drawings; objects are plotted based on plot style definitions.

Current Plot Style Table Settings options

Default Plot Style Table selects the plot style table to attach to new drawings.

Default Plot Style for Layer 0 specifies the default plot style for Layer 0 in new drawings (DefLPlStyle).

Default Plot Style for Objects specifies the default plot style assigned to new objects (DefPlStyle). The list displays BYLAYER, BYBLOCK, Normal styles, and any plot styles defined in the currently-loaded plot style table.

Add or Edit Plot Style Tables displays the Plot Style Manager window; see the StylesManager command.

+OPTIONS COMMAND

Command: +options

Tab index <0>: *(Enter a digit between 0 and 8.)*

COMMAND LINE OPTION

Tab index specifies the tab to display:

Tab Index	Meaning
0	Files tab.
1	Display tab.
2	Open and Save tab.
3	Plotting tab.
4	System tab.
5	User Preferences tab.
6	Drafting tab.
7	Selection tab.
8	Profiles tab.

TIPS

- *Grips* are small squares that appear on an object when the object is selected at the 'Command' prompt. In other Windows applications, grips are known as *handles*.

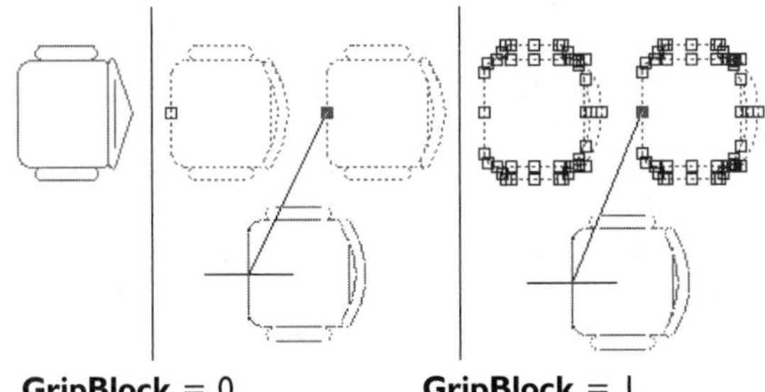

GripBlock = 0. **GripBlock** = 1.

- When an object is first selected, the grips are blue, and are called *unselected* or *cold* grips.
- When a grip is selected, it turns into a solid red square; this is called a *hot* grip.
- Press ESC to turn off unselected grips; press ESC twice to turn off hot grips.
- A larger pickbox makes it easier to select objects, but also easier to select unintended objects accidentally.
- Use **Object Sort Method** if the drawing requires objects to be processed in the order they appear in the drawing, such as for NC (numerically controlled) applications.
- **Plotting** and **PostScript Output** are turned on, by default; setting more sort methods increases processing time.
- The first time you use the **DrawOrder** command, it turns on all object sort method options.
- It's best to choose a crosshair color that is different from your drawing colors.

'Ortho

V. 1.0 Constrains drawing and editing commands to the vertical and horizontal directions only *(short for ORTHOgraphic)*.

Command	Alias	Ctrl+	F-key	Alt+	Status Bar	Tablet
ortho	...	L	F8	...	ORTHO	...

Command: ortho
Enter mode [ON/OFF] <OFF>: *(Enter* **ON** *or* **OFF***.)*

COMMAND LINE OPTIONS

OFF turns off ortho mode.

ON turns on ortho mode.

STATUS BAR OPTIONS

Ortho mode is toggled on and off by clicking **ORTHO** *on the status bar:*

10.1046, 0.9275 , 0.0000 SNAP GRID ORTHO POLAR OSNAP OTRACK LWT MODEL

RELATED COMMANDS

DSettings toggles ortho mode via a dialog box.

Snap rotates the ortho angle.

RELATED SYSTEM VARIABLES

OrthoMode stores the current ortho modes.

SnapAng specifies the rotation angle of the ortho cursor.

TIPS

- Use ortho mode when you want to constrain your drawing and editing to right angles.
- Rotate the angle of ortho with the **Snap** command's **Rotate** option.
- In isoplane mode, ortho mode constrains the cursor to the current isoplane.
- AutoCAD ignores ortho mode when you enter coordinates from the keyboard, and in perspective mode; ortho is also ignored by object snap modes.
- Ortho is not necessarily horizontal or vertical; its orientation is determined by the current UCS and snap alignment.

-OSnap

V. 2.0 Sets and turns on and off object snap modes at the command line (*short for Object SNAP*.

Command	Alias	Ctrl+	F-key	Alt+	Status Bar	Tablet
-osnap	-os	...	F3	...	OSNAP	T15 - U22

*The **OSnap** command displays the Drafting Settings dialog box; see the **DSettings** command.*

Command: -osnap
Current osnap modes: Ext
Enter list of object snap modes: *(Enter one or more modes separated by commas.)*

COMMAND LINE OPTIONS

You only need enter the first three letters as the abbreviation for each option:

APParent snaps to the intersection of two objects that don't physically cross, but appear to intersect on the screen, or would intersect if extended.

CENter snaps to the center point of arcs and circles.

ENDpoint snaps to the endpoint of lines, polylines, traces, and arcs.

EXTension snaps to the extension path of objects.

FROm extends from a point by a given distance.

INSertion snaps to the insertion point of blocks, shapes, and text.

INTersection snaps to the intersection of two objects, or to a self-crossing object, or to objects that would intersect if extended.

MIDpoint snaps to the middle point of lines and arcs.

NEArest snaps to the object nearest to the crosshair cursor.

NODe snaps to a point object.

NONe turns off all object snap modes temporarily.

OFF turns off all object snap modes.

PARallel snaps to a parallel offset.

PERpendicular snaps perpendicularly to objects.

QUAdrant snaps to the quadrant points of circles and arcs.

QUIck snaps to the first object found in the database.

TANgent snaps to the tangent of arcs and circles.

STATUS BAR OPTIONS

*Right-click **OSNAP** on the status bar:*

On turns on previously-set object snap modes.

Off turns off all running object snaps.

Settings displays the Object Snap tab of the Drafting Settings dialog box.

SHORTCUT MENU

Hold down the CTRL key, and then right-click anywhere in the drawing:

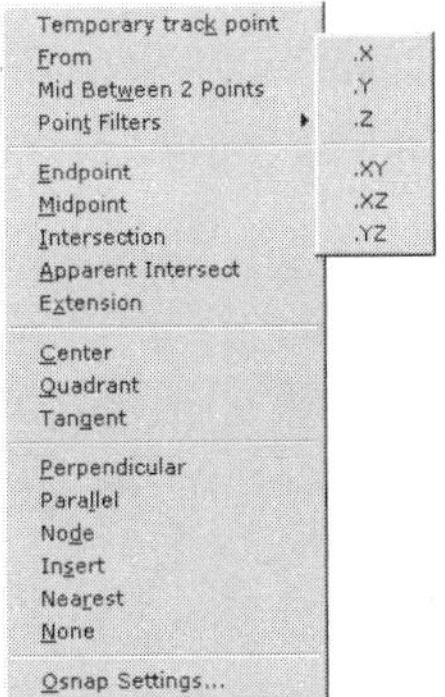

Temporary track point invokes Tracking mode; see the **Tracking** command.

From locates temporary points during drawing and editing commands. AutoCAD prompts:

From point: from
Base point: *(Pick a point.)*
<Offset>: *(Pick another point.)*

Mid Between 2 Points locates a point between two pick points *(new to AutoCAD 2005).*

From point: m2p
First point of mid: *(Pick a point.)*
Second point of mid: *(Pick another point.)*

Point filters invokes point filter modes:

Enter	AutoCAD Requests
.x	**of** *(Pick a point.)* **(need YZ):** *(Enter value for y and z.)*
.y	**of** *(Pick a point.)* **(need XZ):** *(Enter value for x and z.)*
.z	**of** *(Pick a point.)* **(need XY):** *(Enter value for x and y.)*
.xy	**of** *(Pick a point.)* **(need Z):** *(Enter value for z.)*
.xz	**of** *(Pick a point.)* **(need Y):** *(Enter value for y.)*
.yz	**of** *(Pick a point.)* **(need X):** *(Enter value for x.)*

Osnap Settings displays the Object Snap tab of the Drafting Settings dialog box; see **DSettings** command.

RELATED SYSTEM VARIABLES

Aperture specifies the size of the object snap aperture in pixels.

OsnapNodeLegacy determines whether the NODe object snap snaps to text insertion points *(new toe AutoCAD 2005).*

AutoSnap controls the display of AutoSnap (default = 63).

OsnapCoord overrides object snaps when entering coordinates at 'Command' prompt.

OsnapCoord	Meaning
0	Object snap overrides keyboard.
1	Keyboard overrides object snap settings.
2	Keyboard overrides object snap settings, except during a script.

OsMode stores the current object snap mode(s):

OsMode	Meaning
0	NONe (default).
1	ENDpoint.
2	MIDpoint.
4	CENter.
8	NODe.
16	QUAdrant.
32	INTersection.
64	INSertion.
128	PERpendicular.
256	TANgent.
512	NEArest.
1024	QUIck.
2048	APParent intersection.
4096	EXTension.
8192	PARallel.

TIPS

- The **Aperture** command controls the drawing area through which AutoCAD searches.
- If AutoCAD finds no snap matching the current modes, then the pick point is selected.
- The **m2p** modifier can also be entered as **mtp**.
- The **APPint** and **INT** object snap modes should not be used together.
- To turn on more than one object snap at a time, use a comma to separate mode names:

  Enter list of object snap modes: int,end,qua

- The location of all object snaps:

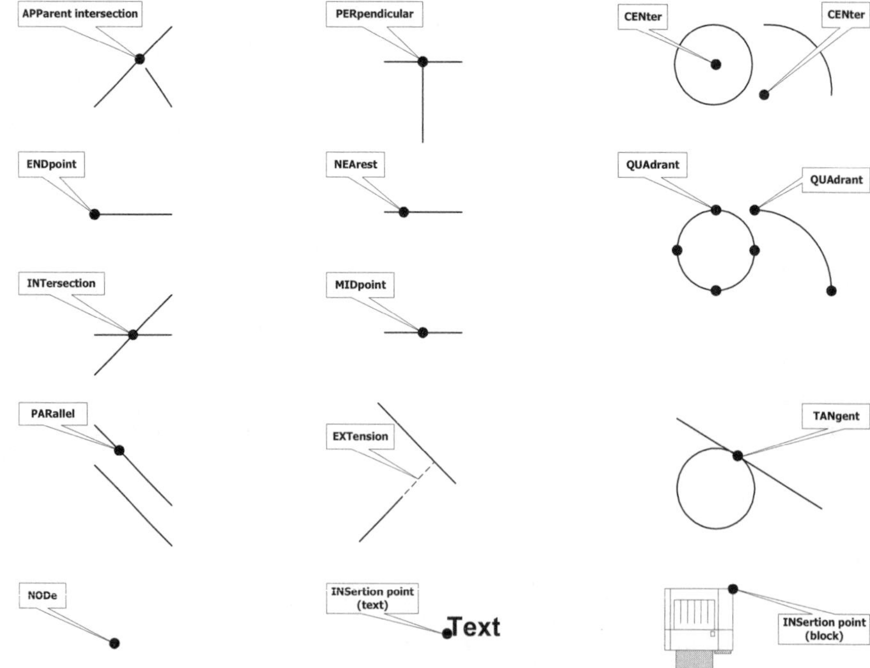

PageSetup

2000 Sets up model and layout views in preparation for plotting drawings. In AutoCAD 2005, this command's **Modify** button performs many of the functions of the Plot dialog box in earlier releases of AutoCAD.

Command	Alias	Ctrl+	F-key	Alt+	Menu Bar	Tablet
pagesetup	...	...	...	**FG**	**File** ↳**Page Setup**	**V25**

Command: pagesetup

Displays dialog box (new user interface in AutoCAD 2005):

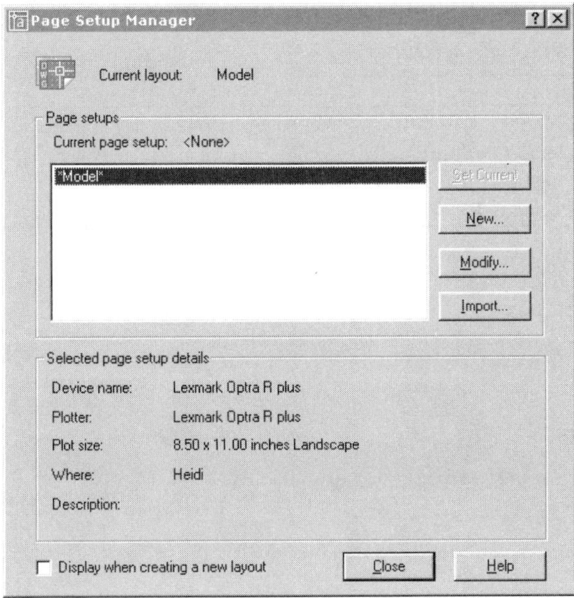

DIALOG BOX OPTIONS

Set Current selects the page setup to be used by the drawing.

New displays the New Page Setup dialog box.

Modify displays the Page Setup dialog box.

Import displays the Select Page Setup From File dialog box; select a .dwg, .dwt, or dxf file, and then click **Open**; displays the Import Page Setups dialog box.

☐ **Display when creating a new layout** displays this dialog box the first time you click a layout tab in new drawings.

New Page Setup dialog box

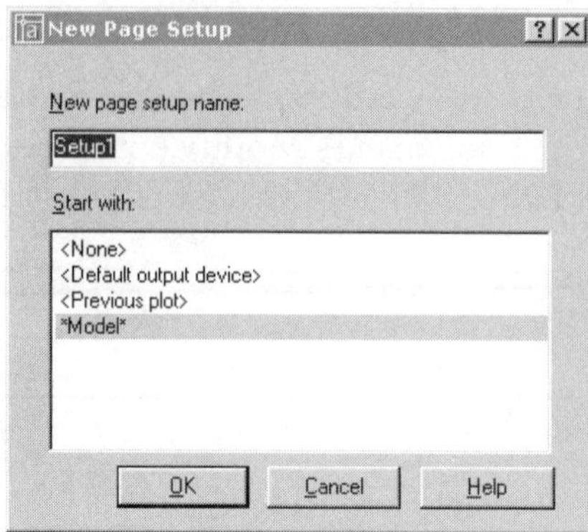

New Page Setup Name specifies the name of the setup.

Start With selects the default settings:

Start With	Meaning
<None>	No page setup used as template.
<Default Output Device>	Selects plotter specified as default output device by Options dialog box.
<Previous Plot>	Uses settings from previous plot job.

OK displays the Page Setup dialog box.

Page Setup dialog box

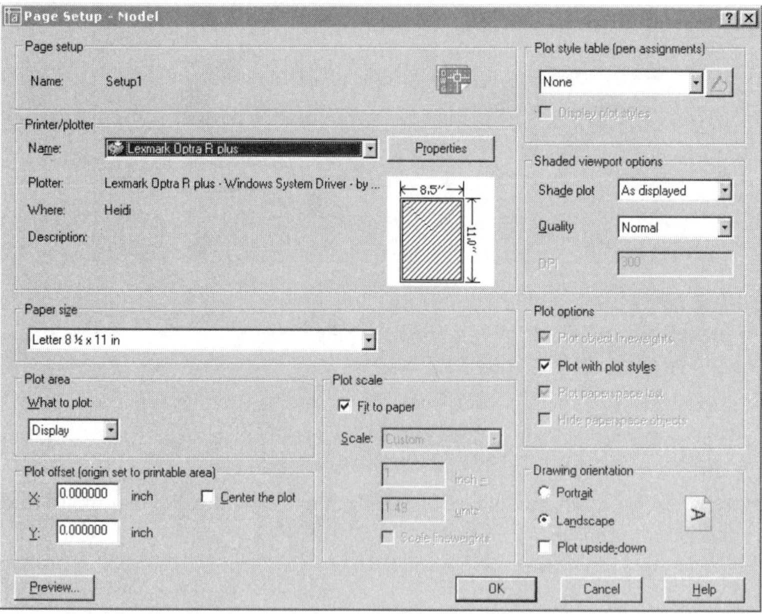

See ***Plot*** *command.*

Import Page Setups dialog box

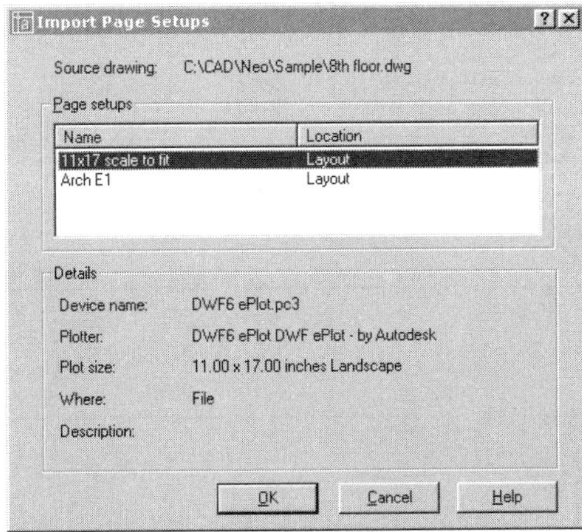

Name lists the names of page setups found in the drawing, if any.

Location indicates the location of the page setups: model or layout.

OK returns to the Page Setup Manager dialog box.

SHORTCUT MENU

Right-click a page setup name:

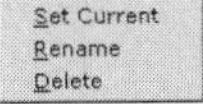

Set Current selects the page setup to use for this drawing.

Rename changes the name of the page setup.

Delete erases the page setup from the drawing, without warning.

RELATED COMMANDS

Layout creates new layouts.

LayoutWizard creates layouts and page setups.

Plot plots drawings based on the settings of the **PageSetup** command.

TIPS

- The sheet set icon shows when this command is activated from the Sheet Set Manager window.
- Right-click Model or Layout tabs, and then select **Page Setup Manager**.

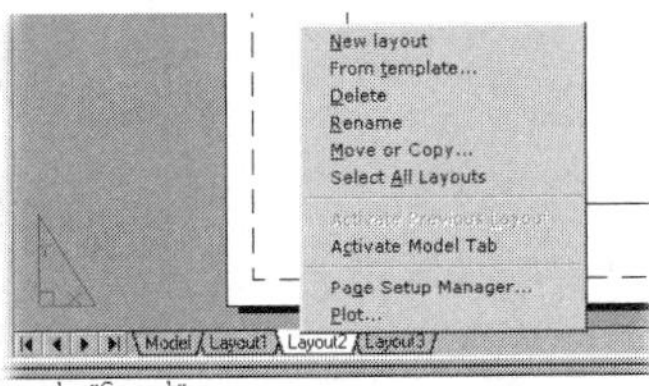

- Page setups cannot be applied to entire sheet sets after they are created.

'Pan

V. 1.0 Moves the view in the current viewport.

Commands	Aliases	Ctrl+	F-key	Alt+	Menu Bar	Tablet
pan	**p**	...	...	**VPT**	**View**	**N11-**
	rtpan				↳**Pan**	**P11**
					↳**Realtime**	
-pan	**-p**			**VPP**	**View**	
					↳**Pan**	
					↳**Point**	

Command: pan

Press Esc or Enter to exit, or right-click to display shortcut menu. *(Move cursor to pan, and then press* ESC *to exit command.)*

Enters real-time panning mode, and displays hand cursor:

Drag the hand cursor to pan the drawing in the viewport.

Press ENTER *or* ESC *to return to the 'Command' prompt.*

SHORTCUT MENU OPTIONS

During real-time pan mode, right-click the drawing to display shortcut menu:

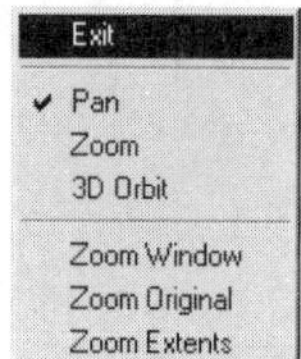

Exit exits real-time pan mode; returns to the 'Command' prompt.

Pan switches to real-time pan mode, when in real-time zoom mode.
Zoom switches to real-time zoom mode; see the **Zoom** command.
3D Orbit switches to 3D orbit mode; see the **3dOrbit** command.

Zoom Window prompts you, 'Press pick button and drag to specify zoom window.'
Zoom Original returns to the view when you first started the **Pan** command.
Zoom Extents displays the entire drawing.

-PAN Command

Command: -pan
Displacement: *(Pick a point, or enter x,y-coordinates.)*
Second point: *(Pick another point, or enter x,y-coordinates.)*

Before panning to the left:

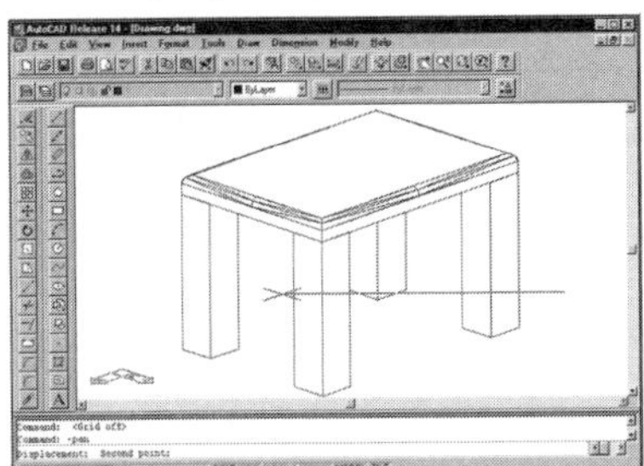

After panning to the left:

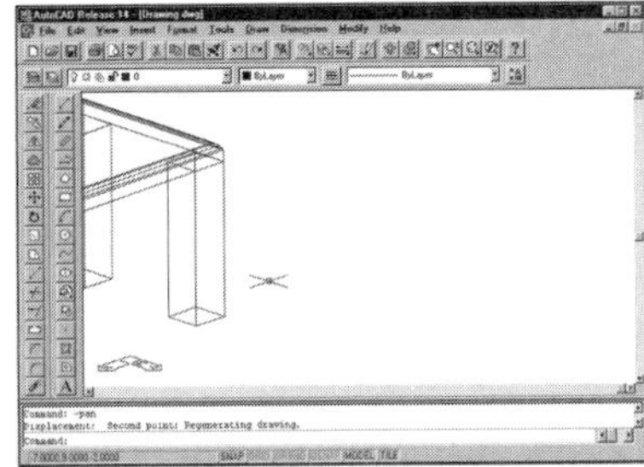

COMMAND LINE OPTIONS

ENTER or **ESC** exits real-time panning mode.

Displacement specifies the distance and direction to pan the view.

Second point pans to this point.

RELATED COMMANDS

DsViewer displays the Aerial View window, which pans in an independent window.

RegenAuto determines how regenerations are handled.

View saves and restores named views.

Zoom pans with the Dynamic option.

3dOrbit pans during perspective mode.

RELATED SYSTEM VARIABLES

MButtonPan determines the action of a mouse's third button or wheel.

RtDisplay toggles the display of raster and OLE images during realtime panning.

TIPS

- You can pan each viewport independently.
- Use **'Pan** to start drawing objects in one area of the drawing, pan over, and then continue working in another area of the drawing.
- You cannot use transparent pan under the following conditions: paper space, perspective mode, **VPoint**, **DView**, another **Pan** command, **View**, **Zoom**, or **3dOrbit**.
- When the drawing no longer moves during real-time panning, you have reached the panning limit; AutoCAD changes the hand icon to show the limit (use **Regen** to increase the limit):

- As an alternative, use the horizontal and vertical scroll bars to pan the drawing.
- With **MButtonPan** set to 1, pan by holding down the wheelbutton and moving the mouse.

PartiaLoad

2000 Loads additional views and layers of a partially-loaded drawing; this command works only with drawings that have been partially loaded.

Commands	Alias	Ctrl+	F-key	Alt+	Menu Bar	Tablet
partiaload	...	...	...	**FR**	**File** ↳**Partial Load**	
-partiaload						

Command: partiaload

If current drawing was not been partially opened (via the ***Open*** *command's* ***Partial Open*** *option), AutoCAD complains, 'Command not allowed unless the drawing has been partially opened.'*

Otherwise, displays dialog box:

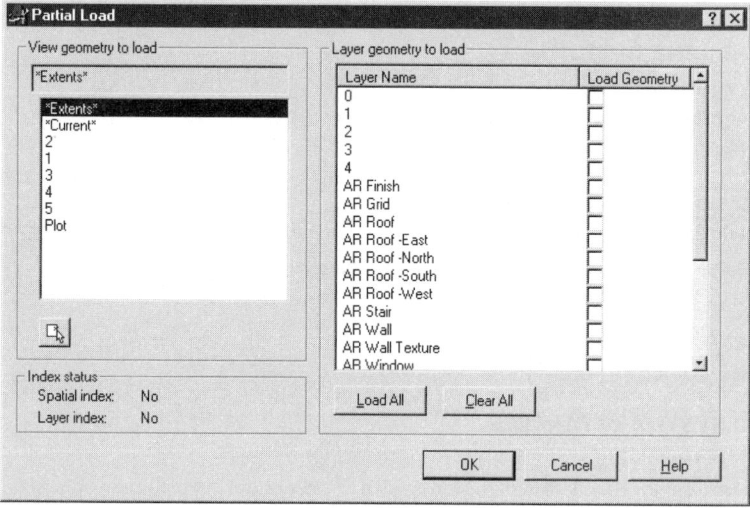

DIALOG BOX OPTIONS

View geometry to load selects the model space views to load. (Paper space views are not available for partial loading.)

Pick a Window button dismisses the dialog box temporarily to allow you to specify an area that becomes the view to load. AutoCAD prompts you:

Specify first corner: *(Pick a point.)*

Specify opposite corner: *(Pick a point.)*

When the Partial Load dialog box returns, *New View* is listed by the View Geometry to Load list.

Layer Geometry to Load options

Load Geometry selects the layers to load.

Load All loads all layers.

Clear All deselects all layer names.

-PARTIALOAD Command

Command: -partiaload

Specify first corner or [View]: *(Pick a point, or type* **V***.)*

Enter view to load or [?] <*Extents*>: *(Enter a view name, or type* **?***.)*

Enter layers to load or [?] <none>: *(Enter a layer name, press* ENTER *for no layers, or type* **?***.)*

COMMAND LINE OPTIONS

Specify first corner specifies a corner to create a new view.

Specify opposite corner specifies the second corner; causes geometry in the new view to be loaded into the drawing.

View options

Enter view to load loads into the drawing the geometry found in the view.

? lists the names of views in the drawing:

Enter view name(s) to list <*>: *(Enter a view name, or type* *****.)*

Sample display:

```
Saved model space views:
View name"1"
"Plot"
```

Extents loads all geometry into the drawing.

Enter layers to load loads geometry on the layers in the drawing.

? lists the names of layers in the drawing:

Enter layers to list <*>: *(Press* ENTER*.)*

Sample display:

```
Layer names:
"0"
"3"
```

None loads no layers.

RELATED COMMANDS

Open partially opens a drawing via a dialog box.

PartialOpen partially opens a drawing via the command line.

RELATED SYSTEM VARIABLE

FullOpen indicates whether the drawing is fully or partially opened.

-PartialOpen

2000 Opens a drawing and loads selected layers and views.

Command	Alias	F-key	Alt+	Menu Bar	Tablet
-partialopen	partialopen	...	...	...	...

Command: -partialopen
Enter name of drawing to open <filename.dwg>: *(Enter a file name.)*
Enter view to load or [?]<*Extents*>: *(Enter a view name, or type* **?***.)*
Enter layers to load or [?]<none>: *(Enter a layer name, or type* **?***.)*
Unload all Xrefs on open? [Yes/No] <N>: *(Type* **Y** *or* **N***.)*

COMMAND LINE OPTIONS

Enter name of drawing to open specifies the name of the DWG file.

Enter view to load loads the geometry found in the view into the drawing.

? lists the names of the views in the drawing.

Extents loads all geometry into the drawing.

Enter layers to load loads geometry on the layers into the drawing.

? lists the names of layers in the drawing.

None loads no layers.

Unload all Xrefs on open:

Yes does not load externally-referenced drawings.

No loads all externally-referenced drawings.

RELATED COMMANDS

Open displays the Select Drawing dialog box; includes the Partial Open option.

PartiaLoad loads additional views or layers of a partially-opened drawing.

RELATED SYSTEM VARIABLES

FullOpen indicates whether the drawing is fully or partially opened.

TIPS

- **PartialOpen** is an alias for the **-PartialOpen** command.
- As an alternative, you may use the the **Partial Open** option of the **Open** command's Select File dialog box.

PasteAsHyperlink

2000 Pastes object as hyperlinks in drawings; works only if the Clipboard contains appropriate data.

Command	Alias	Ctrl+	F-key	Alt+	Menu Bar	Tablet
pasteashyperlink	...	...	...	EH	Edit ↳Paste as Hyperlink	...

Command: pasteashyperlink
Select objects: *(Select one or more objects.)*
Select objects: *(Press* ENTER.*)*

COMMAND LINE OPTION

Select objects selects the objects to which the hyperlink will be pasted.

RELATED COMMANDS

CopyClip copies a hyperlink to the Clipboard.

Hyperlink adds hyperlinks to selected objects.

SelectURL highlights all objects with hyperlinks.

RELATED SYSTEM VARIABLES

None.

PasteBlock

2000 Pastes objects as blocks in drawings; works only when the Clipboard contains AutoCAD objects.

Command	Alias	Ctrl+	F-key	Alt+	Menu Bar	Tablet
pasteblock	...	**Shift+V**	...	**EK**	**Edit** ↳**Paste as Block**	...

Command: pasteblock
Specify insertion point: *(Pick a point.)*

COMMAND LINE OPTION

Specify insertion point specifies the position for the block.

RELATED COMMANDS

Insert inserts blocks in the drawing.

CopyClip copies a block to the Clipboard.

RELATED SYSTEM VARIABLES

None.

TIPS

- This command does not work when the Clipboard contains data that cannot be pasted as a block.
- This command pastes any AutoCAD object as a block, generating a generic block name similar to "A$C65D94228."
- If the Clipboard contains a block, the block is nested by this command.
- As of AutoCAD 2004, you can use the CTRL+SHIFT+V shortcut.

PasteClip

Rel.13 Places an object from the Clipboard in the drawing (*short for PASTE CLIPboard*).

Command	Alias	Ctrl+	F-key	Alt+	Menu Bar	Tablet
pasteclip	...	V	...	EP	Edit ↳Paste	U13

Command: pasteclip

When the Clipboard contains AutoCAD objects, the following prompt is displayed:

Specify insertion point: *(Pick a point.)*

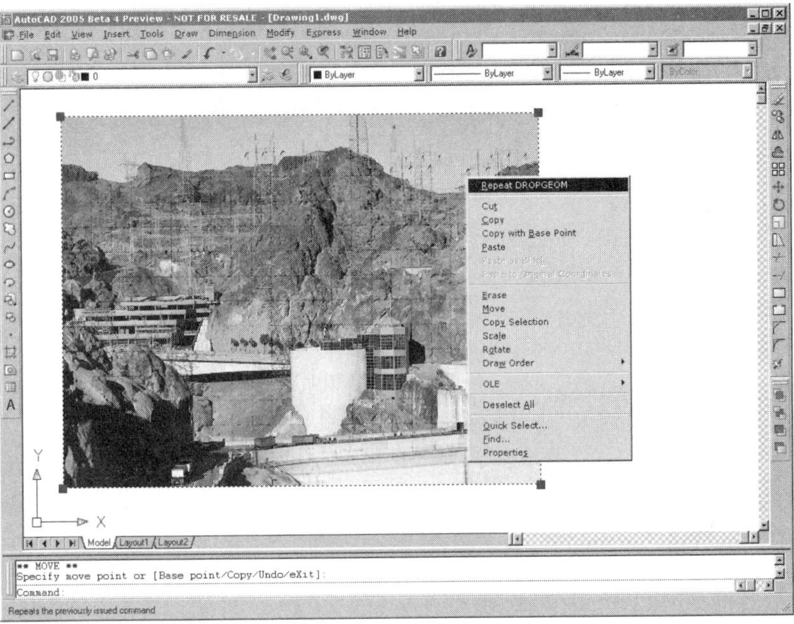

Photograph: Hoover Dam, Nevada

COMMAND LINE OPTION

Specify insertion point specifies the lower-left corner of the pasted object.

SHORTCUT MENU OPTIONS

Right-click pasted objects to display shortcut menu. Not all options appear with all pasted objects.

Cut erases the object from the drawing, and places it on the Clipboard.

Copy copies the object to the Clipboard.

Copy with Basepoint copies to Clipboard, along with insertion point data.

Paste pastes the object in the drawing.

Paste as Block pastes the object as a block.

Paste to Original Coordinates pastes the object to the same insertion point.

OLE displays submenu:

Open opens the object in its source application; does not work with static ActiveX objects.

Reset resets the object to its original settings.

Text Size displays the Text Size dialog box; available only when the object contains text.

Convert converts (or, in most cases, does not convert) the object to another format; displays dialog box:

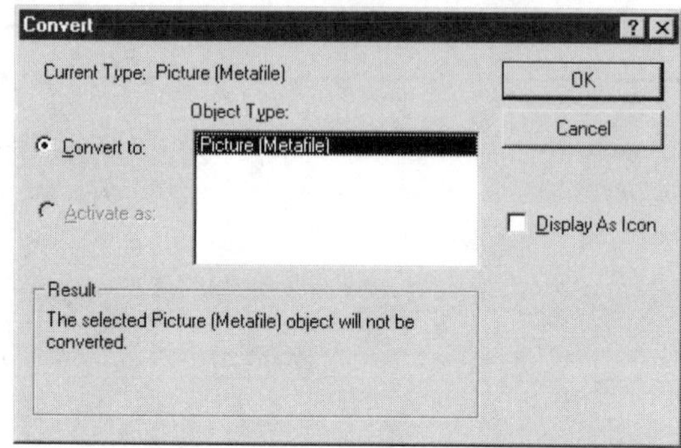

Properties displays the OLE Properties dialog box; see the **OleScale** command.

RELATED COMMANDS

CopyClip copies selected objects to the Clipboard.

Insert inserts an AutoCAD drawing in the drawing.

InsertObj inserts an OLE object in the drawing.

PasteSpec places the Clipboard object as pasted or linked objects.

RELATED SYSTEM VARIABLE

OleHide toggles the display of the OLE object (1 = off).

TIPS

- The **PasteClip** command places all objects in the upper-left corner of the current viewport, unless they are AutoCAD objects.
- Graphical objects are placed in the drawing as OLE objects.
- Text is usually — but not always — placed in the drawing as Mtext objects.
- Use the **PasteSpec** command to paste objects as an AutoCAD block.

Removed Command

PcxIn was removed from AutoCAD Release 14. Use the **ImageAttach** command instead.

PasteOrig

2000 Pastes AutoCAD objects from the Clipboard into the current drawing at the objects' original locations (*short for PASTE ORIGinal*).

Command	Alias	Ctrl+	F-key	Alt+	Menu Bar	Tablet
pasteorig	...	...	...	**ED**	**Edit** **↳Paste to Original Coordinates**	...

Command: pasteorig

COMMAND LINE OPTIONS

None.

RELATED COMMANDS

CopyClip copies a drawing to the Clipboard.

Insert inserts an AutoCAD drawing in the drawing.

PasteBlock pastes AutoCAD objects as a block at a user-specified insertion point.

RELATED SYSTEM VARIABLES

None.

TIPS

- Use this command to copy objects from one drawing to another.
- This command cannot be used to paste objects into the drawing from which they originate.

'PasteSpec

Rel.13 Pastes Clipboard objects in the current drawing as embedded, linked, pasted, or AutoCAD objects (*short for PASTE SPECial*).

Command	Alias	Ctrl+	F-key	Alt+	Menu Bar	Tablet
pastespec	**pa**	...	...	**ES**	**Edit** ↳**Paste Special**	...

Command: pastespec

Displays dialog box:

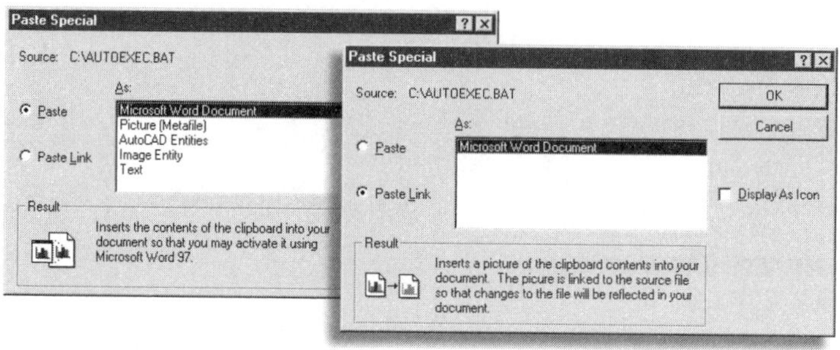

Paste (left) and Paste Link (right).

DIALOG BOX OPTIONS

⊙**Paste** pastes the object as an embedded object.

○**Paste Link** pastes the object as a linked object.

☐ **Display as Icon** displays the object as an icon from the originating application.

Change Icon allows you to select the icon; displays dialog box:

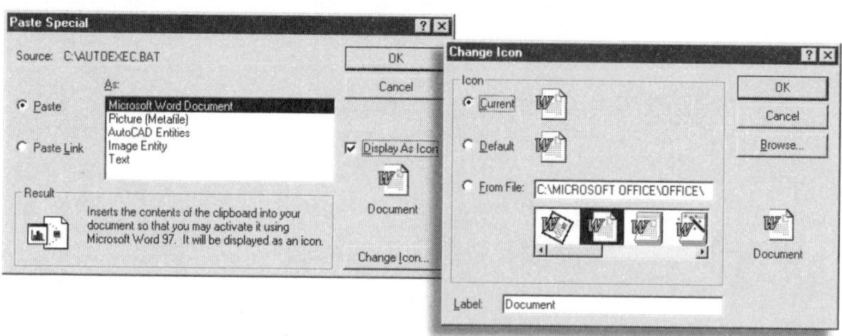

RELATED COMMANDS

CopyClip copies the drawing to the Clipboard.

InsertObj inserts an OLE object in the drawing.

OleLinks edits the OLE link data.

PasteClip places the Clipboard object as a pasted object.

RELATED SYSTEM VARIABLE

OleHide toggles the display of OLE objects (1 = off).

PcInWizard

2000 Converts PCP and PC2 plot configurations to PC3 format.

Command	Alias	Ctrl+	F-key	Alt+	Menu Bar	Tablet
pcinwizard	...	...	...	**TZI**	**Tools** ↳**Wizard** ↳**Import R14 Plot Settings**	...

Command: pcinwizard

Displays dialog box.

DIALOG BOX OPTIONS

Introduction page

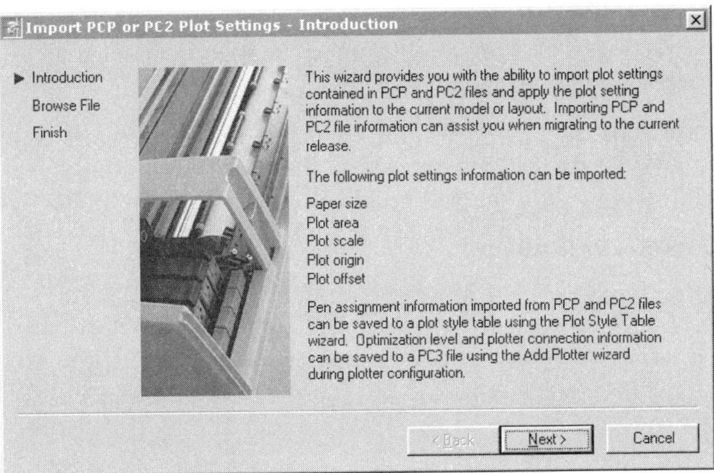

Back returns to the previous step.

Next continues to the next step.

Cancel cancels the wizard.

Browse File page

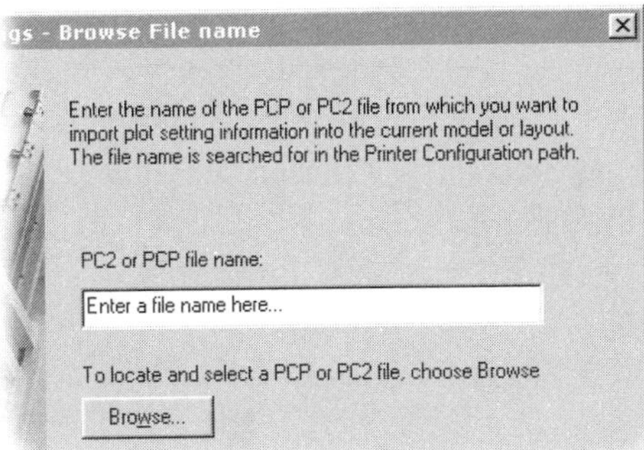

PCP or PC2 filename specifies the *.pcp* or *pc2* name.

Browse displays the **Import** dialog box.

Finish page

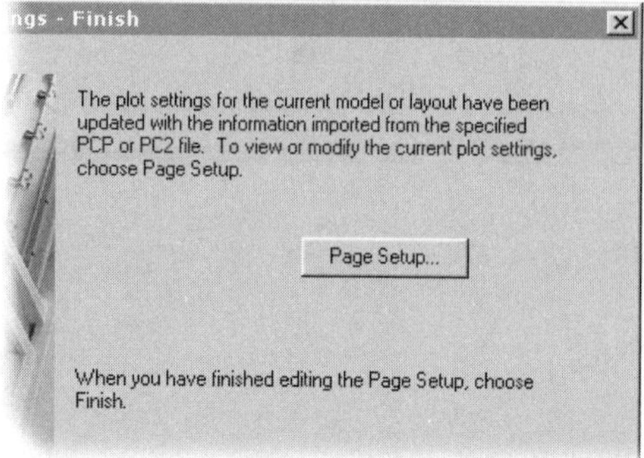

Page Setup displays the Page Setup dialog box; see the **PageSetup** command.

Finish completes the importation process.

RELATED COMMANDS

PageSetup creates and modifies page setups.

Plot uses PC3 files to control the plotter configuration.

RELATED SYSTEM VARIABLES

None.

TIPS

- This command imports *.pcp* and *.pc2* files, and applies them to the current layout or model tab.
- You only use this command when have you created *.pcp* or *.pc2* files with earlier versions of AutoCAD.
- PCP is short for "plotter configuration parameters":

  .pcp files are used by AutoCAD Release 13.
  .pc2 files are used by AutoCAD Release 14.
  .pc3 files are used by AutoCAD 2000 through 2005.
- AutoCAD imports the following information from *.pcp* and *.pc2* files: paper size, plot area, plot scale, plot origin, and plot offset.
- To import color-pen mapping, use the **Plot Style Table** wizard, run by the *StyShWiz.Exe* program.
- To import the optimization level and plotter connection, use the **Add-a-Plotter** wizard, run by the *AddPlWiz.Exe* program.

PEdit

V. 2.1 Edits 2D polylines, 3D polylines, or 3D meshes — depending on the selected object *(short for Polyline EDIT)*.

Command	Alias	Ctrl+	F-key	Alt+	Menu Bar	Tablet
pedit	**pe**	...	...	**MOP**	**Modify** **↳Object** **↳Polyline**	**Y17**

Command: pedit

Options vary, depending on whether a 2D polyline, 3D polyline, or polymesh is picked.

2D Polyline Operations

Select polyline or [Multiple]: *(Pick a 2D polyline, or type* **M.***)*

Enter an option [Open/Join/Width/Edit vertex/Fit/Spline/Decurve/Ltype gen/Undo]: *(Enter an option.)*

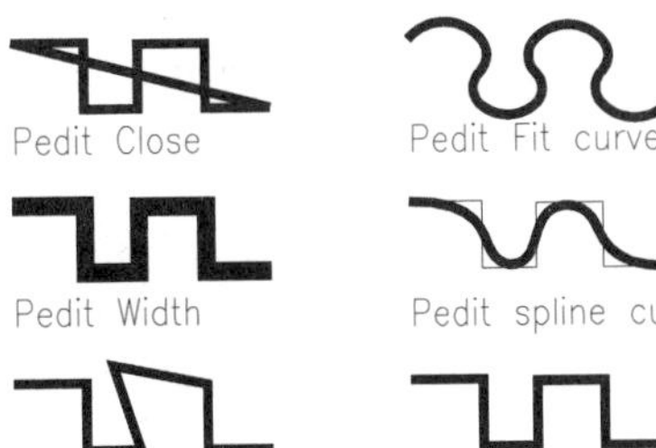

COMMAND LINE OPTIONS

Multiple selects more than one polyline to edit.

Close closes an open polyline by joining the two endpoints with a single segment.

Decurve reverses the effects of a Fit or Spline operation.

Edit vertex options are listed below.

Fit fits a curve to the tangent points of each vertex.

Ltype gen specifies the linetype generation style.

Join adds other polylines to the current polyline.

Open opens a closed polyline by removing the last segment.

Spline fits a splined curve along the polyline.

Undo undoes the most-recent **PEdit** operation.

Width changes the width of the entire polyline.

eXit exits the **PEdit** command.

Edit Vertex options
Enter a vertex editing option
[Next/Previous/Break/Insert/Move/Regen/Straighten/Tangent/Width/eXit]
<N>: *(Enter an option.)*

Break removes a segment or breaks the polyline at a vertex:

Next moves the x-marker to the next vertex.

Previous moves the x-marker to the previous vertex.

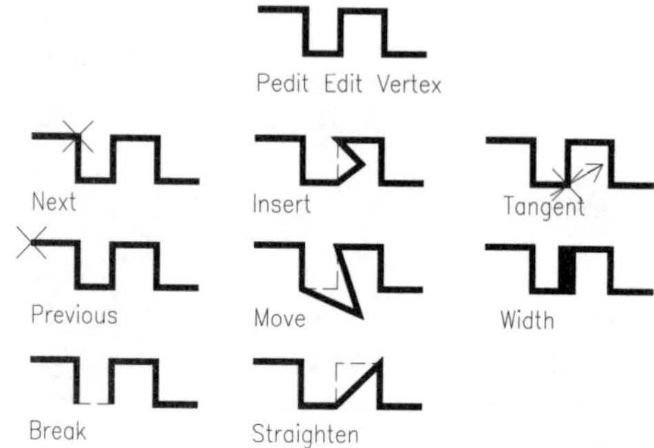

Go performs the break.

eXit exits the **Break** sub-submenu.

Insert inserts another vertex.

Move relocates a vertex; more easily accomplished with grips editing.

Next moves the x-marker to the next vertex.

Previous moves the x-marker to the previous vertex.

Regen regenerates the screen to show the effect of the **PEdit** command.

Straighten draws a straight segment between two vertices:

Next moves the x-marker to the next vertex.

Previous moves the x-marker to the previous vertex.

Go performs the straightening.

eXit exits the Straighten sub-submenu.

Tangent shows the tangent to the current vertex.

Width changes the width of a segment.

eXit exits the Edit vertex submenu.

3D Polyline Operations

Command: pedit

Select polyline or [Multiple]: *(Pick a 3D polyline, or type* **M.***)*

Enter an option [Open/Edit vertex/Spline curve/Decurve/Undo]: *(Enter an option.)*

COMMAND LINE OPTIONS

Multiple selects more than one 3D polyline to edit.

Close closes an open polyline.

Decurve reverses the effects of a Fit-curve or Spline-curve operation.

Open removes the last segment of a closed polyline.

Spline curve fits a splined curve along the polyline.

Undo undoes the most-recent **PEdit** operation.

eXit exits the **PEdit** command.

Edit Vertex options

Enter a vertex editing option

[Next/Previous/Break/Insert/Move/Regen/Straighten/eXit] <N>: *(Enter an option.)*

Break removes a segment or breaks the polyline at a vertex.

- **Next** moves the x-marker to the next vertex.
- **Previous** moves the x-marker to the previous vertex.
- **Go** performs the break.
- **eXit** exits the Break sub-submenu.

Insert inserts another vertex.

Move relocates a vertex; more easily accomplished with grips editing.

Next moves the x-marker to the next vertex.

Previous moves the x-marker to the previous vertex.

Regen regenerates the screen to show the effect of PEdit options.

Straighten draws a straight segment between two vertices:

- **Next** moves the x-marker to the next vertex.
- **Previous** moves the x-marker to the previous vertex.
- **Go** performs the straightening.
- **eXit** exits the Straighten sub-submenu.

eXit exits the Edit-vertex submenu.

Select polyline or [Multiple]: *(Pick a 3D mesh, or type* **M.***)*

Enter an option [Edit vertex/Smooth surface/Desmooth/Mclose/Nclose/ Undo]: *(Enter an option.)*

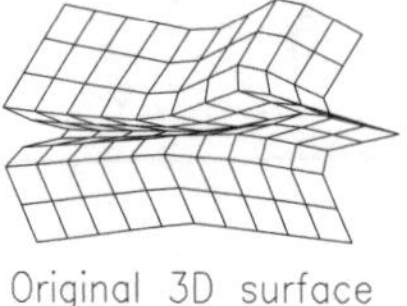
Original 3D surface

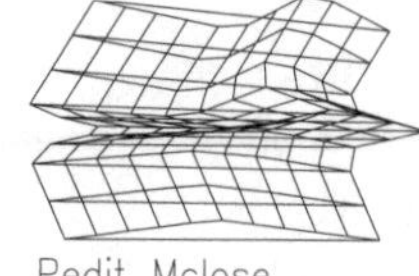
Pedit Mclose

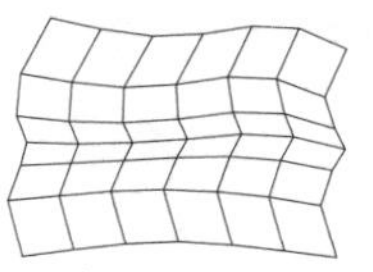
Pedit Smooth surface

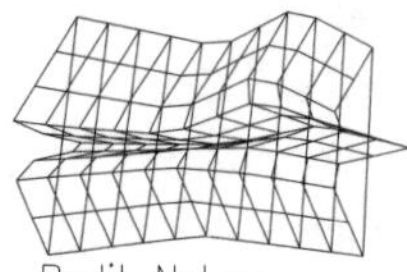
Pedit Nclose

COMMAND LINE OPTIONS

Multiple selects more than one 3D mesh to edit.

Desmooth reverses the effect of the **Smooth** surface options.

Mclose closes the mesh in the m-direction.

Mopen opens the mesh in the m-direction.

Nclose closes the mesh in the n-direction.

Nopen opens the mesh in the n-direction.

Smooth surface smooths the mesh with a Bezier-spline.

Undo undoes the most recent **PEdit** operation.

eXit exits the **PEdit** command.

Edit Vertex options

Current vertex (0,0).

Enter an option [Next/Previous/Left/Right/Up/Down/Move/REgen/eXit] <N>: *(Enter an option.)*

Down moves the x-marker down the mesh by one vertex.

Left moves the x-marker along the mesh by one vertex left.

Move relocates the vertex to a new position; more easily accomplished with grips editing.

Next moves the x-marker along the mesh to the next vertex.

Previous moves the x-marker along the mesh to the previous vertex.

REgen regenerates the drawing to show the effects of the **PEdit** command.

Right moves the x-marker along the mesh by one vertex right.

Up moves the x-marker up the mesh by one vertex.

eXit exits the Edit-vertex submenu.

RELATED COMMANDS

Break breaks a 2D polyline at any position.

Chamfer chamfers all vertices of a 2D polyline.

Convert converts older polylines to the new lwpolyline format.

EdgeSurf draws a 3D mesh.

Fillet fillets all vertices of a 2D polyline.

PLine draws a 2D polyline.

RevSurf draws a 3D surface of revolution mesh.

RuleSurf draws a 3D ruled surface mesh.

TabSurf draws a 3D tabulated surface mesh.

3D draws a 3D surface objects.

3dPoly draws a 3D polyline.

RELATED SYSTEM VARIABLES

SplFrame determines the visibility of a polyline spline frame:

SplFrame	Meaning
0	Do not display control frame (default).
1	Display control frame.

SplineSegs specifies the number (-32768 to 32767) of lines or arcs used to draw a splined polyline (default = 8); when number negative, arc segments are used, when positive, line segments.

SplineType determines the Bezier-spline smoothing for 2D and 3D polylines:

SplineType	Meaning
5	Quadratic Bezier spline.
6	Cubic Bezier spline.

SurfType determines the smoothing using the **Smooth** option:

SurfType	Meaning
5	Quadratic Bezier spline.
6	Cubic Bezier spline.
7	Bezier surface.

TIPS

- During vertex editing, the space bar, the ENTER key, right-clicking, and mouse button #2 moves the x-marker. The marker is moved to the next or previous vertex, whichever was used last.
- It's easier to edit the position of vertices with grips editing, than use this command.

PFace

Rel.11 Draws multisided 3D meshes; intended for use by AutoLISP and ARx programs (*short for Poly FACE*).

Command	Alias	Ctrl+	F-key	Alt+	Menu Bar	Tablet
pface	...	...	...	...	...	...

Command: pface

Specify location for vertex 1: *(Pick point 1.)*

Specify location for vertex 2 or <define faces>: *(Pick point 2, or press ENTER.)*

Face 1, vertex 1:

Enter a vertex number or [Color/Layer]: *(Type a number, or enter an option.)*

Face 1, vertex 2:

Enter a vertex number or [Color/Layer] <next face>: *(Type a number, or enter an option.)*

...and so on.

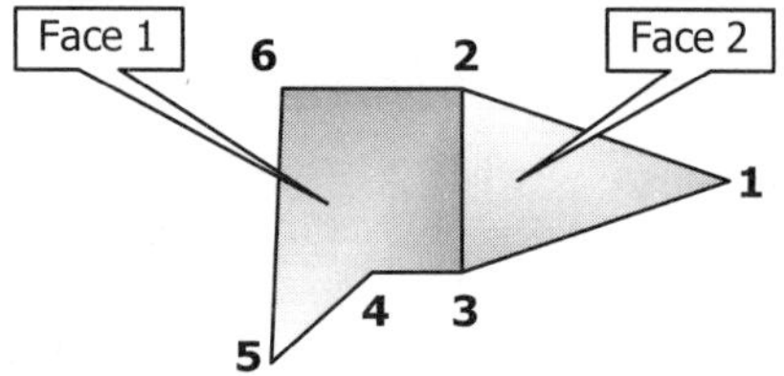

COMMAND LINE OPTIONS

Vertex defines the location of a vertex.

Face defines the faces, based on vertices.

Color gives the face a different color.

Layer places the face on a different layer.

RELATED COMMANDS

3dFace draws three- and four-sided 3D meshes.

3dMesh draws a 3D mesh with polyfaces.

RELATED SYSTEM VARIABLES

PFaceVMax specifies the maximum number of vertices per polyface (default = 4).

SplFrame controls the display of invisible faces (default = 0, not displayed).

TIPS

- **3dFace** creates 3- and 4-sided meshes, while **PFace** creates meshes of an arbitrary number of sides, and allows you to control the layer and the color of each face.
- The maximum number of vertices in the m- and n-direction is 256 when entered from the keyboard, and 32767 vertices when entered from a DXF file or created by programming.
- Make an edge invisible by entering a negative number for the beginning vertex of the edge.
- **PFace** is meant for programmers; to draw 3D surface objects, use these commands instead: **3d**, **3dMesh**, **RevSurf**, **RuleSurf**, **EdgeSurf**, or **TabSurf**.

Plan

Rel.10 Displays the plan view of the WCS, the current UCS, or a named UCS.

Command	Alias	Ctrl+	F-key	Alt+	Menu Bar	Tablet
plan	...	...	...	**V3P**	**View** ↳**3D Views** ↳**Plan View**	**N3**

Command: plan
Enter an option [Current ucs/Ucs/World] <Current>: *(Enter an option.)*

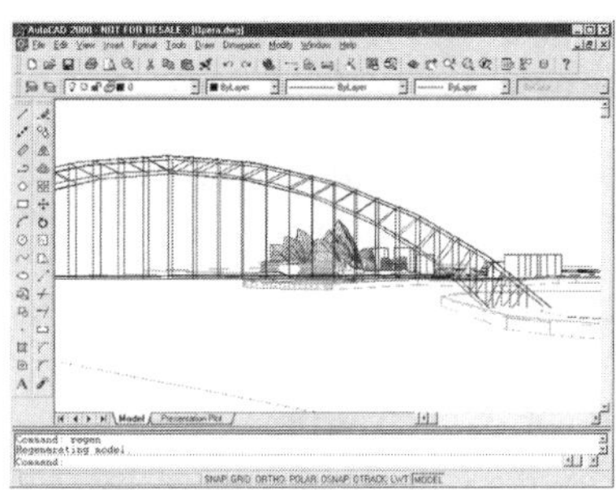

Example of a 3D view.

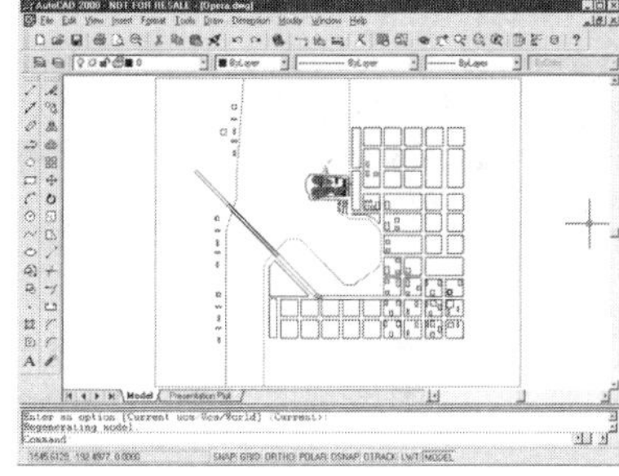

*After using the **Plan World** command.*

COMMAND LINE OPTIONS

Current UCS shows the plan view of the current UCS.

Ucs shows the plan view of the named UCS.

World shows the plan view of the WCS.

RELATED COMMANDS

UCS creates new UCS views.

VPoint changes the viewpoint of 3D drawings.

RELATED SYSTEM VARIABLES

UcsFollow displays the plan view for UCS or WCS automatically.

ViewDir contains the x,y,z coordinates of the current view direction.

TIPS

- Entering **VPoint 0,0,0** is an alternative to the **Plan** command.
- The **Plan** command turns off perspective mode and clipping planes.
- **Plan** does not work in paper space; AutoCAD complains, "** Command only valid in Model space **."
- The **Plan** command is an excellent method for turning off perspective mode.

PLine

V. 1.4 Draws complex 2D lines made of straight and curved sections of constant and variable width; treated as a single object (*short for Poly LINE*).

Command	Alias	Ctrl+	F-key	Alt+	Menu Bar	Tablet
pline	**pl**	...	...	**DP**	**Draw** **↳Polyline**	**N10**

Command: pline

Specify start point: *(Pick a point.)*

Current line-width is 0.0000

Specify next point or [Arc/Halfwidth/Length/Undo/Width]: *(Pick a point, or enter an option.)*

Specify next point or [Arc/Close/Halfwidth/Length/Undo/Width]: *(Pick a point, or enter an option.)*

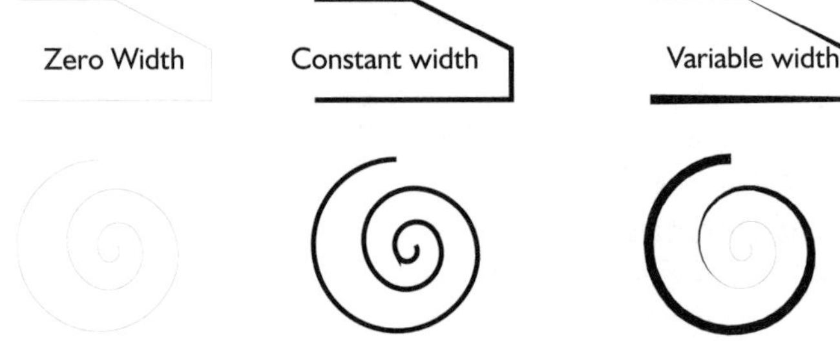

Examples of polylines created with the ***PLine*** *command.*

COMMAND LINE OPTIONS

Specify start point indicates the start of the polyline.

Close closes the polyline with a line segment.

Halfwidth indicates the half-width of the polyline.

Length draws a polyline tangent to the last segment.

Undo erases the last-drawn segment.

Width indicates the width of the polyline.

Endpoint of line indicates the endpoint of the polyline.

Arc options

Specify endpoint of arc or [Angle/CEnter/CLose/Direction/Halfwidth/Line/Radius/Second pt/Undo/ Width]: *(Enter an option.)*

Endpoint of arc indicates the arc's endpoint.

Angle indicates the arc's included angle.

CEnter indicates the arc's center point.

CLose uses an arc to close a polyline.

Direction indicates the arc's starting direction.

Halfwidth indicates the arc's halfwidth.

Line switches back to the menu for drawing lines.

Radius indicates the arc's radius.

Second pt draws a three-point arc.

Undo erases the last drawn arc segment.

Width indicates the width of the arc.

RELATED COMMANDS

Boundary draws a polyline boundary.

Donut draws solid-filled circles as polyline arcs.

Ellipse draws ellipses as polyline arcs when **PEllipse** = 1.

Explode reduces a polyline to lines and arcs with zero width.

Fillet fillets polyline vertices with a radius.

PEdit edits the polyline's vertices, widths, and smoothness.

Polygon draws polygons as polylines of up to 1024 sides.

Rectang draws a rectangle out of a polyline.

Sketch draws polyline sketches, when **SkPoly** = 1.

Xplode explodes a group of polylines into line and arcs of zero width.

3dPoly draws 3D polylines.

RELATED SYSTEM VARIABLES

PLineGen specifies the style of linetype generation:

PlineGen	Meaning
0	Vertex to vertex (default).
1	End to end (compatible with Release 12).

PLineType controls the conversion of old (pre-R14) polylines and the creation of lwpolyline objects by the **PLine** command:

PLineType	Meaning
0	Old polylines not converted; old-format polylines created.
1	Not converted; lwpolylines created.
2	Polylines in pre-R14 drawings converted on open; **PLine** command creates lwpolyline objects (default).

PLineWid sets the current width of polyline (default = 0.0).

TIPS

- **Boundary** uses a polyline to outline a region automatically; use the **List** command to find its area.
- If you cannot see a linetype on a polyline, change system variable **PlineGen** to 1; this regenerates the linetype from one end of the polyline to the other.
- If the angle between a joined polyline and polyarc is less than 28 degrees, the transition is chamfered; at greater than 28 degrees, the transition is not chamfered.
- Use the object snap modes **INTersection** or **ENDpoint** to snap to the vertices of a polyline.

Plot

V.1.0 Creates hardcopies of drawings on vector, raster, and PostScript plotters; also plots to files on disk.

Commands	Alias	Ctrl+	F-key	Alt+	Menu Bar	Tablet
plot	**print**	**P**	**...**	**FP**	**File** ↳**Plot**	**W25**
-plot						

Command: plot

Displays dialog box (user interface changed with AutoCAD 2005):

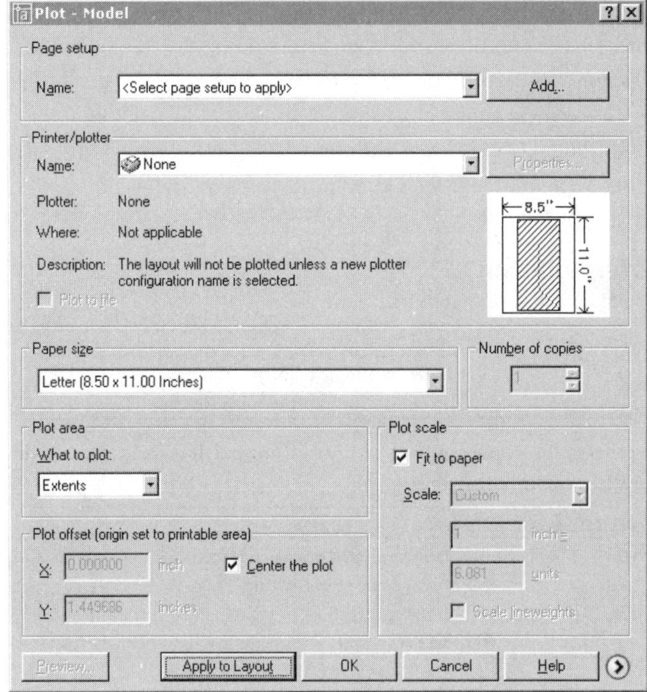

DIALOG BOX OPTIONS

Printer/Plotter Configuration options

Name selects a system printer, or a *.pc3* named plotter configuration.

Properties displays the Plotter Configuration Editor dialog box; see **PlotterManager** command.

Paper Size options

Paper size displays paper sizes supported by the selected output device.

Plot Area options

What to plot:

- **Layout** plots all parts of the drawing within the margins of the specified paper size; the origin is calculated from 0,0 in the layout (not available in model space).
- **Display** plots all parts of the drawing within the current viewport.
- **Extents** plots the entire drawing.
- **Limits** plots the drawing in the Model tab within a rectangle defined by the **Limits** command.
- **View** plots a view saved with the **View** command (not available when no views are named).
- **Window** plots all parts of the drawing within a picked rectangle.

Plot Offset options

Center the plot centers the plot on the paper.

X specifies the plot origin offset from x = 0.

Y specifies the plot origin offset from y = 0.

☑ **Center the plot** centers the plot between the page edge or printer margins (not available when the Layout option is selected, above).

Plot Scale options

☑ **Fit to paper** scales the drawing to fit within the plotter's margins (not available when the Layout option selected).

Scale specifies the plotting scale; ignored by layouts.

Custom specifies user-defined scale.

Inch specifies the number of inches on plotted page.

Units specifies number of units in drawing.

☐ **Scale lineweights** scales lineweights proportionately to the plot scale.

Preview displays the preview window; see **Preview** command.

Apply to layout applies setting to layout.

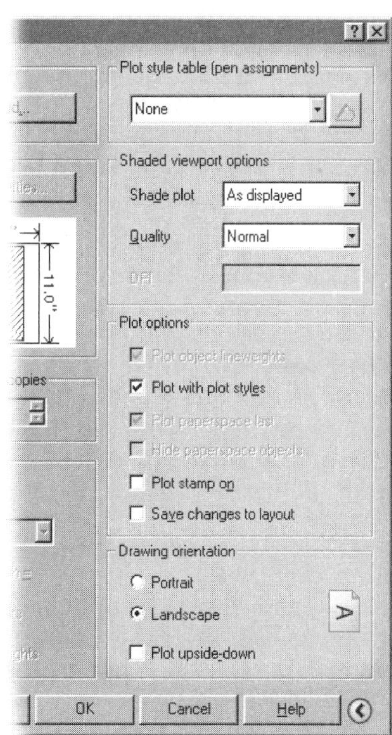

Expanded dialog box

Plot Style Table (Pen Assignments) options

Drop list selects a *.ctb* plot style table file; see the **PlotStyle** command.

New displays the Add Color-Dependent Plot Style Table wizard; see **StylesManager** command.

Edit button displays the Plot Style Table Editor dialog box; see **StylesManager** command.

Shaded Viewport Options options

*Available only in model space; in layouts, right-click the viewport border, and then select **Properties**.*

Shadeplot plots the viewport as-displayed, wireframe, hidden, or rendered.

Quality sets print quality as draft (wireframe), preview (quarter-resolution, up to 150dpi), normal (half resolution, up to 300dpi), presentation (maximum, up to 600dpi), maximum (maximum resolution), or custom (as specified by DPI text box).

DPI specifies the resolution for the output device; DPI is short for "dots for inch."

Plot Options options

☑ **Plot object lineweights** plots objects with lineweights (available only when plot style table turned off).

☑ **Plot with plot styles** plots objects using the plot styles.

☑ **Plot paperspace last** plots the model space objects first, followed by paper space objects (available in layouts only).

☐ **Hide paperspace objects** removes hidden lines before plotting.

Draw Orientation options

⊙ **Portrait** positions the paper upright (vertically).

○ **Landscape** positions the paper wide (horizontally).

☐ **Plot Upside Down** plots the drawing upside down.

-PLOT Command

Command: -plot

Detailed plot configuration? [Yes/No] <No>: *(Type **Y** or **N**.)*

Brief Plot Configuration options

Enter a layout name or [?] <Model>: *(Enter a layout name, or type **?**.)*

Enter a page setup name <>: *(Enter a page setup name, or press* ENTER *for none.)*

Enter an output device name or [?] <default printer>: *(Enter a printer name, or type **?** for a list of printers.)*

Write the plot to a file [Yes/No] <N>: *(Type **Y** or **N**.)*

Save changes to page setup [Yes/No]? <N> *(Type **Y** or **N**.)*

Proceed with plot [Yes/No] <Y>: *(Type **Y** or **N**.)*

Detailed Plot Configuration options

Enter a layout name or [?] <Model>: *(Enter a layout name, or type **?**.)*

Enter an output device name or [?] <default printer>: *(Enter a printer name, or type **?** for a list of printers.)*

Enter paper size or [?] <Letter 8 ½ x 11 in>: *(Enter a paper size, or type **?**.)*

Enter paper units [Inches/Millimeters] <Inches>: *(Type **I** or **M**.)*

Enter drawing orientation [Portrait/Landscape] <Landscape>: *(Type **P** or **L**.)*

Plot upside down? [Yes/No] <No>: *(Type **Y** or **N**.)*

Enter plot area [Display/Extents/Limits/View/Window] <Display>: *(Enter an option.)*

Enter plot scale (Plotted Inches=Drawing Units) or [Fit] <Fit>: *(Enter a scale factor, or type **F**.)*

Enter plot offset (x,y) or [Center] <0.00,0.00>: *(Enter an offset, or type **C**.)*

Plot with plot styles? [Yes/No] <Yes>: *(Type **Y** or **N**.)*

Enter plot style table name or [?] (enter . for none) <>: *(Enter a name, or enter an option.)*

Plot with lineweights? [Yes/No] <Yes>: *(Type* **Y** *or* **N***.)*
Enter shade plot setting [As displayed/Wireframe/Hidden/Rendered] <As displayed>: *(Enter an option.)*

Write the plot to a file [Yes/No] <N>: *(Type* **Y** *or* **N***.)*
Save changes to page setup? Or set shade plot quality? [Yes/No/Quality] <N>: *(Enter an option.)*
Proceed with plot [Yes/No] <Y>: *(Type* **Y** *or* **N***.)*

COMMAND OPTIONS

Enter a layout name specifies the name of a layout; enter **Model** for model view or **?** for a list of layout names.

Enter an output device name specifies the name of a printer, or enter ? for a list of print devices.

Enter paper size specifies the name of a paper size, or enter ? for a list of paper size supported by the print device.

Enter paper units specifies Inches (imperial) or Millimeters (metric) units.

Enter drawing orientation specifies Portrait (vertical) or Landscape (horizontal) orientation.

Plot upside down:

Yes plots the drawing upside down.

No plots the drawing normally.

Enter plot area specifies that the current Display, Extents of the drawing, Limits defined by the **Limits** command, View name, or window area be plotted.

Enter plot scale specifies the scale using the Plotted Inches=Drawing Units format; or enter F to fit the drawing to the page

Enter plot offset specifies the distance between the lower left corner of the paper and the drawing; or enter C to center the drawing on the page.

Plot with plot styles:

Yes prompts you to specify the name of a plot style.

No ignores plot styles, and uses color-dependent styles.

Plot with lineweights:

Yes uses lineweight settings to draw thicker lines.

No ignores lineweights.

Enter shade plot setting specifies whether the plot should be rendered As displayed, Wireframe, Hidden lines removed, or Rendered.

Write the plot to a file:

Yes sends the plot to a file.

No sends the plot to the plot device.

Save changes to page setup:

Yes saves the settings you specified during the previous set of questions.

No does not save the settings.

Quality specifies the quality of the plot.

Proceed with plot:

Yes plots the drawing.

No cancels the plot.

Quality options
Enter shade plot quality [Draft/Preview/Normal/pResentation/Maximum/Custom] <Normal>: *(Select an option.)*

Draft, Preview, Normal, pResentation, and Maximum apply preset dpi (dots per inch) settings to the plot. The higher the dpi, the better the quality, but the slower the plot.

Custom allows you to specify the dpi setting; default = 150dpi.

RELATED COMMANDS

PageSetup selects one or more plotter devices.

Preview goes directly to the plot preview screen.

Publish plots multi-page drawings.

PublishToWeb plots drawings to HTML format.

PsOut saves drawings in *.eps* format.

ViewPlotDetails reports on sucusseful and unsuccessful plots (*new to AutoCAD 2005*).

RELATED SYSTEM VARIABLES

BackgroundPlot toggles background plotting (*new to AutoCAD 2005*).

BgrndPlotTimeout specifies the time AutoCAD waits for plotter to respond (*new to AutoCAD 2005*).

TextFill toggles the filling of TrueType fonts.

TextQlty specifies the quality of TrueType fonts.

RELATED FILES

***.pc3** holds plotter configuration parameter files.

***.plt** holds plot files created with this command.

RELATED EXTERNAL PROGRAMS

BatchPlot.exe plots large numbers of drawings unattended.

TIPS

- AutoCAD R12 replaced **PrPlot** with the **Plot** command. R13 removed **-p** freeplot (plotting without using up a network license). AutoCAD 2005 removed the partial preview.
- AutoCAD cannot plot perspective views to-scale, only to-fit.
- Background plotting (*new to AutoCAD 2005*) displays an icon in the tray; click it for options.

- You must wait for one background plot to complete before beginning the next one.
- Turn off **BackgroundPlot** when plotting drawings with scripts.

'PlotStamp

2000i Adds information about drawings to hardcopy plots.

Commands	Alias	Ctrl+	F-key	Alt+	Menu Bar	Tablet
plotstamp	...	...	...	FPS	File ↳Plot ↳Settings	...
-plotstamp						

Command: plotstamp

Displays dialog box:

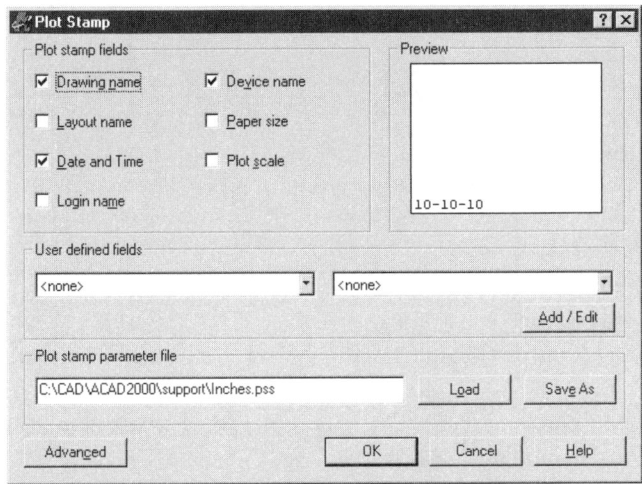

DIALOG BOX OPTIONS

Plot Stamp Fields options

☑ **Drawing name** adds the path and name of the drawing.

☐ **Layout name** adds the layout name.

☑ **Date and time** adds the date (short format) and time of the plot.

☐ **Login name** adds the Windows login name, as stored in the **LogInName** system variable.

☑ **Device name** adds the plotting device's name.

☐ **Paper size** adds the paper size, as currently configured.

☐ **Plot scale** adds the plot scale.

User Defined Fields options

Add/Edit displays the User Defined Fields dialog box.

Plot Stamp Parameter File options

Load/Save displays the Plotstamp Parameter File Name dialog box for opening (or saving) *.pss* files.

Advanced displays the Advanced Options dialog box.

Advanced Options dialog box

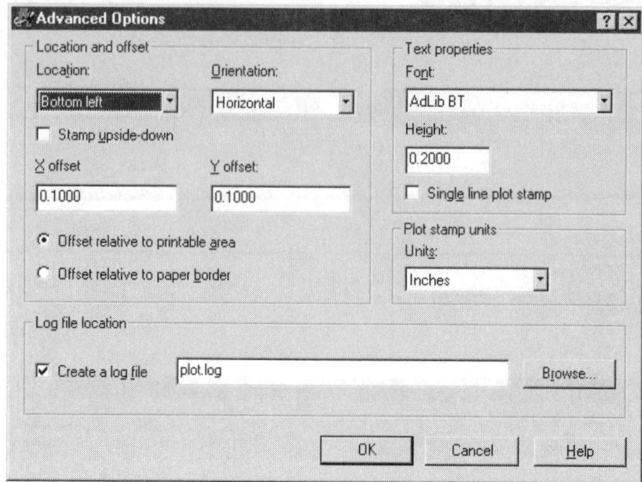

Location and Offset options

Location specifies the location of the plot stamp: Top Left, Bottom Left (default), Bottom Right, or Top Right — relative to the orientation of the drawing on the plotted page.

Stamp upside down:

☑ Draws the plot stamp upside down.

☐ Draws the plot stamp rightside up.

Orientation rotates the plot stamp relative to the page: Horizontal or Vertical.

X offset specifies the distance from the corner of the printable area or page; default = 0.1.

Y offset specifies the distance from the corner of the printable area or page; default = 0.1

⊙ **Offset relative to printable area.**

○ **Offset relative to paper border.**

Text Properties options

Font selects the font for the plot stamp text.

Height specifies the height of the plot stamp text.

Single line plot stamp:

☑ Plotstamp text is placed on a single line.

☐ Plotstamp text is placed on two lines.

Plot Stamp Units option

Units selects either inches, millimeters, or pixels.

Log File Location options

☑ **Create a log file** saves the plot stamp text to a text file.

Browse displays the Log File Name dialog box.

-PLOTSTAMP Command

Command: -plotstamp

Current plot stamp settings:

Displays the current setting of nearly 20 plot stamp settings.

Enter an option [On/OFF/Fields/User Fields/Log file/LOCation/Text properties/UNits]: *(Enter an option.)*

COMMAND LINE OPTIONS

On turns on plot stamping.

OFF turns off plot stamping.

Fields specifies plot stamp data: drawing name, layout name, date and time, login name, plot device name, paper size, plot scale, comment, write to log file, log file path, location, orientation, offset, offset relative to, units, font, text height, and stamp on single line.

User fields specifies two user-defined fields.

Log file writes the plotstamp data to a file instead of the drawing.

LOCation specifies the location and orientation of the plotstamp.

Text properties specifies the font name and height.

UNits specifies the units of measurement: inches, millimeters, or pixels.

RELATED COMMAND

Plot plots the drawing.

RELATED FILES

***.log** is the plotstamp log file; stored in ASCII text format.

***.pss** holds plotstamp parameter file; stored in binary format.

TIPS

- When the options of the **Plot Stamp** dialog box are grayed out, or when the **-Plotstamp** command reports 'Current plot stamp file or directory is read only,' this means that the *Inches.pss* or *Mm.pss* file in the *Support* folder is read-only. To change, in Explorer: (1) right-click the file, (2) select **Properties**, (3) uncheck **Read-only**, and (4) click **OK**.
- You can access the Plot Stamp dialog box via the **Plot** command's More Options dialog box:

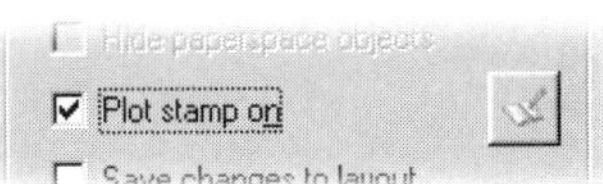

- *Caution!* Too large of an offset value positions the plotstamp text beyond the plotter's printable area, which causes the text to be cut off. To prevent this, use the **Offset relative to printable area** option.

PlotStyle

2000 Selects and assigns plot styles to objects.

Commands	Alias	Ctrl+	F-key	Alt+	Menu Bar	Tablet
plotstyle	...	...	...	OY	Format ↳Plot Style	...
-plotstyle						

Command: plotstyle

When no objects are selected, displays the Current Plot Style dialog box.

When one or more objects are selected, displays the Select Plot Style dialog box.

DIALOG BOX OPTIONS

Plot styles lists the available plot styles.

Active plot style table lists the available plot style tables.

Editor displays the Plot Style Table Editor dialog box; see the StylesManager command.

OK accepts the changes, and closes the dialog box.

Cancel cancels the changes, and closes the dialog box.

Select Plot Style dialog box

Displayed when one or more objects are selected:

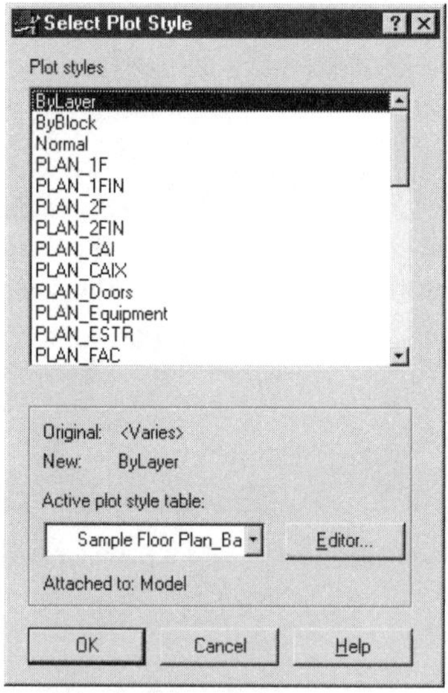

Plot styles lists the plot styles available in the drawing.

Active plot style table selects the plot style table to attach to the current drawing.

Editor displays the Plot Style Table Editor dialog box; see the **StylesManager** command.

Current Plot Style dialog box

Displayed when no objects are selected:

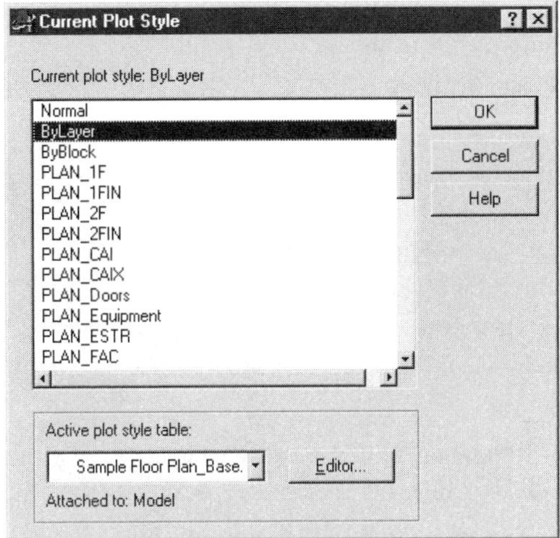

Plot styles lists the plot styles available in the drawing.

Active plot style table selects the plot style table to attach to the current drawing.

Editor displays the Plot Style Table Editor dialog box; see the **StylesManager** command.

-PLOTSTYLE Command

Command: -plotstyle
Current plot style is "Default"
Enter an option [?/Current] : *(Type* **?** *or* **C**.*)*
Set current plot style : *(Enter a name.)*
Current plot style is "*plotstylename*"
Enter an option [?/Current] : *(Press* **ENTER**.*)*

COMMAND LINE OPTIONS

Current prompts you to change plot styles.

Set current plot style specifies the name of the plot style.

ENTER exits the command.

? displays plot style names; sample output:

```
Plot Styles:
-----------------
ByLayer
ByBlock
Normal
ARCH_Dimensions for stairway
Current plot style is "PLAN_ESTR"
```

RELATED COMMANDS

Plot plots drawings with plot styles.

PageSetup attaches plot style tables to layouts.

StylesManager modifies plot style tables.

Layer specifies plot styles for layers.

RELATED SYSTEM VARIABLES

CPlotStyle specifies the plot style for new objects; defined values include "ByLayer," "ByBlock," "Normal," and "User Defined."

DefLPlStyle specifies the plot style name for layer 0.

DefPlStyle specifies the default plot style for new objects.

PStyleMode indicates whether the drawing is in Color-Dependent or Named Plot Style mode.

PStylePolicy determines whether object color properties are associated with plot style.

TIPS

- This command does not operate until a plot style table has been created for the drawing; the **PlotStyle** command displays this message box:

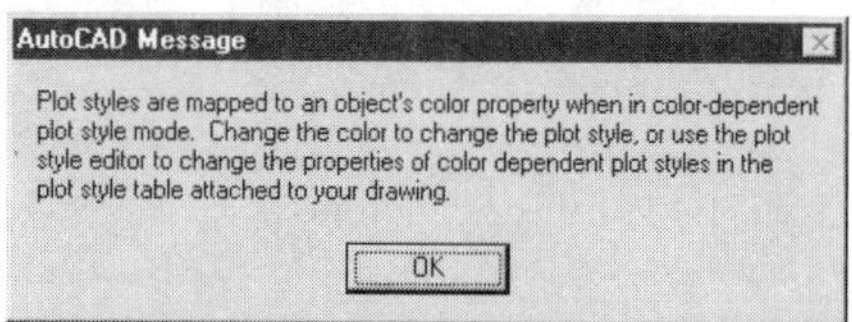

- A plot style can be assigned to any object and to any layer.
- A plot style can override the following plot settings: color, dithering, gray scale, pen assignment, screening, linetype, lineweight, end style, join style, and fill style.
- Plot styles are useful for plotting the same layout in different ways, such as emphasizing objects using different lineweights or colors in each plot.
- Plot style tables can be attached to the Model tab and layout tabs, and attach different plot style tables to layouts, to create different looks for plots.

Quick Start Tutorial

Applying Plot Styles

Before you can apply plot styles to a new drawing, you must turn on plot styles; notice that **Plot Style Control** in the **Object Properties** toolbar is grayed out.

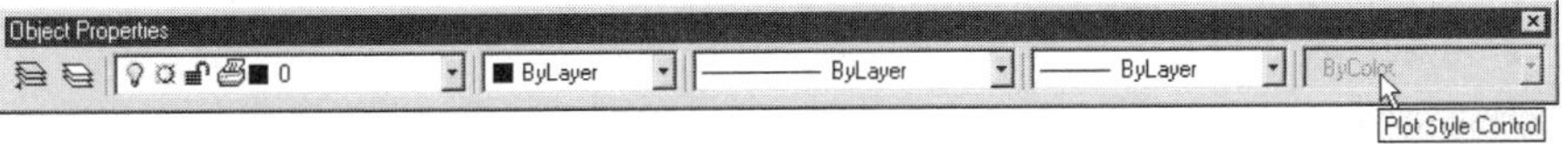

Follow these steps to turn on plot style:

Step 1

From the menu bar, select **Tools | Options | Plotting**.

Step 2

In the **Options** dialog box, click **Use named plot styles** to turn on the feature.

Step 3

In the **Default** plot style list, select any plot style — *except* **Default R14 pen assignments.stb**, because it turns off plot styles.

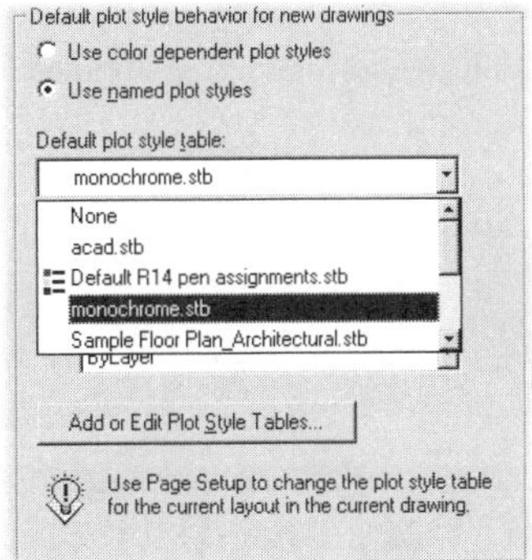

Step 4

Click **OK** to dismiss the Options dialog box. Notice that the **Plot Style Control** in the **Object Properties** toolbar is now available for use.

PlotterManager

2000 Displays the Plotters window, the Add-A-Plotter wizard, and PC3 configuration Editor.

Command	Alias	Ctrl+	F-key	Alt+	Menu Bar	Tablet
plottermanager	...	...	...	**FM**	**File** ↳**Plotter Manager**	**Y24**

Command: plottermanager

Displays window:

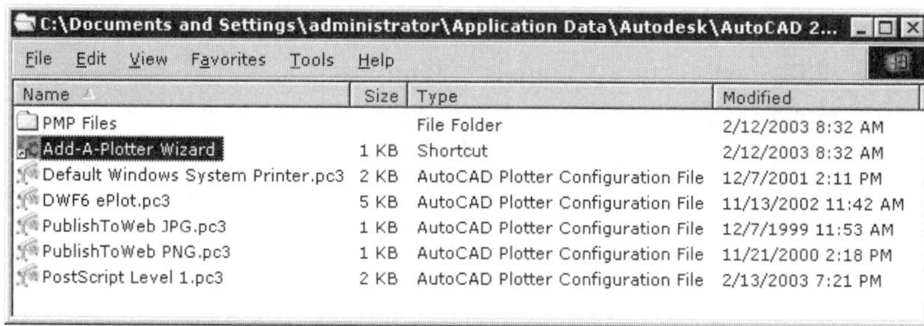

WINDOW OPTIONS

Add-a-Plotter Wizard adds a plotter configuration; double-click to display the **Add Plotter** wizard.

***.pc3** specifies parameters for creating plotted output; double-click to display the Plotter Configuration Editor; see the **StylesManager** command.

ADD PLOTTER WIZARD

The following steps show how to create a PC3 plotter configuration for plotting drawings to an EPS (encapsulated PostScript) file; the steps are similar to creating .pc3 files for other types of plotters:

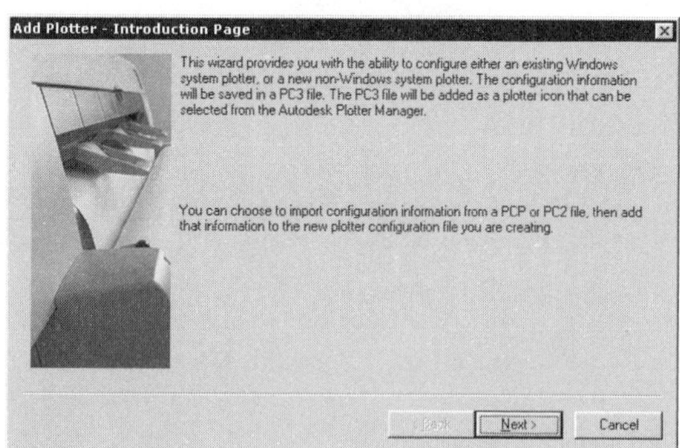

Next displays the next dialog box.

Cancel dismisses the dialog box.

Begin page

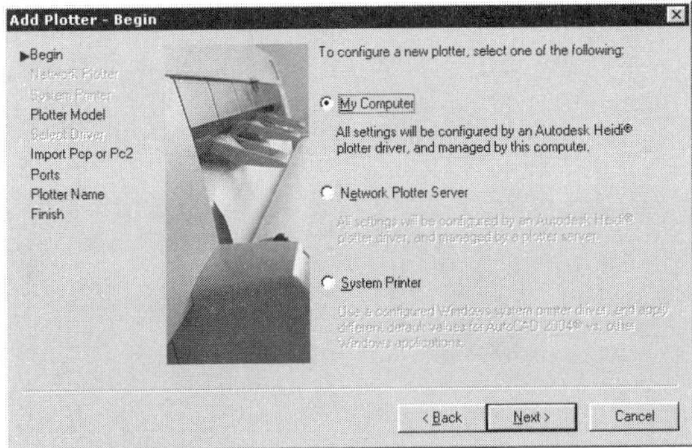

My Computer configures printers and plotters connected to your computer.

Network Plotter Server configures printers and plotters connected to other computers on network.

System Printer configures the default Windows printer.

Plotter Model page

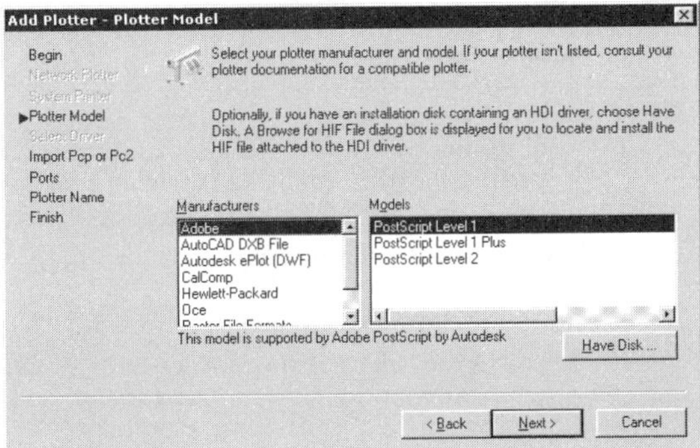

Manufacturer selects a brand name of plotter.

Model selects a specific model number.

Have Disk selects plotter drivers provided by the manufacturer; displays the Open dialog box.

Import Pcp or Pc2 page

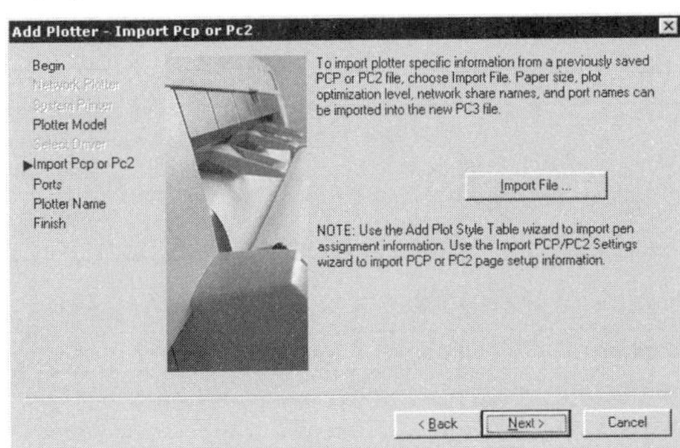

Import File imports *.pcp* and *.pc2* plotter configuration files from earlier versions of AutoCAD.

Ports page

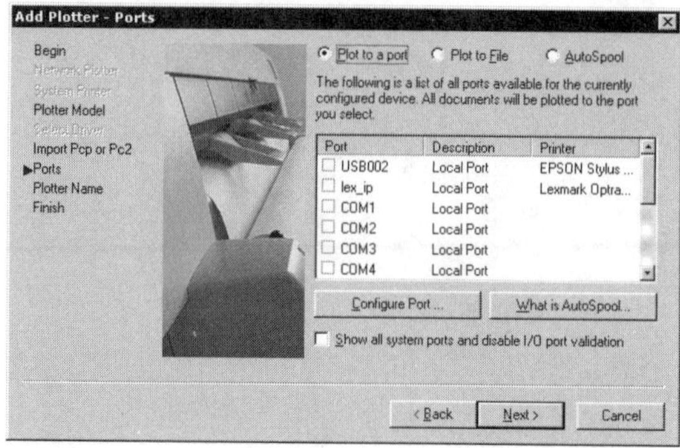

Plot to a port sends the plot to an output port.

Plot to File plots the drawing to a file with a user-definable filename; default filename is the same as the drawing name with the *.plt* extension; PostScript plot files are given the *.eps* extension.

AutoSpool plots the drawing to a file with a filename generated by AutoCAD, and then executes the command specified in the Option dialog box's Files tab. Enter the name of the program that should process the AutoSpool file in **Print Spool Executable**. You may include these DOS command-line arguments:

Argument	Meaning
%s	Substitutes path, spool filename, and extension.
%d	Specifies the path, AutoCAD drawing name, and extension.
%c	Specifies the description for the device.
%m	Returns the plotter model.
%n	Specifies the plotter name.
%p	Specifies the plotter number.
%h	Returns the height of the plot area in plot units.
%w	Returns the width of the plot area in plot units.
%i	Specifies the first letter of the plot units.
%l	Specifies the login name (***LogInName*** *system variable*).
%u	Specifies the user name.
%e	Specifies the equal sign (=).
%%	Specifies the percent sign (%).

Port options

Port lists the virtual ports defined by the Windows operating system:

Port	Meaning
USB	Universal serial bus.
COM	Serial port.
LPT	Parallel port.
HDI	Autodesk's Heidi Device Interface.

Description describes the type of port:

Description	Meaning
Local Port	Printer is connected to your computer.
Network Port	Printer is accessible through the network.

Printers describes the brand name of the printer connected to the port.

Configure Port displays the Configure Port dialog box; allows you to specify parameters specific to the port, such as timeout and protocol.

What is AutoSpool? displays an explanatory dialog box:

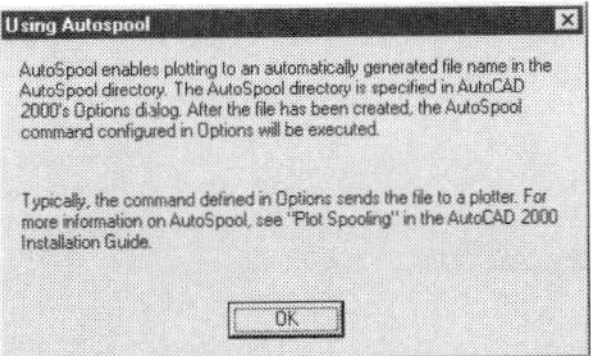

☐ **Show all system ports and disable I/O port validation** prevents AutoCAD from checking whether the port is valid.

Plotter Name page

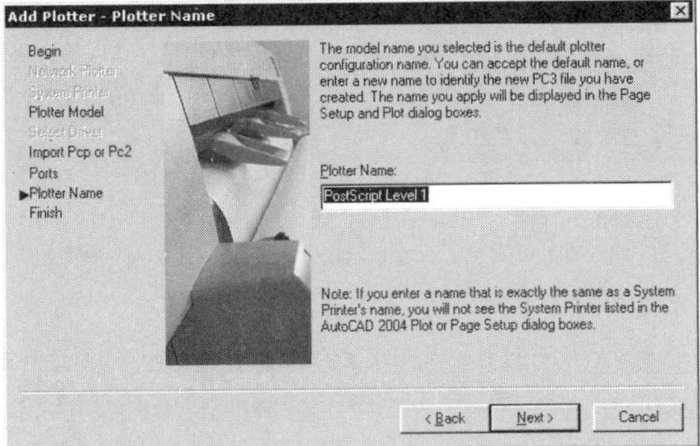

Plotter name Specifies a user-defined name for the plotter configuration; you may have many different configurations for a single plotter.

Finish page

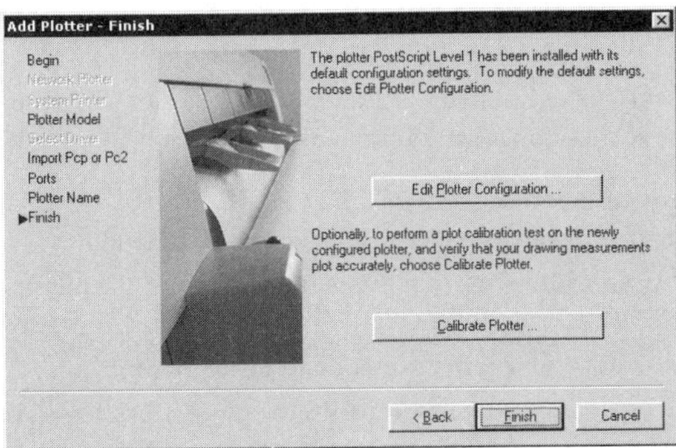

Edit Plotter Configuration displays the Edit Plotter Configuration dialog box.

Calibrate Plotter displays the Calibrate Plotter wizard.

Plotter Configuration Editor dialog boxes

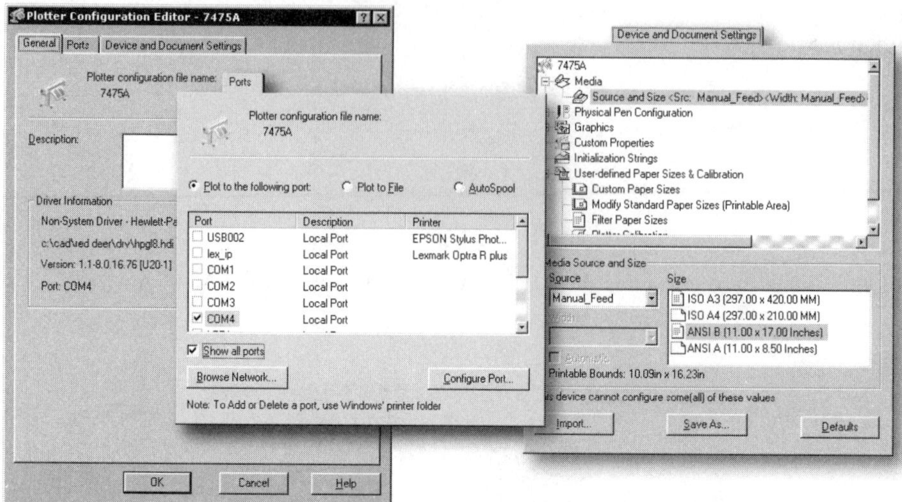

General tab

Description allows you to provide a detailed description of the plotter configuration.

Ports tab

Plot to the following port sends the plot to an output port.

Plot to File plots the drawing to a file with a user-definable filename; default filename is the same as the drawing name, with a *.plt* extension; PostScript plot files are given the *.eps* extension.

AutoSpool plots the drawing to a file with a filename generated by AutoCAD, and then executes the command specified in the Option dialog box's **Files** tab.

Show all ports lists all ports on the computer.

Browse Network displays the Browse for Printer dialog box; selects a printer on the network.

Configure Port displays the Configure Port dialog box; allows you to specify the parameters specific to the port, such as timeout and protocol.

Device and Document Settings tab

Media specifies the paper source, paper size, type, and destination.

Physical Pen Configuration (for pen plotters only) specifies the physical pens in the pen plotter.

Graphics specifies settings for plotting vector and raster graphics and TrueType fonts.

Custom Properties specifies settings specific to the plotter, printer, or other output device.

Initialization Strings (for non-system plotters only) specifies the control codes for pre-initialization, post-initialization, and termination.

Calibration Files and Paper Sizes calibrates the plotter by specifying the *.pmp* file; adds and modifies custom paper sizes; see the Calibrate Plotter wizard.

Import imports *.pcp* and *.pc2* plotter configuration files from earlier versions of AutoCAD.

Save As saves the plotter configuration data to a *.pc3* file.

Defaults resets the plotter configuration settings to the previously-saved values.

CALIBRATE PLOTTER WIZARD

Before using this wizard, use AutoCAD to draw and plot a rectangle — say, 11 inches by 8 inches — then use this wizard to check the accuracy of the plotted drawing:

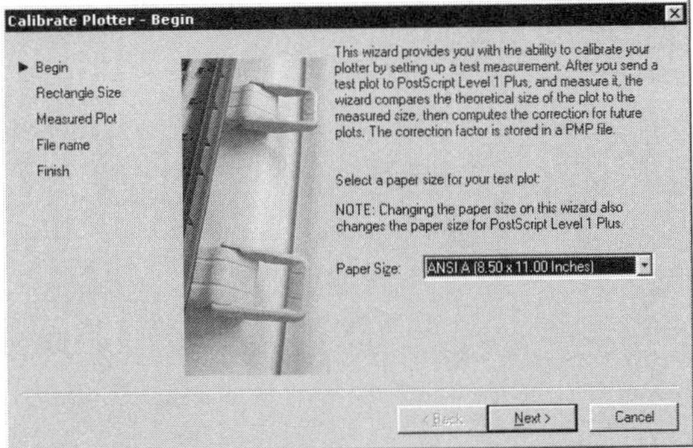

Paper Size selects the size of media or paper.

Next displays the next dialog box.

Cancel dismisses the dialog box.

Rectangle Size page

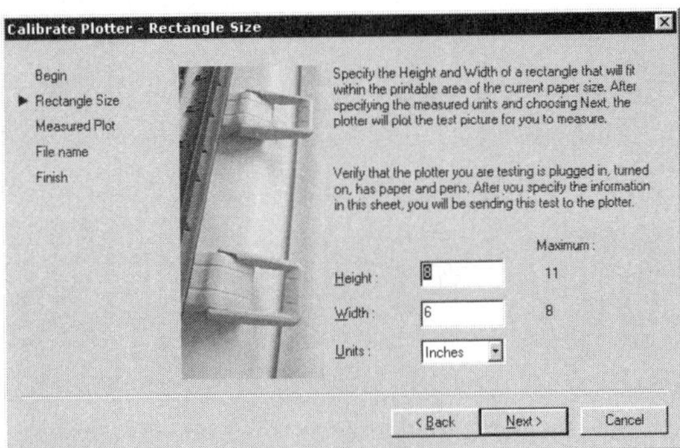

Length specifies the length of the calibration rectangle.

Width specifies the width of the calibration rectangle.

Units selects imperial inches or metric millimeters.

Measured Plot page

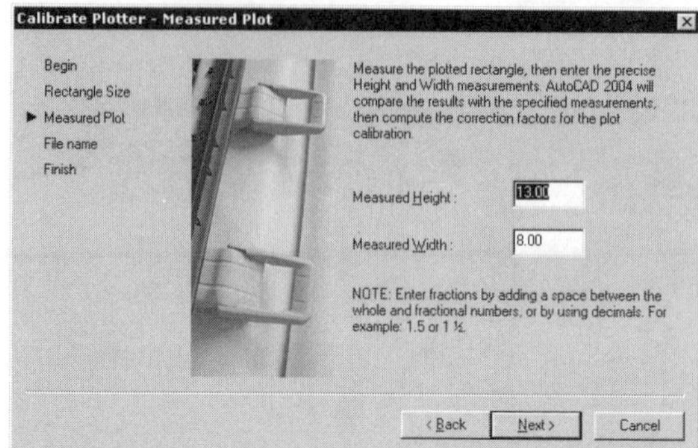

Measured Length specifies the length of the rectangle, which is plotted by AutoCAD, and then measured by you.

Measured Width specifies the width of the rectangle, which is plotted by AutoCAD, and then measured by you.

File Name page

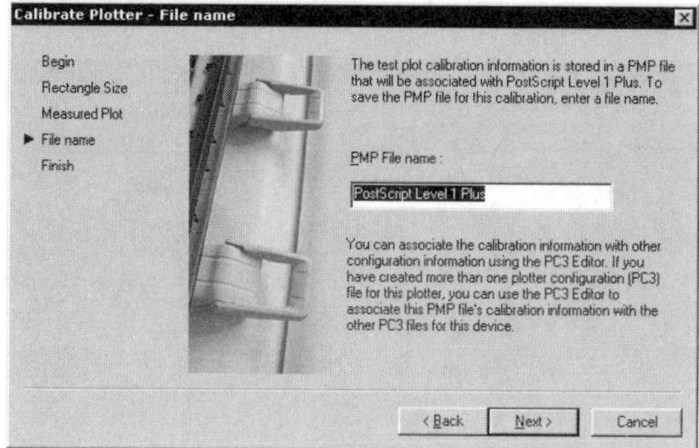

PMP File name specifies the name of the file in which to store the calibration data.

Finish page

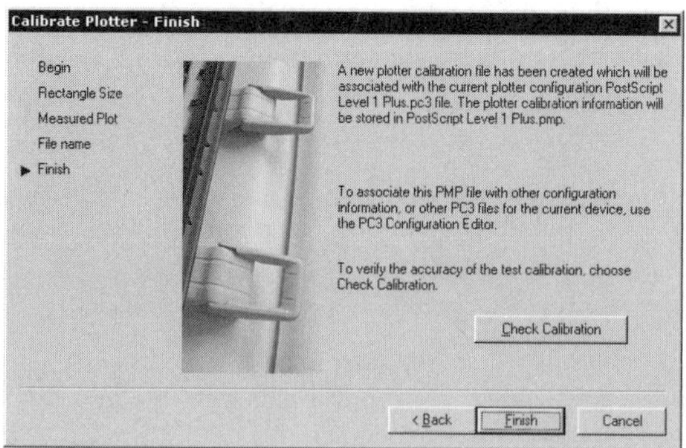

Check Calibration returns to the Calibrate Plotter - Rectangle Size dialog box.

Finish returns to the Add Plotter - Finish dialog box.

RELATED COMMANDS

Plot plots drawings with plot styles.

PageSetup attaches plot style tables to layouts.

StylesManager modifies plot style tables.

AddPlWiz.Exe runs the Add Plotter wizard.

RELATED SYSTEM VARIABLES

PaperUpdate toggles the display of a warning before AutoCAD plots a layout with a paper size different from the size specified by the plotter configuration file:

PaperUpdate	Meaning
0	Displays warning dialog box (default).
1	Changes paper size to match the size specified by the plotter configuration file.

PlotId (*Obsolete*) holds the current plotter configuration ID number.

PlotRotMode controls the orientation of plots:

PlotRotMode	Meaning
0	Aligns rotation icon with media at the lower left for 0 degrees; calculates x and y origin offsets relative to lower-left corner.
1	Aligns the lower-left corner of plotting area with lower-left corner of the paper.
2	Same as 0, except x and y origin offsets relative to the rotated origin position.

Plotter (*Obsolete*) holds the current plotter name.

TIPS

- You can create and edit *.pc3* plotter configuration files without AutoCAD. From the Start button on the Windows toolbar, select **Settings | Control Panel | Autodesk Plotter Manager**.

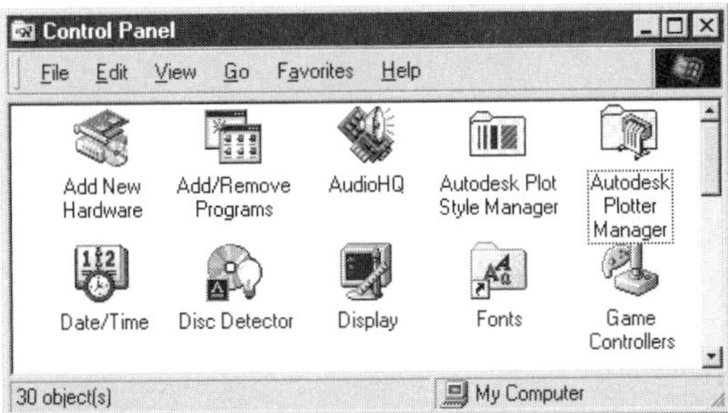

- The *dwf classic.pc3* file specifies parameters for creating *.dwf* files via the Release 14-compatible **DwfOut** command.
- The *dwf eplot.pc3* file specifies parameters for creating *.dwf* files via the **Plot** command.

PngOut

2004 Exports the current view in PNG raster format.

Command	Alias	Ctrl+	F-key	Alt+	Menu Bar	Tablet
pngout	...	...	...	...	...	...

Command: pngout

Displays Create Raster File dialog box. Specify a filename, and then click ***Save****.*

Select objects or <all objects and viewports>: *(Select objects, or press* ENTER *to select all objects and viewports.)*

COMMAND LINE OPTIONS

Select objects selects specific objects.

All objects and viewports selects all objects and all viewports, whether in model space or in layout mode.

RELATED COMMANDS

BmpOut exports drawings in BMP (bitmap) format.

Image places raster images in the drawing.

JpgOut exports drawings in JPEG (joint photographic expert group) format.

TifOut exports drawings in TIFF (tagged image file format) format.

TIPS

- The rendering effects of the **ShadEdge** command are preserved, but not those of the **Render** command.
- PNG files are used as a royalty-free alternative to JPEG files.
- PNG is short for "portable network graphics."

Point

V. 1.0 Draws a 3D point.

Command	Alias	Ctrl+	F-key	Alt+	Menu Bar	Tablet
point	po	...	...	DO	Draw ↳Point	O9

Command: point
Current point modes: PDMODE=0 PDSIZE=0.0000
Specify a point: *(Pick a point.)*

COMMAND LINE OPTION

Point positions a point, or enters a 2D or 3D coordinate.

RELATED COMMANDS

DdPType displays a dialog box for selecting PsMode and PdSize.

Regen regenerates the display to see the new point mode or size.

RELATED SYSTEM VARIABLES

PDMode determines the appearance of a point:

0 1 2 3 4

32 33 34 35 36

64 65 66 67 68

96 97 98 99 100

PDSize determines the size of a point:

PdSize	Meaning
0	5% of height of the **ScreenSize** system variable (default).
1	No display.
-10	*(Negative)* Ten percent of the viewport size.
10	*(Positive)* Ten pixels in size.

TIPS

- The size and shape of the point is determined by **PdSize** and **PdMode**; changing these changes the appearance and size of all points in the drawing with the next regeneration.
- Entering only x,y coordinates places the point at the z coordinate of the current elevation; setting **Thickness** to a value draws points as lines in 3D space.
- Prefix coordinates with * (such as *1,2,3) to place points in the WCS, rather than in the UCS.
- Use the object snap mode **NODe** to snap to points.

Polygon

V. 2.5 Draws 2D polygons of between three to 1024 sides.

Command	Alias	Ctrl+	F-key	Alt+	Menu Bar	Tablet
polygon	pol	...	...	DY	Draw ↳Polygon	P10

Command: polygon
Enter number of sides <4>: *(Enter a number.)*
Specify center of polygon or [Edge]: *(Pick a point, or type* **E***.)*
Enter an option [Inscribed in circle/Circumscribed about circle] <I>: *(Type* **I** *or* **C***.)*
Specify radius of circle: *(Enter a value, or pick two points.)*

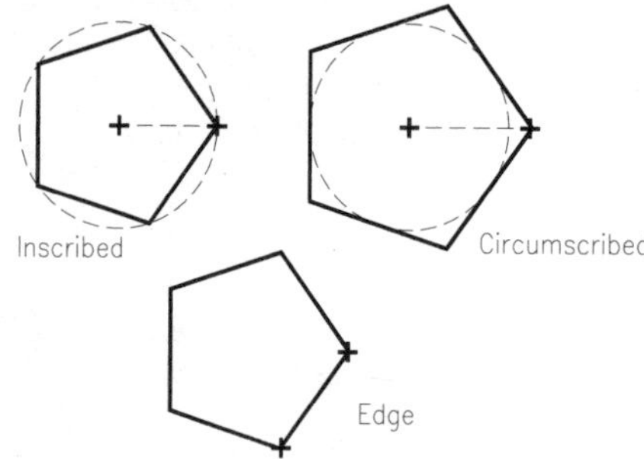

COMMAND LINE OPTIONS

Center of polygon indicates the center point of the polygon; then:

C (*Circumscribed*) fits the polygon outside a circle.

I (*Inscribed*) fits the polygon inside a circle.

Edge draws the polygon based on the length of one edge.

RELATED COMMANDS

PEdit edits polylines, include polygons.

Rectang draws squares and rectangles.

RELATED SYSTEM VARIABLE

PolySides specifiers the most-recently entered number of sides (default = 4).

TIPS

- Polygons are drawn as a polyline; use **PEdit** to edit the polygon, such as its width.
- The pick point determines the polygon's first vertex; polygons are drawn counterclockwise.
- Use the system variable **PolySides** to preset the default number of polygon sides.
- Use the **Snap** command to place the polygon precisely; use **INTersection** or **ENDpoint** object snap modes to snap to the polygon's vertices.

Removed Command

Preferences was removed from AutoCAD 2000; it was replaced by **Options**.

Preview

Rel.13 Displays plot preview; bypasses the **Plot** command.

Command	Alias	Ctrl+	F-key	Alt+	Menu Bar	Tablet
preview	**pre**	...	...	**FV**	**File** ↳**Plot Preview**	**X24**

Command: preview

Press ESC or ENTER to exit, or right-click to display shortcut menu.

Displays preview screen (new user interface in AutoCAD 2005):

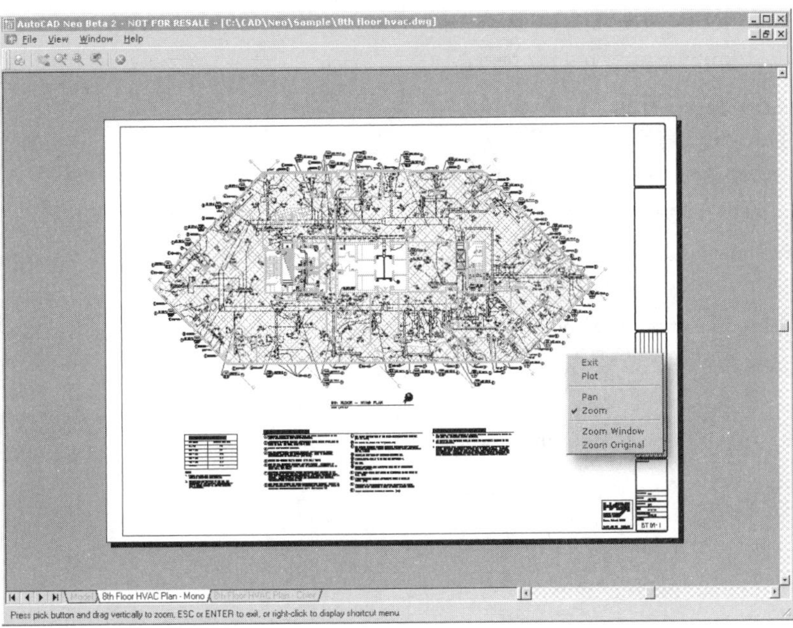

COMMAND LINE OPTION

ESC returns to the drawing window.

TOOLBAR BUTTONS

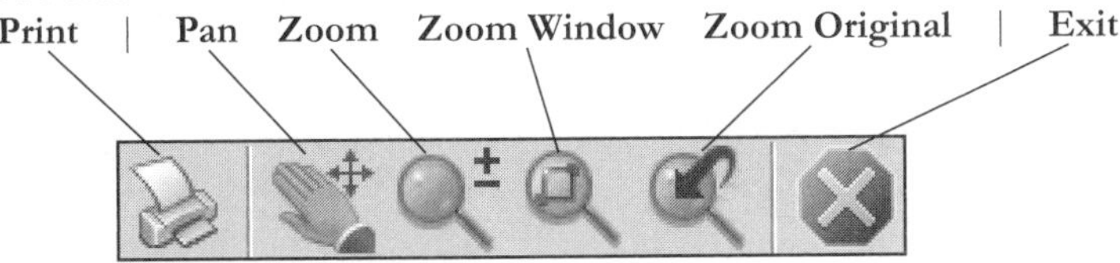

SHORTCUT MENU OPTIONS

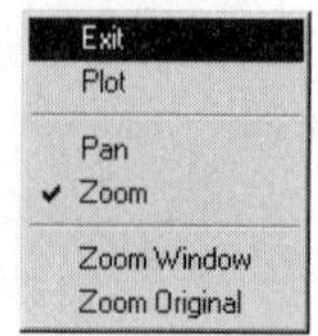

Exit exits preview mode.

Plot plots the drawing; goes immediately to plotting, bypassing the plot dialog box.

Pan pans the preview image in real time.

Zoom enlarges and makes smaller the preview image in real time.

Zoom Window zooms into a windowed area.

Zoom Original returns to the orignal size.

RELATED COMMANDS

Plot plots the drawing.

PageSetup enables plot preview once a plotter is assigned to the layout.

TIPS

- This command does not operate when no plotter is assigned; use the **PageSetup** command to assign a plotter to the Model and layout tabs.
- Press ESC to exit preview mode.
- Partial preview mode was removed from AutoCAD 2005.
- To zoom or pan, select the mode, and then hold down the left mouse button. The image zooms or pans as you move the mouse.

'Properties / 'PropertiesClose

2000 Opens and closes the Properties window for examining and modifying properties of selected objects.

Command	Aliases	Ctrl+	F-key	Alt+	Menu Bar	Tablet
properties	**ch** **props**	**1**	**...**	**TP**	**Tools** **↳Properties**	**Y12-Y13**
	ddchprop **ddmodify** **mo**			**MP**	**Modify** **↳Properties**	
propertiesclose	**prclose**	**1**	**...**	**TP**	**Tools** **↳Properties**	**...**

Command: properties

Displays window. When no objects are selected:

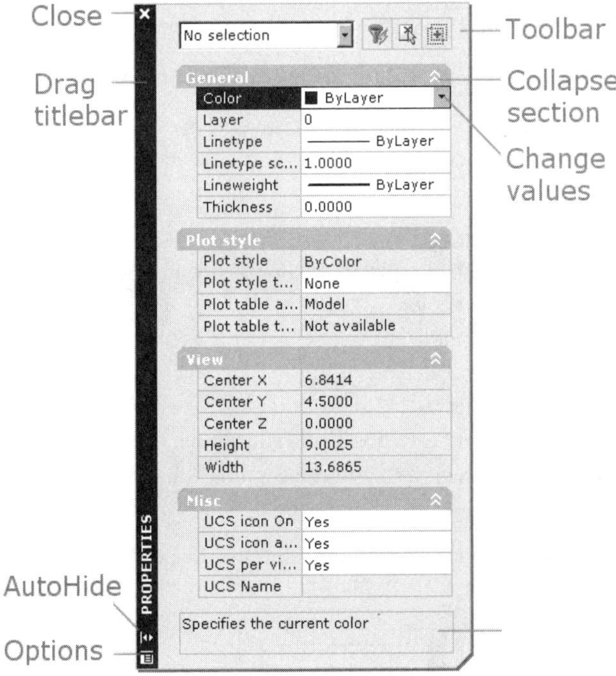

Command: properties

Closes window.

TOOLBAR OPTIONS

Selection lists the selected objects.

Quick Select displays the Quick Select dialog box; see the **QSelect** command.

Select Objects prompts, 'Select objects:'.

Toggle Value of PickAdd Variable controls selections when the SHIFT key is held down:

Off adds objects to the selection set.

On replaces objects in the selection set.

OPTIONS MENU

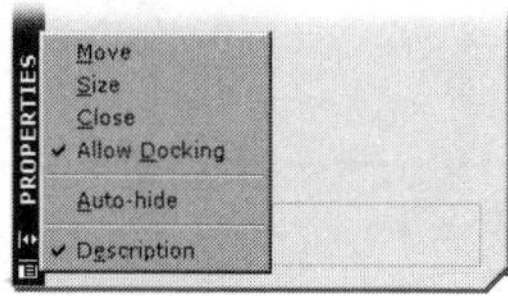

Move moves the window.

Size changes the size of the window.

Close closes the window.

Allow Docking toggles whether the window can be docked at the side of AutoCAD.

Auto-hide toggles whether the window reduces its size when the cursor is elsewhere.

Description toggles the display of the description box.

RELATED COMMANDS

ChProp changes an object's color, layer, linetype, and thickness.

Style creates and changes text styles.

RELATED SYSTEM VARIABLES

OpmState reports whether the Properties window is open.

CeColor specifies the current color.

CeLtScale specifies the current linetype scale factor.

CeLtype specifies the current linetype.

CLayer specifies the current layer.

Elevation specifies the current elevation in the z-direction.

Thickness specifies the current thickness in the z-direction.

TIPS

- Use the **Selection** list to count objects in the drawing. Use CTRL+A to select all objects, and then click the **Selection** list:

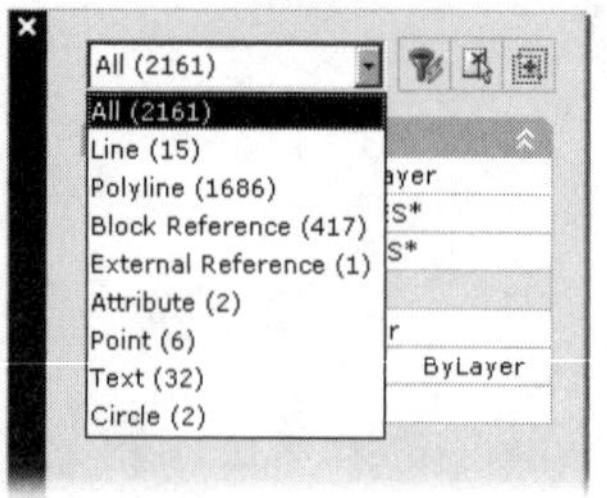

- When one or more objects are selected, the items displayed by the **Properties** window vary; here are three examples:

Arc

General	
Color	ByLayer
Layer	0
Linetype	ByLayer
Linetype sc...	1.0000
Plot style	ByColor
Lineweight	ByLayer
Hyperlink	
Thickness	0.0000
Geometry	
Start X	7.6060
Start Y	5.7092
Start Z	0.0000
Center X	5.9022
Center Y	5.6458
Center Z	0.0000
End X	4.6729
End Y	6.8272
End Z	0.0000
Radius	1.7050
Start angle	2
End angle	136
Total angle	134
Arc length	3.9876
Area	2.3539
Normal X	0.0000
Normal Y	0.0000
Normal Z	1.0000

Mtext

General	
Color	ByLayer
Layer	0
Linetype	ByLayer
Linetype sc...	1.0000
Plot style	ByColor
Lineweight	ByLayer
Hyperlink	
Text	
Contents	tHE iLLUSTRATED ...
Style	Standard
Justify	Top left
Direction	By style
Width	4.3141
Height	0.2000
Rotation	0
Line space ...	1.0000
Line space ...	At least
Geometry	
Position X	5.4490
Position Y	4.3859
Position Z	0.0000

Linear Dimension

General	
Color	ByLayer
Layer	0
Linetype	ByLayer
Linetype sc...	1.0000
Plot style	ByColor
Lineweight	ByLayer
Hyperlink	
Associative	No
Misc	
Dim style	Standard
Lines & Arrows	
Arrow 1	Closed filled
Arrow 2	Closed filled
Arrow size	0.1800
Dim line LW	ByBlock
Ext line LW	ByBlock
Dim line 1	On
Dim line 2	On
Dim line co...	ByBlock
Dim line ext	0.0000
Ext line 1	On
Ext line 2	On
Ext line color	ByBlock
Ext line ext	0.1800
Ext line off...	0.0625
Text	
Fractional t...	Horizontal
Text color	ByBlock
Text height	0.1800
Text offset	0.0900
Text outsid...	Off
Text pos hor	Centered
Text pos vert	Centered
Text style	Standard
Text inside...	Off
Text positi...	7.7601
Text positi...	4.1692
Text rotation	0
Measurem...	4.2798
Text overri...	
Fit	
Dim line fo...	Off
Dim line in...	On
Dim scale ...	1.0000
Fit	Best fit
Text inside	Off
Text move...	Keep dim line wit...
Primary Units	
Decimal se...	.
Dim prefix	

- As an alternative to entering the **PropertiesClose** command, you can click the **x** button on the **Properties** window title bar:

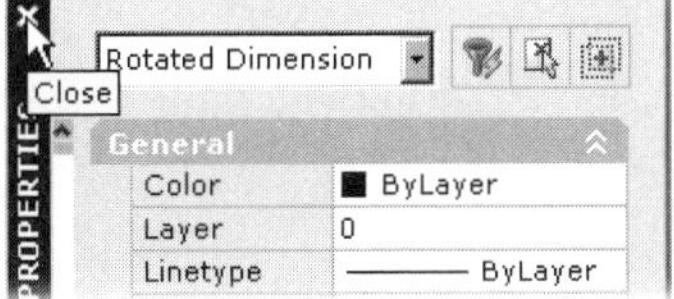

- Double-click the title bar to dock and undock the window within the AutoCAD window.
- The **Properties** window can be dragged larger and smaller by its edges.
- When an item is displayed by gray text, it cannot be modified.
- Bodies, 3D solids, and 2D regions cannot be edited beyond the items in the **General** section.

Removed Command

The **PsDrag** command was discontinued with AutoCAD 2000i. It has no replacement.

PSetupIn

2000 Imports user-defined page setups into the current drawing layout (*short for Page SETUP IN*).

Commands	Alias	Ctrl+	F-key	Alt+	Menu Bar	Tablet
psetupin	...	...	...	...	...	...
-psetupin						

Command: psetupin

Displays the Select File dialog box. Select a .dwg, .dwt, or .dxf file, and then click **Open***.*

AutoCAD displays dialog box:

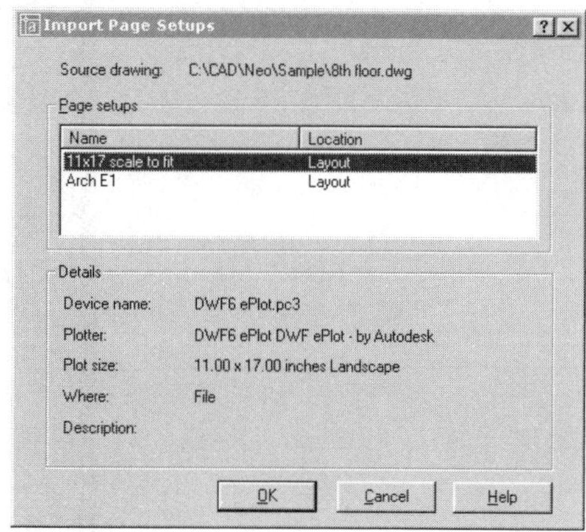

DIALOG BOX OPTIONS

Name lists the names of page setups.

Location lists the location of the setups.

OK closes the dialog box, and loads the page setup.

-PSETUPIN Command

Command: -psetupin

Enter filename: *(Enter .dwg file name.)*

Enter user defined page setup to import or [?]: *(Enter a name, or type* **?***.)*

COMMAND LINE OPTIONS

Enter filename enters the name of a drawing file.

Enter user defined page setup to import enters the name of a page setup.

? lists the names of page setups in the drawing.

RELATED COMMANDS

PageSetup creates a new page setup configuration.

Plot plots the drawing.

PsFill

Rel.12 Fills 2D polyline outlines with raster PostScript patterns (*short for PostScript FILL; undocumented command*).

Command	Alias	Ctrl+	F-key	Alt+	Menu Bar	Tablet
psfill	...	...	...	...	...	...

Command: psfill
Select polyline: *(Pick an outline.)*
Enter PostScript fill pattern name (. = none) or [?] <.>: *(Enter pattern name, or type* **?***.)*

COMMAND LINE OPTIONS

Select polyline selects the closed polyline to fill.

PostScript pattern specifies the name of the fill pattern.

. (dot) selects no fill pattern.

? lists the available fill patterns.

***** specifies that the pattern should not be outlined with a polyline.

RELATED COMMANDS

BHatch fills an area with vector hatch patterns and solid colors.

PsOut exports drawings as PostScript files.

RELATED SYSTEM VARIABLE

PSQuality specifies the display options for PostScript images:

PsQuality	Meaning
75	*(Positive)* Displays filled image at 75dpi (default).
0	Displays bounding box and filename; no image.
-75	*(Negative)* Displays image outline at 75dpi; no fill.

TIPS

- These fill patterns are available:

Grayscale	RGBcolor
AIlogo	Lineargray
Radialgray	Square
Waffle	Zigzag
Stars	Brick
Specks	

- You cannot see fill patterns in AutoCAD drawings until they are exported with the **PsOut** command.

Removed Command

PsIn command was removed with AutoCAD 2000i; there is no replacement.

PsOut

Rel.12 Exports the current drawing as an encapsulated PostScript file (*undocumented command*).

Command	Alias	Ctrl+	F-key	Alt+	Menu Bar	Tablet
psout	...	...	...	**FE**	**File**	...
				↳EPS	**↳Export**	
					↳Encapsulated PS	

Command: psout

Displays the ***Create PostScript File*** *dialog box.*

Select **Tools | Options** *to display dialog box:*

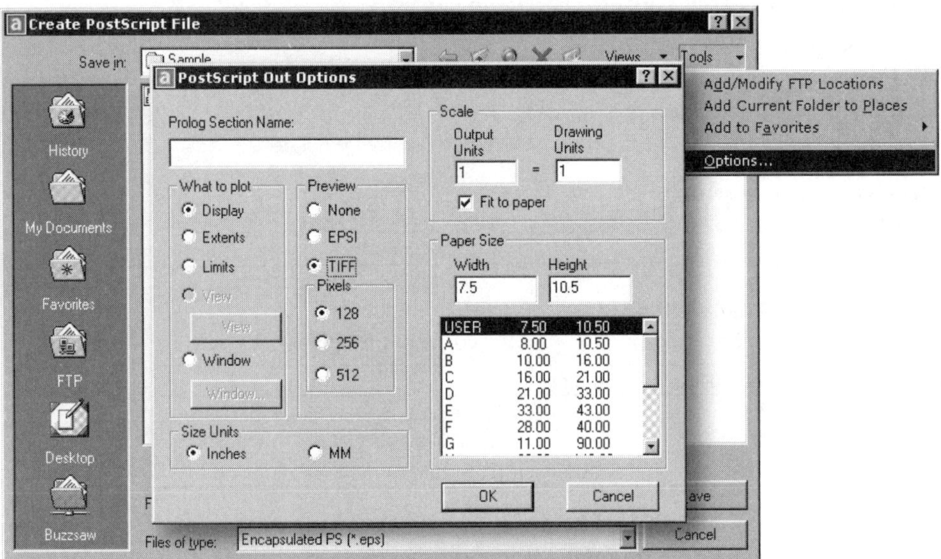

DIALOG BOX OPTIONS

Prolog Section Name (*optional*) specifies the name of the prolog section, which is read from the *acad.psf* file and customizes the PostScript output.

What to Plot options

⊙**Display** selects the current display in the current viewport.

○**Extents** selects the drawing extents.

○**Limits** selects the drawing limits.

○**View** selects a named view.

○**Window** picks two corners of a window.

Preview options

⦿**None** specifies no preview image (default).

○**EPSI** specifies Macintosh preview image format.

○**TIFF** specifies Tagged Image File Format.

Pixels options

128 specifies a preview image size of 128x128 pixels (default).

256 specifies a preview image size of 256x256 pixels.

512 specifies a preview image size of 512x512 pixels.

Size Units options

Inches specifies the plot parameters in inches.

MM specifies the plot parameters in millimeters.

Scale options

Output Units scales the output units.

Drawing Units specifies the drawing units.

Fit to Paper forces the drawing to fit paper size.

Paper Size options

Width enters a width for the output size.

Height enters a height for the output size.

RELATED COMMAND

Plot exports the drawing in a variety of formats, including raster EPS.

RELATED SYSTEM VARIABLE

PSProlog specifies the PostScript prologue information.

RELATED FILE

*.*eps* extension of file produced by **PsOut**.

TIPS

- The "screen preview image" is only used for screen display purposes, since graphics software generally cannot display PostScript graphic files.
- When you select the **Window** option, AutoCAD prompts you for the window corners *after* you finish selecting options.
- Although Autodesk recommends using the smallest screen preview image size (128x128), even the largest preview image (512x512) has a minimal effect on file size and screen display time.
- Some software programs, such as those from Microsoft, might reject an *.eps* file when the preview image is larger than 128x128.
- The screen preview image size has no effect on the quality of the PostScript output.
- If you're not sure which screen preview format to use, select TIFF.
- AutoCAD no longer imports PostScript files.

PSpace

Rel.11 Switches from model space to paper space/layout mode (*short for Paper SPACE*).

Command	Alias	Ctrl+	F-key	Alt+	Menu Bar	Tablet
pspace	ps	...	...	...	...	L5

Command: pspace

COMMAND LINE OPTIONS

None.

RELATED COMMANDS

MSpace switches from paper space to model space.

MView creates viewports in paper space.

UcsIcon toggles the display of the paper space icon.

Zoom scales paper space relative to model space with the XP option.

RELATED SYSTEM VARIABLES

MaxActVP specifies the maximum number of viewports displaying an image.

PsLtScale specifies the linetype scale relative to paper space.

TileMode allows paper space when set to 0.

TIPS

- Use paper space to layout multiple views of a single drawing.
- You can switch to paper space by clicking the word **MODEL** on the status bar; switch to model space by clicking the word **PAPER**.

4.0679, 0.4813, 0.0000 SNAP GRID ORTHO POLAR OSNAP OTRACK LWT MODEL

4.0679, 0.4813, 0.0000 SNAP GRID ORTHO POLAR OSNAP OTRACK LWT PAPER

- When a drawing is in paper space, AutoCAD displays **PAPER** on the status line and the paper space icon:

W

X

- Paper space is known as "drawing composition" in some other CAD packages.

Publish

2004 Outputs multiple layout sheets from one or more drawings as a single, multi-page *.dwf* file or hardcopy plot.

Command	Alias	Ctrl+	F-key	Alt+	Menu Bar	Tablet
publish	...	...	...	**FH**	**File** ⇘**Publish**	...
-publish						

Command: publish

Displays dialog box (user interface changed in AutoCAD 2005):

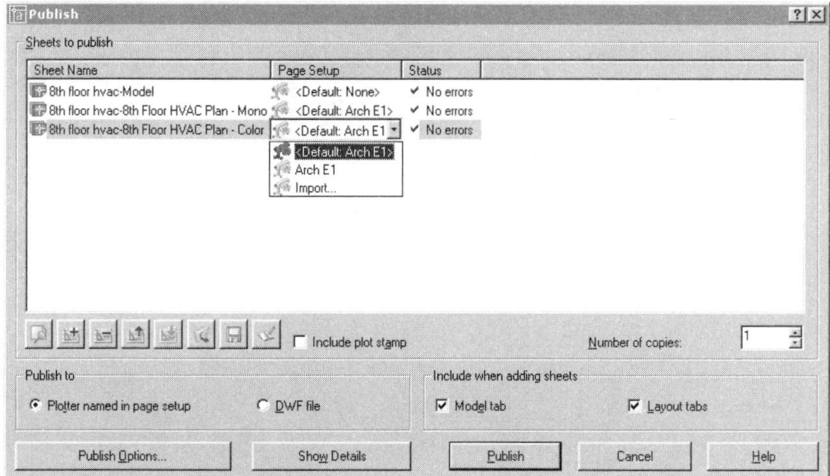

DIALOG BOX OPTIONS

List of Drawing Sheets options

Sheet Name concantenates the drawing name and the layout name with a dash (-); edit sheet names with the Rename option.

Page Setup lists the named page setup for each layout; click to select other setups.

Status displays a message as layouts are published.

☐ **Include plot stamp** adds plot stamp data to the edge of each plot; see the **PlotStamp** command. (Click the Plot Stamp Settings button to access the Plot Stamp dialog box.)

Number of copies allows you make multi-copy prints (available only when plotting).

Publish To options

⊙**Plotters named in page setups** generates hardcopy plots, or to-file plots.

○**DWF file** generates multi-sheet *.dwf* files.

Include when Adding Sheets options

Model Tab determines whether the model layout is included with the list of sheets.

Layout Tabs includes all layouts.

Publish Options displays the Publish Options dialog box.

Show Details expands the dialog box to provide added information about the selected sheet.

Publish generates the *.dwf* file or plot; displays Now Plotting dialog box.

Publish Options dialog box

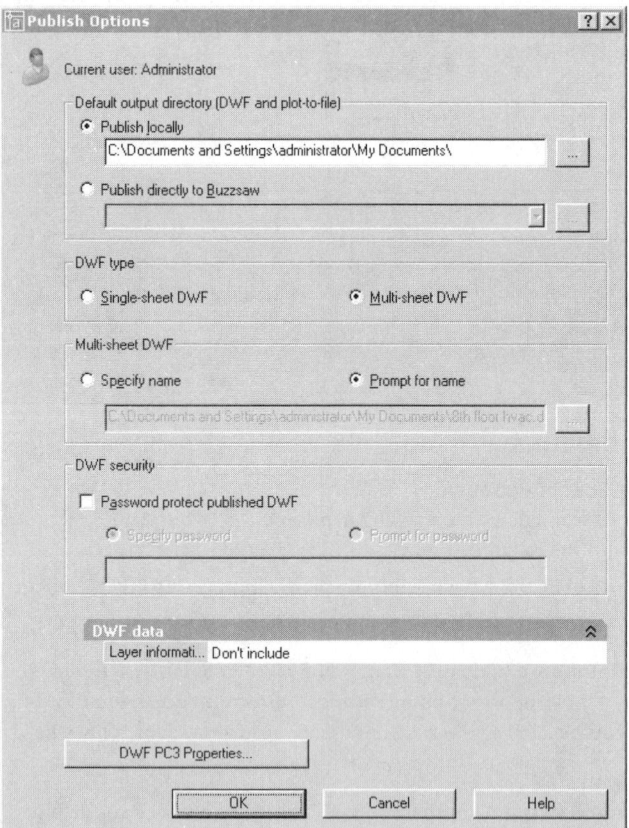

Default Output Directory options

⊙**Publish locally** saves the *.dwf* and *.plt* plot files in a folder.

... displays the Select a Folder for Generated Files dialog box; select a folder, and click **Select**.

○ **Publish Directly to Buzzsaw** saves the files at Autodesk's Buzzsaw Web site.

... displays the Add a New Buzzsaw Location dialog box; enter data, and then click **OK**.

DWF Type options

○ **Single-Sheet DWF** generates one *.dwf* file for each sheet.

⊙ **Multi-Sheet DWF** generates one *.dwf* file for all sheets.

Multi-Sheet DWF Creation options

○ **Specify Name** specifies file and folder names for the multisheet *.dwf* file.

... displays the Select DWF File dialog box; select a file name, and then click **Select**.

⊙ **Prompt for Name** prompts you for the file and folder names during publishing.

DWF Security options

☐ **Password Protect Published DWF** applies a password to help prevent unauthorized access to *.dwf* files.

⊙ **Specify Password** specifies the password.

Caution! Passwords cannot be recovered; passwords are case-sensitive, and may consist of letters, numbers, punctuation, and non-ASCII characters.

○ **Prompt for Password** prompts you for the password during publishing.

DWF Data option

Layer information toggles whether layers are included in *.dwf* files.

TOOLBAR OPTIONS

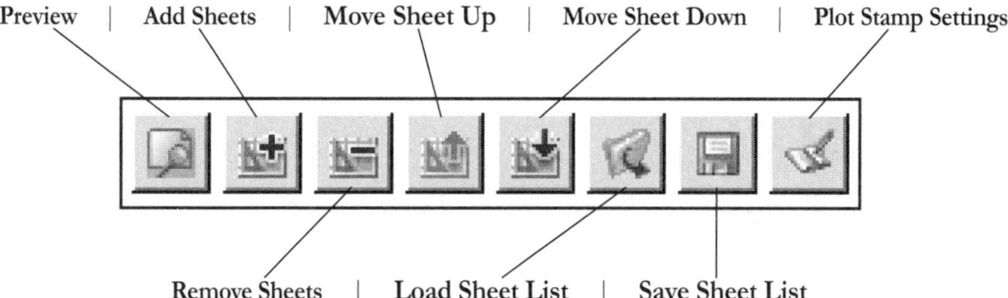

Preview displays the plot preview window; see the **Preview** command.

Add Sheets displays the Select Drawings dialog box; choose one or more drawings, and then click **Select**.

Remove Sheets removes selected sheets from the list, wtihout warning.

Move Sheet Up moves the selected sheets up the list.

Move Sheet Down moves the selected sheets down the list.

Load Sheet List displays the Load List of Sheets dialog box; select a *.dsd* drawing set description or *.bp3* batch plot list file, and then click **Load**.

Save Sheet List displays the Save List As dialog box; enter a file name, and then click **Save**.

Plot Stamp Settings displays the Plot Stamp dialog box; see **PlotStamp** command.

SHORTCUT MENU

Right-click in the Sheets to Publish area:

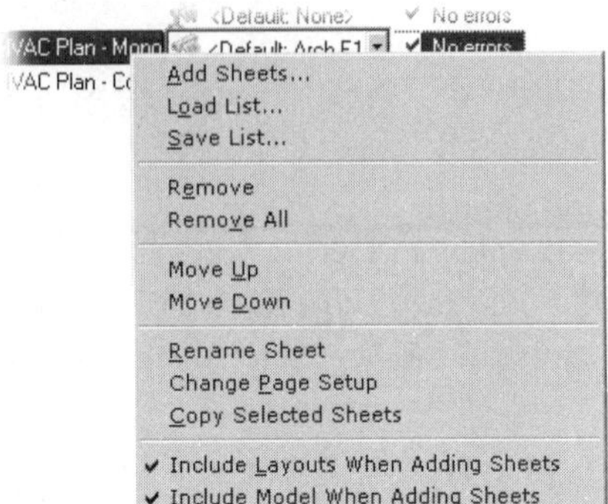

Add Sheets displays the Select Drawings dialog box; choose one or more drawings, and then click **Select**.

Load List displays the Load List of Sheets dialog box; select a *.dsd* drawing set description or *.bp3* batch plot list file, and then click **Load**.

Save List displays the Save List As dialog box; enter a file name, and then click **Save**.

Remove removes selected sheets from the list, wtihout warning.

Remove All removes all sheets.

Move Up moves the selected sheets up the list.

Move Down moves the selected sheets down the list.

Rename Sheet renames the selected sheet.

Change Page Setup displays the Page Setup drop list.

Copy Selected Sheets copies selected sheets, and adds the -Copy*n* suffix to the sheet name.

Include Layouts when Adding Sheets includes all layouts in the sheet set.

Include Model when Adding Sheets includes model space with the layouts.

PUBLISHING COMPLETE DIALOG BOX

After processing is complete, displays dialog box:

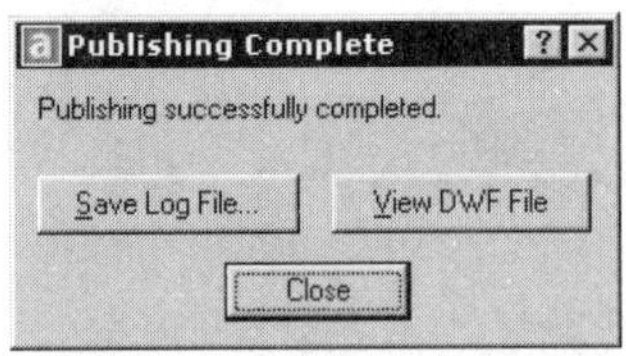

Save Log File saves the log file as a *.csv* (comma separated values) file, which can be opened as a spreadsheet.

View DWF File opens the published file in Autodesk Express Viewer, installed with AutoCAD ; this button is unavailable when **Publish** encounters errors during processing.

PublishToWeb

2000i Exports drawings as DWF, JPG, and PNG images embedded in preformatted Web pages.

Command	Alias	Ctrl+	F-key	Alt+	Menu Bar	Tablet
publishtoweb	...	...	...	**FW**	**File** **↳Publish to Web**	**X25**

Command: publishtoweb

Displays wizard.

WIZARD OPTIONS

Begin page

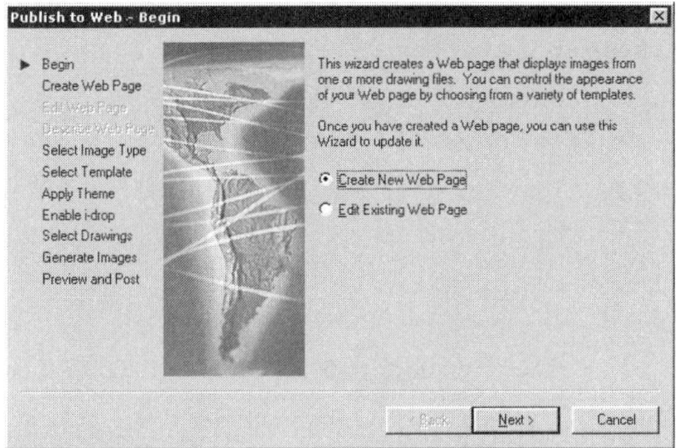

⊙**Create New Web Page** guides you through the steps in creating new Web pages from drawings.

○**Edit Existing Web Page** guides you through the steps in editing existing Web pages.

Create Web Page page

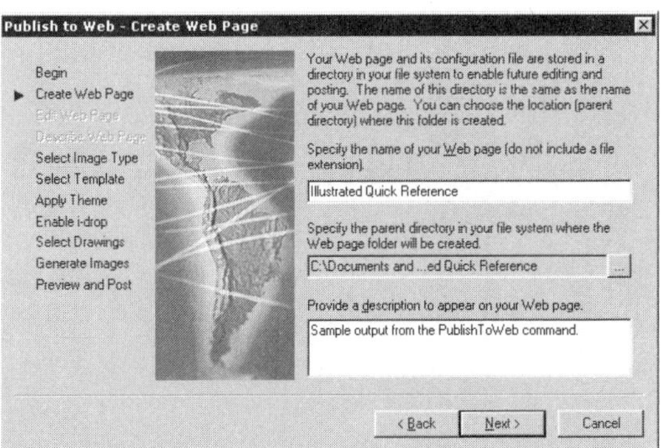

Specify the name of your Web page requires you to enter a file name. (AutoCAD uses the name for the files making up the Web page, which allows you later to edit the Web page.) The name also appears at the top of the Web page.

Provide a description to appear on your Web page specifies a description that appears below the name on the Web page.

Select Image Type page

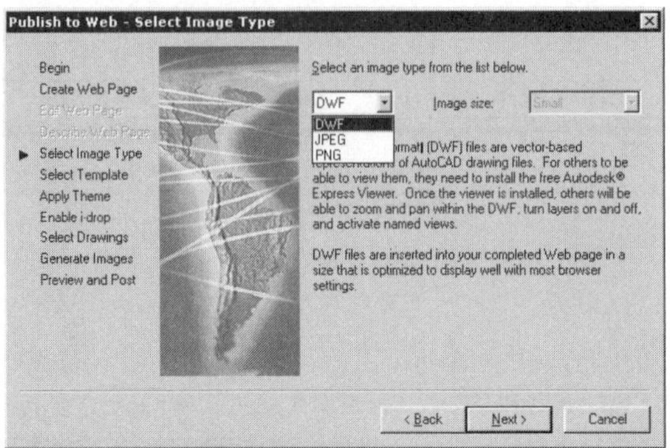

Select an image type from the list below

DWF (drawing Web format) is a vector format that displays cleanly, and can be zoomed and panned; not all Web browsers can display DWF.

JPEG (joint photographic experts group) is a raster format that all Web browsers display; may create *artifacts* (details that don't exist).

PNG (portable network graphics) is a raster format that does not suffer the artifact problem; some older Web browsers do not display PNG.

Image size selects a size of raster image (available for JPEG and PNG only):

Image Size	Resolution	Approximate PNG File Size
Small	789 x 610	60KB
Medium	1009 x 780	90KB
Large	1302 x 1006	130KB
Extra Large	1576 x 1218	170KB

Select Template page

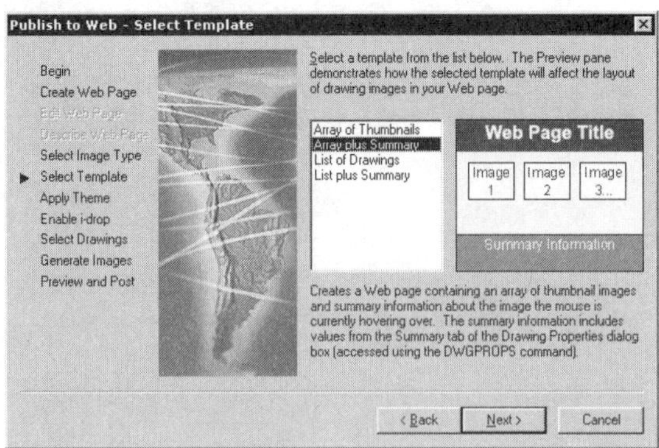

Select a template from the list below selects one of the pre-designed formats for the Web page.

Apply Theme page

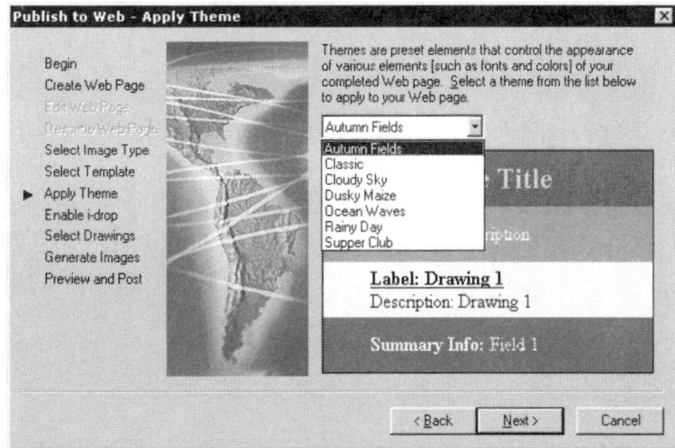

Select a theme from the list below selects one of the pre-designed themes (colors and fonts) for the Web page.

Enable i-drop page

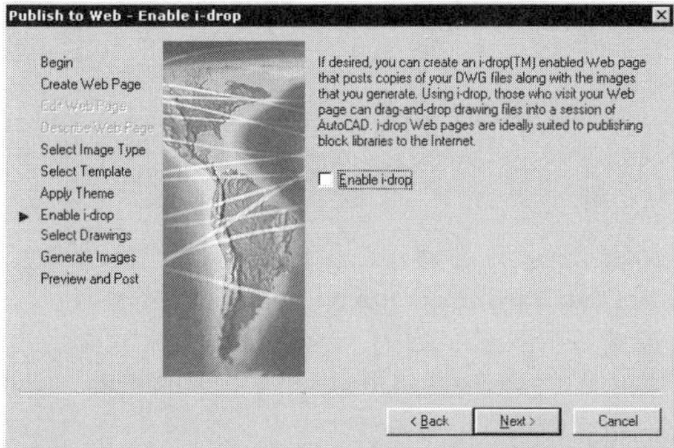

☐ **Enable i-drop** adds i-drop capability to the Web page.

Select Drawings page

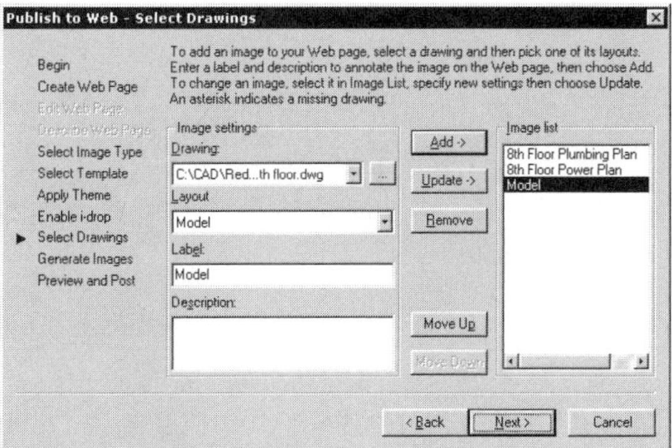

Image Settings options

Drawing selects the drawing; the current drawing is the default.

Layout selects the name of a layout, or Model space.

Label specifies a name, such as the filename or a more descriptive name.

Description specifies a description that appears with the drawing on the Web page.

Buttons

Add adds the image setting to the image list.

Update changes the image setting in the image list.

Remove removes the image setting from the image list.

Move up moves the image setting up the image list.

Move down moves the image setting down the image list.

Generate Images page

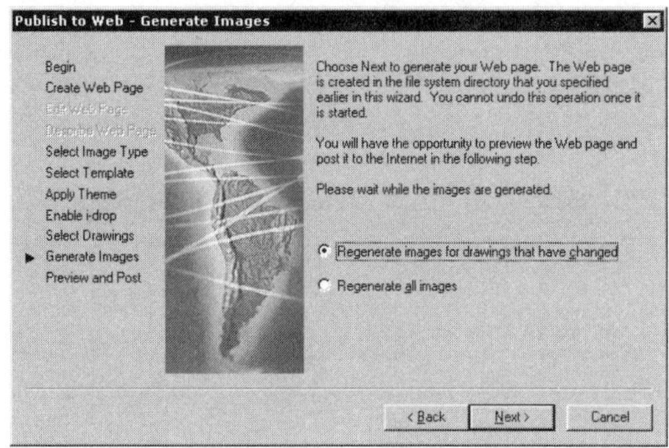

⊙**Regenerate images for drawings that have changed** updates images for those drawings that have been edited.

○**Regenerate all images** regenerates all images from all drawings; ensures all are up to date.

Preview and Post page

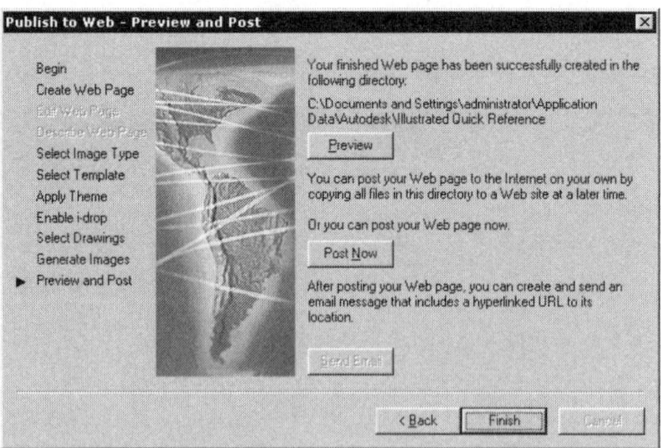

Preview launches the Web browser to preview the resulting Web page.

Post Now uploads the files (HTML, JPEG, PNG, DWF, and so on) to the Web site.

Send Email sends an email to alert others of the posted Web page.

Examples of resulting Web pages:

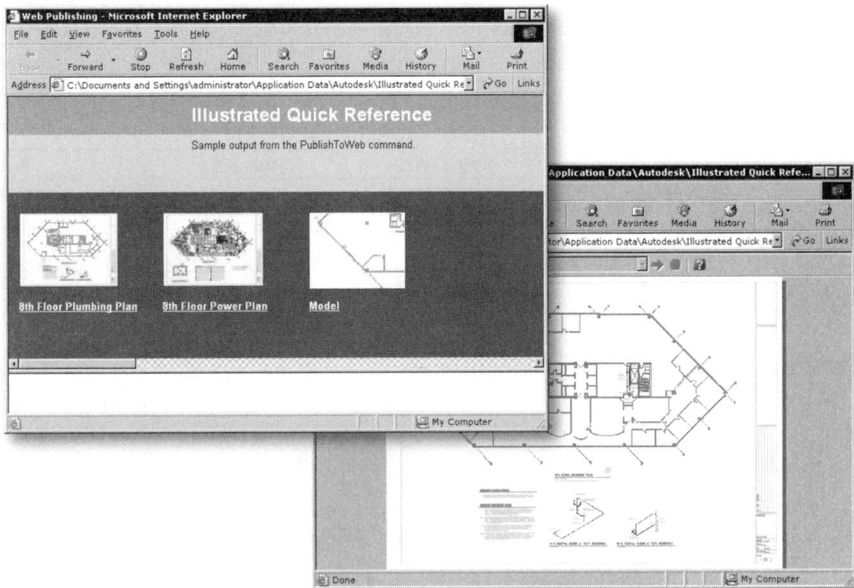

RELATED COMMANDS

Publish exports drawings as multi-sheet *.dwf* files.

Plot exports drawings as *.dwf* files via the ePlot option.

Hyperlink places hyperlinks in drawings.

RELATED FILES

.ptw are PublishToWeb parameter files, stored in tab-delimited ASCII file.

.js are JavaScript files.

.jpg are joint photographic experts group (raster image) files.

.png are portable network graphics (raster image) files.

.dwf are drawing Web format (vector image) files.

TIPS

- Use the **Regenerate all images** option, unless you have an exceptionally slow computer or large number of drawings to process. The **Generate Images** step can take a long time.
- After you click **Preview** to view the Web page (and after AutoCAD launches the Web browser), click the **Back** button to make changes, if the result is not to your liking.
- The **Post Now** option works only if you have correctly set up the FTP (file transfer protocol) parameters. If so, you can have AutoCAD upload the HTML files directly to a Web site. If not, use a separate FTP program to upload the files from the *\windows\applications data\autodesk* folder.
- You can customize the themes and templates by editing the *acwebpublish.css* (themes) and *acwebpublish.xml* (templates) files.

Purge

V. 2.1 Removes unused, named objects from the drawing: blocks, dimension styles, layers, linetypes, plot styles, shapes, text styles, table styles, application ID tables, and multiline styles.

Commands	Alias	Ctrl+	F-key	Alt+	Menu Bar	Tablet
purge	**pu**	...	...	**FUP**	**File** ↳**Drawing Utilities** ↳**Purge**	**X25**
-purge						

Command: purge

Displays dialog box:

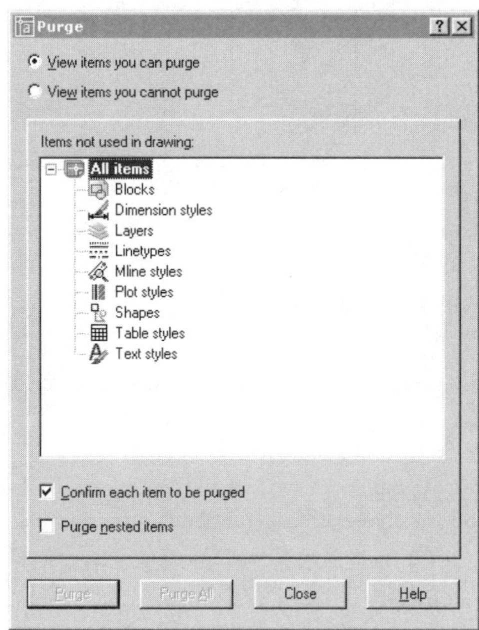

DIALOG BOX OPTIONS

- **View items you can purge** lists objects that can be purged from the drawing.
- **View items you cannot purge** lists objects that cannot be purged from the drawing.

☑ **Confirm each item to be purged** displays a confirmation dialog box for each object being purged.

☐ **Purge nested items** purges nested objects, such as unused blocks within unused blocks.

-PURGE Command

Command: -purge

Enter type of unused objects to purge [Blocks/Dimstyles/LAyers/LTypes/Plotstyles/SHapes/textSTyles/Mlinestyles/Tablestyles/Regapps/All]: *(Enter an option.)*

Sample response:

No unreferenced blocks found.

Purge layer DOORWINS? <N> **y**

Purge layer TEXT? <N> **y**

Purge linetype CENTER? <N> **y**

Purge linetype CENTER2? <N> **y**

No unreferenced text styles found.

No unreferenced shape files found.

No unreferenced dimension styles found.

COMMAND LINE OPTIONS

Blocks purges named but unused (not inserted) blocks.

Dimstyles purges unused dimension styles.

LAyers purges unused layers.

LTypes purges unused linetypes.

Plotstyle purges unused plot styles.

SHapes purges unused shape files.

textSTyles purges unused text styles.

Mlinestyles purges unused multiline styles.

Tablestyles purges unused table styles *(new to AutoCAD 2005).*

Regapps purges unused application ID tables of registered applications.

All purges drawing of all named objects, if possible.

RELATED COMMAND

WBlock writes the current drawing to disk with the * option, and removes spurious information from the drawing.

TIPS

- As of AutoCAD Release 13, this command can be used at any time; it no longer must be the first command after a drawing is loaded.
- It may be necessary to use the **Purge** command several times; follow each purge with the **Close** command, then open the drawing, and purge again. Repeat until the **Purge** command reports nothing to purge.
- As of AutoCAD 2005, it can also purge table styles. Also, STyle option renamed **textSTyle**, and APpids renamed **Regapps**.

 QDim

2000 Dimensions entire objects (with continuous, baseline, ordinate, radius, diameter, and staggered dimensions) using just three picks (*short for Quick DIMensioning*).

Command	Alias	Ctrl+	F-key	Alt+	Menu Bar	Tablet
qdim	...	...	...	**NQ**	**Dimension ↳QDIM**	**W1**

Command: qdim
Select geometry to dimension: *(Select one or more objects; press* CTRL+A *to select the entire drawing.)*
Select geometry to dimension: *(Press* ENTER *to end object selection.)*
Specify dimension line position, or
[Continuous/Staggered/Baseline/Ordinate/Radius/Diameter/datumPoint/Edit/seTtings] <Continuous>: *(Enter an option.)*

COMMAND LINE OPTIONS

Select geometry to dimension selects objects to dimension.

Specify dimension line position specifies the location of the dimension line.

Continuous draws continuous dimensions.

Staggered draws staggered dimensions.

Baseline draws baseline dimensions.

Ordinate draws ordinate dimensions relative to the UCS origin.

Radius draws radial dimensions; prompts 'Specify dimension line position:'.

Diameter draws diameter dimensions; prompts 'Specify dimension line position:'.

datamPoint sets a new datum point for ordinate and baseline dimensions; prompts 'Select new datum point:'.

Edit options

Indicate dimension point to remove, or [Add/eXit] <eXit>: *(Select a dimension, point, or enter an option.)*

Indicate dimension to remove selects the dimension point to remove from the dimension.

Add adds a dimension point to the dimension.

eXit returns to dimension drawing mode.

seTtings options

Associative dimension priority [Endpoint/Intersection] <Endpoint>: *(Enter an option.)*

Endpoint applies associative dimensions to endpoints over intersections.

Intersection applies associative dimensions to intersections over endpoints.

RELATED COMMANDS

DimStyle creates dimension styles, which specify the look of a dimension.

Dim*xxx* draws other kinds of dimensions.

QLeader draws leaders.

RELATED SYSTEM VARIABLE

DimStyle specifies the current dimension style.

TIPS

- Example of continuous dimensions:

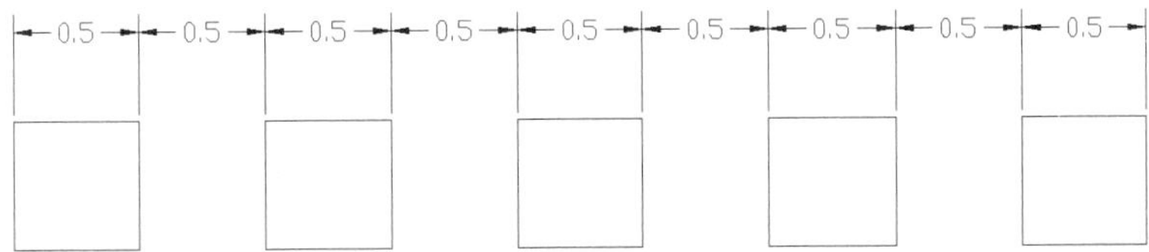

- Example of staggered dimensions:

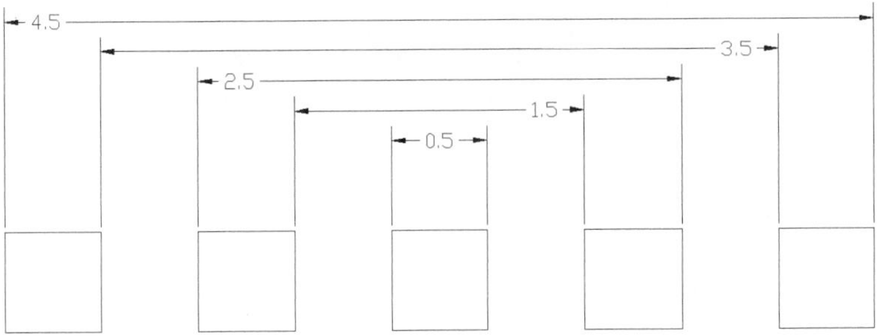

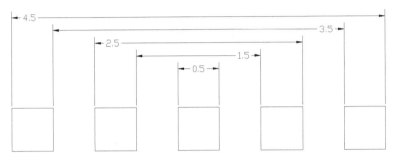

- Example of ordinate dimensions:

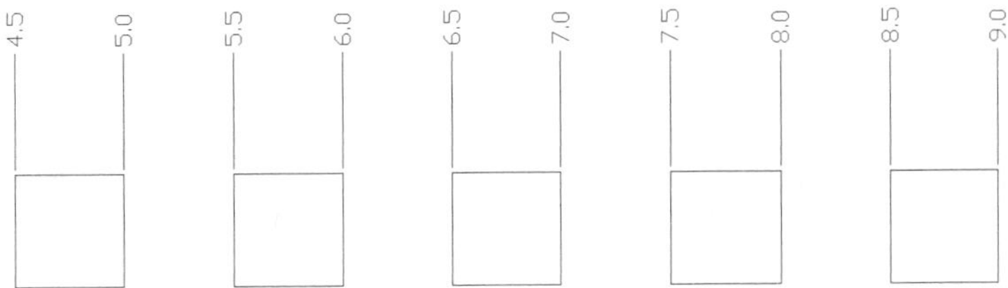

- As of AutoCAD 2004, dimensions created with **QDim** are fully associative.
- This command sometimes fails. AutoCAD complains, "Invalid number of dimension points found."

QLeader

Rel.14 Draws leaders; dialog box specifies options for custom leaders and annotations (*short for Quick LEADER*).

Command	Alias	Ctrl+	F-key	Alt+	Menu Bar	Tablet
qleader	le	...	...	NE	Dimension ↳ Leader	W2

Command: qleader

Specify first leader point, or [Settings] <Settings>: *(Pick point 1 for the arrowhead, or type* **S***.)*

When **S** *is entered, displays dialog box:*

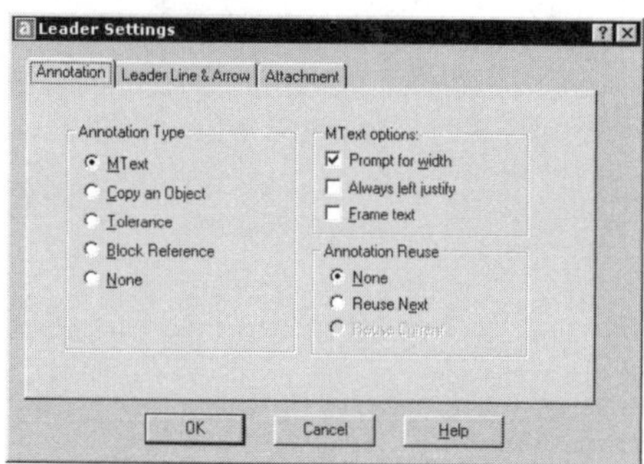

Click **OK** *to continue with the command; the prompts vary, depending on the options selected in the dialog box:*

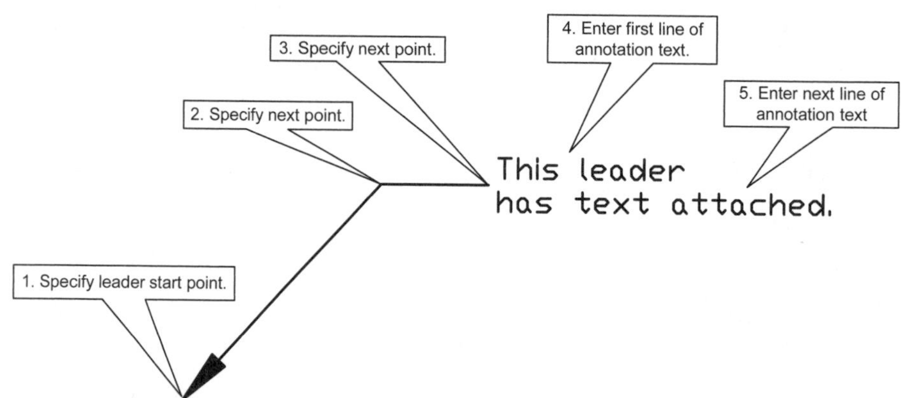

Specify next point: *(Pick point 2 for the leader shoulder.)*

Specify next point: *(Pick point 3, or press* **ENTER** *for text options.)*

Specify text width <0.0000>: *(Enter a value, 4.)*

Enter first line of annotation text <Mtext>: *(Enter text, 5,or press* **ENTER** *for mtext editor.)*

Enter next line of annotation text: *(Press* **ENTER** *to end command.)*

COMMAND LINE OPTIONS

Specify first leader point picks the location for the leader's arrowhead; press ENTER to display tabbed dialog box.

Specify next point picks the vertices of the leader; press ENTER to end the leader line.

Specify text width specifies the width of the bounding box for the leader text.

Enter first line of annotation text specifies the text for leader annotation; press ENTER twice to end.

Enter next line of annotation text specifies more text; press ENTER once to end.

DIALOG BOX OPTIONS

Annotation/Format tab

Annotation Type options

⦿**MText** prompts you to enter text for the annotation.

○**Copy an Object** attaches any object in the drawing as an annotation.

○**Tolerance** prompts you to select tolerance symbols for the annotation.

○**Block Reference** prompts you to select a block for the annotation.

○**None** attaches no annotation.

MText Options options

☑ **Prompt for width** displays the 'Specify text width' prompt.

☐ **Always left justify** forces the text to be left-justified, even when the leader is drawn to the right.

☐ **Frame text** places a rectangle around the text.

Annotation Reuse options

⦿**None** does not retain annotation for next leader.

○**Reuse Next** remembers the current annotation for the next leader.

○**Reuse Current** uses the last annotation for the current leader.

Leader Line & Arrow tab

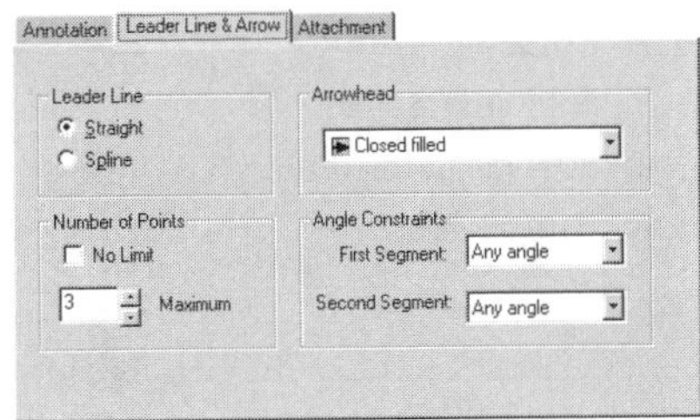

Leader Line options

⦿**Straight** draws the leader with straight lines.

○**Spline** draws the leader as a spline curve.

Number of Points options

☐ **No limit** keeps prompting for leader vertex points until you press ENTER.

Maximum stops the command prompting for leader vertex points; default=3.

Arrowhead option

Arrowhead selects the type of arrowhead, including Closed filled (default), None, and User Arrow.

Angle Constraints options

First Segment selects from Any angle (user-specified), Horizontal (0 degrees), 90, 45, 30, or 15-degree leader line, first segment.

Second Segment selects from Any angle (user-specified), Horizontal, 90, 45, 30, or 15-degree leader line, second segment.

Attachment tab

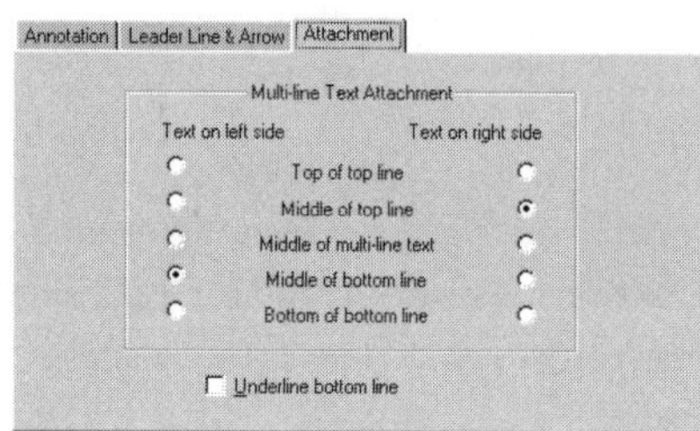

Multiline Text Attachment options

Left Side positions the annotation at one of several locations relative to the last leader segment, when the annotation is located to the left of the leader:

O Top of top line.

O Middle of top line.

⊙ Middle of multi-line text.

O Middle of bottom line (default).

O Bottom of bottom line.

Right Side positions the annotation at one of several locations relative to the last leader segment, when the annotation is located to the right of the leader.

☐ **Underline bottom line** underlines the last line of leader text.

RELATED COMMANDS

DdEdit edits leader text; see the **MText** command.

DimStyle sets dimension variables, including leaders.

Leader draws leaders without dialog boxes.

TIPS

- The **QLeader** command draws leaders, just like the **Leader** command in AutoCAD Release 13 and 14. The difference is that it brings up a triple-tab dialog box for setting the leader options.
- Some options have interesting possibilities, such as using any object in the drawing in place of the leader text.

QNew

2004 Starts new drawings based on a default template file (*short for Quick NEW*).

Command	Alias	Ctrl+	F-key	Alt+	Menu Bar	Tablet
qnew	...	...	...	...	...	...

Command: qnew

*Display depends on **Startup** option in General Options section of the **System** tab (Options dialog box). See the **New** command.*

RELATED COMMANDS

New starts new drawings.

SaveAs saves drawings in *.dwg* or *.dwt* formats; creates template files.

RELATED SYSTEM VARIABLES

DbMod indicates whether the current drawing has changed since being loaded.

DwgPrefix indicates the path to the current drawing.

DwgName indicates the name of the current drawing.

FileDia displays file prompts at the 'Command' prompt.

RELATED FILES

wizard.ini holds the names and descriptions of template files.

***.dwt** are template files stored in *.dwg* format.

TIPS

- The **New** and **QNew** commands operate in an identical manner, except when you specify a default template drawing in the **Option** command's **Files** tab:

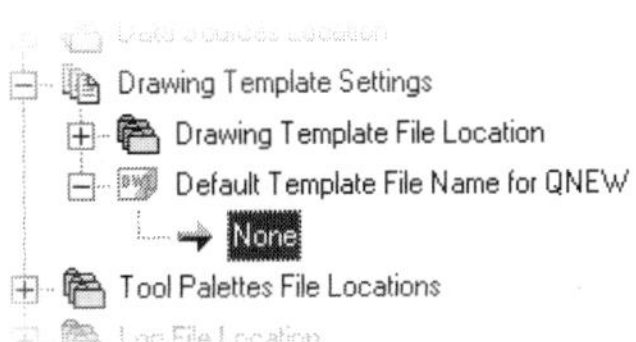

- The toolbar icon executes the **QNew** command, while the **File | New** menu pick executes the **New** command.

QSave

Rel.12 Saves the current drawing without requesting a file name (*short for Quick SAVE*).

Command	Alias	Ctrl+	F-key	Alt+	Menu Bar	Tablet
qsave	...	**S**	...	**FS**	**File**	**U24-**
					⇘Save	**U25**

Command: qsave

If the drawing has never been saved, displays the Drawing Save As dialog box.

COMMAND LINE OPTIONS

None.

RELATED COMMANDS

Quit ends AutoCAD, with or without saving the drawing.

Save saves the drawing, after requesting the filename.

RELATED SYSTEM VARIABLES

DBMod indicates whether the drawing has changed since it was loaded.

DwgName specifies the current drawing filename (default = *drawing1*).

DwgTitled specifies the status of drawing's filename:

DwgTiled	Meaning
0	Drawing is named *drawing1* (default).
1	Drawing was given another name.

TIPS

- When the drawing is unnamed, the **QSave** command displays the Save Drawing As dialog box to request a file name; see the **SaveAs** command.
- When the drawing file, its folder, or drive (such as a CD-ROM drive) is marked read-only, use the **SaveAs** command to save the drawing by another file name, or to another folder or drive.

QSelect

2000 Creates selection sets of objects based on their properties (*short for Quick SELECT*).

Command	Alias	Ctrl+	F-key	Alt+	Menu Bar	Tablet
qselect	...	...	...	**TQ**	**Tools ↳Quick Select**	**X9**

Command: qselect

Displays dialog box:

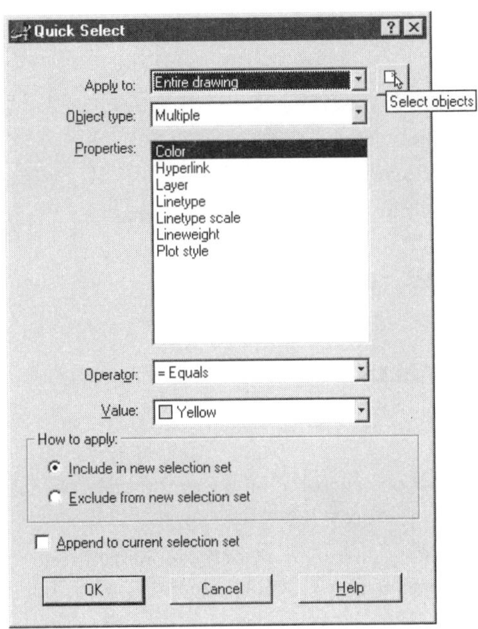

DIALOG BOX OPTIONS

Apply to applies the selection criteria to the entire drawing or current selection set; click Select Objects to create a selection set.

Select Objects allows you to select objects; AutoCAD prompts: 'Select objects: '. Right-click or press ENTER to return to this dialog box.

Object type lists the objects in the selection set; allows you to narrow the selection criteria to specific types of objects (default = Multiple).

Properties lists the properties valid for the selected object types; when you select more than one object type, only the properties in common are listed.

Operator lists logical operators available for the selected property; operators include:

Operator		Meaning
=	**Equals**	Selects objects equal to the property.
<>	**Not Equal To**	Selects objects different from the property.
>	**Greater Than**	Selects objects greater than the property.
<	**Less Than**	Selects objects less than the property.
*	**Wildcard Match**	Selects objects with matching text.

Value specifies the property value for the filter. If known values for the selected property are available, Value becomes a list from which you can choose a value. Otherwise, enter a value.

How to Apply options

⊙**Include in new selection set** creates a new selection set.

○**Exclude from new selection set** inverts the selection set, excluding all objects that match the selection criteria.

Additional options

Append to current selection set

☑ adds to the current selection set.

☐ replaces the current selection set.

RELATED COMMANDS

Select selects objects on the command line.

Filter runs a more sophisticated version of the **QSelect** command.

RELATED SYSTEM VARIABLES

None.

TIPS

- Use this command to select objects based on their properties; use the **Select** command to select objects based on their location in the drawing.
- You may select objects before entering the **QSelect** command, and then add or remove objects from the selection set with the **Quick Select** dialog box's options.
- Since this command is not transparent, you cannot use it within other commands; instead, use the **'Filter** command.
- As of AutoCAD 2005, this command also selects block insertions and tables.
- This command works with the properties of proxy objects created by ObjectARX applications.

'QText

V. 2.0 Displays lines of text as rectangular boxes (*short for Quick TEXT*).

Command	Alias	Ctrl+	F-key	Alt+	Menu Bar	Tablet
qtext	...	...	...	...	...	...

Command: qtext
Enter mode [ON/OFF] <OFF>: *(Enter* **ON** *or* **OFF***.)*

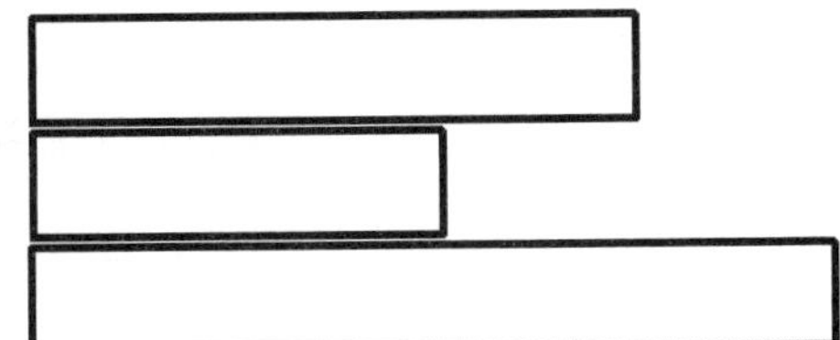

Normal text (at left) and quick text after regeneration (at right).

COMMAND LINE OPTIONS

ON turns on quick text after the next **Regenall** command.

OFF turns off quick text after the next **Regenall** command.

RELATED COMMAND

RegenAll regenerates the screen, which makes quick text take effect.

RELATED SYSTEM VARIABLE

QTextMode holds the current state of quick text mode.

TIPS

- A regeneration is required before AutoCAD displays text in quick outline form:

  Command: regenall
  Regenerating model.

- To reduce regen time, use **QText** to turn lines of text into rectangles, which redraw faster.
- The length of a qtext box does not necessarily match the actual length of text.
- Turning on **QText** affects text during plotting; qtext blocks are plotted as rectangles.
- To find invisible text (such as text made of spaces), turn on **QText**, thaw all layers, zoom to extents, and use the **RegenAll** command.
- The rectangles displayed by this command are affected by lineweights.

Quit

V. 1.0 Exits AutoCAD without saving changes made to the drawing after the most recent **QSave** or **SaveAs** command.

Command	Alias	Ctrl+	F-key	Alt+	Menu Bar	Tablet
quit	exit	Q	...	FX	File	Y25
				F4	↳Exit	

Command: quit

Displays dialog box:

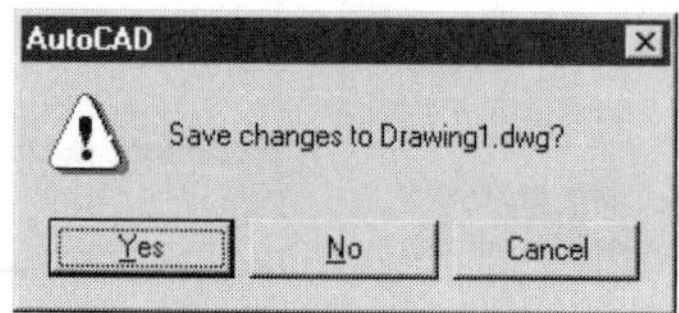

DIALOG BOX OPTIONS

Yes saves changes before leaving AutoCAD.

No does not save changes.

Cancel does not quit AutoCAD; returns to drawing.

RELATED COMMANDS

Close closes the current drawing.

SaveAs saves the drawing by another name and to another folder or drive.

RELATED SYSTEM VARIABLE

DBMod indicates whether the drawing has changed since it was loaded.

RELATED FILES

.dwg are AutoCAD drawing files.

.bak are backup files.

.bkn are additional backup files.

TIPS

- You can change a drawing, yet preserve its original format:
  1. Use the **SaveAs** command to save the drawing by another name.
  2. Use the **Quit** command to preserve the drawing in its most recently saved state.
- Even if you accidently save over a drawing, you can recover the previous version — *if* you remembered to set up AutoCAD to save backup files (see the **Options** command):
  1. Use Windows Explorer to rename the drawing file.
  2. Use Windows Explorer to rename the *.bak* (backup) extension to *.dwg*.
- You cannot save changes to a read-only drawing with the **Quit** command; use the **SaveAs** command instead.

R14PenWizard

2000 Helps create color-dependent plot style tables (*undocumented command*).

Command	Alias	Ctrl+	F-key	Alt+	Menu Bar	Tablet
r14penwizard	...	...	...	TZD	**Tools** ↳**Wizard** ↳**Add Color-Dependent Plot Style Table**	...

*You can also access this command through the Windows Control Panel's **Autodesk Plot Style Manager**. (This command makes the **Plot** command compatible with versions of AutoCAD prior to 2000.)*

Command: r14penwizard

Displays Add Color-Dependent Plot Style Table wizard.

DIALOG BOX OPTIONS

Begin page

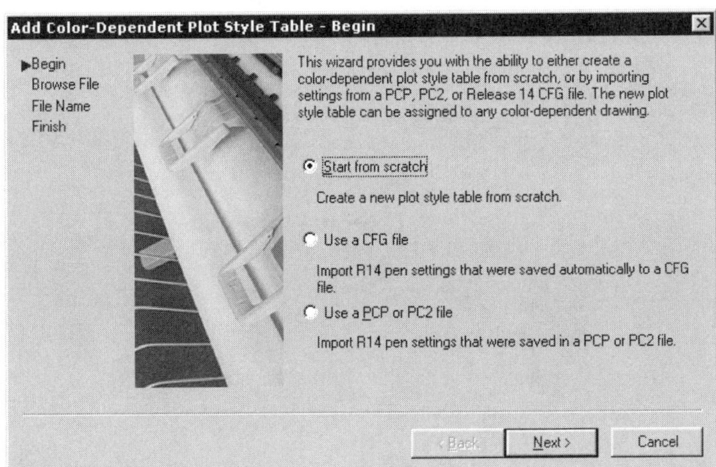

⊙**Start from scratch** creates a new color-dependent plot style table file.

○**Use a CFG file** converts the plotter pen settings stored in AutoCAD Release 14 *acad.cfg* files.

○**Use a PCP or PC2 file** converts the plotter pen settings stored in the plotter configuration parameter *.pcp* and *.pc2* files of earlier versions of AutoCAD.

Buttons

Back goes back to the previous dialog box.

Next moves to the next dialog box.

Cancel exits the wizard.

File Name page

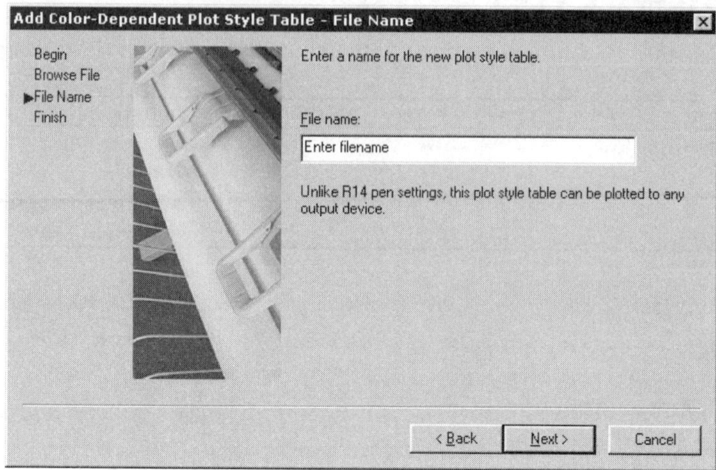

File name specifies name of the file in which to store the new color-dependent plot style table.

Finish page

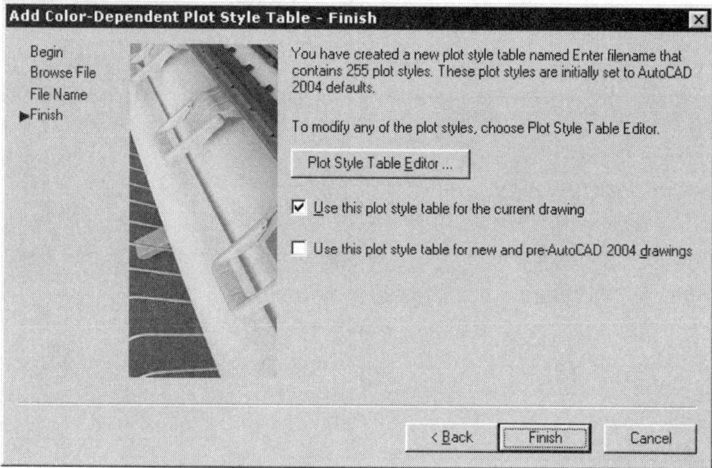

Plot Style Table Editor displays the Plot Style Table Editor dialog box; see **StylesManager** command.

☑ **Use this plot style table for the current drawing** applies the plot style table to the current drawing.

☐ **Use this plot style table for new and pre-AutoCAD 2004 drawings** applies the plot style table to all new drawings and drawings created by versions of AutoCAD prior to 2004.

RELATED COMMANDS

PcInWizard imports *.pcp* and *.pc2* configuration plot files into the current layout.

PlotterManager accesses the Add Plotter wizard and Assign Plot Style wizard.

Plot plots the drawing.

StylesManager displays the Plot Style Table Editor dialog box.

Ray

Rel.13 Draws semi-infinite construction lines.

Command	Alias	Ctrl+	F-key	Alt+	Menu Bar	Tablet
ray	...	...	...	DR	Draw ↳Ray	K10

Command: ray
Specify start point: *(Pick a point.)*
Specify through point: : *(Press* **ENTER** *to end the command.)*

Start point.

COMMAND LINE OPTIONS

Start point specifies the starting point of the ray.

Through point specifies the point through which the ray passes.

RELATED COMMANDS

Properties modifies rays.

Line draws finite line segments.

XLine draws infinite construction lines.

TIPS

- The *ray* object is semi-infinite in length.
- A ray is a "construction line"; it displays and plots, but does not affect the extents.
- A ray has all the properties of a line (including color, layer, and linetype), and can be used as a cutting edge for the **Trim** command.

Removed Command

RConfig — render configuration — was removed from Release 14. It is replaced by **Render**.

Recover

Rel.12 Recovers damaged drawings without user intervention.

Command	Alias	Ctrl+	F-key	Alt+	Menu Bar	Tablet
recover	...	...	...	**FUR**	**File** ↳**Drawing Utilities** ↳**Recover**	...

Command: recover

Displays the Select File dialog box. Select a .dwg file, and then click **Open.**

Sample output:

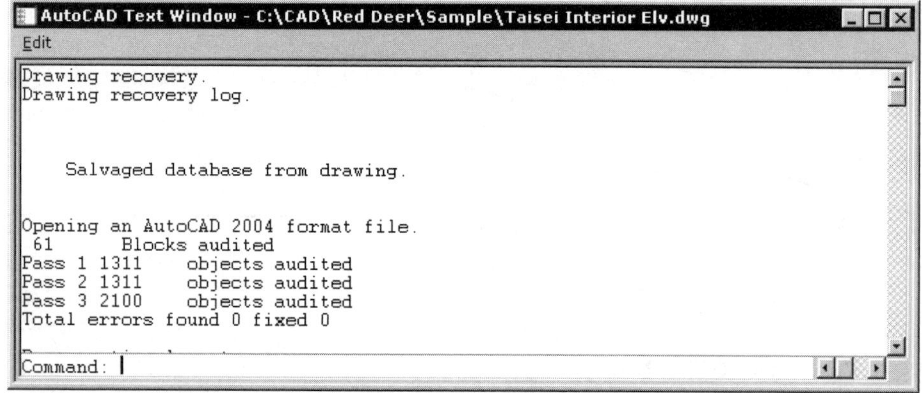

COMMAND LINE OPTIONS

None.

RELATED COMMAND

Audit checks a drawing for integrity.

TIPS

- The **Recover** command does not ask permission to repair damaged parts of the drawing file; use the **Audit** command if you want control over the repair process.
- The **Quit** command discards changes made by the **Recover** command.
- If the **Recover** and **Audit** commands do not fix the problem, try using the **DxfOut** command, followed by the **DxfIn** command.

Rectang

Rel.12 Draws squares and rectangles with a variety of options.

Command	Aliases	Ctrl+	F-key	Alt+	Menu Bar	Tablet
rectang	**rec** **rectangle**	...	...	**DG**	**Draw** ↳**Rectangle**	**Q10**

Command: rectang
Specify first corner point or [Chamfer/Elevation/Fillet/Thickness/Width]: : *(Pick a point, or enter an option)*
Specify other corner point or [Dimensions]: *(Pick another point.)*

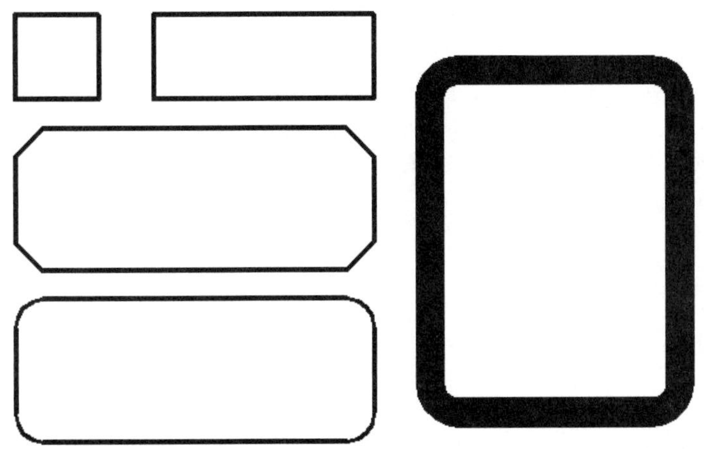

Square and rectangles drawn with the ***Rectangle*** *command's* ***Chamfer, Fillet,*** *and* ***Width*** *options.*

COMMAND LINE OPTIONS

Specify first corner point picks the first corner of the rectangle.

Specify other corner point picks the opposite corner of the rectangle.

Chamfer options
Specify first chamfer distance for rectangles <0.0000>: : *(Enter a value.)*
Specify second chamfer distance for rectangles <0.0000>: *(Enter a value.)*

First chamfer distance for rectangles sets the first chamfer distance for all four corners.

Second chamfer distance for rectangles sets the second chamfer distance for all four corners.

Dimensions options
Specify length for rectangles <0.0>: *(Enter length.)*
Specify width for rectangles <0.0>: *(Enter width.)*
Specify other corner point or [Dimensions]: *(Pick a point.)*

Specify length for rectangles specifies the length along the x axis.

Specify width for rectangles specifies the length along the y axis.

Specify other corner point specifies the orientation of rectangle.

Elevation option
Specify the elevation for rectangles <0.0000>: *(Enter a value.)*

Elevation for rectangles sets the elevation (*height of the rectangle in the z direction*).

Fillet option
Specify fillet radius for rectangles <1.0000>: *(Enter a value.)*

Fillet radius for rectangles sets the fillet radius for all four corners of the rectangle.

Thickness option
Specify thickness for rectangles <0.0000>: *(Enter a value.)*

Thickness for rectangles sets the thickness of the rectangle's sides in the z direction.

Width option
Specify line width for rectangles <0.0000>: *(Enter a value.)*

Width for rectangles sets the width of all segments of the rectangle's four sides.

RELATED COMMANDS

Donut draws solid-filled circles with a polyline.

Ellipse draws ellipsis with a polyline, when **PEllipse** = 1.

PEdit edits polylines, including rectangles.

PLine draws polylines and polyline arcs.

Polygon draws a polygon — 3 to 1024 sides — from a polyline.

RELATED SYSTEM VARIABLE

PlineWid saves the current polyline width.

TIPS

- Rectangles are drawn from polylines; use the **PEdit** command to change the rectangle, such as the width of the polyline.
- The values you set for the **Chamfer**, **Elevation**, **Fillet**, **Thickness**, and **Width** options become the default for the next execution of the **Rectangle** command.
- The pick point determines the location of the rectangle's first vertex; rectangles are drawn counterclockwise.
- Use the **Snap** command and object snap modes to place the rectangle precisely.
- Use object snap modes **ENDpoint** or **INTersection** to snap to the rectangle's vertices.
- This command ignores the settings in the **ChamferA**, **ChamferB**, **Elevation**, **FilletRad**, **PLineWid**, and **Thickness** system variables.

Redefine

Rel. 9 Restores the meaning of AutoCAD commands disabled by the **Undefine** command.

Command	Alias	Ctrl+	F-key	Alt+	Menu Bar	Tablet
redefine	...	...	...	...	...	...

Command: redefine
Enter command name: *(Enter command name.)*

COMMAND LINE OPTION

Enter command name specifies the name of the AutoCAD command to redefine.

RELATED COMMANDS

All AutoCAD commands can be redefined.

Undefine disables the meaning of AutoCAD commands.

TIPS

- Prefix any command with a . (*period*) to redefine the undefinition temporarily, as in:

  Command: .line

- Prefix any command with an _ (*underscore*) to make an English-language command work in any linguistic version of AutoCAD, as in:

  Command: _line

- You must undefine a command with the **Undefine** command before using the **Redefine** command.

Redo

V. 2.5 Reverses the effect of the most-recent **U** and **Undo** commands.

Command	Alias	Ctrl+	F-key	Alt+	Menu Bar	Tablet
redo	...	Y	...	ER	Edit ↳Redo	U12

Command: redo

COMMAND LINE OPTIONS

None.

RELATED COMMANDS

MRedo redoes more than one undo.

Oops un-erases the most recently-erased objects.

U undoes the most recent AutoCAD command.

Undo undoes the most recent series of AutoCAD commands.

TIPS

- The **Redo** command is limited to reversing a single undo, while the **Undo** and **U** commands undo operations all the way back to the beginning of the editing session.
- The **Redo** command must be used immediately following the **U** or **Undo** command.

'Redraw / 'RedrawAll

V. 1.0 Redraws viewports to clean up the screen.

Command	Alias	Ctrl+	F-key	Alt+	Menu Bar	Tablet
redraw	**r**	...	...	...	...	...
redrawall	**ra**	...	...	**VR**	**View ↳Redraw**	**Q11-R11**

Command: redraw

Reraws the current viewport.

Command: redrawall

Redraws all viewports.

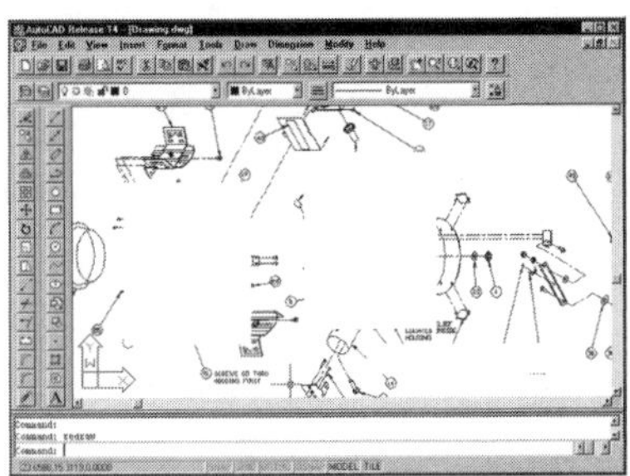

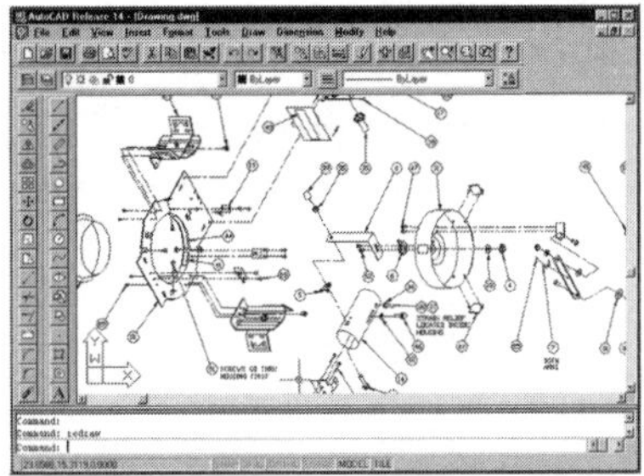

Before redraw, portions of the drawing are "missing" (at left); the drawing is clean after the redraw (at right).

COMMAND LINE OPTION

ESC stops the redraw.

RELATED COMMAND

Regen regenerates the current viewport.

RELATED SYSTEM VARIABLE

SortEnts controls the order of redrawing objects:

SortEnts	Meaning
0	Sorts by the order in the drawing database.
1	Sorts for object selection.
2	Sorts for object snap.
4	Sorts for redraw.
8	Sorts for creating slides.
16	Sorts for regenerations.
32	Sorts for plotting.
64	Sorts for PostScript plotting.

TIPS

- Use this command to clean up the screen after a lot of editing; some commands automatically redraw the screen when they are done.
- **Redraw** does not affect objects on frozen layers.

RefClose

2000 Saves or discards changes made to reference objects (blocks and xrefs) edited in-place (*short for REFerence CLOSE*).

Command	Alias	Ctrl+	F-key	Alt+	Menu Bar	Tablet
refclose	...	...	...	**MBD**	**Modify** **↳In-place Xref and Block Edit** **↳Discard Changes to Reference**	...
				MBS	**Modify** **↳In-place Xref and Block Edit** **↳Save Changes Back to Reference**	

Command: refclose
Enter option [Save/Discard reference changes] <Save>: *(Enter an option.)*

COMMAND LINE OPTIONS

Save saves the editing changes made to the block or externally-referenced file.

Discard discards the changes.

RELATED COMMANDS

RefEdit edits blocks and externally-referenced files attached to the current drawing.

Insert inserts a block in the drawing.

XAttach attaches an externally-referenced drawing.

RELATED SYSTEM VARIABLE

RefEditName specifies the filename of the referenced file being edited.

TIPS

- AutoCAD prompts you with a warning dialog box to ensure you really want to discard or save the changes made to the reference:

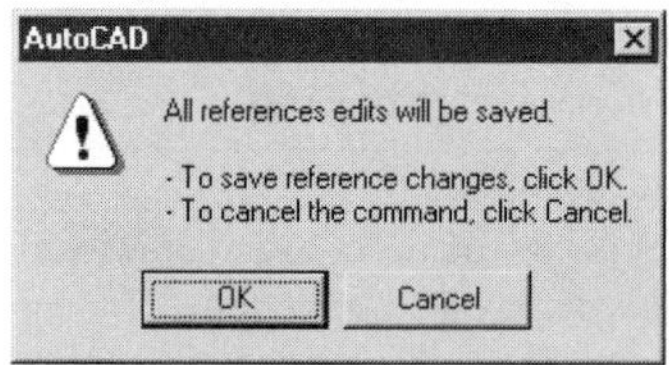

- You can use this command only after the **RefEdit** command; otherwise AutoCAD reports, "** Command not allowed unless a reference is checked out with RefEdit command **."

RefEdit

2000 Edits-in-place blocks and externally-referenced files attached to the current drawing (*short for REFerence EDIT*).

Commands	Alias	Ctrl+	F-key	Alt+	Menu Bar	Tablet
refedit	...	...	...	**MBE**	**Modify** ↳**In-place Xref and Block Edit** ↳**Edit Block or Xref**	...
-refedit						

Command: refedit

Select reference: *(Pick an externally-referenced drawing or block.)*

Displays dialog box:

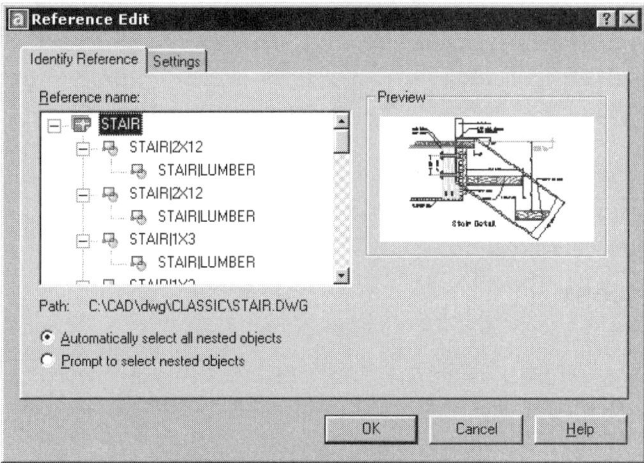

*Click **OK** to continue with the command:*

Select nested objects: *(Pick one or more objects.)*

Use REFCLOSE or the Refedit toolbar to end reference editing session.

Displays RefEdit toolbar.

COMMAND LINE OPTIONS

Select reference selects an externally-referenced drawing or an inserted block for editing.

Select nested objects selects objects within the reference — this becomes the selection set of objects that you may edit; you may select all nested objects with the All option (with the exception of OLE objects and objects inserted with the **MInsert** command, which cannot be refedited).

DIALOG BOX OPTIONS

Identify Reference tab

Reference name lists a tree of the selected reference object and its nested references; a single reference can be edited at a time.

Preview displays a preview image of the selected reference.

⊙**Automatically select all nested objects** selects all nested objects.

○**Prompt to select nested objects** prompts you. 'Select nested objects.'

Settings tab

Identify Reference | Settings

☑ Create unique layer, style, and block names

☐ Display attribute definitions for editing

☑ Lock objects not in working set

Create unique layer, style, block names

- ☑ Prefixes layer and symbol names of extracted objects with **$*n*$**.
- ☐ Retains the names of layers and symbols, as in the reference.

Display attribute definitions for editing option is available only when an xref contains attributes:

- ☑ Makes non-constant attributes invisible; attribute definitions can be edited.
- ☐ Attribute definitions cannot be edited.

Caution! When edited attributes are saved back to the block reference, the attributes of the original reference are not changed; instead, the modified attribute definitions come into effect with the next insertion of the block.

Lock objects not in working set options:

- ☑ Locks unselected objects in a manner similar to locked layers.
- ☐ Objects are not locked.

TOOLBAR OPTIONS

This toolbar appears automatically after you select nested objects to edit:

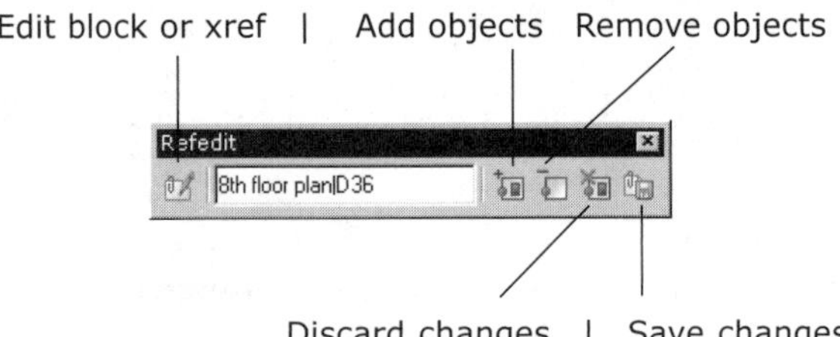

Edit block or xref executes the **RefEdit** command.

Add objects to working set executes the **RefSet Add** command.

Remove objects from working set executes the **RefSet Remove** command.

Discard changes to reference executes the **RefClose Discard** command.

Save back changes to reference executes the **RefClose Save** command.

-REFEDIT Command

Command: -refedit

Select reference: *(Pick an externally-referenced drawing, or a block.)*

Select nesting level [Ok/Next] <Next>: *(Type* **O** *or* **N**.*)*

Enter object selection method [All/Nested] <All>: *(Type* **A** *or* **N**.*)*

Display attribute definitions [Yes/No] <No>: *(Type* **Y** *or* **N**.*)*

Use REFCLOSE or the Refedit toolbar to end reference editing session.

Displays ***Refedit*** *toolbar.*

COMMAND LINE OPTIONS

Select reference selects an externally-referenced drawing or an inserted block for editing.

Select nesting level selects objects within the reference — this becomes the selection set of objects that you may edit; you may select all nested objects with All (with the exception of OLE objects and objects inserted with the **MInsert** command, which cannot be refedited).

Enter object selection method:

All selects all objects.

Nested selects only nested objects.

Display attribute definitions:

Yes makes non-constant attributes invisible; attribute definitions can be edited.

No means attribute definitions cannot be edited.

RELATED COMMANDS

RefSet adds and removes objects from a working set.

RefClose saves or discards editing changes to the reference.

RELATED SYSTEM VARIABLES

RefEditName stores the name of the externally-referenced file or block being edited.

XEdit determines whether the current drawing can be edited while being referenced by another drawing.

XFadeCtl specifies the amount of fading for objects not being edited in place.

TIPS

- In layouts, the drawing must be in model mode for you to be able to select an xref.
- OLE objects and objects inserted with the **MInsert** command cannot be refedited.
- AutoCAD identifies the "working set" as those objects that you have selected to edit in-place.
- Objects *not* in a working set cannot be selected.

RefSet

2000 Adds and removes objects from working sets (*short for* REFerence SET).

Command	Alias	Ctrl+	F-key	Alt+	Menu Bar	Tablet
refset	...	...	...	**MBA**	**Modify** ↳**In-place Xref and Block Edit** ↳**Add Objects to Working Set**	...
				MBR	**Modify** ↳**In-place Xref and Block Edit** ↳**Remove Objects from Working Set**	

Command: refset
Transfer objects between the Refedit working set and host drawing...
Enter an option [Add/Remove] <Add>: *(Type* **A** *or* **R***.)*
Select objects: *(Pick one or more objects.)*

COMMAND LINE OPTIONS

Add prompts you to select objects to add to the working set.

Remove prompts you to select objects to remove from the working set.

Select objects selects the objects to be added or removed.

RELATED COMMANDS

RefEdit edits reference objects in place.

RefClose saves or discards editing changes to the reference.

RELATED SYSTEM VARIABLES

RefEditName stores the name of the xref or block being edited.

XEdit determines whether the current drawing can be edited while being referenced by another drawing.

XFadeCtl specifies the amount of fading for objects not being edited in place.

TIPS

- The purpose of this command is to add objects to — or remove them from — the "working set" of objects, while you are performing in-place editing of a block or an externally-referenced drawing.
- You cannot add or remove objects on locked layers. AutoCAD complains, '** *n* selected objects are on a locked layer.'

Regen / RegenAll

V. 1.0 Regenerates viewports to update the drawing.

Command	Alias	Ctrl+	F-key	Alt+	Menu Bar	Tablet
regen	**re**	...	...	**VG**	**View** ↳**Regen**	**J1**
regenall	**rea**	...	...	**VA**	**View** ↳**Regen All**	**K1**

Command: regen

Regenerating model.

Regenerates the current viewport.

Command: regenall

Regenerating model.

Regenerates all viewports.

COMMAND LINE OPTION

ESC cancels the regeneration.

RELATED COMMANDS

Redraw cleans up the current viewport quickly.

RegenAuto checks with you before doing most regenerations.

ViewRes controls whether zooms and pans are performed at redraw speed.

RELATED SYSTEM VARIABLES

RegenMode toggles automatic regeneration.

WhipArc determines how circles and arcs are displayed:

WhipArc	Meaning
0	Circles and arcs displayed as vectors.
1	Circles and arcs displayed as true circles and arcs.

TIPS

- Some commands automatically force a regeneration of the screen; other commands queue the regeneration.
- The **Regen** command reindexes the drawing database for better display and object selection performance.
- To save on regeneration time, freeze layers you are not working with, apply **QText** to turn text into rectangles, and place hatching on its own layer.
- Use the **RegenAll** command to regenerate all viewports; introduced with Release 10.

'RegenAuto

V. 1.2 Prompts before performing regenerations, when turned off (*short for REGENeration AUTOmatic*).

Command	Alias	Ctrl+	F-key	Alt+	Menu Bar	Tablet
regenauto	...	...	...	...	...	...

Command: regenauto
Enter mode [ON/OFF] <ON>: *(Enter* **ON** *or* **OFF**.*)*

COMMAND LINE OPTIONS

OFF turns on "About to regen, proceed?" message.

ON turns off "About to regen, proceed?" message.

RELATED COMMANDS

Regen forces a regeneration in the current viewport.

RegenAll forces a regeneration in all viewports.

RELATED SYSTEM VARIABLES

Expert suppresses "About to regen, proceed?" message when value is greater than 0.

RegenMode specifies the current setting of automatic regeneration:

RegenMode	Meaning
0	Off.
1	On (default).

TIPS

- If a regeneration is caused by a transparent command, AutoCAD delays it and responds with the message, "Regen queued."
- When off, results in the following prompt:

  Command: regen
  About to regen, proceed? <Y>: *(Press* **ENTER**.*)*
- AutoCAD Release 12 reduces the number of regenerations by expanding the virtual screen from 16 bits to 32 bits.

Region

Rel.11 Creates 2D regions from closed objects.

Command	Alias	Ctrl+	F-key	Alt+	Menu Bar	Tablet
region	reg	...	...	DN	Draw ↳Region	R9

Command: region
Select objects: *(Select one or more closed objects.)*
Select objects: *(Press* **ENTER** *to end object selection.)*
1 loop extracted.
1 region created.

COMMAND LINE OPTION

Select objects selects objects to convert to a region; AutoCAD discards unsuitable objects.

RELATED COMMANDS

All drawing commands.

RELATED SYSTEM VARIABLE

DelObj toggles whether objects are deleted during the conversion by the **Region** command.

TIPS

- The **Region** command converts closed line sets, closed 2D and planar 3D polylines, and closed curves.
- The **Region** command rejects open objects, intersections, and self-intersecting curves.
- The resulting region is unpredictable when more than two curves share an endpoint.
- Polylines with width lose their width when converted to a region.
- Island are "holes" in regions.

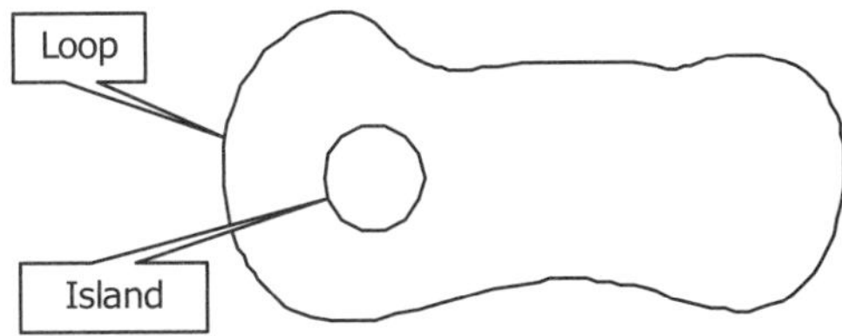

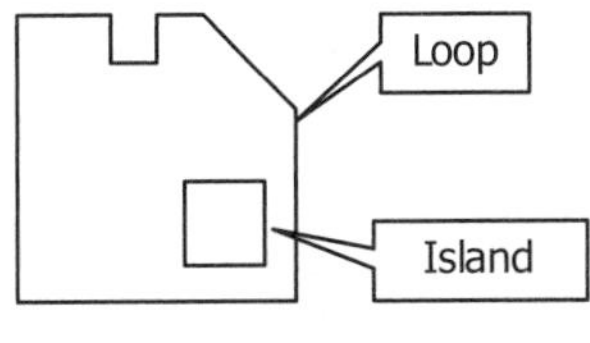

DEFINITIONS

Curve — an object made of circles, ellipses, splines, and joined circular and elliptical arcs.

Island — a closed shape fully within (*not touching or intersecting*) another closed shape.

Loop — a closed shape made of closed polylines, closed lines, and curves.

Region — a 2D closed area defined as a ShapeManager object.

Reinit

Rel.12 Reinitializes digitizers and input-output ports, and reloads the *acad.pgp* file (*short for* REINITialize).

Command	Alias	Ctrl+	F-key	Alt+	Menu Bar	Tablet
reinit	...	...	...	...	...	...

Command: reinit

Displays dialog box:

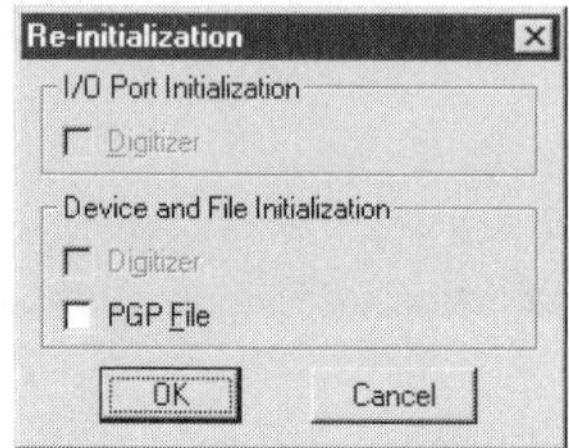

DIALOG BOX OPTIONS

I/O Port Initialization option

☐ **Digitizer** reinitializes ports connected to the digitizer; grayed out if no digitizer is configured.

Device and File Initialization options

☐ **Digitizer** reinitializes the digitizer driver; grayed out if no digitizer is configured.

☐ **PGP File** reloads the *acad.pgp* file.

RELATED COMMAND

MenuLoad reloads menu files.

RELATED SYSTEM VARIABLE

Re-Init reinitializes via system variable settings.

RELATED FILES

acad.pgp is the program parameters file in *\autocad 2005\support* folder.

***.hdi** are device drivers specific to AutoCAD.

TIPS

- AutoCAD allows you to connect both the digitizer and the plotter to the same port, since you do not need the digitizer during plotting; use the **Reinit** command to reinitialize the digitizer after plotting.
- AutoCAD automatically reinitializes all ports and reloads the *acad.pgp* file each time a drawing is loaded.

Rename

V. 2.1 Changes the names of blocks, dimension styles, layers, linetypes, plot styles, text styles, table styles, UCS names, views, and viewports.

Commands	Aliases	Ctrl+	F-key	Alt+	Menu Bar	Tablet
rename	**ren**	...	...	**OR**	**Format** ⇨**Rename**	**V1**
-rename	**-ren**					

Command: rename

Displays dialog box:

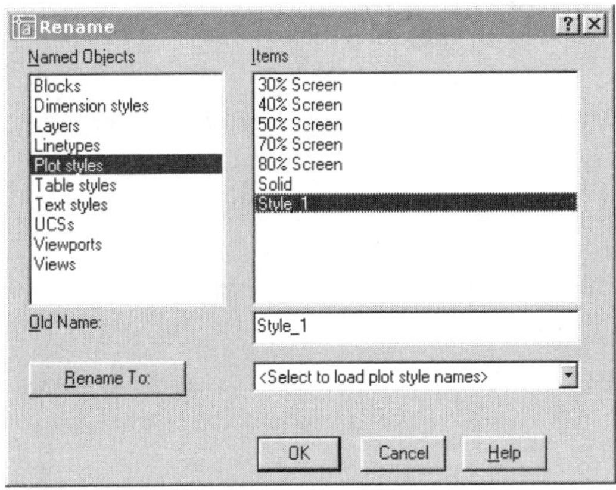

DIALOG BOX OPTIONS

Named Objects lists named objects in the drawing.

Items lists the names of named objects in the current drawing.

Old Name displays the current name of an object to be renamed.

Rename to allows you to enter a new name for the object.

-RENAME Command

Command: -rename

Enter object type to rename
[Block/Dimstyle/LAyer/LType/Plotstyle/Style/Tablestyle/Ucs/VIew/VPort]:
(Enter an option.)

Example usage:

[Block/Dimstyle/LAyer/LType/Plotstyle/Style/Tablestyle/Ucs/VIew/VPort]: b
Enter old block name: diode-20
Enter new block name: diode-02

COMMAND LINE OPTIONS

Block changes the names of blocks.

Dimstyle changes the names of dimension styles.

LAyer changes the names of layers.

LType changes the names of linetypes.

Plotstyle changes the names of plot styles.

Tablestyle changes the names of table styles *(new to AutoCAD 2005)*.

textStyle changes the names of text styles.

Ucs changes the names of UCS configurations.

VIew changes the names of view configurations.

VPort changes the names of viewport configurations.

RELATED SYSTEM VARIABLES

CeLType specifies the name of the current linetype.

CLayer specifies the name of the current layer.

CTableStyle specifies the name of the current table style.

DimStyle specifies the name of the current dimension style.

InsName specifies the name of the current block.

TextStyle specifies the name of the current text style.

UcsName specifies the name of the current UCS view.

TIPS

- You cannot rename layer "0", dimstyle "Standard," anonymous blocks, groups, or linetype "Continuous."
- To rename a group of similar names, use ***** (the wildcard for "all") and **?** (the wildcard for a single character).
- Names can be up to 255 characters in length.
- The **Properties** command does *not* allow you to rename blocks.
- The **PlotStyle** option is available only when plot styles are attached to the drawing.
- In AutoCAD 2004, the **textStyle** option was renamed **Style.** The **Tablestyle** option was added with AutoCAD 2005.
- You cannot use this command during **RefEdit**.

Render

Rel.12 Creates renderings of 3D objects.

Command	Alias	Ctrl+	F-key	Alt+	Menu Bar	Tablet
render	**rr**	...	...	**VER**	**View** ↳**Render** ↳**Render**	**M1**

Command: render

Displays dialog box:

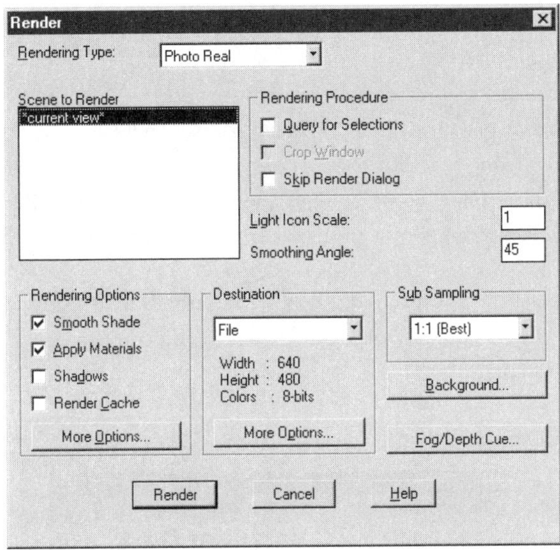

DIALOG BOX OPTIONS

Rendering Type selects between basic Render, Photo Real, or Photo Raytrace; also lists installed third-party renderers.

Scene to Render lists the names of scenes defined by the **Scene** command; default = *Current view*.

Rendering Procedure options

Query For Selections

☑ Prompts you to select the objects to render; unselected objects appear in wireframe in the rendering only when you select the Merge option from the Background option.

☐ Renders all objects in the current viewport.

Crop Window options:

☑ Prompts you to select a windowed area to render.

☐ Renders the entire current viewport.

☐ **Skip Render Dialog** does not display the Render dialog box the next time you use the **Render** command.

Light icon scale sizes light blocks Overhead, Direct, and Sh_Spot.

Smoothing angle converts edges to smooth curves. For example, when the angle between two surfaces is greater than the default of 45 degrees, AutoCAD renders an edge; when it is less than 45 degrees, the edge is smoothed to a curve.

Rendering Options options

☑ **Smooth shade** smooths the edges of multifaced surfaces.

☑ **Apply materials** applies surface materials defined by the **RMat** command.

☐ **Shadows** generates shadows when Photo Real and Photo Raytrace rendering modes are selected.

☐ **Render cache** caches the objects to help speed rendering.

More options displays the Render Options dialog box, which varies according to the type of rendering selected.

Destination options

- **Viewport** displays the rendering in the current viewport.
- **Render Window** displays the rendering in a separate window.
- **File** saves the rendering to a file on disk; does not display the rendering on screen.

More Options displays File Output Configuration dialog box, when **File** is selected.

Sub Sampling renders a fraction of pixels; ranges from 1:1 for best quality (default) to 8:1 for fastest.

Background displays the Background dialog box; see the **Background** command.

Fog/Depth Cue displays the Fog dialog box; see the **Fog** command.

Render renders the scene.

Render Options dialog box

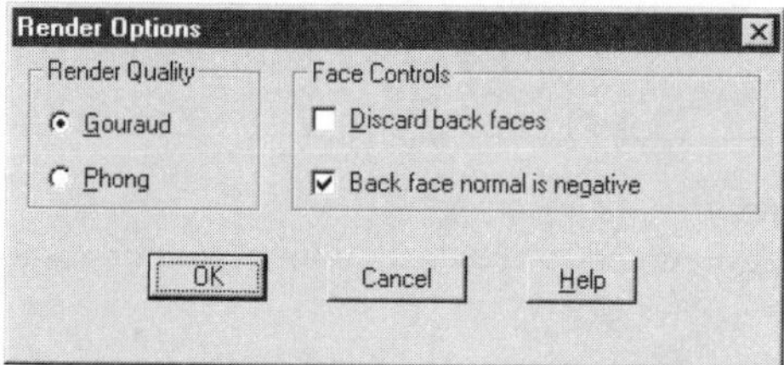

Render Quality options

⊙ **Gouraud** calculates light intensity at each vertex; faster speed.

○ **Phong** calculates light intensity at each pixel; higher quality.

Face Controls options

☐ **Discard back faces** speeds up rendering by ignoring the backs of objects.

☑ **Back face normal is negative** may create odd looking objects at times; turn off.

Photo Real Render Options dialog box

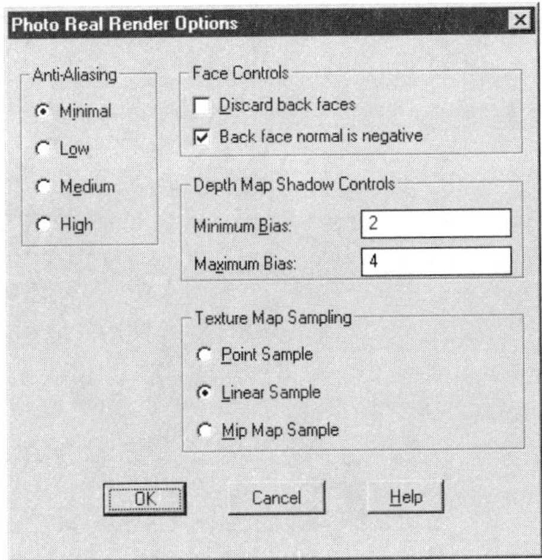

Anti-Aliasing options

⊙ **Minimal** renders with analytical horizontal anti-aliasing; fastest rendering.

○ **Low** renders with four samples per pixel.

○ **Medium** renders with nine samples per pixel.

○ **High** renders with 16 samples per pixel; best quality.

Face Controls options

☐ **Discard back faces** speeds up rendering by ignoring the backs of objects.

☑ **Back face normal is negative** may create odd looking objects at times; turn off.

Depth Map Shadow Controls options

Minimum bias adjusts the shadow map bias to prevent self-shadows and detached shadows; default = 2.0; ranges from 2.0 to 20.0.

Maximum bias limits the shadow map bias to 10 more than minimum bias; default = 4.0.

Texture Map Sampling options

- **Point sample** renders the nearest pixel within a bitmap.
- **Linear sample** averages the four neighbor pixels pyramidically (default).
- **Mip map sample** averages pixels with the *mip* method, which pyramidically averages a square sample area.

Photo Raytrace Render Options dialog box

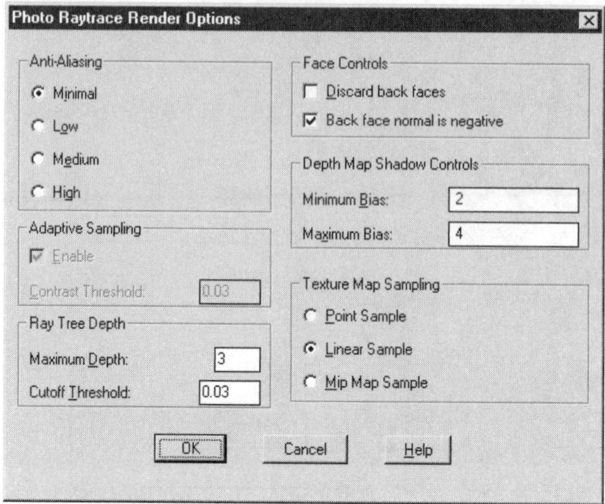

Anti-Aliasing options

⊙**Minimal** renders with analytical horizontal anti-aliasing; fastest rendering but lowest quality.

O**Low** renders with four samples per pixel.

O**Medium** renders with nine samples per pixel.

O**High** renders with 16 samples per pixel; best quality but slowest rendering.

Adaptive Sampling options

☑ **Enable** toggles adaptive sampling; available when minimal anti-aliasing is turned off.

Contrast Threshold specifies the sensitivity of adaptive sampling; larger values increase rendering speed, but might reduce image quality; default = 0.03; ranges from 0.0 to 1.0.

Ray Tree Depth options

Maximum Depth limits the ray tree depth to track reflected and refracted rays; default = 3.

Cutoff Threshold defines the percentage bounce cutoff; default = 0.03 means 3%.

Face Controls options

☐ **Discard back faces** speeds up rendering by ignoring the backs of objects.

☑ **Back face normal is negative** may create odd looking objects at times; turn off.

Depth Map Shadow Controls options

Minimum bias adjusts shadow map bias to prevent self-shadows and detached shadows; default = 2.0; ranges from 2.0 to 20.0.

Maximum bias limits shadow map bias to 10 more than minimum bias; default = 4.0.

Texture Map Sampling options

O**Point sample** renders the nearest pixel within a bitmap.

⊙**Linear sample** averages the four neighbor pixels (default).

O**Mip map sample** averages pixels with the *mip* method, which pyramidically averages a square sample area.

RELATED COMMANDS

All rendering-related commands.

Hide removes hidden lines from wireframe view.

Replay displays a BMP, TIFF, or Targa raster (bitmap) file in the current viewport.

RPref sets up options for the **Render** command.

SaveImg saves the image in the current viewport to a raster file.

ShadEdge performs real-time shading of 3D objects.

TIPS

- If you do not place a light or define a scene, the **Render** command uses the current view and ambient light.
- If you do not select a light or scene, the **Render** command renders all objects using all lights and the current view.
- If you set up the **Render** command to skip the dialog box, use the **RPref** command to set rendering options.
- When outputting to a file, you have the following image format options: BMP, TGA, PCX, PostScript, and TIFF.
- You cannot create a rendering in paper space.
- For smoother renderings, change **FacetRes** to 2.

Removed Command

RenderUnload was removed from Release 14. Use **Arx** to unload *Render.Arx* instead.

Quick Start Tutorial

Your First Rendering

- *Basic Rendering*

Step 1
Create a 3D drawing or select a 3D sample drawing.

Step 2
Start the **Render** command, click the **Render** button, and wait a few seconds.

- *Advanced Rendering*

Step 1
Create a 3D drawing or select a 3D sample drawing.

Step 2
Use the **MatLib** command to load material definitions into the drawing.

Step 3
Using the **RMat** command to assign materials to colors, layers, and/or objects.

Step 4
Use the **Light** command to place and aim lights: point, spot, and distant.

Step 5
Use the **Scene** command to collect lights and a named view into named scenes.

Step 6
Render the named scene with the **Render** command.

Step 7
Use the **SaveImg** command to save the rendering to a TIFF, Targa, or GIF image on disk.

Step 8
View the saved rendering file with the **Replay** command.

RENDERING EFFECTS

Original wireframe drawing:

Basic rendering; most options turned off:

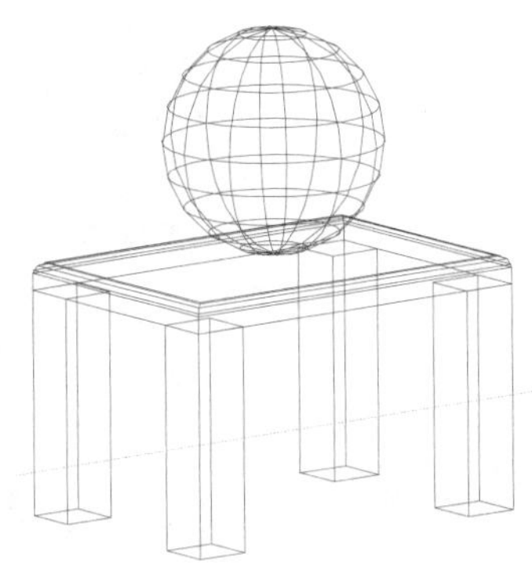

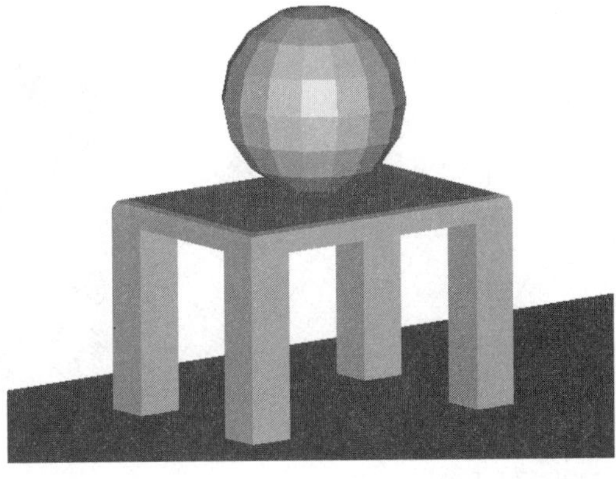

Smooth Shading on:

Attach Materials on; requires Photo Real or Raytrace mode:

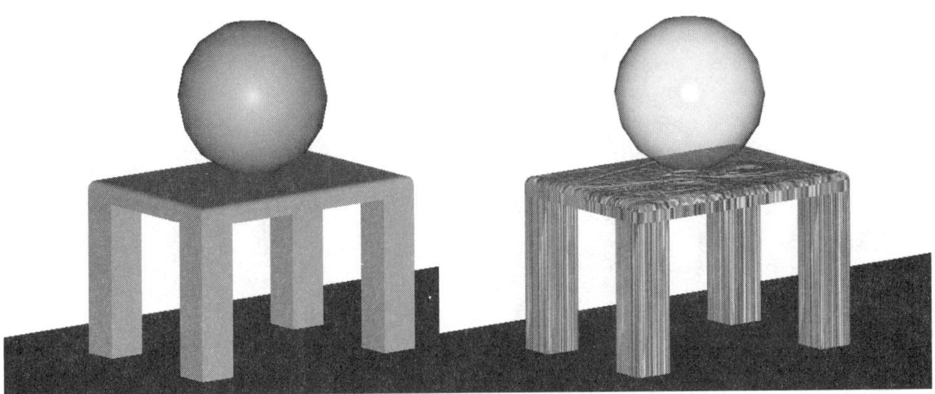

Background set to **Gradient**:

Background set to **Image**:

Fog set to white: **Fog** set to black:

Shadows turned on: **Shadow Volumes** turned off:

Sub Sampling set to **4:1**:

RendScr

Rel.12 Redisplays the most-recent rendering (*short for RENDer SCReen).*

Command	Alias	Ctrl+	F-key	Alt+	Menu Bar	Tablet
rendscr	...	...	...	...	...	...

Command: rendscr

COMMAND LINE OPTIONS

None.

RELATED COMMAND

Render creates a rendering of the current 3D viewport.

RELATED SYSTEM VARIABLES

None.

TIP

- This command appears to not work under Windows. Autodesk notes that this command is for "systems with a nonwindowing single monitor display that is configured for full-screen rendering."

Replay

Rel.12 Displays BMP, TIFF, and Targa images in the current viewport.

Command	Alias	Ctrl+	F-key	Alt+	Menu Bar	Tablet
replay	...	...	...	TDV	Tools ↳Display Image ↳View	V8

Command: replay

Displays the Select File dialog box. Select a file, and then click ***Open****.*

Displays dialog box:

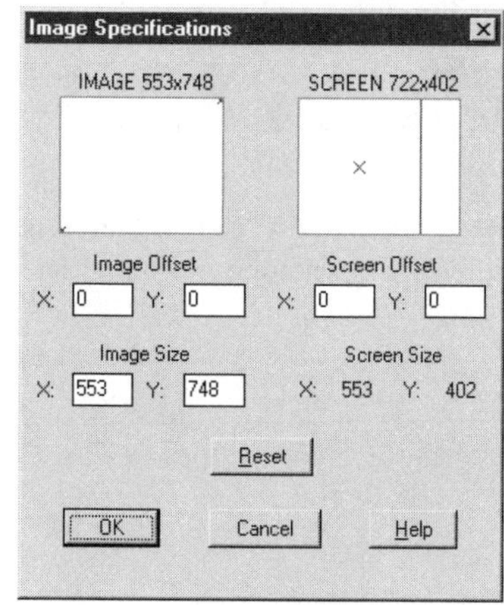

DIALOG BOX OPTIONS

Image selects the displayed area by clicking on the image tile.

Image Offset specifies the x,y coordinates of the image's lower-left corner.

Image Size specifies the size of the image in pixels.

Screen specifies the maximum size of the image.

Screen Offset specifies the x,y coordinates of the image's lower-left corner.

Reset restores the default values.

RELATED COMMANDS

Import displays a dialog box to loading some raster and vector files.

SaveImg saves a rendering as a BMP, TIFF, or Targa raster file.

RELATED FILES

****.bmp*** are Windows bitmap files.

****.tif*** are RGBA TIFF files, up to 32-bits in color depth.

****.tga*** are RGBA Targa v2.0 files, up to 32-bits in color depth.

In \textures folder provided with AutoCAD:

****.tga*** contains many Targa-format images.

'Resume

V. 2.0 Resumes script files paused by an error, or by pressing the BACKSPACE or ESC keys.

Command	Alias	Ctrl+	F-key	Alt+	Menu Bar	Tablet
resume	...	...	...	...	...	...

Command: resume

COMMAND LINE OPTIONS

BACKSPACE pauses the script file.

ESC stops the script file.

RELATED COMMANDS

RScript reruns the current script file.

Script loads and runs a script file.

RevCloud

2004 Draws revision clouds, and converts objects into revision clouds.

Command	Alias	Ctrl+	F-key	Alt+	Menu Bar	Tablet
revcloud	...	...	...	DU	Draw ↳Revision Cloud	...

Command: revcloud
Minimum arc length: 0.5000 Maximum arc length: 0.5000 Style: Normal
Specify start point or [Arc length/Object/Style] <Object>: s
Select arc style [Normal/Calligraphy] <Normal>: c
Arc style = Calligraphy
Specify start point or [Arc length/Object/Style] <Object>: *(Pick a point, or enter an option.)*
Guide crosshairs along cloud path... *(Move cursor to create cloud.)*
Revision cloud finished.

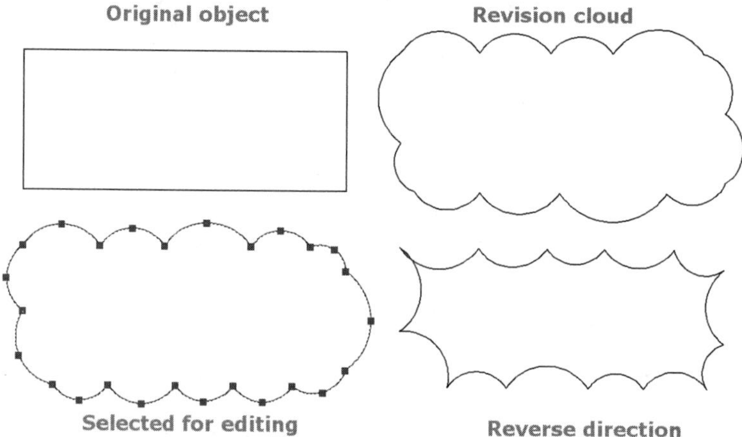

COMMAND LINE OPTIONS

Specify start point specifies the starting point of the cloud.

Guide crosshairs along cloud path indicates the outline of the cloud.

Arc Length options
Specify minimum length of arc <0.5000>: *(Enter a minimum value.)*
Specify maximum length of arc <1.0000>: *(Enter a maximum value.)*

Specify minimum length of arc specifies the minimum arc length.

Specify maximum length of arc specifies the maximum arc length.

Object options
Select object: *(Select one object to convert to a revision cloud.)*
Reverse direction [Yes/No] <No>: *(Type **Y** or **N**.)*

Select object selects the object to convert into a revision cloud.

Reverse direction turns the revision cloud inside-out.

Style options *(new to AutoCAD 2005)*
Select arc style [Normal/Calligraphy] <Normal>: *(Type* **N** *or* **C**.*)*

Normal draws clouds with uniform-width arcs.

Calligraphy draws clouds from variable-width polyarcs.

RELATED SYSTEM VARIABLE

DimScale affects the size and width of the arcs.

TIPS

- When the cursor reaches the start point, the revision cloud closes automatically.
- The revision cloud is drawn as a polyline.
- To edit the revision cloud, select it, and then move the grips.
- The arc length is not available from a system variable because it is stored in the Windows registry.
- The arc length is multiplied by the value stored in the **DimScale** system variable.

Revolve

Rel.11 Creates 3D solid objects by revolving closed 2D objects about an axis.

Command	Alias	Ctrl+	F-key	Alt+	Menu Bar	Tablet
revolve	**rev**	...	...	**DIR**	**Draw** ↳**Solids** ↳**Revolve**	**Q7**

Command: revolve

Current wire frame density: ISOLINES=4

Select objects: *(Select one or more closed objects.)*

Select objects: *(Press* **ENTER** *to end object selection.)*

Specify start point for axis of revolution or define axis by [Object/X (axis)/Y (axis)]: *(Pick a point, or enter an option.)*

Specify angle of revolution <360>: *(Enter a value.)*

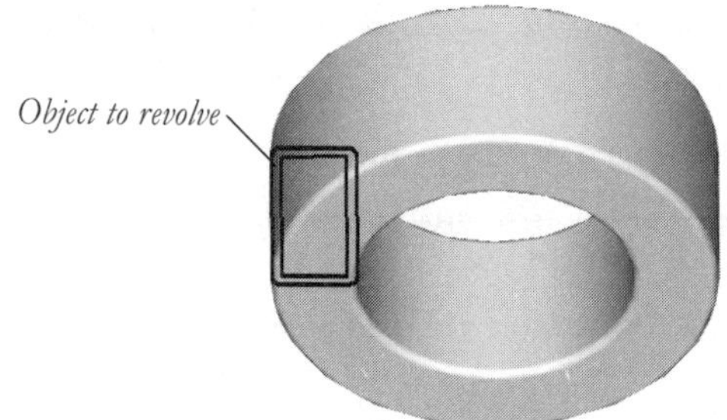

The two rectangles (left) were used to create the revolved object.

COMMAND LINE OPTIONS

Select objects selects a closed object to revolve: closed polyline, circle, ellipse, donut, polygon, closed spline, or a region.

Axis of revolution options

Specify start point for axis of revolution indicates the axis of revolution; you must specify the endpoint.

Object selects the object that determines the axis of revolution.

X uses the positive x axis as the axis of revolution.

Y uses the positive y axis as the axis of revolution.

Specify angle of rotation specifies the amount of rotation; full circle = 360 degrees.

RELATED COMMANDS

Extrude extrudes 2D objects into a 3D solid model.

Rotate rotates open and closed objects, forming a 3D surface.

TIPS

- **Revolve** works with just one object at a time.
- This command does not work with open objects or self-intersecting polylines.

RevSurf

Rel.10 Generates 3D surfaces of revolution defined by a path curve and an axis (*short for REVolved SURFace*).

Command	Alias	Ctrl+	F-key	Alt+	Menu Bar	Tablet
revsurf	...	...	...	DFS	**Draw** ↳**Surfaces** ↳**Revolved Surface**	O8

Command: revsurf
Current wire frame density: SURFTAB1=6 SURFTAB2=6
Select object to revolve: *(Select an object.)*
Select object that defines the axis of revolution: *(Select an object.)*
Specify start angle <0>: *(Enter a value.)*
Specify included angle (+=ccw, -=cw) <360>: *(Enter a value.)*

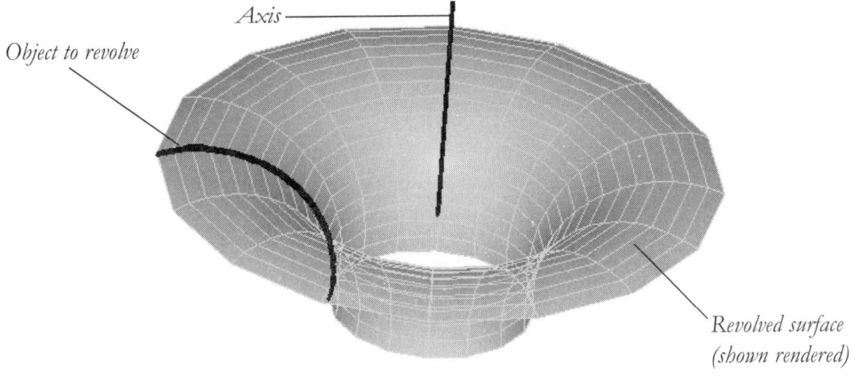

COMMAND LINE OPTIONS

Select object to revolve selects the single object that will be revolved about an axis.

Select object that defines the axis of revolution selects the axis object.

Start angle specifies the starting angle.

Included angle specifies the angle to rotate about the axis.

RELATED COMMANDS

PEdit edits revolved surfaces.

Revolve revolves a 2D closed object into a 3D solid.

RELATED SYSTEM VARIABLES

SurfTab1 specifies the mesh density in the m-direction (default = 6).

SurfTab2 specifies the mesh density in the n-direction (default = 6).

TIPS

- Unlike the **Revolve** command, **RevSurf** works with open and closed objects.
- If a multi-segment polyline is the axis of revolution, the rotation axis is defined as the vector pointing from the first vertex to the last vertex, ignoring the intermediate vertices.

RMat

Rel.13 Applies material definitions to colors, layers, and objects; used by the **Render** command (*short for Render MATerials*).

Command	Alias	Ctrl+	F-key	Alt+	Menu Bar	Tablet
rmat	...	...	...	**VEM**	**View** ↳**Render** ↳**Materials**	**P1**

Command: rmat

Displays dialog box:

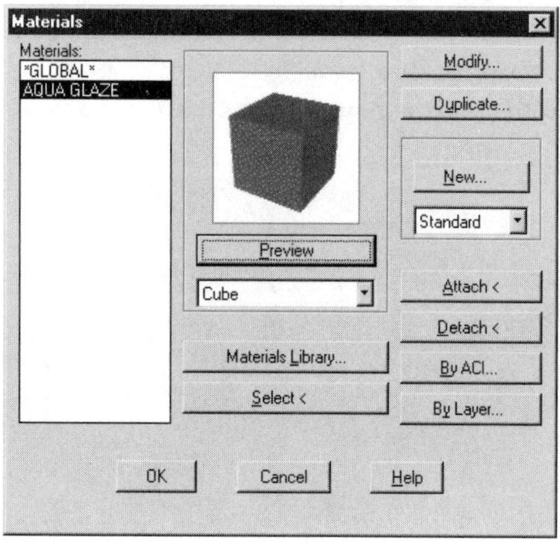

DIALOG BOX OPTIONS

Materials lists the names of materials loaded into drawing by the **MatLib** command.

Preview previews the material mapped to a sphere or cube.

Materials Library displays the Materials Library dialog box; see the **MatLib** command.

Select selects the objects to which to attach the material definition.

Modify edits a material definition; displays a different Modify Material dialog box for standard, granite, marble and wood-based materials; see the New option.

Duplicate duplicates a material definition so that you can edit it; displays a different Modify Material dialog box for standard, granite, marble and wood-based materials; see the New option.

New creates a new material definition; displays a different dialog box for the four types of materials: standard, granite, marble, wood.

Attach selects the objects to which to attach the material definition.

Detach selects the objects from which to detach the material definition.

By ACI attaches material via ACI number; displays dialog box:

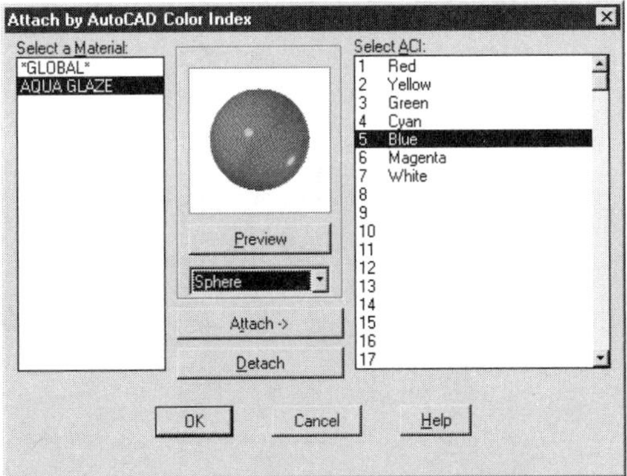

By Layer attaches material to a layer name; displays dialog box:

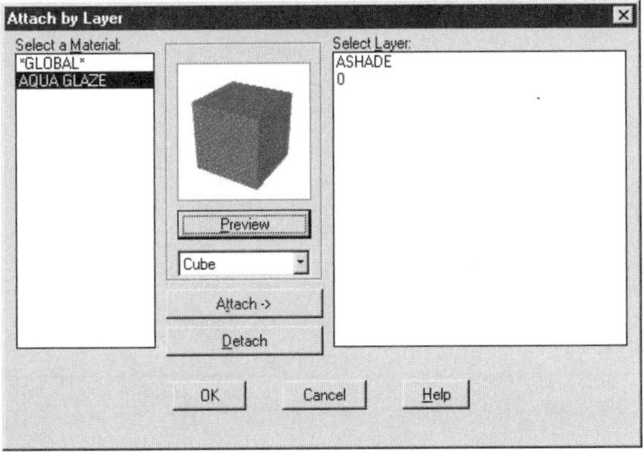

New Material options

Material Name names the material.

Value adjusts the value of selected Attributes between 0 and a larger number.

Color options

☐ **By ACI** makes the material's base color the same as the object's.

Lock locks the ambient and reflective colors to the base color.

Mirror allows mirrored reflection.

Color System selects RGB or HLS color system:

- **HLS** specifies color by levels of Hue, Luminescence, and Saturation.
- **RGB** specifies color by levels of Red, Green, Blue.

Bitmap Blend determines the degree the bitmap is rendered.

File Name specifies the bitmap's filename.

Adjust Bitmap displays the Adjust Material Bitmap Placement dialog box.

Find File selects the bitmap file; displays the Bitmap File dialog box.

New Standard Material dialog box

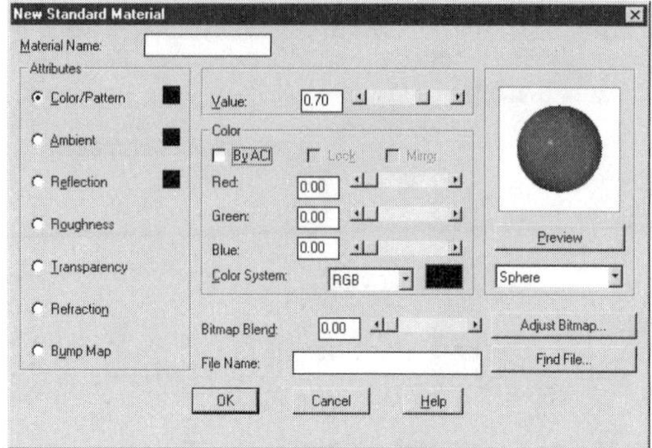

Attributes options

⊙**Color/Pattern** sets the base color of the material.

○**Ambient** sets the ambient color shown in shadowed areas.

○**Reflection** sets the highlight color of the material.

○**Roughness** sets the size of the highlighted area; the higher the value of the roughness, the larger the highlight area.

○**Transparency** sets the amount of transparency of the material:

Transparency	Meaning
0.0	No transparency.
0.1 *through* **9.9**	Increasing transparency with edge fall-off.
1.0	Perfect transparency.

○**Refraction** controls refraction of the material; Photo Raytrace rendering only.

○**Bump Map** attaches a bitmap to the material.

New Granite Material dialog box

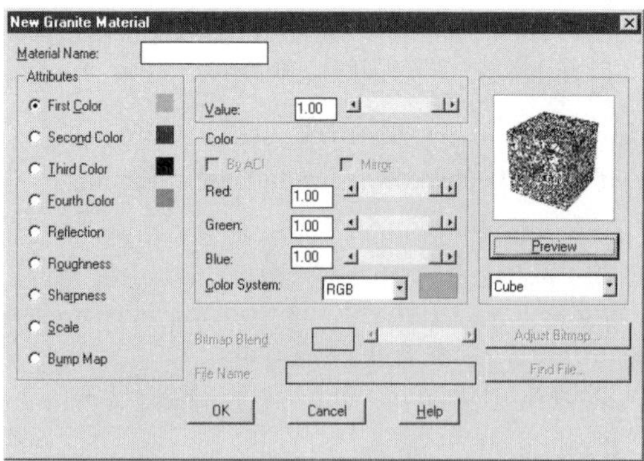

Attributes options

⊙**First Color** sets the base color of the granite material.

○**Second Color** sets the second color of the granite material.

○ **Third Color** sets the third color of the granite material.

○ **Fourth Color** sets the fourth color of the granite material.

○ **Reflection** sets the highlight color of the material.

○ **Roughness** sets the size of the highlighted area; the higher the value of the roughness, the larger the highlighted area.

○ **Sharpness** sets the sharpness of the four colors of the material:

Sharpness	Meaning
0.0	Complete blurring of colors.
0.1 *through* **9.9**	Increasing sharpness of colors.
1.0	All four colors perfectly distinct.

○ **Scale** sizes the material relative to its attached object.

○ **Bump Map** attaches a bitmap to the material.

New Marble Material dialog box

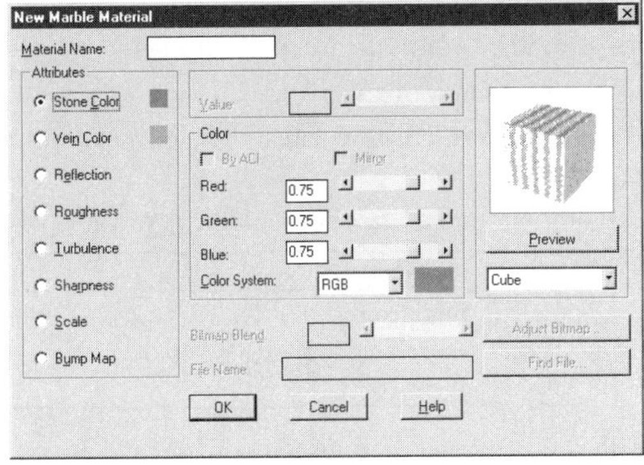

Attributes options

⦿ **Stone Color** sets the base color of the marble material.

○ **Vein Color** sets the secondary color of the marble material.

○ **Reflection** sets the highlight color of the material.

○ **Roughness** sets the size of the highlighted area; the higher the value of the roughness, the larger the highlighted area.

○ **Turbulence** determines the swirling of the vein color.

○ **Sharpness** sets the sharpness of the four colors of the material:

Sharpness	Meaning
0.0	Complete blurring of colors.
0.1 *through* **9.9**	Increasing sharpness of colors
1.0	All colors perfectly distinct.

New Wood Material dialog box

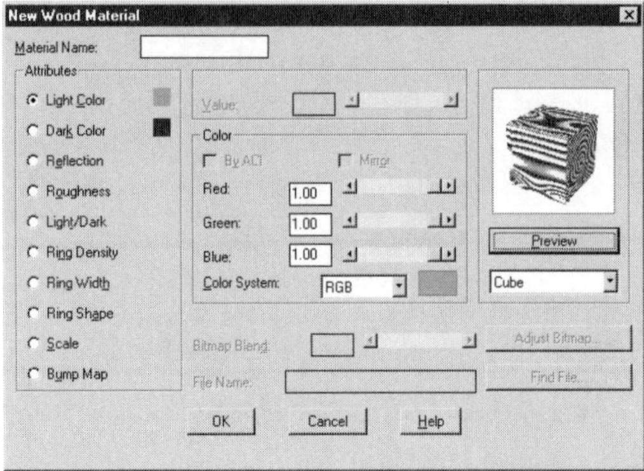

Attributes options

⊙ **Light Color** sets the base color of the wood material.

○ **Dark Color** sets the secondary color of the wood material.

○ **Reflection** sets the highlight color of the material.

○ **Roughness** sets the size of the highlighted area; the higher the value of the roughness, the larger the highlight area.

○ **Light/Dark** sets the amount of contrast between the two colors of the material:

Light/Dark	Meaning
0.0	Dark color.
0.1 *through* **9.9**	Increasing lightness of color.
1.0	Light color.

○ **Ring Density** specifies the number of rings in the material definition.

○ **Ring Width** specifies the variation in ring width; 0 = narrow rings; 1.0 = wide rings.

○ **Ring Shape** specifies 0 = circular rings; 1.0 = irregular rings.

○ **Scale** sizes the material relative to its attached object.

○ **Bump Map** attaches a bitmap to the material.

RELATED COMMANDS

MatLib loads material definitions into the drawing.

Render renders the drawing using material definitions.

TIPS

- You must use the commands in this order:

    1. **MatLib** loads materials into the drawing.
    2. **RLib** attaches materials to objects.
    3. **Render** with **Apply Materials** turned on.

- The **By ACI** option lets you attach a material definition to all objects of one color.
- The **By Layer** option lets you attach a material definition to all objects on one layer.

RMLin

2000i Inserts redline markup files from Autodesk's Volo products (*short for Redline Markup Language INput*).

Commands	Alias	Ctrl+	F-key	Alt+	Menu Bar	Tablet
rmlin	...	...	...	**IU**	**Insert** ↳**Markup**	...
-rmlin						

Command: rmlin

Displays Insert Markup dialog box.

Select a .rml file, and then click ***OK****.*

-RMLIN Command

Command: -rmlin

Enter name of RML file: *(Enter the name of file.)*

COMMAND LINE OPTION

Enter name of RML file specifies the name of the redline markup file to import.

RELATED COMMANDS

Markup displays *.dwf* files containing markups created by Composer software.

OpenDwfMarkup opens marked-up *.dwf* files from Composer.

DxfIn inserts a *.dxf* file, which can contain redline markups from other software.

TIPS

- When an *.rml* file is inserted in the drawing, AutoCAD places it on a new layer called "_Markup_." This layer has the following properties:

Property	Value
Name	Markup.
Color	Red.
Locked/unlocked	Locked.

- Redlines or markup are comments added to drawings; redlines are not normally part of the drawing. Redlines consists of notes, circles, arrows, hyperlinks, and so on.
- This command is superceeded by the **Markup** command, as of AutoCAD 2005.

Rotate

V. 2.5 Rotates objects about an axis perpendicular to the current UCS.

Command	Alias	Ctrl+	F-key	Alt+	Menu Bar	Tablet
rotate	ro	...	...	MR	Modify ↳Rotate	V20

Command: rotate
Current positive angle in UCS: ANGDIR=counterclockwise ANGBASE=0
Select objects: *(Select one or more objects.)*
Select objects: *(Press* **ENTER** *to end object selection.)*
Specify base point: *(Pick a point.)*
Specify rotation angle or [Reference]: *(Enter a value, or type* **R***.)*

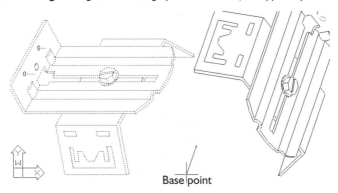

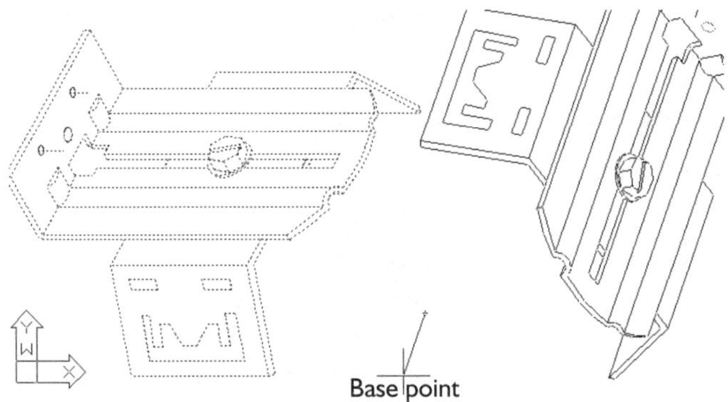

COMMAND LINE OPTIONS

Select objects selects the objects to be rotated.

Specify base point picks the point about which the objects will be rotated.

Specify rotation angle specifies the angle by which the objects will be rotated.

Reference allows you to specify the current rotation angle and new rotation angle.

RELATED COMMANDS

Rotate3D rotates objects in 3D space.

UCS rotates the coordinate system.

Snap rotates the cursor.

TIPS

- AutoCAD rotates the selected object(s) about the base point.
- At the 'Specify rotation angle' prompt, you can show the rotation by moving the cursor. AutoCAD dynamically displays the new rotated position as you move the cursor.
- Use object snap modes, such as **INTersection**, to position the base point and rotation angle(s) accurately.

Rotate3D

Rel.11 Rotates objects about any axis in 3D space.

Command	Alias	Ctrl+	F-key	Alt+	Menu Bar	Tablet
rotate3d	...	...	...	M3R	Modify ↳3D Operation ↳Rotate 3D	W22

Command: rotate3d
Current positive angle: ANGDIR=counterclockwise ANGBASE=0
Select objects: *(Select one or more objects.)*
Select objects: *(Press* **ENTER** *to end object selection.)*
Specify first point on axis or define axis by [Object/Last/View/Xaxis/Yaxis/Zaxis/2points]: *(Pick a point, or enter an option.)*
Specify second point on axis: *(Pick another point.)*
Specify rotation angle or [Reference]: *(Enter a value, or type* **R**.*)*

COMMAND LINE OPTIONS

Select objects selects the objects to be rotated.

Specify rotation angle rotates objects by a specified angle: relative rotation.

Reference specifies the starting and ending angle: absolute rotation.

Define Axis By options

Object selects object to specify the rotation axis.

Last selects the previous axis.

View specifies that the current view direction is the rotation axis.

Xaxis specifies that the x axis is the rotation axis.

Yaxis specifies that the y axis is the rotation axis.

Zaxis specifies that the z axis is the rotation axis.

2 points defines two points on the rotation axis.

RELATED COMMANDS

Align rotates, moves, and scales objects in 3D space.

Mirror3d mirrors objects in 3D space.

Rotate rotates objects in 2D space.

RPref

Rel.12 Specifies options for the **Render** command (*short for Render PREFerences*).

Command	Alias	Ctrl+	F-key	Alt+	Menu Bar	Tablet
rpref	**rpr**	...	...	**VEP**	**View** ↳**Render** ↳**Preferences**	**R2**

Command: rpref

Displays dialog box:

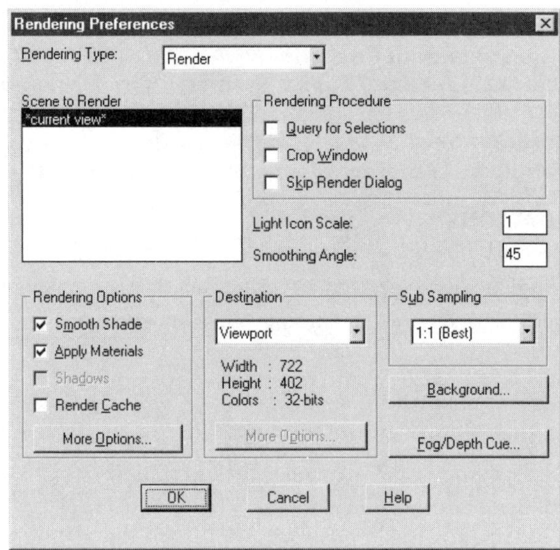

DIALOG BOX OPTIONS

*Options are identical to the **Render** command, with the exception of the **OK** button replacing the **Render** button.*

RELATED COMMANDS

All rendering-related commands.

Hide removes hidden lines from wireframe view.

Render renders the scene.

ShadEdge produces real-time shading of 3D objects.

3dOrbit creates perspective view.

TIP

- If you set up the **Render** command to skip the dialog box, use the **RPref** command to set rendering options.

'RScript

V. 2.0 Repeats script files (*short for Repeat SCRIPT*).

Command	Alias	Ctrl+	F-key	Alt+	Menu Bar	Tablet
rscript	...	...	...	...	...	...

Command: rscript

COMMAND LINE OPTIONS

None.

RELATED COMMANDS

Resume resumes a script file after being interrupted.

Script loads and runs a script file.

TIP

- Placed at the end of a script file, this command causes the script file to run repeatedly until cancelled with **BACKSPACE** or **ESC**.

RuleSurf

Rel.10 Draws 3D ruled surfaces between two objects (*short for RULEd SURFace*).

Command	Alias	Ctrl+	F-key	Alt+	Menu Bar	Tablet
rulesurf	...	...	...	**DFR**	**Draw** **⇘Surfaces** **⇘Ruled Surface**	**Q8**

Command: rulesurf
Select first defining curve: *(Pick an object.)*
Select second defining curve: *(Pick another object.)*

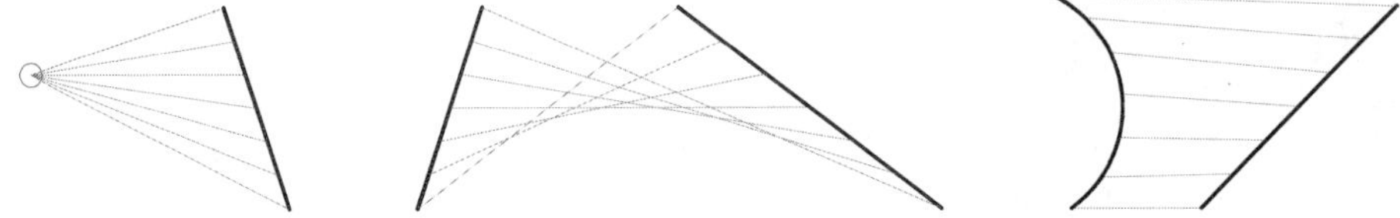

Ruled surfaces created between: point and line; line and line; arc and line.

COMMAND LINE OPTIONS

Select first defining curve selects the first object for the ruled surface.

Select second defining curve selects the second object for the ruled surface.

RELATED COMMANDS

EdgeSurf draws a 3D surface bounded by four edges.

RevSurf draws a 3D surface of revolution.

TabSurf draws a 3D tabulated surface.

RELATED SYSTEM VARIABLE

SurfTab1 determines the number of faces drawn.

TIPS

- **RuleSurf** uses these objects as the boundary curve: points, lines, arcs, circles, polylines, and 3D polylines.
- Pick order is important: the ruled surface is drawn starting at the endpoint nearest to the pick point. This can result in a twisted surface, as illustrated by the line-line example above.
- Both boundaries must either be closed or open; the exception is using the point object.
- The **RuleSurf** command begins drawing its mesh as follows:

Object	RuleSurf
Open object	From the object's endpoint closest to your pick.
Circle	From the zero-degree quadrant.
Closed polyline	From the last vertex.

- **RuleSurf** draws with a circle in the opposite direction of a closed polyline.

Save / SaveAs

V. 2.0 Saves the current drawing to disk, after prompting for a filename.

Commands	Alias	Ctrl+	F-key	Alt+	Menu Bar	Tablet
save	**saveurl**	...	...	...	...	...
saveas	...	**Shift+S**	...	**FA**	**File** ↳**Save As**	**V24**

Command: save *or* saveas

Displays dialog box.

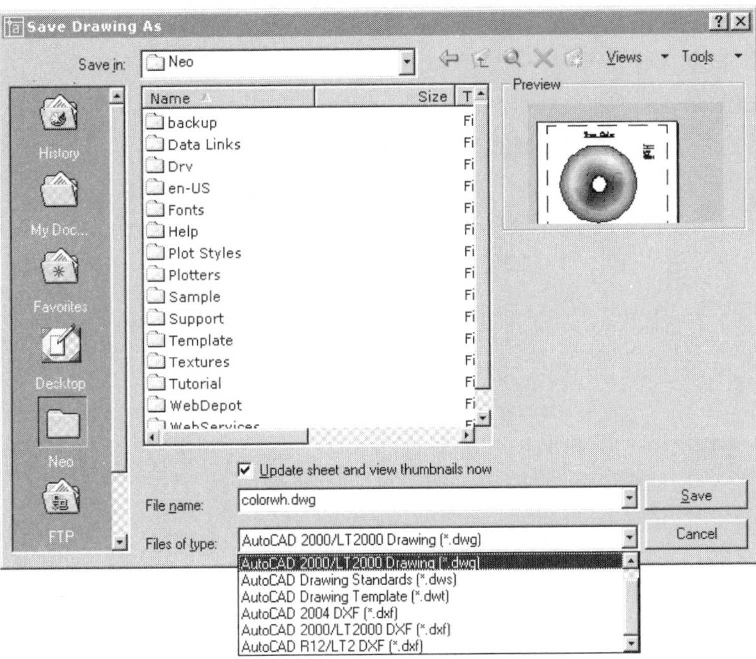

DIALOG BOX OPTIONS

Save in selects the folder, hard drive, or network drive in which to save the drawing.

☑ **Update sheet and view thumbnails now** *(new to AutoCAD 2005).*

File name names the drawing; maximum = 255 characters; default=*drawing1.dwg.*

Save saves the drawing, and then returns to AutoCAD.

Cancel does not save the drawing.

When a drawing of the same name already exists in the same folder, displays dialog box:

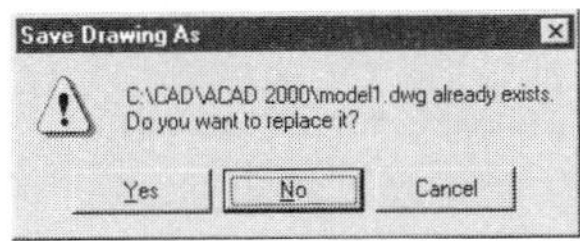

Files of type saves the drawing in a variety of formats:

Save as type	Meaning
AutoCAD 2004 DWG	Saves drawings in native format.
AutoCAD 2000/LT 2000	Saves drawings in AutoCAD 2000/2000i/ 2002 format.
Drawing Standards	Saves drawings as CAD standards.
Drawing Template	Saves drawings in the *template* folder.
AutoCAD 2004 DXF	Exports drawings in 2004 DXF format.
AutoCAD 2000/LT 2000 DXF	Exports drawings in 2000/2000i/2002 DXF format.
R12, LT 2 DXF	Exports drawings in R12 DXF format.

Saveas Options dialog box

From the toolbar, select ***Tools | Options:***

DWG Options tab

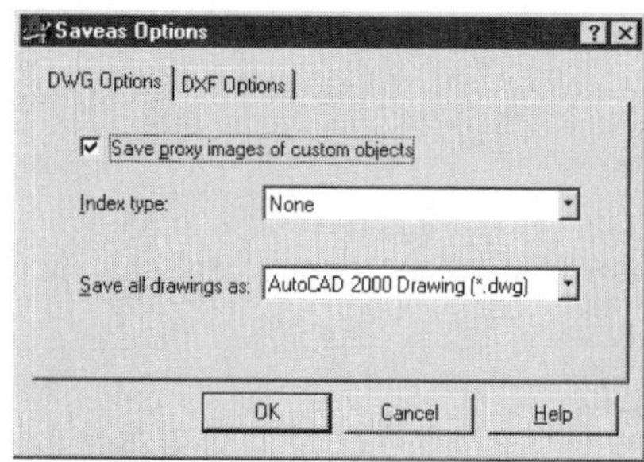

Save proxy images of custom objects option:

☑ Saves an image of the custom objects. in the drawing file.

☐ Creates a frame around each custom object, instead of the image.

Index type saves indices with drawings; useful only for xrefed and partially-loaded drawings:

Index type	Meaning
None	Creates no indices (default).
Layer	Loads layers that are on and thawed.
Spatial	Loads only the visible portion of a clipped xref.
Layer and Spatial	Combines the above two options.

Save all drawings as selects the default format for saving drawings.

DXF Options tab

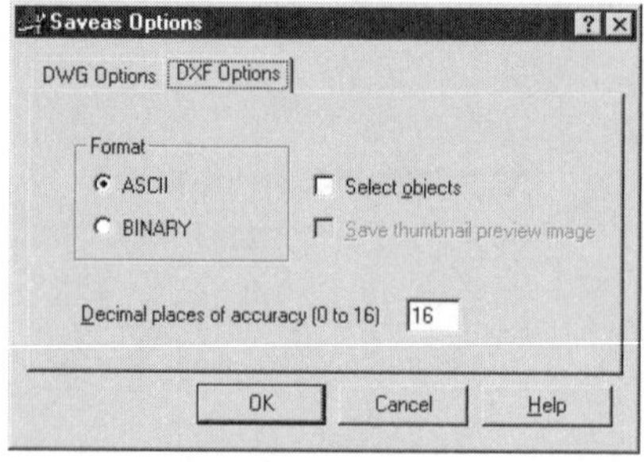

Format options

⦿ **ASCII** saves the drawing in ASCII format, which can be read by humans – as well as common text editors – but takes up more disk space.

○ **BINARY** saves the drawing in binary format, which takes up less disk space, but cannot be read by some software programs.

Additional options

☐ **Select objects** allows you to save selected objects to the DXF file; AutoCAD prompts you, 'Select objects:'.

☐ **Save thumbnail preview image** saves a thumbnail image of the drawing.

Decimal places of accuracy specifies the number of decimal places for real numbers: 0 to 16.

Template Description dialog box

Displayed when drawing is saved as .dwt template file:

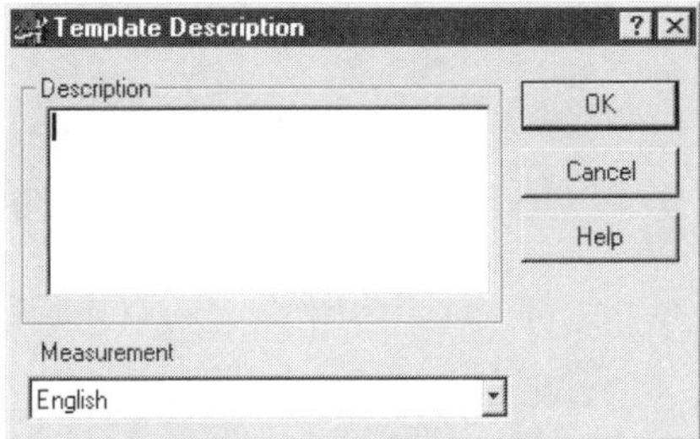

Description describes the template file; stored in *wizard.ini*.

Measurement selects English or metric as the default measurement system.

RELATED COMMANDS

Export saves the drawing in a variety of raster and vector formats.

Plot saves the drawing in even more raster and vector formats.

Quit exits AutoCAD without saving the drawing.

QSave saves the drawing without prompting for a name.

RELATED SYSTEM VARIABLES

DBMod indicates that the drawing was modified since being opened.

DwgName specifies the name of the drawing; *drawing1.dwg* when unnamed.

TIPS

- The **Save** and **SaveAs** commands perform exactly the same function.
- Use these commands to save the drawing by another name.
- When a drawing is opened as read-only, use these commands to save the drawing to another filename.
- To save drawings without seeing the Save Drawing As dialog box, use the **QSave** command.
- As of AutoCAD 2004, you can use CTRL+SHIFT+S to display the Save Drawing As dialog box.

Removed Command

SaveAsR12 was removed from AutoCAD Release 14. Use **SaveAs** instead, which saves drawings in earlier formats.

SaveImg

Rel.12 Saves the current viewport image — raster and vector — as BMP, TIFF, or Targa images on disk (*short for Save Image*).

Command	Alias	Ctrl+	F-key	Alt+	Menu Bar	Tablet
saveimg	...	...	...	TIS	Tools ↳Display Image ↳Save	V7

Command: saveimg

Displays dialog box:

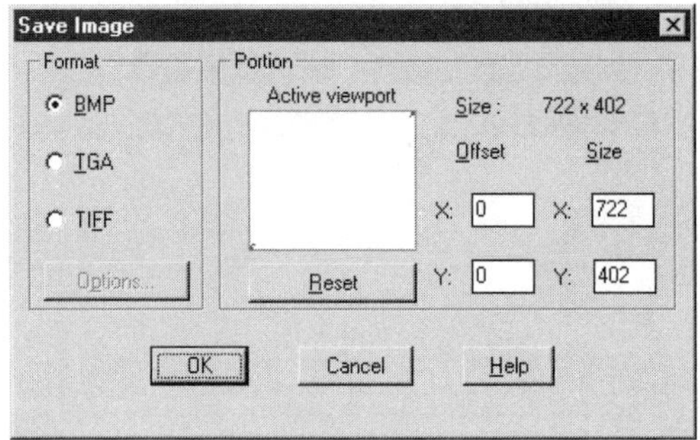

DIALOG BOX OPTIONS

Format options

⊙**BMP** saves in Windows bitmap format.

O**TGA** saves in Targa format.

O**TIFF** saves in Tagged image file format.

Options displays the Options dialog box.

Portion options

Active viewport selects the area of the current viewport to save.

Offset specifies the lower-left x,y coordinates of the image area (default = 0,0).

Size specifies the upper-right x,y coordinates of the image area.

TGA Options dialog box

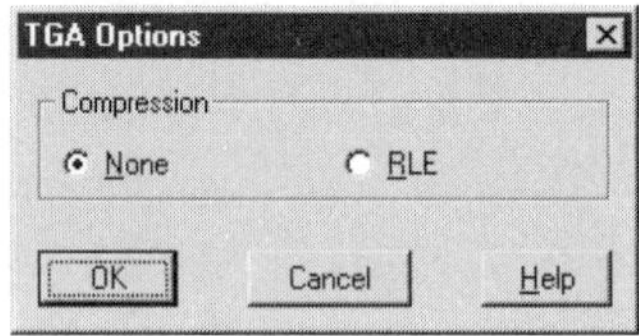

Compression options

⊙ **None** saves *.tga* files with no file compression.

O **RLE** saves *.tga* files with run-length encoded compression.

TIFF Options dialog box

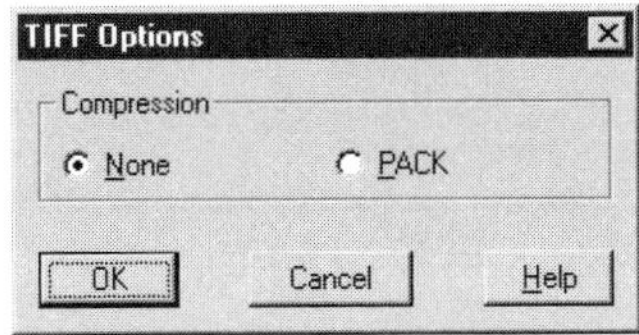

Compression options

⊙**None** saves TIFF files with no compression.

○**PACK** saves TIFF files with pack bits compression.

RELATED COMMANDS

Replay displays a raster image in the current viewport.

JpgOut saves the drawing in JPEG raster format.

PngOut saves the drawing in PNG raster format.

TifOut saves the drawing in TIFF raster format.

WmfOut saves the drawing in WMF vector format.

PRT SCR saves the entire screen to the Clipboard.

TIPS

- The **SaveImg** command saves files in these formats:

Format	Specification
BMP	24-bit Windows bitmap.
TGA	32-bit RGBA TrueVision v2.0.
TIFF	32-bit RGBA tagged image file.

- The GIF format, which is used to display images on the Internet and was found in AutoCAD Release 12 and 13, was replaced by the BMP format in AutoCAD Release 14.
- After using the **SaveImg** command to create images of the drawing, use the **Replay** command to display them in AutoCAD.

Removed Command

SaveUrl was removed from AutoCAD 2000; it was replaced by **Save** and **SaveAs**.

Scale

V. 2.5 Changes the size of selected objects, to make them smaller or larger.

Command	Alias	Ctrl+	F-key	Alt+	Menu Bar	Tablet
scale	**sc**	...	...	**ML**	**Modify** ↳**Scale**	**V21**

Command: scale
Select objects: *(Select one or more objects.)*
Select objects: *(Press* **ENTER** *to end object selection.)*
Specify base point: *(Pick a point.)*
Specify a scale factor or [Reference]: *(Enter scale factor, or type* **R***.)*

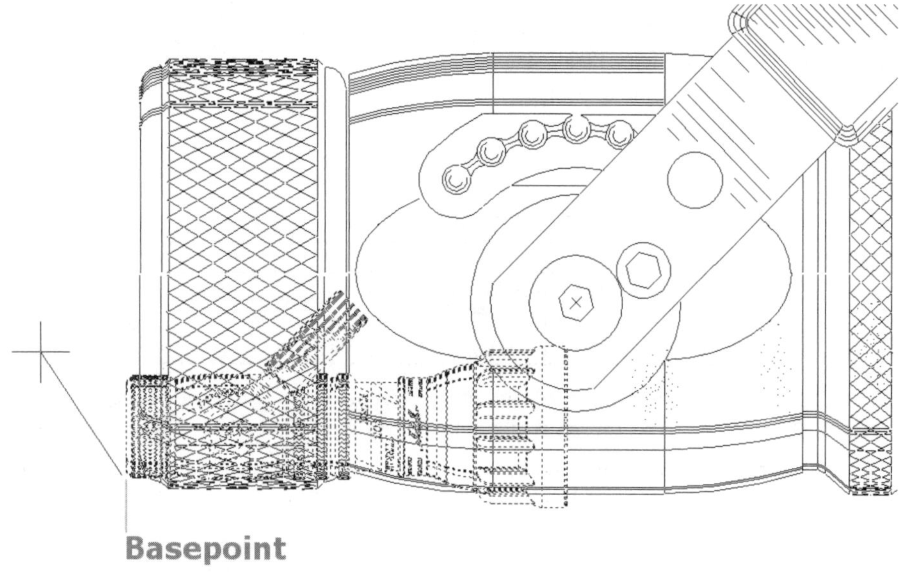

COMMAND LINE OPTIONS

Base point specifies the point from which scaling takes place.

Reference requests a distance, followed by a new distance.

Scale factor indicates the scale factor, which applies equally in the x and y directions.

Scale Factor	Meaning
> **1.0**	Enlarges object(s).
= **1.0**	Makes no change.
> **0.0** *and* < **1.0**	Reduces object(s).
= **0.0** *or negative*	Illegal values.

Reference options
Specify reference length <1>: *(Enter a base length.)*
Specify new length: *(Enter a new length.)*

Specify reference length specifies a baseline length; you can enter a value, or pick two points.

Specify new length specifies the new length that results in determining the scale factor.

RELATED COMMANDS

Align scales an object in 3D space.

Insert allows a block to be scaled independently in the x, y, and z directions.

Plot allows a drawing to be plotted at any scale.

TIPS

- You can interactively scale the object by moving the cursor to "grow" the object larger and smaller.
- Scale factors larger than 1.0 grow the object; smaller than 1.0 shrink the object.
- This command changes the size in all dimensions (x,y,z) equally; to change an object in just one dimension, use the **Stretch** command.
- The **Reference** option is useful for scaling a raster image to match the size of a drawing, or for changing the units in a drawing. For example, 1:12 changes decimal feet into inches; 2.54:1 changes cm to inches.

ScaleText

2002 Changes the height of text objects relative to their insertion points.

Command	Alias	Ctrl+	F-key	Alt+	Menu Bar	Tablet
scaletext	...	...	...	**MOTS**	**Modify** ↳**Object** ↳**Text** ↳**Scale**	...

Command: scaletext
Select objects: *(Select one or more objects.)*
Select objects: *(Press* **ENTER** *to end object selection.)*
Enter a base point option for scaling [Existing/Left/Center/Middle/Right/TL/TC/TR/ML/MC/MR/BL/BC/BR] <Existing>: *(Enter an option, or press* **ENTER** *to keep insertion point as is.)*
Specify new height or [Match object/Scale factor] <0.2000>: *(Enter a value, or an option.)*

COMMAND LINE OPTIONS

Enter a base point option for scaling specifies the point from which scaling takes place.

Existing uses the existing insertion point for each text object.

Specify new height specifies the new height of the text.

Match object matches the height of another text object.

Scale factor scales the text by a factor:

Scale Factor	Meaning
> 1.0	Enlarges text.
1.0	Makes no change in size.
> 0.0 *and* **< 1.0**	Reduces text.
0.0 *or negative*	Illegal values.

Match Object options
Select a text object with the desired height: *(Pick another text object.)*
Height=*nnn*

Select a text object with the desired height selects another text object to match its height.

Scale Factor options
Specify scale factor or [Reference] <2.0000>: *(Enter a value, or type* **R***.)*

Specify scale factor scales the text by a factor; see **Scale** command.

Reference supplies a reference value, followed by a new value.

RELATED COMMANDS

Justify changes the justification of the text.

Properties changes all aspects of the text.

Style creates text styles.

TIP

- *Caution!* Scaling text larger may make it overlap nearby text.

Scene

Rel.12 Collects lights and a view into a named scene for renderings.

Command	Alias	Ctrl+	F-key	Alt+	Menu Bar	Tablet
scene	...	...	...	VES	View ↳Render ↳Scene	N1

Command: scene

Displays dialog box:

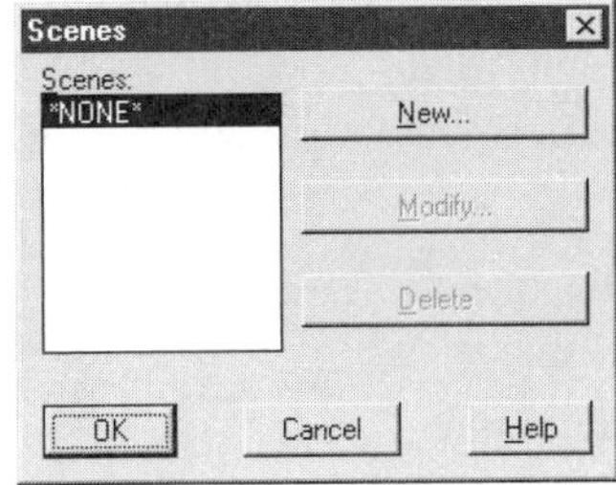

DIALOG BOX OPTIONS

New creates new named scenes; displays the New Scene dialog box.

Modify changes existing scene definitions; displays the Modify Scene dialog box.

Delete deletes scenes from drawings; displays dialog box:

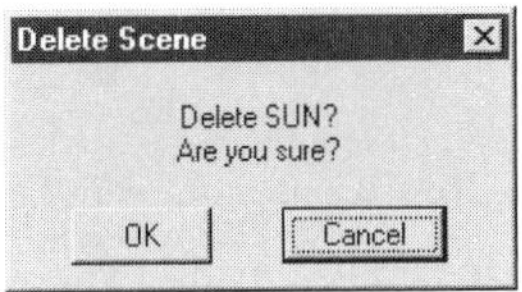

New Scene dialog box

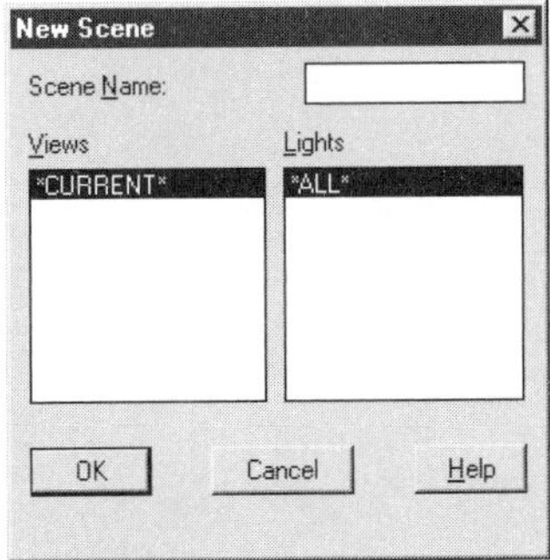

Scene Name specifies a name for the scene; maximum = 8 characters.

Views selects one named view or *CURRENT*, the current view.

Lights selects zero or more lights, or *ALL*, all lights.

Modify Scene dialog box

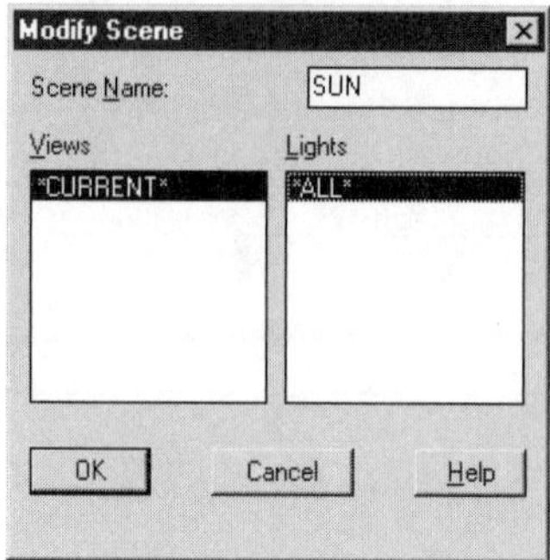

Scene Name changes the name of the scene; maximum = 8 characters.

Views selects a named view.

Lights selects one or more lights.

RELATED COMMANDS

Light places lights in the drawing for the **Scene** command.

View creates named views.

Render uses scenes to create renderings.

RPref specifies options for renderings.

View creates named views for the **Scene** command.

TIPS

- If the *CURRENT* view is selected, it is the view current at rendering time, not the current view at the time the scene was created.
- If *ALL* lights are selected, it is all the lights available at rendering time, not all the lights at the time the scene was created.
- If you select no lights in a scene, **Render** uses ambient light.
- Scene parameters are stored as extended entity data in invisible blocks on the ASHADE layer.

'Script

V. 1.4 Runs ASCII files containing sequences of AutoCAD commands and options.

Command	Alias	Ctrl+	F-key	Alt+	Menu Bar	Tablet
script	scr	...	...	TR	Tools ↳Run Script	V9

Command: script

Displays the Select Script File dialog box. Select an .scr file, and then click ***Open****.*

Script file begins running as soon as it is loaded.

COMMAND LINE OPTIONS

BACKSPACE interrupts the script.

ESC stops the script.

RELATED COMMANDS

Delay specifies the delay in milliseconds; pauses the script before executing the next command.

Resume resumes a script after it is interrupted.

RScript repeats script files.

TIPS

- Since the **Script** command is transparent, it can be used during another command.
- Use the **/s** command-line switch to run a script when AutoCAD starts.
- Prefix the **VSlide** command with * to preload it; this results in a faster slide show:

```
*vslide
```

- You can make a script file more flexible — such as pausing for user input, branching with conditionals — by inserting AutoLISP functions.
- Dialog boxes, toolbars, menus, and tool palettes cannot be controlled by scripts. Toolbar and menu macros can contain scripts, however.

Section

Rel.11 Creates 2D region objects from the intersection of a plane and a 3D solid.

Command	Alias	Ctrl+	F-key	Alt+	Menu Bar	Tablet
section	sec	...	...	DIE	Draw ↳Solid ↳Section	...

Command: section
Select objects: *(Select one or more objects.)*
Select objects: *(Press* **ENTER** *to end object selection.)*
Specify first point on section plane by [Object/Zaxis/View/XY/YZ/ZX/ 3points]: *(Pick a point, or enter an option.)*

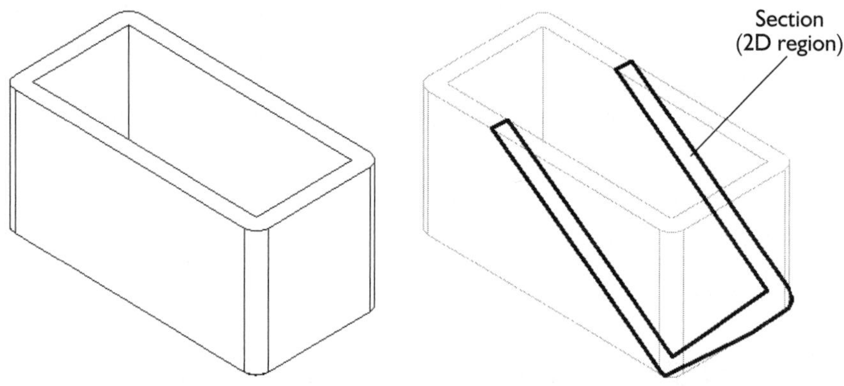

COMMAND LINE OPTIONS

Select objects selects the 3D solid objects to be sectioned.

Object aligns the section plane with a selected object: circle, ellipse, arc, elliptical arc, 2D spline, or 2D polyline.

Zaxis specifies the normal (*z axis*) to the section plane.

View uses the current view plane as the section plane.

XY uses the x,y plane of the current view.

YZ uses the y,z plane of the current view.

ZX uses the z,x plane of the current view.

3 points picks three points to specify the section plane.

RELATED COMMAND

Slice cuts a slice out of a solid model, creating another 3D solid.

TIPS

- Section regions are placed on the current layer, not the object's layer.
- Regions are ignored.
- One cutting plane is required for each selected solid.

SecurityOptions

2004 Sets passwords and digital signatures for securing drawings against unauthorized access.

Command	Alias	Ctrl+	F-key	Alt+	Menu Bar	Tablet
securityoptions	...	...	...	...	...	...

Command: securityoptions

Displays dialog box:

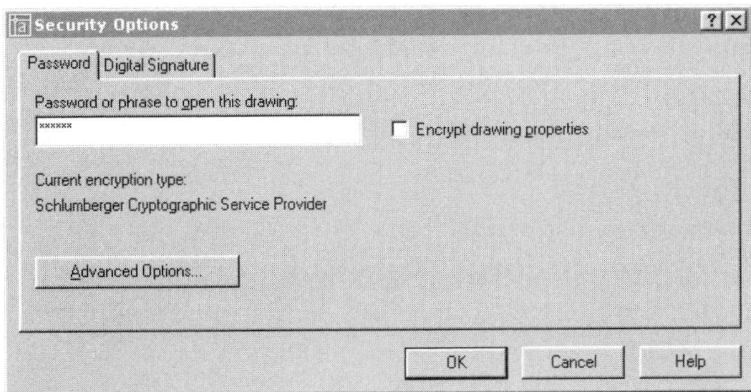

DIALOG BOX OPTIONS

Password tab

Password or phrase to open this drawing assigns a password to lock the drawing.

☐ **Encrypt drawing properties** encrypts drawing properties data; see the **DwgProps** command.

Advanced Options displays the Advanced Options dialog box; available only after a password is entered.

OK displays the Confirm Password dialog box:

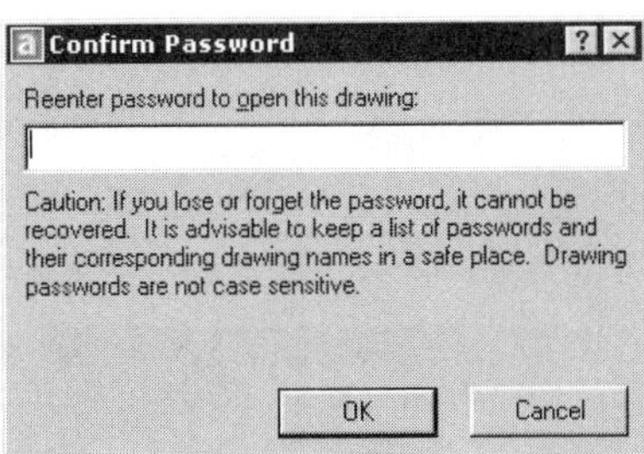

Digital Signature tab

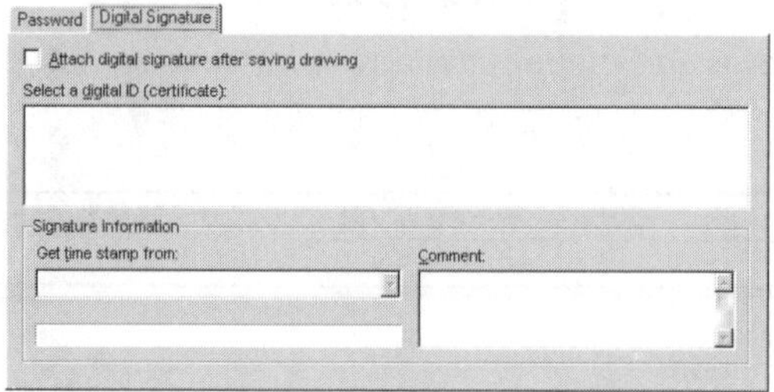

☐ **Attach digital signature after saving drawing** attaches the digital signature the next time the drawing is saved.

Select a digital ID (certificate) lists digital signature services.

Signature information options

Get time stamp from lists time servers.

Comment provides additional comments to be included with the digital signature.

Advanced Options dialog box

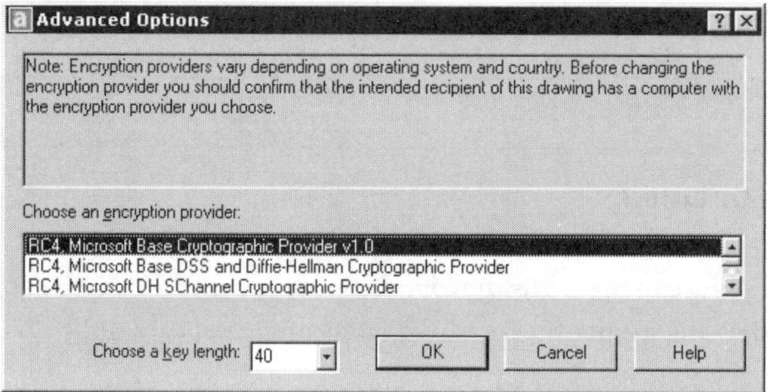

Choose an encryption provider selects from several encryption types.

Choose a key length selects 40, 48, or 56 bits of encryption.

RELATED COMMAND

SigValidate displays information about the digital signatures attached to drawings.

TIPS

- When opening drawings protected by passwords, AutoCAD displays the **Password** dialog box for entering the password:

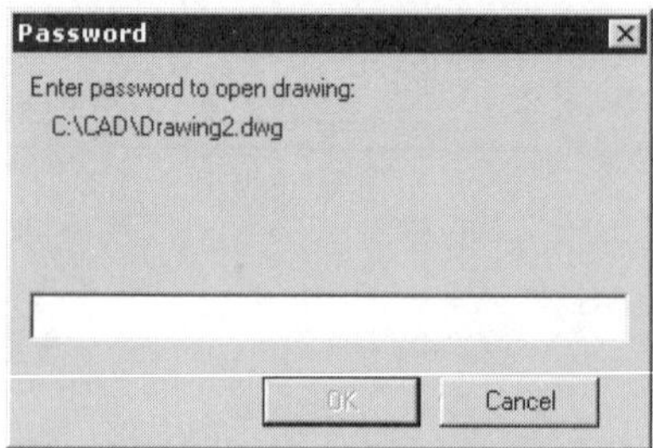

- When the password entered is correct, AutoCAD opens the drawing.
- When the password entered is incorrect, AutoCAD complains:

- To remove the password, in the **Security Options** dialog box, remove the password, and then click **OK**.
- The first time you assign a digital signature to the drawing, you are prompted to obtain one:

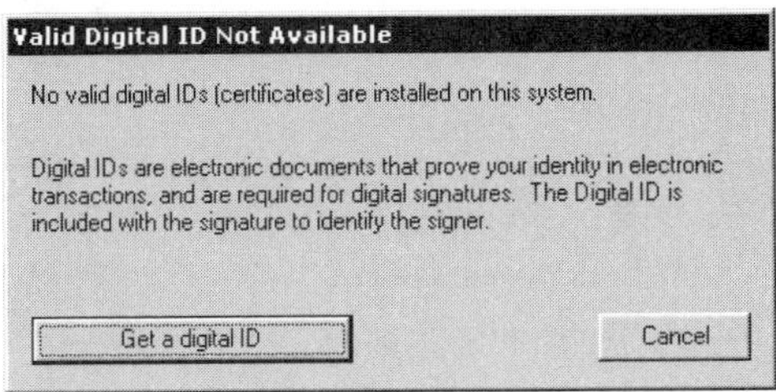

- *Caution!* If the password is lost, the drawing is not recoverable. Autodesk recommends that you (1) save a backup copy of the drawing not protected with any password; and (2) maintain a list of password names stored in a secure location.
- *Caution!* The encryption provider must be the same on the receiving computers; otherwise the drawings cannot be opened.
- To remove this feature from AutoCAD, install with the **Custom** option, and then unselect **Drawing Encryption**.

Select

V. 2.5 Creates a selection set of objects based on their location in the drawing.

Command	Alias	Ctrl+	F-key	Alt+	Menu Bar	Tablet
select	...	A	...	...	...	...

Command: select
Select objects: *(Select one or more objects, or enter an option.)*
Select objects: *(Press* **ENTER** *to end object selection.)*

COMMAND LINE & TOOLBAR OPTIONS

pick selects a single object.

AU switches from pick mode to C or W mode, depending on whether an object is found at the initial pick point (short for AUtomatic).

ALL selects all objects in the drawing; press CTRL+A to select all objects.

BOX goes into C or W mode, depending on how the cursor moves.

C selects objects in and crossing the selection box (Crossing).

CP selects all objects inside and crossing the selection polygon (Crossing Polygon).

F selects all objects crossing a selection polyline (Fence).

G selects objects contained in a named group (Group).

M sakes multiple selections before AutoCAD scans the drawing; saves time in a large drawing (Multiple).

SI selects only a single set of objects before terminating the **Select** command (SIngle).

W selects all objects inside the selection box (Window).

WP selects objects inside the selection polygon (Windowed Polygon).

L selects the last-drawn object still visible on the screen (Last).

P selects previously-selected objects (Previous).

R removes objects from the selection set (Remove).

U removes the most-recently added selected objects (Undo).

A continues to add objects after using the R option (Add).

ENTER exits the **Select** command.

ESC aborts the **Select** command.

RELATED COMMANDS

All commands that prompt 'Select objects'.

Filter specifies objects to be added to the selection set.

QSelect selects objects via a dialog box interface based on their properties.

RELATED SYSTEM VARIABLES

PickAdd controls how objects are added to a selection set.

PickAuto controls automatic windowing at the 'Select objects' prompt.

PickDrag controls the method of creating a selection box.

PickFirst controls the command-object selection order.

TIPS

- Objects selected by the **Select** command become the Previous selection set and may selected by the first subsequent 'Select objects:' prompt by responding P for "previous."
- To view a list of all selection options, enter any non-valid text at the prompt, such as "asdf":

  Command: select
  Select objects: asdf
  Invalid selection
  Expects a point or
  Window/Last/Crossing/BOX/ALL/Fence/WPolygon/CPolygon/Group/Add/Remove/Multiple/Previous/Undo/AUto/SIngle
  Select objects:

- Pressing CTRL+A selects all objects in the drawing without needing the **Select** command.
- **All** selects all objects in the drawing, except those on frozen and locked layers.
- **Crossing** selects objects within and crossing the selection rectangle.
- **Window** selects objects within the selection rectangle.
- **Fence** selects objects crossing the selection polyline.
- **CPolygon** selects objects within and crossing the selection polygon.
- **WPolygon** selects objects within the selection polygon.

SelectURL

Rel.14 Highlights all objects and areas that have hyperlinks attached to them (*undocumented command*).

Command	Alias	Ctrl+	F-key	Alt+	Menu Bar	Tablet
selecturl	...	...	...	...	...	...

Command: selecturl

Highlights hyperlined objects and areas with grips.

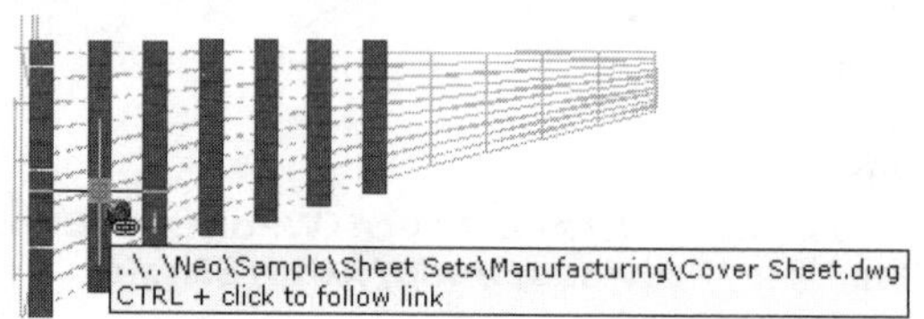

If there are no hyperlinks in the drawing, AutoCAD reports:

No objects with hyperlinks found.

COMMAND LINE OPTIONS

None.

RELATED COMMANDS

AttachURL attaches URLs to objects and areas.

Hyperlink attaches URLs to objects via a dialog box.

TIPS

- Examples of URLs include:

http://www.autodeskpress.com	Autodesk Press Web site.
news://adesknews.autodesk.com	Autodesk news server.
ftp://ftp.autodesk.com	Autodesk FTP server.
http://www.upfrontezine.com	Author Ralph Grabowski's Web site.

- Do not delete layer URLLAYER.
- The URL is stored as follows:

Attachment	URL
One object	Stored as xdata (extended entity data).
Multiple objects	Stored as xdata in each object.
Area	Stored as xdata in a rectangle object on layer URLLAYER.

DEFINITIONS

DWF — short for "drawing Web format," Autodesk's file format for displaying drawings on the Internet.

URL — short for "uniform resource locator," the universal file naming convention.

'SetIDropHandler

2004 Specifies how i-drop objects should be handed in drawings.

Command	Alias	Ctrl+	F-key	Alt+	Menu Bar	Tablet
'setidrophandler	...	...	...	...	...	...

Command: setidrophandler

Displays dialog box:

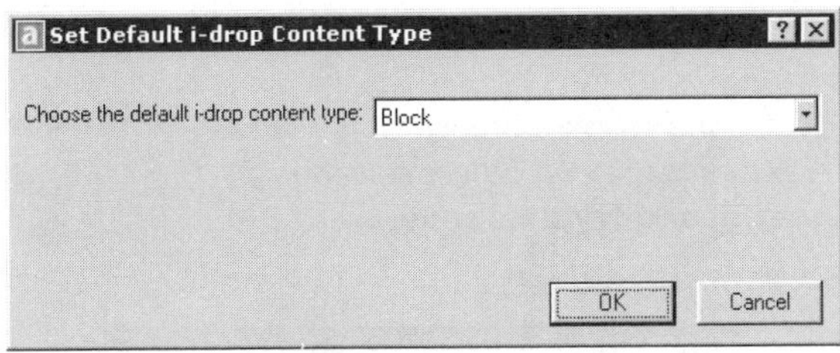

DIALOG BOX OPTION

Choose the default i-drop content type specifies how i-drop objects are handled when placed in drawings.

TIPS

- When i-drop content is dragged from Web pages into drawings, AutoCAD treats the objects as blocks.
- At time of writing this book, the only option is "Block", so this command serves no purpose.
- For more information on i-drop (short for "Internet drag'n drop") technology, see www.autodesk.com/developidrop.

SetUV

Rel.14 Provides positioning control over materials mapped onto objects for rendering.

Command	Alias	Ctrl+	F-key	Alt+	Menu Bar	Tablet
setuv	...	...	...	VEA	View ↳Render ↳Mapping	R1

Command: setuv
Select objects: *(Select one or more objects.)*
Select objects: *(Press* **ENTER** *to end object selection.)*
Updating the Render geometry database...

Displays dialog box:

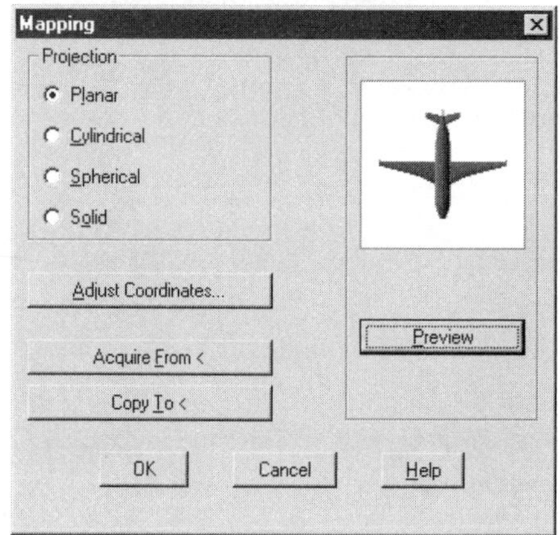

DIALOG BOX OPTIONS

Projection options

⊙**Planar** specifies a plane for projecting the bitmap onto the selected object.

○**Cylindrical** specifies an axis of the cylindrical coordinate system and wrap line for projecting the bitmap onto the selected object.

○**Spherical** specifies the polar axis of the spherical coordinate system and wrap line for projecting the bitmap onto the selected object.

○**Solid** adjusts the coordinates to shift marble, granite, or wood materials.

Additional options

Adjust Coordinates displays different dialog boxes, depending on the setting of Projection.

Acquire From dismisses the dialog box temporarily, allowing you to select a mapping object already in the drawing.

Copy To dismisses the dialog box temporarily, allowing you to select the objects to which to apply the mapping.

Adjust Planar Coordinates dialog box

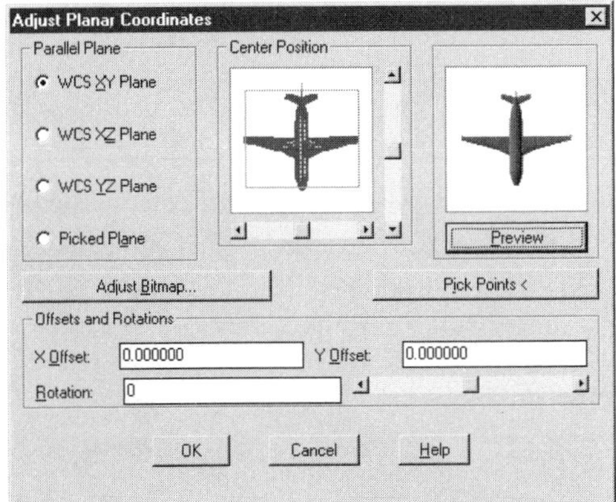

Parallel Plane selects a WCS reference, or picks a plane with the **Pick Points** radio button.

Center Position shows a parallel projection of the selected object's mesh onto the current parallel plane:

Position	Meaning
Blue	Current projection square.
Blue tick mark	Top of the projection square.
Green	Projection square's left edge.

Adjust Bitmap displays the Adjust Object Bitmap Placement dialog box.

Pick Points specifies a projection plane by picking points in the drawing.

Offset changes the x and y offset of the map.

Rotation changes the rotation angle of the map.

Adjust Cylindrical Coordinates dialog box

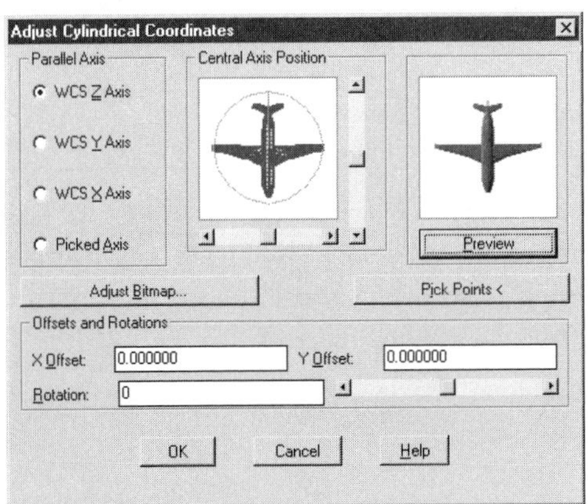

Parallel Axis selects a WCS reference axis of the WCS, or picks an axis with the Pick Points radio button.

Central Axis Position displays a parallel projection of the object's mesh onto a plane perpendicular to the current axis.

Position	Meaning
Blue circle	Projection axis.
Green radius	Wrap line.

Adjust Bitmap displays the Adjust Object Bitmap Placement dialog box.

Pick Points specifies a projection axis by picking points in the drawing.

Offset changes the x and y offset of the map.

Rotation changes the rotation angle of the map.

Adjust Spherical Coordinates dialog box

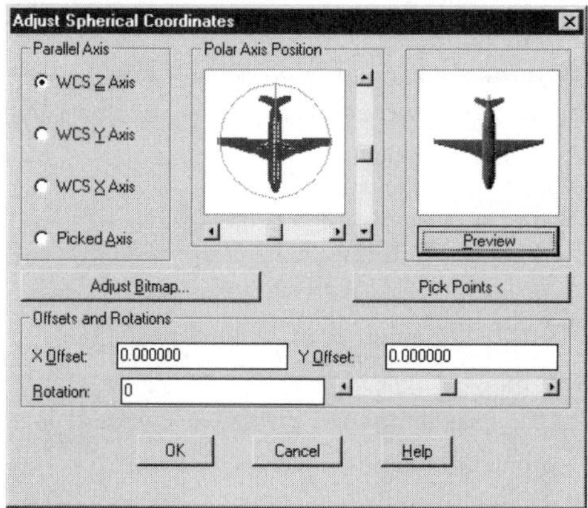

Parallel Axis selects one of the three perpendicular axes of the WCS, or picks an axis with the Pick Points button.

Polar Axis Position displays a parallel projection of the object's mesh onto a plane perpendicular to the current axis.

Position	Meaning
Blue circle	Projection axis.
Green radius	Wrap line.

Adjust Bitmap displays the Adjust Object Bitmap Placement dialog box.

Pick Points specifies a projection axis by picking points in the drawing.

Offset changes the x and y offset of the map.

Rotation changes the rotation angle of the map.

Adjust Solid Coordinates dialog box

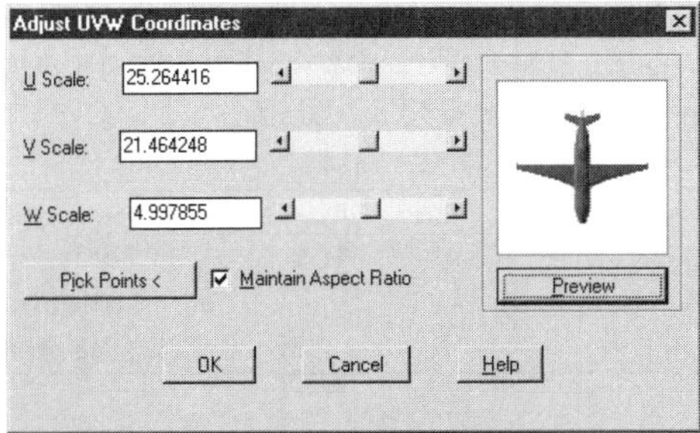

U Scale, V Scale, W Scale sets 3D projection coordinates.

Pick Points specifies a projection axis by picking points in the drawing.

Adjust Object Bitmap Placement dialog box

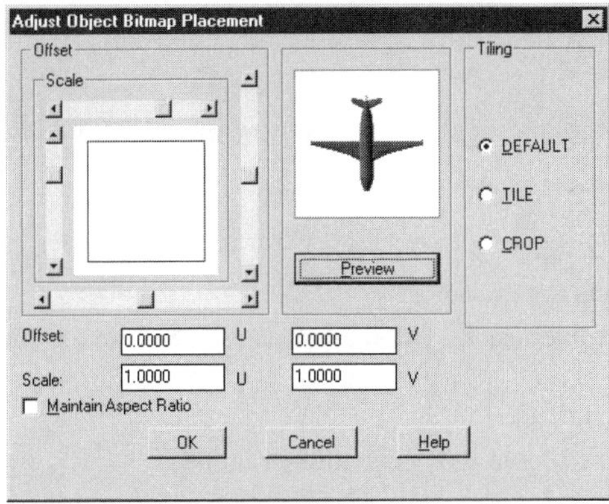

Offset sets the u and v offset distances via slider bars interactively.

Scale sets the u and v scale of the bitmap interactively.

Tiling options

⊙ **Default** means something or other.

○ **Tile** tiles the bitmap.

○ **Crop** does not tile the bitmap.

TIPS

- The upper-left corner pixel of the bitmap defines the transparent color; to have no transparent colors in the bitmap, deliberately make that pixel a unique color.
- **Crop** tiling turns on transparent mode: all pixels have the same color as the upper-left pixel, which is treated as transparent.
- UV (and sometimes W) are equivalent to the x,y,z coordinates; the letters U, V, W are used, however, because they are independent of the x,y,z coordinates.

'SetVar

V. 2.5 Lists the settings of system variables; allows you to change variables that are not read-only (*short for SET VARiable*).

Command	Alias	Ctrl+	F-key	Alt+	Menu Bar	Tablet
setvar	**set**	...	...	**TQV**	**Tools** ↳**Inquiry** ↳**Set Variable**	**U10**

Command: setvar
Enter variable name or [?]: *(Enter a name, or type* **?***.)*

COMMAND LINE OPTIONS

Enter variable name indicates the system variable name you want to access.

? lists the names and settings of system variables.

TIPS

- See Appendix D for the complete list of all system variables found in AutoCAD.
- Example usage of this command:

  Command: setvar
  Enter variable name or [?]: visretain
  Enter new value for VISRETAIN <0>: 1

- Almost all system variables can be entered without the **SetVar** command. For example:

  Command: visretain
  New value for VISRETAIN <0>: 1

- System variables marked "read only" cannot be changed:

  Command: _pkser
  _PKSER = "341-35000000" (read only)

Removed Command

Shade was removed from AutoCAD 2000; it was replaced by **ShadeMode**.

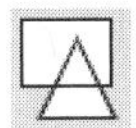

ShadeMode

2000 Generates renderings of 3D models quickly in a variety of modes (*replaces the Shade command*).

Command	Alias	Ctrl+	F-key	Alt+	Menu Bar	Tablet
shademode	**shade**	**...**	**...**	**VS**	**View** ↳**Shade**	**N2**

Command: shademode

Current mode: 2D wireframe

Enter option [2D wireframe/3D wireframe/Hidden/Flat/Gouraud/ fLat+edges/gOuraud+edges] <2D wireframe>: *(Enter an option.)*

COMMAND LINE OPTIONS

2D wireframe displays wireframe models in 2D space.

3D wireframe displays wireframe models in 3D space.

Hidden removes hidden faces.

Flat displays flat shaded faces.

fLat+edges displays flat shaded faces, with outlined faces of the background color.

Gouraud displays smooth shaded faces.

gOuraud+edges displays smooth shaded faces, with outlined faces of the background color.

RELATED COMMANDS

Hide performs a true hidden-line removal of 3D drawings.

MSlide saves a rendered view as an SLD-format slides on disk.

MView performs a hidden-line view of individual viewports during plotting.

Plot performs a hidden-line view during plotting.

Render performs a more realistic rendering.

3dOrbit performs hidden-line removal of perspective views.

TIPS

- As an alternative to the **ShadeMode** command, the **Render** command creates high-quality renderings of 3D drawings, but takes somewhat longer to do so.
- The smaller the viewport, the faster the shading.

This drawing is displayed in **ShadeMode**'s default **2D wireframe** mode; notice the standard 2D UCS icon.

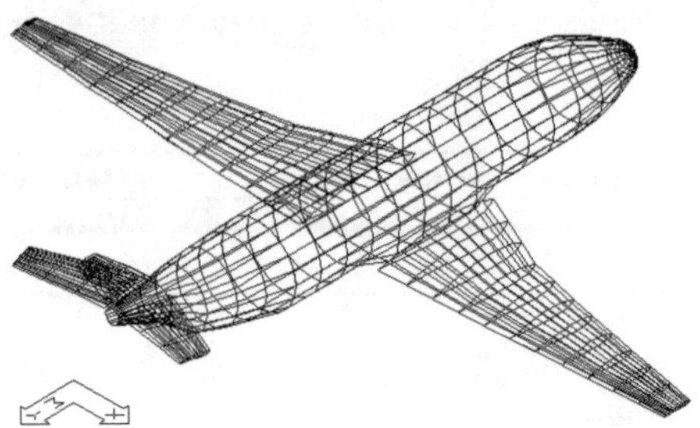

This drawing is displayed in **3D wireframe** mode; notice the 3D UCS icon.

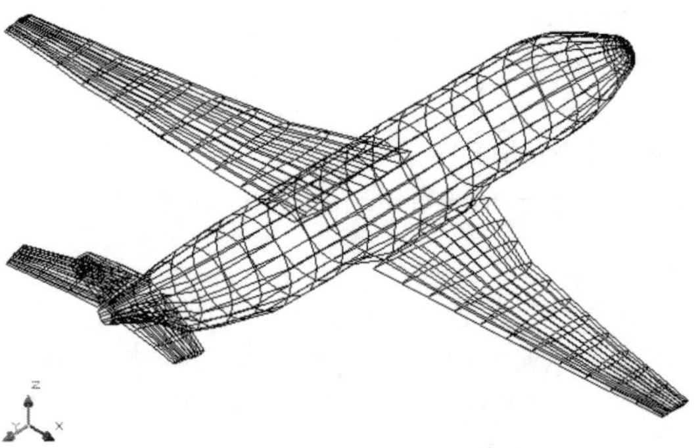

This drawing is displayed in **Hidden** mode; notice that faces hidden by other faces are not displayed:

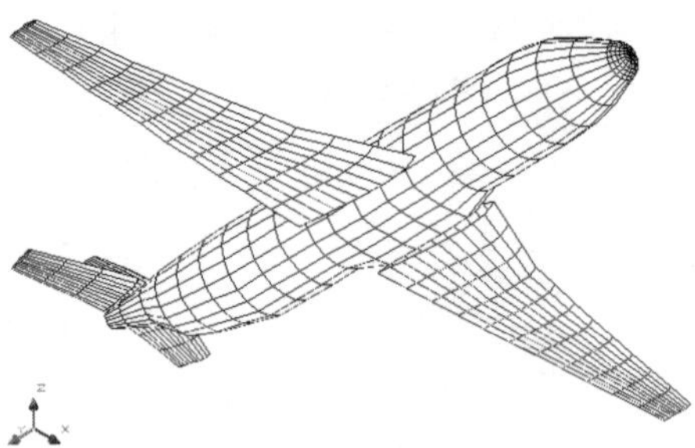

This drawing is displayed in **Flat** mode; notice that each face is filled with a shade of gray.

This drawing is displayed **Gouraud** mode; notice that the faces are smoothed.

This drawing is displayed in **fLat+edges** mode; notice that each face is outlined by the background color (*white*).

This drawing is displayed in **gOuraud+edges** mode; notice the outlining of each face.

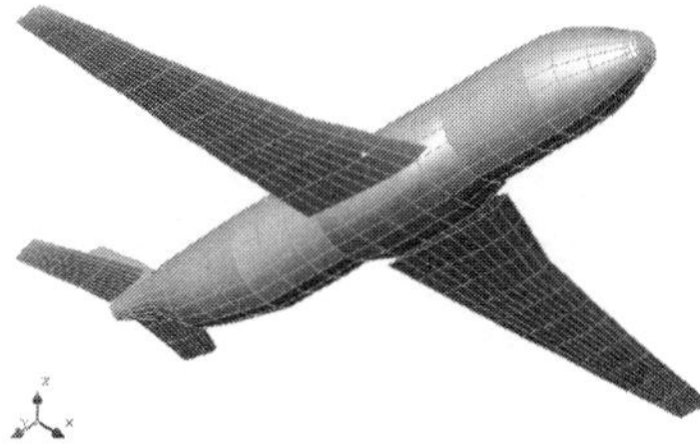

Shape

V. 1.0 Inserts shapes into drawings.

Command	Alias	Ctrl+	F-key	Alt+	Menu Bar	Tablet
shape	...	...	...	...	...	...

Command: shape
Enter shape name or [?]: *(Enter the name, or type* **?***.)*
Specify insertion point: *(Pick a point.)*
Specify height <1>: *(Specify the height.)*
Specify rotation angle <0>: *(Specify the angle.)*

COMMAND LINE OPTIONS

Enter shape name indicates the name of the shape to insert.

? lists the names of currently-loaded shapes.

Specify insertion point indicates the insertion point of the shape.

Specify height specifies the height of the shape.

Specify rotation angle specifies the rotation angle of the shape.

RELATED COMMANDS

Load loads SHX-format shape files into the drawings.

Insert inserts blocks into drawings.

Style loads *.shx* font files into drawings.

RELATED SYSTEM VARIABLE

ShpName specifies the default *.shp* file name.

TIPS

- Shapes are used to define the text and symbols found in complex linetypes.

Some electronic shapes.

- Shapes were an early alternative to blocks, but now are used primarily with complex linetypes. They take up extemely small amounts of memory, but are difficult to create.
- Shapes are defined by *.shp* files, which must be compiled into *.shx* files before they can be loaded by the **Load** command.
- In addition, shapes must be loaded by the **Load** command before they can be inserted with the **Shape** command.
- Compile an *.shx* file into an *shp* file with the **Compile** command.

Sheetset / SheetsetHide

2005 Displays and hides the Sheetset Manager window.

Command	Alias	Ctrl+	F-key	Alt+	Menu Bar	Tablet
sheetset	**ssm**	**4**	...	**TM**	**Tools** ↳**Sheetset Manager**	...
sheetsethide	...	**4**	...	**TM**	**Tools** ↳✓**Sheetset Manager**	...

Command: sheetset

Displays the Sheet Set Manager window:

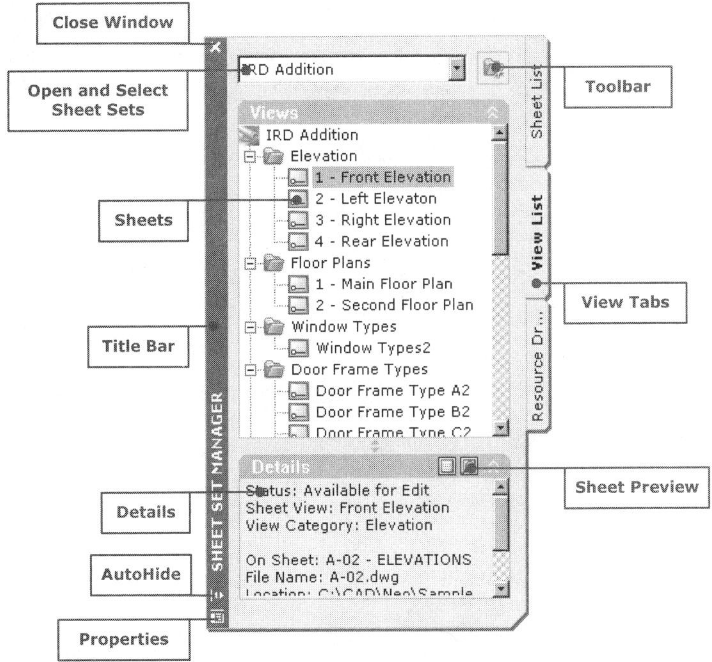

Command: sheetsethide

Closes the window:

WINDOW OPTIONS

x closes the window; alternatively, use the **SheetsetClose** command.

Toolbar changes, depending on the tab selected.

Title bar docks against the side of AutoCAD, or can be dragged around.

View Tabs changes the view between Sheet List, View List, and Resource Drawings views.

Views lists sheets and sheet set names; right-click for shortcut menus.

- **Sheet List** displays names of sheets, which are organized into subsets.
- **View List** displays views for each sheet set; views can be organized into categories.
- **Resource Drawings** displays folders and files used by the sheet set

Open and select sheet sets displays a droplist listing recently-opened sheet sets.

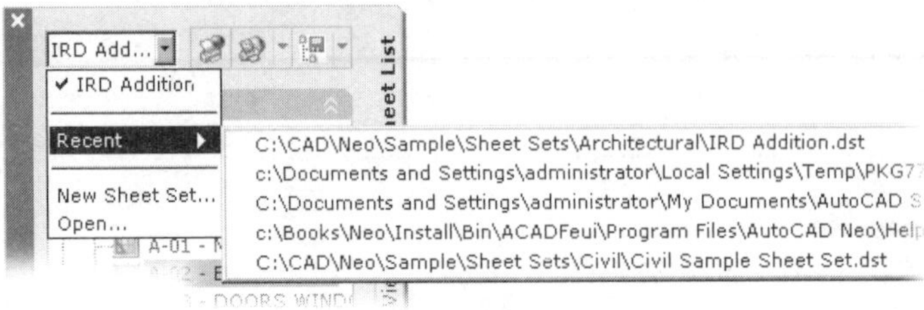

- **Recent** lists recently-opened *.dst* sheet set data files.
- **New Sheet Set** starts the Create Sheet Set wizard; see **NewSheetset** command.
- **Open** displays the Open Sheetset dialog box; select a *.dst* file, and then click **Open**. See the **NewSheetset** command.

Details lists information about selected sheets and sheet sets.

Sheet Preview shows preview images of selected sheets.

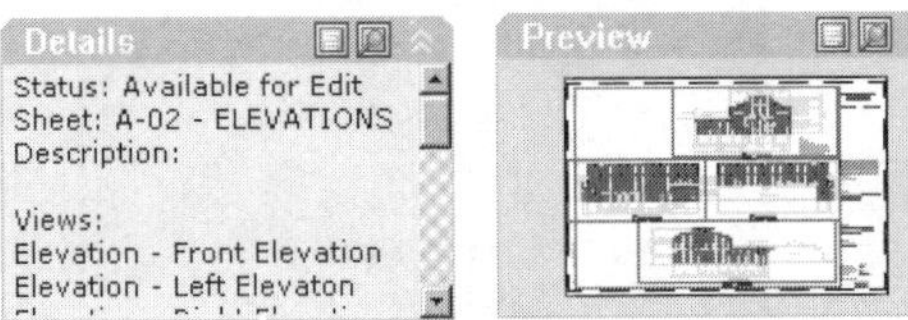

AutoHide reduces window to titlebar when cursor not over the window.

Properties displays a shortcut menu.

SHEET LIST VIEW

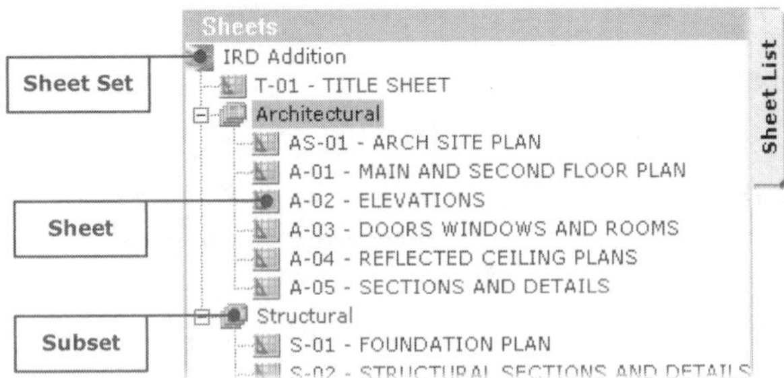

Sheetset shortcut menus

 *Right-click a sheet set name:*

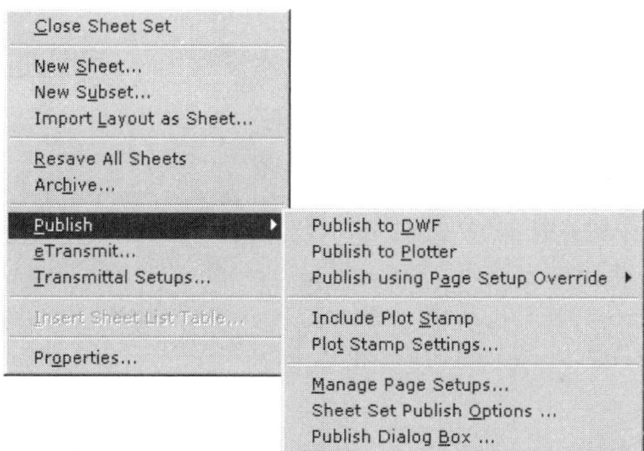

Close Sheet Set closes the entire sheet set; the use Open option to open another sheet set.

New Sheet displays the New Sheet dialog box; enter a number and name, and then click **OK**.
New Subset dispays the Subsheet Properties dialog box; enter a name, and then click **OK.**
Import Layout as Sheet dispays the Import Layout as Sheet dialog box; select a drawing and layouts, and then click **OK**.

Resave All Sheets saves sheet set infomation stored in each drawing file.
Archive displays the Archive as Sheet Set dialog box; see **Archive** command.

Publish plots the sheet set as *.dwf* files or to a plotter; see **Publish** command.
eTransmit collates the sheet set for transmittal by CD or email; see **eTransmit** command.
Transmittal Setups displays the Transmittal Setup dialog box; see **eTransmit** command.

Insert Sheet List Table creates a table listing all sheets in the sheet set, and adds it to a sheet.

Properties displays the Sheet Set Properties dialog box; make changes, and then click **OK**.

Subset shortcut menu

 *Right-click a subset name:*

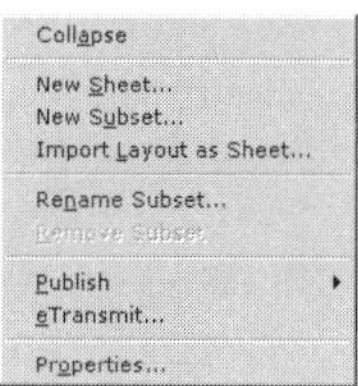

Collapse hides the names of sheets.
Rename Subset changes the name of the subset.
Remove Subset erases the subset; available only when subset contains no sheets.
Properties displays the Subset Properties dialog box; make changes, and then click **OK**.

Sheet shortcut menu

 *Right-click a sheet name:*

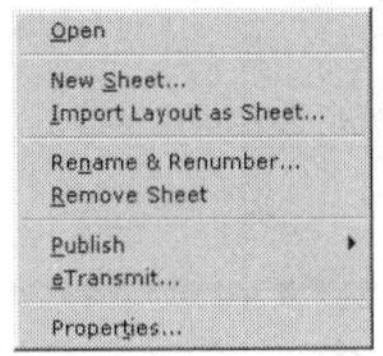

Open opens the drawing in a new window.

Rename & Renumber displays the Rename & Renumber Sheet dialog box.

Remove Sheet removes the sheet from the set; does *not* erase the drawing or layout.

Properties displays the Sheet Properties dialog box; make changes, and then click **OK**.

Sheet List toolbar

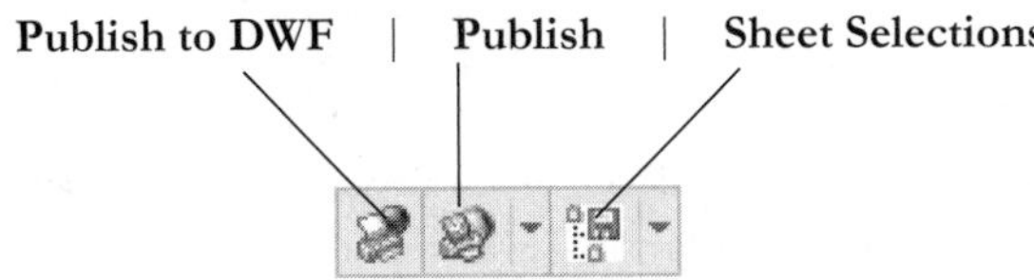

VIEW LIST VIEW

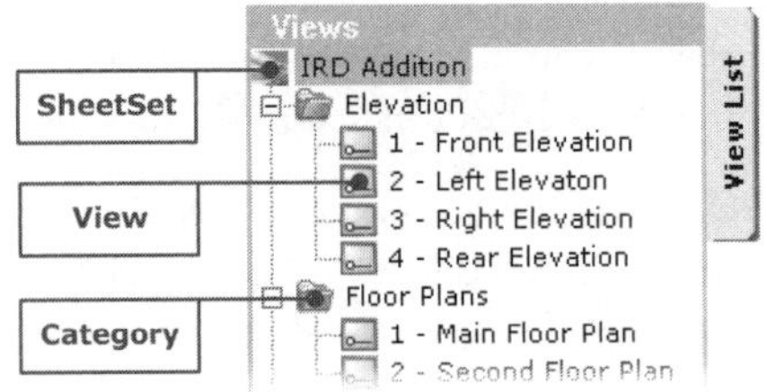

Sheetset shortcut menu

 *Right-click a sheet set name:*

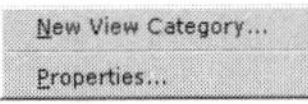

New View Category displays the View Category dialog box; enter info, and then click **OK**.

Properties displays the Sheetset Properties dialog box; make changes, and then click **OK**.

Category shortcut menu

 *Right-click a category name:*

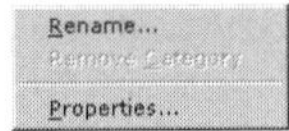

Rename changes the name of the category.

Remove Category erases the category; available only when category is empty.

Properties displays the View Category dialog box; make changes, and then click **OK**.

View shortcut menu

 *Right-click a view name:*

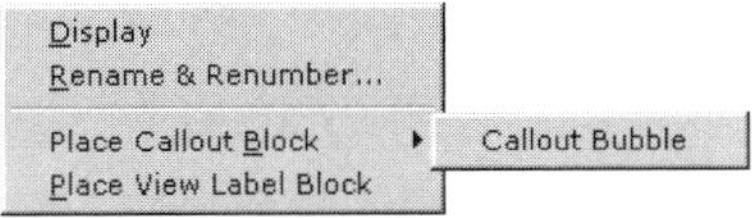

Open displays the drawing in a new window.

Rename & Renumber displays the Rename & Renumber View dialog box.

Place Callout Block

- **Callout Buble** places "Callout Bubble" block in drawings, with view number (1) and sheet number (A-02) field text filled in; see the **Insert** command.

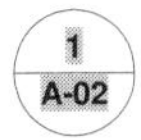

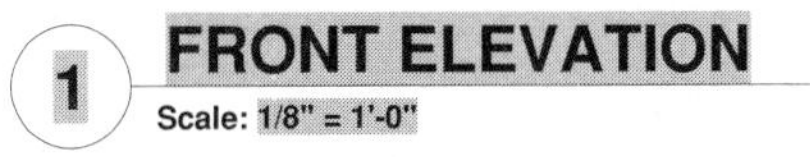

Place View Label Block places "Drawing Name" block in drawings, with view number (1), view name (Front Elevation), and viewport scale filled in as field text. (Gray background indicates field text; see the **Field** command.)

View List toolbar

 **New View Category** displays the View Category dialog box.

RESOURCE DRAWINGS VIEW

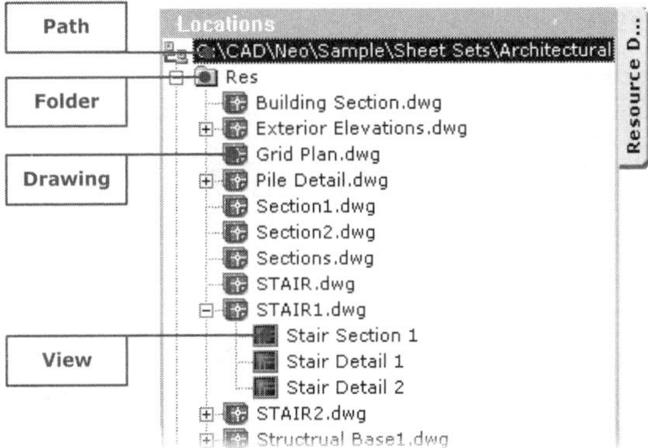

Path shortcut menu

 *Right-click a path name:*

Expand
Add New Location...
Remove Location

Expand displays folders, drawings, and views.

Add New Location displays the Browse for Folder dialog box; select a folder, and then click **Open**.

Remove Location deletes a location, with no warning.

There is no shortcut menu for folders.

Drawing shortcut menu

 *Right-click a drawing name:*

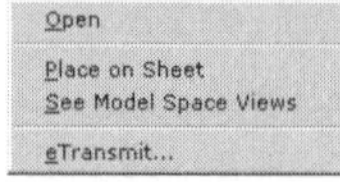

Open opens the drawing in a new window.

Place on Sheet places the drawing in the current sheet; works only in layout mode.

See Model Space Views displays model space view icons, if any.

View shortcut menu

 *Right-click a view name:*

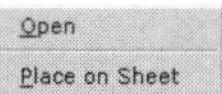

Open opens the view in a new window.

Place on Sheet places the drawing in the current sheet; works only in layout mode.

Resource Drawings toolbar

 **Add New Location** displays the Browse for Folder dialog box.

DIALOG BOX OPTIONS

New Sheet / Subset dialog box

Similar dialog boxes are used for properties of subsets and new sheets.

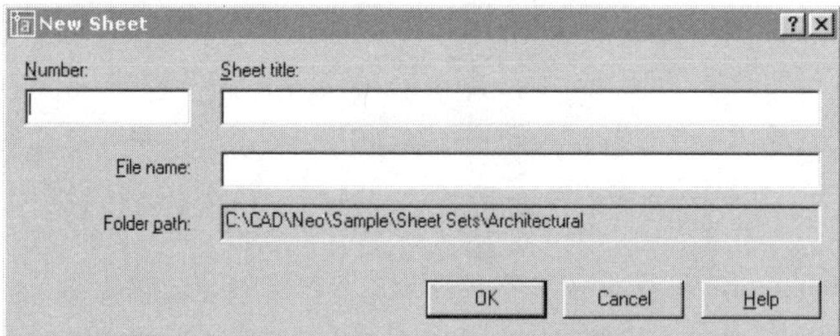

Number specifies the order number of the sheet; allows sheets to be placed in correct order.

Sheet Title names the sheet, usually the name of the related layout tab.

File Name specifies the name given the sheet file; usually consists of the sheet number and sheet title.

Subset Sheet Properties dialog box

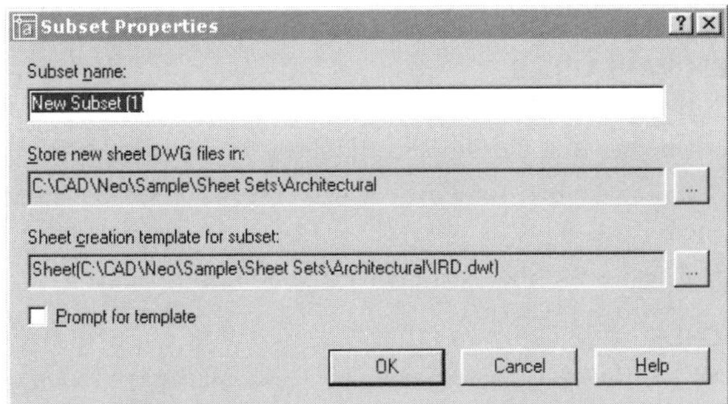

Subset name specifies the name of the subset.

Store new sheet DWG files in selects the folder for storing drawing files.

... displays Browser for Folder dialog box; select a folder, and then click **Open**.

Sheet Creation Template for Subset specifies the name of the *.dwt* template file to use.

... displays Browser for Folder dialog box; select a folder, and then click **Open**.

☐**Prompt for template** prompts for the template name later.

Import Layouts as Sheets dialog box

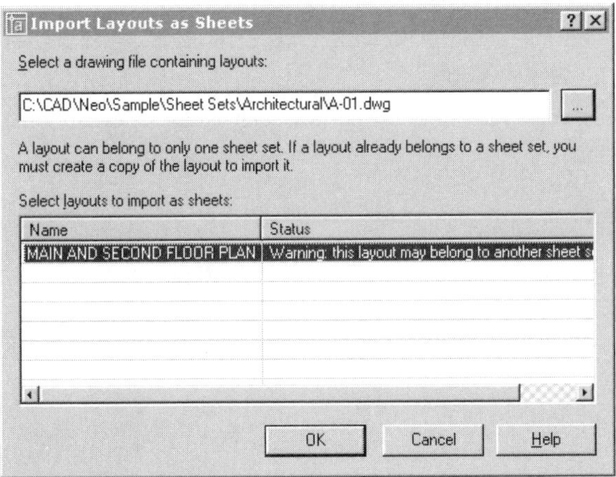

Select a drawing file containing layouts specifies the name of a drawing file.

... displays Browser for Folder dialog box; select a folder, and then click **Open**.

Select layouts to import as sheets lists the names of layouts found in the drawing file.

RELATED COMMANDS

NewSheetset creates new sheet sets.

OpenSheetset opens a specific sheet set.

RELATED SYSTEM VARIABLES

SsFound records the path and file name of current sheet set.

SsLocate toggles whether AutoCAD opens the sheet sets associated with drawings.

SsmAutoOpen toggles whether AutoCAD displays the Sheet Set Manager window when drawings are opened.

SsmState reports whether the Sheet Set Manager window is open.

STARTUP SWITCHES

/nossm (no Sheet Set Manager) prevents the Sheet Set Manager window from opening at startup.

/set loads a specified *.dst* sheet set data file at startup.

RELATED FILES

.dst are sheet set data files.

TIPS

- The purpose of sheet sets is to create a hierarchy of drawings, arranged in order; as well, drawings can be organized into subsets (sheets) and categories (views).
- *SSM* is short for "sheet set manager."
- Sheet sets can be created with the **NewSheetset** command, either based on existing drawings or on another sheet set.
- Sheet sets, subsets, and sheets have properties, such as titles, descriptions, file paths, and custom properties.

Shell

V. 2.5 Starts a new instance of the Windows command interpreter; runs external commands defined in *acad.pgp*.

Command	Alias	Ctrl+	F-key	Alt+	Menu Bar	Tablet
shell	**sh**	...	...	...	...	...

Command: shell
OS Command: *(Enter a command.)*

Opens window:

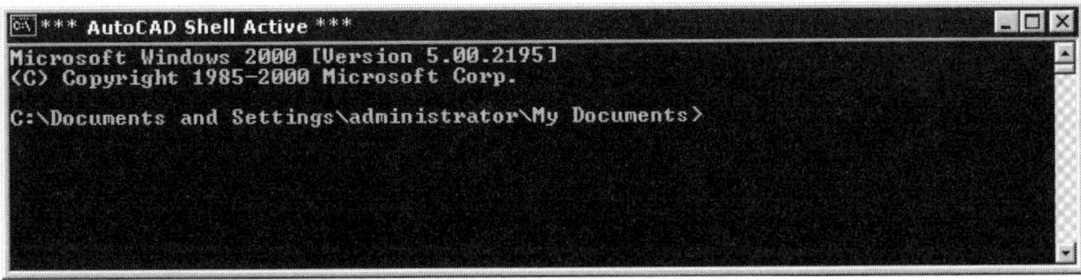

COMMAND LINE OPTIONS

OS Command specifies an operating system command, and then closes the new instance of the Windows command interpreter.

ENTER remains in the operating system's command mode for more than one command.

Exit returns to AutoCAD from the OS command mode.

RELATED COMMAND

Quit exits AutoCAD.

RELATED FILE

acad.pgp is the external command definition file.

ShowMat

Rel.13 Lists the rendering material attached to objects (*short for SHOW MATerial*).

Command	Alias	Ctrl+	F-key	Alt+	Menu Bar	Tablet
showmat	...	...	...	...	...	...

Command: showmat
Select object: *(Select one object.)*

Sample response:

Material BRONZE is explicitly attached to the object.

COMMAND LINE OPTION

Select object selects the object to examine.

RELATED COMMANDS

MatLib loads the material definitions into the drawing.

RMat attaches the materials to objects, colors, and layers.

TIP

- When no material is attached to the model, AutoCAD reports, "Material *GLOBAL* is attached by default or by block."

SigValidate

2004 Displays digital signature information stored in drawings.

Command	Alias	Ctrl+	F-key	Alt+	Menu Bar	Tablet
sigvalidate	...	...	...	...	...	...

Command: sigvalidate

Displays dialog box:

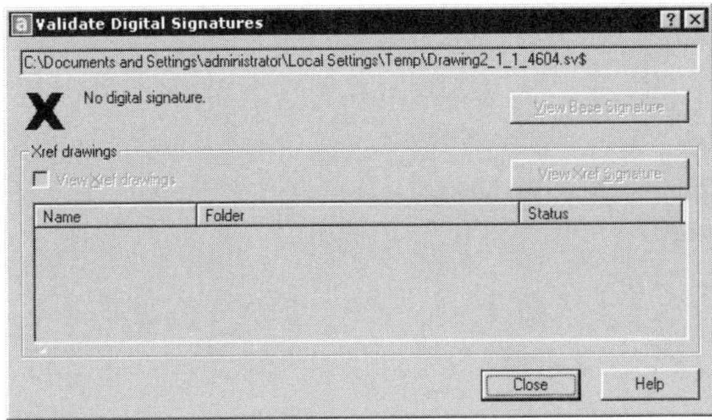

DIALOG BOX OPTIONS

View Base Signature displays the Digital Signature Contents dialog box.

Xref Drawings options

View Xref Drawings

☑ Lists xref drawings attached to this drawing.

☐ Does not list xrefs.

View Xref Signature displays the Digital Signatures Contents dialog box for the selected xref; available only when the selected xref has a digital signature.

RELATED COMMAND

SecurityOptions attaches digital signatures to drawings.

TIPS

- Digital signatures validate the authenticity of drawings.
- Digital signatures indicate whether the drawing changed since being signed.
- Autodesk notes that digital signatures become invalid for these reasons: the drawing was corrupted when the digital signature was attached; the drawing was corrupted in transit; and the digital ID is no longer valid.
- This command requires that the drawing be saved first; if not, AutoCAD reports, "This drawing has been modified. The last saved version will be validated."
- If this command does not work, it was removed from AutoCAD; re-install with the Custom option, and then select Drawing Encryption.

Sketch

V. 1.4 Allows freehand drawing as a series of lines or polylines.

Command	Alias	Ctrl+	F-key	Alt+	Menu Bar	Tablet
sketch	...	...	...	...	...	...

Command: sketch
Record increment <0.1000>: *(Enter a value.)*
Sketch. Pen eXit Quit Record Erase Connect .

Click pick button to begin drawing:

<Pen down>

Click pick button again to stop drawing:

<Pen up>

Press ENTER *to record sketching and exit **Sketch**:*

nnn **lines recorded.**

COMMAND LINE OPTIONS

Commands can be invoked by mouse and digitizer buttons:

Connect connects to the last drawing segment (*as an alternative, press button #6)*.

Erase erases temporary segments as the cursor moves over them (*button #5*).

eXit records the temporary segments, and exits the **Sketch** command (*button #3 or* SPACEBAR *or* ENTER).

Pen lifts and lowers the pen (*pick button #1*).

Quit discards temporary segments, and exits the **Sketch** command (*button #4 or* ESC).

Record records the temporary segments as permanent (*button #2*).

. (*Period*) connects the last segment to the current point (*button #1*).

RELATED COMMANDS

Line draws line segments.

PLine draws polyline and polyline arc segments.

RELATED SYSTEM VARIABLES

SketchInc specifies the current recording increment for the **Sketch** command (default = 0.1).

SKPoly controls the type of sketches recorded:

SkPoly	Meaning
0	Record sketches as lines (default).
1	Record sketches as polylines.

TIPS

- During the **Sketch** command, the definitions of the pointing device's buttons change to:

Button	Meaning	Keystroke
0	Raises and lowers the ***p***en	P
1	Draws line to the current ***point***	.
2	***R***ecords the sketch	R
3	Records the sketch and e**X**its	X *or* **ENTER** *or* **SPACEBAR**
4	Discards the sketch and ***Q***uits	Q *or* **ESC**
5	*E*rases the sketch	E
6	*C*onnects to last-drawn segment	C

- Only the first several button commands are available on a mouse.
- Pull-down menus are unavailable during the **Sketch** command.

Slice

Rel.11 Cuts 3D solids with planes, creating one or two 3D solids.

Command	Alias	Ctrl+	F-key	Alt+	Menu Bar	Tablet
slice	sl	...	...	DIL	Draw ↳Solids ↳Slice	...

Command: slice

Select objects: *(Select one or more solid objects.)*

Select objects: *(Press* **ENTER** *to end object selection.)*

Specify first point on slicing plane by [Object/Zaxis/View/XY/YZ/ZX/3points] <3points>: *(Pick a point, or enter an option.)*

Specify a point on desired side of the plane or [keep Both sides]: *(Pick a point, or type* **B***.)*

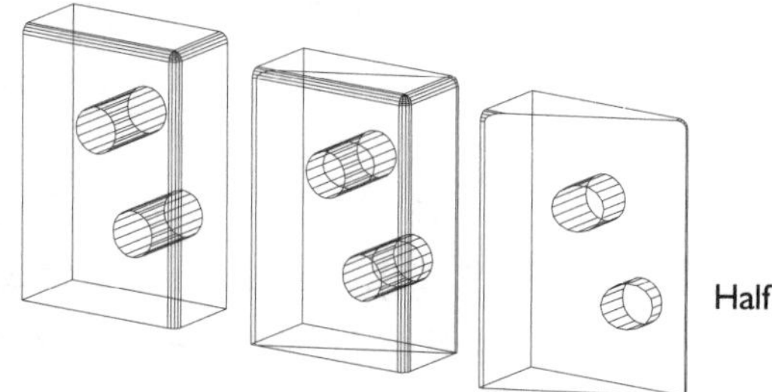

COMMAND LINE OPTIONS

Select objects selects the 3D solid model to slice.

Slicing plane options

Object aligns the cutting plane with a circle, ellipse, arc, elliptical arc, 2D spline, or 2D polyline.

View aligns the cutting plane with the viewing plane.

XY aligns the cutting plane with the x,y plane of the current UCS.

YZ aligns the cutting plane with the y,x plane of the current UCS.

Zaxis aligns the cutting plane with two normal points.

ZX aligns the cutting plane with the z,x plane of the current UCS.

3 points aligns the cutting plane with three points.

Additional options

keep Both sides retains both halves of the cut solid model.

Specify a point on the desired side of the plane retains either half of the cut solid model.

TIPS

- This command cannot slice a 2D region, a 3D wireframe model, or other 2D shapes.
- It is helpful to use object snap when specifying points.

'Snap

V. 1.0 Sets the drawing "resolution," the origin for the grid and hatches, isometric mode, and the angle of the grid, hatches, and orthographic mode.

Command	Alias	Ctrl+	F-key	Alt+	Menu Bar	Tablet
snap	sn	B	F9	...	...	...

Command: snap

Specify snap spacing or [ON/OFF/Aspect/Rotate/Style/Type]<0.5>: *(Enter a value, or an option.)*

COMMAND LINE OPTIONS

Snap spacings sets the snap increment.

Aspect sets separate x and y increments.

OFF turns off snap.

ON turns on snap.

Rotate rotates the crosshairs for snap and grid.

Style switches between standard and isometric style.

Type switches between grid or polar snap.

Aspect options

Specify horizontal spacing <0.5>: *(Enter a value.)*

Specify vertical spacing <0.5>: *(Enter a value.)*

Specify horizontal spacing specifies the spacing between snap points along the x-direction.

Specify vertical spacing specifies the spacing between snap points along the y-direction.

Rotate options

Specify base point <0.0,0.0>: *(Enter a value.)*

Specify rotation angle <0>: *(Enter a value.)*

Specify base point specifies the point from which snap increments are determined.

Specify rotation angle rotates the snap about the base point.

Style options

Enter snap grid style [Standard/Isometric] <S>: *(Type* **S** *or* **I**.*)*

Specify vertical spacing <0.5>: *(Enter a value.)*

Standard specifies rectangular snap.

Isometric specifies isometric snap.

Specify vertical spacing specifies the spacing between snap points along the y-direction.

Type options

Enter snap type [Polar/Grid] <Grid>: *(Type* **P** *or* **G**.*)*

Polar specifies polar snap.

Grid specifies rectangular snap.

STATUS BAR OPTION

Click **SNAP** *to turn snap on and off; right-click for shortcut menu:*

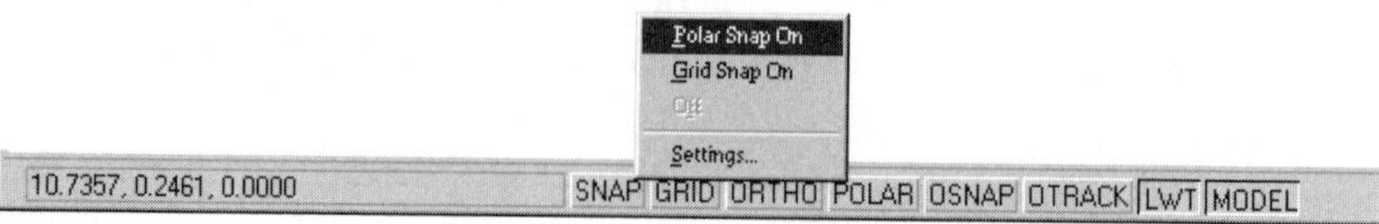

Polar Snap On turns on polar snap.

Grid Snap On turns on snap.

Off turns off snap.

Settings displays the Snap and Grid tab of the Drafting Settings dialog box.

RELATED COMMANDS

DSettings sets snap values via a dialog box.

Grid turns on the grid.

Isoplane switches to a different isometric drawing plane.

RELATED SYSTEM VARIABLES

SnapAng specifies the current angle of the snap rotation.

SnapBase specifies the base point of the snap rotation.

SnapIsoPair specifies the current isometric plane setting.

SnapMode determines whether snap is on.

SnapStyl determines the style of snap.

SnapUnit specifies the current snap increment in the x and y directions.

TIPS

- Setting the snap is setting the cursor resolution. For example, setting a snap distance of 0.1 means that when you move the cursor, it jumps in 0.1 increments. You can, however, still type in numerical values of greater resolution, such as 0.1234.
- There is no snap in the z-direction.
- The **Aspect** option is not available when the **Style** option is set to **Isometric**; you may, however, rotate the isometric grid.
- The options of the **Snap** command affect several other commands:

Snap Option	Command	Effect
Style	Ellipse	Adds **Isocircle** option to the **Ellipse** command.
SnapBase	Hatch	Relocates the origin of the hatch.
Rotate	Hatch	Rotates the hatching angle.
Rotate	Grid	Rotates the grid display.
Rotate	Ortho	Rotates the ortho angle.

- You can toggle snap mode by clicking the word **SNAP** on the status bar.

SolDraw

Rel.13 Creates profiles and sections in viewports created with the **SolView** command (*short for SOLids DRAWing*).

Command	Alias	Ctrl+	F-key	Alt+	Menu Bar	Tablet
soldraw	...	...	...	**DIUD**	**Draw** ↳**Solids** ↳**Setup** ↳**Drawing**	...

Command: soldraw
Select viewports to draw ..
Select objects: *(Select one or more viewports.)*
Select objects: *(Press* **ENTER** *to end object selection.)*

COMMAND LINE OPTION

Select objects selects a viewport; must be a floating viewport in model space (**Tilemode**=0).

RELATED COMMANDS

SolProf creates profile images of 3D solids.

SolView creates floating viewports.

TIPS

- The **SolView** command must be used before this command.
- This command performs the following actions:
  1. Creates visible and hidden lines representing the silhouette and edges of solids in the viewport.
  2. Projects to a plane perpendicular to the viewing direction.
  3. Generates silhouettes and edges for all 3D solids and portions of solids behind the cutting plane.
  4. Crosshatches sectional views.
- Existing profiles and sections in the selected viewport are erased.
- All layers — except the ones needed to display the profile or section — are frozen in each viewport.
- The following layers are used by **SolDraw**, **SolProf**, and **SolView**: *viewname*-**VIS**, *viewname*-**HID**, and *viewname*-**HAT**.
- Hatching uses the values set in system variables **HpName**, **HpScale**, and **HpAng**.

Solid

V. 1.0 Draws solid-filled triangles and quadrilaterals; does *not* create 3D solids.

Command	Alias	Ctrl+	F-key	Alt+	Menu Bar	Tablet
solid	**so**	...	...	**DF2**	**Draw** ↳**Surfaces** ↳**2D Solid**	**L8**

Command: solid
Specify first point: *(Pick a point.)*
Specify second point: *(Pick a point.)*
Specify third point: *(Pick a point.)*
Specify fourth point or <exit>: *(Pick a point, or press* ENTER *to draw triangle.)*

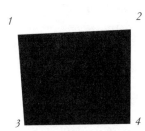

Three-point solid.

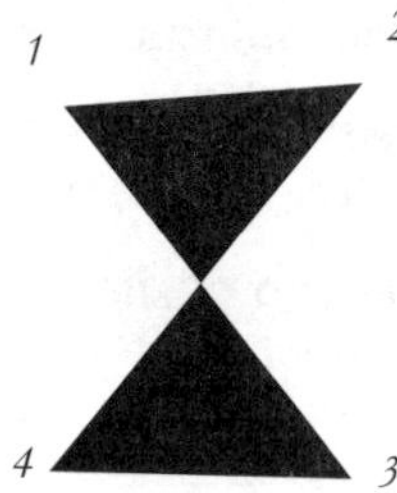

Pick order makes a difference for four-point solids.

COMMAND LINE OPTIONS

First point picks the first corner.

Second point picks the second corner.

Third point picks the third corner.

Fourth point picks the fourth corner; alternatively, press ENTER to draw triangle.

ENTER draws quadilaterials; ends the **Solid** command.

RELATED COMMANDS

Fill turns object fill off and on.

BHatch fills any shape with a solid fill pattern.

Trace draws lines with width.

PLine draws polylines and polyline arcs with width.

RELATED SYSTEM VARIABLE

FillMode determines whether solids are displayed filled or outlined.

 SolidEdit

2000 Edits the faces and edges of 3D solids (*short for SOLids EDITor*).

Command	Alias	Ctrl+	F-key	Alt+	Menu Bar	Tablet
solidedit	...	...	...	MN	Modify ⤷Solids Editing	...

Command: solidedit
Solids editing automatic checking: SOLIDCHECK=1
Enter a solids editing option [Face/Edge/Body/Undo/eXit] <eXit>: *(Enter an option.)*

COMMAND LINE OPTIONS

Undo undoes the last editing actions, one at a time, up to the start of the **SolidEdit** command.

eXit exits body mode.

Face options
Enter a face editing option
[Extrude/Move/Rotate/Offset/Taper/Delete/Copy/coLor/Undo/eXit] <eXit>:
(Enter an option.)

Extrude extrudes one or more faces to the specified distance, or along a path; a positive value extrudes the face in the direction of its normal.

Move moves one or more faces the specified distance.

Rotate rotates one or more faces about an axis by a specified angle.

Offset offsets one or more faces by the specified distance, or through a specified point; a positive value increases the size of the solid, while a negative value decreases the size.

Taper tapers one or more faces by a specified angle; a positive angle tapers in, while negative angle tapers out.

Delete removes the selected faces; also removes attached chamfers and fillets.

Copy copies the selected faces as a 3D region or a 3D body object.

Color changes the color of the selected faces.

Edge options
Enter an edge editing option [Copy/coLor/Undo/eXit] <eXit>: *(Enter an option.)*

Copy copies the selected 3D edges as a line, arc, circle, ellipse, or spline.

coLor changes the color of the selected edges.

Body options
Enter a body editing option
[Imprint/seParate solids/Shell/cLean/Check/Undo/eXit] <eXit>: *(Enter an option.)*

Imprint imprints a selection set of arcs, circles, lines, 2D and 3D polylines, ellipses, splines, regions, bodies, and 3D solids on the face of a 3D solid.

seParate solids separates 3D solids into independent 3D solid objects; the solid objects must have disjointed volumes; this option does *not* separate 3D solids that were joined by a Boolean editing command, such as **Intersect**, **Subtract**, and **Union**.

Shell creates a hollow, thin-walled solid of specified thickness; a positive thickness creates a shell toward the inside of the solid, while a negative value creates the shell on the outside of the solid.

cLean removes redundant edges and vertices, imprints, unused geometry, shared edges, and shared vertices.

Check checks whether the object is a 3D solid; duplicates the function of the **SolidCheck** system variable.

RELATED SYSTEM VARIABLE

SolidCheck toggles solid validation on and off (default = on).

RELATED COMMANDS

All commands related to creating and editing 3D solid models.

TIP

- When working with the **SolidEdit** command, you can select a face, an edge, or an internal point on a face, or use the **CP** (*crossing polygon*), **CW** (*crossing window*), **F** (*fence*) object selection options.

SolProf

Rel.13 Creates profile images of 3D solids (*short for SOLid PROFile*).

Command	Alias	Ctrl+	F-key	Alt+	Menu Bar	Tablet
solprof	...	...	...	**DIUP**	**Draw** ↳**Solids** ↳**Setup** ↳**Profile**	...

Command: solprof
Select objects: *(Select one or more objects.)*
Select objects: *(Press* **ENTER** *to end object selection.)*
Display hidden profile lines on separate layer? [Yes/No] <Y>: *(Type* **Y** *or* **N***.)*
Project profile lines onto a plane? [Yes/No] <Y>: *(Type* **Y** *or* **N***.)*
Delete tangential edges? [Yes/No] <Y>: *(Type* **Y** *or* **N***.)*
n **solids selected.**

COMMAND LINE OPTIONS

Select objects selects the objects to profile.

Display hidden profile lines on separate layer?

No specifies that all profile lines are visible; a block is created for the profile lines for every selected solid.

Yes generates just two blocks: one for visible lines and one for hidden lines.

Project profile lines onto a plane?

No creates profile lines with 3D objects.

Yes creates profile lines with 2D objects.

Delete tangential edges?

No does not display *tangential edges*, the transition line between two tangent faces.

Yes displays tangential edges.

RELATED COMMANDS

SolDraw creates profiles and sections in viewports.

SolView creates floating viewports.

TIPS

- The **SolView** command must be used before the **SolProf** command.
- Solids that share a common volume can produce dangling edges, if you generate profiles with hidden lines. To avoid this, first use the **Union** command.
- AutoCAD must display a layout, and a modelspace viewport must be active before you can use the **SolProf** command.

SolView

Rel.13 Creates floating viewports in preparation for the **SolDraw** and **SolProf** commands (*short for SOLid VIEWs*).

Command	Alias	Ctrl+	F-key	Alt+	Menu Bar	Tablet
solview	...	...	...	DIUV	Draw ↳Solids ↳Setup ↳View	...

Command: solview
Enter an option [Ucs/Ortho/Auxiliary/Section]: *(Enter an option.)*

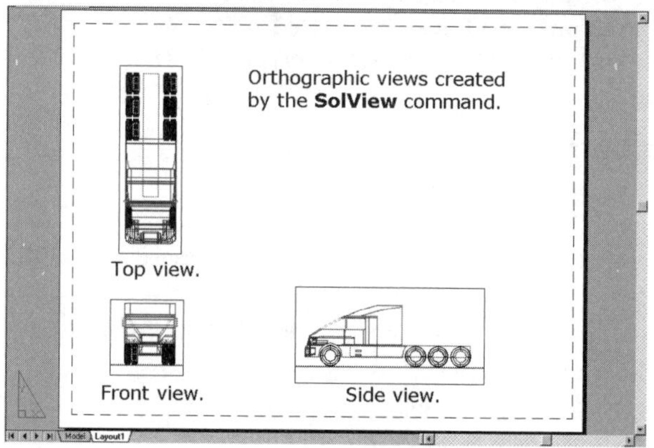

COMMAND LINE OPTIONS

Ucs options
Enter an option [Named/World/?/Current] <Current>: *(Enter an option.)*

Named creates the profile view using the x,y plane of a named UCS.

World creates the profile view using the x,y plane of the WCS.

? lists the names of existing UCSs.

Current creates the profile view using the x,y-plane of the current UCS.

Ortho options
Specify side of viewport to project: *(Pick a point.)*
Specify view center: *(Pick a point.)*
Enter view name: *(Enter a name.)*

Pick side of viewport to project selects the edge of one viewport.

View center picks the center of the view.

Clip picks two corners for a clipped view.

View name names the view.

Auxiliary options
Specify first point of inclined plane: *(Pick a point.)*
Specify second point of inclined plane: *(Pick a point.)*
Specify side to view from: *(Pick a point.)*
Specify view center: *(Pick a point.)*
Enter view name: *(Enter a name.)*

Inclined plane's 1st point picks the first point.

Inclined plane's 2nd point picks the second point.

Side to view from determines the view side.

Section options
Specify first point of cutting plane: *(Pick a point.)*
Specify second point of cutting plane: *(Pick a point.)*
Specify side to view from: *(Pick a point.)*
Enter view scale <5.9759>: *(Pick a point.)*
Specify view center <specify viewport>: *(Pick a point.)*
Specify first corner of viewport: *(Pick a point.)*
Specify opposite corner of viewport: *(Pick a point.)*
Enter view name: *(Enter a name.)*

Cutting plane 1st point picks the first point.

Cutting plane 2nd point picks the second point.

Side to view from determines the view side.

Viewscale specifies the scale of the new view.

eXit exits the command.

RELATED COMMANDS

SolDraw creates profiles and sections in viewports.

SolProf creates profile images of 3D solids.

TIPS

- This command creates orthographic, auxiliary, and sectional views.
- This command must be used before the **SolDraw** command, because it creates the layers required by **SolDraw**.
- This command is useful for creating layouts for the **SolProf** command.
- The layers created by this command have the following prefixes:

Layer Name	View
viewname-**VIS**	Visible lines.
viewname-**HID**	Hidden lines.
viewname-**DIM**	Dimensions.
viewname-**HAT**	Hatch patterns for sectional views.

- *Caution!* Autodesk warns that "The information stored on these layers is deleted and updated when you run **SolDraw**. Do not place permanent drawing information on these layers."

'SpaceTrans

2002 Converts distances between model and space units (*short for SPACE TRANSlation*).

Command	Alias	Ctrl+	F-key	Alt+	Menu Bar	Tablet
spacetrans	...	...	...	...	...	...

Command: spacetrans

This command does not operate in Model tab. In model view of a layout tab:

Specify paper space distance <1.000>: *(Enter a value.)*

In paper space of a layout tab:

Select a viewport: *(Pick a point.)*

Specify model space distance <1.000>: *(Enter a value.)*

COMMAND LINE OPTIONS

Specify paper space distance specifies the paper space length to be converted to its model space equivalent, usually the scale factor.

Select a viewport selects a paper space viewport.

Specify model space distance specifies the model space length to be converted to its paper space equivalent, usually the scale factor.

RELATED COMMANDS

Text places text in the drawing.

SolProf creates profile images of 3D solids.

TIPS

- This command is meant to be used transparently during another command, and not at the 'Command:' prompt.
- The purpose of this command is to convert lengths from model or paper space to an equivalent in the other space.
- Autodesk recommends using this command for converting model space text heights into paper space text heights.

'Spell

Rel.13 Checks the spelling of text in the drawing.

Command	Alias	Ctrl+	F-key	Alt+	Menu Bar	Tablet
spell	**sp**	...	...	**TE**	**Tools** ↳**Spelling**	**T10**

Command: spell
Select objects: *(Select one or more text objects.)*
Select objects: *(Press* **ENTER** *to end object selection.)*

When unrecognized text is found, displays dialog box:

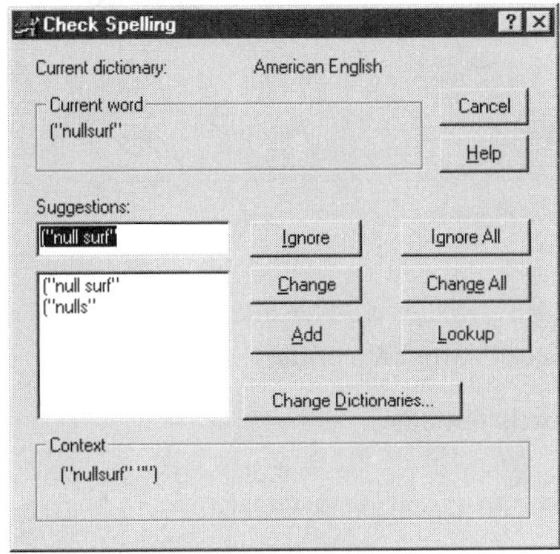

When selected text is recognized, or when spelling check is complete, displays dialog box:

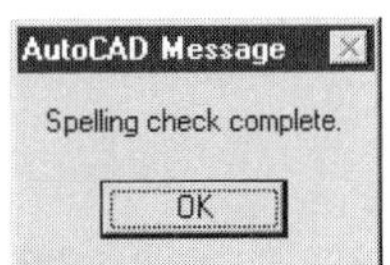

DIALOG BOX OPTIONS

Ignore ignores the word, and goes on to the next word.

Ignore All ignores all words with this spelling.

Change changes the word to the suggested spelling.

Change All changes all words with this spelling.

Add adds the word to the user (custom) dictionary.

Lookup checks the spelling of the word in the Suggestions box.

Change dictionaries selects a different dictionary; displays the Change Dictionaries dialog box.

Change Dictionaries dialog box

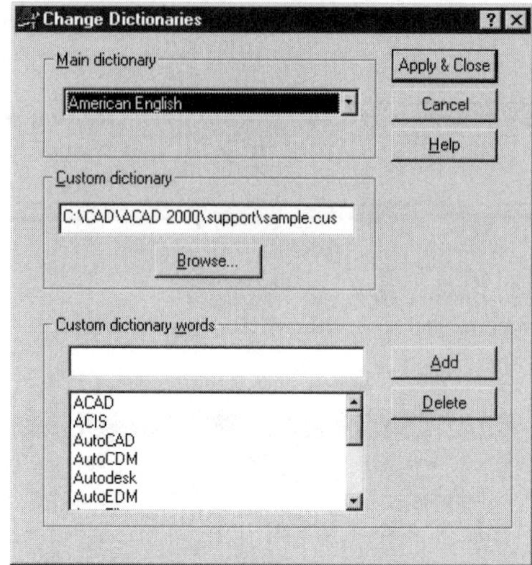

Main dictionary selects a language for the dictionary.

Custom dictionary options

Directory specifies the drive, folder, and filename of the custom dictionary.

Browse displays the **Select Custom Dictionary** dialog box.

Custom Dictionary Words options:

Add adds a word.

Delete removes a custom word from the dictionary.

RELATED COMMANDS

DdEdit edits text.

MText places paragraph text.

Text places lines of text in the drawing.

RELATED SYSTEM VARIABLES

DctCust specifies the name of the custom spelling dictionary.

DctMain specifies the name of the main spelling dictionary.

RELATED FILES

enu.dct is the dictionary word file.

.cus are the custom dictionary files.

TIPS

- A spell checker does *not* check your spelling; words that are spelled correctly but used incorrectly (such as *its* and *it's*) are not flagged. Rather, a spell checker looks for words that it does not recognize, which are words not in its dictionary file.
- As of AutoCAD 2002, the **Spell** command also checks words in blocks.

Sphere

Rel.11 Draws 3D spheres as solid models.

Command	Alias	Ctrl+	F-key	Alt+	Menu Bar	Tablet
sphere	...	...	...	DIS	**Draw** ↳**Solids** ↳**Sphere**	K7

Command: sphere
Current wire frame density: ISOLINES=4
Specify center of sphere <0,0,0>: *(Pick a point.)*
Specify radius of sphere or [Diameter]: *(Enter a value, or type* **D***.)*

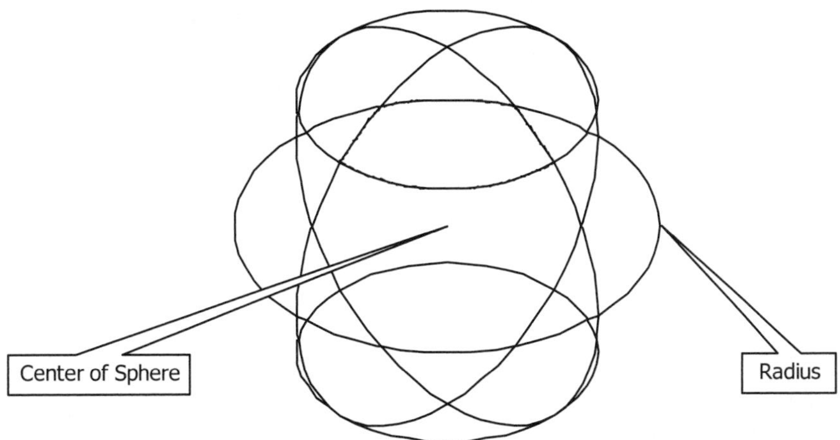

COMMAND LINE OPTIONS

Specify center of sphere locates the center point of the sphere.

Diameter specifies the diameter of the sphere.

Radius specifies the radius of the sphere.

RELATED COMMANDS

Box draws solid boxes.

Cone draws solid cones.

Cylinder draws solid cylinders.

Torus draws solid tori.

Wedge draws solid wedges.

Ai_Sphere draws surface meshed spheres.

RELATED SYSTEM VARIABLES

DispSilh specifies the silhouette display of 3D solids:

DispSilh	Meaning
0	Off.
1	On.

IsoLines specifies the number of tessellation lines that define the surface of the 3D solid:

IsoLines	Meaning
0	Minimum.
4	Default.
2047	Maximum.

TIPS

- The **Sphere** command places the sphere's central axis parallel to the z axis of the current UCS, with the latitudinal isolines parallel to the x,y plane.
- Notice the effect of the **DispSilh** system variable, which toggles the silhouette display of 3D solids, after executing the **Hide** command:

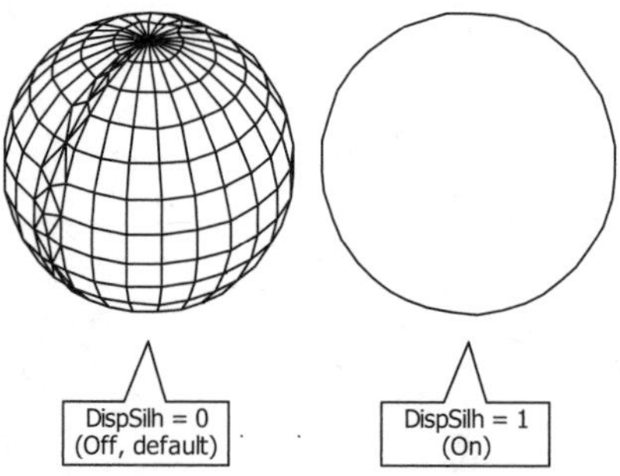

- Notice the effect of the **IsoLines** system variable, which controls the number of tessellation lines used to define the surface of a 3D solid:

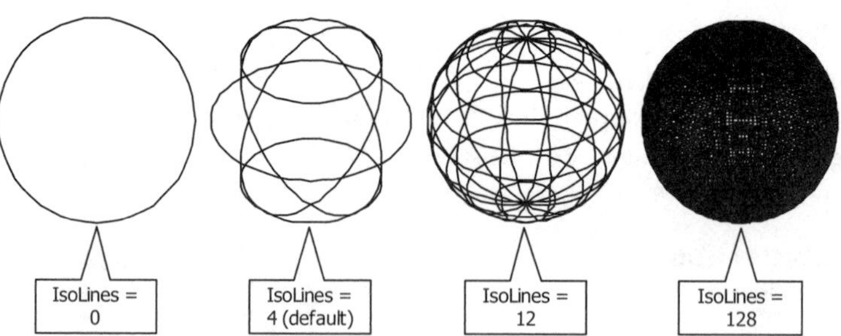

- You must use the **Regen** command after changing the **DispSilh** and **IsoLines** system variables.

Spline

Rel.13 Draws NURBS (*Non-Uniform Rational Bezier Spline*) curves.

Command	Alias	Ctrl+	F-key	Alt+	Menu Bar	Tablet
spline	**spl**	...	...	**DS**	**Draw** ↳**Spline**	**L9**

Command: spline
Specify first point or [Object]: *(Pick a point, or type* **O***.)*
Specify next point: *(Pick a point.)*
Specify next point or [Close/Fit tolerance] <start tangent>: *(Pick a point, or enter an option.)*
Specify next point or [Close/Fit tolerance] <start tangent>: *(Press* ENTER *to define tangency points.)*
Specify start tangent: *(Pick a point, or press* ENTER*.)*
Specify end tangent: *(Pick a point, or press* ENTER*..)*

Open spline

Closed spline

COMMAND LINE OPTIONS

Specify first point picks the starting point of the spline.
Object converts 2D and 3D splined polylines into a NURBS spline.
Specify next point picks the next tangent point.
Close closes the spline at the start point.
Fit changes the spline tolerance; 0 = curve passes through fit points.
Specify start tangent specifies the tangency of the starting point of the spline.
Specify end tangent specifies the tangency of the endpoint of the spline.

RELATED COMMANDS

PEdit edits splined polylines.
PLine draws splined polylines.
SplinEdit edits NURBS splines.

RELATED SYSTEM VARIABLE

DelObj toggles whether the original polyline is deleted with the **Object** option.

TIP

- A closed spline has the same start and end tangent.

SplinEdit

Rel.13 Edits NURBS splines.

Command	Alias	Ctrl+	F-key	Alt+	Menu Bar	Tablet
splinedit	**spe**	...	...	...	**Modify** ↳**Spline**	**Y18**

Command: splinedit
Select spline: *(Select one spline object.)*
Enter an option [Fit data/Close/Move vertex/Refine/rEverse/Undo]: *(Enter an option.)*

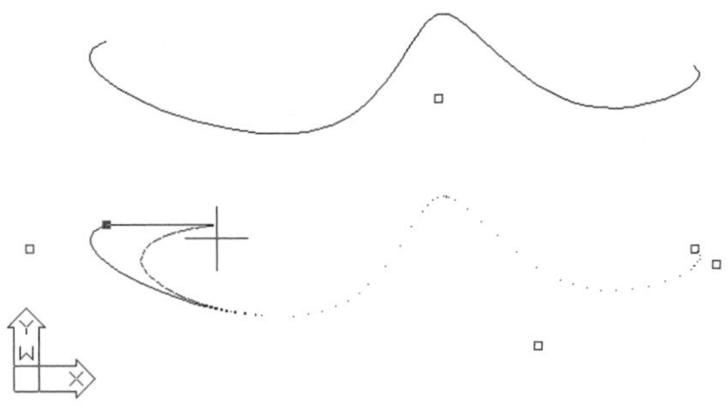

COMMAND LINE OPTIONS

Close closes the spline, if open.

Move vertex moves a control vertex.

Open opens the spline, if closed.

Refine adds a control point; changes the spline's order or weight.

rEverse reverses the spline's direction.

Undo undoes the most-recent edit change.

eXit exits the command.

Fit Data options
Enter a fit data option [Add/Close/Delete/Move/Purge/Tangents/toLerance/eXit] <eXit>: *(Enter an option.)*

Add adds fit points.

Close closes the spline, if open.

Delete removes fit points.

Move moves fit points.

Open opens the spline, if closed.

Purge removes fit point data from the drawing.

Tangents edits the start and end tangents.

toLerance refits the spline with the new tolerance value.

eXit exits suboptions.

RELATED COMMANDS

PEdit edits a splined polyline.

PLine creates polylines, including splined polylines.

Spline draws a NURBS spline.

TIPS

- The spline loses its fit data when you use these **SplinEdit** command options:

  Refine

  Fit Purge

  Fit Tolerance followed by **Fit Move**

  Fit Tolerance followed by **Fit Open** or **Fit Close**.
- The maximum order for a spline is 26; once the order has been elevated, it cannot be reduced.
- The larger the weight, the closer the spline is to the control point.
- This command automatically converts a spline-fit polyline into a spline object, even if you do not edit the polyline.

 Standards

2002 Loads standards into the current drawing.

Command	Alias	Ctrl+	F-key	Alt+	Menu Bar	Tablet
standards	...	...	...	TSC	Tools ↳CAD Standards ↳Configure	...

Command: standards

Displays tabbed dialog box:

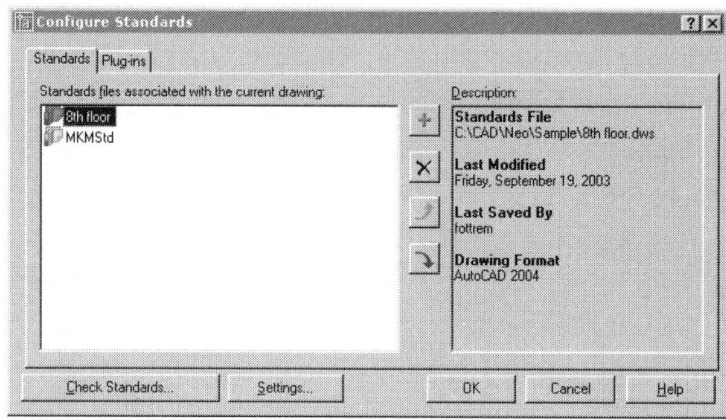

DIALOG BOX OPTIONS

Add adds a standards file (F3); displays the Select Standards File dialog box.

Remove Standards removes the selected standards file (DEL).

Move up moves a standards file higher in the list.

Move down moves a standards file lower in the list.

Check Standards displays the Check Standards dialog box; see the **CheckStandards** command.

Settings displays CAD Standards Settings dialog box (*new to AutoCAD 2005*). See **CheckStandards**.

Plug-ins tab

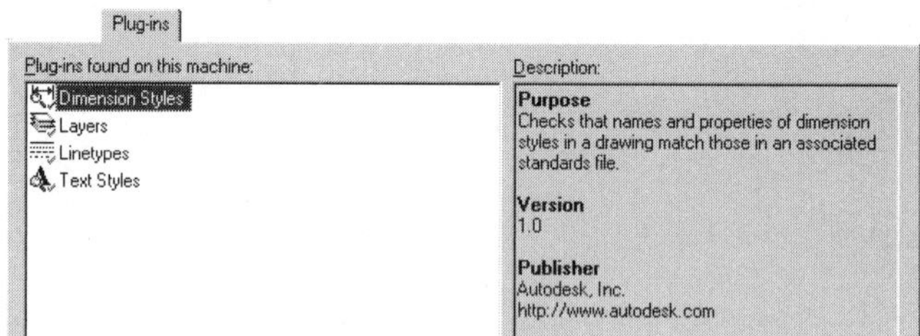

Displays information about each item in the CAD standard.

RELATED COMMAND

CheckStandards checks the drawing against standards loaded by the **Standards** command.

RELATED FILE

***.dws** are the CAD Standards files, stored in DWG format.

Stats

Rel.12 Lists statistics of the most-recent rendering (*short for STATisticS*).

Command	Alias	Ctrl+	F-key	Alt+	Menu Bar	Tablet
stats	**...**	**...**	**...**	**VET**	**View** **↳Render** **↳Statistics**	**...**

Command: stats

Displays dialog box:

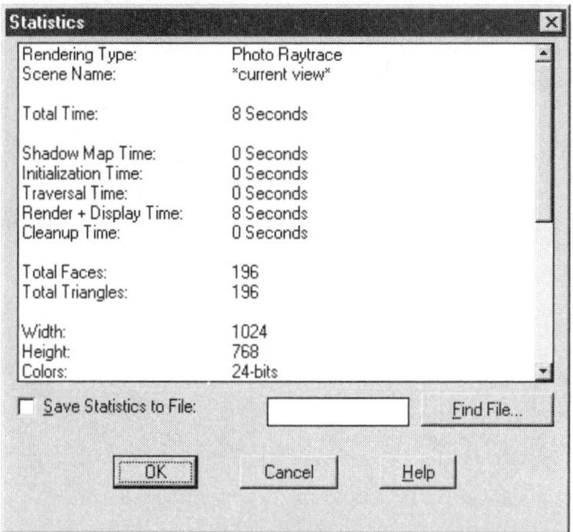

DIALOG BOX OPTION

☐ **Save Statistics to File** saves the rendering statistics to the file specified in the adjacent text entry box.

RELATED AUTOCAD COMMAND

Render creates renderings.

TIP

- The dialog box is blank if no rendering has taken place in the drawing.

DEFINITIONS

Scene Name — name of the currently-selected scene; when no scene is current, displays "(none)."

Last Rendering Type — name of the currently-selected renderer; default is AutoCAD Render.

Rendering Time — time required to the create the most recent rendering; reported in HH:MM:SS (hours, minutes, seconds) format.

Total Faces — number of faces processed in the most recent rendering; a single 3D object consists of many faces.

Total Triangles — number of triangles processed in the most recent rendering; a rectangular face is typically divided into two triangles.

'Status

V. 1.0 Displays information about the current drawing and environment.

Command	Alias	Ctrl+	F-key	Alt+	Menu Bar	Tablet
status	...	...	...	**TYS**	**Tools** ↳**Inquiry** ↳**Status**	...

Command: status

Example output:

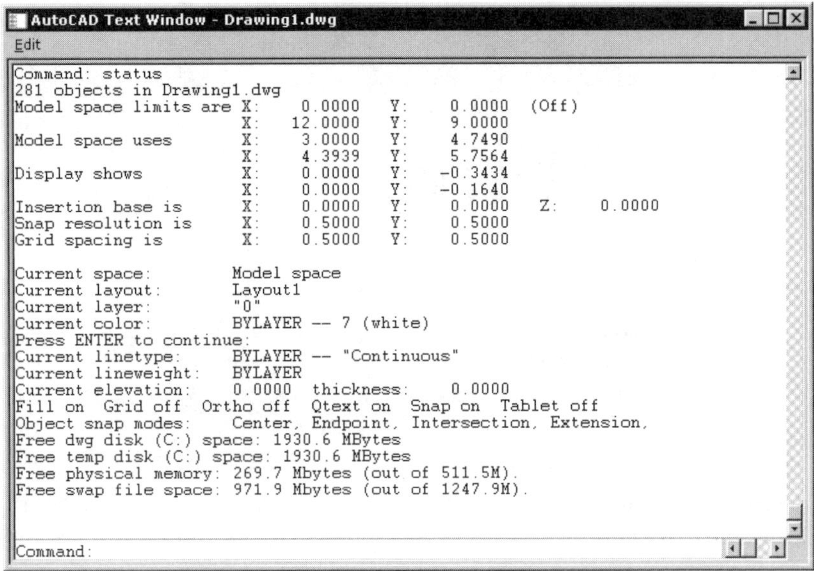

COMMAND LINE OPTIONS

ENTER continues the listing.

F2 returns to the graphics screen.

RELATED COMMANDS

DbList lists information about all the objects in the drawing.

List lists information about the selected objects.

Stats lists information about the most recent rendering.

DEFINITIONS

Model Space limits, Paper Space limits

– the x,y coordinates stored in the **LimMin** and **LimMax** system variables; 'Off' indicates limits checking is turned off (**LimCheck**).

Model Space uses, Paper Space uses

– the x,y coordinates of the lower-left and upper-right extents of objects in the drawing; 'Over' indicates drawing extents exceed the drawing limits.

Display shows

– the x,y coordinates of the lower-left and upper-right corners of the current display.

Insertion base is

– the x,y,z coordinates stored in system variable **InsBase**.

Snap resolution is

Grid spacing is

– the snap and grid settings, as stored in the **SnapUnit** and **GridUnit** system variables.

Current space

– the indication of whether whether model space or paper space is current.

Current layout

– the name of the current layout.

Current layer

Current color

Current linetype

Current lineweight

Current plot style

Current elevation

Thickness

– the current values for the layer name, color, linetype name, elevation, and thickness, as stored in system variables **CLayer**, **CeColor**, **CeLType**, **CeLweight, CPlotSytle, Elevation**, and **Thickness**.

Fill

Grid

Ortho

Qtext

Snap

Tablet

– the current settings for the fill, grid, ortho, qtext, snap, and tablet modes, as stored in the system variables **FillMode**, **GridMode**, **OrthoMode**, **TextMode**, **SnapMode**, and **TabMode**.

Object Snap modes

– the currently-set object modes, as stored in the system variable **OsMode**.

Free disk (dwg + temp = C)

– the amount of free disk space on the drive storing AutoCAD's temporary files, as held by by system variable **TempPrefix**.

Free physical memory

– the amount of free RAM.

Free swap file space

– the amount of free space in AutoCAD's swap file on disk.

StlOut

Rel.12 Exports 3D solids and bodies in binary or ASCII SLA format (*short for STereoLithography OUTput*).

Command	Alias	Ctrl+	F-key	Alt+	Menu Bar	Tablet
stlout	...	...	...	**FE**	**File**	...
				↳STL	**↳Export**	
					↳Lithography	

Command: stlout
Select a single solid for STL output...
Select objects: *(Select one or more solid objects.)*
Select objects: *(Press* **ENTER** *to end object selection.)*
Create a binary STL file? [Yes/No] <Y>: *(Type* **Y** *or* **N**.*)*

*Displays the Create STL File dialog box; enter a name, and then click **Save**.*

COMMAND LINE OPTIONS

Select objects selects a single 3D solid object.

Y creates a binary-format *.sla* file.

N creates an ASCII-format *.sla* file.

RELATED COMMANDS

All solid modeling commands.

AcisOut exports 3D solid models to an ASCII SAT-format ACIS file.

AmeConvert converts AME v2.x solid models into ACIS models.

RELATED SYSTEM VARIABLE

FaceTRes determines the resolution of triangulating solid models.

RELATED FILE

***.stl** is the the SLA-compatible file with STL extension created by this command.

TIPS

- The solid model must lie in the positive x,y,z-octant of the WCS.
- The **StlOut** command exports a single 3D solid; it does not export regions or any other AutoCAD object.
- Even though this command prompts you twice to 'Select objects', selecting more than one solid causes AutoCAD to complain, "Only one solid per file permitted."
- The resulting *.stl* file cannot be imported back into AutoCAD.

DEFINITIONS

STL — stereolithography data file, which consists of a faceted representation of the model.

SLA — stereoLithography Apparatus.

Stretch

V. 2.5 Stretches objects to lengthen, shorten, or distort them.

Command	Alias	Ctrl+	F-key	Alt+	Menu Bar	Tablet
stretch	**s**	...	...	**MH**	**Modify** ↳**Stretch**	**V22**

Command: stretch
Select objects to stretch by crossing-window or crossing-polygon...
Select objects: *(Type* **C** *or* **CP***.)*
Specify first corner: *(Pick a point.)*
Specify opposite corner: *(Pick a point.)*
Select objects: *(Press* **ENTER** *to end object selection.)*
Specify base point or displacement: *(Pick a point.)*
Specify second point of displacement or <use first point as displacement>: *(Pick a point.)*

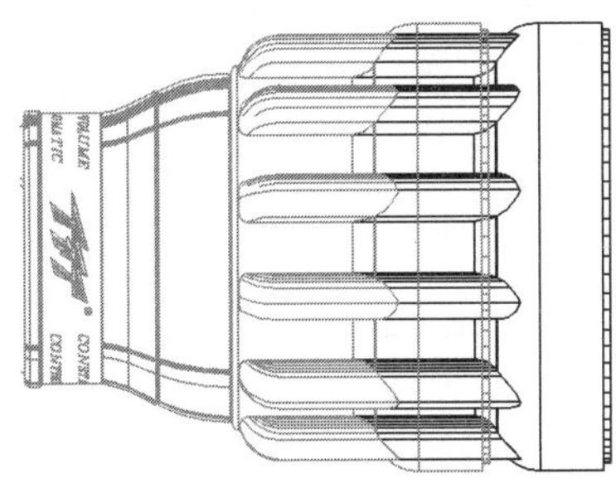

COMMAND LINE OPTIONS

First corner selects object; must be CPolygon or Crossing object selection.

Select objects selects other objects using any selection mode.

Base point indicates the starting point for stretching.

Second point stretches the object larger or smaller.

RELATED COMMANDS

Change changes the size of lines, circles, text, blocks, and arcs.

Scale increases or decreases the size of any object.

TIPS

- The effect of the **Stretch** command is not always obvious; be prepared to use the **Undo** command.
- The first time you select objects for the **Stretch** command, you must use **Crossing** or **CPolygon** object selection; objects entirely within the selection window are moved.
- The **Stretch** command will not move a hatch pattern unless the hatch's origin is included in the selection set.
- Use the **Stretch** command to update associative dimensions automatically by including the dimension's endpoints in the selection set.

'Style

V. 2.0 Creates and modifies text styles, which define the properties of fonts.

Commands	Aliases	Ctrl+	F-key	Alt+	Menu Bar	Tablet
style	**st**	...	...	**OS**	**Format**	**U2**
	ddstyle				**↳Text Style**	
-style						

Command: style

Displays dialog box:

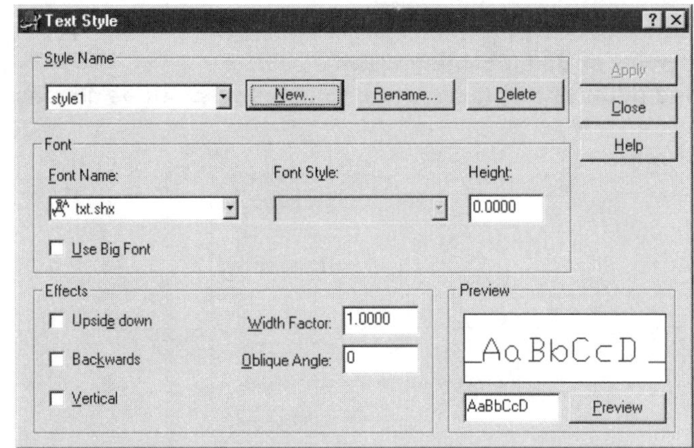

DIALOG BOX OPTIONS

Style Name options

Style Name selects an existing text style.

New creates new text styles; displays dialog box:

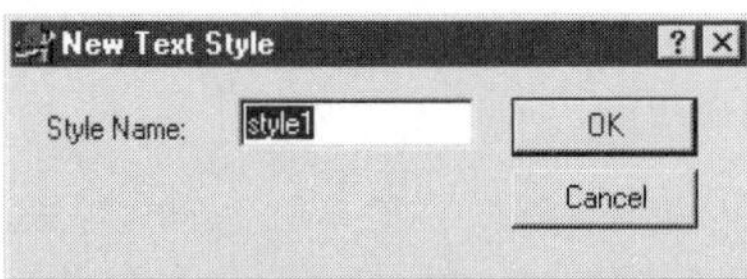

Rename renames existing text styles; displays dialog box:

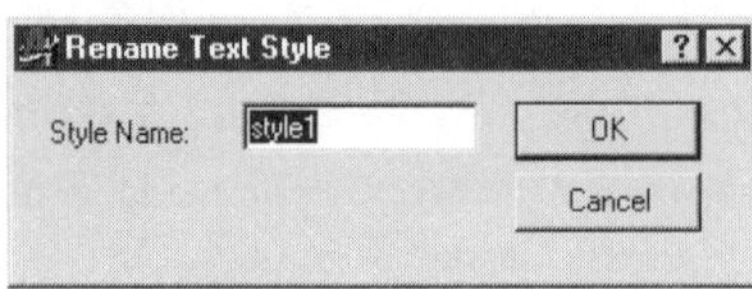

Delete deletes text styles.

Font options

Font Name specifies the names of AutoCAD SHX and TrueType TTF fonts.

Font Style selects from available font styles, such as **Bold**, *Italic*, and ***Bold Italic***.

Height specifies the text height.

☐ **Use Big Font** specifies the use of a big font file, typically for Asian alphabets.

Effects options (not available for all fonts)

☐ **Upside Down** draws text upside down:

☐ **Backwards** draws text backwards:

☐ **Vertical** draws text vertically.

Width Factor changes the width of characters.

Width factor = 0.5 (left) and 2.0 (right):

Oblique Angle slants characters forward or backward:

Oblique angle = 30 (left) and -30 (right):

Preview previews the effects on the style.

Buttons

Apply applies the changes to the style.

Close closes the dialog box; in some cases, you can click the **Close** button before clicking the **Apply** button; then AutoCAD displays the following warning dialog box:

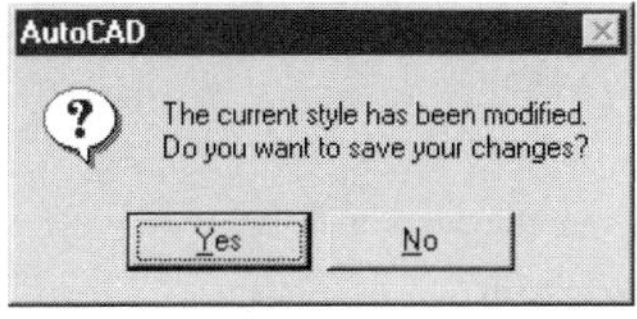

-STYLE Command

Command: -style

Enter name of text style or [?] <STANDARD>: *(Enter a name, or type* **?***.)*

Specify full font name or font filename (TTF or SHX) <txt>: *(Enter a name.)*

Specify height of text <0.0000>: *(Enter a value.)*

Specify width factor <1.0000>: *(Enter a value.)*

Specify obliquing angle <0>: *(Enter a value.)*

Display text backwards? [Yes/No] <N>: *(Type* **Y** *or* **N***.)*

Display text upside-down? [Yes/No] <N>: *(Type* **Y** *or* **N***.)*

Vertical? <N> *(Type* **Y** *or* **N***.)*

"STANDARD" is now the current text style.

COMMAND LINE OPTIONS

Enter name of text style names the text style; maximum = 31 characters (default = "STANDARD").

? lists the names of styles already defined in the drawing.

Specify full font name or font filename names the font file (SHX or TTF) from which the style is defined (default = *txt.shx*).

Specify height of text specifies the height of the text (default = 0 units).

Specify width factor specifies the width factor of the text (default = 1.00).

Specify obliquing angle specifies the obliquing angle or slant of the text (default = 0 degrees).

Display text backwards

- **Yes** prints text backwards — mirror writing.
- **No** prints text forwards.

Display text upside-down

- **Yes** prints text upside-down.
- **No** prints text rightside-up (default).

Vertical

- **Yes** prints text vertically; not available for all fonts.
- **No** prints text horizontally (default).

RELATED COMMANDS

Change changes the style assigned to selected text.

Purge removes any unused text style definitions.

Rename renames a text style.

MText places paragraph text.

Text places a single line of text.

RELATED SYSTEM VARIABLES

TextStyle specifies the current text style.

TextSize specifies the current text height.

RELATED FILES

***.shp** is Autodesk's format for vector source fonts.

***.shx is** Autodesk's format for compiled vector fonts; stored in *\autocad 2005\fonts* folder.

***.ttf** are TrueType font files; stored in *\windows\fonts* folder.

TIPS

- The **Style** command affects the font used with the **Text** and **MText**, as well as dimension and table styles.
- A **Width Factor** of 0.85 fits in 15% more text without sacrificing legibility.
- An **Obliquing Angle** of +15% can sometimes enhance the look of a font.
- The **Obliquing Angle** can be positive (forward slanting) or negative (backward slanting).
- To use PostScript fonts in the drawing, use the **Compile** command to convert PostScript *.pfb* files into *.shx* format.
- You can use any TrueType font with AutoCAD.
- Some of the fonts included with AutoCAD:

- A text height of 0 lets you specify the text height while you are adding text with the **Text**, **DText**, and **MText** commands. If you specify a text height other than 0 in the **Style** command, that height is always used with that particular style name.

StylesManager

2000 Displays the Plot Styles window.

Command	Alias	Ctrl+	F-key	Alt+	Menu Bar	Tablet
stylesmanager	...	...	...	**FY**	**File** ⇘**Plot Style Manager**	**Y24**

You can also access this command through the Windows Control Panel: ***Autodesk Plot Style Manager.***

Command: stylesmanager

Displays window:

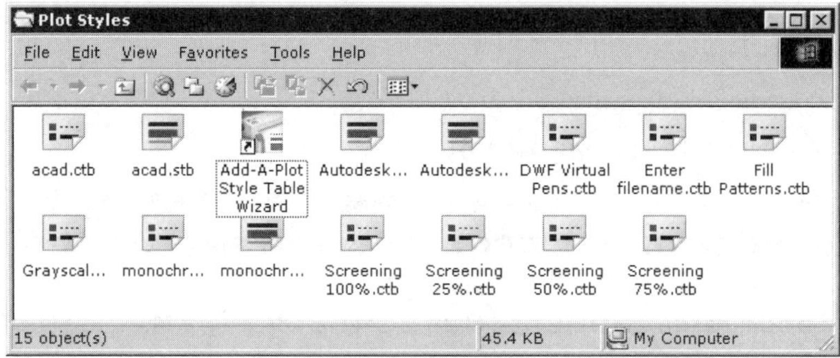

WINDOW OPTIONS

***.ctb** (color-table based) opens the Plot Style Table Editor dialog box.

*** .stb** (style-table based) opens the Plot Style Table Editor dialog box, as well.

Add-A-Plot Style Table Wizard opens Add Plot Style Table wizard; see the R14PenWizard command.

DEFINITIONS

StylesManager — the program that modifies plot styles stored in plot style tables.

Plot styles — when drawings have a plot style table attached to their model and layout tabs, any changes to the plot style changes the object using the plot style. Plot styles can be assigned by object or by layer. See the **PlotStyle** and **Layer** commands.

Color-table based (.cbt) — assigns colors to objects and layers, as used by older releases of AutoCAD. For example, the color of the object specifies the width of pen. This style of controlling the plot is now called "color dependent."

Style-table based (.stb) — newer releases of AutoCAD can control every aspect of the plot through "plot styles." By changing the *.stb* file attached to a layout tab, you immediately change the plot style for all objects and layers in the layout. This, for example, allows a quick change from monochrome to color plotting. A single drawing file can contain multiple plot styles.

DIALOG BOX OPTIONS

General tab

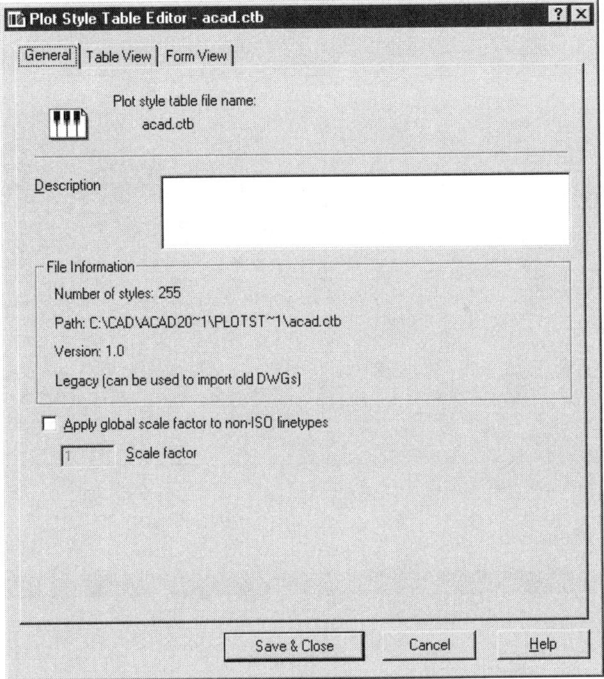

Description allows you to describe the plot style.

☐ **Apply global scale factor to non-ISO linetypes and fill patterns** applies the scale factor to all non-ISO linetypes and hatch patterns in the plot.

Scale factor specifies the scale factor.

Save and Close saves the changes, and closes the dialog box.

Cancel cancels the changes, and closes the dialog box.

CBT File Table View tab

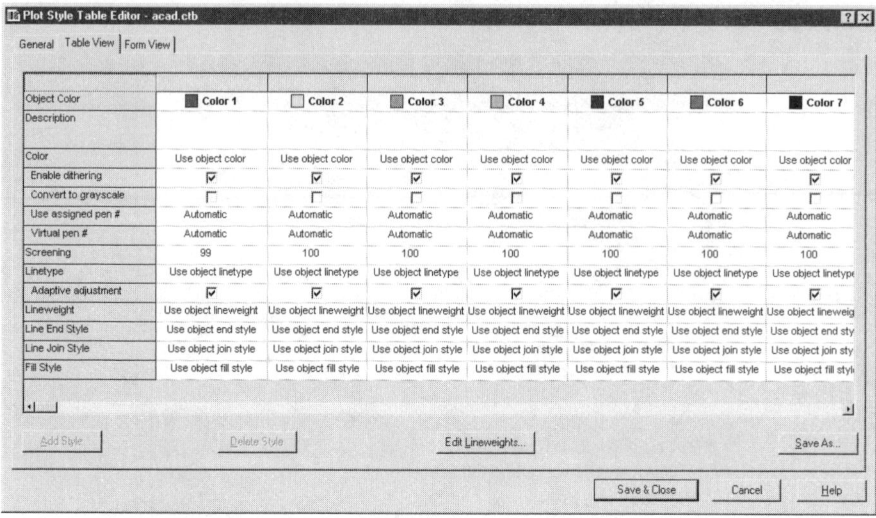

See the following pages for options.

CBT File Form View tab

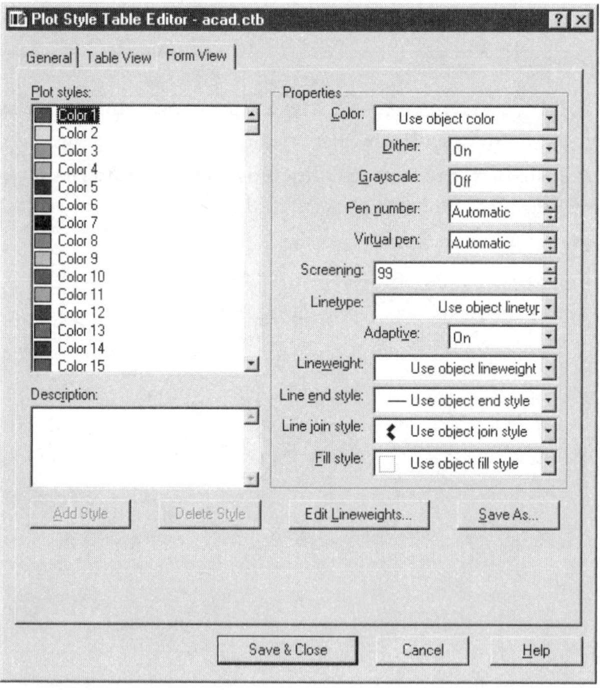

See the following pages for options.

STB File Table View tab

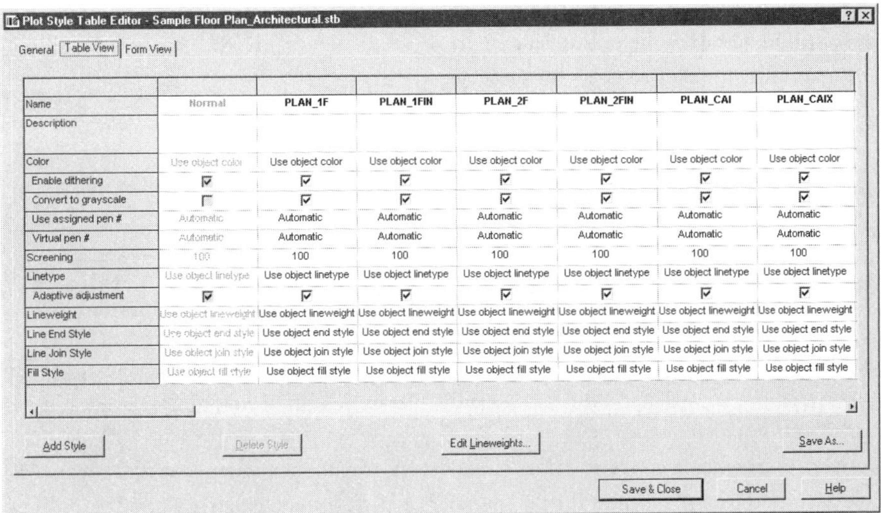

See the following pages for options.

STB File Form View tab

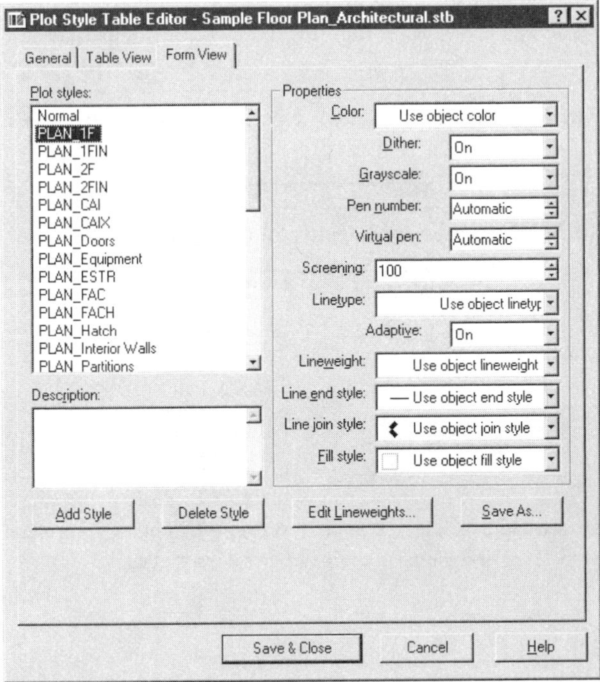

See the following pages for options.

Table View and Form View options are identical

Object color specifies the color of the object.

Description describes the plot style.

Color specifies the plotted color for the objects.

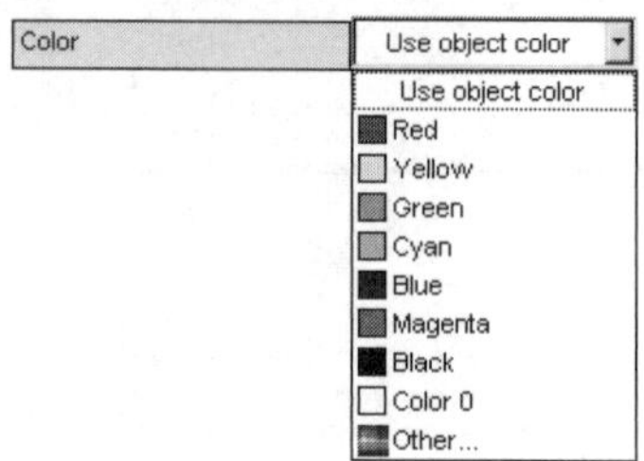

Enable dithering toggles dithering, if the plotter supports dithering, to generate more colors than the plotter is capable of; this setting should be turned off for plotting vectors, and turned on for plotting renderings.

Convert to grayscale converts colors to shades of gray, if the plotter supports gray scale.

Use assigned pen # specifies the pen number of pen plotters; range is 1 to 32.

Virtual pen number specifies the virtual pen number (default = Automatic); range is 1 to 255. A value of 0 (or Automatic) tells AutoCAD to assign virtual pens from ACI (AutoCAD Color Index); this setting is meant for non-pen plotters that can make use of virtual pens.

Screening specifies the intensity of plotted objects; range is 0 (plotted "white") to 100 (full density):

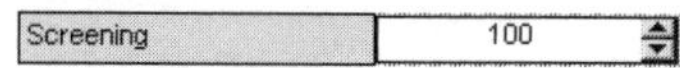

Linetype specifies the linetype with which to plot the objects:

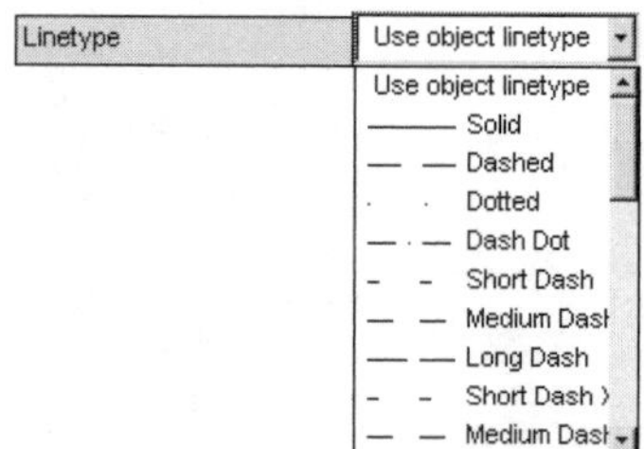

Adaptive adjustment adjusts linetype scale to prevent the linetype from ending in the middle of its pattern; keep off when the plotted linetype scale is crucial.

Lineweight specifies how wide lines are plotted, in millimeters:

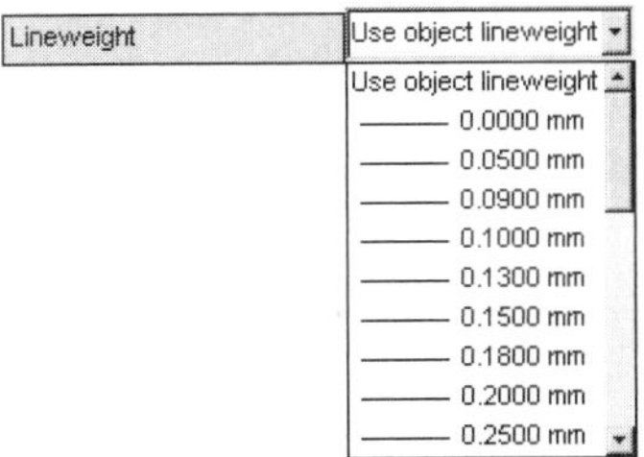

Line end style specifies how the ends of lines are plotted:

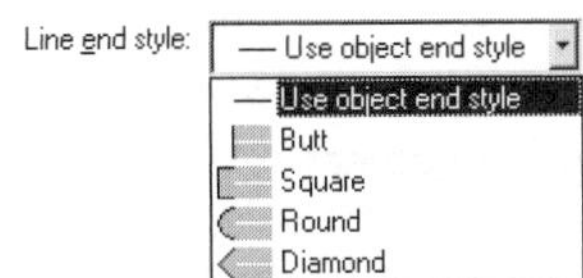

Line join style specifies how the intersections of lines are plotted:

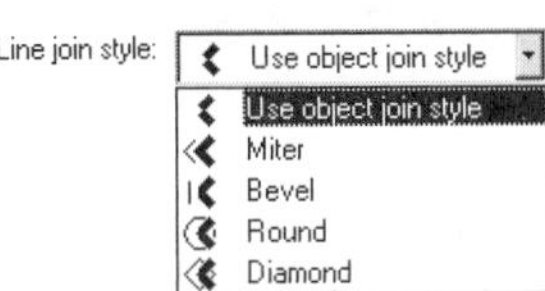

Fill style specifies how objects are filled:

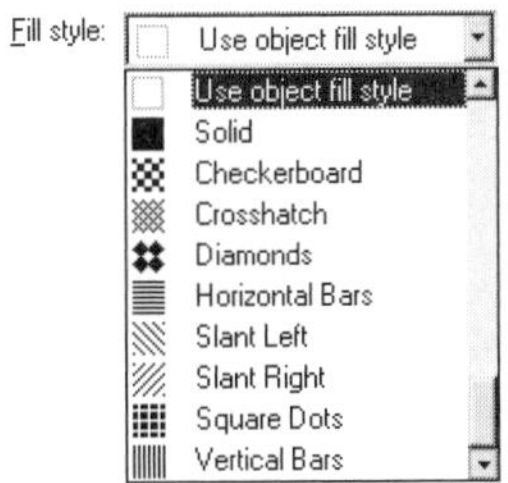

Add Style adds a plot style.

Delete Style removes a plot style.

Edit Lineweights displays the Edit Lineweights dialog box.

Save As displays the Save As dialog box.

TIPS

- To configure a printer or plotter for *virtual pens*:
    1. In the **Device and Document Settings** tab, open the PC3 Editor dialog box.
    2. In the **Vector Graphics** section, select **255 Virtual Pens** from **Color Depth**.
- **CTB** is short for "color-dependent based" style table, which is compatible with earlier versions of AutoCAD.
- **STB** is short for "style-table based."
- You can attach a different *.stb* file to each layout and to the model tab in a drawing.

Subtract

Rel.11 Removes the volume of one set of 3D models or 2D regions from another.

Command	Alias	Ctrl+	F-key	Alt+	Menu Bar	Tablet
subtract	**su**	...	...	**MNS**	**Modify** ↳**Solids Editing** ↳**Subtract**	**X16**

Command: subtract
Select objects: *(Select one or more solid objects.)*
Select objects: *(Press* **ENTER** *to end object selection.)*
1 solid selected.
Objects to subtract from them...
Select objects: *(Select one or more solid objects.)*
Select objects: *(Press* **ENTER** *to end object selection.)*
1 solid selected.

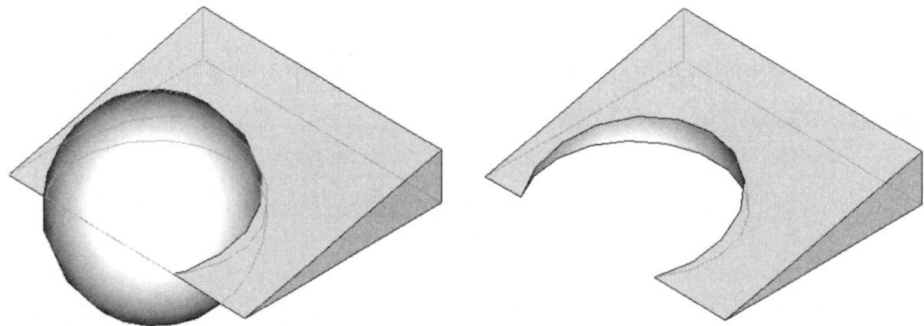

Original objects (left); sphere subtracted from wedge (right).

COMMAND LINE OPTION

Select objects selects the objects to be subtracted.

RELATED COMMANDS

Intersect removes all but the intersection of two solid volumes.

Union joins two solids.

TIPS

- AutoCAD subtracts the objects you select *second* from the objects you select *first*.
- You can use this commands on 3D solids and 2D regions.
- To subtract one region from another, both must lie in the same plane.
- When one solid is fully inside another, AutoCAD performs the subtraction, but reports, "Null solid created — deleted."

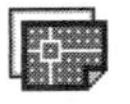

SysWindows

Rel.13 Controls multiple windows (*short for SYStem WINDOWS*).

Command	Alias	Ctrl+	F-key	Alt+	Menu Bar	Tablet
syswindows	...	F6	...	W	**Windows**	...
					↳*varies*	

Command: syswindows
Enter an option [Cascade/tile Horizontal/tile Vertical/Arrange icons]: *(Enter an option.)*

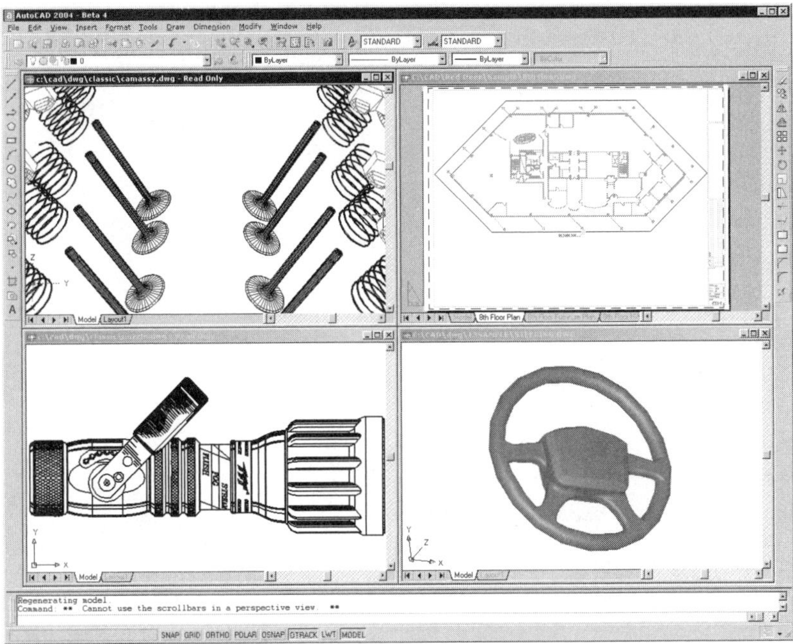

COMMAND LINE OPTIONS

Arrange icons arranges icons in an orderly fashion.

Cascade cascades the window.

tileHorizontal tiles the window horizontally.

tileVertical tiles the window vertically.

TITLE BAR OPTIONS

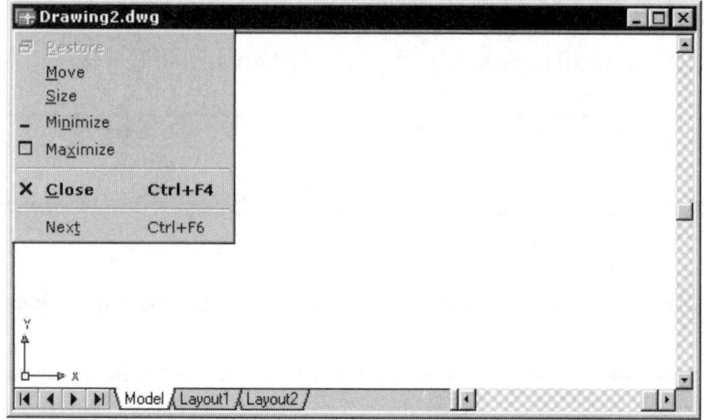

Restore restores the window to its "windowized" size.

Move moves the window.

Size resizes the window.

Minimize minimizes the window.

Maximize maximizes the window.

Close closes the window.

Next switches the focus to the next window.

RELATED COMMANDS

Close closes the current window.

CloseAll closes all windows.

Open opens one or more drawings, each in its own window.

MView creates paper space viewports in a window.

Vports creates model space viewports in a window.

TIPS

- The **SysWindows** command had no practical effect until AutoCAD 2000, because AutoCAD Release 13 and 14 supported a single window only.
- Window control icons:

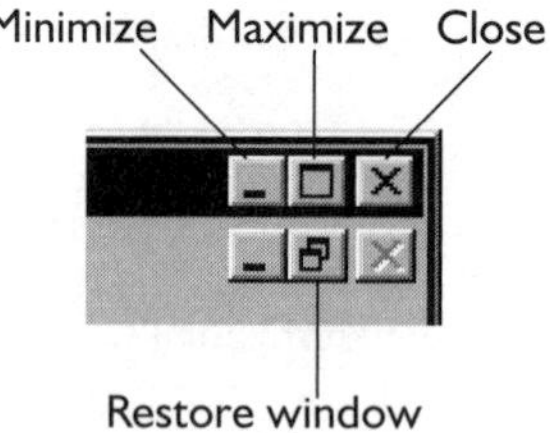

- Press **CTRL+F6** to switch quickly between currently-loaded drawings.
- Press **CTRL+F4** to close the current drawing.

Table

2005 Inserts formatted tables in drawings.

Command	Alias	Ctrl+	F-key	Alt+	Menu Bar	Tablet
table	**tb**	...	...	**DA**	**Draw** ⇘**Table**	...
-table						

Command: table

Displays dialog box:

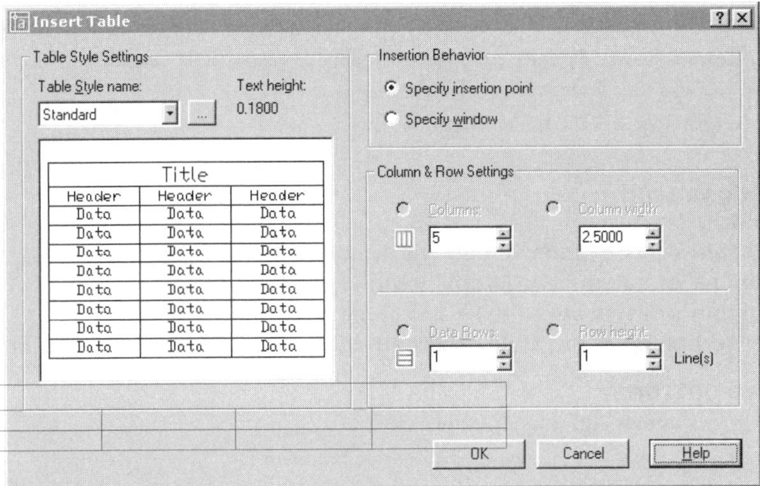

DIALOG BOX OPTIONS

Table Style Settings options

Table Style Name selects from a list of pre-formatted styles; new drawings contain just one style, named Standard.

... displays the Table Style dialog box to edit or create new styles; see **TableStyle** command.

Insertion Behavior options

⊙**Specify insertion point** locates the table by its upper-left corner; after you click **OK**, AutoCAD prompts:

Specify insertion point: *(Pick a point.)*

AutoCAD inserts the table in the drawing (default = 2 rows x 5 columns):

Title row
Header row
Cell

○**Specify window** fits the table to the window; after you click **OK**, AutoCAD prompts:

Specify first corner: *(Pick point 1.)*

Specify second corner: *(Pick point 2.)*

AutoCAD places the table at the first pick point, and fits it to the second pick point.

Pick Point 1 (insertion point)

Pick Point 2 (lower right corner)

Column & Row Settings options

- **Columns** specifies the initial number of columns.
- **Column width** specifies the initial width of columns.
- **Data Rows** specifies the initial number of "data" rows; the style determines whether the header and title rows are also placed.
- **Row height** specifies the initial height of rows.

-TABLE Command

Command: -table
Current table style: Standard Cell width: 2.5000 Cell height: 1 line(s)
Enter number of columns or [Auto] <5>: *(Type* **A***, or enter a number.)*
Enter number of rows or [Auto] <1>: *(Type* **A***, or enter a number.)*
Specify insertion point or [Style/Width/Height]: *(Pick a point, or enter an option.)*

COMMAND OPTIONS

Auto creates columns and rows to fit the table.

Style selects the table style name; see **TableStyle** command.

Width specifies the initial width of the table.

Height specifies the initial height of the table.

RELATED COMMANDS

MatchCell matches the style of one cell to other cells.

TableEdit edits text in table cells.

TableExport exports tables in CSV format.

TableStyle defines table styles.

RELATED SYSTEM VARIABLE

CTableStyle specifies the current table style name.

TIPS

- To copy tables between drawings, using DesignCenter.
- Whether the table extends up or down from the insertion point depends on its style.
- The minimum width of a column is one character; the minimum height of a row is the height of a character + margins.

TablEdit

2005 Edits the text in cells; other methods edit the table itself.

Command	Alias	Ctrl+	F-key	Alt+	Menu Bar	Tablet
tabledit	...	...	...	...	...	...

As an alternative to using the **TablEdit** *command, double-click a cell to edit its content.*

Command: tabledit

Pick a table cell: *(Pick inside a cell.)*

AutoCAD highlights the cell, and then displays the mtext editor (see the **MText** *command):*

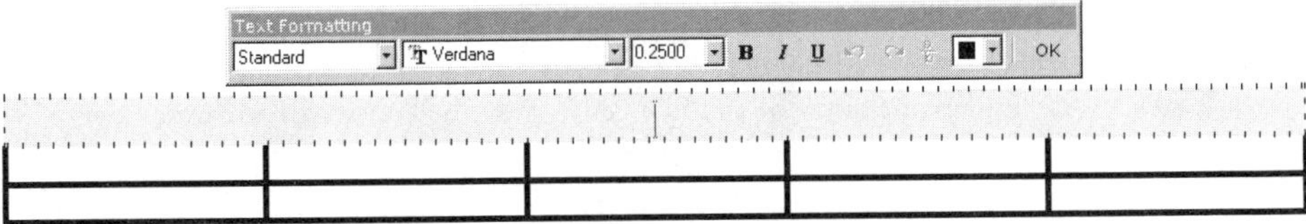

To edit other cells, press the arrow keys; to move to the "next" cell, press **Tab**.

ADDITIONAL CELL EDITING OPTIONS

Click the center of a cell, right-click, and then select options from the shortcut menu.

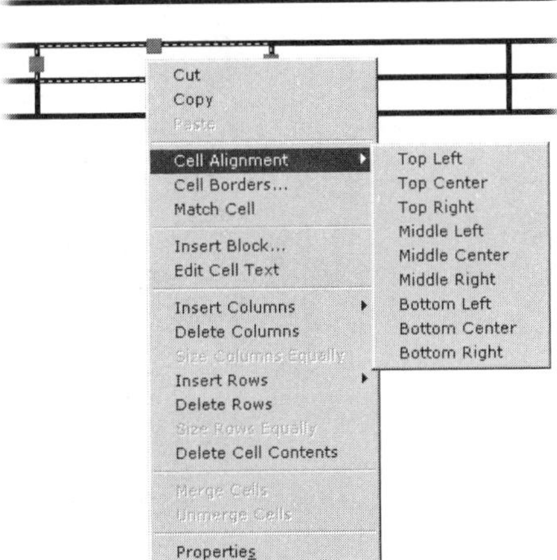

Cell Formating options

Cell Alignment displays a submenu of alignment options.

Cell Borders displays the Cell Border Properties dialog box; change the thickness and color of cell border lines, and then click **OK**.

Match Cell copies the formatting of the current cell to other cells; see the **MatchCell** command.

Cell Editing options

Insert Block displays the Insert a Block in a Table Cell dialog box; select a block, set its options, and then click **OK**. Only one block can be inserted per cell.

Edit Cell Text displays the mtext editor.

Row and Column options

Insert Columns inserts one column to the left or right of the current cell; does not add columns to the title row.

Delete Columns deletes the current column without warning.

Size Columns Equally (*available only when two or more cells are selected*) makes selected columns the same width.

Insert Rows inserts one row above or below the current cell.

Delete Rows deletes the current row without warning.

Size Rows Equally (*available only when two or more cells are selected*) makes selected rows the same height.

Delete Cell Contents erases text and blocks from the selected cell without warning.

Cell Merging options

Merge Cells (*available only when two or more cells are selected*) joins the selected cells into a single cell. *Warning!* Only the text or block of the first cell is kept; the content of the other cells is erased.

Unmerge Cells (*available only when a previously merged cell is selected*) splits merged cells back to the original number of cells; erased cell contents are not returned.

TABLE EDITING

Select a table, right-click, and then select an option from the shortcut menu:

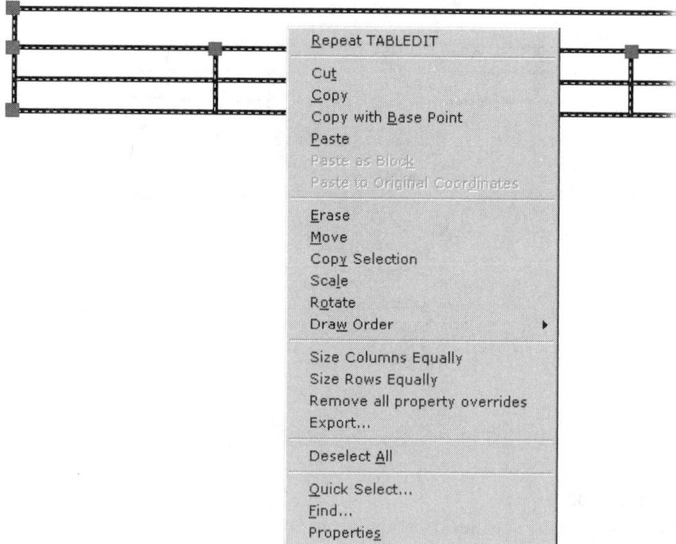

Size Columns Equally returns all columns to their original width.

Size Rows Equally adjusts the height of all rows to the height of the tallest row.

Remove All Property Overrides returns the table to its original style settings.

Export exports the table in CSV format; see the **TableExport** command.

Grips Editing of Tables and Cells

Select a table, a cell, or several cells, and then use grips editing:

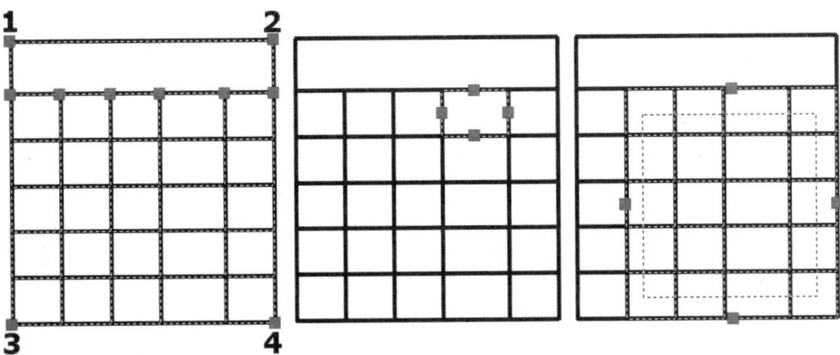

Table grips editing

Grip 1 enlarges and reduces the entire table.

Grip 2 widens and narrows all columns.

Grip 3 increases and decreases the height of all rows.

Grip 4 enlarges and/or reduces the entire table, depending on how you move the cursor.

Other grips widen and narrow individual columns.

Cell grips editing

Grips on vertical borders change the width of the cell(s).

Grips on horizontal borders change the height of the cell(s).

Properties Editing of Tables and Cells

Select one or more cells or tables, right-click, and then select ***Properties*** *from the shortcut menu:*

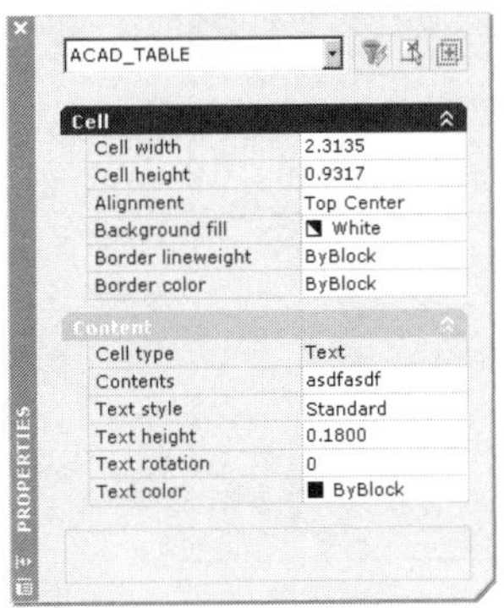

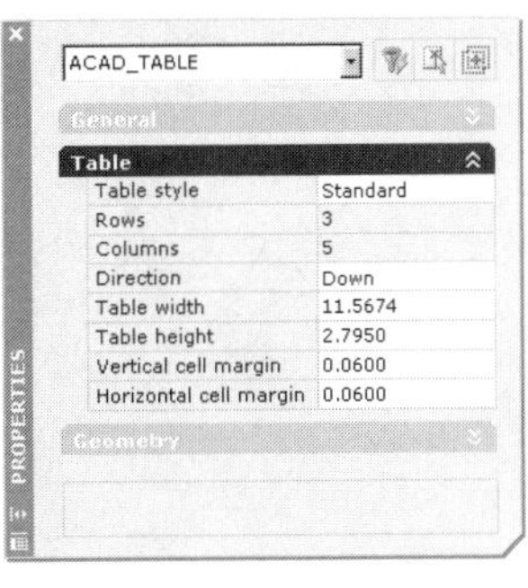

DIALOG BOXES

Cell Border Properties dialog box

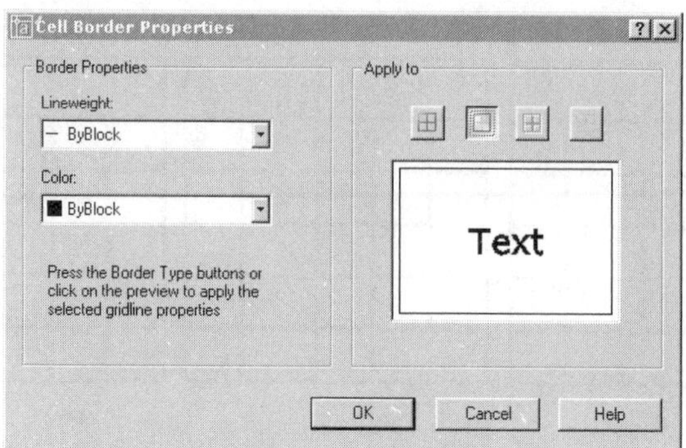

Lineweights selects the width of cell borders; see **Lineweight** command.

Color selects the color of cell borders; see the **Color** command.

To change the background color of cells, use the ***Properties*** *command, and then change the Background Fill.*

Apply to changes the borders:

All Borders applies properties to all cell borders.

Outside Borders applies properties to outer borders.

Inside Borders applies properties to inner borders.

No Borders applies properties to none of the borders.

Insert a Block in a Table Cell dialog box

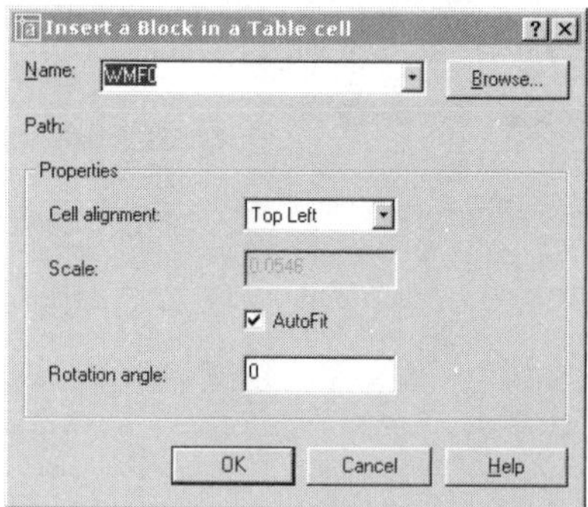

Name specifies the name of the block; the droplist shows the names of blocks defined in the drawing.

Browse displays the Select Drawing File dialog box; select a *.dwg* or *.dxf* file, and then click **Open**.

Cell Alignment positions the block in the cell: top, middle, bottom, left, center, or right.

Scale specifies the scale factor; not available when AutoFit is turned on.

☑**AutoFit** scales the block to fit the cell.

Rotation Angle rotates the block.

TableExport

2005 Exports tables as CSV data files.

Command	Alias	Ctrl+	F-key	Alt+	Menu Bar	Tablet
tableexport	...	...	...	...	...	...

Command: tableexport
Select a table: *(Select one table.)*

Displays the Export Data dialog box.

*Enter a file name, select a folder, and then click **Save**.*

TIPS

- *CSV* is short for "comma-separated values." Each table row is one line of data, and each cell is separated by a comma. This format can be read by spreadsheet and database programs.
- To import tables from Microsoft's Office XP spreadsheet, copy and paste it into the drawing. AutoCAD converts the spreadsheet into a table object.
- To import tables from other brands of spreadsheet software:

  1. In the spreadsheet, copy the data to the Clipboard with CTRL+C.
  2. In AutoCAD, use the **Edit | Paste Special** command to paste the data:

Format	Meaning
Picture (Metafile)	Pasted as an OLE object.
Device Independent Bitmap	Pasted as an OLE object.
AutoCAD Entities	Pasted as a table object.
Image Entity	Pasted as a raster image.
Text	Pasted as mtext.
Unicode Text	Pasted as mtext.

- AutoCAD cannot import tables from CSV files.
- To copy tables between drawings, use DesignCenter.

TableStyle

2005 Creates and edits table styles.

Command	Alias	Ctrl+	F-key	Alt+	Menu Bar	Tablet
tablestyle	...	...	...	**FB**	**Format** ↳**Table Style**	...

Command: tablestyle

Displays dialog box:

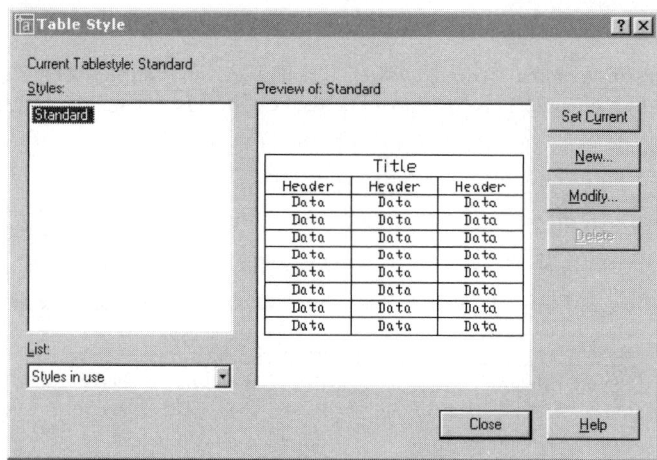

DIALOG BOX OPTIONS

Styles lists the styles defined in the drawing; every new drawing has the "Standard" style.

List filters the style names:

- **All** lists all styles.
- **Styles in use** lists the styles being used by tables in the drawing.

Set Current sets the selected style as current.

New creates new table styles based on the settings of the current style; displays Create New Table Style dialog box.

Modify changes the settings of the table style; displays the Modify Table Style dialog box.

Delete erases the style from the drawing; unavailable if the "Standard" table style is selected.

Create New Table Style dialog box

New Style Name specifies the name of the new table style.

Start With selects the existing style upon which to base the new style.

Continue displays the New Table Style dialog box (identical to the Modify Table Style dialog box).

Modify Table Style dialog box

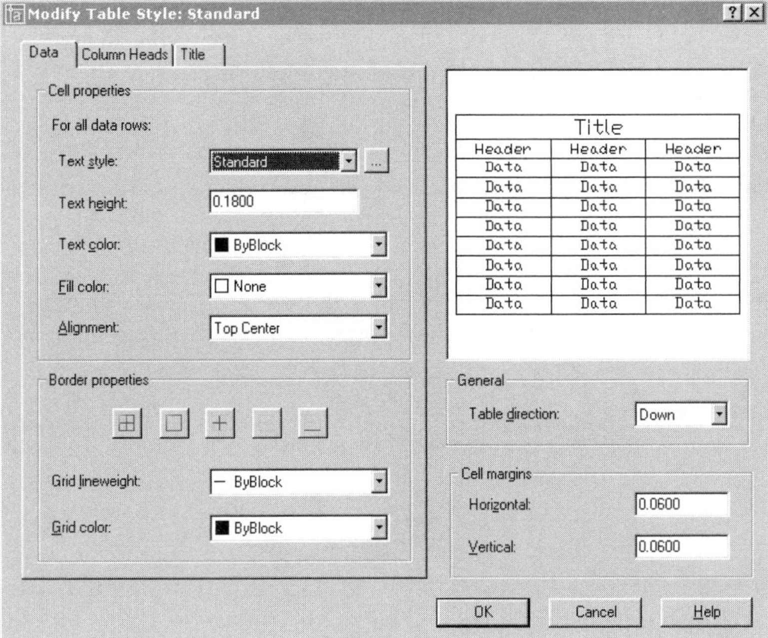

General options

Table Direction determines how the table grows relative to the insertion point:

- **Down** (insertion point is at upper-left corner).
- **Up** (insertion point is at lower-left corner).

Cell Margins options

Horizontal specifies the horizontal margin between cell borders and text (default = 0.06).

Vertical specifies the vertical margin (default = 0.06).

Data tab

Data is also known as "cells."

Cell Properties options

Text Style selects a predefined text style.

... displays the Text Style dialog box; see the **Style** command.

Text Height specifies the height of the text; unavailable if the height is defined by text style.

Text Color specifies the color of the text; select **Select Color** for additional colors (see **Color** command).

Fill Color specifies the background color of the cell.

Alignment determines the justification of the text within the cell: top, middle, bottom, left, center, or right.

Border Properties options

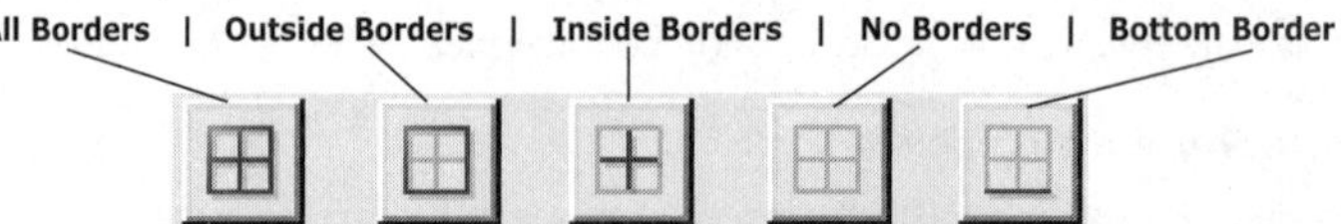

Grid Lineweight selects the width of cell borders; see the **Lineweight** command.

Grid Color selects the color of cell borders; see the **Color** command.

Column Heads tab

Identical to the Data tab, except for this option:

☑**Include Header Row** determines whether tables have a column head row; use this row to provide titles to columns.

Title options

Identical to the Data tab, except for this option:

☑**Include Title Row** determines whether tables have a title row; use this row to provide a title to the table.

SHORTCUT MENU

Right-click a style name to display shortcut menu:

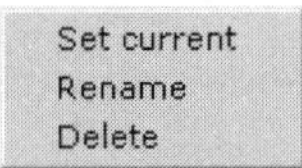

Set Current sets the selected style as current.

Rename changes the name of the style.

Delete erases the style from the drawing.

RELATED COMMANDS

Table places tables in drawings.

MatchCell matches the style of one cell to other cells.

TableEdit edits text in table cells.

Rename renames table styles.

Purge purges unused table styles from drawings.

DesignCenter shares tables styles between drawings.

Properties changes table style names, and overrides style settings.

RELATED SYSTEM VARIABLE

CTableStyle specifies the current table style name.

TIPS

- To copy table styles from other drawings, copy a table with CTRL+C, and then paste it into the drawing with CTRL+V. Erase the table; the style remains in the drawing.
- To format indiviual cells, select the cell(s), right-click, and then choose Properties.
- To revert a table to its original style, select the table, right-click, and then select **Remove All Property Overrides.**

'Tablet

V. 1.0 Configures and calibrates digitizing tablets, and toggles tablet mode.

Command	Alias	Ctrl+	F-key	Alt+	Menu Bar	Tablet
tablet	ta	T	F4	TT	Tools ↳Tablet	X7

Command: tablet

When no tablet is configured, AutoCAD complains, 'Your pointing device cannot be used as a tablet.'

When a digitizing tablet is configured, AutoCAD continues:

Enter an option [ON/OFF/CAL/CFG]: *(Enter an option.)*

COMMAND LINE OPTIONS

CAL calibrates the coordinates for the tablet.

CFG configures the menu areas on the tablet.

OFF turns off the tablet's digitizing mode.

ON turns on the tablet's digitizing mode.

RELATED SYSTEM VARIABLE

TabMode toggles use of the tablet:

TabMode	Meaning
0	Tablet mode disabled.
1	Tablet mode enabled.

RELATED FILES

acad.mnu is the menu source code that defines the functions of tablet menu areas.

tablet.dwg is an AutoCAD drawing of the printed template overlay.

TIPS

- To change the tablet overlay, edit the *tablet.dwg* file, and then plot it to fit your digitizer.
- **Tablet** does not work if a digitizer has not been configured with the **Options** command.
- AutoCAD supports up to four independent menu areas; macros are specified by the ***TABLET1 through ***TABLET4 sections of the *acad.mnu* menu file.
- Menu areas may be skewed, but corners must form a right angle.
- Projective transformation is a limited form of "rubber sheeting": straight lines remain straight, but not necessarily parallel.

DEFINITIONS

Affine Transformation — requires three pick points; sets an arbitrary linear 2D transformation with independent x,y scaling and skewing.

Orthogonal Transformation — requires two pick points; sets the translation; the scaling and rotation angles remain uniform.

Residual Error — largest where mapping is least accurate.

Outcome of Fit — reports on the results of transformation types:

Outcome	Meaning
Exact	Enough points to transform data.
Success	More than enough points to transform data.
Impossible	Not enough points to transform data.
Failure	Too many colinear and coincident points.
Cancelled	Fitting cancelled during projective transformation.

Projective Transformation — maps a perspective projection from one plane to another plane.

RMS Error — specifies root mean square error; smaller is better; measures closeness of fit.

Standard Deviation —when near zero, residual error at each point is roughly the same.

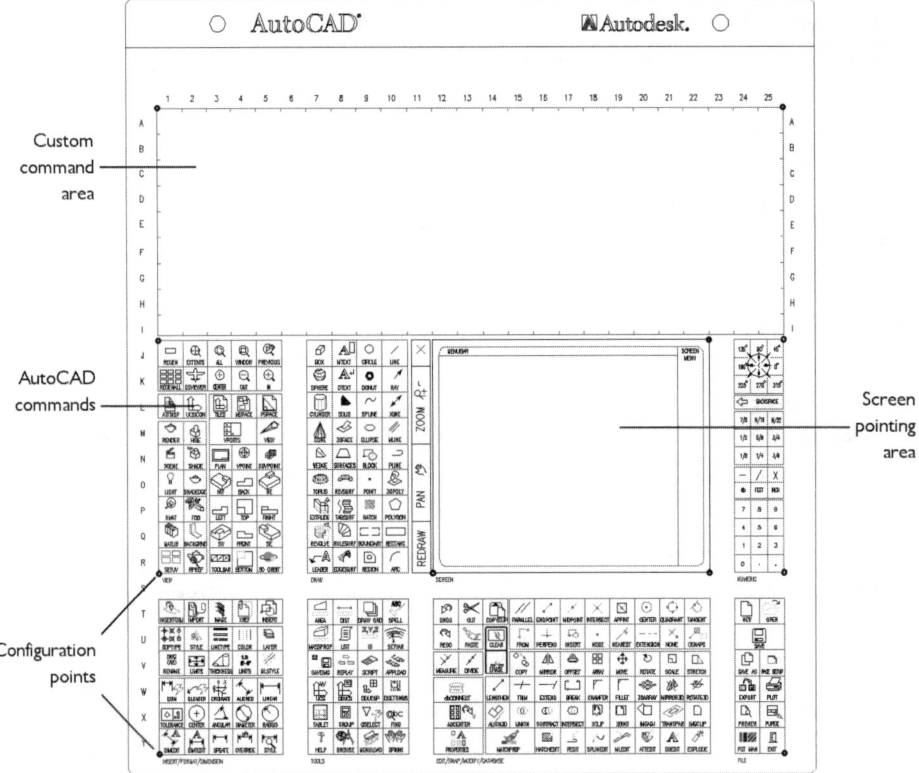

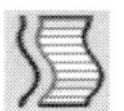

TabSurf

Rel.10 Draws tabulated surfaces as 3D meshes, defined by path curves and direction vectors (*short for TABulated SURFace*).

Command	Alias	Ctrl+	F-key	Alt+	Menu Bar	Tablet
tabsurf	...	...	...	**DFT**	**Draw** ↳**Surfaces** ↳**Tabulated Surface**	**P8**

Command: tabsurf
Select object for path curve: *(Select an object.)*
Select object for direction vector: *(Select an object.)*

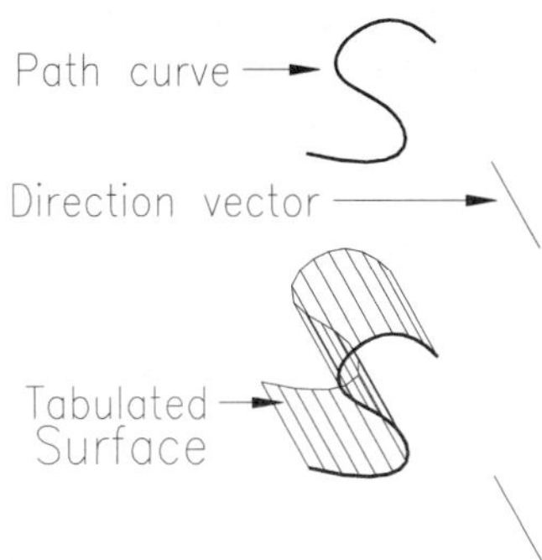

COMMAND LINE OPTIONS

Select object for path curve selects the object that defines the tabulation path.

Select object for direction vector selects the vector that defines the tabulation direction.

RELATED COMMANDS

Edge changes the visibility of 3D face edges.

Explode reduces tabulated surface to 3D faces.

PEdit edits 3D meshes, such as tabulated surfaces.

EdgeSurf draws 3D mesh surfaces between boundaries.

RevSurf draws revolved 3D mesh surfaces around an axis.

RuleSurf draws 3D mesh surfaces between open or closed boundaries.

RELATED SYSTEM VARIABLE

SurfTab1 defines the number of tesselations drawn by **TabSurf** in the *n*-direction.

TIPS

- The path curve can be an open or closed object: line, 2D polyline, 3D polyline, arc, circle, or ellipse.
- The direction in which you draw the direction vector determines the direction of the extrusion; the length of the vector determines the thickness of the extrusion.
- The number of *m*-direction tabulations is always 1, and lies along the direction vector.
- The number of *n*-direction tabulations is determined by system variable **SurfTab1** (default = 6) along curves only.

Text

V. 1.0 Places text, one line at a time, in drawings.

Command	Alias	Ctrl+	F-key	Alt+	Menu Bar	Tablet
text	**dtext**	...	...	...	**Draw**	**K8**
	dt				**↳Text**	
					↳Single Line Text	

Command: text
Current text style: "Standard" Text height: 0.2000
Specify start point of text or [Justify/Style]: *(Pick a point, or enter an option.)*
Specify height <0.2000>: *(Enter a value.)*
Specify rotation angle of text <0>: *(Enter a value.)*
Enter text: *(Enter text, and then press* **ENTER***.)*
Enter text: *(Press* **ENTER** *to end the command.)*

COMMAND LINE OPTIONS

Specify start point of text indicates the starting point of the text.

ENTER continues text one line below the previously-placed text.

Specify height indicates the height of the text; this prompt does not appear wen the style sets the height to a value other than 0.

Specify rotation angle of text indicates the rotation angle of the text.

Enter text specifies the text; press ENTER twice to end the command.

Justify options

Enter an option [Align/Fit/Center/Middle/Right/TL/TC/TR/ML/MC/MR/BL/BC/BR]: *(Enter an option.)*

Align aligns the text between two points with adjusted text height.

Fit fits the text between two points with fixed text height.

Center centers the text along the baseline.

Middle centers the text horizontally and vertically.

Right right-justifies the text.

TL top-left justification.

TC top-center justification.

TR top-right justification.

ML middle-left justification.

MC middle-center justification.

MR middle-right justification.

BL bottom-left justification.

BC bottom-center justification.

BR bottom-right justification.

Style options

Enter style name or [?] <Standard>: *(Enter a name, or type* **?***.)*

Style name indicates a different style name.

? lists the currently-loaded styles.

SPECIAL SYMBOLS

%%c draws diameter symbol: Ø.

%%d draws degree symbol: °.

%%o starts and stops overlining.

%%p draws the plus-minus symbol: ±.

%%u starts and stops underlining.

%%% draws the percent symbol: %.

RELATED COMMANDS

DdEdit edits text.

Change changes the text height, rotation, style, and content.

Properties changes all aspects of text.

Style creates new text styles.

MText places paragraph text in drawings.

RELATED SYSTEM VARIABLES

TextSize is the current height of text.

TextStyle is the current style of text.

ShpName is the default shape name

TIPS

- Use the **Text** command to place text easily in many locations in the drawing. It displays text on screen as you type.
- You can erase text by pressing the BACKSPACE key while at the 'Text' prompt.
- *Warning!* The spacing between lines of text does not match the current snap spacing.
- Transparent commands do not work during the **Text** command.
- You can enter any justification mode at the 'Start point' prompt.
- The 'Enter text' prompt repeats until cancelled with ENTER.

The dot indicates the text insertion point.

TextToFront

2005 Places all text and dimensions visually on top of overlapping objects.

Command	Alias	Ctrl+	F-key	Alt+	Menu Bar	Tablet
texttofront	...	...	...	...	**Tools** **↳Draw Order** **↳Bring Text and Dimenions to Front**	...

Command: texttofront
Bring to front [Text/Dimensions/Both] <Both>: *(Enter an option.)*
nnn **object(s) brought to front.**

COMMAND LINE OPTIONS

Text brings all text to the front.

Dimensions brings all dimensions to the front.

Both brings both text and dimensions to the front.

RELATED COMMANDS

DrawOrder controls the display order of selected objects.

BHatch controls the display of hatch and fill patterns relative to their boundaries and other objects.

TIPS

- You must apply this command separately to model space and each layout.
- Newer objects are drawn on top of older objects in drawings; to change this behavior, see the **DrawOrderCtrl** system variable.

'TextScr

V. 2.1 Switches from the AutoCAD window to the Text window (*short for TEXT SCReen*).

Command	Alias	Ctrl+	F-key	Alt+	Menu Bar	Tablet
textscr	...	...	**F2**	**VLT**	**View** ↳**Display** ↳**Text Window**	...

Command: textscr

Displays the **Text** *window:*

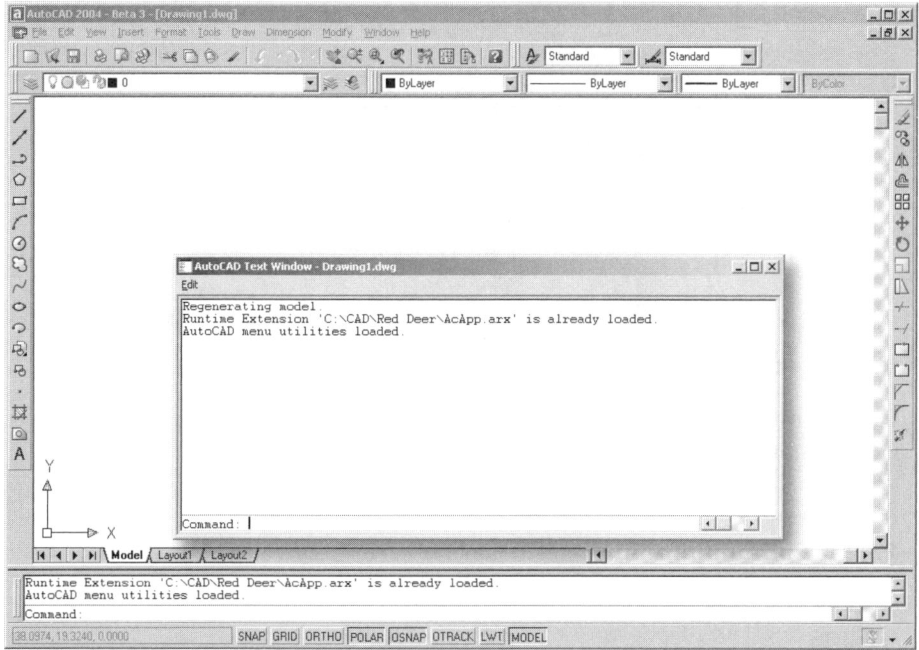

EDIT MENU OPTIONS

Paste to CmdLine pastes text from the Clipboard to the command line; available only when the Clipboard contains text.

Copy copies selected text to the Clipboard.

Copy History copies all text to the Clipboard.

Paste pastes text from the Clipboard into the Text window; available only when the Clipboard contains text.

Options displays Options dialog box; see the **Options** command.

COMMAND LINE OPTIONS

Command window navigation:

Key	Meaning
←	Moves the cursor left by one character.
→	Moves the cursor right by one character.
↑	Displays the previous line in the command history.
↓	Displays the next line in the command history.
Page Up	Displays the previous screen of text.
Page Down	Moves to the next screen of text.
Home	Moves the cursor to the start of the line.
End	Moves the cursor to the end of the line.
Insert	Toggles insert mode.
Delete	Deletes the character to the right of the cursor.
BACKSPACE	Deletes the character to the left of the cursor.

SHORTCUT MENU OPTIONS

*Right-click the **Text** window:*

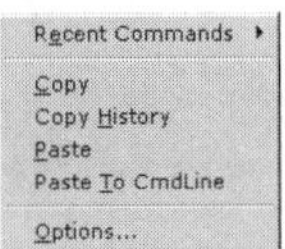

Recent Command displays a list of ten recently-used commands.

Copy copies selected text to the Clipboard.

Copy History copies all text to the Clipboard.

Paste pastes text from the Clipboard into the Text window; available only when Clipboard contains text.

Paste to CmdLine pastes text from the Clipboard to the command line; available only when the Clipboard contains text.

Options displays the **Options** dialog box; see the **Options** command.

RELATED COMMAND

GraphScr switches from the Text window to the AutoCAD drawing window.

RELATED SYSTEM VARIABLE

ScreenMode reports whether the screen is in text or graphics mode:

ScreenMode	Meaning
0	Text screen.
1	Graphics screen.
2	Dual screen displaying both text and graphics.

Removed Command

TiffIn was removed from AutoCAD Release 14. Use the **ImageAttach** command instead.

TifOut

2004 Exports the current viewports in TIFF raster format.

Command	Alias	Ctrl+	F-key	Alt+	Menu Bar	Tablet
tifout	...	...	...	...	...	...

Command: tifout

Displays Create Raster File dialog box. Specify a filename, and the click ***Save****.*

Select objects or <all objects and viewports>: *(Select objects, or press* **ENTER** *to select all objects and viewports.)*

COMMAND LINE OPTIONS

Select objects selects specific objects.

All objects and viewports selects all objects and all viewports, whether in model space or in layout mode.

RELATED COMMANDS

BmpOut exports drawings in BMP (bitmap) format.

Image places raster images in the drawing.

JpgOut exports drawings in JPEG (joint photographic expert group) format.

PngOut exports drawings in PNG (portable network graphics) format.

TIPS

- The rendering effects of the **ShadEdge** command are preserved, but not of the **Render** command.
- TIFF files are commonly used in desktop publishing.
- TIFF is short for "tagged image file format," and was developed by Aldus, the forerunner of Adobe.

'Time

V. 2.5 Displays time-related information about the current drawing.

Command	Alias	Ctrl+	F-key	Alt+	Menu Bar	Tablet
time	...	...	...	TQT	Tools ↳Inquiry ↳Time	...

Command: time

Display/ON/OFF/Reset: *(Enter an option.)*

Example output:

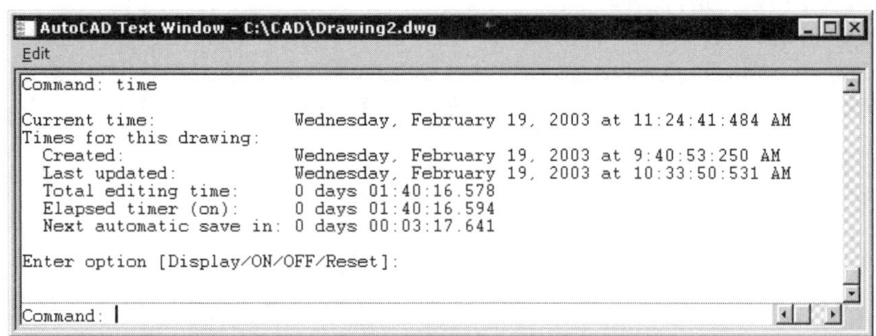

AutoCAD Text Window - C:\CAD\Drawing2.dwg
Edit
Command: time
Current time: Wednesday, February 19, 2003 at 11:24:41:484 AM
Times for this drawing:
Created: Wednesday, February 19, 2003 at 9:40:53:250 AM
Last updated: Wednesday, February 19, 2003 at 10:33:50:531 AM
Total editing time: 0 days 01:40:16.578
Elapsed timer (on): 0 days 01:40:16.594
Next automatic save in: 0 days 00:03:17.641
Enter option [Display/ON/OFF/Reset]:
Command:

COMMAND LINE OPTIONS

Display displays the current time and date.

OFF turns off the user timer.

ON turns on the user timer.

Reset resets the user timer.

RELATED COMMANDS

Status displays information about the current drawing and environment.

Preferences sets the automatic backup time.

RELATED SYSTEM VARIABLES

CDate is the current date and time.

Date is the current date and time in Julian format.

SaveTime is the interval for automatic drawing saves.

TDCreate is the date and time the drawing was created.

TDInDwg is the time the drawing spent in AutoCAD.

TDUpdate is the last date and time the drawing was changed.

TDUsrTimer is the current user timer setting.

TIP

- The time displayed by the **Time** command is only as accurate as your computer's clock; unfortunately, the clock in some personal computers can stray by minutes per week.

Removed Command

Today was removed from AutoCAD 2004; it was replaced by the Communication Center.

TInsert

2005 Inserts a block or a drawing in a table cell (*short for Table INSERT; an undocumented command*).

Command	Alias	Ctrl+	F-key	Alt+	Menu Bar	Tablet
tinsert	...	...	...	...	...	...

Command: tinsert
Pick a table cell: *(Select a single cell in a table.)*

Displays dialog box.

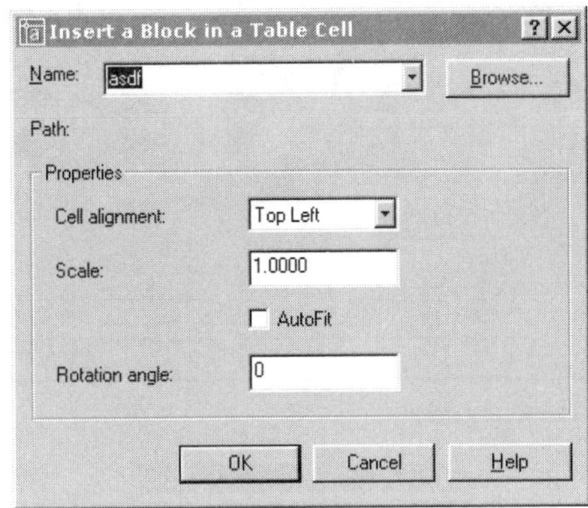

If the block contains attributes, the Enter Attributes dialog box is displayed.

COMMAND LINE OPTION

Pick a table cell specifies the cell in which to place the block.

DIALOG BOX OPTIONs

Name specifies the name of the block; the droplist lists the names of all blocks found in the current drawing.

Browse displays the Select Drawing File dialog box; select a *.dwg* or *.dxf* file, and click **Open**.

Properties options

Cell Alignment specifies position of the block within the cell, from top-left to bottom-right.

Scale sizes the block, and adjusts the cell to fit the block.

☐ **AutoFit** fits the block to the cell.

Rotation Angle specifies the angle at which to place the block.

RELATED COMMANDS

Table creates new tables.

Block creates blocks.

TIPS

- Alternatively, select a cell, right-click, and then choose **Insert Block** from the shortcut menu.
- A cell can contain at most one block.

Tolerance

Rel.13 Places geometric tolerancing symbols and text.

Command	Alias	Ctrl+	F-key	Alt+	Menu Bar	Tablet
tolerance	**tol**	...	...	**NT**	**Dimension** **↳Tolerance**	**X1**

Command: tolerance

Displays dialog box.

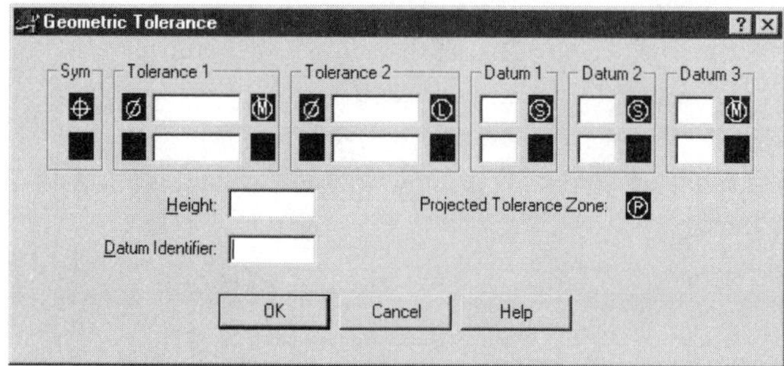

Clicking a symbol displays this dialog box:

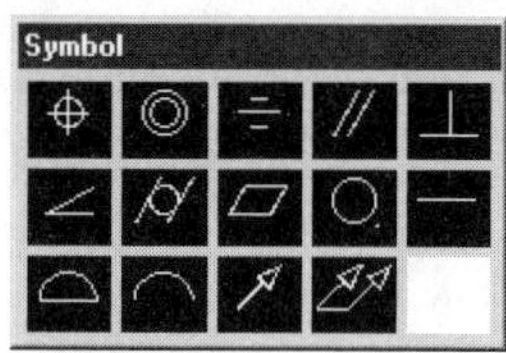

Enter tolerance location: *(Pick a point.)*

DIALOG BOX OPTIONS

Sym specifies the geometric characteristic symbol.

Tolerance specifies the first tolerance value.

Dia specifies the places optional Ø (diameter) symbol.

Value specifies the tolerance value.

Datum specifies the datum reference.

Height specifies the projected tolerance zone value.

Projected Tolerance Zone places the projected tolerance zone symbol.

Datum Identifier creates the datum identifier symbol, such as -A-.

MC displays Material Condition dialog box.

Material Condition dialog box

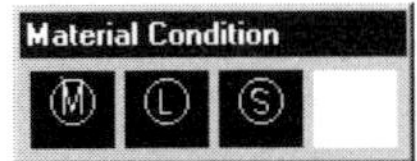

Ⓜ specifies maximum material condition.

Ⓛ specifies least material condition.

Ⓢ specifies regardless of feature size.

RELATED FILES

gdt.shp is the tolerance symbol definition source file.

gdt.shx is the compiled tolerance symbol file.

TIP

- You can use the **DdEdit** command to edit tolerance symbols and feature control frames.

DEFINITIONS

Datum — a theoretically-exact geometric reference that establishes the tolerance zone for the feature. These objects can be used as a datum: point, line, plane, cylinder, and other geometry.

Material Condition — symbols that modify the geometric characteristics and tolerance values (modifiers for features that vary in size).

Projected Tolerance Zone — the height of the fixed perpendicular part's extended portion; changes the tolerance to positional tolerance.

Tolerance — the amount of variance from perfect form.

Orientation Symbols

⌖ Position.

◎ Concentricity and coaxiality.

⌯ Symmetry.

// Parallelism.

⊥ Perpendicularity.

∠ Angularity.

Form Symbols

⌭ Cylindricity.

⏥ Flatness.

○ Circularity and roundness.

— Straightness.

Profile Symbols

⌓ Profile of the surface.

⌒ Profile of the line.

↗ Circular runout.

⌰ Total runout.

-Toolbar

Rel.13 Displays and hides toolbars via the command line.

Command	Alias	Ctrl+	F-key	Alt+	Menu Bar	Tablet
-toolbar	...	...	...	...	...	...

Note: ***Toolbar*** *and* ***TbConfig*** *are aliases for the* ***Customize*** *command.*

Command: -toolbar
Enter toolbar name or [ALL]: *(Enter a name, or type* **ALL***.)*
Enter an option [Show/Hide]: *(Type* **S** *or* **H***.)*

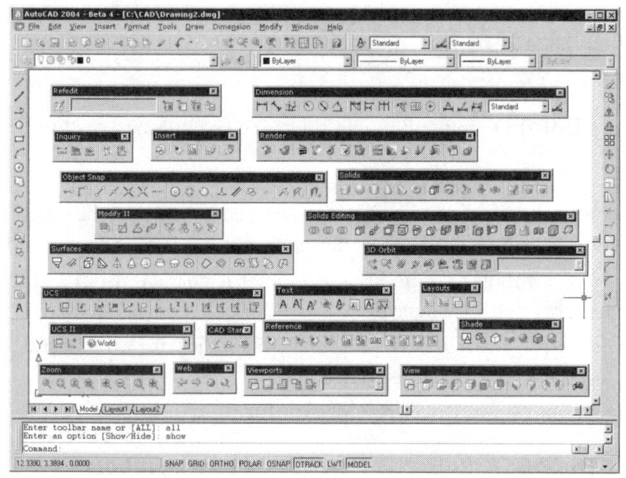

Opening all of AutoCAD's toolbars with the ***-Toolbar All Show*** *command.*

COMMAND LINE OPTIONS

Toolbar name specifies the name of the toolbar.

ALL applies the command to all toolbars; must be entered in all capital letters.

Show displays the toolbar.

Hide dismisses the toolbar.

RELATED COMMANDS

Customize customizes toolbars via a dialog box.

MenuLoad loads a partial menu file, including toolbar definitions.

Tablet configures the tablet.

RELATED SYSTEM VARIABLE

ToolTips toggles the display of tooltips.

RELATED FILES

****.mnc*** are compiled menu files.

****.mns*** are AutoCAD source menu files.

****.bmp*** are bitmap files that define custom icon buttons.

ToolPalettes / ToolPalettesClose

2004 Displays and closes the Tool Palettes window.

Command	Alias	Ctrl+	F-key	Alt+	Menu Bar	Tablet
toolpalettes	...	3	...	TP	**Tools** ↳**Tool Palettes Window**	...
toolpalettesclose	...	3	...	TP	**Tools** ↳**Tool Palettes Window**	...

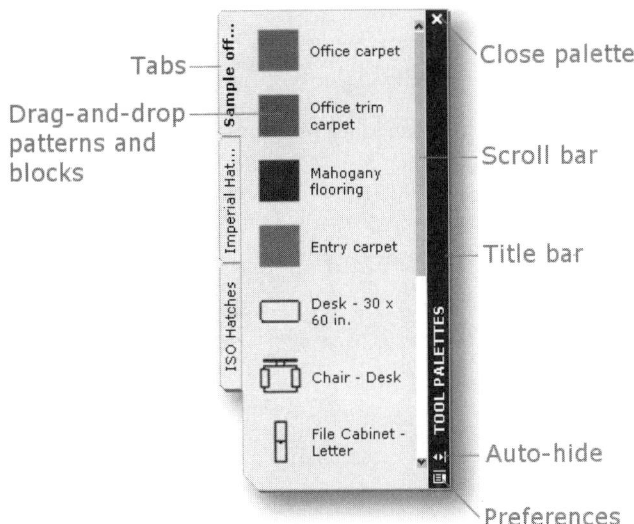

Command: toolpalettesclose

Closes the window.

WINDOW OPTIONS

Tabs selects various sets of palettes.

x closes the palette window; alternatively, press CTRL+3.

Title bar drags around the window.

Auto-hide collapses the window when the cursor moves away.

Preferences displays a shortcut menu.

PREFERENCES MENU

Right-click the palette, or click the ***Preferences*** *button:*

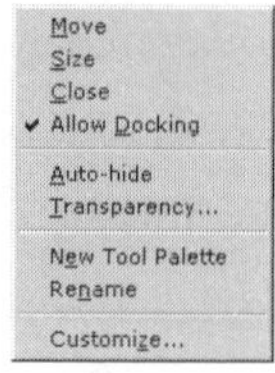

Move moves the window.

Size changes the size of the window.

Close closes the window.

Allow Docking toggles whether the window can be docked at the side of AutoCAD.

Auto-hide toggles whether the window reduces its size when the cursor is elsewhere.

Transparency displays the Transparency dialog box.

New Tool Palette creates a new tab, and then prompts you for a name:

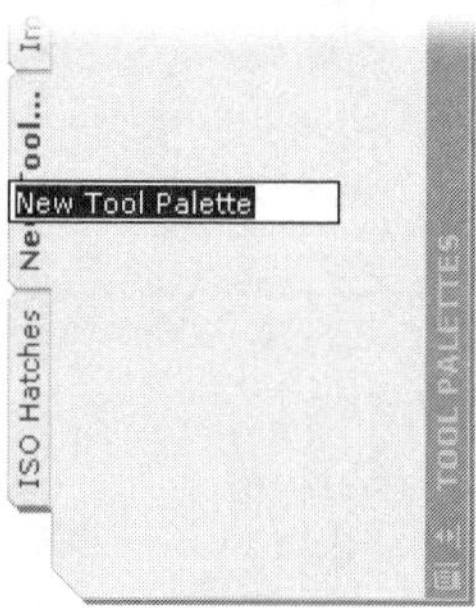

Rename renames the tab.

Customize displays the Tool Palettes tabs of the Customize dialog box; see the **Customize** command.

SHORTCUT MENU

Right-click a tab:

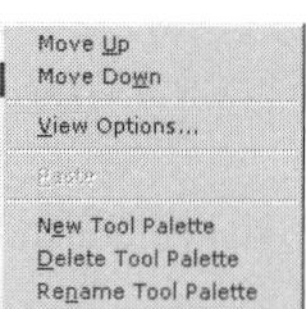

Move up moves the tab up one position.

Move down moves the tab down one position.

View Options displays the View Options dialog box.

New Tool Palette creates a new tab, and then prompts you for a name.

Delete Tool Palette removes the tab, after prompting you for confirmation.

Rename Tool Palette renames the tab.

Transparency dialog box

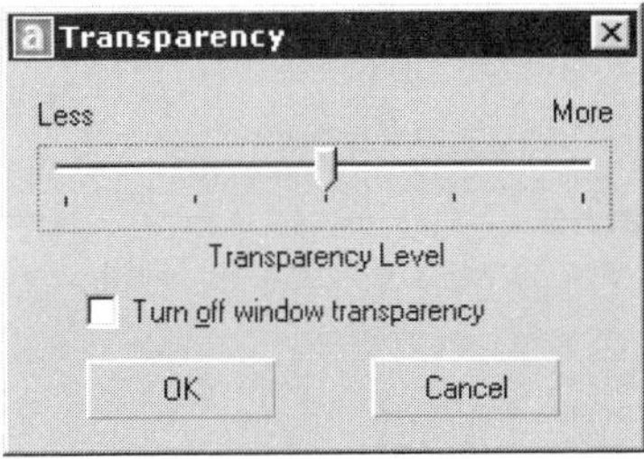

Transparency Level changes the transparency of the palette from none (Less) to More; the figure illustrates maximum transparency.

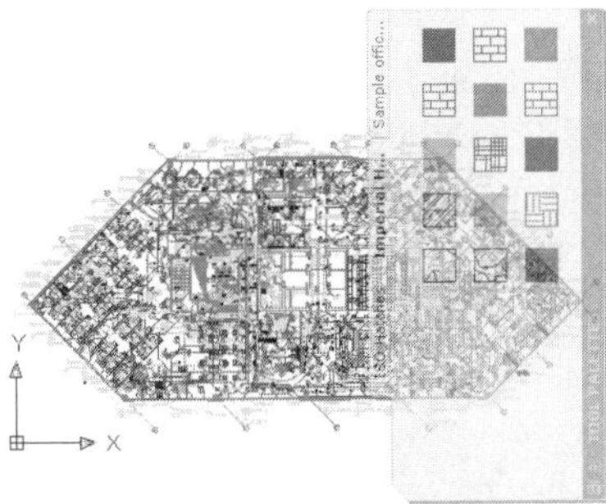

☐ **Turn off window transparency** makes the palette opaque.

View Options dialog box

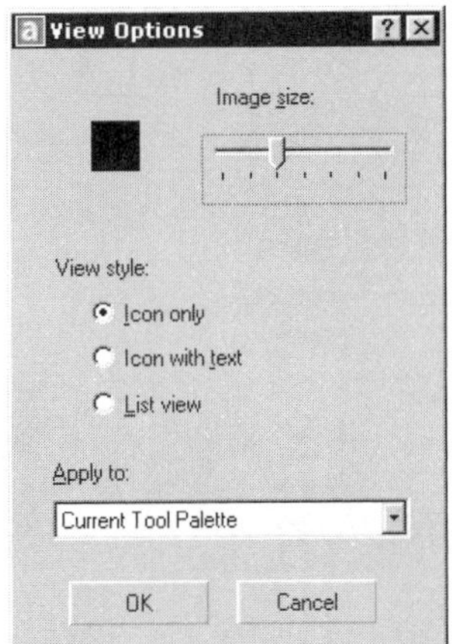

Image Size changes the size of icons from small (14 pixels square) to large (54 pixels).

View style toggles the display of icons and text:

⊙ **Icon only** displays icons only.

○ **Icon with text** displays icons and text.

○ **List view** displays small icons with text.

Apply to determines if changes apply to the current palette, or to all palettes.

RELATED COMMANDS

Customize exports and imports Tool Palettes, and creates groups of palettes.

DesignCenter displays the content of drawings.

ToolPalettesClose closes the Tool Palette window.

RELATED SYSTEM VARIABLES

PaletteOpaque determines whether the Tool Palette window can be transparent.

TpState notes whether the Tool Palettes window is open.

RELATED FILES

***.xtp** stores the content of Tool Palette windows.

TIPS

- To share the content of Tool Palettes:
  1. From the menu bar, select **Tools | Customize**.
  2. In the **Tool Palette** tab, click **Export** to save your tool palettes to disk.
  3. Alternatively, click **Import** to access Tool Palettes created by others.
- To bring content from the DesignCenter into the Tool Palette window:
  1. In **DesignCenter**, right-click an item.
  2. From the shortcut menu, select **Create Tool Palette**.

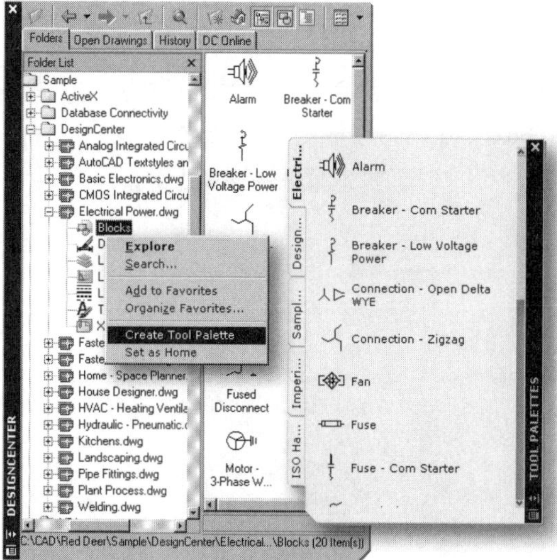

- The *.xtp* files store content in XML (extended markup language) format.
- As of AutoCAD 2005, you can drag objects (hold down the right mouse button while dragging) and toolbar buttons (have the Customize dialog box open) onto palettes. Program code (macros, scripts, and so on) can also be stored on palettes by editing existing tools.
- The Customize dialog box allows you to create groups of palettes (*new to AutoCAD 2005*).

Torus

Rel.11 Draws 3D tori as solid models.

Command	Alias	Ctrl+	F-key	Alt+	Menu Bar	Tablet
torus	**tor**	...	...	**DIT**	**Draw** ↳**Solids** ↳**Torus**	**O7**

Command: torus

Current wire frame density: ISOLINES=4

Specify center of torus <0,0,0>: *(Pick point 1.)*

Specify radius of torus or [Diameter]: *(Specify the radius, 2, or type* **D***.)*

Specify radius of tube or [Diameter]: *(Specify the radius, 3, or type* **D***.)*

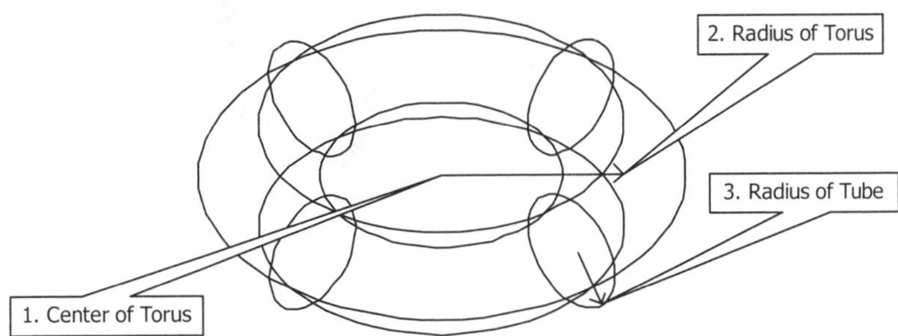

COMMAND LINE OPTIONS

Center of torus indicates the center of the torus.

Diameter indicates the diameter of the torus and the tube.

Radius indicates the radius of the torus and the tube.

RELATED COMMANDS

Ai_Torus creates a torus from 3D polyfaces.

Cone draws solid cones.

Cylinder draws solid cylinders.

Sphere draws solid spheres.

TIPS

- When the torus radius is negative, the tube radius must be a larger positive number. A negative torus radius creates a football shape. Specify a tube diameter larger than the torus diameter to create a hole-less torus.

Trace

V. 1.0 Draws line segments with width.

Command	Alias	Ctrl+	F-key	Alt+	Menu Bar	Tablet
trace	...	...	...	...	...	...

Command: trace
Specify trace width <0.050>: *(Enter a value.)*
Specify start point: *(Pick a point.)*
Specify next point: *(Pick another point.)*
Specify next point: *(Press* **ENTER** *to end the command.)*

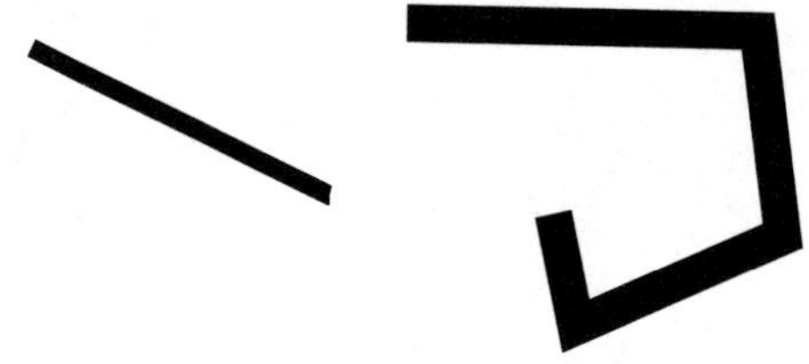

COMMAND LINE OPTIONS

Trace width specifies the width of the trace.

Start point picks the starting point.

Next point picks the next vertex.

ENTER exits the **Trace** command.

RELATED COMMANDS

Line draws lines with zero width.

MLine draws up to 16 parallel lines.

PLine draws polylines and polyline arcs with varying widths.

LWeight gives every object a width.

RELATED SYSTEM VARIABLES

FillMode toggles display of fill or outline traces (default = 1, on).

TraceWid specifies the current width of the trace (default = 0.05).

TIPS

- This command has largely been replaced by the **PLine** command.
- Traces are drawn along the centerline of the pick points.
- During drawing, the display of a trace segment is delayed by one pick point.
- During the drawing of traces, you cannot back up, because an **Undo** option is missing; if you require this feature, draw wide lines with the **PLine** command, setting the **Width** option.
- There is no option for controlling joints (always bevelled) or endcapping (always square); if you require these features, draw wide lines with the **MLine** command, setting the solid fill, endcap, and joint options with the **MlStyle** command.

Tracking

Rel.14 Locates x and y points visually, relative to other points in the command sequence; *not* a command, but a command modifier.

Modifer	Alias	Ctrl+	F-key	Alt+	Menu Bar	Tablet
tracking	tk track	...	...	...	...	T15

Example usage:

Command: line

Specify first point: *(Pick a point.)*

Specify next point or [Undo]: tk

Enters tracking mode:

First tracking point: *(Pick a point.)*

Next point (Press ENTER to end tracking): *(Pick a point.)*

Next point (Press ENTER to end tracking): *(Press* **ENTER** *to end tracking.)*

Exits tracking mode:

Specify next point or [Undo]: *(Pick a point.)*

COMMAND LINE OPTIONS

First tracking point picks the first tracking point.

Next point picks the next tracking point.

ENTER exits tracking mode.

RELATED KEYBOARD MODIFIERS

Direct distance entry specifies an angle and relative distance to the next point.

from locates an offset point.

m2p finds a point midway between two picked points.

TIPS

- **Tracking** is not a command, but a command option modifier.
- **Tracking** can be used in conjunction with direct distance entry.
- In tracking mode, AutoCAD automatically turns on **Ortho** mode to constrain the cursor vertically and horizontally.
- If you start tracking in the x direction, the next tracking direction is y, and vice versa.
- You can use tracking as many times as you need to at the 'Specify first point' and 'Specify next point' prompts.

Transparency

Rel.14 Toggles the transparency of pixels in raster images.

Command	Alias	Ctrl+	F-key	Alt+	Menu Bar	Tablet
transparency	...	...	...	**MOIT**	**Modify** ↳**Object** ↳**Image** ↳**Transparency**	**X21**

Command: transparency
Select image(s): *(Select one or more images inserted with the* **Image** *command.)*
Select image(s): *(Press* ENTER *to end object selection.)*
Enter transparency mode [ON/OFF] <OFF>: *(Type* **ON** *or* **OFF***.)*

COMMAND LINE OPTIONS

Select image(s) selects the objects whose transparency to change.

ON makes the background pixels transparent.

OFF makes the background pixels opaque.

RELATED COMMANDS

ImageAttach attaches a raster image as an externally-referenced file.

ImageAdjust changes the brightness, contrast, and fading of a raster image.

TIPS

- This command works only with raster images placed in drawings with the **Image** command.
- This command is meant for use with raster file formats that allow transparent pixels.
- When on, transparent pixels allow graphics under the image to show through.
- For this command to work correctly, you must first specify the color of the pixels to be made transparent, by using a raster editor, such as PaintShop Pro. When saving the image, specify the color to be designated as transparent.

TraySettings

2004 Specifies settings for the Communications Center, located at the right end of the status bar (the "tray").

Command	Alias	Ctrl+	F-key	Alt+	Menu Bar	Tablet
traysettings	...	...	...	...	...	...

You can also access this dialog box by clicking the Status Bar Menu icon at the end of the status bar.

Command: traysettings

Displays dialog box:

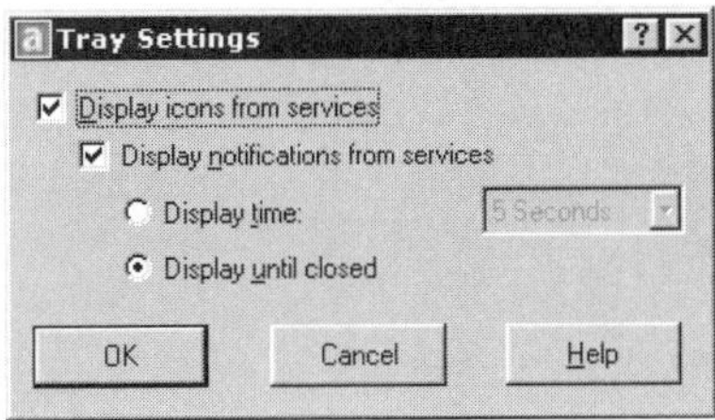

DIALOG BOX OPTIONS

Display icons from services:

☑ Displays "tray" at right end of the status line; see below.

☐ Turns off the tray.

Display notification from services:

☑ Displays balloon notifications from a variety of services; see below.

☐ Turns off notifications:

○ **Display time** specifies the duration that a notification balloon is displayed.

⊙ **Display until closed** specifies that the notification balloon is displayed until closed by user.

RELATED SYSTEM VARIABLES

TrayIcons toggles the display of the tray on the status bar.

TrayNotify toggles whether notification balloons are displayed.

TrayTimeOut determines the length of time that notification balloons are displayed

TIP

- Examples of notification balloons: click underlined links for more information, or click **x** to dismiss.

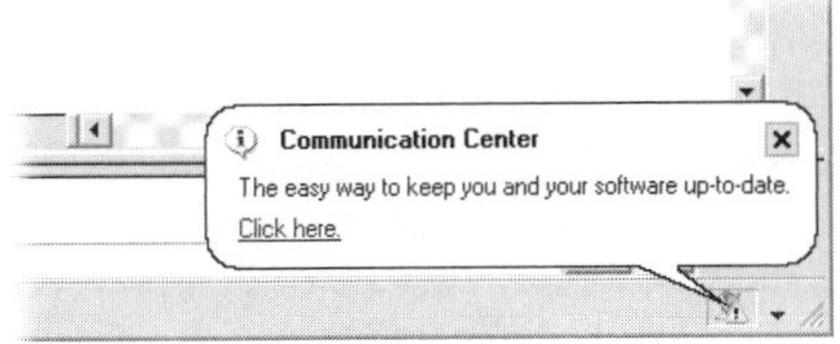

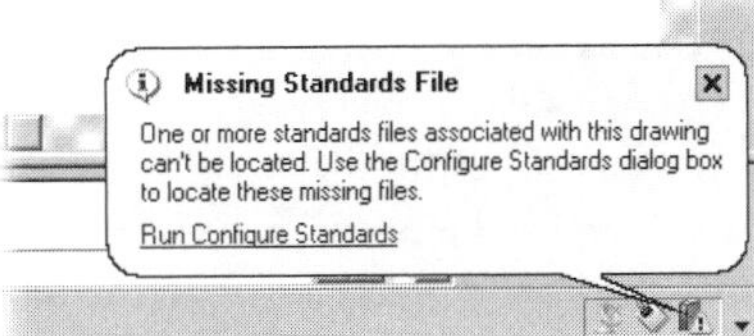

'TreeStat

Rel.12 Displays the status of the drawing's spatial index, including the number and depth of nodes (*short for TREE STATistics*).

Command	Alias	Ctrl+	F-key	Alt+	Menu Bar	Tablet
treestat	...	...	...	...	...	...

Command: treestat

Sample output:

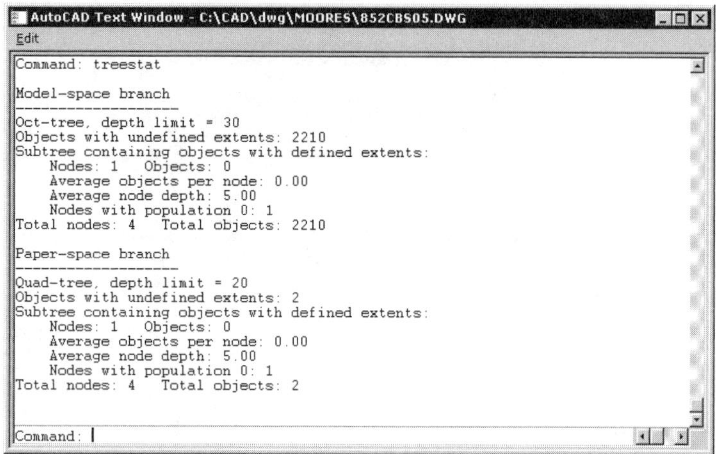

RELATED SYSTEM VARIABLES

TreeDepth specifies the size of the tree-structured spatial index in *xxyy* format:

Depth	Meaning
xx	Number of model space nodes (default = 30).
yy	Number of paper space nodes (default = 20).
-xx	2D drawing.
+xx	3D drawing (default).

TreeMax is the maximum number of nodes (default = 10,000,000).

TIPS

- Better performance occurs with fewer objects per oct-tree node. When redraws and object selection seem slow, increase the value of **TreeDepth**.
- Each node consumes 80 bytes of memory.

DEFINITIONS

Oct Tree — the model space branch of the spatial index, where all objects are either 2D or 3D. *Oct* comes from the eight volumes in the x,y,z coordinate system of 3D space.

Quad Tree — the paper space branch of the spatial index, where all objects are two-dimensional. *Quad* comes from the four areas in the x,y coordinate system of 2D space.

Spatial Index — objects indexed by oct-region to record their position in 3D space; has a tree structure with two primary branches: oct tree and quad tree. Objects are attached to *nodes*; each node is a branch of the *tree*.

Trim

V. 2.5 Trims lines, arcs, circles, 2D polylines, and hatches to real or projected cutting lines and views.

Command	Alias	Ctrl+	F-key	Alt+	Menu Bar	Tablet
trim	tr	...	...	MT	Modify ↳Trim	W15

Command: trim
Current settings: Projection=UCS Edge=None
Select cutting edges ...
Select objects: *(Select one or more objects.)*
Select objects: *(Press* ENTER *to end object selection.)*
Select object to trim or [Project/Edge/Undo]: *(Select object, or enter an option.)*
Select object to trim or [Project/Edge/Undo]: *(Press* ENTER *to end command.)*

COMMAND LINE OPTIONS

Select objects selects the cutting edges.

Select object to trim picks the object at the trim end.

Undo untrims the last trim action.

Edge options

Enter an implied edge extension mode [Extend/No extend] <No extend>: *(Type* **E** *or* **N**.*)*

Extend extends the cutting edge to trim object.

No extend trims only at an actual cutting edge.

Project options

Enter a projection option [None/Ucs/View] <Ucs>: *(Enter an option.)*

None uses only objects as cutting edge.

Ucs trims at the x,y plane of the current UCS.

View trims at the current view plane.

RELATED COMMANDS

Change changes the size of lines, arcs and circles.

Extend lengthens lines, arcs and polylines.

Lengthen lengthens open objects.

RELATED SYSTEM VARIABLES

EdgeMode determines whether this command projects cutting edges.

ProjMode determines how cutting edges are projected in 3D space.

TIPS

- As of AutoCAD 2005, this command also trims hatches.
- You can select all objects in the drawing by pressing ENTER at the 'Select objects' prompt.

U

V. 2.5 Undoes the most recent AutoCAD command (*short for Undo*).

Command	Alias	Ctrl+	F-key	Alt+	Menu Bar	Tablet
u	...	Z	...	EU	Edit ↳Undo	T12

Command: u

COMMAND LINE OPTIONS

None.

RELATED COMMANDS

Oops unerases the most-recently erased object.

Quit exits the drawing, undoing all changes.

Redo reverses the most-recent undo, if the prior command was **U** or **Undo**.

Undo allows more sophisticated control over undo than **U**.

RELATED SYSTEM VARIABLE

None.

TIPS

- The **U** command is convenient for stepping back through the design process, undoing one command at a time.
- The **U** command is the same as the **Undo 1** command; for greater control over the undo process, use the **Undo** command.
- The **Redo** command redoes the most-recent undo only; use **MRedo** otherwise.
- The **Quit** command, followed by the **Open** command, restores the drawing to its original state, if not already saved.
- Because the undo mechanism creates a mirror drawing file on disk, disable the **Undo** command with system variable **UndoCtl** (set to 0) when your computer is low on disk space.
- Commands that involve writing to file, plotting, and some display functions (such as **Render**, **Shade**, and **Hide**) are not undone.

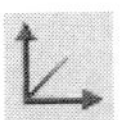

Ucs

Rel.10 Defines new coordinate planes, and restores existing UCSs (*short for User-defined Coordinate System*).

Command	Alias	Ctrl+	F-key	Alt+	Menu Bar	Tablet
ucs	...	...	...	**TW**	**Tools** ↳**New UCS**	**W7**
				TH	**Tools** ↳**Orthographic UCS**	...

Command: ucs
Current ucs name: *TOP*
Enter an option [New/Move/orthoGraphic/Prev/Restore/Save/Del/Apply/?/World] <World>: *(Enter an option.)*

COMMAND LINE OPTIONS

New creates a new user-defined coordinate system.

Move moves the UCS along the z axis.

orthoGraphic selects a standard orthographic UCS: top, bottom, front, back, left, and right.

Prev restores the previous UCS orientation.

Restore restores a named UCS.

Save saves the current UCS by name.

Del deletes the name of a saved UCS.

Apply applies the UCS setting to a selected viewport, or all active viewports.

? lists the names of saved UCS orientations.

World aligns the UCS with the WCS.

New options

Specify origin of new UCS or [ZAxis/3point/OBject/Face/View/X/Y/Z] <0,0,0>: *(Pick a point, press* **ENTER**, *or enter an option.)*

For compatibility with earlier versions of AutoCAD, you may enter any of these options at the earlier 'Enter an option' prompt.

Specify origin of new UCS moves the UCS to a new origin point.

ZAxis aligns the UCS with a new origin and z axis.

3point aligns the UCS with a point on the positive x-axis and positive x,y plane.

OBject aligns the UCS with a selected object.

Face aligns the UCS with the face of a 3D solid object.

View aligns the UCS with the current view.

X rotates the UCS about the x axis.

Y rotates the UCS about the y axis.

Z rotates the UCS about the z axis.

RELATED TOOLBARS

UCS toolbar

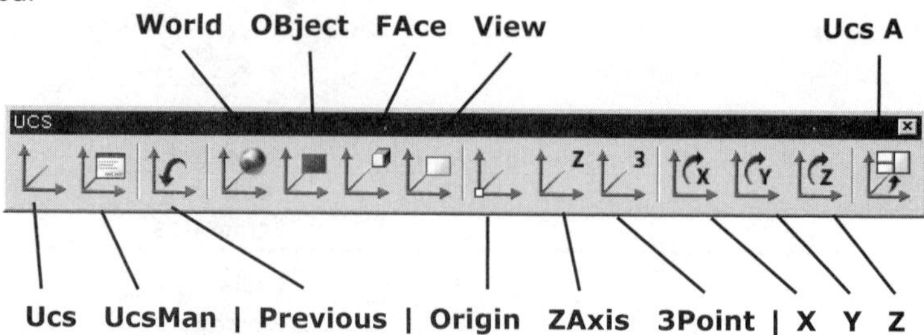

UCS II toolbar

RELATED SYSTEM VARIABLES

UcsAxisAng specifies the default rotation angle when the UCS is rotated around an axis using the **X**, **Y**, or **Z** options of this command

UcsBase specifies the UCS that defines the origin and orientation of orthographic UCS settings.

UcsFollow automatically aligns the UCS with the view of a newly activated viewport.

UcsIcon determines visibility and location of the UCS icon:

UcsIcon	Meaning
0	UCS icon not displayed.
1	UCS icon displayed in lower-right corner.
2	UCS icon displayed at the UCS origin, when possible.
3	UCS icon displayed at UCS always (default).

UcsOrg specifies the WCS coordinates of UCS icon (default = 0,0,0).

UcsOrtho specifies whether the related UCS is automatically displayed when an orthographic view is restored.

UcsView specifies whether the current UCS is saved when a view is created with the **View** command.

UcsVp specifies that the UCS reflects the UCS of the currently-active viewport.

UcsXdir specifies the X direction of the current UCS (default = 1,0,0).

UcsYdir specifies the Y direction of the current UCS (default = 0,1,0).

WorldUcs correlates the WCS to the UCS:

WorldUcs	Meaning
0	Current UCS is WCS.
1	UCS is same as WCS (default).

RELATED COMMANDS

UcsMan modifies the UCS via a dialog box.

UcsIcon controls the visibility of the UCS icon.

Plan changes the view to the plan view of the current UCS.

TIPS

- Use the **UCS** command to draw objects at odd angles in 3D space.
- Although you can create UCSs in paper space, you cannot use 3D viewing commands in paper space.
- UCSs can be aligned with these objects: points, lines, traces, 2D polylines, solids, arcs, circles, texts, shapes, dimensions, attribute definitions, 3D faces, and block references.
- UCSs do *not* align with these objects: mlines, rays, xlines, 3D polylines, splines, ellipses, leaders, viewports, 3D solids, 3D meshes, and regions.

DEFINITIONS

UCS — user-defined 2D coordinate system oriented in 3D space; sets a working plane, orients 2D objects, defines the extrusion direction, and the axis of rotation.

WCS — world coordinate system; the default 3D x,y,z coordinate system.

UcsIcon

Rel.10 Controls the location and display of the UCS icon.

Command	Alias	Ctrl+	F-key	Alt+	Menu Bar	Tablet
ucsicon	...	...	...	VLU	View ↳Display ↳UCS Icon	L2

Command: ucsicon

Enter an option [ON/OFF/All/Noorigin/ORigin/Properties] <ON>: *(Enter an option.)*

COMMAND LINE OPTIONS

All applies changes of this command to viewports.

Noorigin displays the UCS icon always in lower-left corner.

OFF turns off the display of the UCS icon.

ON turns on the display of the UCS icon.

ORigin displays the UCS icon at the current UCS origin.

Properties displays the UCS Icon dialog box.

DIALOG BOX OPTIONS

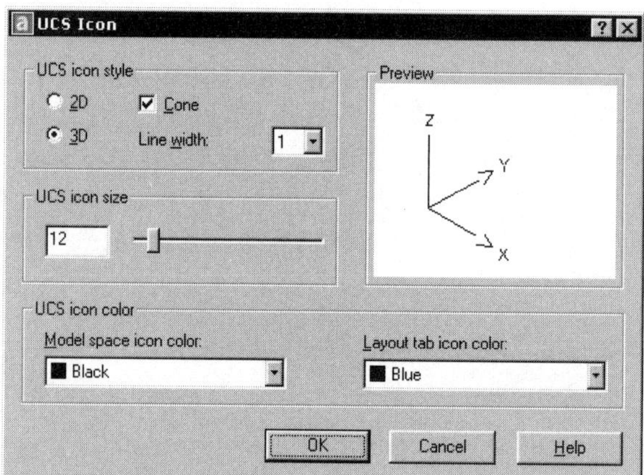

UCS icon style options

○ **2D** displays flat UCS icon.

⊙ **3D** displays tripod UCS icon.

Cone option (available only with 3D style):

☑ Displays arrowheads as 3D cones.

☐ Displays plain arrowheads.

Line width changes the line width from 1 to 2 to 3 pixels.

UCS icon size option

Slide bar changes the icon size from 5 to 95 pixels.

UCS icon color

Model space icon color selects the icon's color in model space.

Layout tab icon color selects the icon's color in layout (paper space).

RELATED SYSTEM VARIABLE

UcsIcon determines the visibility and location of UCS icon:

UcsIcon	Meaning
0	UCS icon not displayed.
1	UCS icon displayed in lower-right corner.
2	UCS icon displayed at the UCS origin, when possible.
3	UCS icon displayed at UCS always (default).

RELATED COMMAND

UCS creates and controls user-defined coordinate systems.

TIPS

- The UCS icon varies, depending on the current viewpoint relative to the active UCS. Below are variations on the 2D icon styles:

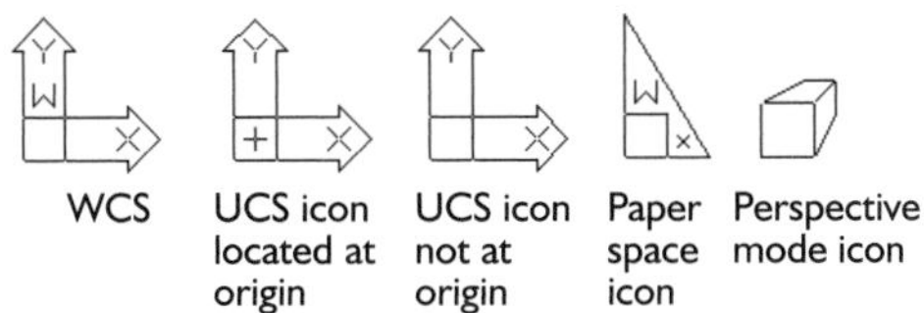

- When AutoCAD switches from 2D wireframe mode to one of the **ShadeMode** command's 3D options, the UCS icon changes to a rendered 3D icon:

- There is generally no need for the UCS icon in 2D drafting, and it can be safely turned off.

UcsMan

2000 Displays the UCS dialog box.

Commands	Aliases	Ctrl+	F-key	Alt+	Menu Bar	Tablet
ucsman	**dducs**	...	...	...	**Tools** ⇨ **Named UCS**	**W8**
	dducsp				**Tools** ⇨ **Orthographic UCS** ⇨ **Preset**	**W9**
+ucsman						

Command: ucsman

Displays dialog box.

DIALOG BOX OPTIONS

Named UCSs tab

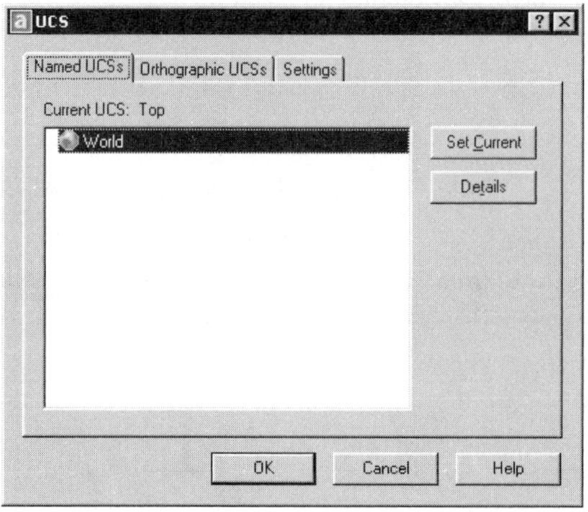

Named UCSs lists the names of the AutoCAD-generated and user-defined coordinate systems of the current viewport in the active drawing; the arrowhead points to the current UCS.

Set Current restores the selected UCS.

Details displays the UCS Details dialog box:

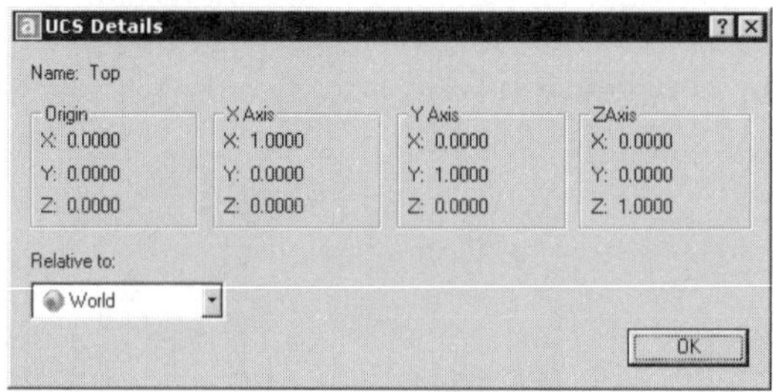

Orthographic UCSs tab

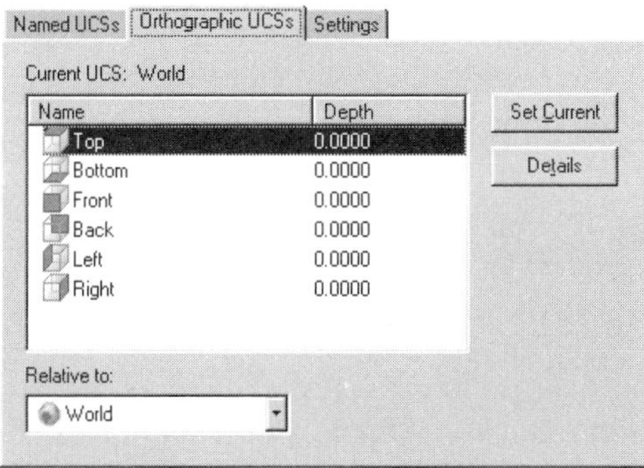

Name lists the six standard orthographic UCS views: top, bottom, front, back, left, and right.

Depth specifies the height of the UCS above the x,y plane.

Relative to specifies the orientation of the selected UCS relative to WCS or to a customized UCS.

Set Current activates the selected UCS.

Details displays the UCS Details dialog box.

Settings tab

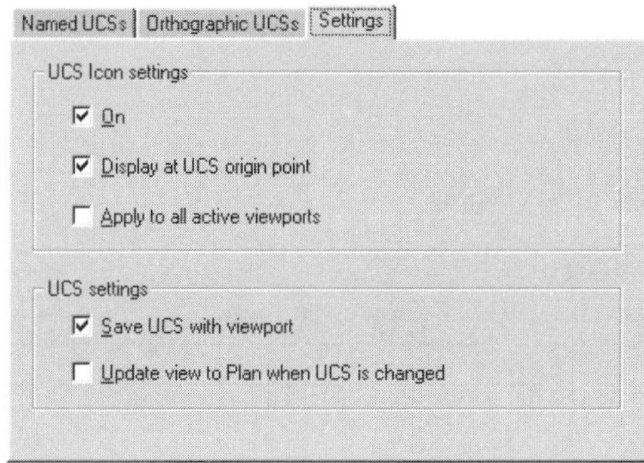

UCS icon settings options

☑ **On** displays the UCS icon in the current viewport; each viewport can display the UCS icon independently.

Display at UCS origin point:

☑ Displays the UCS icon at the origin of the current UCS.

☐ Displays the UCS icon at the lower-left corner of the viewport.

Apply to all active viewports:

☑ Applies these UCS icon settings to all active viewports in the current drawing.

☐ Applies to the current viewport only.

UCS settings options

Save UCS with viewport:

☑ Saves the UCS setting with the viewport.

☐ Current viewport determines UCS settings.

☑ **Update view to Plan when UCS is changed** restores plan view when the UCS changes.

SHORTCUT MENU OPTIONS

Right-click the list in the ***Named UCSs*** *tab:*

Set Current
Rename
Delete
Details...

Set Current sets the selected UCS as active.

Rename renames the selected UCS; you cannot rename the World UCS.

Delete erases the selected UCS; you cannot delete the World UCS.

Details displays the UCS Details dialog box.

Right-click the list in the ***Orthographic UCS*** *tab:*

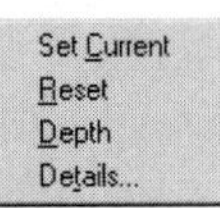

Set Current sets the selected UCS as active.

Reset restores the origin of the selected UCS.

Depth moves the UCS in the z direction.

Details displays the UCS Details dialog box.

+UCSMAN Command

Command: +ucsman

Tab index <0>: *(Type* **1**, **2**, *or* **3**.*)*

COMMAND LINE OPTION

Tab index displays the tab related to the tab number:

Index	Meaning
0	Named UCS tab.
1	Orthographic UCS tab.
2	Settings tab.

RELATED SYSTEM VARIABLES

See ***UCS*** *command.*

RELATED COMMANDS

UCS displays the UCS options at the command line.

UcsIcon changes the display of the UCS icon.

Plan displays the plan view of the WCS or a UCS.

TIPS

- Functions of this command were formerly carried out by the **DdUcs** and **DdUcsP** commands.
- **Unnamed** is the first entry, when the current UCS is unnamed.
- **World** is the default for new drawings; it cannot be renamed or deleted.
- **Previous** is the previous UCS; you can move back through several previous UCSs.

Undefine

Rel. 9 Makes an AutoCAD command unavailable.

Command	Alias	Ctrl+	F-key	Alt+	Menu Bar	Tablet
undefine	...	...	...	...		

Command: undefine
Enter command name: *(Enter name.)*

Example usage:
Command: undefine
Enter command name: line
Command: line
Unknown command. Type ? for list of commands.
Command: .line
From point:

COMMAND OPTIONS

Enter command name specifies the name of the command to make unavailable.

. *(period)* is the prefix for undefined commands to redefine them temporarily.

RELATED COMMAND

Redefine redefines an AutoCAD command.

TIPS

- This command allows AutoLISP and ObjectARX to override native AutoCAD commands.
- Commands created by programs cannot be undefined, including the following programming interfaces:

  AutoLISP and Visual LISP.
  ObjectARx.
  Visual Basic for Applications.
  External commands.
  Aliases.

- In menu macros written with international language versions of AutoCAD, precede command names with an underscore character (_) to translate the command name into English automatically.

V. 2.5 Undoes the effect of the previous command(s).

Command	Alias	Ctrl+	F-key	Alt+	Menu Bar	Tablet
undo	...	...	...	...	...	...

Command: undo

Enter the number of operations to undo or [Auto/Control/BEgin/End/Mark/Back] <1>: *(Enter a number, or an option.)*

COMMAND LINE OPTIONS

Auto treats a menu macro as a single command.

Control limits the options of the **Undo** command.

BEgin groups a sequence of operations (formerly the Group option).

End ends the group option.

Mark sets a marker.

Back undoes back to the marker.

number indicates the number of commands to undo.

Control options

Enter an UNDO control option [All/None/One] <All>:

All toggles on full undo.

None turns off the undo feature.

One limits the **Undo** command to a single undo.

RELATED COMMANDS

Oops unerases the most-recently erased object.

Quit leaves the drawing without saving changes.

Redo undoes the most recent undo.

MRedo undoes multiple undoes.

U undoes a single step.

RELATED SYSTEM VARIABLES

UndoCtl determines the state of undo control:

UndoCtrl	Meaning
0	Undo disabled.
1	Undo enabled.
2	Undo limited to one command.
4	Auto-group mode.
8	Group currently active.

UndoMarks specifies the number of undo marks placed in the **Undo** control stream.

TIPS

- Since the undo mechanism creates a mirror drawing file on disk, disable the **Undo** command with system variable **UndoCtl** (set it to 0) when your computer is low on disk space.
- There are some commands that cannot be undone, such as **Save** and **Plot**.

Union

2000 Joins two or more solids and regions into a single body.

Command	Alias	Ctrl+	F-key	Alt+	Menu Bar	Tablet
union	uni	...	...	MNU	Modify ↳Solids Editing ↳Union	X15

Command: union
Select objects: *(Select one or more solid objects.)*
Select objects: *(Select one or more solid objects.)*
Select objects: *(Press* **ENTER** *to end command.)*

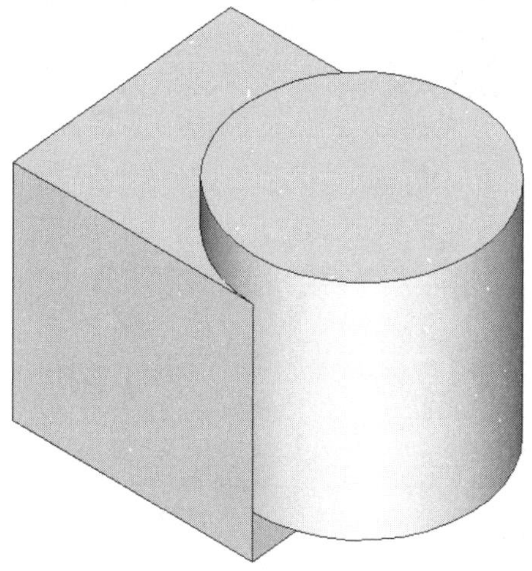

A box and a cylinder unioned into a single object.

COMMAND LINE OPTION

Select objects selects the objects to join into a single object; you must select at least two solid objects.

RELATED COMMANDS

Intersect creates a solid model from the intersection of two objects.

Subtract creates a solid model by subtracting one object from another.

TIPS

- You must select at least two solid or coplanar region objects.
- The two objects need not overlap for this command to operate.

'Units

V. 1.4 Controls the display and format of coordinates and angles, as well as the orientation direction of angles.

Commands	Aliases	Ctrl+	F-key	Alt+	Menu Bar	Tablet
units	**un** **ddunits**	...	...	**OU**	**Format** ↳**Units**	**V4**
-units	**-un**					

command: units

Displays dialog box:

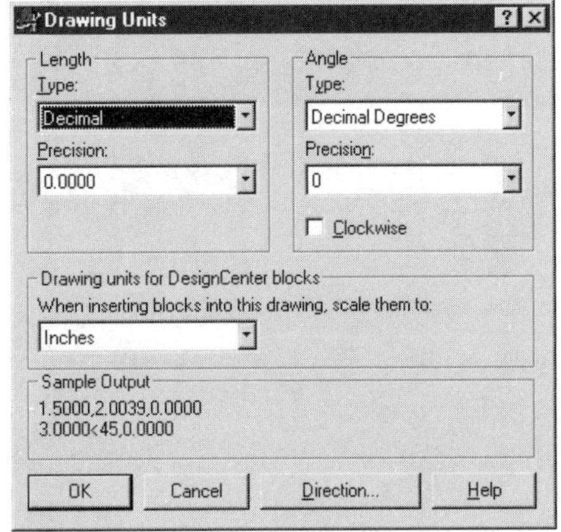

DIALOG BOX OPTIONS

Length options

Type sets the format for units of linear measure displayed by AutoCAD: Architectural, Decimal, Engineering, Fractional, or Scientific.

Precision specifies the number of decimal places or fractional accuracy.

Angle options

Type sets the current angle format.

Precision sets the precision for the current angle format.

☐ **Clockwise** calculates positive angles in the clockwise direction.

Drawing units for DesignCenter blocks specifies the units when blocks are inserted from the DesignCenter.

Direction displays the Direction Control dialog box.

Direction Control dialog box

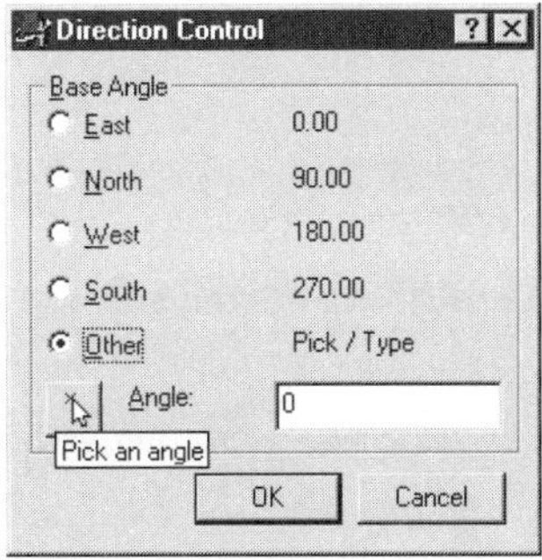

Base Angle options

○ **East** sets the base angle to 0 degrees (default).

○ **North** sets the base angle to 90 degrees.

○ **West** sets the base angle to 180 degrees.

○ **South** sets the base angle to 270 degrees.

⊙ **Other** turns on the Angle option.

Angle sets the base angle to any direction.

Pick an angle dismisses the dialog box temporarily, and allows you to define the base angle by picking two points in the drawing; AutoCAD prompts you 'Pick angle' and 'Specify second point:'.

-UNITS Command

Command: -units

Report formats: (Examples)

1. Scientific	**1.55E+01**
2. Decimal	**15.50**
3. Engineering	**1'-3.50"**
4. Architectural	**1'-3 1/2"**
5. Fractional	**15 1/2**

With the exception of Engineering and Architectural formats, these formats can be used with any basic unit of measurement. For example, Decimal mode is perfect for metric units as well as decimal English units.

Enter choice, 1 to 5 <2>: *(Enter a value.)*

Enter number of digits to right of decimal point (0 to 8) <4>: *(Enter a value.)*

Systems of angle measure: (Examples)

1. Decimal degrees	**45.0000**
2. Degrees/minutes/seconds	**45d0'0"**
3. Grads	**50.0000g**
4. Radians	**0.7854r**
5. Surveyor's units	**N 45d0'0" E**

Enter choice, 1 to 5 <1>: *(Enter a value.)*

Enter number of fractional places for display of angles (0 to 8) <0>: *(Enter a value.)*

Direction for angle 0:

East 3 o'clock = 0

North 12 o'clock = 90

West 9 o'clock = 180

South 6 o'clock = 270

Enter direction for angle 0 <0>: *(Enter a value.)*

Measure angles clockwise? [Yes/No] <N> *(Enter* **Y** *or* **N***.)*

COMMAND LINE OPTIONS

Report formats selects scientific, decimal, engineering, architectural, or fractional format for length display.

Number of digits to right of decimal point specifies the number of decimal places between 0 and 8.

Systems of angle measure selects decimal degrees, degrees/minutes/seconds, grads, radians, or surveyor's units for angle display.

Denominator of smallest fraction to display specifies the denominator of fraction displays, such as 1/2 or 1/128.

Number of fractional places for display of angles specifies the number of decimal places between 0 and 8.

Direction for angle 0 selects the direction for 0 degrees from east, north, west, or south.

Do you want angles measured clockwise?

Yes measures angles clockwise.

No measures angles counterclockwise.

RELATED SYSTEM VARIABLES

InsUnits specifies the drawing units for blocks dragged from the DesignCenter:

InsUnits	Meaning
0	Unitless
1	Inches
2	Feet
3	Miles
4	Millimeters
5	Centimeters
6	Meters
7	Kilometers
8	Microinches
9	Mils
10	Yards; 3 feet
11	Angstroms; 0.1 nanometers
12	Nanometers; 10E-9 meters
13	Microns; 10E-6 meters
14	Decimeters; 0.1 meter
15	Decameters; 10 meters
16	Hectometers; 100 meters
17	Gigameters; 10E9 meters
18	Astronomical Units; 149.597E8 kilometers
19	Light Years; 9.4605E9 kilometers
20	Parsecs; 3.26 light years

AngBase specifies the direction of zero degrees.

AngDir specifies the direction of angle measurement.

AUnits specifies the units of angles.

AuPrec specifies the displayed precision of angles.

InsUnitsDefSource specifies source units to be used.

InsUnitsDefTarget specifies target units to be used.

LUnits specifies the units of measurement.

LuPrec specifies the displayed precision of coordinates.

UnitMode toggles the type of display units.

RELATED COMMAND

New sets up drawings with Imperial or metric units.

TIPS

- Because **'Units** is a transparent command, you can change units during another command.
- The 'Direction Angle' prompt lets AutoCAD start the angle measurement from any direction.
- AutoCAD accepts the following notations for angle input:

Notation	Meaning
<	Specify an angle based on current units setting.
<<	Bypass angle translation set by **Units** command to use 0-angle-is-east direction and decimal degrees.
<<<	Bypass angle translation; use angle units set by the **Units** command and 0-angle-is-east direction.

- The system variable **UnitMode** forces AutoCAD to display units in the same manner that you enter them.
- Do not use a suffix — such as 'r' or 'g' — for angles entered as radians or grads; instead, use the **Units** command to set angle measurement to radians and grads.
- The **Drawing units for DesignCenter blocks** option is for inserting blocks from AutoCAD DesignCenter, and especially when the block was created in other units.
- To prevent blocks from being scaled when dragged from the DesignCenter window, select **Unitless**.

UpdateField

2005 Forces the updates of selected fields.

Command	Alias	Ctrl+	F-key	Alt+	Menu Bar	Tablet
updatefield	...	...	...	TT	**Tools ↳Update Fields**	...

Command: updatefield
Select objects: *(Select one or more fields.)*
Select objects: *(Press* **ENTER** *to end field selection.)*
***n* field(s) found**
***n* field(s) updated**

COMMAND LINE OPTION

Select objects selects one or more fields; press CTRL+A to select all field text in the drawing.

RELATED COMMANDS

Field places field text, which is automatically updated as its value changes.

Find finds fields.

DdEdit edits the properties of field text.

RELATED SYSTEM VARIABLE

FieldEval determines when fields are updated (default = 31, all on):

FieldEval	Meaning
0	Fields are not updated (static).
1	Fields are updated when drawing is opened.
2	Fields are updated when drawing is saved.
4	Field are updated when plotted.
8	Fields are updated with the **eTransmit** command.
16	Fields are updated when drawing is regenerated.

TIP

- This command forces individual fields to update. Depending on the setting of the **FieldEval** system variable, all fields are updated automatically.

UpdateThumbsNow

2005 Forces the update of preview images in the Sheet Set Manager window.

Command	Alias	Ctrl+	F-key	Alt+	Menu Bar	Tablet
updatethumbsnow	...	...	...	...	...	...

Command: updatethumbsnow

COMMAND LINE OPTIONS

None.

RELATED COMMAND

SheetSet controls sheet sets.

RELATED SYSTEM VARIABLE

UpdateThumbnail determines which thumbnails updated (default = 15):

FieldEval	Meaning
0	Thumbnail previews not updated.
1	Model view thumbnail updated.
2	Sheet view thumbnails updated.
4	Sheet thumbnails updated.
8	Thumbnails updateded when sheets and views created, modified, and restored.
16	Thumbnails updated when drawing saved.

TIP

- This command controls when preview images are updated, required when the drawing changes.

VbaIde

2000 Displays the Visual Basic window (*short for Visual Basic for Applications Integrated Development Environment*).

Command	Alias	Alt+	F-key	Alt+	Menu Bar	Tablet
vbaide	...	**F11**	...	**TMB**	**Tools** ↳**Macro** ↳**Visual Basic Editor**	...

Command: vbaide

Displays window:

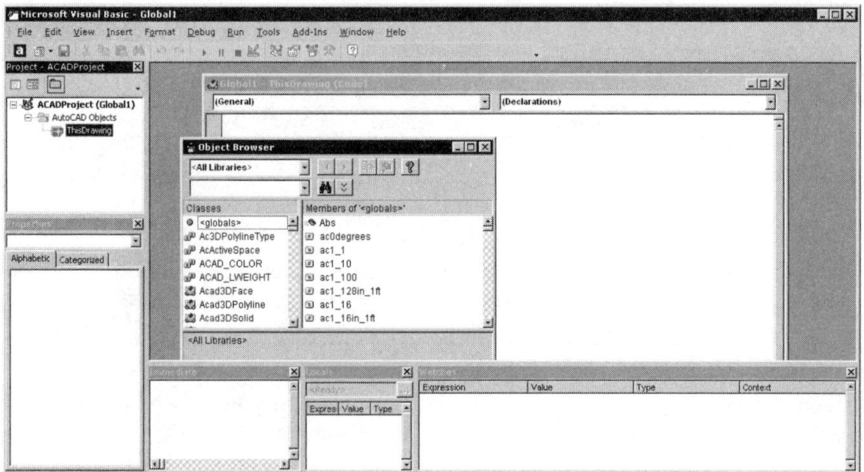

MENU BAR OPTION

Select **Help | Microsoft Visual Basic Help** *for assistance in using this VBA IDE window.*

RELATED COMMANDS

VbaLoad loads a VBA project; displays the Open VBA Project dialog box.

VbaMan displays the VBA Manager dialog box.

VbaRun displays the Macros dialog box.

VbaStmt executes a VBA expression at the command line.

VbaUnload unloads a VBA project.

RELATED SYSTEM VARIABLES

None.

TIPS

- *VBA* is short for "Visual Basic for Applications," a macro programming language common to a number of Windows applications; it is based on the Visual Basic programming language.
- AutoCAD contains sample VBA projects in the *\autocad 2005\sample\vba* folder.
- For more information about Visual Basic for Applications, read the *ActiveX and VBA Developer's Guide* included with AutoCAD.
- Because many viruses can be spread via VBA macros, I strongly recommend that your computer have real-time virus protection to prevent infection.
- Because loading a macro from other sources into AutoCAD can also expose your computer to malicious viruses, AutoCAD displays the following warning:

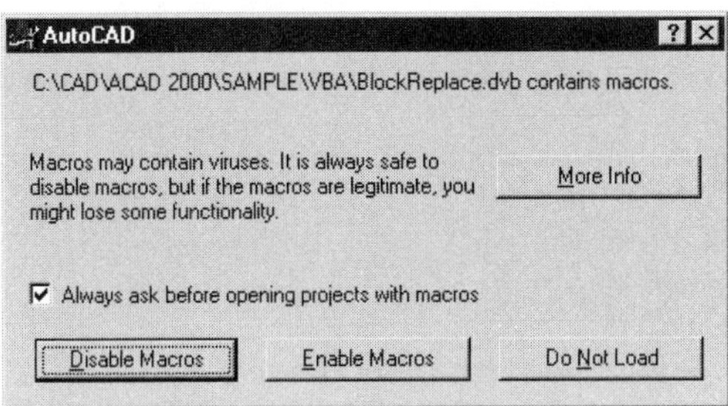

More Info displays AutoCAD's on-line help window.

Always ask before opening projects with macros:

☑ Displays this dialog box.

☐ Prevents this dialog box from being displayed; to turn back on, use the **VbaRun** command's Options dialog box; see **VbaRun** command.

Disable Macros loads a VBA project file, but disables macros; you can view, edit, and save the macros. To renable the macros, close the project, and then open it again with Enable Macros.

Enable Macros loads the project file with macros enabled.

Do Not Load prevents the project file from being loaded.

VbaLoad

2000 Loads a VBA project into AutoCAD (*short for Visual Basic for Applications LOAD*).

Commands	Alias	Ctrl+	F-key	Alt+	Menu Bar	Tablet
vbaload	...	...	...	TML	Tools ↳Macro ↳Load Project	...
-vbaload						

Command: vbaload

Displays the Open VBA Project dialog box; select a .dvd file, and then click ***Open****.*

When the project contains macros, the AutoCAD dialog box is displayed; see the ***VbaIde*** *command.*

-VBALOAD Command

Command: -vbaload

Open VBA project <*projectname*>: *(Enter project name.)*

COMMAND LINE OPTION

Open VBA project specifies the project path and filename.

RELATED COMMANDS

Vbaide displays the Visual Basic for Applications development environment window.

VbaMan displays the VBA Manager dialog box.

VbaRun displays the Macros dialog box.

VbaStmt executes a VBA expression at the command line.

VbaUnload unloads a VBA project.

RELATED SYSTEM VARIABLES

None.

TIPS

- You may load one or more VBA projects; there is no practical limit to the number.
- To unload a VBA project, use the **VbaUnload** command.
- This command does not load embedded VBA projects; these projects are automatically loaded with the drawing.
- When the project contains macros, AutoCAD displays the AutoCAD dialog box to warn you about protection against macro viruses; see the **Vbaide** command.
- For more information about Visual Basic for Applications, read the *ActiveX and VBA Developer's Guide* included with AutoCAD.

VbaMan

2000 Displays the VBA Manager dialog box (*short for Visual Basic for Applications MANager*).

Command	Alias	Ctrl+	F-key	Alt+	Menu Bar	Tablet
vbaman	...	...	...	**TMV**	**Tools** **↳Macro** **↳VBA Manager**	...

Command: vbaman

Displays dialog box:

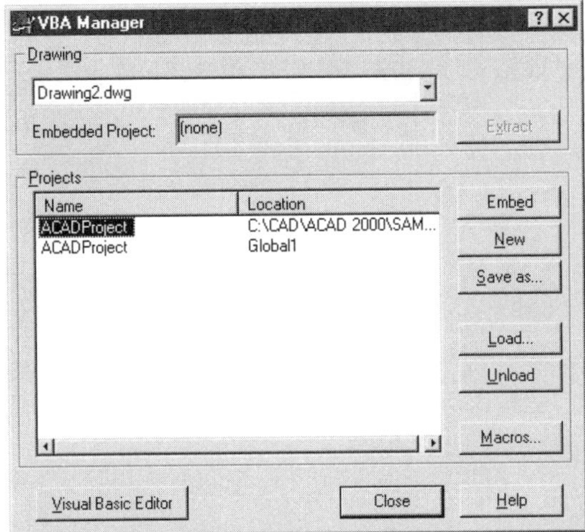

DIALOG BOX OPTIONS

Drawing options

Drawing lists the names of drawings currently loaded in AutoCAD.

Embedded Project specifies the name of the embedded project.

Extract moves the embedded project from the drawing to a global project file.

Projects options

Embed embeds the project in the drawing.

New creates a new project; default name = Global *n*.

Save as saves a global project.

Load displays the Open VBA Project dialog box; see the **VbaLoad** command.

Unload unloads the global project.

Macros displays the Macros dialog box; see the **VbaRun** command.

Additional options

Visual Basic Editor displays the Visual Basic Editor; see the **VbaIde** command.

VbaRun

2000 Displays the Macros dialog box (*short for Visual Basic for Applications* RUN).

Command	Alias	Alt+	F-key	Alt+	Menu Bar	Tablet
vbarun	...	F8	...	TMM	Tools ↳Macro ↳Macros	...
-vbarun						

Command: vbarun

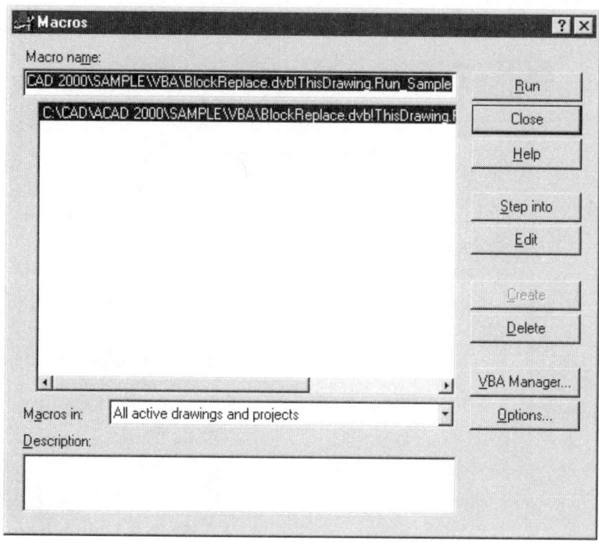

DIALOG BOX OPTIONS

Macro name specifies the name of the macro; enter a name or select one from the list.

Macros in specifies the projects and drawings containing macros from all active drawings and projects, all active drawings, all active projects, and any single drawing or project currently open in AutoCAD.

Description describes the macro; you may modify the description.

Buttons

Run runs the macro.

Close closes the dialog box.

Help displays context-sensitive on-line help.

Step into displays the Visual Basic Editor, and executes the macro, pausing at the first executable line of code.

Edit displays the Visual Basic Editor with the macro; see the **VbaIde** command.

Create displays the Visual Basic Editor with an empty procedure.

Delete erases the selected macro.

VBA Manager displays the VBA Manager dialog box; see the **VbaMan** command.

Options displays the VBA Options dialog box.

VBA Options dialog box

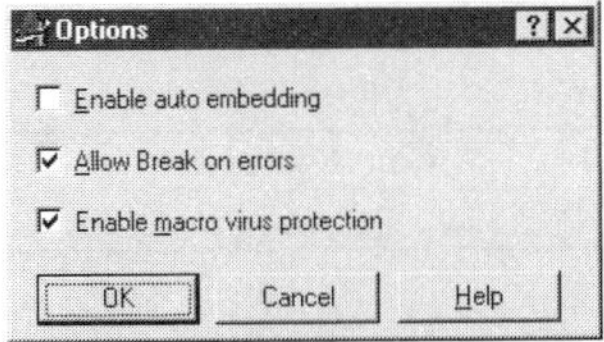

☐ **Enable auto embedding** creates an embedded VBA project for all drawings when you open the drawing:

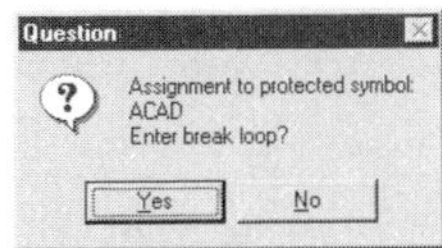

Allow Break on errors:

☑ Stops the macro, and displays the **Visual Basic Editor** with the code, showing the error in the macro.

☐ Displays an error message, and stops the macro.

☑ **Enable macro virus protection** enables virus protection, which displays a dialog box when VBA macros are loaded; see the **VbaIde** command.

-VBARUN Command

Command: -vbarun

Macro name: *(Enter macro name.)*

COMMAND LINE OPTION

Macro name treats a menu macro as a single command.

RELATED COMMANDS

Vbaide displays the Visual Basic for Applications integrated development environment window.

VbaLoad loads a VBA project; displays the Open VBA Project dialog box.

VbaMan displays the VBA Manager dialog box.

VbaStmt executes a VBA expression at the command line.

VbaUnload unloads a VBA project.

RELATED SYSTEM VARIABLES

None.

TIPS

- A *macro* is an executable subroutine; each project can contain one or more macros.
- When the macro's name is not unique among loaded projects, include the module's project and names in this format: **Project.Module.Macro**
- When the macro is not yet loaded, include the *.dvb* file name using this format: **Filenamedvb!Project.Module.Macro**

VbaStmt

2000 Executes a single line VBA expression at the command prompt (*short for Visual Basic for Applications StaTeMenT*).

Command	Alias	Ctrl+	F-key	Alt+	Menu Bar	Tablet
vbastmt	...	...	...	...	...	...

Command: vbastmt
Statement: *(Enter VBA statement.)*

COMMAND LINE OPTION

Statement specifies the VBA statement for AutoCAD to execute.

RELATED COMMANDS

Vbaide displays the Visual Basic for Applications integrated development environment window.

VbaLoad loads a VBA project; displays the Open VBA Project dialog box.

VbaMan displays the VBA Manager dialog box.

VbaRun displays the Macros dialog box.

VbaUnload unloads a VBA project.

RELATED SYSTEM VARIABLES

None.

TIPS

- A VBA *statement* is a complete instruction containing keywords, operators, variables, constants, and expressions.
- A VBA *macro* is an executable subroutine.
- At the 'Statement' prompt, enter a single line of code; use the colon (:) to separate multiple statements on the single line.

VbaUnload

2000 Unloads a VBA project (*short for Visual Basic for Applications* UNLOAD).

Command	Alias	Ctrl+	F-key	Alt+	Menu Bar	Tablet
vbaunload	...	...	...	...	...	...

Command: vbaunload
Unload VBA Project: *(Enter project name.)*

COMMAND LINE OPTION

Unload VBA Project specifies the name of the VBA project to unload.

ENTER unloads the active global project.

RELATED COMMANDS

VbaIde displays the Visual Basic for Applications integrated development environment window.

VbaLoad loads a VBA project; displays the Open VBA Project dialog box.

VbaMan displays the VBA Manager dialog box.

VbaRun displays the Macros dialog box.

VbaStmt executes a VBA expression at the command line.

RELATED SYSTEM VARIABLES

None.

TIPS

- When you do not enter a project name, AutoCAD unloads the active global project.
- To load a VBA project, use the **VbaLoad** command.

View

V. 2.0 Saves and displays view by name in the current viewport; creates view categories for sheet sets.

Commands	Aliases	Ctrl+	F-key	Alt+	Menu Bar	Tablet
view	**v**	**...**	**...**	**VN**	**View** **↳Named Views**	**M5**
+view						
-view	**-v**			**V3**	**View** **↳3D Views**	**O3-Q5**

Command: view

Displays dialog box:

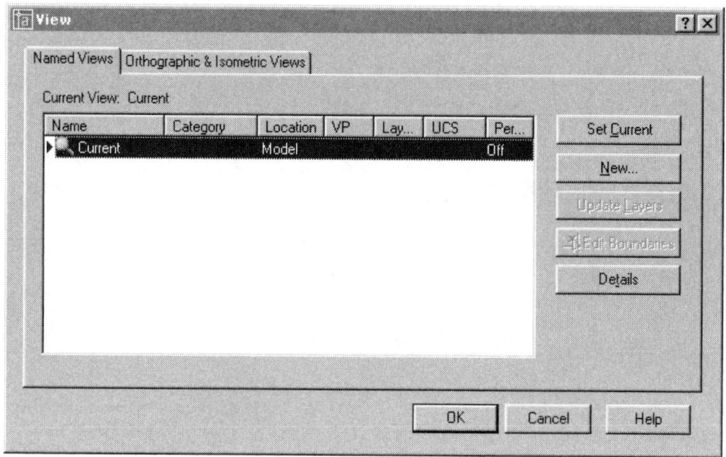

DIALOG BOX OPTIONS

Named Views tab

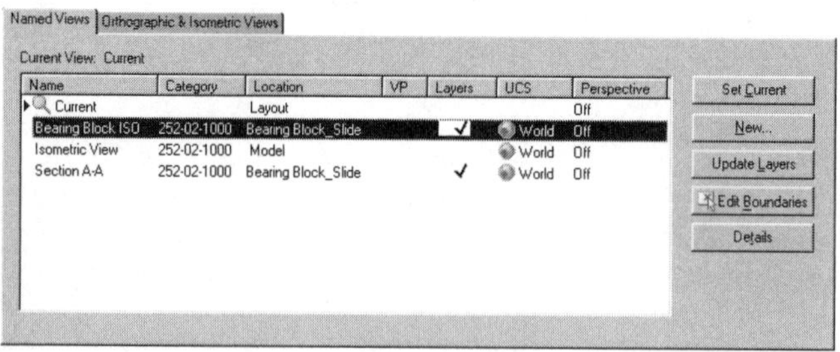

Name lists the names of saved views in the current drawing.

Category appears on the View List tab in the Sheet Set Manager *(new to AutoCAD 2005)*.

Location specifies the model space or the name of a layout tab.

VP indicates whether the view is associated with a viewport on a sheet in a sheet set *(new to AutoCAD 2005).*

Layers indicates whether layer visibility settings are saved with the view *(new to AutoCAD 2005).*

UCS names the UCS saved with the view.

Perspective indicates whether the view was saved in perspective view, was clipped, or neither.

Buttons

Set Current restores the named view; alternatively, double-click the view name.

New displays the New View dialog box.

Update Layers updates view to match layer visibility settings of the current viewport *(new to AutoCAD 2005).*

Edit Boundaries highlights the view, by graying out the area outside of the view *(new to AutoCAD 2005).*

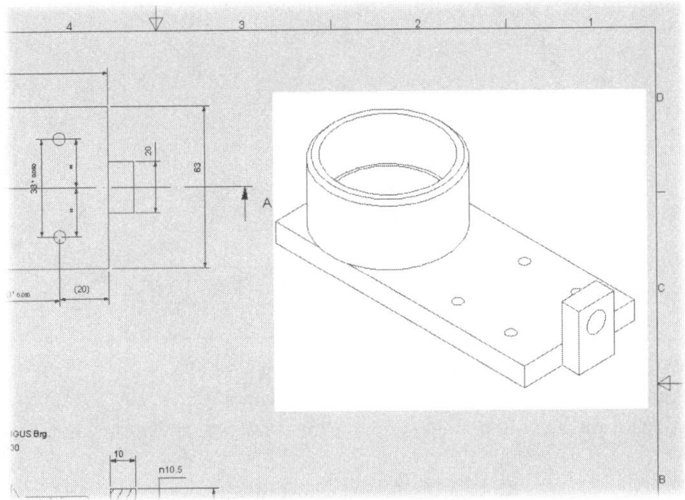

AutoCAD clears the dialog box, and prompts:

Specify first corner: *(Pick a point.)*

Specify opposite corner: *(Pick another point.)*

Specify first corner (or press ENTER to accept): *(Press* **ENTER** *to return to the dialog box.)*

Details displays the View Details dialog box.

Orthographic & Isometric Views tab

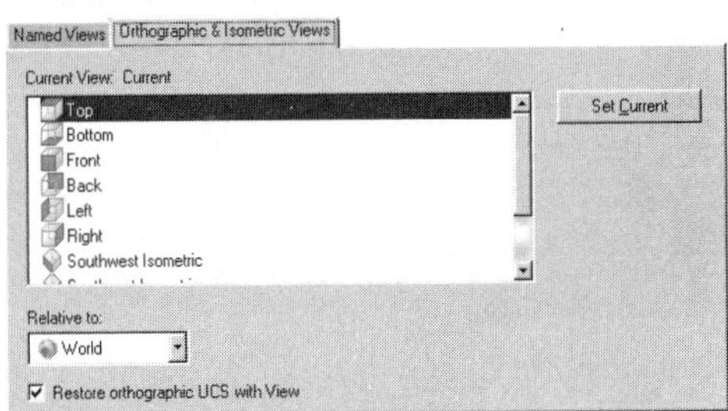

Names of standard orthographic and isometric views are listed; an arrowhead points to the current view.

Relative to sets the selected view relative to:

- World coordinate system.
- A named UCS, if any.

☑ **Restore orthographic UCS with View** restores the associated UCS.

Set Current sets the selected view; after clicking **OK**, AutoCAD automatically zooms to the extents of the view.

New View dialog box

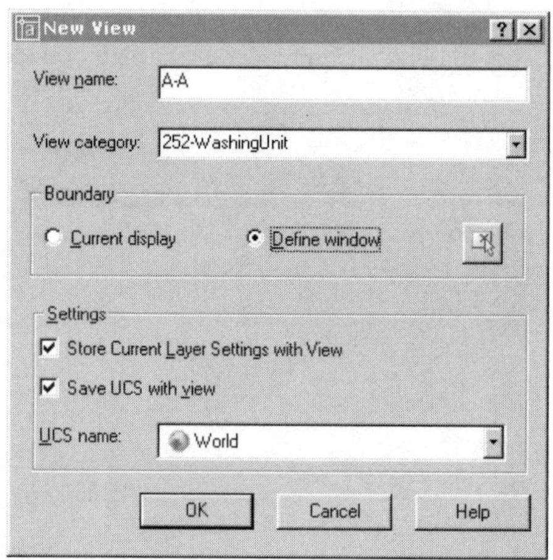

View name specifies the view name; up to 255 characters long.

View Category specifies the default prefix for named views *(new to AutoCAD 2005)*. In the Sheet Set Manager, select the View List tab to see the names of views, under categories:

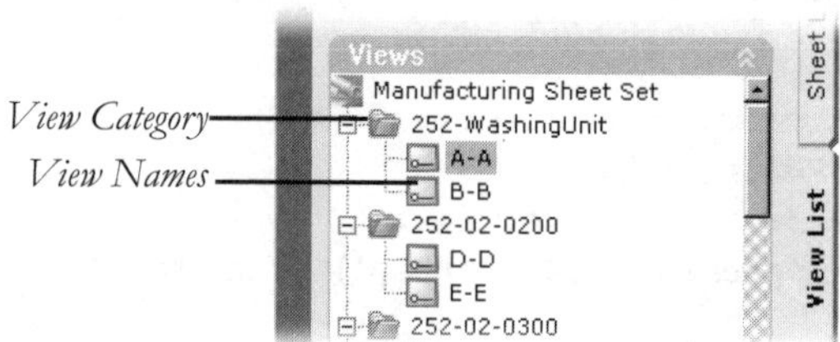

Boundary options

○ **Current display** stores the current viewport as the named view.

⊙ **Define window** stores a windowed area as the named view.

Define View Window dismisses the dialog box temporarily, and prompts you to pick two corners that define the view.

Settings options

☑ **Save Current Layer Settings with View** toggles whether layer settings (freeze, lock, plot, and so on) are stored with the view.

☑ **Save UCS with view** toggles the option to save a UCS with the named view.

UCS name selects the name of a UCS to store with the named view.

View Details dialog box

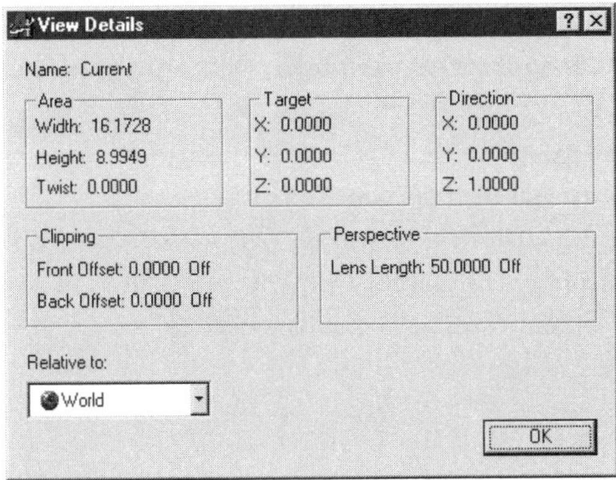

Relative to sets the selected view relative to the WCS or a named UCS.

SHORTCUT MENU OPTIONS

Right-click the list in the ***Named Views*** *tab:*

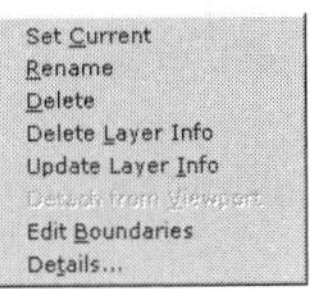

Set Current sets the selected view as active.

Rename renames the selected view; you cannot rename the Current view.

Delete erases the selected view; you cannot delete the Current view.

Delete Layer Info removes layer visibility info stored with the view *(new to AutoCAD 2005).*

Update Layer Info updates the view to match the layer visibility settings of the current viewport *(new to AutoCAD 2005).*

Detach from Viewport breaks the link between the view and its viewport in the sheet set *(new to AutoCAD 2005).*

Edit Boundaries highlights the view, by graying out the area outside of the view *(new to AutoCAD 2005).*

Details displays the View Details dialog box.

+VIEW Command

Command: +view

Tab index <0>: *(Type* **0** *or* **1***.)*

COMMAND LINE OPTIONS

Tab index specifies the tab of the **View** dialog box to display:

Index	Meaning
0	Named Views tab (default).
1	Orthographic & Isometric Views tab.

-VIEW Command

Command: -view

Enter an option [?/Categorize/lAyer state/Orthographic/Delete/Restore/Save/Ucs/Window]: *(Enter an option.)*

COMMAND LINE OPTIONS

? lists the names of views saved in the current drawing.

Categorize provides the default category prefix *(new to AutoCAD 2005).*

lAyer state saves the current layer settings with the view *(new to AutoCAD 2005).*

Orthographic restores predefined orthographic views.

Delete deletes a named view.

Restore restores a named view.

Save saves the current view with a name.

Ucs saves the current UCS with the view.

Window saves a windowed view with a name.

Orthographic options

Enter an option [Top/Bottom/Front/BAck/Left/Right] <Top>: *(Enter an option.)*

Select Viewport for view: *(Pick a viewport.)*

Regenerating model.

Enter an option selects a standard orthographic view for the current viewport: Top, Bottom, Front, BAck, Left, or Right.

Select Viewport for view selects the viewport — in either Model or Layout tab — in which to apply the orthographic view.

RELATED COMMANDS

Rename changes the names of views via a dialog box.

UCS creates and displays user-defined coordinate systems.

PartialLoad loads portions of drawings based on view names.

Open opens drawings and optionally starts with a named view.

Plot plots named views.

SheetSet uses named views.

RELATED SYSTEM VARIABLES

DefaultViewCategory specifies the default name for categories (*new to AutoCAD 2005*).

ViewCtr specifies the coordinates of the center of the view.

ViewSize specifies the height of the view.

TIPS

- Name views in your drawing to move quickly from one detail to another.
- The **Plot** command plots named views of a drawing.
- Objects outside of the window created by the **Window** option may be displayed, but are not plotted.
- You create separate views in model and paper space; when listing named views with ?, AutoCAD indicates an 'M' or 'P' next to the view name.
- As of AutoCAD 2005, this command creates views for sheets.

ViewPlotDetails

2005 Displays a report of plotting errors and successes.

Command	Alias	Ctrl+	F-key	Alt+	Menu Bar	Tablet
viewplotdetails	...	...	...	**FB**	**File** ⇘**View Plot Details**	...

Command: viewplotdetails

Displays dialog box:

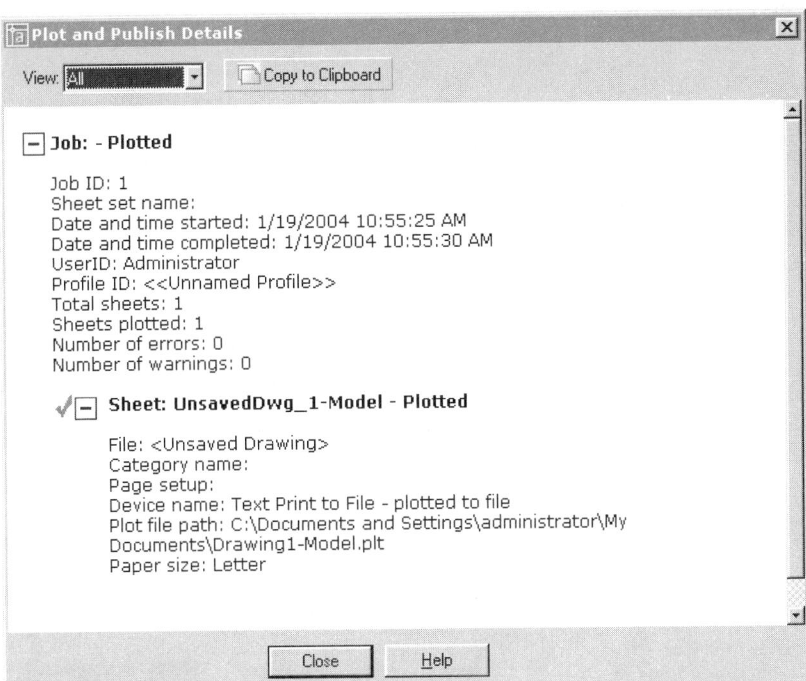

DIALOG BOX OPTIONS

Green checkmark indicates successful plot.

Red x warns of plotting error.

View determines whether all messages are displays, or just errors:

- All messages.
- Errors only.

Copy to Clipboard copies selected text to Clipboard.

[−] Collapses text under heading.

[+] Expands text under heading.

SHORTCUT MENU

The ***Cut*** *and* ***Paste*** *options are unavailable.*

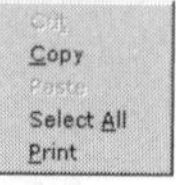

Copy copies selected text to the Clipboard.

Select All selects all text in the dialog box; use the cursor to select portions of text.

Print displays Print dialog box; select a printer and its options, and then click **Print**. *All* text in the dialog box is printed, not just selected text.

RELATED COMMANDS

Plot plots drawings.

TraySettings toggles the display of icons in the tray.

TIPS

- When a plot is completed, AutoCAD displays a balloon. Clicking the blue underlined text displays the dialog box.

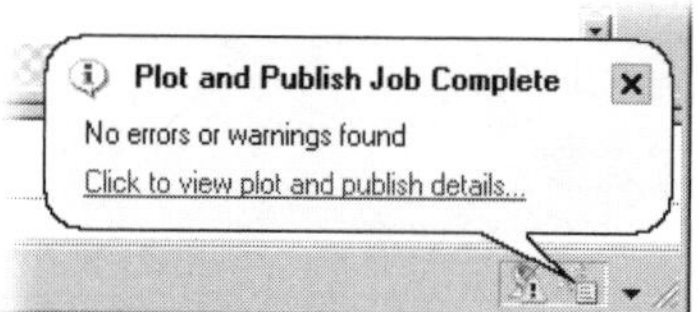

- You can also access this dialog box by right-clicking the Plotting icon in the tray, and then selecting **View Plot Details**.

ViewRes

V. 2.5 Controls the roundness of curved objects; determines whether zooms and pans are performed as redraws or regens (*short for VIEW RESolution*).

Command	Alias	Ctrl+	F-key	Alt+	Menu Bar	Tablet
viewres	...	...	...	...	...	...

Command: viewres
Do you want fast zooms? [Yes/No] <Y>: *(Type* **Y** *or* **N***.)*
Enter circle zoom percent (1-20000) <100>: *(Enter a value.)*

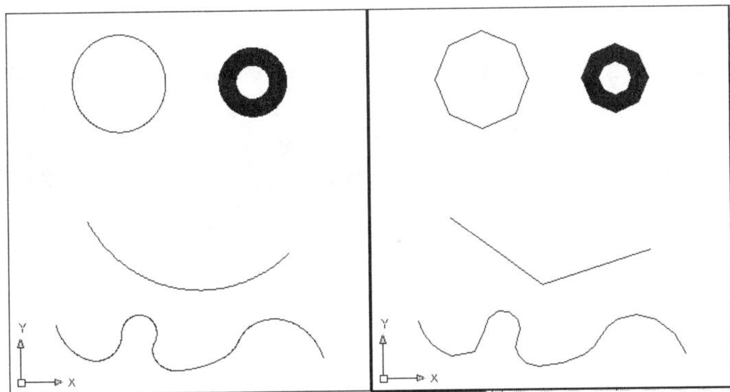

Circle zoom percent = 100 (at left) and 1 (at right).

COMMAND LINE OPTIONS

Do you want fast zooms?

Yes: AutoCAD tries to make every zoom and pan a redraw (faster).

No: Every zoom and pan causes a regeneration (slower).

Enter circle zoom percent specifies that smaller values display faster, but makes circles look less round (see figure); default = 100.

RELATED SYSTEM VARIABLE

WhipArc toggles the display of circles and arcs as vectors or as true, rounded objects.

RELATED COMMAND

RegenAuto determines whether AutoCAD uses redraws or regens.

TIP

- Setting **WhipArc** to 1 is recommended over increasing the value of **ViewRes**.

Removed Command

VlConv was removed from AutoCAD Release 14; use **3dsIn** and **3dsOut** instead.

VLisp

2000 Opens the VLisp integrated development environment (*short for Visual LISP*).

Commands	Alias	Ctrl+	F-key	Alt+	Menu Bar	Tablet
vlisp	**vlide**	...	...	**TSV**	**Tools** **⇘AutoLISP** **⇘Visual LISP Editor**	...

Command: vlisp

Displays window:

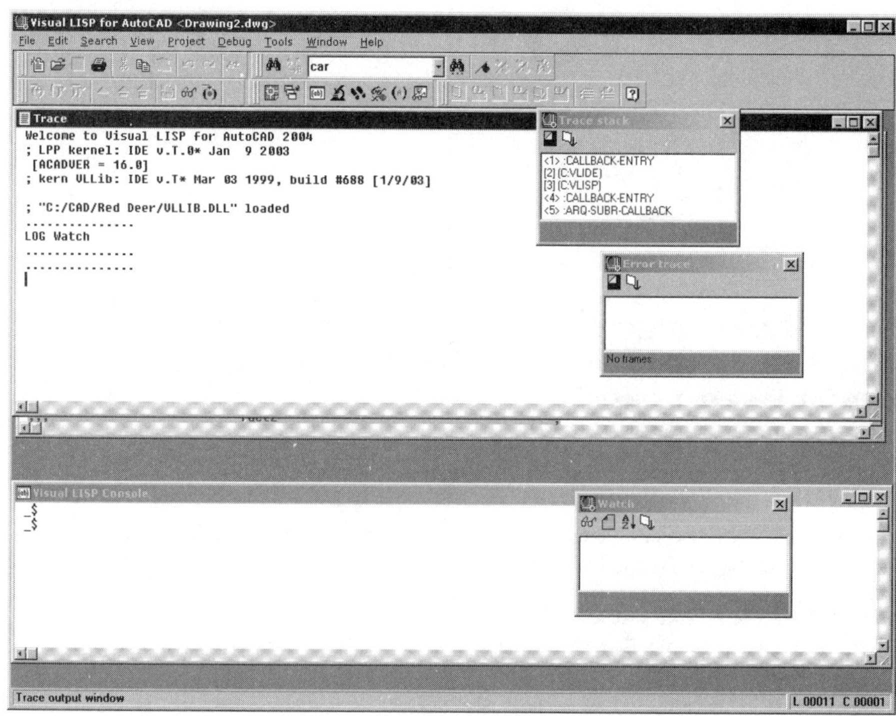

MENU BAR OPTION

*Select **Help | Visual LISP Help Topics** for assistance in using this VLISP window.*

RELATED COMMAND

AppLoad loads Visual LISP applications, as well as programs written in AutoLISP and other APIs.

TIP

- Sample VLisp code can be found in the *autocad 2005**sample**vlisp* folder.

VpClip

2000 Clips a layout viewport (*short for ViewPort CLIPping*).

Command	Alias	Ctrl+	F-key	Alt+	Menu Bar	Tablet
vpclip	...	...	...	...	**Modify** ↳**Clip** ↳**Viewport**	...

Command: vpclip
Select viewport to clip: *(Pick a viewport.)*
Select clipping object or [Polygonal] <Polygonal>: *(Select an object, or type* **P**.*)*

The selected viewport disappears, and is replaced by the new clipped viewport.

COMMAND LINE OPTIONS

Select viewport to clip selects the viewport that will be clipped.

Select clipping object selects the object that defines the clipping boundary: closed polyline, circle, ellipse, closed spline, or region.

Polygonal options
Specify start point: *(Pick a point.)*
Specify next point or [Arc/Close/Length/Undo]: *(Pick a point, or enter an option.)*
Specify next point or [Arc/Close/Length/Undo]: *(Type* **C** *to close.)*

Specify start point specifies the starting point for the polygon.

Arc draws an arc segment; see the **Arc** command.

Close closes the polygon.

Length draws a straight segment of specified length.

Undo undoes the previous polygon segment.

RELATED COMMAND

Mview creates rectangular and polygonal viewports in paper space.

TIPS

- This command does not operate in Model tab.
- An example of clipped viewports:

VpLayer

Rel. 11 Controls the visibility of layers in viewports, when a layout tab other than Model is selected (*short for ViewPort LAYER*).

Command	Alias	Ctrl+	F-key	Alt+	Menu Bar	Tablet
vplayer	...	...	...	...	...	...

Command: vplayer
Enter an option [?/Freeze/Thaw/Reset/Newfrz/Vpvisdflt]: *(Enter an option.)*
Select a viewport: *(Pick a viewport.)*

COMMAND LINE OPTIONS

Freeze indicates the names of layers to freeze in this viewport.

Newfrz creates new layers that are frozen in all newly-created viewports (short for NEW FReeZe).

Reset resets the state of layers based on the Vpvisdflt settings.

Thaw indicates the names of layers to thaw in this viewport.

Vpvisdflt determines which layers will be frozen in a newly-created viewport and default visibility in existing viewports (short for ViewPort VISibility DeFauLT).

? lists the layers frozen in the current viewport.

RELATED COMMANDS

Layer creates and controls layers in all viewports.

MView creates and joins viewports when tilemode is off.

RELATED SYSTEM VARIABLE

TileMode controls whether viewports are tiled (model) or overlapping (layouts).

VpMax / VpMin

2005 Maximizes or minimizes the selected viewport in the AutoCAD window (*short for ViewPortMAXimize*).

Command	Alias	Ctrl+	F-key	Alt+	Menu Bar	Tablet
vpmax	...	...	...	...	...	...
vpmin						

Command: vpmax

When the layout contains more than one viewport, AutoCAD prompts:

Select a viewport to maximize: *(Select a viewport.)*

AutoCAD maximizes the viewport to fill the entire AutoCAD window, and switches to model space for editing.

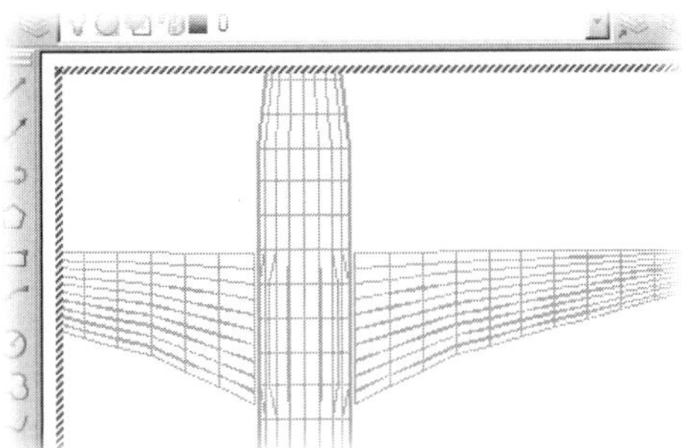

The red dashed border indicates you are editing in vpmax mode.

COMMAND LINE OPTION

Select a viewport to maximze selects the viewport.

RELATED COMMAND

VPorts creates viewports.

RELATED SYSTEM VARIABLE

VpMaximizedState determines whether the viewport is maximized.

TIPS

- Use the **VpMin** command to return to the layout tab, and restore the viewport.
- When the viewport is maximized, AutoCAD displays the Minimize Viewport icon on the status bar. Click the arrows to move from one viewport to the next.

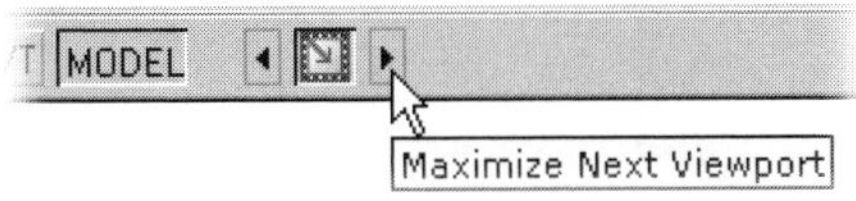

VPoint

<u>**V. 2.1**</u> Changes the viewpoint of 3D drawings (*short for ViewPOINT*).

Command	Alias	Ctrl+	F-key	Alt+	Menu Bar	Tablet
vpoint	**-vp**	...	...	**V3V**	**View** ↳**3D Views** ↳**VPOINT**	**N4**

Command: vpoint
Current view direction: VIEWDIR=0.0000,0.0000,1.0000
Specify a view point or [Rotate] <display compass and tripod>: *(Enter an option, or press* **ENTER** *for the compass-tripod.)*

COMMAND LINE OPTIONS

Specify a view point indicates the new 3D viewpoint by coordinates.

Rotate indicates the new 3D viewpoint by angle.

ENTER brings up visual guides (see figure below).

RELATED COMMANDS

DdVpoint adjusts the viewpoint via a dialog box.

DView changes the viewpoint of 3D objects, and allows perspective mode.

3dOrbit rotates the viewpoint in real-time.

RELATED SYSTEM VARIABLES

VpointX is the x-coordinate of the current 3D view.

VpointY is the y-coordinate of the current 3D view.

VpointZ is the z-coordinate of the current 3D view.

WorldView determines whether **VPoint** coordinates are in WCS or UCS.

TIPS

- The *compass* represents the globe, flattened to two dimensions:

  The north pole (0, 0, z) is in the center.
  The equator (x, y, 0) is the inner circle
  The south pole (0, 0, -z) is the outer circle.

- As the cursor is moved on the compass, the *axis tripod* rotates showing the 3D view direction.
- To select the view direction, pick a location on the globe and press the pick button.

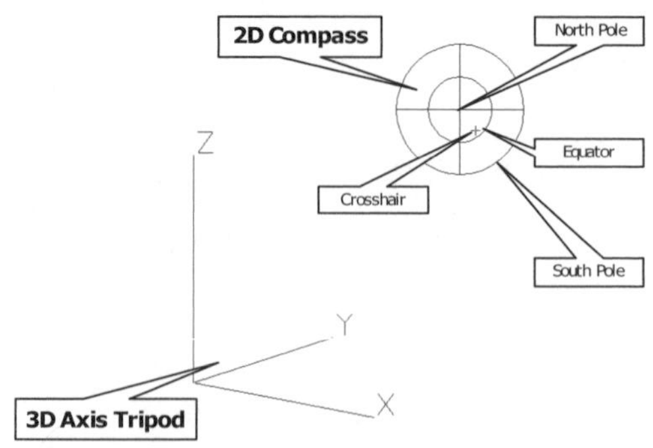

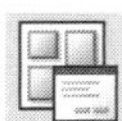

VPorts

Rel. 10 Creates viewports of the current drawing, when **TileMode** is on (*short for ViewPORTS*).

Commands	Alias	Ctrl+	F-key	Alt+	Menu Bar	Tablet
vports	**viewports**	**R**	**...**	**VV**	**View** ↳**Viewports**	**M3-4**
+vports						
-vports				**VV1**	**View** ↳**Viewports** ↳**1 Viewport**	

Command: vports

Displays tabbed dialog box.

DIALOG BOX OPTIONS

New Viewports tab

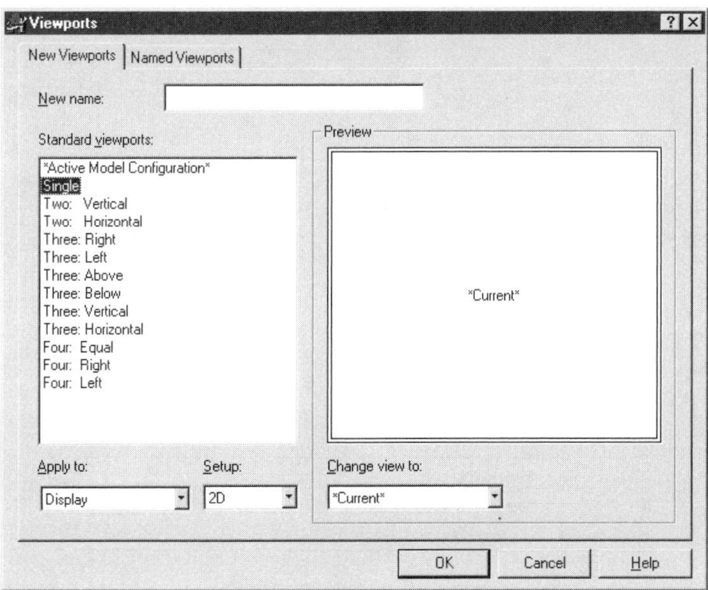

In model space

New name specifies the name for the viewport configuration; can be up to 255 characters long.

Standard viewports lists the available viewport configurations.

Preview displays a preview of the viewport configuration.

Apply to applies the viewport configuration to:

- Display.
- Current Viewport.

Setup selects 2D or 3D configuration; the 3D option applies orthogonal views, such as top, left, and front.

Change view to selects the type of view; in 3D mode, selects a standard orthoganal view.

In paper space

Viewport spacing specifies the spacing between the floating viewports; default = 0 units.

Named Viewports tab

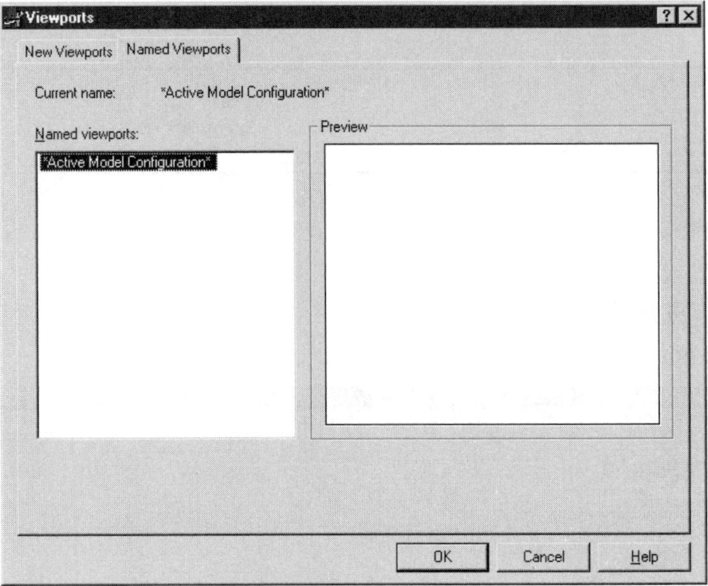

Named viewports lists the names of saved viewport configurations.

+VPORTS Command

Command: +vports

Tab index <0>: *(Type* **0** *or* **1***.)*

Tab index specifies the tab to display:

Index	Meaning
0	New Viewports tab (default).
1	Named Viewports tab.

-VPORTS Command

Command: -vports

*In layout mode, displays **MView** command prompts. In model space, prompts:*

Enter an option [Save/Restore/Delete/Join/SIngle/?/2/3/4] <3>: *(Enter an option.)*

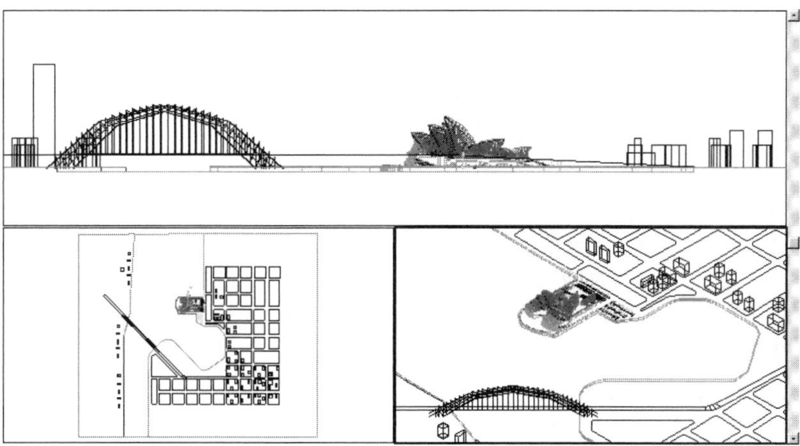

COMMAND LINE OPTIONS

In model space

Save saves the settings of a viewport by name.

Restore restores a viewport definition.

Delete deletes a viewport definition.

Join joins two viewports together as one when they form a rectangle.

SIngle joins all viewports into a single viewport.

? lists the names of saved viewport configurations.

4 divides the current viewport into four.

2 options

Horizontal creates one viewport over another.

Vertical creates one viewport beside another (default).

3 options

Horizontal creates three viewports over each other.

Vertical creates three viewports beside each other.

Above creates one viewport overtop of two viewports.

Below creates one viewport below two viewports.

Left creates one viewport left of two viewports.

Right creates one viewport right of two viewports (default).

In paper space
Specify corner of viewport or [ON/OFF/Fit/Hideplot/Lock/Object/Polygonal/ Restore/2/3/4]<Fit>: *(Enter an option.)*

ON turns on the viewport; the objects in the viewport become visible.

OFF turns off the viewport; the objects in the viewport become invisible.

Fit creates one viewport that fills the display area.

Hideplot removes hidden lines when plotting in layout mode.

Lock locks the viewport, so that no editing can take place.

Object converts a closed polyline, ellipse, spline, region, or circle into a viewport.

Polygonal creates an non-rectangular viewport

Other options are identical to those displayed in model tab.

RELATED COMMANDS

MView creates viewports in paper space.

RedrawAll redraws all viewports.

RegenAll regenerates all viewports.

VpClip clips a viewport.

RELATED SYSTEM VARIABLES

CvPort is the current viewport number.

MaxActVp limits the maximum number of active viewports.

TileMode controls whether viewports can be overlapped or tiled.

TIPS

- The **Join** option joins two viewports only when they form a rectangle.
- You can restore saved viewport arrangements in paper space using the **MView** command.
- Many display-related commands (such as **Redraw** and **Grid)** affect the current viewport only.

VSlide

V. 2.0 Displays slide files in the current viewport (*short for View SLIDE*).

Command	Alias	Ctrl+	F-key	Alt+	Menu Bar	Tablet
vslide	...	...	...	...	...	...

Command: vslide

Displays Select Slide File dialog box. Select an .sld file, and then click ***Open.***

COMMAND LINE OPTIONS

None.

RELATED COMMANDS

MSlide creates slide files of the current viewport.

Redraw erases slides from the screen.

RELATED AUTODESK PROGRAM

slidelib.exe creates an SLB-format library file of a group of slide files.

RELATED AUTOCAD FILES

.sld stores individual slide files.

.slb stores a library of slide files.

TIP

- The following applies when **FileDia** is set to 0, or when using this command in a script:

  For faster viewing of a series of slides, placing an asterisk preceding the **VSlide** command preloads the *.sld* slide file, as in:

  Command: *vslide filename

  Use the following format to display a specific slide stored in an SLB slide library file:

  Command: vslide
  Slide file: acad.slb(slidefilename)

WBlock

V. 1.4 Writes blocks or entire drawings to disk (*short for Write BLOCK*).

Commands	Aliases	Ctrl+	F-key	Alt+	Menu Bar	Tablet
wblock	**w**	...	...	**FE** ↳**DWG**	**File** ↳**Export** ↳**Block**	...
-wblock	**-w**					

Command: wblock

Displays dialog box:

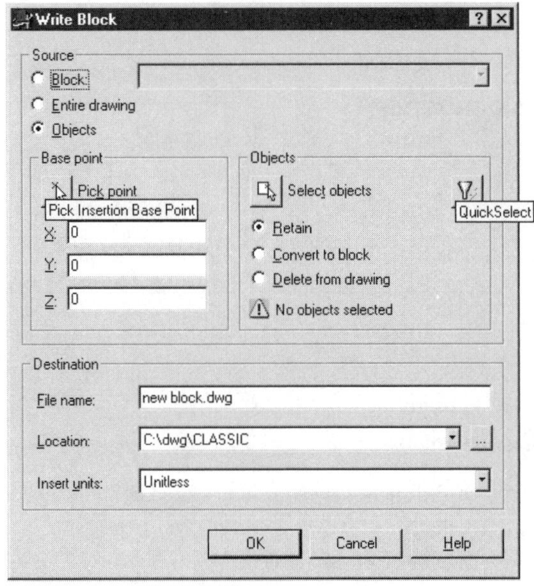

DIALOG BOX OPTIONS

Source options

○ **Block** specifies the name of the block to save as a *.dwg* file.

○ **Entire drawing** selects the current drawing to save as a *.dwg* file.

⊙ **Objects** specifies the objects from the drawing to save as a *.dwg* file.

Base point options

Pick Insertion Base Point dismisses the dialog box temporarily to allow you to select the insertion base point.

X specifies the x coordinate of the insertion point.

Y specifies the y coordinate of the insertion point.

Z specifies the z coordinate of the insertion point.

Objects options

Select Objects dismisses the dialog box temporarily to allow you to select one or more objects.

Quick Select displays the Quick Select dialog box; see the **QSelect** command.

⊙ **Retain** retains the selected objects in the current drawing after saving them as a drawing file.

○ **Convert to block** converts the selected objects to a block in the drawing, after saving them as a *.dwg* file; names the block under File name in the Destination section.

○ **Delete from drawing** deletes the selected objects from the drawing, after saving them as a *.dwg* file.

Destination options

File name specifies the file name for the block or objects.

Location specifies the path.

... displays the Browse for Folder dialog box.

Insert units specifies the units when the *.dwg* file is inserted as a block.

-WBLOCK Command

Command: -wblock

*Displays the Create Drawing File dialog box. Name the file, and then click **Save**.*

Enter name of existing block or
[= (block=output file)/* (whole drawing)] <define new drawing>: *(Enter a name, or use = and * options, or press **ENTER**.)*

COMMAND LINE OPTIONS

Enter name of existing block specifies the name of a current block in the drawing.

= (*equals*) writes block to a *.dwg* file, using the block's name as the file name.

* (*asterisk*) writes the entire drawing to a *.dwg* file.

ENTER creates a *.dwg* drawing file on disk of the selected objects. (The selected objects are erased from the drawing; use **Oops** to bring them back.)

RELATED COMMANDS

Block creates a block of a group of objects.

Insert inserts a block or another drawing into the drawing.

RELATED SYSTEM VARIABLES

None.

TIPS

- Use the **WBlock** command to extract blocks from the drawing and store them on a disk drive. This allows the creation of a block library.
- Support for the DesignXML (extended markup language) format was withdrawn from AutoCAD 2004.

Wedge

Rel.11 Draws 3D wedges as solid models.

Command	Alias	Ctrl+	F-key	Alt+	Menu Bar	Tablet
wedge	**we**	...	...	**DIW**	**Draw** ↳**Solids** ↳**Wedge**	**N7**

Command: wedge
Specify first corner of wedge or [CEnter] <0,0,0>: *(Pick a point, or type* **CE***.)*
Specify corner or [Cube/Length]: *(Pick a point, or enter an option.)*
Specify height: *(Pick a point.)*

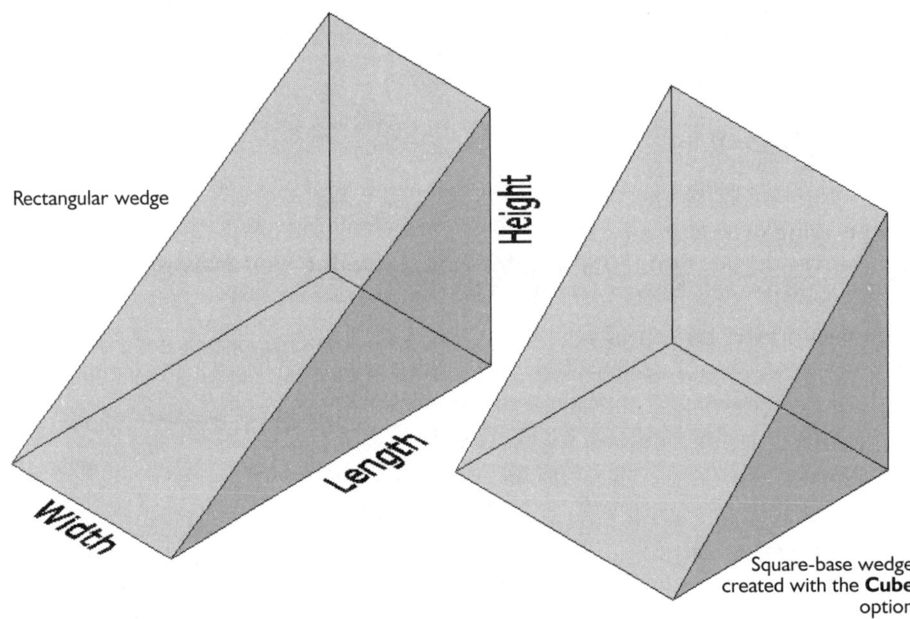

COMMAND LINE OPTIONS

Corner specifies the lower-left corner of the wedge.

CEnter draws the wedge's base about the center of the sloped face.

Cube draws cubic wedges.

Length specifies the length, width, and height of the wedge.

CEnter options
Specify center of wedge <0,0,0>: *(Pick a point.)*
Specify opposite corner or [Cube/Length]: *(Pick a point, or enter an option.)*
Specify height: *(Pick a point.)*

Specify center of wedge indicates the center of the wedge's inclined face.

Specify opposite corner indicates the distance from the midpoint to one corner.

Cube options

Specify corner or [Cube/Length]: *(Type* **C***.)*

Specify length: *(Specify the length.)*

Specify length indicates the length of all three sides.

Length options

Specify corner or [Cube/Length]: *(Type* **L***.)*

Specify length: *(Specify the length.)*

Specify width: *(Specify the width.)*

Specify height: *(Specify the height.)*

Specify length indicates the length parallel to the x-axis.

Specify width indicates the width parallel to the y-axis.

Specify height indicates the height parallel to the z-axis.

RELATED COMMANDS

Ai_Wedge draws wedges as 3D surface models.

Box draws solid boxes.

Cone draws solid cones.

Cylinder draws solid cylinders.

Sphere draws solid spheres.

Torus draws solid tori.

RELATED SYSTEM VARIABLES

None.

TIPS

- *Length* means size in the x-direction.
- *Width* means size in the y-direction.
- *Height* means size in the z-direction.
- Use negative values for length, width, and height to draw the wedge in the negative x, y, and z directions.
- The **IsoLines** system variable has no effect on wedges.

WhoHas

2000 Determines which computer has drawings open.

Command	Alias	Ctrl+	F-key	Alt+	Menu Bar	Tablet
whohas	...	...	...	...	...	...

Command: whohas

Displays the Select Drawing to Query dialog box. Select a drawing file, and then click ***Open****.*

When the drawing is open, reports:

Owner: ralphg

Computer's Name : HEATHER

Time Acessed : Monday, March 3, 2006 11:27:28 AM

When the drawing is not open, reports:

User: unknown.

COMMAND LINE OPTIONS

None.

RELATED COMMANDS

Open opens drawings.

XAttach attaches drawings that can be opened by other users.

TIP

- This command is meant for use over networks as a convenient way to find out which users are editing specific drawings.

WipeOut

2004 Fills areas with the background color to "wipe out" portions of drawings.

Command	Alias	Ctrl+	F-key	Alt+	Menu Bar	Tablet
wipeout	...	...	...	...	...	...

Command: wipeout
Specify first point or [Frames/Polyline] <Polyline>: *(Pick a point, or enter an option.)*
Specify next point: *(Pick a point.)*
Specify next point or [Undo]: *(Pick a point, or type* **U***.)*
Specify next point or [Close/Undo]: *(Pick a point, or type* **C** *or* **U***.)*
Specify next point or [Close/Undo]: *(Type* **C** *to end the command.)*

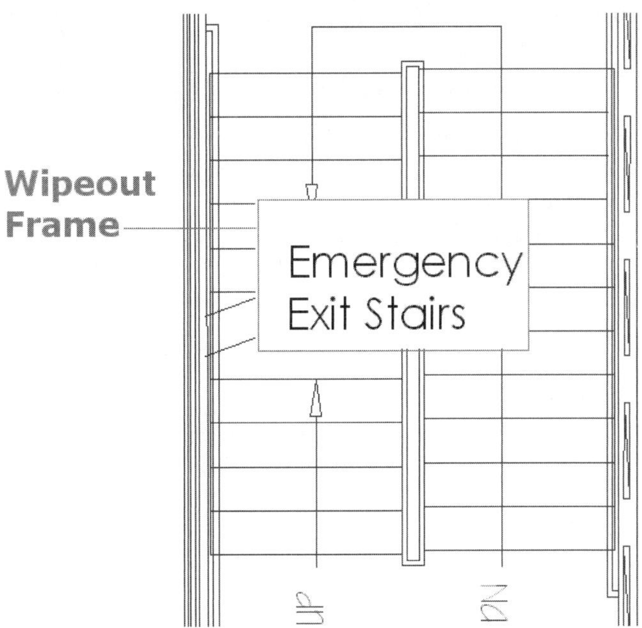

COMMAND LINE OPTIONS

Specify first point specifies the starting point of the polygon.

Undo undoes the last segment.

Close closes the polygon.

Frames options
Enter mode [ON/OFF] <ON>: *(Type* **ON** *or* **OFF***.)*

ON turns on the wipeout boundary polygon.

OFF turns off the boundary polygon.

Polyline options
Select a closed polyline: *(Pick a closed polyline.)*
Erase polyline? [Yes/No] <No>: *(Type* **Y** *or* **N**.*)*

Select a closed polyline picks a polyline that forms the wipeout boundary.

Erase polyline?

Yes erases the polyline.

No leaves the polyline in place.

RELATED COMMAND

DrawOrder displays overlapping objects in a different order.

TIPS

- Wipeout boundaries can be edited with grips editing.
- When the **Frames** option is turned off, wipeouts cannot be edited.
- The **Frames** option applies to all wipeouts in the drawing; you cannot turn frames on and off for individual wipeouts.
- The wipeout consists of an image object drawn with the background color.
- To create a rectangular wipeout frame, use the **Rectangle** command.
- To make text appear above the wipeout, use the **DrawOrder** command to move the text to the **Front** — after creating the wipeout.
- To create a wipeout under text, you may find it easier to use the Background Mask option of the **MText** command.

WmfIn

Rel.12 Imports *.wmf* and *.clp* files (*short for Windows MetaFile IN*).

Command	Alias	Ctrl+	F-key	Alt+	Menu Bar	Tablet
wmfin	...	...	...	**IW**	**Insert** ↳**Windows Metafile**	...

Command: wmfin

Displays the Import WMF dialog box. Select a file, and then click ***Open****.*

Specify insertion point or [Scale/X/Y/Z/Rotate/PScale/PX/PY/PZ/PRotate]: *(Pick a point, or enter an option.)*

Enter X scale factor, specify opposite corner, or [Corner/XYZ] <1>: *(Specify a value, pick a point, or enter an option.)*

Enter Y scale factor <use X scale factor>: *(Specify a value, or press* ENTER*.)*

Specify rotation angle <0>: *(Specify a value, or press* ENTER*.)*

COMMAND LINE OPTIONS

Insertion point picks the insertion point of the lower-left corner of the WMF image.

X scale factor scales the WMF image in the x direction (default = 1).

Corner scales the WMF image in the x and y directions.

XYZ scales the image in the x, y, and z directions.

Y scale factor scales the image in the y direction (default = x scale).

Rotation angle rotates the image (default = 0).

RELATED COMMANDS

WmfOpts controls the importation of *.wmf* files.

WmfOut exports selected objects in WMF format.

RELATED FILES

**.clp* are Windows Clipboard files.

**.wmf* are Windows Metafiles.

TIPS

- The WMF image is placed as a block with the name **WMF0**; subsequent placements of *.wmf* files increases the number by one: **WMF1**, **WMF2**, and so on.
- Exploding the WMF*n* block results in polylines; even circles, arcs, and text are converted to polylines; solid-filled areas are exploded into solid triangles.
- The *.clp* files are created by the Windows Clipboard. After using CTRL+C to copy objects to the Clipboard, you can open the Clipboard Viewer, and then save the image as a *.clp* file.
- The CLP support is undocumented by Autodesk; the **WmfOpts** command has no effect on imported *.clp* files.
- The *.clp* file is pasted as a block; when exploded, constituent parts are 2D polylines.
- WMF is short for "Windows meta file," a vector format that Microsoft based on CGM (computer graphics metafile).

WmfOpts

Rel.12 Controls the importation of *.wmf* files (*short for Windows Meta File OPTionS*).

Command	Alias	Ctrl+	F-key	Alt+	Menu Bar	Tablet
wmfopts	...	...	...	**IWP**	**Insert** **↳Windows Metafile** **↳Options**	...

Command: wmfopts

Displays dialog box:

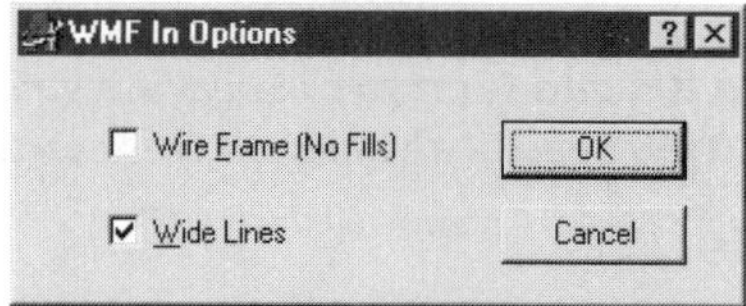

DIALOG BOX OPTIONS

Wire Frame

☑ Displays the WMF images with lines only, no filled areas (default).

☐ Displays area fills.

Wide Lines

☑ Displays lines with width (default).

☐ Displays lines with a width of zero.

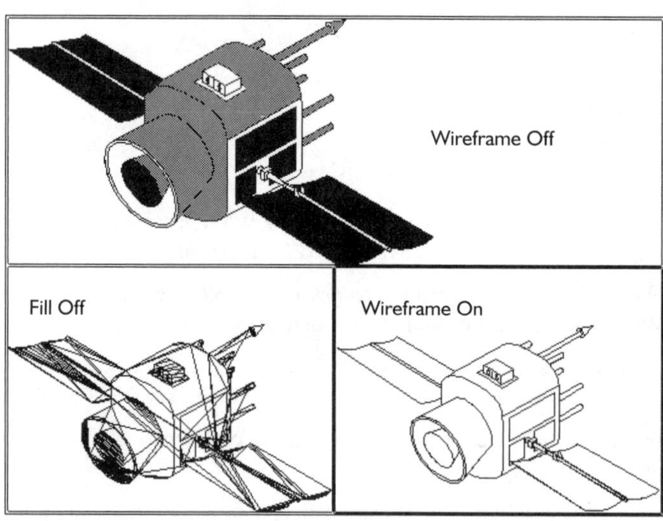

RELATED COMMANDS

WmfIn imports *.wmf* files.

WmfOut exports selected objects in WMF format.

WmfOut

Rel.12 Exports selected objects in WMF format (*short for Windows MetaFile OUTput*).

Command	Alias	Ctrl+	F-key	Alt+	Menu Bar	Tablet
wmfout	...	...	...	**FE**	**File**	...
				↳WMF	**↳Export**	
					↳Metafile	

Command: wmfout

Displays the Create WMF File dialog box. Enter a file name, and then click ***Save****.*

Select objects: *(Select one or more objects.)*

Select objects: *(Press* **ENTER** *to end object selection.)*

COMMAND LINE OPTION

Select objects selects the objects to export. Press CTRL+A to select all objects visible in the current viewport.

RELATED SYSTEM VARIABLE

WmfBkgnd toggles the background color of exported *.wmf* files:

WmfBkgnd	Meaning
0	Transparent background.
1	AutoCAD background color.

WmfForegnd switches the foreground and background colors of exported *.wmf* files as required, such that:

WmfForegnd	Meaning
0	Foreground is darker than background color.
1	Background is darker than foreground color.

RELATED COMMANDS

WmfOpts controls the importation of *.wmf* files.

WmfIn imports files in WMF format.

CopyClip copies selected objects to the Clipboard in several formats, including *.wmf*, also called "picture" format.

TIPS

- *.wmf* files created by AutoCAD are resolution-dependent; small circles and arcs lose their roundness.
- The **All** selection does not select all objects in the drawing; instead, the **WmfOut** command selects all objects *visible* in the current viewport.

XAttach

Rel.14 Attaches externally-referenced drawings to the current drawing (*short for eXternal reference ATTACH*).

Command	Alias	Ctrl+	F-key	Alt+	Menu Bar	Tablet
xattach	xa	...	...	IX	Insert ↳External Reference	...

Command: xattach

Displays the Select File to Attach dialog box. Select a file, and click **Open**.

Displays dialog box:

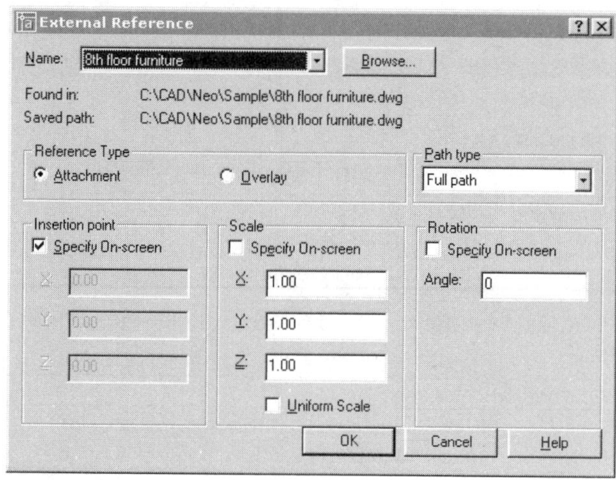

After you click **OK**, *AutoCAD confirms at the command line:*

Attach Xref FILENAME: C:\filename.dwg
FILENAME loaded.

DIALOG BOX OPTIONS

Name specifies the file name of the external *.dwg* file to be attached; the drop list shows the names of currently-attached xrefs (externally-referenced files).

Browse displays the Select File To Attach dialog box.

Retain Path

☑ Saves the xref's filename and full path in the *.dwg* file.

☐ Saves only the filename of the xref.

Reference Type options

⊙ **Attachment** attaches the xref.

○ **Overlay** overlays the xref.

Insertion Point options

☑ **Specify On-screen** specifies the insertion point of the xref in the drawing.

*After you click **OK** to dismiss the dialog box, AutoCAD prompts:*

Specify insertion point or [Scale/X/Y/Z/Rotate/PScale/PX/PY/PZ/PRotate]: *(Pick a point, or enter an option.)*

Scale sets the scale factor for the x, y, and z axes.

X sets the x-scale factor.

Y sets the y-scale factor.

Z sets the z-scale factor.

Rotate specifies the rotation angle.

PScale presets the scale factor for the x, y, and z axes.

PX presets the x-scale factor.

PY presets the y-scale factor.

PZ presets the z-scale factor.

PRotate presets the rotation angle.

Scale options

☐ **Specify On-screen** specifies the scale of the xref in the drawing.

X sets the x-scale factor.

Y sets the y-scale factor.

Z sets the z-scale factor.

☐ **Uniform scale** sets the Y and Z factors equal to X (*new to AutoCAD 2005*).

*After you click **OK** to dismiss the dialog box, AutoCAD prompts:*

Enter X scale factor, specify opposite corner, or [Corner/XYZ] <1>: *(Enter a value, or enter an option.)*

Enter Y scale factor <use X scale factor>: *(Enter a value, or press **ENTER**.)*

X scale factor scales the xref in the x direction.

Corner indicates the x,y scale factor by picking two points of a rectangle.

XYZ specifies the scale factor in the x, y, and z directions.

Y scale factor scales the xref in the y direction.

Rotation Angle options

☐ **Specify On-screen** specifies the rotation of the xref in the drawing.

*After you click **OK** to dismiss the dialog box, AutoCAD prompts:*

Specify rotation angle <0>: *(Enter a value, or press **ENTER**.)*

Rotation angle specifies the rotation angle of the xref.

RELATED SYSTEM VARIABLES

XRefType determines whether xrefs are attached or overlaid, by default (*new to 2005*).

XEdit determines whether the drawing may be edited in-place, when being referenced by another drawing.

XLoadPath stores the path of temporary copies of demand-loaded xref drawings.

XRefCtl controls whether *.xlg* external reference log files are written.

ProjectName holds the project name for the current drawing (default = "").

DemandLoad specifies if and when AutoCAD demand-loads a third-party application when a drawing contains custom objects created by the application:

DemandLoad	Meaning
0	Turns off demand loading.
1	Loads application when drawings contain proxy objects.
2	Loads application when the application's commands are invoked.
3	Loads application when drawings contain proxy objects, or when the application's commands are invoked.

IdxCtl controls the creation of layer and spatial indices:

IndexCtl	Meaning
0	Creates no indices (default).
1	Creates layer index.
2	Creates spatial index.
3	Creates both layer and spatial indices.

VisRetain specifies how the layer settings — on-off, freeze-thaw, color, and linetype — in xref drawings are defined by the current drawing:

VisRetain	Meaning
0	Xref layer definition in the current drawing takes precedence.
1	Settings for xref-dependent layers take precedence over xref layer definition in the current drawing.

XLoadCtl controls the loading of xref drawings:

XLoadCtl	Meaning
0	Loads the entire xref drawing.
1	Demand loading; xref is opened.
2	Demand loading; copy of the xref is opened.

RELATED COMMANDS

RefEdit edits xref drawings.

XBind binds portions of xref drawings to the current drawing.

XClip clips the display of xrefs.

XRef controls xrefs.

TIP

- When AutoCAD cannot find an xref, it searches in the following order:

  The folder of the current drawing.
  The project search paths defined in the Options dialog box's Files tab and the **ProjectName** system variable.
  The support search paths defined in the Options dialog box's Files tab.
  The Start In folder specified in the shortcut that launched AutoCAD.

XBind

Rel.11 Binds portions of externally-referenced drawings to the current drawing (*short for eXternal BINDing*).

Commands	Aliases	Ctrl+	F-key	Alt+	Menu Bar	Tablet
xbind	**xb**	...	...	**MOEB**	**Modify** ↳**Object** ↳**External Reference** ↳**Bind**	**X19**
-xbind	**-xb**					

Command: xbind

Displays dialog box:

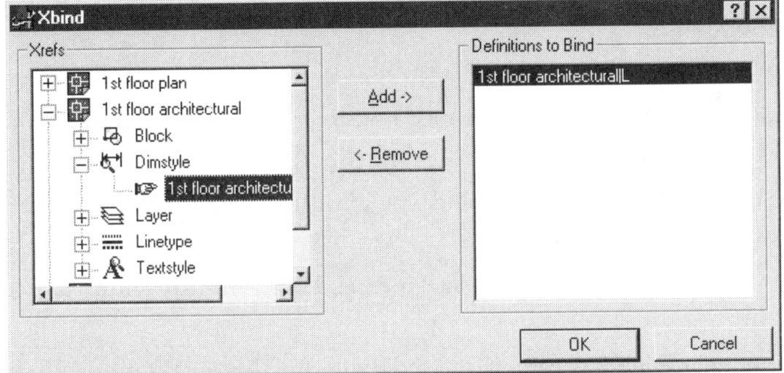

DIALOG BOX OPTIONS

Xrefs lists xrefs (externally-referenced drawings), along with their bindable objects: blocks, dimension styles, layer names, linetypes, and text styles.

Definitions to Bind lists definitions that will be bound.

Buttons

Add adds a definition to the binding list.

Remove removes a definition from the binding list.

-XBIND Command

Command: -xbind

Enter symbol type to bind [Block/Dimstyle/LAyer/LType/Style]: *(Enter an option.)*

Enter dependent name(s): *(Enter one or more names, separated by commas.)*

COMMAND LINE OPTIONS

Block binds blocks to the current drawing.

Dimstyle binds dimension styles to the current drawing.

LAyer binds layer names to the current drawing.

LType binds linetype definitions to the current drawing.

Style binds text styles to the current drawing.

Enter dependent names specifies the named objects to bind.

RELATED COMMANDS

RefEdit edits xref drawings.

XRef controls xrefs.

TIPS

- The **XBind** command lets you copy named objects from another drawing to the current drawing.
- Before you can use the **XBind** command, you must first use the **XAttach** command to attach an xref to the current drawing.
- Blocks, dimension styles, layer names, linetypes, and text styles are known as "dependent symbols."
- When a dependent symbol is part of an xrefed drawing, AutoCAD uses a vertical bar (|) to separate the xref name from the symbol name, as in *filename* | *layername*.
- After you use the **XBind** command, AutoCAD replaces the vertical bar with **0**, as in *filename***0***layername*. The second time you bind that layer from that drawing, **XBind** increases the digit, as in *filename***1***layername*.
- When the **XBind** command binds a layer with a linetype (other than Continuous), it automatically binds the linetype.
- When the **XBind** command binds a block — with a nested block, dimension style, layer, linetype, text style, and/or reference to another xref — it automatically binds those objects as well.

XClip

Rel.12 Clips portions of blocks and externally-referenced drawings (*short for eXternal CLIP; formerly the* ***XRefClip*** *command*).

Command	Alias	Ctrl+	F-key	Alt+	Menu Bar	Tablet
xclip	**xc**	**...**	**...**	**...**	**Modify** **↳Clip** **↳Xref**	**X18**

Command: xclip
Select objects: *(Select one or more blocks or xrefs.)*
Select objects: *(Press* **ENTER** *to end object selection.)*
Enter clipping option
[ON/OFF/Clipdepth/Delete/generate Polyline/New boundary] <New>: *(Enter an option.)*

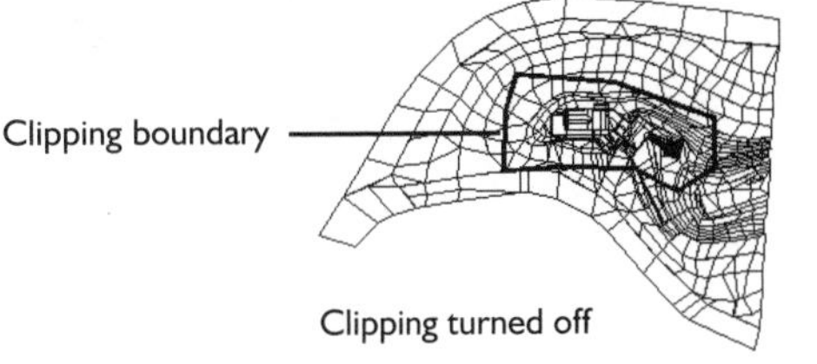

Clipping turned off

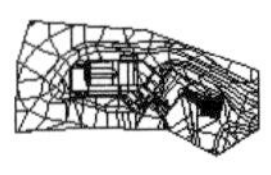

Clipping turned on

COMMAND LINE OPTIONS

Select objects selects the xref or block, *not* the clipping polyline.

ON turns on clipped display.

OFF turns off clipped display; displays all of the xref or block.

Clipdepth sets front and back clipping planes for 3D xrefs and blocks.

Delete erases the clipping boundary.

generate Polyline extracts the existing boundary as a polyline.

New boundary places a new rectangular or irregular polygon clipping boundary, or creates an irregular clipping boundary from an existing polyline.

RELATED COMMANDS

XBind bind parts of the xrefs drawing to the current drawing.

Xref controls xrefs.

RELATED SYSTEM VARIABLE

XClipFrame toggles the display of the clipping boundary.

TIPS

- While the old **XRefClip** command could not create an irregularly-clipped xref, the new **XClip** command creates arbitrary clipping boundaries.
- The **XClip** command works for both blocks and xrefs.
- A spline-fit polyline results in a curved clip boundary, but a curve-fit polyline does not.

XLine

Rel.13 Places infinitely long construction lines in drawings.

Command	Alias	Ctrl+	F-key	Alt+	Menu Bar	Tablet
xline	xl	...	...	DT	Draw ↳Construction Line	L10

Command: xline

Specify a point or [Hor/Ver/Ang/Bisect/Offset]: *(Pick a point, or enter an option.)*

Through point: *(Pick a point.)*

Through point: *(Press* ENTER *to end the command.)*

COMMAND LINE OPTIONS

Specify a point picks the midpoint for the xline.

Through point picks another point through which the xline passes.

Ang places the construction line at an angle.

Bisect bisects an angle with the construction line.

From point places the construction line through a point.

Hor places a horizontal construction line.

Offset places the construction line parallel to another object.

Ver places a vertical construction line.

ENTER exits the command.

Angle options

Enter angle of xline <0> or [Reference]: *(Enter an angle, or type* **R***.)*

Enter angle of xline specifies the angle of the xline relative to the x-axis.

Reference specifies the angle relative to two points.

Bisect options

Specify angle vertex point: *(Pick a point.)*

Specify angle start point: *(Pick a point.)*

Specify angle end point: *(Pick a point.)*

Specify angle vertex point specifies the vertex of the angle.

Specify angle start point specifies the angle start point.

Specify angle end point specifies the angle end point.

Offset options
Specify offset distance or [Through] <1.0000>: *(Enter a distance, or type* **T***.)*
Select a line object: *(Select a line, xline, ray, or polyline line segment.)*
Specify side to offset: *(Pick a point.)*

Specify offset distance specifies the distance between xlines.

Through picks a point through which the xline should pass.

Select a line object selects the line, xline, ray, or polyline to offset.

Specify side to offset specifies the offset side.

RELATED COMMANDS

Properties modifies characteristics of xline and ray objects.

Ray places semi-infinite construction lines.

RELATED SYSTEM VARIABLE

OffsetDist specifies the current offset distance.

TIPS

- Use xlines to find the bisectors of triangles (using **MIDpoint** object snap), or to create intersection snap points (using **INTersection** object snap).

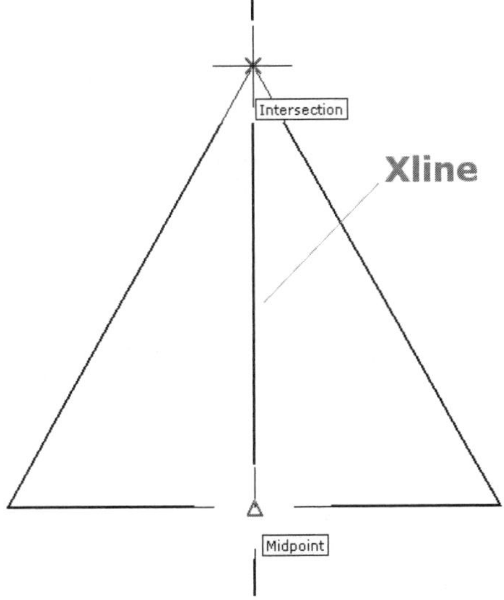

- Ray and xline construction lines are plotted; they do not affect the extents.

XOpen

2004 Opens externally-referenced drawings in new windows (*short for eXternal OPEN*).

Command	Alias	Ctrl+	F-key	Alt+	Menu Bar	Tablet
xopen	...	...	...	...	...	...

Command: xopen

Select xref: *(Select an externally-referenced drawing.)*

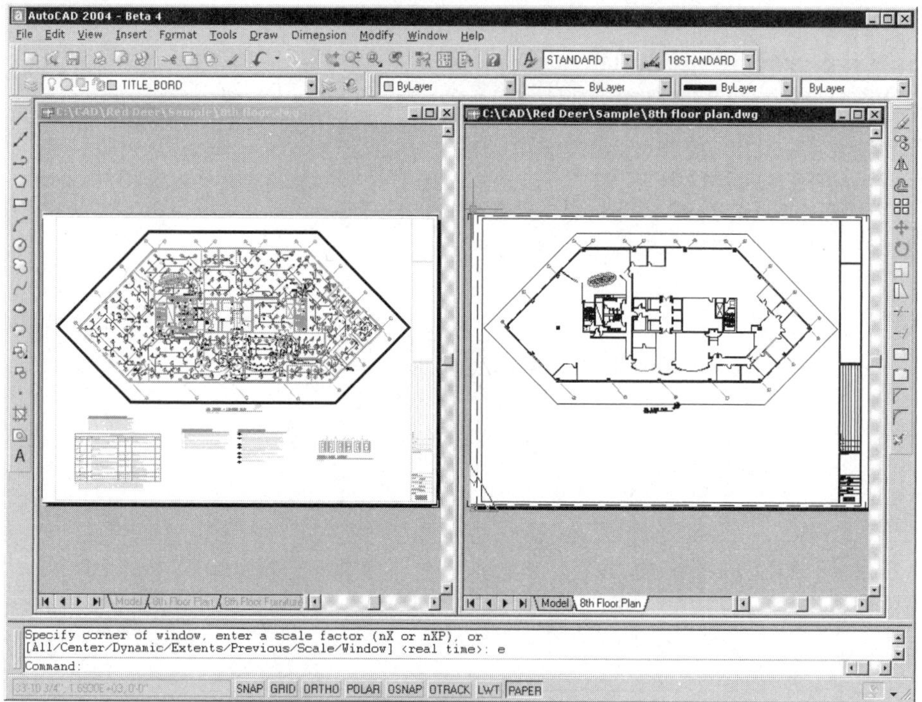

At left: original drawing.
At right: "xopened" drawing in separate window.

COMMAND LINE OPTION

Select xrefs selects the xref to open; the xref must be inserted in the current drawing.

TIPS

- This command does not work with blocks.
- When you select a non-xref object, AutoCAD complains, 'Object is not an Xref.'

Xplode

Rel.12 Explodes complex objects into simpler objects, with user control (*short for eXPLODE*).

Command	Alias	Ctrl+	F-key	Alt+	Menu Bar	Tablet
xplode	**xp**	...	...	...	...	...

Command: xplode
Select objects to XPlode.
Select objects: *(Select one or more objects.)*
Select objects: *(Press* **ENTER** *to end object selection.)*
Enter an option [Individually/Globally] <Globally>: *(Type* **I** *or* **G**.*)*
Enter an option [All/Color/LAyer/LType/Inherit from parent block/Explode] <Explode>: *(Enter an option.)*

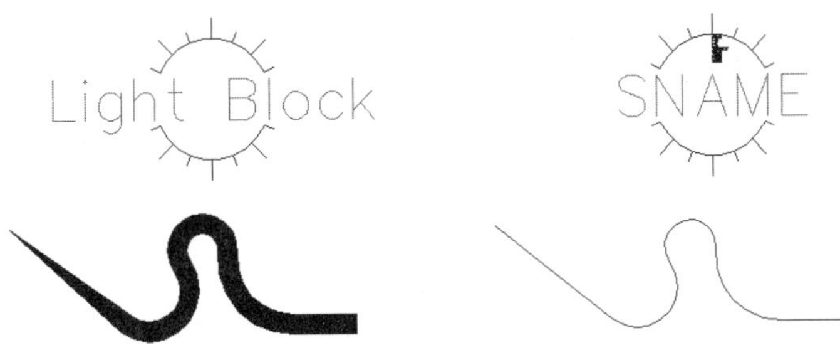

Block and polyline (at left); exploded (at right).

COMMAND LINE OPTIONS

Select objects selects objects to be exploded.

Individually allows you to specify options for each selected object.

Globally applies options to all selected objects.

All sets the color, layer, linetype and lineweight of exploded objects.

Color sets the color of objects after they are exploded: red, yellow, green, cyan, blue, magenta, white, bylayer, byblock, or any color number.

LWeight specifies a lineweight.

LAyer sets the layer for the exploded objects.

LType specifies any loaded linetype name for the exploded objects.

Inherit from parent block assigns the color, linetype, lineweight, and layer of the exploded objects, based on the original object.

Explode reduces the complex object into its components.

RELATED COMMANDS

Explode explodes the object without options.

U reverses the explosion.

RELATED SYSTEM VARIABLES

None.

TIPS

- Examples of complex objects include blocks and polylines; examples of simple objects include lines, circles, and arcs.
- Mirrored blocks can be exploded.
- The **LWeight** option is not displayed when the lineweight is off.
- The 'Enter an option [Individually/Globally]:' prompt appears only when more than one valid object is selected for explosion.
- Specifying **BYLayer** for the color or linetype means that the exploded objects take on the color or linetype of the object's original layer.
- Specifying **BYBlock** for the color or linetype means that the exploded objects take on the color or linetype of the original object.
- The default layer is the current layer, not the exploded object's original layer.
- The **XPlode** command breaks down complex objects as follows:

Object	Exploded into
Attributes	Attribute values are deleted; displays attribute definitions.
Block	Component objects.
Leaders	Line segments, splines, mtext, and tolerance objects; arrowheads become solids or blocks.
Mtext	Text.
Multiline	Line and arc segments.
Polyface mesh	Point, line, or 3D faces.
Region	Lines, arcs, and splines.
Table	Lines.
2D polyline	Line and arc segments; width and tangency are lost.
3D polyline	Line segments.
3D solids	Planar surfaces become regions; nonplanar surfaces become bodies.
3D bodies	Single-surface body, regions, or curves.

- The **Inherit** option works only when the parts were originally drawn with color, linetype, and lineweight set to BYBLOCK, and drawn on layer 0.

XRef

Rel.11 Controls externally-referenced drawings in the current drawing (*short for eXternal REFerence*).

Commands	Aliases	Ctrl+	F-key	Alt+	Menu Bar	Tablet
xref	**xr**	...	...	**IR**	**Insert** ⇘**Xref Manager**	**T4**
-xref	**-xr**					

Command: xref

Displays dialog box:

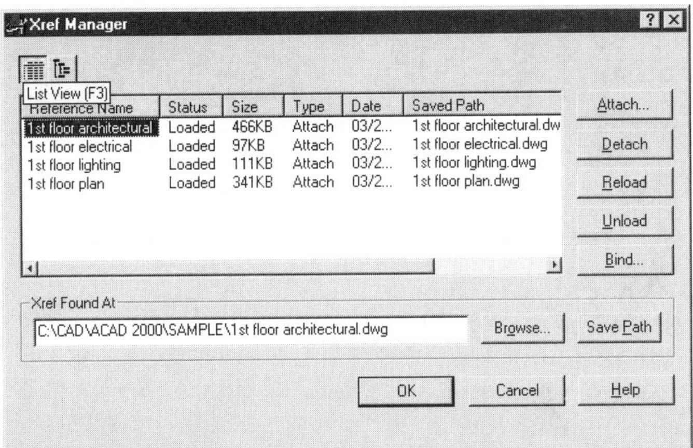

DIALOG BOX OPTIONS

Buttons

Attach attaches a drawing as an xref (externally-referenced drawing); displays the Attach Xref dialog box; see the **XAttach** command.

Detach detaches xrefs.

Reload reloads and displays the most-recently saved version of the xref.

Unload unloads the xref; does not remove it permanently; rather it does not display the xref.

Bind binds named objects — blocks, dimension styles, layer names, linetypes, and text styles — to the current drawing; displays the Bind Xrefs dialog box; see the **XBind** command.

Additional options

Xref Found At displays the path to the xref file.

Browse selects a path or file name; displays the Select New Path dialog box.

Save Path saves the path displayed by the Xref Found At option.

-XREF Command

Command: -xref

Enter an option [?/Bind/Detach/Path/Unload/Reload/Overlay/Attach] <Attach>: *(Enter an option.)*

COMMAND LINE OPTIONS

? lists the names of xref files.

Bind makes the xref drawing part of the current drawing.

Detach removes xref files.

Path respecifies paths to xref files.

Unload unloads xref files.

Reload updates the xref files.

Overlay overlays the xref files.

Attach attaches another drawing to the current drawing.

RELATED TOOLBAR ICONS

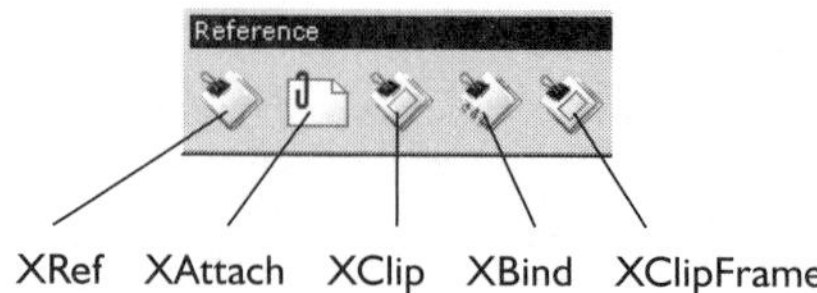

RELATED COMMANDS

Insert adds another drawing to the current drawing.

RefEdit edits xrefs.

XBind binds parts of xrefs to the current drawing.

XClip clip portions of xrefs.

RELATED SYSTEM VARIABLES

See the ***XAttach*** *command.*

TIPS

- *Caution!* Nested xrefs cannot be unloaded.
- Tree view shows how the xrefs are nested (press **F4**):

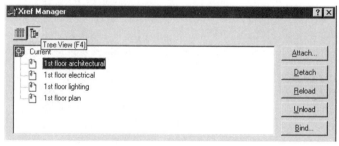

- Press **F3** to return to list view.

'Zoom

V. 1.0 Makes drawings larger or smaller in the current viewport.

Command	Aliases	Ctrl+	F-key	Alt+	Menu Bar	Tablet
zoom	z	...	...	VZ	View	K11
	rtzoom				↳Zoom	

Command: zoom

Specify corner of window, enter a scale factor (nX or nXP), or [All/Center/Dynamic/Extents/Previous/Scale/Window/Object] <real time>:
(Pick a point, enter an option, or press ENTER*.)*

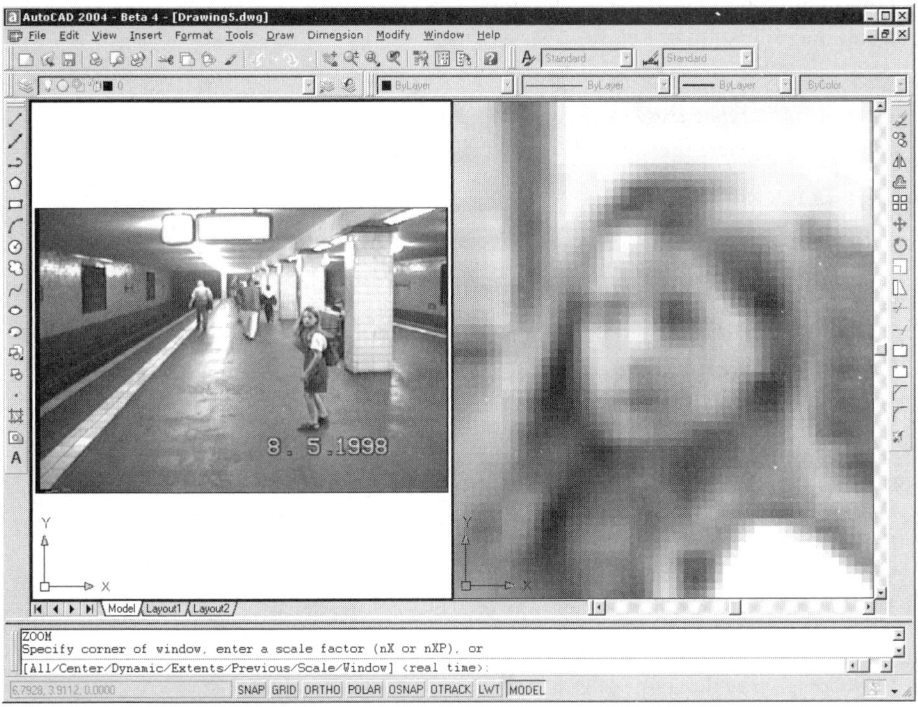

Photograph: Subway station in former East German portion of Berlin.

COMMAND LINE OPTIONS

(pick a point) begins the Window option.

realtime press ENTER to start real-time zoom.

ENTER *or* ESC ends real-time zoom.

All displays the drawing limits or extents, whichever is greater.

Dynamic brings up the dynamic zoom view.

Extents displays the current drawing extents.

Previous displays the previous view generated by the **Pan**, **View**, or **Zoom** commands.

Vmax displays the current virtual screen limits (short for Virtual MAXimum; undocumented).

Window indicates the two corners of the new view.

Object zooms to the extents of selected objects (*new to AutoCAD 2005*).

Center options
Specify center point: *(Pick a point.)*
Enter magnification or height <>: *(Enter a value.)*

Center point indicates the center point of the new view.

Enter magnification or height indicates a magnification value or height of view.

Left options (undocumented)
Lower left corner point: *(Pick a point.)*
Enter magnification or height <>: *(Enter a value.)*

Lower left corner point indicates the lower-left corner of the new view.

Enter magnification or height indicates a magnification value or height of view.

Scale(X/XP) options

*n***X** displays a new view as a factor of the current view.

*n***XP** displays a paper space view as a factor of model space.

RIGHT-CLICK OPTIONS

During real-time zoom, right-click in the drawing:

Exit real-time zoom mode: Exit
Real-time pan: Pan
Real-time zoom: ✓ Zoom
3D orbit mode: 3D Orbit
Zoom window: Zoom Window
Original view: Zoom Original
Zoom to drawing extents: Zoom Extents

RELATED COMMANDS

DsViewer displays Aerial View window, which zooms and pans.

Pan moves the view.

View saves zoomed views by name.

3dZoom performs real-time zooms in perspective viewing mode.

RELATED SYSTEM VARIABLES

ViewCtr specifies the coordinates of the current view's center point.

ViewSize specifies the height of the current view.

TIPS

- A magnification of 1x leaves the drawing unchanged; a magnification of 2x enlarges objects (zooms in), while 0.5x makes objects smaller (zooms out)
- Transparent zoom is *not* possible during the **VPoint**, **Pan**, **DView**, and **View** commands.
- As of AutoCAD 2005, you can zoom to the extents of selected objects.

3D

Rel.11 Draws 3D surface primitives with polygon meshes (*short for three Dimensions*).

Command	Alias	Ctrl+	F-key	Alt+	Menu Bar	Tablet
3d	...	...	...	**DF3**	**Draw** ↳**Surfaces** ↳**3D Surfaces**	**N8**

Command: 3d
Enter an option
[Box/Cone/DIsh/DOme/Mesh/Pyramid/Sphere/Torus/Wedge]: *(Enter an option.)*

*See the **Ai_** commands for details, such as **Ai_Box** and **Ai_Wedge**.*

*Selecting **Draw | Surfaces | 3D Surfaces** from the menu bar displays the dialog box:*

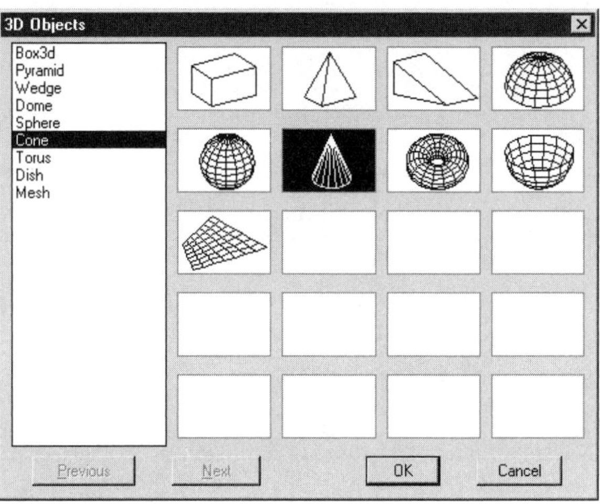

ICON MENU OPTIONS

Box draws 3D boxes and cubes.

Cone draws cones.

DIsh draws dishes, the bottom half of spheres.

Dome draws domes, the top half of spheres.

Mesh draws a 3D mesh.

Pyramid draws a variety of pyramids.

Sphere draws spheres.

Torus draws tori (3D donut) shapes.

Wedge draws wedges.

RELATED COMMANDS

Ai_Box draws 3D surface boxes and cubes.

Ai_Cone draws 3D surface cones.

Ai_Dish draws 3D surface dishes.

Ai_Dome draws 3D surface domes.

Ai_Mesh draws 3D meshes.

Ai_Pyramid draws 3D surface pyramids.

Ai_Sphere draws 3D surface spheres.

Ai_Torus draws 3D surface tori.

Ai_Wedge draws 3D surface wedges.

Box draws 3D solid boxes and cubes.

Cone draws 3D solid cones.

Cylinder draws 3D solid cylinders.

Sphere draws 3D solid spheres.

Torus draws 3D solid tori.

Wedge draws 3D solid wedges.

TIPS

- The **3D** command creates 3D objects made of 3D polygon meshes, and *not* of 3D solids.
- To draw a cylinder with endcaps, apply thickness to a circle.
- You *cannot* perform Boolean operations on 3D surface models.
- To convert a 3D solid model to a 3D surface model, export the drawing with the **3dsOut** command, then import with the **3dsIn** command.
- Use the **Ucs** or **Align** command to place 3D surface models in space; use the **VPoint** and **Dview** commands to view surface models from different 3D viewpoints.
- You can apply the **Hide**, **Shade,** and **Render** commands to 3D surface models.

3dArray

Rel.11 Creates 3D rectangular and polar arrays.

Command	Alias	Ctrl+	F-key	Alt+	Menu Bar	Tablet
3darray	...	...	...	M33	**Modify**	W20
					↳**3D Operation**	
					↳**3D Array**	

Command: 3darray
Select objects: *(Select one or more objects.)*
Select objects: *(Press* **ENTER** *to end object selection.)*
Enter the type of array [Rectangular/Polar] <R>: *(Type* **R** *or* **P**.*)*

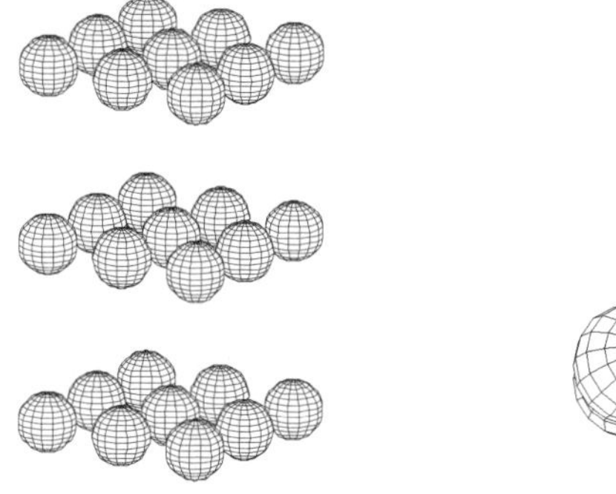

3D rectangular array

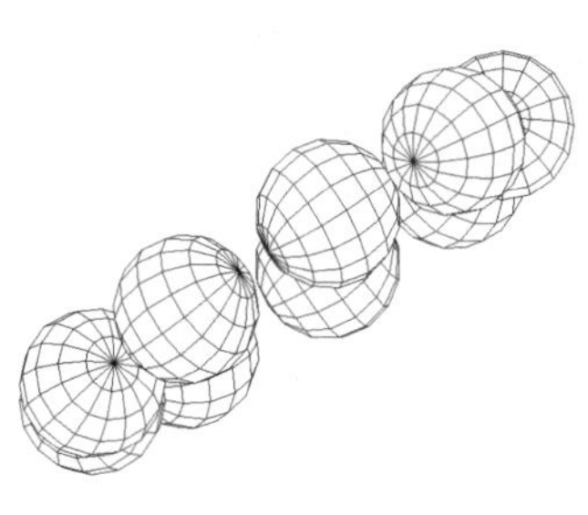

3D Polar array.

COMMAND LINE OPTIONS

Select objects selects the objects to be arrayed.

R creates rectangular 3D arrays.

P creates a polar array in 3D space.

Rectangular Array options
Enter the number of rows (---) <1>: *(Enter a value.)*
Enter the number of columns (|||) <1>: *(Enter a value.)*
Enter the number of levels (...) <1>: *(Enter a value.)*
Specify the distance between rows (---) <1>: *(Enter a value.)*
Specify the distance between columns (|||) <1>: *(Enter a value.)*
Specify the distance between levels (...) <1>: *(Enter a value.)*

Enter the number of rows specifies the number of rows in the x direction.

Enter the number of columns specifies the number of columns in the y direction.

Enter the number of levels specifies the number of levels in the z direction.

Specify the distance between rows specifies the distance between objects in the x direction.

Specify the distance between columns specifies the distance between objects in the y direction.

Specify the distance between levels specifies the distance between objects in the z direction.

Polar Array options

Enter the number of items in the array: *(Enter a number.)*
Specify the angle to fill (+=ccw, -=cw) <360>: *(Enter an angle.)*
Rotate arrayed objects? [Yes/No] <Y>: *(Type* **Y** *or* **N**.*)*
Specify center point of array: *(Pick a point.)*
Specify second point on axis of rotation: *(Pick a point.)*

Enter the number of items specifies the number of objects to array.

Specify the angle to fill specifies the distance along the circumference that objects are arrayed (default = 360 degrees).

Rotate arrayed objects?

Yes objects rotate so that they face the central axis (default).

No objects rotate.

Specify center point of array specifies center point of the array, and one end of the axis.

Specify second point on axis of rotation specifies the other end of the array axis.

ESC interrupts the drawing of arrays.

RELATED COMMANDS

Array creates a rectangular or polar array in 2D space.

Copy creates one or more copies of the selected object.

MInsert creates a rectangular block-array of blocks.

3dClip

2000 Performs real-time front and back clipping (*short for three Dimensional CLIPping*).

Command	Alias	Ctrl+	F-key	Alt+	Menu Bar	Tablet
3dclip	...	...	...	...	...	...

Command: 3dclip

Displays window:

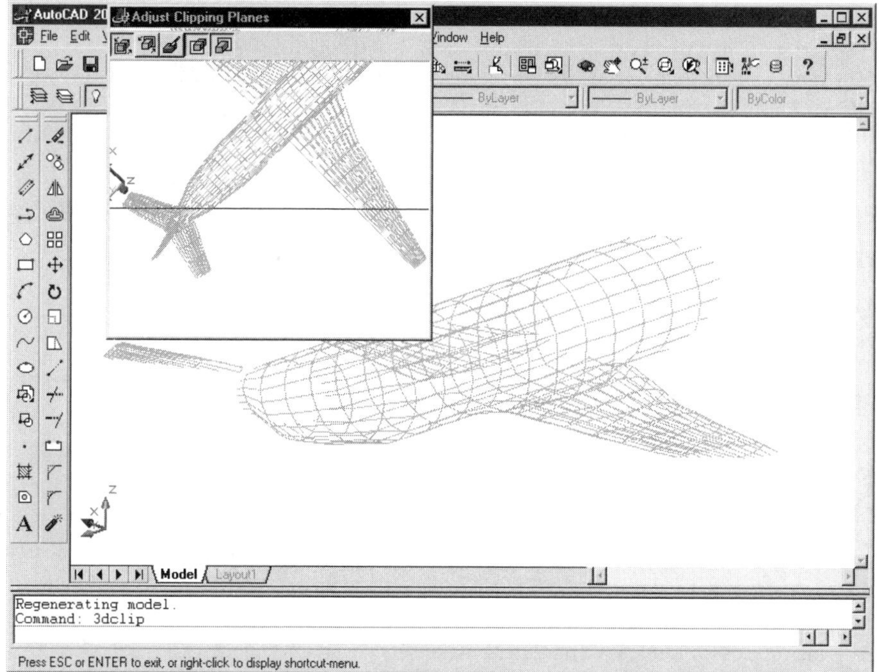

TOOLBAR OPTIONS

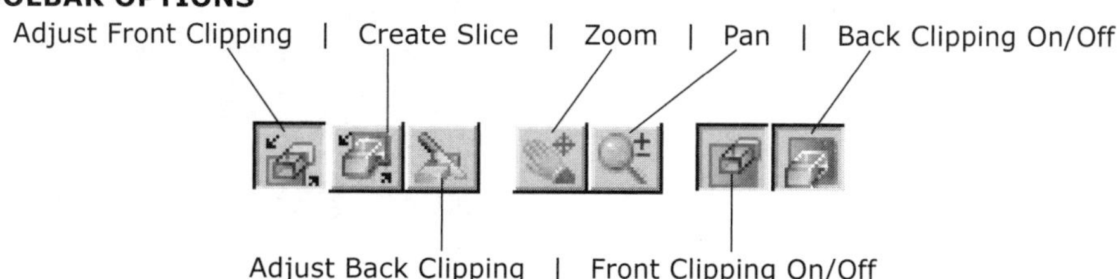

SHORTCUT MENU OPTIONS

Right-click ***Adjust Clipping Planes window:***

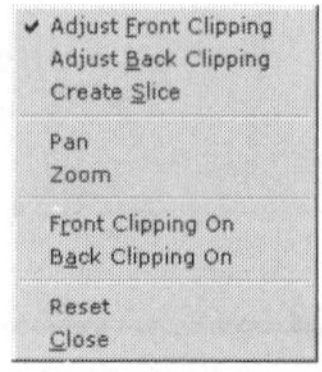

Adjust Front Clipping switches to front clipping mode.

Adjust Back Clipping switches to back clipping mode.

Create Slice switches to slicing — ganged front and back clipping — mode.

Pan moves the model within the window *(new to AutoCAD 2005).*

Zoom enlarges and reduces the view of the model within the window *(new to AutoCAD 2005).*

Front Clipping On toggles on and off front clipping.

Back Clipping On toggles on and off back clipping.

Reset removes clipping planes, and returns view to original state *(new to AutoCAD 2005).*

Close closes the Adjust Clipping Planes window.

RELATED COMMANDS

Vpoint creates a static 3D viewpoint.

3dCOrbit places the drawing into real-time, 3D, continuous orbit mode.

3dDistance performs real-time 3D forward and backward panning.

3dOrbit provides real-time 3D viewing of the drawing.

3dPan performs real-time 3D sideways panning.

3dSwivel tilts the 3D view.

3dZoom performs real-time 3D zooming.

TIP

- Use this command to remove objects in the front of a 3D scene, or to expose the interior of a 3D model, such as this airplane.

3dConfig

2004 Configures display characteristics of your computer's 3D graphics boards.

Command	Alias	Ctrl+	F-key	Alt+	Menu Bar	Tablet
3dconfig	...	...	...	...	...	...

Command: 3dconfig
Configure: 3DCONFIG
Enter option [Adaptive degradation/Dynamic tessellation/Render options/ Geometry/acceLeration/eXit] <Adaptive degradation>: *(Enter an option, or type* **V** *for undocumented adVanced options.)*

COMMAND LINE OPTIONS

Adaptive degradation allows AutoCAD to switch to lower-quality rendering in order to maintain display speed.

Dynamic tessellation specifies the smoothness of faceted objects in 3D drawings.

Render options enhances the display of lights, materials, textures, and transparency in 3D views.

Geometry specifies the display of isolines and back faces in 3D drawings.

acceLeration specifies hardware or software acceleration.

adVanced specifies redraw on window expose; cache viewport draw geometry; display lists; fast hidden line only; and pixel deviation (*undocumented options*).

eXit exits the command.

Adaptive degradation options

Enter mode [ON/OFF] <ON>: *(Enter* **ON** *or* **OFF***.)*

ON or **OFF** turns on or off adaptive degradation.

When on, the following options become available:

Current display options: Wireframe Bounding Box
Enter option [Flat shaded/Wireframe/Bounding box/Maintain speed fps/ eXit] <Flat shaded>: *(Enter an option.)*

Flat shaded degrades the 3D display to flat shading.

Wireframe degrades the 3D display to wireframe.

Bounding box degrades the 3D display, showing each object as a rectangular box.

Maintain speed fps specifies the speed at which to display, in frames per second; range is 5fps to 60fps.

Dynamic tessellation options

Enter option [Surface tessellation/Curve tessellation/Tessellations to cache/ eXit] <Surface tessellations>: *(Enter an option.)*

Surface tessellation specifies the number of tessellation lines to display on surfaces; ranges from 0 to 100 (default = 88).

Curve tessellation specifies the number of tessellation lines to display on curved surfaces; ranges from 0 to 100 (default = 88).

Tessellations to cache specifies the number of tessellations to cache; ranges from 1 to 3 (default = 3).

Render options
Enter mode [ON/OFF] <ON>: *(Enter* **ON** *or* **OFF***.)*
Enter option [Lights/Materials/eXit] <Lights>: *(Enter an option.)*

ON or **OFF** turns on or off render options.

When Render mode is on, the following options become available:

Lights determines whether assigned lights (with the **Light** command) or the default global light are used.

Materials determines whether assigned materials (with the **RMat** command) or default global material are used.

When Materials mode is on, the following options become available:

Configure: Textures
Enter mode [ON/OFF] <ON>: *(Enter* **ON** *or* **OFF***.)*
Configure: Transparency
Enter mode [Low/Medium/High] <Low>: *(Enter an option.)*

Textures determines whether assigned textures (with the **RMat** and **SetUV** commands) are used.

Transparency determines the quality of transparency:

- **Low** applies screen-door effect for faster speed (default).
- **Medium** applies blending effect for moderate speed.
- **High** applies blending and extra processing for slower speed.

Geometry options
Enter option [Isolines on top/Discard backfaces] <Isolines on top>: *(Enter an option.)*

Isolines on top

On displays isolines on front and back faces in all shading modes (except hidden).

Off hides isolines on back faces.

Discard backfaces

On discards back faces to enhance performance.

Off displays back faces.

acceLeration options
Enter option [Hardware/Software/eXit] <Hardware>: *(Enter an option.)*

Hardware configures graphics card to perform 3D display rendering, the faster option.

Software configures 3D display through software, if the graphics board is unable to.

When Hardware option selected:

Enter option [Driver name/Geometry acceleration/Antialias lines/eXit] <Driver name>:

Driver name allows you to select a specific driver for the graphics board.

Geometry acceleration allows a more precise display, if turned on and if supported by the graphics board.

Antialias lines uses anti-aliasing to drawn smoother lines.

RELATED COMMAND

3dOrbit provides real-time 3D viewing of the drawing.

 3dCOrbit/Distance/Pan/Swivel/Zoom

2000 This collection of commands is a subset of **3dOrbit**.

Command	Alias	Ctrl+	F-key	Alt+	Menu Bar	Tablet
3dcorbit	...	...	...	...	...	...

Command: 3dcorbit

Places the 3D view into continuous orbiting mode.

Command: 3ddistance

Interactively changes the 3D viewing distance.

Command: 3dpan

Interactively pans the 3D view.

Command: 3dswivel

Interactively twists the 3D view.

Command: 3dzoom

Interactively zooms the 3D view.

COMMAND LINE OPTIONS

ESC exits the command.

ENTER exits the command.

RELATED COMMANDS

Vpoint creates a static 3D viewpoint.

3dClip performs real-time 3D front and back clipping.

3dOrbit performs real-time 3D viewing of the drawing.

3dViewCtr specifies the center point for 3D orbiting views.

TIPS

- To set the 3D model to continuous orbit mode, drag the cursor across the drawing.
- You can use **3dPan** and **3dZoom** commands when the current drawing is in perspective mode; when you try to use the regular **Pan** and **Zoom** command, AutoCAD complains, '** That command may not be invoked in a perspective view **.'

3dFace

V. 2.6 Draws 3D faces with three or four corners.

Command	Alias	Ctrl+	F-key	Alt+	Menu Bar	Tablet
3dface	**3f**	...	...	**DFF**	**Draw** **↳Surfaces** **↳3D Face**	**M8**

Command: 3dface

Specify first point or [Invisible]: *(Pick a point, or type* **I** *followed by the point.)*

Specify second point or [Invisible]: *(Pick a point, or type* **I** *followed by the point.)*

Specify third point or [Invisible] <exit>: *(Pick a point, or type* **I** *followed by the point.)*

Specify fourth point or [Invisible] <create three-sided face>: *(Pick a point, or type* **I** *followed by the point.)*

Specify third point or [Invisible] <exit>: *(Press* **ENTER** *to exit the command.)*

COMMAND LINE OPTIONS

First point picks the first corner of the face.

Second point picks the second corner of the face.

Third point picks the third corner of the face.

Fourth point picks the fourth corner of the face; or press **ENTER** to create a triangular face.

Invisible makes the edge invisible.

RELATED COMMANDS

3D draws a 3D object: box, cone, dome, dish, pyramid, sphere, torus, or wedge.

Properties modifies the 3d face, including visibility of edges.

Edge changes the visibility of the edges of 3D faces.

EdgeSurf draws 3D surfaces made of 3D meshes.

PEdit edits 3D meshes.

PFace draws generalized 3D meshes.

RELATED SYSTEM VARIABLE

SplFrame controls the visibility of edges.

TIPS

- A 3D face is the same as a 2D solid, except that each corner can have a different z coordinate.
- Unlike the procedure for **Solid**, corner coordinates are entered in natural order.
- The **i** (short for invisible) suffix must be entered before object snap modes, point filters, and corner coordinates.
- Invisible 3D faces, where all four edges are invisible, do not appear in wireframe views; they hide objects behind them in hidden-line mode, and are rendered in shaded views.
- You can use the **Properties** and **Edge** commands to change the visibility of 3D face edges.
- 3D faces cannot be extruded.

Rel.10 Draws open polygon 3D meshes.

Command	Alias	Ctrl+	F-key	Alt+	Menu Bar	Tablet
3dmesh	...	...	...	DFM	Draw ↳Surfaces ↳3D Mesh	...

Command: 3dmesh
Enter size of mesh in M direction: *(Enter a value.)*
Enter size of mesh in N direction: *(Enter a value.)*
Specify location for vertex (0, 0): *(Pick a point.)*
Specify location for vertex (0, 1): *(Pick a point.)*
Specify location for vertex (1, 0): *(Pick a point.)*
Specify location for vertex (1, 1): *(Pick a point.)*

COMMAND LINE OPTIONS

Enter size of mesh in M direction specifies the m-direction mesh size (between 2 and 256).

Enter size of mesh in N direction specifies the n-direction mesh size (between 2 and 256).

Specify location for vertex (*m*,*n*) specifies a 2D or 3D coordinate for each vertex.

RELATED COMMANDS

3D draws a variety of 3D objects.

3dFace draws a 3D face with three or four corners.

Explode explodes a 3D mesh into individual 3D faces.

PEdit edits a 3D mesh.

PFace draws a generalized 3D face.

Xplode explodes a group of 3D meshes.

TIPS

- It is more convenient to use the **EdgeSurf**, **RevSurf**, **RuleSurf**, and **TabSurf** commands than the **3dMesh** command. The **3dMesh** command is meant for use by AutoLISP and other programs.
- The range of values for the m- and n-mesh size is 2 to 256.
- The number of vertices = **m** x **n**.
- The first vertex is (0,0). The vertices can be at any point in space.
- The coordinates for each vertex in row **m** must be entered before starting on vertices in row **m+1**.
- Use the **PEdit** command to close the mesh, since it is always created open.
- The **SurfU** and **SurfV** system variables do not affect the 3D mesh object.

3dOrbit

2000 Provides interactive 3D viewing of the drawing.

Command	Alias	Ctrl+	F-key	Alt+	Menu Bar	Tablet
3dorbit	orbit	...	...	VB	View 3D Orbit	R5

Command: 3dorbit

Press ESC or ENTER to exit, or right-click to display shortcut-menu. *(Move the cursor to change the viewpoint, and then press* **ENTER** *to exit the command.)*

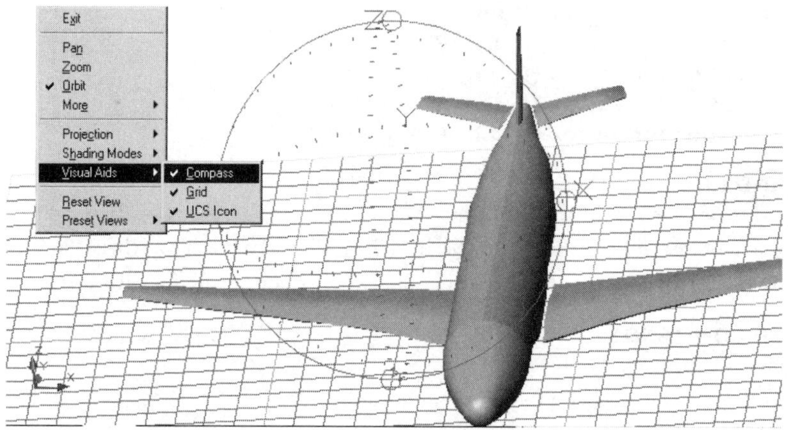

COMMAND LINE OPTIONS

ESC exits the command.

ENTER exits the command.

SHORTCUT MENU OPTIONS

Right-click the drawing:

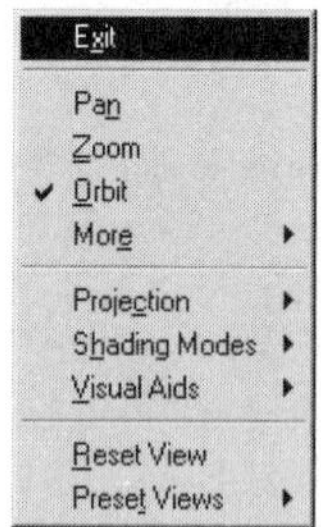

Exit exits the command.

Pan pans the view in real-time; see the **3dPan** command.

Zoom zooms the view in real-time; see the **3dZoom** command.

Orbit rotates the view in real-time.

More displays a submenu with additional view controls.

Projection displays a submenu for toggling between parallel and perspective projection.

Shading Modes displays a submenu for selecting the type of wireframe or shaded modes; see the **ShadeMode** command.

Visual Aids displays a submenu of visual aids for navigating in 3D space.

Reset View resets to the original view when you first began the command.

Preset Views displays a submenu of standard views.

More options

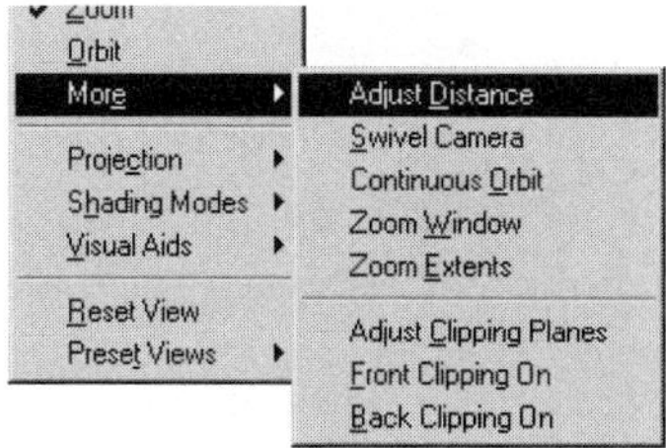

Adjust Distance moves the view closer and further away; see the **3dDistance** command.

Swivel Camera rotates the view; see the **3dSwivel** command.

Continuous Orbit continuously rotates the model; see the **3dCOrbit** command.

Zoom Window performs a windowed zoom; see the **Zoom** command.

Zoom Extents displays the entire drawing.

Adjust Clipping Planes sets the front and back clipping planes; see the **3dClip** command.

Front Clipping On toggles the front clipping plane.

Back Clipping On toggles the back clipping plane.

Projection options

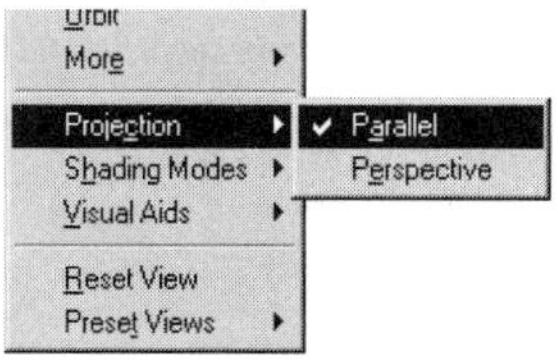

Parallel displays the view in orthogonal view.

Perspective displays the view in one-point perspective view; some commands do not work in perspective mode, such as the **Pan** and **Zoom** commands.

Shading Modes options (see **ShadeMode** command)

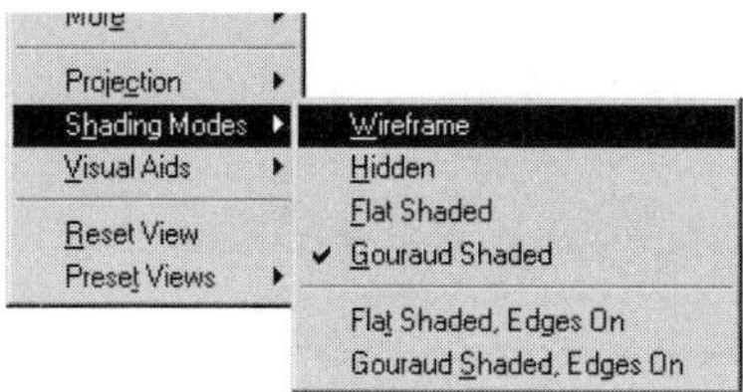

Wireframe displays the model in wireframe mode.

Hidden removes hidden lines from the view.

Flat Shaded flat-shades the model.

Gouraud Shaded smooth-shades the model.

Flat Shaded, Edges On flat-shades the model, and outlines faces with the background color.

Gouraud Shaded, Edges On smooth-shades the model, and outlines faces with the background color.

Visual Aids options

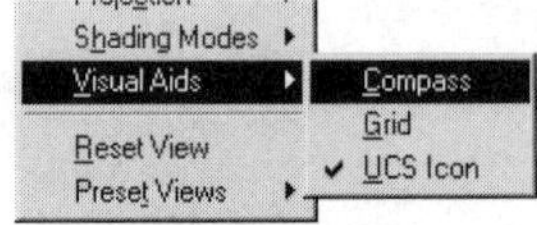

Compass toggles the display of the compass to help you navigate in 3D space.

Grid toggles the display of the grid as lines, to help you see the x,y-plane; see the **Grid** command.

UCS Icon toggles the display of the 3D UCS icon; see the **UcsIcon** command.

Preset Views options

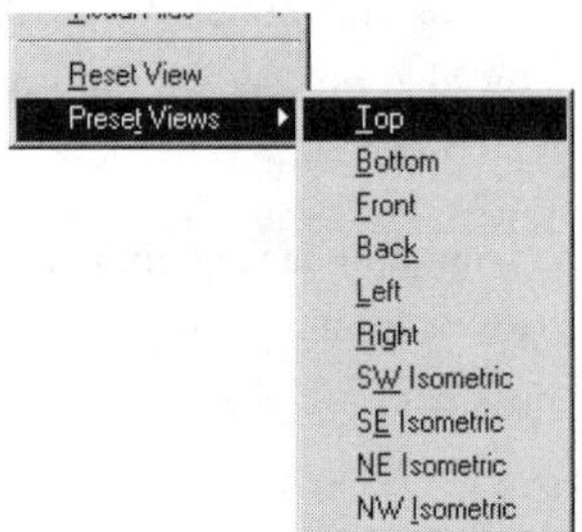

Top, Bottom, Front, Back, Left, Right displays the standard orthogonal views.

SW Isometric, SE Isometric, NW Isometric, NW Isometric displays the standard isometric views.

RELATED COMMANDS

Vpoint creates a static 3D viewpoint.

3dClip performs real-time 3D front and back clipping.

3dCOrbit places the drawing in real-time, 3D, continuous orbit mode.

3dDistance performs real-time 3D forward and backward panning.

3dPan performs real-time 3D sideways panning.

3dSwivel tilts the 3D view.

3dZoom performs real-time 3D zooming.

TIP

- This command is meant to replace the **DView** command.

3dOrbitCtr

2004 Specifies the center point for 3D orbiting views.

Command	Alias	Ctrl+	F-key	Alt+	Menu Bar	Tablet
3dorbitctr	...	...	...	...	...	...

Command: 3dorbitctr
Specify orbit center point: *(Pick a point.)*

COMMAND LINE OPTION

Specify orbit center point specifies the center of rotation.

RELATED COMMANDS

3dClip performs real-time 3D front and back clipping.

3dCOrbit places the drawing in real-time, 3D, continuous orbit mode.

3dDistance performs real-time 3D forward and backward panning.

3dOrbit provides real-time 3D viewing of the drawing.

3dSwivel tilts the 3D view.

3dZoom performs real-time 3D zooming.

3dPoly

Rel.10 Draws 3D polylines (*short for 3D POLYline*).

Command	Alias	Ctrl+	F-key	Alt+	Menu Bar	Tablet
3dpoly	...	...	...	**D3**	**Draw** ↳**3D Polyline**	**O10**

Command: 3dpoly
Specify start point of polyline: *(Pick a point.)*
Specify endpoint of line or [Undo]: *(Pick a point, or type* **U***.)*
Specify endpoint of line or [Undo]: *(Pick a point, or type* **U***.)*
Specify endpoint of line or [Close/Undo]: *(Pick a point, type* **U** *or* **C***, or press* ENTER *to end the command.)*

COMMAND LINE OPTIONS

Specify start point indicates the starting point of the 3D polyline.

Close joins the last endpoint with the start point.

Undo erases the last-drawn segment.

Specify endpoint of line indicates the endpoint of the current segment.

ENTER ends the command.

RELATED COMMANDS

Explode reduces a 3D polyline into lines and arcs.

PEdit edits 3D polylines.

PLine draws 2D polylines.

TIPS

- Because 3D polylines are made of straight lines, you can use the **PEdit** command to spline-fit the polyline.
- 3D polylines do not support linetypes and widths.
- You may use lineweights to fatten up a 3D polyline.
- Use the .xy point filter to place pick points with object snaps, and then specify the z coordinate.

3dsIn

Rel.13 Imports *.3ds* files created by 3D Studio and other applications (*short for 3D Studio IN*).

Command	Alias	Ctrl+	F-key	Alt+	Menu Bar	Tablet
3dsin	...	...	...	**I3**	**Insert** ↳**3D Studio**	...

Command: 3dsin

Displays the 3D Studio File Import dialog box. Select a .3ds file, and then click ***Open****.*

AutoCAD displays dialog box:

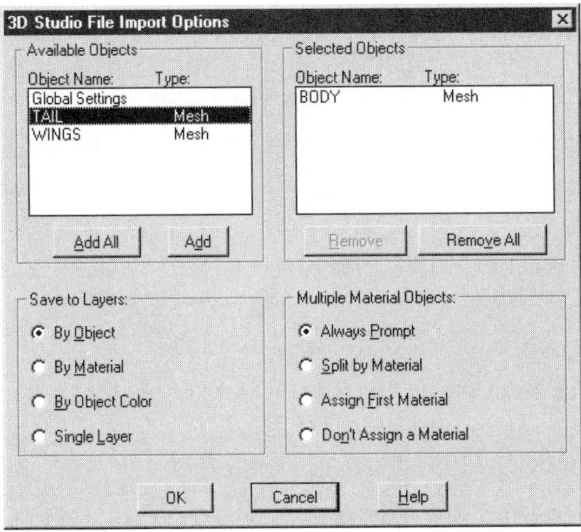

DIALOG BOX OPTIONS

Available and Selected Objects options

Object Name names the object.

Type specifies the type of object.

Add adds the object to the Selected Objects list.

Add All adds all objects to the Selected Objects list.

Remove removes the object from Selected Objects list.

Remove All removes all objects from Selected Objects list.

Save to Layers options

⊙**By Object** places each object on its own layer.

○**By Material** places the objects on layers named after materials.

○**By Object Color** places the objects on layers named "Color*nn*."

○**Single Layer** places all objects on layer "AvLayer."

Multiple Material Objects options

⊙**Always Prompt** prompts for each material.

○**Split by Material** splits objects with more than one material into multiple objects, each with one material.

○**Assign First Material** assigns the first material to the entire object.

○**Don't Assign to a Material** removes all 3D Studio material definitions.

RELATED COMMAND

3dsOut exports drawing as a 3DS file.

RELATED FILES

.3ds are 3D Studio files.

.tga are converted bitmap and animation files.

TIPS

- You are limited to selecting a maximum of 70 3D Studio objects.
- Conflicting object names are truncated and given a sequence number.
- The **By Object** option gives the AutoCAD layer the name of the object.
- The **By Object Color** option places all objects on layer "ColorNone" when no colors are defined in the 3DS file.
- 3D Studio assigns materials to faces, elements, and objects; AutoCAD assigns materials only to objects, colors, and layers.
- 3D Studio bitmaps are converted to *.tga* (Targa format) bitmaps.
- Only the first frame of animation files (CEL, CLI, FLC, and IFL) is converted to Targa bitmap files.
- Converted *.tga* files are saved to the *.3ds* file's folder.
- 3D Studio "ambient lights" lose their color.
- 3D Studio "omni lights" become point lights in AutoCAD.
- 3D Studio "cameras" become named views in AutoCAD.

3dsOut

Rel.13 Exports AutoCAD drawings as 3D Studio files (*short for 3D Studio OUT*).

Command	Alias	Ctrl+	F-key	Alt+	Menu Bar	Tablet
3dsout	...	...	...	**FE**	**File**	...
				↳3DS	**↳Export**	
					↳3D Studio	

Command: 3dsout
Select objects: *(Select one or more objects.)*
Select objects: *(Press* **ENTER** *to end object selection.)*

Displays 3D Studio File Export dialog box. Enter a file name, and then click ***Save****.*

AutoCAD displays dialog box:

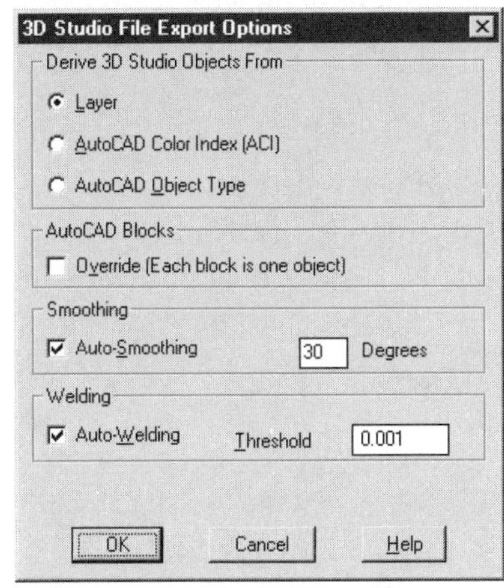

After you select options, the ***3dsOut*** *command exports selected objects:*

Generating objects
Writing preamble
Converting and writing material definitions
UCSVIEW = 1 UCS will be saved with view
Collecting geometry
Converting object BODY
Unifying normals
Assigning smoothing
3D Studio file output completed

COMMAND LINE OPTION

Select objects selects objects to export; note that **3dsOut** exports only objects with a surface.

DIALOG BOX OPTIONS

Derive 3D Studio Objects From options

⊙**Layer** exports all objects on AutoCAD layers to single 3D Studio objects (default).

○**ACI** exports all objects of an ACI color to single 3D Studio objects.

○**Object Type** exports all objects of an AutoCAD object type to single 3D Studio objects.

AutoCAD Blocks option

☐ **Override** specifies that each AutoCAD block becomes a single 3D Studio object; overrides the Derive 3D Studio Objects From options.

Smoothing options

Auto-Smoothing

☑ Creates a 3D Studio smoothing group (default).

☐ Does not assign smoothing to new 3D Studio objects.

Degrees smooths the face normals, when the angle between two face normals is equal to or less than this value (default = 30 degrees).

Welding options

Auto-Welding

☑ Creates a 3D Studio welded vertex (default).

☐ Does not created welded vertices upon export.

Threshold welds two vertices into a single vertex when their interdistance is less than or equal to this value (default = 0.001).

RELATED AUTOCAD COMMAND

3dsIn imports 3DS files into drawings.

RELATED FILE

****.3ds*** are 3D studio files.

TIPS

- AutoCAD objects with 0 thickness are not exported, with the exception of circles, polygons, and polyface meshes.
- Solids and 3D faces must have at least three vertices.
- 3D solids and bodies are converted to meshes.
- AutoCAD blocks are exploded unless the **Override** option is turned on.
- The weld threshold distance ranges from 0.00 000 001 to 99,999,999.
- AutoCAD named views become 3D Studio "cameras."
- AutoCAD point lights become 3D Studio "omni lights."

APPENDIX A
Express Tools

The following commands are included with the Express Tools package.

Command	Description
AliasEdit	Edits the aliases stored in *acad.pgp*.
AlignSpace	Aligns model space objects, whether in different viewports or with objects in paper space.
ArcText	Places text along an arc.
AttIn	Imports attribute data.
AttOut	Quickly extracts attributes in tab-delimited format.
BExtend	Extends open objects to objects in blocks and xrefs.
BlockReplace	Replaces all inserts of one block with another.
BlockToXref	Convert blocks to xrefs.
BreakLine	Creates the break-line symbol.
BScale	Scales blocks from their insertion points.
BTrim	Trims to objects nested in blocks and external references.
Burst	Explodes blocks, converts attributes to text.
ChSpace	Moves objects between model and paper space.
ChUrls	**DdEdit**-like editor for hyperlinks (URL addresses).
ClipIt	Adds arcs, circles, and polylines to the **XClip** command.
CopyM	**Copy** command with repeat, divide, measure, and array options.
CopyToLayer	Copies objects to other layers.
DimEx	Exports dimension styles to an ASCII file.
DimIm	Imports dimension style files created with **DimEx**.
EditTime	Pauses the timer when not editing (*new to AutoCAD 2005*).
EtBug	Sends bug reports to Autodesk.
ExOffset	Adds options to the **Offset** command.
ExPlan	Adds options to the **Plan** command.
Flatten	Reduces 3D drawings to 2D (*new to AutoCAD 2005*).
FS	Selects objects that touch the selected object.
FullScreen	Toggles between full-screen and regular window.
GetSel	Selects objects based on layer and type.
ImageApp	Specifies the external image editor.
ImageEdit	Launches the image editor to edit selected images.
LayCur	Changes the layer of selected objects to the current layer.
LayDel	Deletes layers from drawings — permanently.
LayFrz	Freezes the layers of selected objects.
LayIso	Isolates layers of selected objects (all other layers are frozen).
LayLck	Locks the layers of selected objects.
LayMch	Changes the layer of selected objects to that of a selected object.
LayMrg	Merges two layers; removes the first layer from the drawing.
LayOff	Turns off layers of selected objects.
LayOn	Turns on all layers.
LayoutMerge	Places objects from layouts onto one layout.
LayThw	Thaws all layers.
LayUlk	Unlocks layer of selected object.
LayVpi	Isolates object's layer in viewport.

Command	Description
LayWalk	Isolates each layer in sequential order.
LMan	Saves and restores layer settings.
Lsp	AutoLISP function searching utility.
LspSurf	LISP file viewer.
MkLtype	Creates linetypes from selected objects.
MkShape	Creates shapes from selected objects.
MoCoRo	Moves, copies, rotates, and scales objects.
MoveBak	Moves *.bak* files to specified folders.
MStretch	Stretches with multiple selection windows.
NCopy	Copies objects nested inside blocks and xrefs.
OverKill	Removes overlapping duplicate objects (*new to AutoCAD 2005*).
Plt2Dwg	Imports HPGL files into the drawings.
Propulate	Updates, lists, and clears drawing properties.
PsBScale	Sets and updates the scale of blocks relative to paper space.
PsTScale	Sets text height relative to paper space.
QlAttach	Associates leaders with annotation objects.
QlAttachSet	Associates leaders with annotations.
QlDetachSet	Dissassociates leaders from annotations.
QQuit	Closes all drawings, and then exits AutoCAD.
ReDir	Changes paths for xrefs, images, shapes, and fonts.
RepUrls	Replaces hyperlinks.
Revert	Closes the drawing, and re-opens the original.
RText	Inserts and edits remote text objects.
RtUcs	Changes UCSs in real time.
SaveAll	Saves all drawings.
ShowUrls	Lists URLs in a dialog box.
Shp2blk	Converts from a shape definition to a block definition.
SuperHatch	Uses images, blocks, external references, or wipeouts as hatch patterns.
SysvDlg	Launches an editor for system variables.
TCase	Changes text between Sentence, lower, UPPER, Title, and tOGGLE cASE.
TCircle	Surrounds text and multiline text with circles, slots, and rectangles.
TCount	Prefixes text with sequential numbers.
TextFit	Fits text between points.
TextMask	Places masks behind selected text.
TextUnmask	Removes masks from behind text.
TFrames	Toggles the frames surrounding images and wipeouts.
TJust	Justifies text created with the **MText** and **AttDef** commands.
TOrient	Re-orients text, multiline text, and block attributes.
TScale	Scales text, multiline text, attributes, and attribute definitions.
TSpaceInvaders	Finds and selects text with overlapping objects.
Txt2Mtxt	Converts single-line to multiline text.
TxtExp	Explodes selected text into polylines.
VpScale	Lists the scale of the selected viewports.
VpSynch	Synchronizes viewports with a master viewport.
XData	Attaches xdata to objects.
XdList	Lists xdata attached to objects.
XList	Displays properties of objects nested in blocks and xref.

APPENDIX B

External Programms

These programs run under Windows from outside of AutoCAD.

AcSignApply.exe

Applies digital signatures to drawings (*short for AutoCad SIGNnature APPLY*).

Start | Programs | Autodesk | AutoCAD 2005 | Attach Digital Signatures

Displays dialog box:

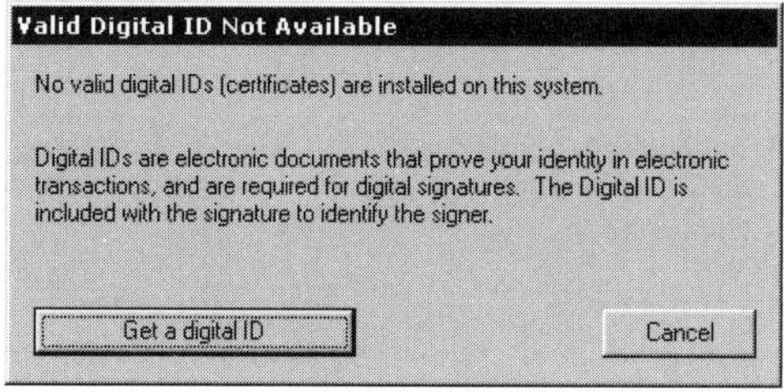

DIALOG BOX OPTIONS

Get a digital ID attempts to contact servers to download digital IDs (identification).

Cancel cancels the program.

RELATED AUTOCAD COMMANDS

SecurityOptions applies passwords and digital signatures to drawings.

SigValidate checks whether the drawing has a valid digital signature.

TIPS

- Digital signatures validate the authenticity of drawings.
- Digital signatures indicate whether the drawing was changed since signed.
- Autodesk notes that digital signatures become invalid for these reasons: the drawing was corrupted when the digital signature was attached, the drawing was corrupted in transit, and the digital ID is no longer valid.

AddPlWiz.exe

Runs the Add Plotter wizard (*short for ADD PLotter WIZARD).*

Start | Settings | Control Panel | Autodesk Plotter Manager

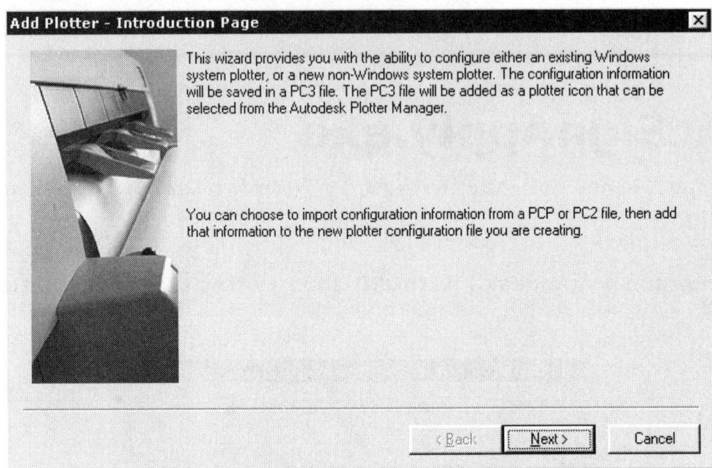

DIALOG BOX OPTIONS

Next goes to the next dialog box.

Cancel cancels the program.

RELATED AUTOCAD COMMANDS

PlotterManager opens the Plotters window, which includes the Add-a-Plotter wizard.

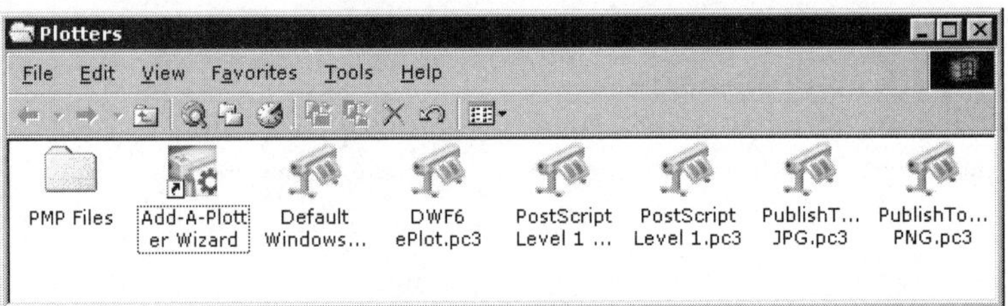

RELATED FILES

**.pc3* are AutoCAD plotter configuration files, third generation.

**.pmp* are AutoCAD plotter model parameter files.

TIPS

- See the **PlotterManager** command for details on using this program.
- You can create and edit *.pc3* plotter configuration files without AutoCAD. From the **Start** button on the Windows toolbar, select **Settings | Control Panel | Autodesk Plotter Manager**.

AdMigrator.exe

Migrates settings from older releases to AutoCAD 2005 (*short for AutoDesk MIGRATOR).*

Start | Programs | Autodesk | AutoCAD 2005 | Migrate Custom Settings

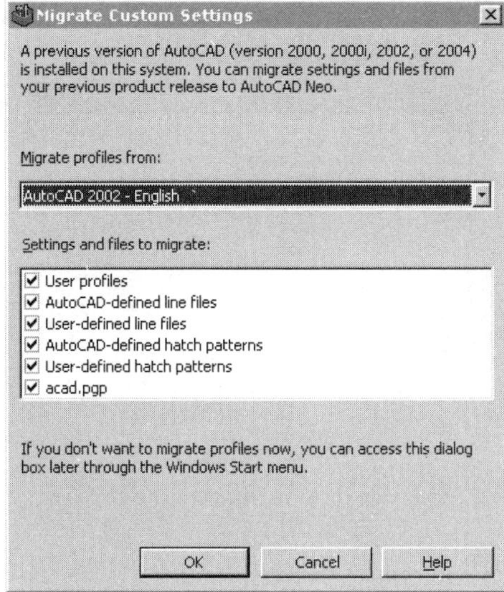

DIALOG BOX OPTIONS

Migrate profiles from selects the previous release of AutoCAD installed on your computer.

Settings and files to migrate selects the settings to migrate to AutoCAD 2005.

TIPS

- This program works with AutoCAD 2000, 2000i, 2002, and 2004.
- The following migration programs are also available from Autodesk's Web site:

  Autodesk Batch Drawing Converter.

  AutoLISP Compatibility Analyzer.

  Menu and Toolbar Porter.

  Command Alias (PGP) Porter.

  ScriptPro.

  Layer State Converter.

 AdRefMan.exe

Allows you to change the path to external files, such as text fonts, images, plot configurations, and xrefs *(short for AutoDesk REFerence MANager)*.

Start | Programs | Autodesk | AutoCAD 2005 | Reference Manager

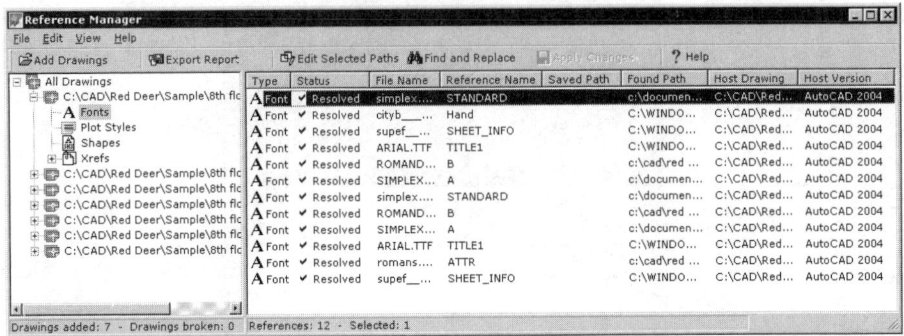

MENU BAR OPTIONS

File menu

Add Drawings (CTRL+O) adds one or more drawings to the list.

Remove Drawings removes selected drawings from the list.

Export Report creates a report in *.csv* (comma separated values) format.

Apply Changes applies the changed paths to drawings.

Exit (ALT+F4) exits the program.

Edit menu

Select All (CTRL+A) selects all drawings.

Unselect All unselects all drawings.

Invert Selection reverses the selection: unselected drawings become selected.

Find Selected Paths (F2) displays the Edit Selected Paths dialog box.

Find and Replace (F3) displays the Find and Replace Selected Paths dialog box.

View menu

List by Drawing (CTRL+D) lists by files: drawings and xrefs.

List by Reference Type (CTRL+R) lists by objects: fonts, plot styles, shapes, and xrefs.

Options displays the Options dialog box.

Help menu

Help (F1) displays online help for this program.

About displays the About Autodesk Reference Manager dialog box.

TOOLBAR OPTIONS

Add Drawings selects the drawings to be checked; displays the Add Xrefs dialog box when a drawing contains xrefs.

Export Report creates a report in *.csv* (comma separated values) format of the findings by this program; displays the Export Report dialog box.

Edit Selected Paths displays the Edit Selected Paths dialog box; enter a new path, and click **OK**.

Find and Replace displays the Find and Replace Selected Paths dialog box.

Apply Changes applies the changed paths to drawings; displays the Summary dialog box.

Help displays online help for this program.

Options dialog box

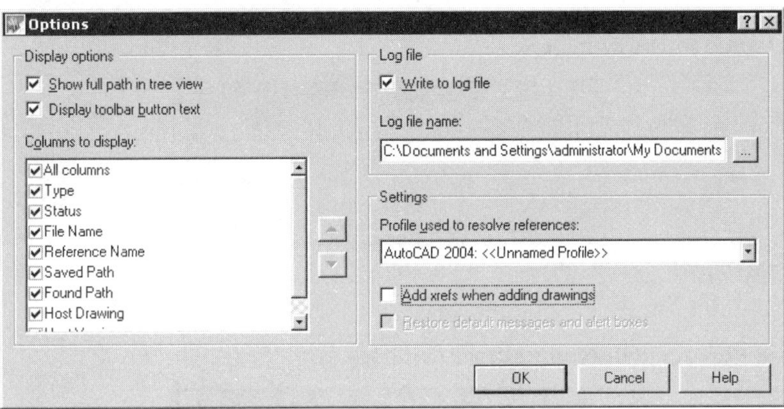

Display Options options

☑ **Show full path in tree view** determines whether the full path is displayed or just the file name.

☑ **Display toolbar button text** toggles the display of text next to toolbar buttons.

Columns to display determines which data is displayed for each drawing.

Log File options

☑ **Write to log file** determines whether the results of this session are written to a *.txt* file.

Log file name specifies the name of the log file.

... displays the Browse for Folder dialog box.

Settings options

Profile used to resolve references selects a profile (created with the Profile tab in AutoCAD's **Options** command), which specifies paths.

☐ **Add xrefs when adding drawings** loads all nested xref drawing files.

☐ **Restore default messages and alert boxes** turns on all dialog boxes that have been turned off with the "Don't display this message again" option.

Add Xrefs dialog box

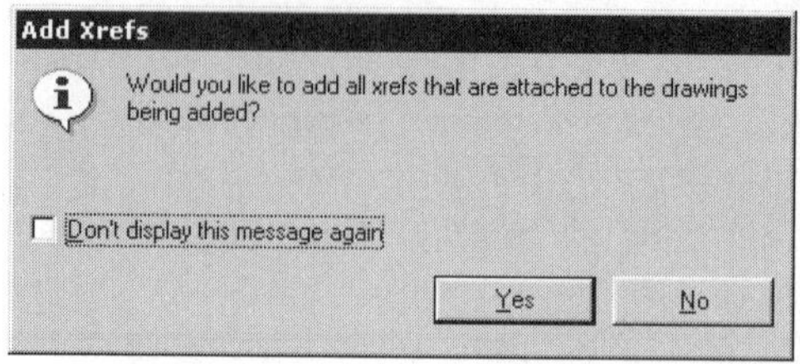

☐ **Don't display this message again** suppresses further displays of this dialog box.

Yes loads all externally-referenced drawings (recommended).

No does not load xrefs.

Edit Selected Paths dialog box

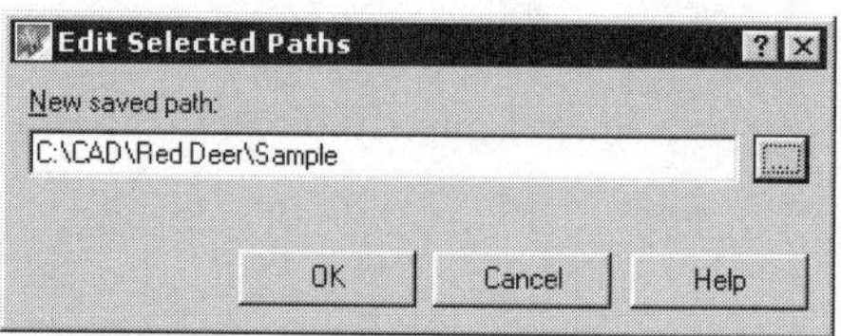

New saved path specifies the path to the object.

... displays the Browse for Folder dialog box.

Find and Replace Selected Paths dialog box

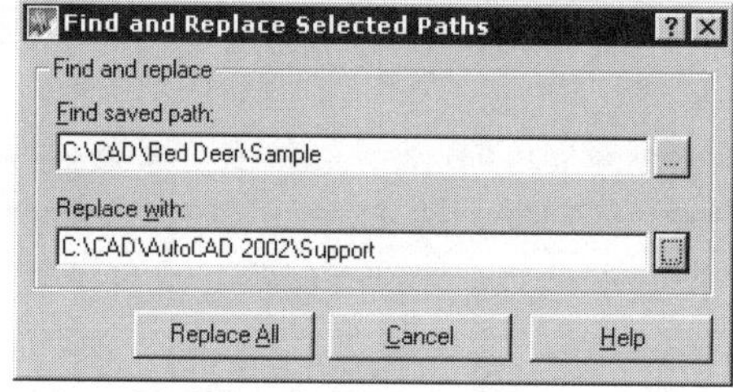

Find saved path specifies the current path to the object.

Replace with specifies the new path to the object.

... displays the Browse for Folder dialog box.

Replace All replaces the old path with the new path.

Summary dialog box

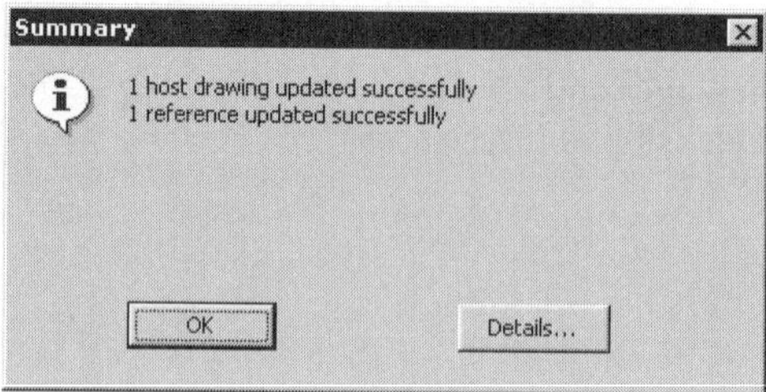

OK dismisses dialog box.

Details displays the Details dialog box:

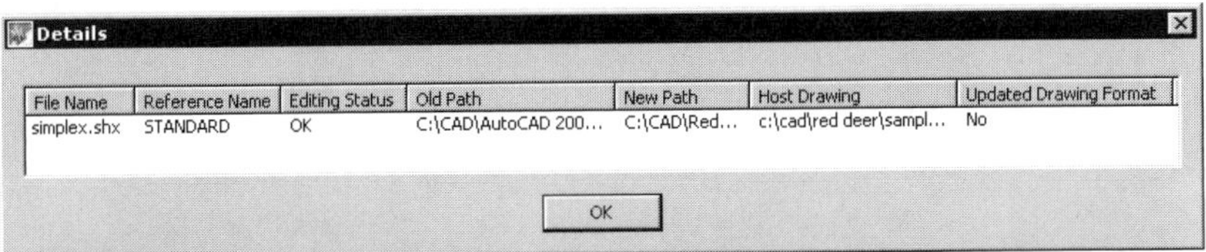

RELATED AUTOCAD COMMANDS

Image attaches raster images to drawings.

Options specifies the default paths to external files.

Style defines the look of *.shx* and *.ttf* fonts.

Shapes loads *.shx* shape files into drawings.

XRef attaches drawings as external references.

TIPS

- When drawings are moved, they may need to reference different folders and disk drives, which means that saved paths need to be updated. This program allows you to modify the paths without opening each drawing file in AutoCAD.
- This program does not work with the following references:

  Text fonts not associated with a text style.

  OLE links.

  Hyperlinks.

  Database file links.

  Xrefs linked to URLs on the Web.
- This program cannot work with drawings open in AutoCAD, or with file attributes set to read-only.

BatchPlot.exe

Launches AutoCAD to batch plot drawings.

 **| BatchPlot.exe**

Displays program:

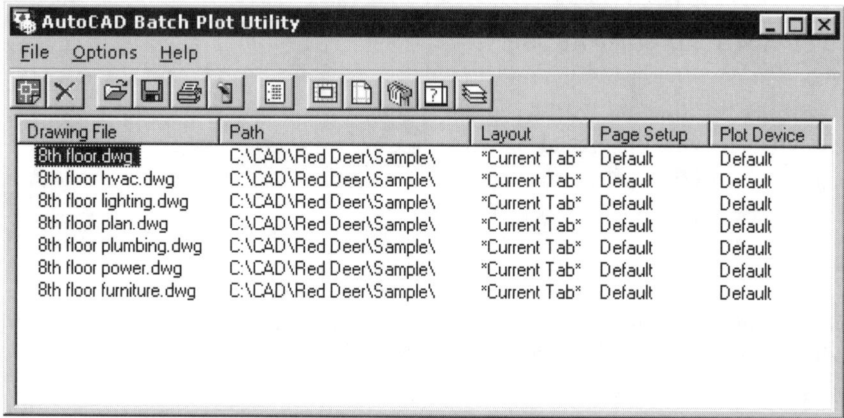

MENU BAR & TOOLBAR OPTIONS

File menu

 **Add Drawing** adds drawings to be plotted; displays the Add Drawing File dialog box.

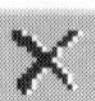 **Remove** removes selected drawings from the list.

New List clears drawings from the list.

 **Open List** (CTRTL+O) opens *.bp3* files, which specify lists of drawings to plot.

Append List prompts you to select additional *.bp3* files, which adds drawings to the list.

 **Save List** (CTRTL+S) saves the list of drawings in *.bp3* files; displays the Save Batch Plot List File dialog box.

 **Plot** plots the drawings using AutoCAD.

 **Plot Test** loads (but does not plot) drawings to check for missing xrefs, fonts, and fonts.

 **Logging** displays the Logging dialog box.

Options menu

Layouts displays the Layouts dialog box.

Page Setups displays the Page Setups dialog box.

Plot Devices displays the Plot Devices dialog box.

Plot Settings displays the Plot Settings tab of the Plot Settings dialog box.

Layers displays the Layers tab of the Plot Settings dialog box.

Help menu

Help displays online help for this program.

About displays the About Batch Plot dialog box.

Logging dialog box

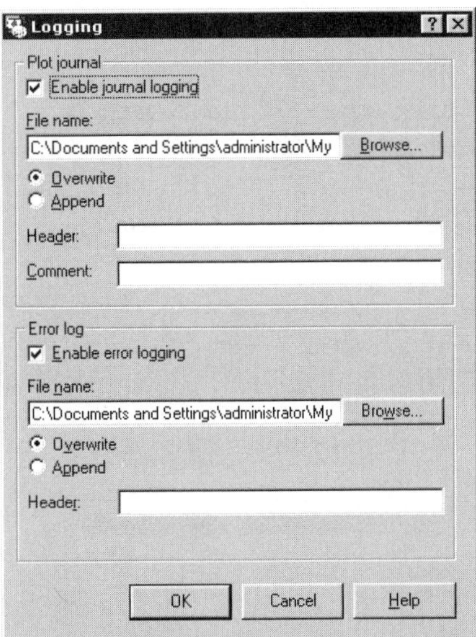

Plot Journal options

☑ **Enable Journal Logging** turns on plot logging.

File Name specifies the name of the log file.

⊙ **Overwrite** overwrites the previous plot file.

○ **Append** adds to the end of an existing plot file.

Header specifies a line of text to be added to the beginning of the file.

Comment specifies additional text to be added to the beginning of the file.

Error Log options

☑**Enable Error Logging** turns on error logging; records problems that occur during batch plotting.

File Name specifies the name of the error log file.

⊙**Overwrite** overwrites the previous error log file.

○**Append** adds error message to the end of a the error log file.

Header specifies a line of text to be added to the beginning of the file.

Layouts dialog box

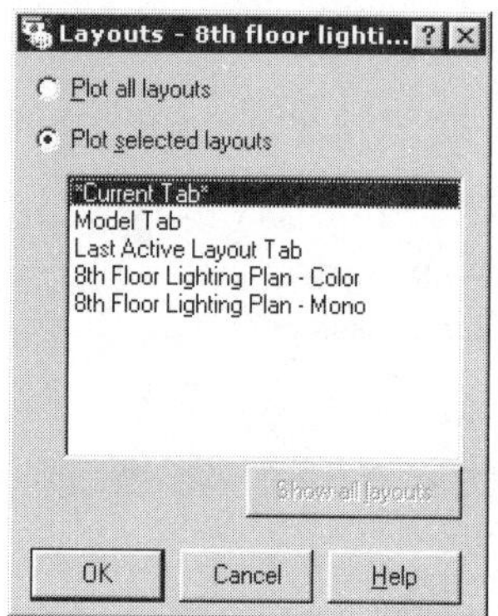

○**Plot All Layouts** instructs AutoCAD to plot all the layouts in the selected drawing.

⊙**Plot Selected Layouts** instructs AutoCAD to plot only selected layouts, as listed below.

- ***Current Tab*** plots the tab current when the drawing was saved.
- **Model Tab** plots the model tab.
- **Last Active Layout Tab** plots the last layout tab active (other than a model tab).

Show All Layouts shows all layout names in the drawing.

Page Setups dialog box

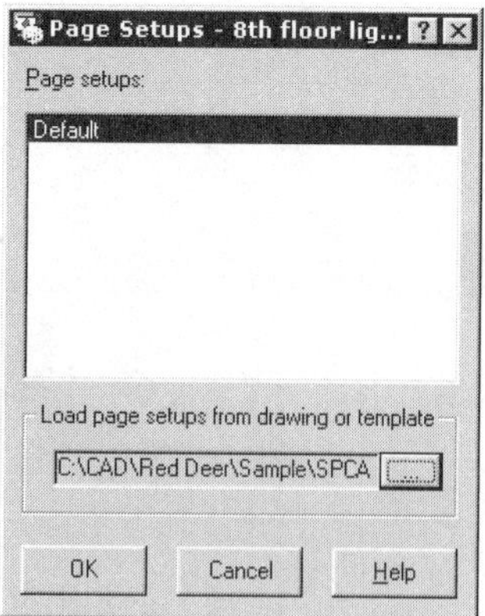

Page Setups lists the page setups in the selected drawing.

Load Page Setups from Drawing or Template options

... displays the Select Drawing dialog box.

Plot Devices dialog box

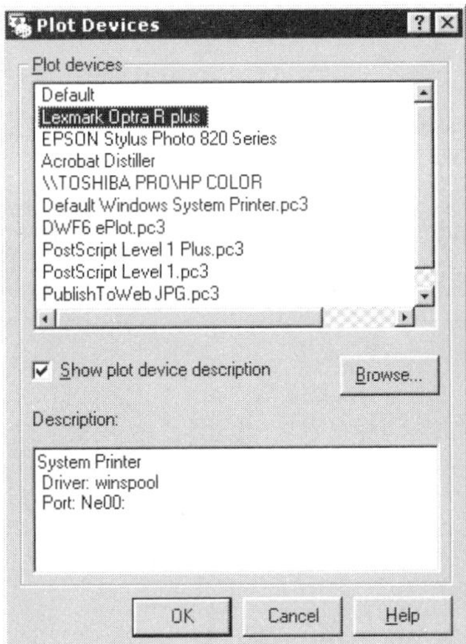

Plot Devices lists the printers and plotters supported by Windows and AutoCAD.

☑**Show Plot Device Description** displays a description of the selected device.

Browse displays the Open Plot Configuration File dialog box; locate a plotter configuration file, and then click **Open**.

Plot Settings dialog box

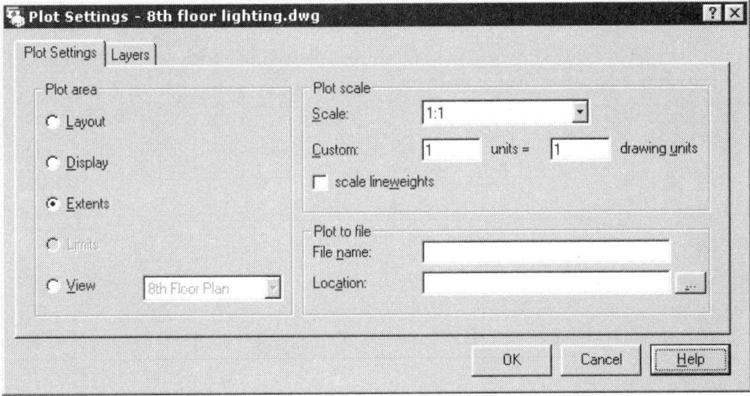

Plot Settings tab

Plot Area options

○**Layout** plots a specified layout.

○**Display** plots the view stored when the drawing was last saved.

⊙**Extents** plots the drawing extents.

○**Limits** plots the drawing limits.

○**Plots** the named view.

Plot Scale options

Plot Scale selects a scale factor; layout tabs default to 1:1 scale, while model tabs default to Scaled to Fit.

Custom allows you to set your own scale factor.

☐**Scale Lineweights** makes lineweights thinner or thicker, according to the scale factor.

Plot to File options

File Name specifies the name of the plot file.

Location selects the path and drive for the plot file.

Layers tab

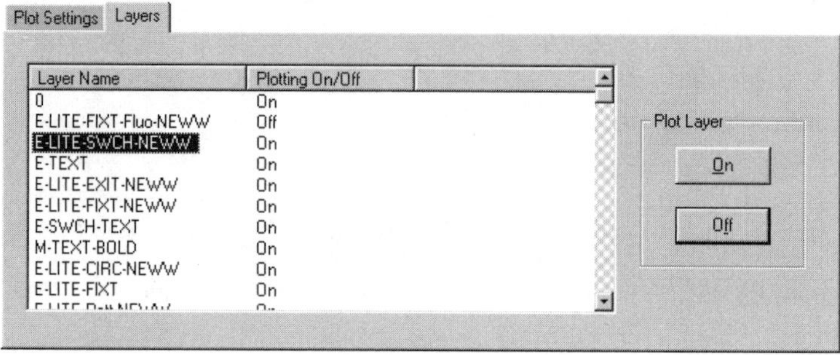

Plot Layer options

On turns on the selected layer for plotting.

Off turns off the selected layer for plotting.

TIP

- After you select **Plot**, each drawing is loaded into AutoCAD for plotting. When the drawing plots successfully, a check mark is displayed next to the drawing name. When the drawing fails to plot, an "x" is displayed.

DwgCheckStandards.exe

Checks if drawing elements meet prescribed standards (*short for DraWinG CHECK STANDARDS*).

Start | Programs | Autodesk | AutoCAD 2005 | Batch Standards Checker

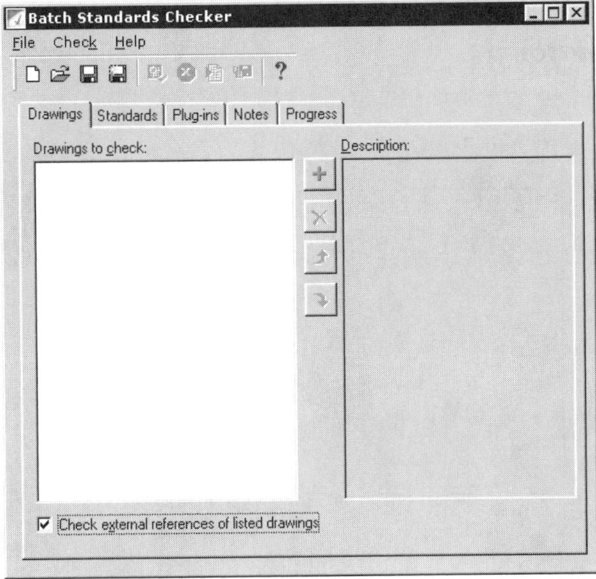

MENU BAR & TOOLBAR OPTIONS

File menu

New Check File (CTRL+N) clears the settings.

Open Check File (CTRL+O) displays the File Open dialog box; select *.chx* file, click **Open**.

Save Check File (CTRL+S) saves the parameters of this program as *.chx* files.

Save As (ALT+S) displays the File Save dialog box; name the file, and click **Save**.

Exit (ALT+F4) exits the program.

Check menu

Start Check (ALT+T) examines the drawing for violations of CAD standards.

Stop Check (ALT+P) halts the standards checking process.

 **View Report** (ALT+V) displays reports in Web browsers.

 **Export Report** (ALT+E) saves reports in HTML format.

Help menu

Help (F1) displays online help for this command.

About displays the About Batch Standards Checker dialog box.

DIALOG BOX OPTIONS

Drawings tab

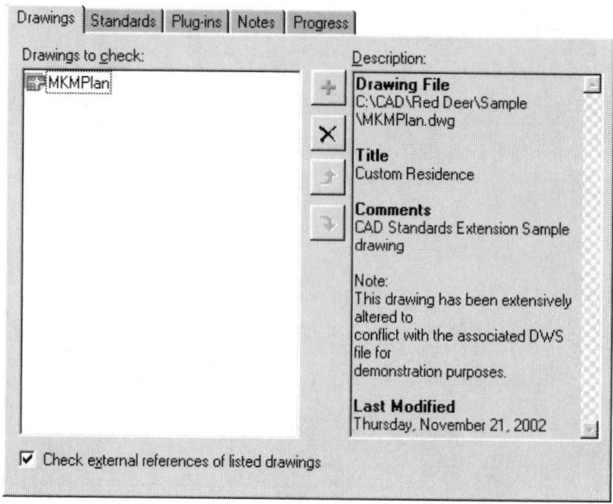

Add Drawing (F3) opens additional drawing files; displays the File Open dialog box.

Remove Drawing (DEL) removes selected drawings from the list.

Move Up (F4) moves the selected drawings higher.

Move Down (F5) moves the selected drawings lower. Drawings are checked in the order in which they appear.

☑ **Check external references of listed drawings** causes this program to check the standards of all attached xrefs. Drawings that cannot be located have an exclamation (!) prefix.

Standards tab

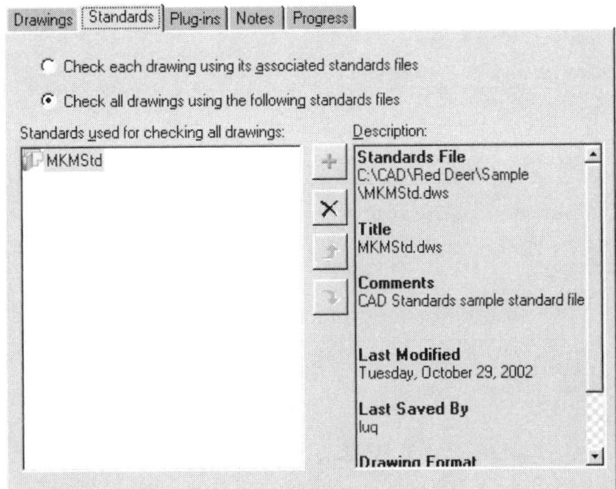

○ **Check each drawing using its associated standards files** checks each drawing against its associated standard.

⊙ **Check all drawings using the following standards files** checks all drawings against a single standard.

Add Standards File (F3) adds additional *.dws* standards list to the list.

Remove Standards File (DEL) removes selected standards from the list.

Move Up (F4) moves the selected standards higher.

Move Down (F5) moves the selected standards lower.

When two or more standards are loaded, sometimes they conflict. In this case, the settings in the higher standards take precedence over lower standards.

Plug-ins tab

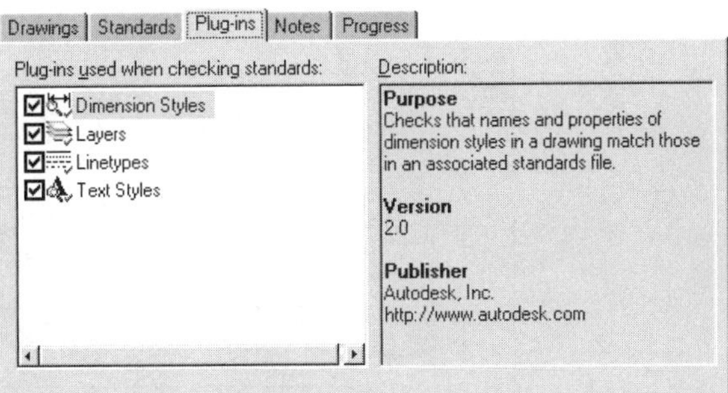

Plug-ins used when checking standards selects the styles to check.

☑ Style is checked.

☐ Style is not checked.

Notes tab

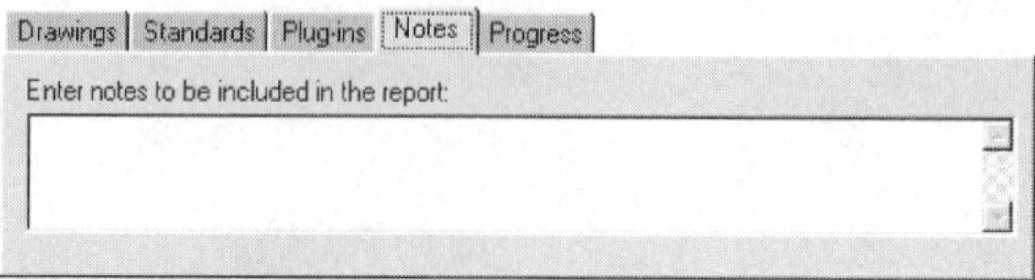

Enter notes to be included in the report provides an area for you to enter text.

Progress tab

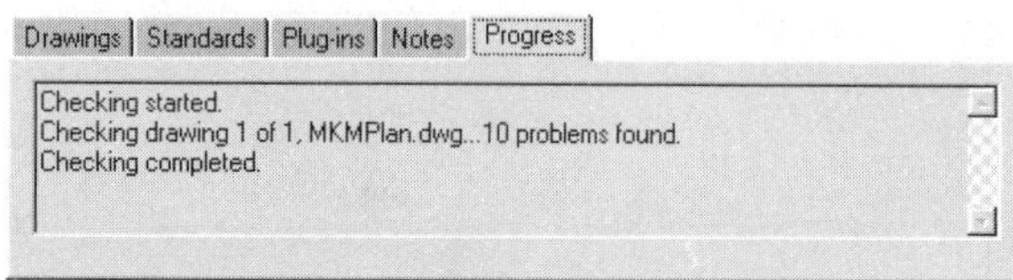

Displays the progress of the standards checking.

RELATED AUTOCAD COMMANDS

CheckStandards checks whether drawings match CAD standards.

Standards applies CAD standards to drawings.

RELATED FILES

.*chx are CAD standards files stored in XML format.

.*dws are CAD Standards files, stored in DWG format.

TIP

- This program produces reports in HTML format, which can be viewed by Web browsers:

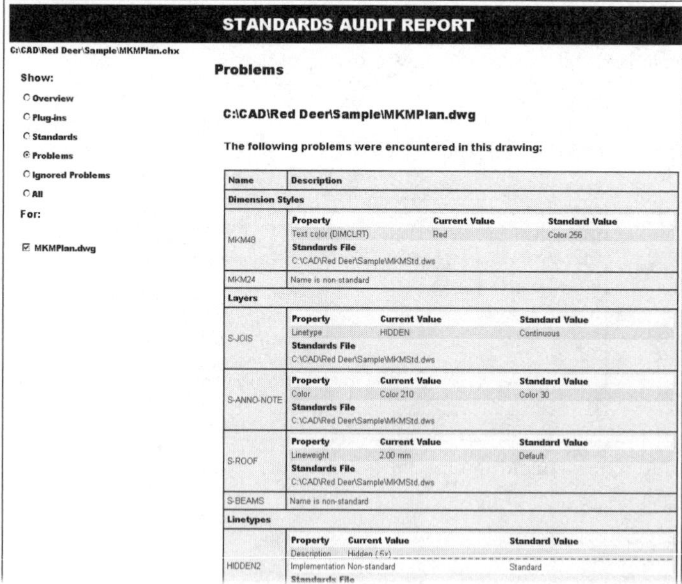

DwfViewer.exe

Displays and plots *.dwf* files.

Start | Programs | Autodesk | Autodesk DWF Viewer

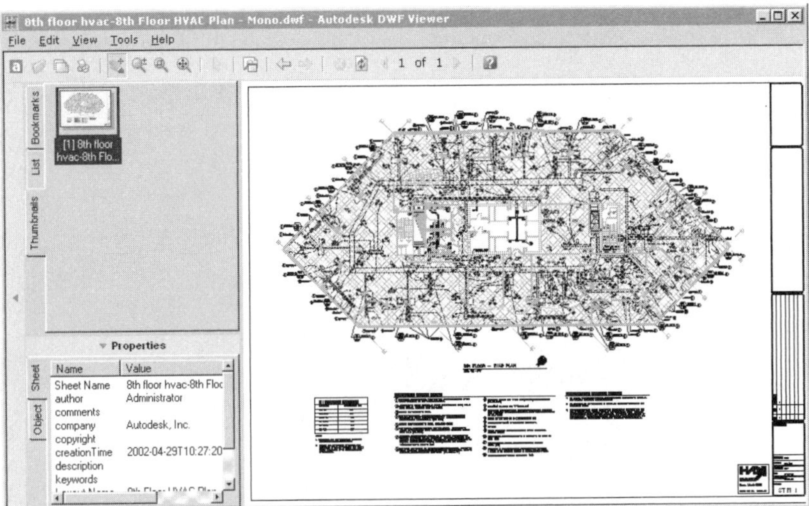

MENU BAR & TOOLBAR OPTIONS

File menu

Open (CTRL+O) opens *.dwf* files; displays the Open File dialog box.

 **Print** (CTRL+P) displays the Print dialog box.

Exit (ALT+F4) exits this program.

Edit menu

 **Copy** (CTRL+C) copies the image to the Clipboard as a bitmap.

View menu

 **Layers** (l) displays the Layers window; allows layers to be turned on and off.

Sheets (S) displays the Sheets windows; allows a specific layout to be displayed.

Views (V) displays the Views window; allows a specific view to be displayed.

Show displays additional viewing options:

- **Hyperlinks** (CTRL+H) toggles the display of hyperlinks embedded in the drawing.
- **Page Tiles** (CTRL+T) illustrates sheets of paper, when Fit to Paper Size option is turned off in the Print dialog box.
- **Paper Background** (CTRL+B) toggles the background color between white and gray.
- **Toolbar** toggles the display of the toolbar.

Tools menu

Pan (*arrow keys*) moves the view.

Zoom (+ *and* -) enlarges and reduces the drawing.

Zoom Rectangle (CTRL+R) enlarges a windowed area of the drawing.

Fit in Window (HOME) changes the view of the drawing to fit to the window.

Options displays the Options dialog box; select colors, and then click **OK**.

Help menu

Contents (F1) displays online help for this program.

Check for Viewer Updates requires an Internet connection.

About displays About Autodesk DWF Viewer dialog box.

Options dialog box

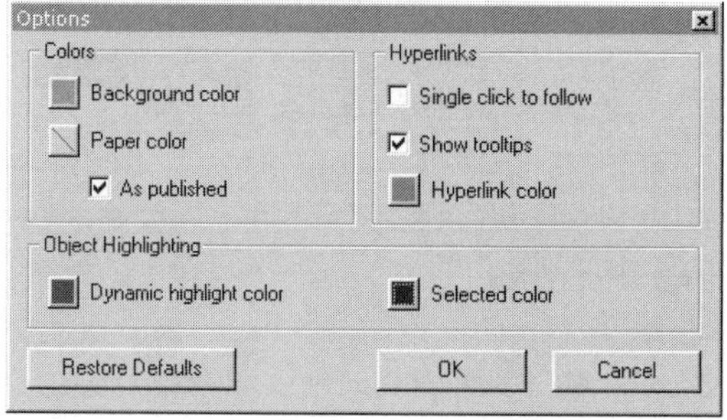

Color options

Background color specifies the color of the background behind the paper.

Paper color specifies the color of the "paper."

☑ **As published** displays the same color as specified by AutoCAD.

Hyperlink options

http://www.wware.com
CTRL + click to follow link

☐ **Single click to follow** activates hyperlink with a single click; otherwise, hold down CTRL key.

☑ **Show tooltips** shows tooltip with hyperlink info when cursor pauses over link.

Hyperlink color specifies color of objects containing hyperlinks.

Object Highlighting options

Dynamic highlight color specifies color of highlighted objects.

Selected color specifies color of selected objects.

Restore Defaults returns colors and settings to their original values.

Print dialog box

Printer options

Name selects a Windows system printer; Express view does not support AutoCAD plotter drivers.

Properties displays the Windows printer Properties dialog box.

Print What options

⊙ **Full page** prints the entire drawing.

○ **Current View** prints the zoomed in view.

Force all geometry to black prints all elements in black; useful for monochrome printers.

Paper Size & Orientation options

Paper Size selects the size of paper from those supported by the selected printer.

Portrait orients the page vertically.

Landscape orients the page horizontally.

Reduce/Enlarge Drawing options

Fit to Paper Size

☑ fits the drawing to the paper size; ignores scale.

☐ tiles the print over several pages.

Print Drawing to Scale plots the drawing to a scale relative to its full size.

Show tiles in page view displays the viewer with dashed blue lines indicating the margins of pages.

Print Range options

All prints all sheets in the drawing.

Current page only prints the currently-displayed page only.

Sheets prints a range of sheets.

Copies options

Number of copies prints one or more copies.

Reverse order prints multiple sheets in reverse order.

Collate prints multiple sheets together.

Layers window

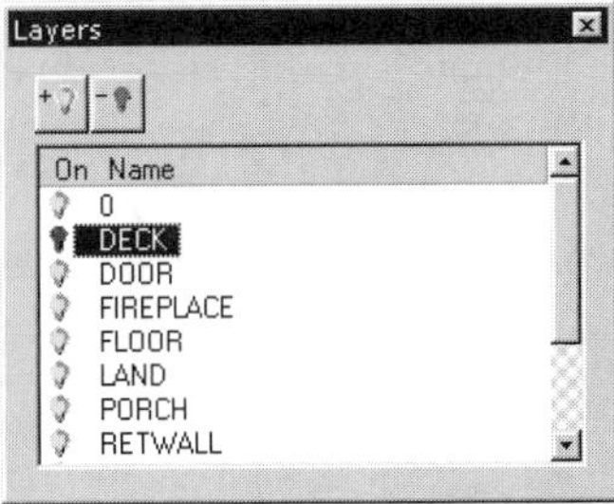

On turns on the selected layer.

Off turns off the selected layer.

x dismisses the window.

Sheets window

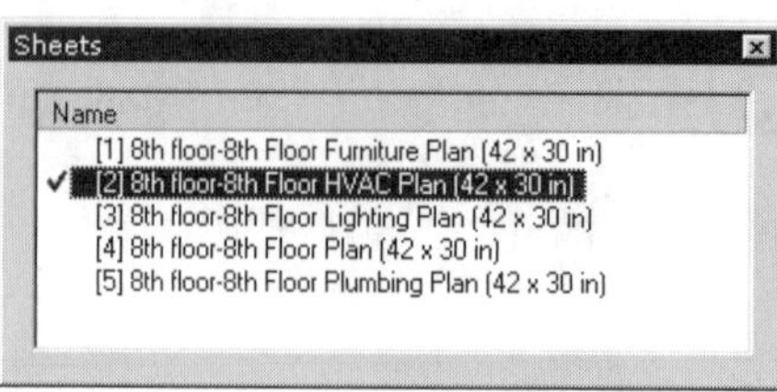

✓ Displays the selected sheet.

x Dismisses the window.

Displays the previous sheet (layout).

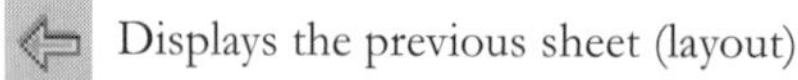

Selects the name of a sheet.

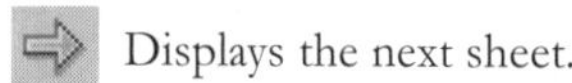

Displays the next sheet.

Stops display of a sheet.

Views window

✓ Displays the selected view.

x Dismisses the window.

RELATED AUTOCAD COMMANDS

Publish creates multi-sheet *.dwf* files.

TIPS

- DWF Viewer displays (and prints) *.dwf* files only; it doesn't handle *.dwg* or *.dxf* drawing files.
- In the Print dialog box:

  Reverse order option is useful when your printer produces pages face up, meaning they are printed in reverse order. Turn on this option to reverse the reverse order, producing a print set in correct order.

  Collate option is useful when you have more than one set of drawings to produce. Turn on this option to print sheets together; the drawback is that printing takes longer.
- When **Page Tiles** is turned on, blue dashed lines show the edges of paper.
- Use a wheelmouse to zoom in and out by rolling the wheel forward and backward; to pan, hold down the wheel while moving the mouse.
- DWF Viewer was known as "Express Viewer" prior to AutoCAD 2005.

SendDmp.exe

Sends error reports to Autodesk (*short for SEND DuMP.*)

AutoCAD | **SendDmp.exe**

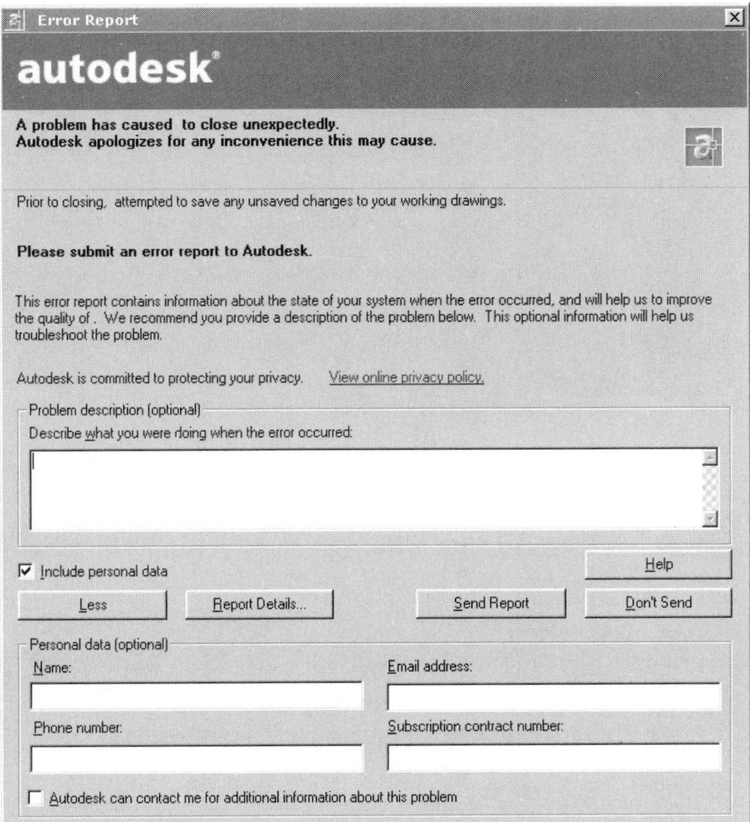

DIALOG BOX OPTIONS

Problem description describes what happened when AutoCAD crashed.

☑ **Include personal data** includes personal information, such as you name and phone number.

Less reduces the size of the dialog box.

Report details displays the Reporting Details dialog box; select a file name, and then click **View file contents.**

Send Report sends the report via email.

Don't Send cancels the report.

TIPS

- This program runs automatically whenever AutoCAD crashes.
- Autodesk receives a file called *dmpuserinfo.xml*, which contains the crash data.

StyShWiz.exe

Runs the Add-a-Plot Style Table wizard (*short for STYle SHeet WIZard.)*

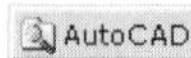

| StyShWiz.exe

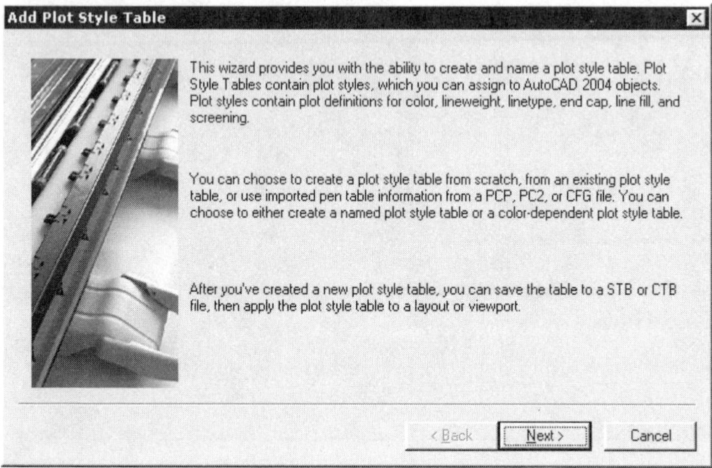

DIALOG BOX OPTIONS

Next goes to the next dialog box.

Cancel cancels the program.

RELATED AUTOCAD COMMANDS

StylesManager opens the Plot Styles window, which includes the Add-a-Plot Style Table wizard.

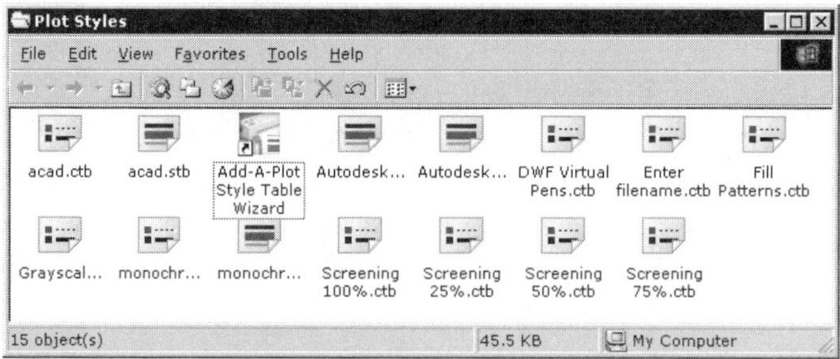

RELATED FILES

***.ctb** are AutoCAD color-table based files.

***.stb** are AutoCAD style-table based files.

TIPS

- See the **StylesManager** command for details on using this program.
- You can create and edit plotter style files without AutoCAD. From the **Start** button on the Windows toolbar, select **Settings | Control Panel | Autodesk Plot Style Manager**.

APPENDIX C

Obsolete & Removed Commands

The following commands have been removed from AutoCAD since v 2.5.

Command	Introduced	Removed	Replacement	Reaction
3Dline	R9	R11	Line	"Line"
AmeLite	R11	R12	Region	"Unknown command"
AscText	R11	R13	MText	"Unknown command"
Ase...	R12	R13	ASE...	"Unknown command"
	(Most R12 ASE commands were combined into ASE commands with R13.)			
Ase...	R13	2000	dbConnect	"Unknown command"
AseUnload	R12	R14	Arx Unload	"Unknown command"
Axis	v1.4	R12	*none*	"Discontinued command"
BMake	R12	2000	Block	Displays **Block Definition** dialog.
CConfig	R13	2000	PlotStyle	"Discontinued command"
Config	R12	R14	Options	Displays **Options** dialog.
DdAttDef	R12	2000	AttDef	Displays **Attribute Def** dialog.
DdAttE	R9	2000	AttEdit	Displays **Edit Attributes** dialog.
DdAttExt	R12	2000	AttExt	Displays **Attribute Ext** dialog.
DdChProp	R12	2000	Properties	Displays **Properties** window.
DdColor	R13	2000	Color	Displays **Select Color** dialog.
DdEModes	R9	R14	Object Properties	"Discontinued command"
DdGrips	R12	2000	Options	Displays **Options** dialog.
DDim	R12	2000	DimStyle	Displays **Dim Style Mgr** dialog.
DdInsert	R12	2000	Insert	Displays **Insert** dialog.
DdLModes	R9	R14	Layer	Displays **Layer Manager** dialog.
DdLType	R9	R14	Linetype	Displays **Linetype Mgr** dialog.
DdModify	R12	2000	Properties	Displays **Properties** window.
DdOSnap	R12	2000	DSettings	Displays **Drafting Settings** dialog.
DdRename	R12	2000	Rename	"Unknown command"
DdRModes	R9	2000	DSettings	Displays **Drafting Settings** dialog.
DdSelect	R12	2000	Options	Displays **Options** dialog.
DdUcs	R10	2000	UcsMan	Displays **UCS** dialog.
DdUcsP	R12	2000	UcsMan	Displays **UCS** dialog.
DdUnits	R12	2000	Units	Displays **Units** dialog.
DdView	R12	2000	View	Displays **View** dialog.
DText	v2.5	2000	Text	Executes **Text** command.
DL, DLine	R11	R13	MLine	"Unknown command"
DwfOut	R14	2004	Publish	Executes **Plot** command.
DwfOutD	R14	2000	DwfOut	"Unknown command"

Command	Introduced	Removed	Replacement	Reaction *(con't)*
End	R11	R13	Quit	"Discontinued command"
EndRep	v1.0	v2.5	Minsert	"Discontinued command"
EndSv	v2.0	v2.5	End	"Discontinued command"
EndToday	2000i	2004	*none*	"Unknown command"
ExpressTools	2000	2002	ExpressTools	Restored in AutoCAD 2004.
Files	v1.4	R14	*Explorer*	"Discontinued command"
FilmRoll	v2.6	R13	*none*	"Unknown command"
FlatLand	R10	R11	*none*	"Cannot set Flatland to that value"
GifIn	R12	R14	ImageAttach	"No longer supported"
HpConfig	R12	2000	PlotStyle	"Discontinued command"
IgesIn, IgesOut	v2.5	R13	*none*	"Discontinued command"
InetCfg	R14	2000	*none*	"Unknown command"
InetHelp	R14	2000	Help	"Unknown command"
InsertUrl	R14	2000	Insert	Displays **Insert** dialog.
ListUrl	R14	2000	QSelect	"Unknown command"
MakePreview	R13	R14	RasterPreview	"Discontinued command"
MeetNow	2000i	2004	*none*	"Unkown command"
OceConfig	R13	2000	PlotStyle	"Discontinued command"
OpenUrl	R14	2000	Open	Displays **Select File** dialog.
OSnap	v2.0	2000	DSettings	Displays **Drafting Settings** dialog.
PcxIn	R12	R14	ImageAttach	"No longer supported"
PsDrag	R12	2000i	*none*	"Unknown command"
PsIn	R12	2000i	*none*	"Unknown command"
Preferences	R11	2000	Options	Displays Options dialog.
PrPlot	v2.1	R12	Plot	"Discontinued command."
QPlot	v1.1	v2.0	SaveImg	"Unknown command"
RConfig	R12	R14	*none*	"Unknown command"
RenderUnload	R12	R14	Arx Unload	"Unknown command"
Repeat	v1.0	v2.5	Minsert	"Discontinued command"
SaveAsR12	R13	R14	SaveAs	"Unknown command"
SaveUrl	R14	2000	SaveAs	Displays **Save Drawing As** dialog.
Snapshot	v2.0	v2.1	Saveimg	"Unknown command"
Sol...	R11	R13	*(AME commands lost their SOL-prefix.)*	
TbConfig	R12	2000i	Customize	Displays **Customize** dialog box.
TiffIn	R12	R14	ImageAttach	"No longer supported"
Today	2000i	2004	*none*	"Unknown command"
Toolbar	R13	2000i	Customize	Displays **Customize** dialog box.
VlConv	R13	R14	3dsIn & 3dsOut	"Unknown command"

APPENDIX D

System Variables

AutoCAD stores information about its current state, the drawing, and the operating system in over 400 *system variables*. These variables help users and programmers — who often work with macros and AutoLISP — determine the state of the AutoCAD system.

CONVENTIONS

The following pages list all documented system variables, plus several more not documented by Autodesk. The listing uses the following conventions:

Bold	System variable is documented in AutoCAD 2005.
Italicized	System variable is not listed by the **SetVar** command or Autodesk documentation.
~~*Strikethru Italic*~~	System variable was removed from AutoCAD.
⌨	System variable must be accessed via the **SetVar** command.
▣	System variable is new to AutoCAD 2005.

COLUMN HEADINGS

Default	Default value, as set in the *acad.dwg* prototype drawing.
R/O	Read-only; cannot be changed by the user or by a program.
Loc	Location where the value of the system variable is saved:

Location	Meaning
ACAD	Set by AutoCAD.
DWG	Saved in current drawing.
REG	Saved in Windows registry for the logged-in user.
...	Not saved.

TIPS

- The **SetVar** command lets you change the value of all variables, except those marked read-only (R/O).
- You can get a list of most system variables at the 'Command:' prompt with the **?** option of the **SetVar** command, as follows:

  Command: setvar
  Variable name or ?: ?
  Variable(s) to list <*>: *(Press* ENTER.*)*

- When a system variable is:

  Stored in the Windows registry, it affects all drawings.
  Stored in the drawing, it affects the current drawing only.
  Not stored, the variable is set when AutoCAD loads. The value of the variable is either read from the operating system, or set to a default value.

Variable	Default	R/O	Loc	Meaning
~~*_LInfo*~~				*Removed from AutoCAD 2004.*
_PkSer	*varies*	R/O	*ACAD*	*Software serial number, such as "117-69999999".*
_Server	*0*	R/O	REG	*Network authorization code.*
_VerNum	*varies*	R/O	REG	*Internal program build number, such as "N.41.101".*

A

Variable	Default	R/O	Loc	Meaning
AcadLspAsDoc	0	...	REG	*acad.lsp* is loaded into: **0** Just the first drawing. **1** Every drawing.
AcadPrefix	*varies*	R/O	...	Paths specified by the AutoCAD search path in Options Files tab. May be controlled by the **ACAD** environment variable.
AcadVer	"16.1"	R/O	...	AutoCAD version number.
AcisOutVer	40	R/O	...	ACIS version number, such as 15, 16, 17, 18, 20, 21, 30, 40, or 70.
AcGiDumpMode	0	...	...	Value of 0 or 1.
AdcState	0	R/O	...	Specifies if **DesignCenter** is active.
AFlags	0	...	...	Attribute display code: **0** No mode specified. **1** Invisible. **2** Constant. **4** Verify. **8** Preset.
AngBase	0	...	DWG	Direction of zero degrees relative to UCS.
AngDir	0	...	DWG	Rotation of positive angles: **0** Clockwise. **1** Counterclockwise.
ApBox	0	...	REG	AutoSnap aperture box cursor: **0** Off. **1** On.
Aperture	10	...	REG	Object snap aperture in pixels: **1** Minimum size. **50** Maximum size.
AssistState	0	R/O	...	Specifies if **Info Palette** is active.
Area	0.0	R/O	...	Area measured by the last **Area**, **List**, or **Dblist** commands.
AttDia	0	...	DWG	Attribute entry interface: **0** Command-line prompts. **1** Dialog box.
AttMode	1	...	DWG	Display of attributes: **0** Off. **1** Normal. **2** On.
AttReq	1	...	REG	Attribute values during insertion are: **0** Default values. **1** Prompt for values.

Variable	Default	R/O	Loc	Meaning
AuditCtl	0	...	REG	Determines creation of *.adt* audit log file: **0** File not created. **1** *.adt* file created.
AUnits	0	...	DWG	Mode of angular units: **0** Decimal degrees. **1** Degrees-minutes-seconds. **2** Grads. **3** Radians. **4** Surveyor's units.
AUPrec	0	...	DWG	Decimal places displayed by angles.
AutoSnap	63	...	REG	Controls AutoSnap display: **0** Turns off all AutoSnap features. **1** Turns on marker. **2** Turns on SnapTip. **4** Turns on magnetic cursor. **8** Turns on polar tracking . **16** Turns on object snap tracking . **32** Turns on tooltips for polar tracking and object snap tracking .
AuxStat	*0*	...	DWG	*-32768 Minimum value.* *32767 Maximum value.*
~~*AxisMode*~~	*0*	...	DWG	*Removed from AutoCAD 2002.*
AxisUnit	*0.0*	...	DWG	*Obsolete system variable.*

B

Variable	Default	R/O	Loc	Meaning
BackZ	0.0	R/O	DWG	Back clipping plane offset.
BackgroundPlot	2	...	REG	Controls background plotting and publishing (ignored during scripts): **0** Plot foreground; publish foreground. **1** Plot background; publish foreground. **2** Plot foreground; publish background. **3** Plot background; publish background.
BgrdPlotTimeout	*20*	...	...	*Background plot timeout; ranges from 0 to 300 secs.*
BindType	0	...	...	When binding an xref or editing an xref, xref names are converted from: **0** **xref|name** to **xref0name**. **1** **xref|name** to **name**.
BlipMode	0	...	DWG	Display of blip marks: **0** Off. **1** On.

C

Variable	Default	R/O	Loc	Meaning
CDate	*varies*	R/O	...	Current date and time in the format YyyyMmDd.HhMmSsDd, such as 20010503.18082328

Variable	Default	R/O	Loc	Meaning
CeColor	"BYLAYER"	...	DWG	Current color.
CeLtScale	1.0	...	DWG	Current linetype scaling factor.
CeLType	"BYLAYER"	...	DWG	Current linetype.
CeLWeight	-1	...	DWG	Current lineweight in millimeters; valid values are 0, 5, 9, 13, 15, 18, 20, 25, 30, 35, 40, 50, 53, 60, 70, 80, 90, 100, 106, 120, 140, 158, 200, and 211, plus the following: **-1** BYLAYER. **-2** BYBLOCK. **-3** DEFAULT as defined by **LwDdefault**.
ChamferA	0.5	...	DWG	First chamfer distance.
ChamferB	0.5	...	DWG	Second chamfer distance.
ChamferC	1.0	...	DWG	Chamfer length.
ChamferD	0	...	DWG	Chamfer angle.
ChamMode	0	...	...	Chamfer input mode: **0** Chamfer by two lengths. **1** Chamfer by length and angle.
CircleRad	0.0	...	...	Most-recent circle radius.
CLayer	"0"	...	DWG	Current layer name.
CleanScreenState	0	R/O	...	Specifies if cleanscreen mode is active.
CmdActive	1	R/O	...	Type of current command: **1** Regular command. **2** Transparent command. **4** Script file. **8** Dialog box. **16** AutoLISP is active .
CmdDia	*1*	...	*REG*	*Replaced by **PlQuiet** in AutoCAD 2000.*
CmdEcho	1	...	...	AutoLISP command display: **0** No command echoing. **1** Command echoing.
CmdNames	*varies*	R/O	...	Current command, such as "SETVAR".
CMLJust	0	...	DWG	Multiline justification mode: **0** Top. **1** Middle. **2** Bottom.
CMLScale	1.0	...	DWG	Scales width of multiline: **-*n*** Flips offsets of multiline. **0** Collapses to single line. *n* Scales by a factor of *n*.
CMLStyle	"STANDARD"	...	DWG	Current multiline style name.
Compass	0	...	...	Toggles display of the 3D compass: **0** Off. **1** On.

Variable	Default	R/O	Loc	Meaning
Coords	1	...	DWG	Coordinate display style: **0** Updated by screen picks. **1** Continuous display. **2** Polar display upon request.
CPlotStyle	"ByColor"	...	DWG	Current plot style; values defined by AutoCAD are: "ByLayer" "ByBlock" "Normal" "User Defined"
CProfile	"<<Unnamed Profile>>"	R/O	REG	Current profile.
CpuTicks	*592020023071334.1*	R/O	...	*Number of CPU ticks.*
CTab	"Model"	R/O	DWG	Current tab.
~~*CurrentProfile*~~	"*<<Unnamed Profile>>*"	...	...	*Removed from AutoCAD 2000; replaced by **CProfile**.*
CursorSize	5	...	REG	Cursor size, in percent of viewport: **1** Minimum size. **100** Full viewport.
CVPort	2	...	DWG	Current viewport number.

D

Variable	Default	R/O	Loc	Meaning
Date	*varies*	R/O	...	Current date in Julian format, such as 2448860.54043252
~~*DBGListAll*~~	*0*	...	*ACAD*	*Removed from AutoCAD 2002.*
DBMod	4	R/O	...	Drawing modified, as follows: **0** No modification since last save. **1** Object database modified. **2** Symbol table modified. **4** Database variable modified. **8** Window modified. **16** View modified.
DbcState	0	R/O	DWG	Specifies if **dbConnect Manager** is active.
DctCust	"d:\acad 2005\support\sample.cus"	...	REG	Name of custom spelling dictionary.
DctMain	"enu"	...	REG	Code for spelling dictionary: **ca** Catalan. **cs** Czech. **da** Danish. **de** German; sharp 's'. **ded** German; double 's'. **ena** English; Australian. **ens** English; British 'ise'. **enu** English; American. **enz** English; British 'ize'. **es** Spanish; unaccented capitals. **esa** Spanish; accented capitals. **fi** Finish. **fr** French; unaccented capitals.

Variable	Default	R/O	Loc	Meaning
				fra French; accented capitals. **it** Italian. **nl** Dutch; primary. **nls** Dutch; secondary. **no** Norwegian; Bokmal. **non** Norwegian; Nynorsk. **pt** Portuguese; Iberian. **ptb** Portuguese; Brazilian. **ru** Russian; infrequent 'io'. **rui** Russian; frequent 'io'. **sv** Swedish.
DefaultViewCategory	""	...	...	Default name for View Category in the **View** command's New View dialog box
DefLPlStyle	"ByColor"	R/O	REG	Default plot style for new layers.
DefPlStyle	"ByColor"	R/O	REG	Default plot style for new objects.
DelObj	1	...	REG	Toggle source objects deletion: **0** Objects deleted. **1** Objects retained.
DemandLoad	3	...	REG	When drawing contains proxy objects: **0** Demand loading turned off. **1** Load app when drawing opened. **2** Load app at first command. **3** Load app when drawing opened or at first command.
DiaStat	1	R/O	...	User exited dialog box by clicking: **0** **Cancel** button. **1** **OK** button.

Dimension Variables

Variable	Default	R/O	Loc	Meaning
DimADec	0	...	DWG	Angular dimension precision: **-1** Use **DimDec** setting (default). **0** Zero decimal places (minimum). **8** Eight decimal places (maximum).
DimAlt	Off	...	DWG	Alternate units: **On** Enabled. **Off** Disabled.
DimAltD	2	...	DWG	Alternate unit decimal places.
DimAltF	25.4	...	DWG	Alternate unit scale factor.
DimAltRnd	0.0	...	DWG	Rounding factor of alternate units.
DimAltTD	2	...	DWG	Tolerance alternate unit decimal places.
DimAltTZ	0	...	DWG	Alternate tolerance units zeros: **0** Zeros not suppressed. **1** All zeros suppressed. **2** Include 0 feet, but suppress 0 inches . **3** Includes 0 inches, but suppress 0 feet. **4** Suppresses leading zeros. **8** Suppresses trailing zeros.

Variable	Default	R/O	Loc	Meaning
DimAltU	2	...	DWG	Alternate units: **1** Scientific. **2** Decimal. **3** Engineering. **4** Architectural; stacked. **5** Fractional; stacked. **6** Architectural. **7** Fractional. **8** Windows desktop units setting.
DimAltZ	0		DWG	Zero suppression for alternate units: **0** Suppress 0 ft and 0 in. **1** Include 0 ft and 0 in. **2** Include 0 ft; suppress 0 in. **3** Suppress 0 ft; include 0 in. **4** Suppress leading 0 in dec dims. **8** Suppress trailing 0 in dec dims. **12** Suppress leading and trailing zeroes.
DimAPost	""	...	DWG	Prefix and suffix for alternate text.
DimAso	On	...	DWG	Toggle associative dimensions: **On** Dimensions are created associative. **Off** Dimensions are not associative.
DimAssoc	2	...	DWG	Controls creation of dimensions: **0** Dimension elements are exploded. **1** Single dimension object, attached to defpoints. **2** Single dimension object, attached to geometric objects.
DimASz	0.18	...	DWG	Arrowhead length.
DimAtFit	3	...	DWG	When insufficient space between extension lines, dimension text and arrows are fitted: **0** Text and arrows outside extension lines. **1** Arrows first outside, then text. **2** Text first outside, then arrows. **3** Either text or arrows, whichever fits better.
DimAUnit	0	...	DWG	Angular dimension format: **0** Decimal degrees. **1** Degrees.Minutes.Seconds. **2** Grad. **3** Radian. **4** Surveyor units.
DimAZin	0	...	DWG	Supress zeros in angular dimensions: **0** Display all leading and trailing zeros. **1** Suppress 0 in front of decimal. **2** Suppress trailing zeros behind decimal. **3** Suppress zeros in front and behind the decimal.

Variable	Default	R/O	Loc	Meaning
DimBlk	""	R/O	DWG	Arrowhead block name: Architectural tick: "Archtick" Box filled: "Boxfilled" Box: "Boxblank" Closed blank: "Closedblank" Closed filled: "" (default) Closed: "Closed" Datum triangle filled: "Datumfilled" Datum triangle: "Datumblank" Dot blanked: "Dotblank" Dot small: "Dotsmall" Dot: "Dot" Integral: "Integral" None: "None" Oblique: "Oblique" Open 30: "Open30" Open: "Open" Origin indication: "Origin" Right-angle: "Open90"
DimBlk1	""	R/O	DWG	Name of first arrowhead's block; uses same list of names as under **DimBlk**. . No arrowhead.
DimBlk2	""	R/O	DWG	Name of second arrowhead's block; uses same list of names as under **DimBlk**. . No arrowhead.
DimCen	0.09	...	DWG	Center mark size: **-*n*** Draws center lines. **0** No center mark or lines drawn. +*n* Draws center marks of length *n*.
DimClrD	0	...	DWG	Dimension line color: **0** BYBLOCK (default) **1** Red. ... **255** Dark gray. **256** BYLAYER.
DimClrE	0	...	DWG	Extension line and leader color.
DimClrT	0	...	DWG	Dimension text color.
DimDec	4	...	DWG	Primary tolerance decimal places.
DimDLE	0.0	...	DWG	Dimension line extension.
DimDLI	0.38	...	DWG	Dimension line continuation increment.
DimDSep	"."	...	DWG	Decimal separator (must be a single char.)
DimExe	0.18	...	DWG	Extension above dimension line.
DimExO	0.0625	...	DWG	Extension line origin offset.
~~DimFit~~	*3*	...	*DWG*	*Obsolete: Autodesk recommends use of **DimATfit** and **DimTMove** instead.*

Variable	Default	R/O	Loc	Meaning
DimFrac	0	...	DWG	Fraction format when **DimLUnit** set to 4 or 5: **0** Horizontal. **1** Diagonal. **2** Not stacked.
DimGap	0.09	...	DWG	Gap from dimension line to text.
DimJust	0	...	DWG	Horizontal text positioning: **0** Center justify. **1** Next to first extension line. **2** Next to second extension line. **3** Above first extension line. **4** Above second extension line.
DimLdrBlk	""	...	DWG	Block name for leader arrowhead; uses same name as **DimBlock**. **.** Supresses display of arrowhead.
DimLFac	1.0	...	DWG	Linear unit scale factor.
DimLim	Off	...	DWG	Generate dimension limits.
DimLUnit	2	...	DWG	Dimension units (except angular); replaces **DimUnit**: **1** Scientific. **2** Decimal. **3** Engineering. **4** Architectural. **5** Fractional. **6** Windows desktop.
DimLwD	-2	...	DWG	Dimension line lineweight; valid values are BYLAYER, BYBLOCK, or an integer multiple of 0.01mm.
DimLwE	-2	...	DWG	Extension lineweight; valid values are BYLAYER, BYBLOCK, or an integer multiple of 0.01mm.
DimPost	""	...	DWG	Default prefix or suffix for dimension text (maximum 13 characters): **""** No suffix. **<>mm** Millimeter suffix. **<>Å** Angstrom suffix.
DimRnd	0.0	...	DWG	Rounding value for dimension distances.
DimSAh	Off	...	DWG	Separate arrowhead blocks: **Off** Use arrowhead defined by **DimBlk**. **On** Use arrowheads defined by **DimBlk1** and **DimBlk2**.
DimScale	1.0	...	DWG	Overall scale factor for dimensions: **0** Value is computed from the scale between current modelspace viewport and paperspace. **>0** Scales text and arrowheads.

Variable	Default	R/O	Loc	Meaning
DimSD1	Off	...	DWG	Suppress first dimension line: **On** First dimension line is suppressed. **Off** Not suppressed.
DimSD2	Off	...	DWG	Suppress second dimension line: **On** Second dimension line is suppressed. **Off** Not suppressed.
DimSE1	Off	...	DWG	Suppress the first extension line: **On** First extension line is suppressed. **Off** Not suppressed.
DimSE2	Off	...	DWG	Suppress the second extension line: **On** Second extension line is suppressed. **Off** Not suppressed.
DimSho	On	...	DWG	Update dimensions while dragging: **On** Dimensions are updated during drag. **Off** Dimensions are updated after drag.
DimSOXD	Off	...	DWG	Suppress dimension lines outside extension lines: **On** Dimension lines not drawn outside extension lines. **Off** Are drawn outside extension lines.
DimStyle	"STANDARD"	R/O	DWG	Current dimension style.
DimTAD	0	...	DWG	Vertical position of dimension text: **0** Centered between extension lines. **1** Above dimension line, except when dimension line not horizontal and **DimTIH** = 1. **2** On side of dimension line farthest from the defining points. **3** Conforms to JIS.
DimTDec	4	...	DWG	Primary tolerance decimal places.
DimTFac	1.0	...	DWG	Tolerance text height scaling factor.
DimTIH	On	...	DWG	Text inside extensions is horizontal: **Off** Text aligned with dimension line. **On** Text is horizontal.
DimTIX	Off	...	DWG	Place text inside extensions: **Off** Text placed inside extension lines, if room. **On** Force text between the extension lines.
DimTM	0.0	...	DWG	Minus tolerance.
DimTMove	0	...	DWG	Determines how dimension text is moved: **0** Dimension line moves with text. **1** Adds a leader when text is moved. **2** Text moves anywhere; no leader.
DimTOFL	Off	...	DWG	Force line inside extension lines: **Off** Dimension lines not drawn when arrowheads are outside. **On** Dimension lines drawn, even when arrowheads are outside.

Variable	Default	R/O	Loc	Meaning
DimTOH	On	...	DWG	Text outside extension lines: **Off** Text aligned with dimension line. **On** Text is horizontal.
DimTol	Off	...	DWG	Generate dimension tolerances: **Off** Tolerances not drawn. **On** Tolerances are drawn.
DimTolJ	1	...	DWG	Tolerance vertical justification: **0** Bottom. **1** Middle. **2** Top.
DimTP	0.0	...	DWG	Plus tolerance.
DimTSz	0.0	...	DWG	Size of oblique tick strokes: **0** Arrowheads. >**0** Oblique strokes.
DimTVP	0.0	...	DWG	Text vertical position when **DimTAD**=0: **1** Turns **DimTAD** on. >**-0.7** *or* <**0.7** Dimension line is split for text.
DimTxSty	"STANDARD"	...	DWG	Dimension text style.
DimTxt	0.18	...	DWG	Text height.
DimTZin	0	...	DWG	Tolerance zero suppression: **0** Suppress 0 ft and 0 in. **1** Include 0 ft and 0 in. **2** Include 0 ft; suppress 0 in. **3** Suppress 0 ft; include 0 in. **4** Suppress leading 0 in decimal dim. **8** Suppress trailing 0 in decimal dim. **12** Suppress leading and trailing zeroes.
~~DimUnit~~	*2*	...	*DWG*	*Obsolete; replaced by **DimLUnit** and **DimFrac**.*
DimUPT	Off	...	DWG	User-positioned text: **Off** Cursor positions dimension line. **On** Cursor also positions text.
DimZIN	0	...	DWG	Suppression of 0 in feet-inches units: **0** Suppress 0 ft and 0 in. **1** Include 0 ft and 0 in. **2** Include 0 ft; suppress 0 in. **3** Suppress 0 ft; include 0 in. **4** Suppress leading 0 in decimal dim. **8** Suppress trailing 0 in decimal dim. **12** Suppress leading and trailing zeroes.
DispSilh	0	...	DWG	Silhouette display of 3D solids: **0** Off. **1** On.
Distance	0.0	R/O	...	Distance measured by last **Dist** command.
~~Dither~~				*Removed from Release 14.*
DonutId	0.5	...	...	Inside radius of donut.
DonutOd	1.0	...	...	Outside radius of donut.

Variable	Default	R/O	Loc	Meaning
DragMode	2	...	REG	Drag mode: **0** No drag. **1** On if requested. **2** Automatic.
DragP1	10	...	REG	Regen drag display.
DragP2	25	...	REG	Fast drag display.
DrawOrderCtrl	3	...	DWG	Determines behavior of draw order: **0** Draw order not restored until next regen or drawing reopened. **1** Normal draw order behavior. **2** Draw order inheritance. **3** Combines options 1 and 2.
DwgCheck	0	...	REG	Toggles checking if drawing was edited by software other than AutoCAD: **0** Supresses dialog box. **1** Displays warning dialog box.
DwgCodePage	*varies*	R/O	DWG	Drawing code page, such as "ANSI_1252".
DwgName	*varies*	R/O	...	Current drawing filename, such as "drawing1.dwg".
DwgPrefix	*varies*	R/O	...	Drawing's drive and folder, such as "d:\acad 2005\".
DwgTitled	0	R/O	...	Drawing filename is: **0** "drawing1.dwg". **1** User-assigned name.
~~DwgWrite~~				*Removed from AuoCAD Release 14.*

E

Variable	Default	R/O	Loc	Meaning
EdgeMode	0	...	REG	Toggle edge mode for **Trim** and **Extend** commands: **0** No extension. **1** Extends cutting edge.
Elevation	0.0	...	DWG	Current elevation, relative to current UCS.
EntExts	*1*	...	...	*Controls how drawing extents are calculated:* ***0*** *Extents calculated every time; slows down AutoCAD but uses less memory.* ***1*** *Extents of every object are cached as a two-byte value (default).* ***2*** *Extents of every object are cached as a four-byte value (fastest but uses more memory).*
EntMods	*0*	R/O	...	*Increments by one each time an object is modified to indicate that an object has been modified since the drawing was opened; value ranges from 0 to 4.29497E9.*
ErrNo	0	...	...	Error number from AutoLISP, ADS, & Arx.
~~ExeDir~~				*Removed from Release 14.*

Variable	Default	R/O	Loc	Meaning
Expert	0	...	...	Suppresses the displays of prompts: **0** Normal prompts. **1** "About to regen, proceed?" and "Really want to turn the current layer off?" **2** "Block already defined. Redefine it?" and "A drawing with this name already exists. Overwrite it?" **3** **Linetype** command messages. **4** **UCS Save** and **VPorts Save**. **5** **DimStyle Save** and **DimOverride.**
ExplMode	1	...	...	Toggles whether **Explode** and **Xplode** commands explode non-uniformly scaled blocks: **0** Does not explode. **1** Explodes.
ExtMax	-1.0E+20, -1.0E+20, -1.0E+20	R/O	DWG	Upper-right coordinate of drawing extents.
ExtMin	1.0E+20, 1.0E+20, 1.0E+20	R/O	DWG	Lower-left coordinate of drawing extents.
ExtNames	1	...	DWG	Format of named objects: **0** Names are limited to 31 characters, and can include A - Z, 0 - 9, dollar ($), underscore (_), and hyphen (-). **1** Names are limited to 255 characters, and can include A - Z, 0 - 9, spaces, and any characters not used by Windows or AutoCAD for special purposes.

F

Variable	Default	R/O	Loc	Meaning
FaceTRatio	0	...	...	Controls the aspect ratio of facets on rouunded 3D bodies: **0** Creates an *n* by 1 mesh. **1** Creates an *n* by *m* mesh.
FaceTRres	0.5000	...	DWG	Adjusts smoothness of shaded and hidden-line objects: **0.01** Minimum value. **2** Recommended value. **10** Maximum value.
~~*FfLimit*~~	...	...	...	*Removed from AutoCAD Release 14.*
FieldDisplay	1	...	REG	Toggles gray background to field text.
FieldEval	31	...	DWG	Determines how fields are updated: **0** Not updated **1** Updated with **Open**. **2** Updated with **Save**. **4** Updated with **Plot**. **8** Updated with **eTransmit**. **16** Updated with regeneration.

Variable	Default	R/O	Loc	Meaning
FileDia	1	...	REG	User interface for file-accessing commands: **0** Command-line prompts. **1** File dialog boxes.
FilletRad	0.5	...	DWG	Current fillet radius.
FillMode	1	...	DWG	Fill of solid objects and hatches: **0** Off. **1** On.
Flatland	*0*	R/O	...	*Obsolete system variable.*
FontAlt	"simplex.shx"	...	REG	Font name that substitutes for missing fonts.
FontMap	"acad.fmp"	...	REG	Name of font mapping file.
Force_Paging	*0*	...	...	***0*** *Minimum (default).* ***4.29497E9*** *Maximum.*
FrontZ	0.0	R/O	DWG	Front clipping plane offset.
FullOpen	1	R/O	...	Drawing is: **0** Partially loaded. **1** Fully open.

G

Variable	Default	R/O	Loc	Meaning
GfAng	0	...	...	Angle of gradient fill; 0 to 360 degrees.
GfClr1	"RGB 000,000,255"	...	...	First gradient color in RGB format.
GfClr2	"RGB 255,255,153"	...	...	Second gradient color in RGB format.
GfClrLum	1.0	...	...	Level of gray in one-color gradients: **0** Black. **1** White.
GfClrState	1	...	...	Specifies type of gradient fill: **0** Two-color. **1** One-color.
GfName	1	...	...	Specifies style of gradient fill: **1** Linear. **2** Cylindrical. **3** Inverted cylindrical. **4** Spherical. **5** Inverted spherical. **6** Hemispherical. **7** Inverted hemispherical. **8** Curved. **9** Inverted curved.
GfAShift	0	...	...	Specifies the origin of the gradient fill: **0** Centered. **1** Shifted up and left.
GlobCheck	*0*	...	...	*Reports statistics on dialog boxes:* ***-1*** *Turn off local language.* ***0*** *Turn off.* ***1*** *Warns if larger than 640x400.* ***2*** *Also reports size in pixels.* ***3*** *Additional information.*

Variable	Default	R/O	Loc	Meaning
GridMode	0	...	DWG	Display of grid: **0** Off. **1** On.
GridUnit	0.5,0.5	...	DWG	X,y-spacing of grid.
GripBlock	0	...	REG	Display of grips in blocks: **0** At block insertion point. **1** Of all objects within block.
GripColor	160	...	REG	ACI color of unselected grips: **1** Minimum color number; red. **160** Default color; blue. **255** Maximum color number.
GripHot	1	...	REG	ACI color of selected grips: **1** Default color, red. **255** Maximum color number.
GripHover	3	...	REG	ACI grip fill color when cursor hovers.
GripLegacy	*0*	...	...	*Value of 0 or 1.*
GripObjLimit	100	...	REG	Grips not displayed when more than this number: **1** Minimum. **32767** Maximum.
Grips	1	...	REG	Display of grips: **0** Off. **1** On
GripSize	3	...	REG	Size of grip box, in pixels: **1** Minimum size. **255** Maximum size.
GripTips	1	...	REG	Determines if grip tips are displayed when the cursor hovers over custom objects: **0** Off. **1** On.

H

Variable	Default	R/O	Loc	Meaning
HaloGap	0	...	DWG	Distance to shorten a haloed line; specified as the percentage of 1".
Handles	1	R/O	...	Obsolete system variable.
HidePrecision	0	...	DWG	Controls the precision of hide calculations: **0** Single precision, less accurate, faster. **1** Double precision, more accurate, but slower (recommended).
HideText	0	...	...	Determines whether text is hidden during the **Hide** command: **0** Text is not hidden nor hides other objects, unless text object has thickness. **1** Text is hidden and hides other objects.
Highlight	1	...	...	Object selection highlighting: **0** Disabled. **1** Enabled.

Variable	Default	R/O	Loc	Meaning
HPAng	0	...	...	Current hatch pattern angle.
HpAssoc	1	...	...	Determines if hatches are associative: **0** Not associative. **1** Associative.
HpBound	1	...	REG	Object created by **BHatch** and **Boundary**: **0** Region. **1** Polyline.
HpDouble	0	...	...	Double hatching: **0** Disabled. **1** Enabled.
HpDrawOrder	3	...	...	Draw order of hatch patterns and fills: **0** None. **1** Behind all other objects. **2** In front of all other objects. **3** Behind the hatch boundary. **4** In front of the hatch boundary.
HpGapTol	0	...	REG	Largest gap allowed in hatch boundary; ranges from 0 to 5000 units.
HpName	"ANSI31"	...	...	Current hatch pattern name **""** No default. **.** Set no default.
HpScale	1.0	...	...	Current hatch scale factor; cannot be zero.
HpSpace	1.0	...	...	Current spacing of user-defined hatching; cannot be zero.
HyperlinkBase	""	...	DWG	Path for relative hyperlinks.

I

Variable	Default	R/O	Loc	Meaning
ImageHlt	0	...	REG	When a raster image is selected: **0** Image frame is highlighted. **1** Entire image is highlighted.
IndexCtl	0	...	DWG	Creates layer and spatial indices: **0** No indices created. **1** Layer index created. **2** Spatial index created. **3** Both indices created.
InetLocation	"www.autodesk.com"	...	REG	Default browser URL.
InsBase	0.0,0.0,0.0	...	DWG	Insertion base point relative to the current UCS for **Insert** and **XRef** commands.
InsName	""	...	...	Current block name: **.** Set to no default.
InsUnits	1	...	...	Drawing units when a block is dragged into drawing from DesignCenter: **0** Unitless. **1** Inches. **2** Feet. **3** Miles. **4** Millimeters.

Variable	Default	R/O	Loc	Meaning
				5 Centimeters. **6** Meters. **7** Kilometers. **8** Microinches. **9** Mils. **10** Yards. **11** Angstroms. **12** Nanometers. **13** Microns. **14** Decimeters. **15** Decameters. **16** Hectometers. **17** Gigameters. **18** Astronomical Units. **19** Light Years. **20** Parsecs.
InsUnitsDefSource	1	...	REG	Source drawing units value; ranges from 0 to 20; see above.
InsUnitsDefTarget	1	...	REG	Target drawing units value; ranges from 0 to 20.
IntersectionColor	257	...	DWG	Color of intersection polylines: **0** Color is byblock. **1-255** AutoCAD color index. **256** Color is bylayer. **257** Color is byentity.
IntersectionDisplay	0	...	DWG	Determines 3D surface intersections during **Hide** command: **0** Does not draw intersections. **1** Draws polylines at intersections.
ISaveBak	1	...	REG	Controls whether *.bak* file is created: **0** No file created. **1** *.bak* backup file created.
ISavePercent	50	...	REG	Percentage of waste in saved *.dwg* file before cleanup occurs: **0** Every save is a full save. **>0** Faster partial saves.
IsoLines	4	...	DWG	Isolines on 3D solids: **0** No isolines; minimum. **16** Good-looking. **2,047** Maximum.

L

Variable	Default	R/O	Loc	Meaning
LastAngle	0	R/O	...	Ending angle of last-drawn arc.
LastPoint	*varies*	...	...	Last-entered point, such as 15,9,56.

Variable	Default	R/O	Loc	Meaning
LastPrompt	""	R/O	...	Last string on the command line; includes user input.
LazyLoad	*0*	...	...	*Toggle: 0 or 1.*
LayoutRegenCtl	2	...	REG	Controls display list for layouts: **0** Display list regen'ed with each tab change. **1** Display list is saved for model tab and last layout tab. **2** Display list is saved for all tabs.
LensLength	50.0	R/O	DWG	Perspective view lens length, in mm.
LimCheck	0	...	DWG	Drawing limits checking: **0** Disabled. **1** Enabled.
LimMax	12.0,9.0	...	DWG	Upper right drawing limits.
LimMin	0.0,0.0	...	DWG	Lower left drawing limits.
LispInit	1	...	REG	AutoLISP functions and variables are: **0** Preserved from drawing to drawing. **1** Valid in current drawing only.
Locale	"enu"	R/O	...	ISO language code; see DctMain.
LocalRootPrefix	"d:\documents and Settings*username*\local settings\appli..."	R/O	REG	Path to folder holding local customizable files.
LogFileMode	0	...	REG	Text window written to *.log* file: **0** No. **1** Yes.
LogFileName	"d:\acad 2005\Drawing1_1_1_0000.log"	R/O	DWG	Filename and path for *.log* file.
LogFilePath	"d:\acad 2005\"	...	REG	Path for the *.log* file.
LogInName	"*username*"	R/O	...	User's login name; max = 30 chars.
~~*LongFName*~~				*Removed from AutoCAD Release 14.*
LTScale	1.0	...	DWG	Current linetype scale factor; cannot be 0.
LUnits	2	...	DWG	Linear units mode: **1** Scientific. **2** Decimal. **3** Engineering. **4** Architectural. **5** Fractional.
LUPrec	4	...	DWG	Decimal places (or inverse of smallest fraction) of linear units.
LwDefault	25	...	REG	Default lineweight, in millimeters; must be one of the following values: 0, 5, 9, 13, 15, 18, 20, 25, 30, 35, 40, 50, 53, 60, 70, 80, 90, 100, 106, 120, 140, 158, 200, or 211.

Variable	Default	R/O	Loc	Meaning
LwDisplay	0	...	DWG	Toggles whether lineweight is displayed; setting is saved separately for Model space and each layout tab. **0** Not displayed. **1** Displayed.
LwUnits	1	...	REG	Determines units for lineweight: **0** Inches. **1** Millimeters.
M				
MacroTrace	*0*	...	...	*Diesel debug mode:* ***0*** *Off.* ***1*** *On.*
MaxActVP	64	...	...	Maximum viewports to regenerate: **2** Minimum. **64** Maximum (increased from 48 in R14).
MaxObjMem	*0*	...	...	*Maximum number of objects in memory; object pager is turned off when value = 0, <0, or 2,147,483,647.*
MaxSort	200	...	REG	Maximum names sorted alphabetically.
MButtonPan	1	...	REG	Determines behavior of wheelmouse: **0** As defined by AutoCAD menu file. **1** Pans when dragging with wheel.
MeasureInit	0	...	REG	Drawing units for default drawings: **0** English. **1** Metric.
Measurement	0	...	DWG	Current drawing units: **0** English. **1** Metric.
MenuCtl	1	...	REG	Submenu display: **0** Only with menu picks. **1** Also with keyboard entry.
MenuEcho	0	...		Menu and prompt echoing: **0** Display all prompts. **1** Suppress menu echoing. **2** Suppress system prompts. **4** Disable **^P** toggle. **8** Display all input-output strings.
MenuName	"acad"	R/O	REG	Current menu filename.
Millisecs	*248206921*	R/O	...	*Number of milliseconds since timing started.*
MirrText	0	...	DWG	Text handling during **Mirror** command: **0** Retain text orientation. **1** Mirror text.
ModeMacro	""	...	...	Invoke Diesel macro.
MsmState	0	R/O	...	Specifies if **Markup Set Manager** is active.

Variable	Default	R/O	Loc	Meaning
MsOleScale	1.0	...	DWG	Determines the size of text-containing OLE objects when pasted in model space: **-1** Scaled by value of **PlotScale**. **0** Scale by value of **DimScale**. **>0** Scale factor.
MTextEd	"Internal"	...	REG	Name of the **MText** editor: **.** Use default editor. **0** Cancel the editing operation. **-1** Use the secondary editor. **"blank"** MTEXT internal editor. **"Internal"** MTEXT internal editor. **"Notepad"** Windows Notepad editor. **":lisped"** Built-in AutoLISP function. *string* Name of editor fewer than 256 characters long using this syntax: *:AutoLISPtextEditorFunction*#*TextEditor*.
MTextFixed	0	...	REG	Specifies mtext editor appearence: **0** Display mtext editor and text at same size and position as object being edited. **1** Display mtext editor at the same location as last used; fixed height text.
MTJigStrings	"abc"	...	REG	Sample text displayed by mtext editor; maximum 10 letters; enter . for no text.
MyDocumentsPrefix	"C:\Documents and Settings*username*\My Documents"	R/O	REG	Path to the *my documents* folder of the currently logged-in user.

N

Variable	Default	R/O	Loc	Meaning
NodeName	*"AC$"*	R/O	REG	*Name of network node; range is one to three characters.*
NoMutt	0	...	...	Suppresses the display of message (a.k.a. muttering) during scripts, LISP, macros: **0** Display prompt, as normal. **1** Suppress muttering.
NwfState	*1*	...	...	*Reports whether New Features Workshop displays when AutoCAD starts.*

O

Variable	Default	R/O	Loc	Meaning
ObscureColor	0	...	DWG	Color of objects obscured by **Hide** command: **0** Invisible. **1 - 255** Color number.
ObscureLtype	0	...	DWG	Linetype of objects obscured by **Hide** command: **0** Invisible. **1** Solid. **2** Dashed.

Variable	Default	R/O	Loc	Meaning
				3 Dotted. **4** Short dash. **5** Medium dash. **6** Long dash. **7** Double short dash. **8** Double medium dash. **9** Double long dash. **10** Medium long dash. **11** Sparse dot.
OffsetDist	1.0	...	...	Current offset distance: <**0** Offsets through a specified point. >**0** Default offset distance.
OffsetGapType	0	...	REG	Determines how to reconnect polyline when individual segments are offset: **0** Extend segments to fill gap. **1** Fill gap with fillet (arc segment). **2** Fill gap with chamfer (line segment).
OleFrame	2	...	DWG	Controls the visibility of the frame around OLE objects: **0** Frame is not displayed and not plotted. **1** Frame is displayed and is plotted. **2** Frame is displayed but is not plotted.
OleHide	0	...	REG	Display and plotting of OLE objects: **0** All OLE objects visible. **1** Visible in paper space only. **2** Visible in model space only. **3** Not visible.
OleQuality	1	...	REG	Quality of display and plotting of embedded OLE objects: **0** Line art quality. **1** Text quality. **2** Graphics quality. **3** Photograph quality. **4** High quality photograph.
OleStartup	0	...	DWG	Loading OLE source application improves plot quality: **0** Do not load OLE source app. **1** Load OLE source app when plotting.
OpmState	*0*	...	...	*Specifies if **Properties** window is active.*
OrthoMode	0	...	DWG	Orthographic mode: **0** Off. **1** On.
OsMode	4133	...	REG	Current object snap mode: **0** NONe. **1** ENDpoint. **2** MIDpoint. **4** CENter. **8** NODe. **16** QUAdrant. **32** INTersection.

Variable	Default	R/O	Loc	Meaning
				64 INSertion. **128** PERpendicular. **256** TANgent. **512** NEARest. **1024** QUIck. **2048** APPint. **4096** EXTension. **8192** PARallel. **16383** All modes on. **16384** Object snap turned off via **OSNAP** on the status bar.
OSnapCoord	2	...	REG	Keyboard overrides object snap: **0** Object snap overrides keyboard. **1** Keyboard overrides object snap. **2** Keyboard overrides object snap, except in scripts.
OSnapHatch	0	...	...	Toggles whether hatches are snapped.
OSnapNodeLegacy	1	...	...	Toggles whether osnap snaps to text insertion points.

P

Variable	Default	R/O	Loc	Meaning
PaletteOpaque	0	...	REG	Determines if palettes can be made transparent: **0** Turned off by user. **1** Turned on by user. **2** Unavailable, but turned on by user. **3** Unavailable, and turned off by user.
PaperUpdate	0	...	REG	Determines how AutoCAD plots a layout with paper size different from plotter's default: **0** Displays a warning dialog box. **1** Changes paper size to that of the plotter configuration file.
PDMode	0	...	DWG	Point display mode: **0** Dot. **1** No display. **2** +-symbol. **3** x-symbol. **4** Short line. **32** Circle. **64** Square.

0 1 2 3 4

32 33 34 35 36

64 65 66 67 68

96 97 98 99 100

Variable	Default	R/O	Loc	Meaning
PDSize	0.0	...	DWG	Point display size, in pixels: **>0** Absolute size. **0** 5% of drawing area height. **<0** Percentage of viewport size.
PEdit Accept	0	...	REG	Suppresses display of the **PEdit** command's "Object selected is not a polyline. Do you want to turn it into one? <Y>" prompt.
PEllipse	0	...	DWG	Toggle ellipse creation: **0** True ellipse. **1** Polyline arcs.
Perimeter	0.0	R/O	...	Perimeter calculated by the last **Area**, **DbList**, and **List** commands.
PFaceVMax	4	R/O	...	Maximum vertices per 3D face.
PHandle	*0*	...	*ACAD*	*Ranges from 0 to 4.29497E9.*
PickAdd	1	...	REG	Effect of **SHIFT** key on selection set: **0** Adds to selection set. **1** Removes from selection set.
PickAuto	1	...	REG	Selection set mode: **0** Single pick mode. **1** Automatic windowing and crossing.
PickBox	3	...	REG	Object selection pickbox size, in pixels: **0** Minimum size. **50** Maximum size.
PickDrag	0	...	REG	Selection window mode: **0** Pick two corners. **1** Pick a corner; drag to second corner.
PickFirst	1	...	REG	Command-selection mode: **0** Enter command first. **1** Select objects first.
PickStyle	1	...	REG	Groups and associative hatches in selection sets: **0** Neither included. **1** Include groups. **2** Include associative hatches. **3** Include both.
Platform	"Microsoft Windows NT Version 5.0 (x86)"	R/O	...	Name of the operating system.
PLineGen	0	...	DWG	Polyline linetype generation: **0** From vertex to vertex. **1** From end to end.
PLineType	2	...	REG	Automatic conversion and creation of 2D polylines by **PLine**: **0** Not converted; creates old-format polylines. **1** Not converted; creates optimized lwpolylines. **2** Polylines in older drawings are converted on open; **PLine** creates optimized lwpolyline objects.

Variable	Default	R/O	Loc	Meaning
PLineWid	0.0	...	DWG	Current polyline width.
PlotId	*""*	...	*REG*	*Obsolete; has no effect in AutoCAD.*
PlotOffset	0	...	REG	Sets the plot offset: **0** Relative to edge of margins. **1** Relative to edge of paper.
PlotRotMode	1	...	DWG	Orientation of plots: **0** Lower left = 0,0. **1** Lower left plotter area = lower left of media. **2** X, y-origin offsets calculated relative to the rotated origin position.
Plotter	*0*	...	*REG*	*Obsolete; has no effect in AutoCAD.*
PlQuiet	0	...	REG	Toggles display during batch plotting and scripts (replaces **CmdDia**): **0** Plot dialog boxes and nonfatal errors are displayed. **1** Nonfatal errors are logged; plot dialog boxes are not displayed.
PolarAddAng	""	...	REG	Contains a list of up to 10 user-defined polar angles; each angle can be up to 25 characters long, each separated with a semicolon (;). For example: 0;15;22.5;45.
PolarAng	90	...	REG	Specifies the increment of polar angle; contrary to Autodesk documentation, you may specify any angle.
PolarDist	0.0	...	REG	The polar snap increment when **SnapStyl** is set to 1 (isometric).
PolarMode	0	...	REG	Settings for polar and object snap tracking: **0** Measure polar angles based on current UCS (absolute), track orthogonally; don't use additional polar tracking angles; and acquire object tracking points automatically. **1** Measure polar angles from selected objects (relative). **2** Use polar tracking settings in object snap tracking. **4** Use additional polar tracking angles (via **PolarAng**). **8** Press **SHIFT** to acquire object snap tracking points.
PolySides	4	...	...	Current number of polygon sides: **3** Minimum sides. **1024** Maximum sides.
Popups	1	R/O	...	Display driver support of AUI: **0** Not available. **1** Available.

Variable	Default	R/O	Loc	Meaning
Product	"AutoCAD"	R/O	ACAD	Name of the software.
Program	"acad"	R/O	ACAD	Name of the software's executable file.
ProjectName	""	...	DWG	Project name of the current drawing; searches for xref and image files.
ProjMode	1	...	REG	Projection mode for **Trim** and **Extend** commands: **0** No projection. **1** Project to x,y-plane of current UCS. **2** Project to view plane.
ProxyGraphics	1	...	REG	Proxy image saved in the drawing: **0** Not saved; displays bounding box. **1** Image saved with drawing.
ProxyNotice	1	...	REG	Display warning message: **0** No. **1** Yes.
ProxyShow	1	...	REG	Display of proxy objects: **0** Not displayed. **1** All displayed. **2** Bounding box displayed.
ProxyWebSearch	0	...	REG	Object enablers are checked: **0** Do not check for object enablers. **1** Check for object enablers if an Internet connection is present.
PsLtScale	1	...	DWG	Paper space linetype scaling: **0** Use model space scale factor. **1** Use viewport scale factor.
PsProlog	*""*	...	*REG*	*PostScript prologue filename.*
PsQuality	*75*	...	*REG*	*Resolution of PostScript display, in pixels:* *<0 Display as outlines; no fill.* *0 Not displayed.* *>0 Display filled.*
PStyleMode	1	...	DWG	Toggles the plot color matching mode of the drawing: **0** Use named plot style tables. **1** Use color-dependent plot style tables.
PStylePolicy	1	...	REG	Determines whether the object color is associated with its plot style: **0** Not associated. **1** Associated.
PsVpScale	0	...	...	Sets the view scale factor (the ratio of units in paper space to the units in newly-created model space viewports) for all newly-created viewports: **0** Scaled to fit.

Variable	Default	R/O	Loc	Meaning
PUcsBase	""	...	DWG	Name of UCS defining the origin and orientation of orthographic UCS settings in paper space only.

Q

Variable	Default	R/O	Loc	Meaning
QAFlags	*0*	...	...	*Quality assurance flags:* ***0*** *Turned off.* ***1*** *The ^C metacharacters in a menu macro cancels grips, just as if user pressed **ESC**.* ***2*** *Long text screen listings do not pause.* ***4*** *Error and warning messages are displayed at the command line, instead of in dialog boxes.* ***128*** *Screen picks are accepted via the AutoLISP (command) function.*
QaUcsLock	*0*	...	...	*Either 0 or 1.*
QTextMode	0	...	DWG	Quick text mode: **0** Off. **1** On.
QueuedRegenMax	*2147483647*	...	...	*Ranges between very large and very small numbers.*

R

Variable	Default	R/O	Loc	Meaning
R14RasterPlot	*0*	...	...	*Either 1 or 0.*
RasterPreview	1	R/O	REG	Preview image: **0** None saved. **1** Saved in BMP format.
RefEditName	""	...	...	The reference filename when it is in reference-editing mode.
RegenMode	1	...	DWG	Regeneration mode: **0** Regen with each view change. **1** Regen only when required.
Re-Init	0	...	...	Reinitialize I/O devices: **1** Digitizer port. *2 Plotter port.* **4** Digitizer. *8 Plotter.* **16** Reload PGP file.
RememberFolders	1	...	REG	Controls path search method: **0** Path specified in desktop AutoCAD icon is default for file dialog boxes. **1** Last path specified by each file dialog box is remembered.
ReportError	1	...	REG	Determines if AutoCAD sends an error report to Autodesk: **0** No error report created. **1** Error report is generated and sent to Autodesk.

Variable	Default	R/O	Loc	Meaning
RoamableRootPrefix	"d:\documents and settings*username*\application aata\aut..."	R/O	REG	Path to root folder where roamable customized files are located.
~~*RIAspect*~~				*Removed from AutoCAD Release 14.*
~~*RIBackG*~~				*Removed from AutoCAD Release 14.*
~~*RIEdge*~~				*Removed from AutoCAD Release 14.*
~~*RIGamit*~~				*Removed from AutoCAD Release 14.*
~~*RIGrey*~~				*Removed from AutoCAD Release 14.*
~~*RIThresh*~~				*Removed from AutoCAD Release 14.*
RTDisplay	1	...	REG	Raster display during real-time zoom and pan: **0** Display the entire raster image. **1** Display raster outline only.

S

Variable	Default	R/O	Loc	Meaning
SaveFile	"auto.sv$"	R/O	REG	Automatic save filename.
SaveFilePath	"d:\temp\"	...	REG	Path for automatic save files.
SaveName	""	R/O	...	Drawing save-as filename.
SaveTime	10	...	REG	Automatic save interval, in minutes: **0** Disable auto save.
ScreenBoxes	0	R/O	ACAD	Maximum number of menu items **0** Screen menu turned off.
ScreenMode	3	R/O	...	State of AutoCAD display screen: **0** Text screen. **1** Graphics screen. **2** Dual-screen display.
ScreenSize	*varies*	R/O	...	Current viewport size, in pixels, such as 719.0000,381.0000.
SDI	0	...	REG	Toggles multiple-document interface (SDI is "single document interface"): **0** Turns on MDI. **1** Turns off MDI (only one drawing may be loaded into AutoCAD). **2** MDI disabled for apps that cannot support MDI; read-only. **3** (R/O) MDI disabled for apps that cannot support MDI, even when **SDI**= 1.
ShadEdge	3	...	DWG	**Shade** style: **0** Shade faces; 256-color shading. **1** Shade faces; edges background color. **2** Hidden-line removal. **3** 16-color shading.
ShadeDif	70	...	DWG	Percent of diffuse to ambient light: **0** Minimum. **100** Maximum.

Variable	Default	R/O	Loc	Meaning
ShortcutMenu	11	...	REG	Toggles availability of shortcut menus: **0** Disables all default, edit, and command shortcut menus. **1** Enables default shortcut menus. **2** Enables edit shortcut menus. **4** Enables command shortcut menus whenever a command is active. **8** Enables command shortcut menus only when command options are available at the command line.
ShpName	""	...	...	Current shape name: **.** Set to no default.
SigWarn	1	...	REG	Determines whether a warning is displayed when a file is opened with a digital signature.
SketchInc	0.1	...	DWG	**Sketch** command's recording increment.
SkPoly	0	...	DWG	Sketch line mode: **0** Record as lines. **1** Record as a polyline.
SnapAng	0	...	DWG	Current rotation angle for snap and grid.
SnapBase	0.0,0.0	...	DWG	Current origin for snap and grid.
SnapIsoPair	0	...	DWG	Current isometric drawing plane: **0** Left isoplane. **1** Top isoplane. **2** Right isoplane.
SnapMode	0	...	DWG	Snap mode: **0** Off. **1** On.
SnapStyl	0	...	DWG	Snap style: **0** Normal. **1** Isometric.
SnapType	0	...	REG	Toggles between standard or polar snap for the current viewport: **0** Standard snap. **1** Polar snap.
SnapUnit	0.5,0.5	...	DWG	X,y-spacing for snap.
SolidCheck	1	...	...	Toggles solid validation: **0** Off. **1** On.
SortEnts	96	...	DWG	Object display sort order: **0** Off. **1** Object selection. **2** Object snap. **4** Redraw. **8** Slide generation. **16** Regeneration. **32** Plot. **64** PostScript output.
SpaceSwitch	*1*	...	...	*Either 1 or 9.*

Variable	Default	R/O	Loc	Meaning
SplFrame	0	...	DWG	Polyline and mesh display: **0** Polyline control frame not displayed; display polygon fit mesh; 3D faces invisible edges not displayed. **1** Polyline control frame displayed; display polygon defining mesh; 3D faces invisible edges displayed.
SplineSegs	8	...	DWG	Number of line segments that define a splined polyline.
SplineType	6	...	DWG	Spline curve type: **5** Quadratic Bezier spline. **6** Cubic Bezier spline.
SsFound	""	...	...	Path and file name of sheet set.
SsLocate	1	...	USER	Determine whether sheet set files are opened with drawing.
SsmAutoOpen	1	...	USER	Determines whether the **Sheet Set Manager** is opened with drawing.
SsmState	0	R/O	...	Reports whether **Sheet Set Manager** is open.
StandardsViolation	2	...	REG	Determines whether alerts are displayed when CAD standards are violated: **0** No alerts. **1** Alert displayed when CAD standard violated. **2** Displays icon on status bar when file is opened with CAD standards, and when non-standard objects are created.
Startup	0	...	REG	Determines which dialog box is displayed by the **New** and **QNew** commands: **0** Displays **Select Template** dialog box. **1** Displays **Startup** and **Create New Drawing** dialog box.
~~*StartupToday*~~				~~*Removed from AutoCAD 2004.*~~
SurfTab1	6	...	DWG	Density of surfaces and meshes: **5** Minimum. **32766** Maximum.
SurfTab2	6	...	DWG	Density of surfaces and meshes: **2** Minimum. **32766** Maximum.
SurfType	6	...	DWG	Pedit surface smoothing: **5** Quadratic Bezier spline. **6** Cubic Bezier spline. **8** Bezier surface.
SurfU	6	...	DWG	Surface density in m-direction: **2** Minimum. **200** Maximum.

Variable	Default	R/O	Loc	Meaning
SurfV	6	...	DWG	Surface density in n-direction: **2** Minimum. **200** Maximum.
SysCodePage	"ANSI_1252"	R/O	...	System code page.

T

Variable	Default	R/O	Loc	Meaning
TabMode	0	...	...	Tablet mode: **0** Off. **1** On.
Target	0.0,0.0,0.0	R/O	DWG	Target in current viewport.
Tbaskbar	*1*	...	...	*Determines whether each drawing appears as a button on the Windows taskbar.*
TbCustomize	*1*	...	...	*Determines whether toolbars can be customized.*
TDCreate	*varies*	R/O	DWG	Time and date drawing created, such as 2448860.54014699.
TDInDwg	*varies*	R/O	DWG	Duration drawing loaded, such as 0.00040625.
TDuCreate	*varies*	R/O	DWG	The universal time and date the drawing was created, such as 2451318.67772165.
TDUpdate	*varies*	R/O	DWG	Time and date of last update, such as 2448860.54014699.
TDUsrTimer	*varies*	R/O	DWG	Decimal time elapsed by user-timer, such as 0.00040694.
TDuUpdate	*varies*	R/O	DWG	The universal time and date of the last save, such as 2451318.67772165.
TempPrefix	"d:\temp"	R/O	...	Path for temporary files set by **Temp** envar.
TextEval	0	...	...	Interpretation of text input: **0** Literal text. **1** Read **(** and **!** as AutoLISP code.
TextFill	1	...	REG	Toggle fill of TrueType fonts: **0** Outline text. **1** Filled text.
TextQlty	50	...	DWG	Resolution of TrueType fonts: **0** Minimum resolution. **100** Maximum resolution (preferred).
TextSize	0.2000	...	DWG	Default height of text (2.5 in metric units).
TextStyle	"Standard"	...	DWG	Default name of text style.
Thickness	0.0000	...	DWG	Default object thickness.
TileMode	1	...	DWG	View mode: **0** Display layout tab. **1** Display model tab.
ToolTips	1	...	REG	Display tooltips: **0** Off. **1** On.

Variable	Default	R/O	Loc	Meaning
TpState	0	R/O	...	Determines if Tool Palettes is open.
TraceWid	0.0500	...	DWG	Current width of traces.
TrackPath	0	...	REG	Determines the display of polar and object snap tracking alignment paths: **0** Display object snap tracking path across the entire viewport. **1** Display object snap tracking path between the alignment point and "From point" to cursor location. **2** Turn off polar tracking path. **3** Turn off polar and object snap tracking paths.
TrayIcons	1	...	REG	Determines if the tray is displayed on the status bar.
TrayNotify	1	...	REG	Determines whether service notifications are displayed by the tray.
TrayTimeout	5	...	REG	Specifies the length of time (in seconds) that tray notificaitons are displayed: **0** Minimium. **10** Maximum.
TreeDepth	3020	...	DWG	Maximum branch depth in *xxyy* format: *xx* Model-space nodes. *yy* Paper-space nodes. *>0* 3D drawing. *<0* 2D drawing.
TreeMax	10000000	...	REG	Limits memory consumption during drawing regeneration.
TrimMode	1	...	REG	Trim toggle for **Chamfer** and **Fillet** commands: **0** Leave selected edges in place. **1** Trim selected edges.
TSpaceFac	1.0	...	...	Mtext line spacing distance; measured as a factor of "normal" text spacing; valid values range from 0.25 to 4.0.
TSpaceType	1	...	...	Type of mtext line spacing: **1** At Least: adjust line spacing based on the height of the tallest character in a line of mtext. **2** Exactly: use the specified line spacing; ignores character height.
TStackAlign	1	...	DWG	Vertical alignment of stacked text (fractions): **0** Bottom aligned. **1** Center aligned. **2** Top aligned.

Variable	Default	R/O	Loc	Meaning
TStackSize	70	...	DWG	Sizes stacked text as a percentage of the current text height: **1** Minimum %. **127** Maximum %.

U

Variable	Default	R/O	Loc	Meaning
UcsAxisAng	90	...	REG	Default angle for rotating the UCS around an axes (via the **UCS** command using the **X**, **Y**, or **Z** options; valid values are limited to: 5, 10, 15, 18, 22.5, 30, 45, 90, or 180.
UcsBase	""	...	DWG	Name of the UCS that defines the origin and orientation of orthographic UCS settings.
UcsFollow	0	...	DWG	New UCS views: **0** No change. **1** Automatically align UCS with new view.
UcsIcon	3	...	DWG	Display of UCS icon: **0** Off. **1** On. **2** Display at UCS origin, if possible. **3** On, and displayed at origin.
UcsName	"World"	R/O	DWG	Name of current UCS view: "" Current UCS is unnamed.
UcsOrg	0.0,0.0,0.0	R/O	DWG	Origin of current UCS relative to WCS.
UcsOrtho	1	...	REG	Determines whether the related orthographic UCS setting is restored automatically: **0** UCS setting remains unchanged when orthographic view is restored. **1** Related orthographic UCS is restored automatically when an orthographic view is restored.
UcsView	1	...	REG	Determines whether the current UCS is saved with a named view: **0** Not saved. **1** Saved.
UcsVp	1	...	DWG	Determines whether the UCS in active viewports remains fixed (locked) or changes (unlocked) to match the UCS of the current viewport: **0** Unlocked. **1** Locked.
UcsXDir	1.0,0.0,0.0	R/O	DWG	X-direction of current UCS relative to WCS.
UcsYDir	0.0,1.0,0.0	R/O	DWG	Y-direction of current UCS relative to WCS.

Variable	Default	R/O	Loc	Meaning
UndoCtl	5	R/O	...	State of undo: **0** Undo disabled. **1** Undo enabled. **2** Undo limited to one command. **4** Auto-group mode. **8** Group currently active.
UndoMarks	0	R/O	...	Current number of undo marks.
UnitMode	0	...	DWG	Units display: **0** As set by **Units** command. **1** As entered by user.
UpdateThumbnail	7	...	DWG	Determines whether thumbnails are created when drawing is saved: **0** Thumbnail previews not updated. **1** Sheet views updated. **2** Model views updated. **4** Sheets updated.
UserI1 *thru* **UserI5**	0	...	DWG	Five user-definable integer variables.
UserR1 *thru* **UserR5**	0.0	...	DWG	Five user-definable real variables.
UserS1 *thru* **UserS5**	""	...	...	Five user-definable string variables.

V

Variable	Default	R/O	Loc	Meaning
ViewCtr	*varies*	R/O	DWG	X,y,z-coordinate of center of current view, such as 6.2,4.5,0.0.
ViewDir	*varies*	R/O	DWG	Current view direction relative to UCS, such as 0.0,0.0,1.0 for plan view.
ViewMode	0	R/O	DWG	Current view mode: **0** Normal view. **1** Perspective mode on. **2** Front clipping on. **4** Back clipping on. **8** UCS-follow on. **16** Front clip not at eye.
ViewSize	9.0	R/O	DWG	Height of current view in drawing units.
ViewTwist	0	R/O	DWG	Twist angle of current view.
VisRetain	1	...	DWG	Determines xref drawing's layer settings — on-off, freeze-thaw, color, and linetype: **0** Xref-dependent layer settings are not saved in the current drawing. **1** Xref-dependent layer settings are saved in current drawing, and take precedence over settings in xrefed drawing next time the current is loaded.

Variable	Default	R/O	Loc	Meaning
VpMaximizedState	0	R/O	...	Specifies whether viewport is maximized by **VpMax** command.
VSMax	*varies*	R/O	DWG	Upper-right corner of virtual screen, such as 37.46,27.00,0.00.
VSMin	*varies*	R/O	DWG	Lower-left corner of virtual screen, such as -24.97,-18.00,0.0.

W

Variable	Default	R/O	Loc	Meaning
WhipArc	0	...	REG	Display of circlular objects: **0** Displayed as connected vectors. **1** Displayed as true circles and arcs.
WhipThread	3	...	REG	Controls multithreaded processing on two CPUs (if present) during drawing redraw and regeneration: **0** Single-threaded calculations. **1** Regenerations multi-threaded. **2** Redraws multi-threaded. **3** Regens and redraws multi-threaded.
WmfBkgnd	1	...	...	Controls background of *.wmf* files: **0** Background is transparent. **1** Background is same as AutoCAD's background color.
WmfForegnd	0	...	...	Controls foreground colors of exported WMFimages: **0** Foreground is darker than background. **1** Foreground is lighter than background.
WorldUcs	1	R/O	...	Matching of WCS with UCS: **0** Current UCS does not match WCS. **1** UCS matches WCS.
WorldView	1	...	DWG	Display during **3dOrbit**, **DView**, and **VPoint** commands: **0** Current UCS. **1** WCS.
WriteStat	1	R/O	...	Indicates whether drawing file is read-only: **0** Drawing file cannot be written to. **1** Drawing file can be written to.

X

Variable	Default	R/O	Loc	Meaning
XClipFrame	0	...	DWG	Visibility of xref clipping boundary: **0** Not visible. **1** Visible.
XEdit	1	...	DWG	Toggles whether drawing can be edited in-place when referenced by another drawing: **0** Cannot in-place refedit. **1** Can in-place refedit.

Variable	Default	R/O	Loc	Meaning
XFadeCtl	50	...	REG	Fades objects not being edited in-place: **0** No fading; minimum value. **90** 90% fading; maximum value.
XLoadCtl	2	...	REG	Controls demand loading: **0** Demand loading turned off; entire drawing is loaded. **1** Demand loading turned on; xref file opened. **2** Demand loading turned on; a *copy* of the xref file is opened.
XLoadPath	"d:\temp"	...	REG	Path for storing temporary copies of demand-loaded xref files.
XRefCtl	0	...	REG	Determines creation of *.xlg* xref log files: **0** File not written. **1** Log file written.
XrefNotify	2	...	REG	Determines if alert is displayed for updated and missing xrefs: **0** No alert displayed. **1** Icon indicates xrefs are attached; a yellow alert indicates missing xrefs. **2** Also displays balloon messages when an xref is modified.
XrefType	*0*	...	...	*Determines whether xrefs are attached or overlaid.*

Z

Variable	Default	R/O	Loc	Meaning
ZoomFactor	60	...	REG	Controls the zoom level via mouse wheel; valid values range between 3 and 100; 15 is recommended.

AutoCAD 2005 Keystrokes & Shortcuts

Object Snap

Modes:

APP	Apparent intersection.
CEN	Center.
END	Endpoint.
EXT	Extension.
FROM	From.
INS	Insertion point.
INT	Intersection.
MID	Midpoint.
M2P	Midpoint between 2 points.
NEA	Nearest.
NOD	Node (point).
PAR	Parallel.
PER	Perpendicular.
QUA	Quadrant.
TAN	Tangent.
QUI	Quick mode.
NON	None.

Setting:

OFF	Turn off object snap.

Selection Sets

(Pick)	Selects one object.
ALL	Selects all objects.
AU	AUtomatic: *(pick)* or BOX.
BOX	Left to right = Crossing; Right to left = Window.
C	Crossing.
CP	Crossing polygon.
F	Fence.
G	Group.
L	Last.
M	Multiple (no highlighting).
P	Previous.
SI	Single selection.
W	Window.
WP	Window polygon.

Selection modes:

A	Add to selection set (default).
R	Remove from selection set.
SHIFT	Remove from selection set.
U	Undo change to selection set.

Special Text Characters

*For **Text** command:*

%%c	Diameter symbol	Ø
%%d	Degree symbol	°
%%o	Overline	
%%%	Percent symbol	%
%%p	Plus-minus symbol	±
%%u	Underline	
%%*nnn*	ASCII character *nnn*	

*For **MText** command:*

\~	Nonbreaking space.
****	Backslash.
\{	Opening brace.
\}	Closing brace.
\C*n*	Sets color *n*.
\F*x***;**	Changes to font file name *x*.
\H*n***;**	Changes text height to *n* units.
\L	Underline.
\l	Turns off underline.
\M+*n*	Multibyte shape number *n*.
\O	Overline.
\o	Turns off overline.
\P	End of paragraph.
\Q*n***;**	Changes obliquing angle to *n*.
\S*n***^***m*	Stacks character *n* over *m*.
\T*n***;**	Changes tracking to *n*.
\U+*n*	Places Unicode character *n*.
\W*n***;**	Changes width factor to *n*.

Color Numbers & Names

0	...	Background color.
1	R	Red.
2	Y	Yellow.
3	G	Green.
4	C	Cyan.
5	B	Blue.
6	M	Magenta.
7	W	White/black.
8-249	...	Other colors.
250-255	...	Shades of grey.
BYLAYER	...	Color from layer.
BYBLOCK	...	Color from block.